D0583895

PLANT ANATOMY

Of Related Interest from the Benjamin/Cummings Series in the Life Sciences

C. J. Avers
Molecular Cell Biology (1986)

M. G. Barbour, J. H. Burk, and W. D. Pitts
Terrestrial Plant Ecology, second edition (1987)

W. M. Becker
The World of the Cell (1986)

N. A. Campbell
Biology (1987)

R. J. Lederer
Ecology and Field Biology (1984)

E. Zeiger and L. Taiz
Plant Physiology (forthcoming in 1989)

PLANT ANATOMY

James D. Mauseth

University of Texas, Austin

1988

The Benjamin/Cummings Publishing Company, Inc.

Menlo Park, California • Reading, Massachusetts
Don Mills, Ontario • Wokingham, U.K. • Amsterdam • Sydney
Singapore • Tokyo • Madrid • Bogota • Santiago • San Juan

Sponsoring editors: Andrew Crowley,
Robin Williams

Associate editor: Langdon Faust

Production editor: Laura Kenney

Text and cover designer: Gary Head

Illustrators: Ginny Mickelson,
Elizabeth Morales-Denney

Production artist: Victoria Philp

Compositor: G&S Typesetters, Inc.

About the cover:

The cover shows a light micrograph of a leaf clearing of *Acalypha* (copperleaf). The leaf was treated with warm sodium hydroxide to make it transparent. Basic fuchsin was added to stain the lignin of the xylem, which appears as bright white lines. The red structure is a plant hair on the surface of the leaf. The photograph was taken by the author with polarized light to make the crystals of calcium oxalate visible. Notice that each larger vein branches to produce smaller veins, but they are also capable of merging with other large veins. The veins are so close together that no leaf cell is far from vascular tissue. The functions of the crystals are not known for certain; they may make the leaf too gritty for most insects to eat, or they may be a means of keeping the level of dissolved, free calcium low.

Copyright © 1988 by The Benjamin/Cummings Publishing Company, Inc. All rights reserved. No part of this publication may be reproduced, stored in a retrieval system, or transmitted in any form or by any means, electronic, mechanical, photo-copying, recording, or otherwise, without the prior written permission of the publisher. Printed in the United States of America. Published simul-taneously in Canada.

Library of Congress Cataloging-in-Publication Data

Mauseth, James D.
    Plant anatomy.
    Benjamin/Cummings (series in the life sciences)
    Bibliography: p.
    Includes index.
    1. Botany—Anatomy. I. Title. II. Series.
QK641.M38   1988   581.4   87-35799

ISBN 0-8053-4570-1

ABCDEFGHIJ-DO-898

The Benjamin/Cummings Publishing Company, Inc.
2727 Sand Hill Road
Menlo Park, California 94025

Preface

Plant anatomy—the actual, physical structure of a plant—should be thought of as a dynamic *process*, not as a static *thing*. The cells, tissues, and organs of any particular plant are the corporal results of the interaction of the plant's metabolism and its environment. If either had been different, then the anatomy would have been different also. Furthermore, the anatomy of a plant is the result of the natural selection that has occurred on all of that plant's ancestors. The varying selection pressures that have acted on plants have produced about 400,000 species. This is a large number, but more could have arisen, and more will arise in the future.

The fact that so many species have evolved indicates that there is not one single, universal optimal anatomy for plants. Instead, different types of structure are advantageous or disadvantageous, functional or nonfunctional, depending on the complete biology of the plant: its microhabitat, the climate, its pollinators, predators, pathogens, and competitors. Because of this, Plant Anatomy—the study of plant structure—must also be dynamic and analytical. It must not be treated either as a science of rote memorization or as observations devoid of interpretations. Certainly there are anatomical principles and generalizations such that a single microscope slide can provide great insight into the total biology of a plant. However, plant biology and plant anatomy are rich and complex enough that it is rewarding to spend time thinking and speculating about the anatomy of a plant and how that anatomy is re-lated to its physiology, growth, reproduction, and survival.

Speculation should be an integral part of Plant Anatomy. It does us no good to simply catalog observations; they must be interpreted. If it is not possible to perform experiments that can confirm or reject an interpretation of the function of a particular structure, then speculation and the construction of hypotheses become especially important. Speculation may lead to multiple hypothetical interpretations of the function of a structure. Obviously, only one can be correct, but the presence of the other hypotheses guides our further studies and ensures that the proper research is done to decide which one most accurately represents reality.

In this book I have tried to emphasize the importance of understanding the mechanisms by which a particular structure affects the total biology of the plant. It is necessary to consider not only how the plant's structure is created by its metabolism but also how it simultaneously acts as the framework and physical environment for that metabolism. Consequently, I have placed less emphasis on extensive terminology. Of course, terminology is important for clear and efficient communication among scientists, but too often the terms tend to take on a life of their own. Terminology is never more important than the concepts for which it is constructed. Because plants are part of an evolutionary continuum, so are plant cells, tissues, and organs. Terminology tends to be discrete and discriminatory, and it is never easily applied to continua. In many cases, it is not possible to give a precise, uncontestable definition for certain terms; and then it is more important to understand why there are ambiguities. It is more important to understand the concept than its terminology.

Electron microscopy, both transmission and scanning, have revolutionized our understanding of many aspects of plant anatomy, yet I have purposefully avoided using many electron micrographs. This book will serve, I believe, as both a textbook for graduate and undergraduate students and as a reference manual for all. Most of the people using it will be working with light microscopy, not electron microscopy, and the illustrations that will be most relevant to their own personal observations will be light micrographs.

Finally, I hope that in this book I have demonstrated that the science of Plant Anatomy is still very much alive and active. With the recent emphasis on functional Plant Anatomy, on the analysis of the interrelationships of plant anatomy with physiology and ecology, Plant Anatomy is now flourishing more than ever. Older observations are now being reinterpreted in light of new theories of biochemistry, metabolism, and survival. As new types of photosynthesis are discovered, the implications of previous observations on leaf structure become clear. As water movement in aquatic plants is studied, new interpretations of hydathodes become tenable. As genetic engineers increase their capacity to modify plants, they must know which structures should be altered to achieve the desired effect. Plant anatomy is such a fundamental component of all aspects of plant biology that it is safe to say that, as long as there are any areas in which plant research is being carried out, the job of the plant anatomist is not finished.

Reviewers

H. J. Arnott, University of Texas, Arlington

Judith Croxdale, University of Wisconsin, Madison

Nancy G. Dengler, University of Toronto

Philip R. Larson, Forestry Sciences Laboratory, Forest Service

Nels R. Lersten, Iowa State University, Ames

Elizabeth Lord, University of California, Riverside

Roger D. Meicenheimer, Miami University

Randy Moore, Baylor University

Karl J. Niklas, Cornell University

Alan R. Orr, University of Northern Iowa, Cedar Falls

Robert C. Romans, Bowling Green State University

Edward L. Schneider, Southwest Texas State University

James L. Seago, Jr., State University of New York, College at Oswego

Wendy Kuhn Silk, University of California, Davis

Brief Contents

Detailed Contents

PART IV
The Primary Vegetative Body of the Plant   199

11 Stem   201

12 Leaf   231

13 Root   269

Introduction

Introduction to Plant Anatomy

Plant anatomy, as a defined, scientific discipline, is one of the oldest fields of botany, having been initiated by the microscopical studies of Nehemiah Grew and Marcello Malpighi in 1671 (Morton, 1981). An advantage to such a long history is that many fundamental aspects of anatomy were discovered, interpreted, and explained long ago. Today researchers can focus on fine details and applications and are able to build on the basic knowledge gained through three centuries of anatomical studies.

At present, just as in the past, one of the goals of anatomical studies is to understand the function of structures rather than simply catalog or describe them. This is not an easy task, because it depends on our interpretation of the physiology, ecology, and other aspects of the plant's biology. For example, evolution based on natural selection is now universally recognized as the primary means of understanding plants: we attempt to understand the effects of a particular structure and then evaluate whether it is selectively advantageous. As a corollary, we accept that plants (and thus their anatomy) evolve and change with time and that nothing is perfectly adapted. The environment changes as the climate cools or warms, becomes drier or more moist, and as mountain ranges rise or erode away, so consequently a

particular anatomy that had been adaptive for a species may no longer be optimum. For these reasons, it is important always to think about the biology of individual species and interpret their anatomy in terms of that biology, keeping in mind as well the ancestry of the species and the ancestry of the anatomy. We must always expect to find a diversity of structures, possibly with one structure grading into another through many intermediate forms. Additionally, we should expect to encounter some structures that are not adaptive in the environmental conditions in which the plant finds itself.

The current approach to plant anatomy, based on natural selection, is not the approach used during much of the history of anatomy. Prior to the postulation of natural selection, it was believed that everything had been created as a series of "types." We are all familiar with the old idea that species had been divinely created, but that concept went further: individual genera and species represented specific divinely conceived types of organisms, just as roots, shoots, and leaves represented divinely conceived types of organs, and parenchyma, collenchyma, and sclerenchyma were types of cells. This idea of types was extended to everything: plant hairs, stomata, nectaries, fruits. Just as there is no reason to expect intermediates between divinely created species, there should be no intermediates between divinely

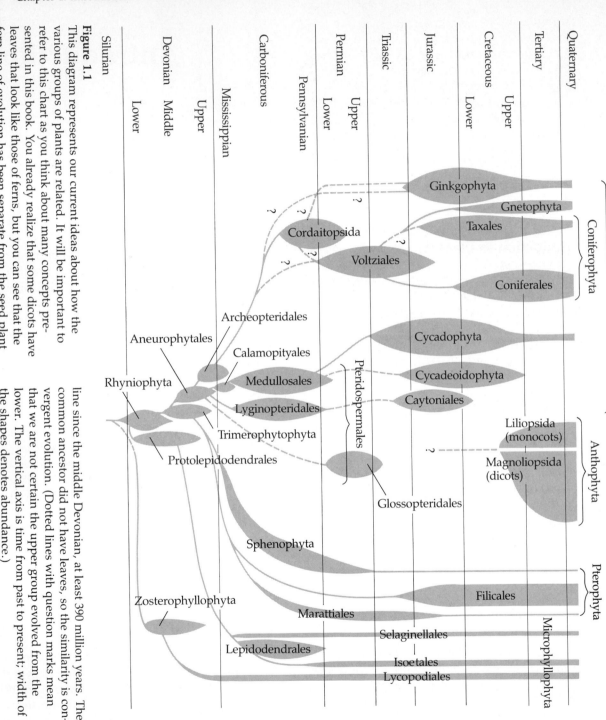

Figure 1.1
This diagram represents our current ideas about how the various groups of plants are related. It will be important to refer to this chart as you think about many concepts presented in this book. You already realize that some dicots have leaves that look like those of ferns, but you can see that the fern line of evolution has been separate from the seed plant line since the middle Devonian, at least 390 million years. The common ancestor did not have leaves, so the similarity is convergent evolution. (Dotted lines with question marks mean that we are not certain the upper group evolved from the lower. The vertical axis is time from past to present; width of the shapes denotes abundance.)

created types. As a consequence, in the nineteenth century there was a tremendous amount of categorizing and classifying in plant anatomy. Because this was a period in which much work was being done, many of these classifications and terms became firmly fixed in the literature of plant anatomy.

It has only been in the present century that the disciplines of physiology, biochemistry, and genetics have become sufficiently advanced that they are able to interact with anatomy in such a way that meaningful explanations of the anatomy can be offered. It is only recently that we have been able to understand fully the structure, function, and selective advantage of many plant tissues and organs. As a consequence, at present, plant anatomy is still in the process of changing over to a foundation of **function** from one of **types**. In this book, the functional approach has been used wherever it was possible, and in those areas still dominated by the concept of types I have tried to explain the problems arising from that approach and I have attempted to describe the studies that must be done in order to understand the function.

An important difference between the two approaches to plant anatomy is that the functional one is dynamic whereas the typological approach is static: once all the types have been identified and classified, there would truly be nothing to do but make long lists of families or genera that have each particular type. With a functional approach, each structure in each species must be individually analyzed; in many cases this is easy: in almost all leaves the stomata function to permit the entry of carbon dioxide while limiting the loss of water. But when stomata occur over the secretory tissue of a nectary, then a new function must be at least considered: the regulation of the release of nectar. Anatomy based on function requires very little memorization, but it demands constant thinking, constant analysis, and a constant consideration for individual structures on individual species and their own particular biologies.

Introduction to Higher Plants

The Major Groups of Plants

The Kingdom Plantae is ancient, having originated more than 550,000,000 years ago. The plants existed exclusively as algae for over one hundred million years, then about 420,000,000 years ago land plants appeared (Fig. 1.1; Stewart, 1983). From that time, the land plants have evolved rapidly and become much more complex than any of the algae, developing diverse types of cells, tissues, and organs. Land plants are classified into the major divisions primarily on the basis of their anatomy: the evolution of new types of spores, roots, leaves, and vascular tissues is considered sufficient justification for distinguishing groups of plants.

For many years, the classification of the Kingdom Plantae into divisions has been stable and almost universally accepted. In this book, the system and names presented by Bold, Alexopoulos, and Delevoryas (1980) will be used (Table 1.1). The plant kingdom is divided informally into the **vascular plants** (also called the **tracheophytes**, those that have vascular tissues specialized for the conduction of water or nutrients) and the **thallophytes** (those that have no vascular tissues: the algae, mosses, and liverworts).

Vascular plants have been quite successful evolutionarily and have diversified along several lines, forming many major groups. The vascular plants are subdivided into two groups. The first contains the **vascular cryptogams**, plants with "hidden" reproduction: Division Psilotophyta (whisk ferns), Division Microphyllophyta (also called Lycophyta, the lycopods), Division Arthrophyta (also called Sphenophyta, the horsetails), and Division Pterophyta (also called Filicophyta, the ferns). The second group of vascular plants contains the **seed plants** which are more often called the **spermatophytes**, or **phanerogams**.

The seed plants formerly had been divided into the "gymnosperms" (plants without

Table 1.1. The Classification of Green Plants

Division	Chlorophyta	Green algae
Division	Bryophyta	Mosses
Division	Hepatophyta	Liverworts
Division	*Rhyniophyta	Early land plants
Division	*Zosterophyllophyta	
Division	Microphyllophyta	Lycopods
Order	*Protolepidodendrales	
Order	*Lepidodendrales	
Order	Lycopodiales	
Order	Selaginellales	
Order	*Pleuromeiales	
Order	Isoetales	
Division	Psilotophyta	Psilotum
Division	Arthrophyta	Sphenophytes
Order	*Sphenophyllales	
Order	Equisetales	
Order	*Calamitales	
Division	*Trimerophytophyta	Ancestors of megaphyllous plants
Division	Pteridophyta	Ferns
Class	*Cladoxylopsida	
Class	Filicopsida	
Division	*Progymnospermophyta	Progymnosperms
Class	Aneurophytopsida	
Class	Archaeopteridopsida	
Class	Protopityopsida	
Division	*Pteridospermophyta	Seed ferns
Order	Lyginopteridales	
Order	Medullosales	
Order	Callistophytales	
Order	Caytoniales	
Order	Corystospermales	
Order	Peltaspermales	
Order	Glossopteridales	
Division	Cycadophyta	Cycads
Division	*Cycadeoidophyta	Cycadeoids

Table 1.1. (*continued*)

Division	Ginkgophyta	Ginkgos
Division	Coniferophyta	
Class	*Cordaitopsida	Cordaites
Class	Coniferopsida	Conifers
Order	*Voltziales	
Order	Coniferales	
Order	Taxales	
Division	Gnetophyta	Gnetum, Ephedra, and Welwitschia
Division	Anthophyta	Flowering plants (angiosperms)
Class	Liliopsida	Monocots
Class	Magnoliopsida	Dicots

*Extinct

flowers and fruits) and the "angiosperms" (those with flowers and fruits). Although these terms are still commonly employed for convenience, the classification is no longer widely accepted as scientifically valid. The reason for the change is that we now believe that the gymnosperms represent at least two separate, ancient lines of evolution. Starting in the extinct groups Division Trimerophytophyta and Division Progymnospermophyta, many new, sophisticated types of anatomy began, such as wood, bark, and seeds (Stewart, 1983; Taylor, 1981). From these two divisions have come the modern, living higher plants that we are all familiar with—Division Cycadophyta (the cycads), Division Coniferophyta (the conifers), and Division Anthophyta (the flowering plants, still very often called the angiosperms)—as well as some not so familiar—Division Ginkgophyta (*Ginkgo* or maidenhair tree) and Division Gnetophyta, with three rather unusual groups, *Gnetum*, *Ephedra*, and *Welwitschia*.

The Division Anthophyta is composed of just two classes. The Class Liliopsida (formerly the Monocotyledonae) contains the monocots (such plants as palms, grasses, orchids, bromeliads, and lilies). The Class Magnoliopsida (formerly the Dicotyledonae) contains the dicots, or broadleaf plants (Cronquist, 1981). The dicots constitute a much larger group than the monocots and contain all of the woody flowering plants and a majority of the herbaceous ones.

Throughout its history, plant anatomy has been concerned with all of the land plants, but to avoid being too voluminous and overwhelming to students, plant anatomy textbooks have always been restricted to considering primarily the seed plants. I will follow that tradition in this book, emphasizing the seed plants primarily and discussing the vascular cryptogams as often as possible but only where they provide especially illustrative examples. The seed plants are not selected as the focus of a plant anatomy text merely for convenience nor because they are so valuable economically, but because they represent several closely related lines of evolution and thus share a common "ground plan," a common fundamental organization. At first glance, and perhaps also at second glance, pine trees, oaks, and onions may seem dramatically different, but actually they are quite similar, and on an anatomical basis they share the same types of tissues. It is estimated that there are about 600,000 species of seed plants, and on this basis we expect at least a little diversity in their structure. In this expectation we are not disappointed: each plant does differ from all others, sometimes subtly, often dramatically, but always for a reason, and the reason is usually logical and understandable.

The Structure of Higher Plants

The ground plan of higher vascular plants can be considered from two points of view, the external organization and the internal organization. If you consider a plant, you realize it is composed of just three basic organs: the shoot axis, the leaf, and the root (Fig. 1.2). The leaves are arranged on the shoot axis in a particular order, usually either a steep or gentle spiral; occasionally they seem to be aligned on the stem in straight lines. The growth in length of the shoot occurs at its extreme apex, which contains a shoot apical meristem. This meristem produces the cells of both the shoot axis and the new leaf primordia. Also, just above each leaf primordium, a group of cells is organized into an axillary meristem which develops into an axillary bud. The axillary buds may become dormant and then remain inactive indefinitely; they may become activated and produce a new leafy branch, a flower, or inflorescence; or they may produce a special shoot that looks very different from the main shoot: a spine, a tendril, or a gland.

Roots, like shoots, grow at their apices by an apical meristem. The root apical meristem differs from the shoot apical meristem because it produces no leaf primordia, but it does produce a protective layer of cells in front of itself (the root cap). Because roots have no leaves, they also have no axillary buds; branch roots arise deep within the root as opposed to superficially.

As for reproduction, major differences occur between the different divisions, and it is not possible to speak so generally. Ferns have the simplest reproductive tissues: they produce spores in relatively exposed sporangia located on the underside of their leaves. In gymnosperms, the sporangia are grouped tightly together and are protected, thus forming cones. In angiosperms, the sporogenous tissues are protected within stamens and carpels, and these in turn are usually surrounded by protective sepals and pollinator-attracting petals, the whole group of structures being called a flower. After pollination, seeds develop within cones or flowers; in angiosperms, the flowers themselves change greatly after pollination as the carpel matures into a fruit.

If we consider plants from an interior point of view, even more uniformity of ground plan is discovered (Figs. 1.3, 1.4, 1.5). Plants have only three fundamental types of cells: parenchyma, with just thin walls, collenchyma, with irregularly thickened walls, and sclerenchyma, with very thick, strong walls. The primary, herbaceous body is always covered by an epidermis, which often contains glands, hairs, and stomata. Interior to this is the cortex, which in many plants is a thin layer composed of green photosynthetic cells in the shoot and colorless starch storage cells in the root. But in plants with enlarged stems such as cacti and other succulents, it is the cortex that is increased in size. Often the outermost layers of cortex, those adjacent to the epidermis, have thickened walls and constitute a collenchymatous or sclerenchymatous hypodermis. The bulk of the cortex is made up of parenchyma, and it may contain many types of specialized cells that synthesize crystals, tannins, latex, mucilage, and other compounds.

Interior to the cortex is the stele, the set of vascular tissues. There are two types: the xylem, which conducts water upward from the root, and the phloem, which conducts minerals and organic compounds throughout the plant. Both of these are complex tissues; that is, each is composed not of just one type of cell but rather several types. The xylem contains both tracheids and vessel members for conduc-

Figure 1.2

(A) This diagram is of a small herbaceous dicot, *Bergia texana;* it has a root system and shoot system. The shoot contains leaves attached to the stem at nodes. Internodes and axillary buds are also visible.

(B) This is a monocot, the grass *Cyanodon dactylis.* Like a dicot it has roots and stems, but the root system is not unitary; instead new adventitious roots are produced at most nodes. The leaves do not have petioles but instead form a sheath around the stem.

tion, and also it has much parenchyma for storing sugar and starch in the times of dormancy. Phloem contains a combination of storage parenchyma and conducting parenchyma, the latter being either sieve tube members and companion cells in angiosperms or sieve cells and albuminous cells in nonangiosperms.

The vascular tissues are usually organized

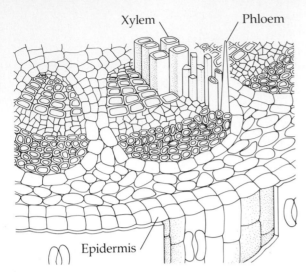

Figure 1.3
In most dicot stems, there is a ring of vascular bundles, each containing xylem and phloem. Inside the ring is pith; outside are cortex and an epidermis.

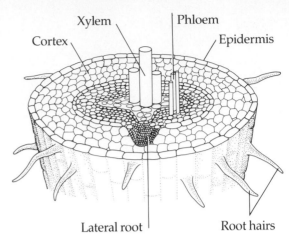

Figure 1.5
The arrangement of tissues within a root is somewhat similar to the arrangement in a stem: outermost is the epidermis, then cortex, phloem, and xylem, but frequently there is no pith.

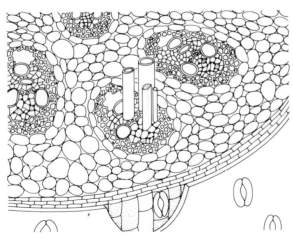

Figure 1.4
A monocot stem also contains vascular bundles but usually in a more complex pattern.

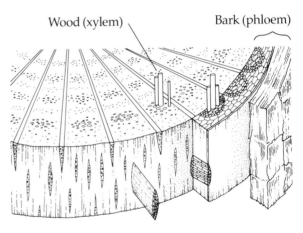

Figure 1.6
Some plants are able to become woody; that is, they undergo secondary growth, as shown in this cross section of a tree trunk. They produce both wood and bark; plants that do not are herbs.

into one or many bundles in both roots and shoots. If several bundles are present, then they are arranged into a ring which surrounds the pith in all seed plants except the monocots; in these latter plants the bundles are organized in complex three-dimensional patterns embedded in a matrix of pithlike tissues.

If the plant is an herb, no further growth occurs, and the plant is said to have only a primary plant body. But if it is a woody species, a layer of meristematic cells (the vascular cambium) arises between the xylem and phloem,

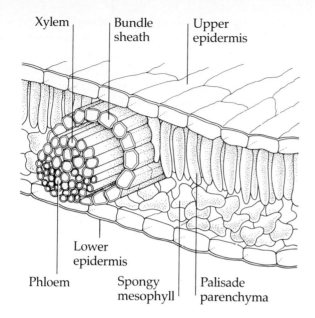

Xylem   Bundle sheath   Upper epidermis

Lower epidermis

Phloem   Spongy mesophyll   Palisade parenchyma

Figure 1.7
This diagram represents the arrangement of tissues that occurs in many leaves. The xylem distributes water to the tissues, the phloem absorbs the sugars produced by photosynthesis in the mesophyll, and the epidermis retains water in the leaf while permitting the absorption of carbon dioxide. Other types of tissue arrangements are theoretically possible, and some other types actually do occur, but this particular pattern is by far the most common.

and it produces secondary xylem (wood) to the interior and secondary phloem (bark) to the exterior (Fig. 1.6). In the outer regions of these stems and roots, other layers of cells become meristematic and function as cork cambia, producing layers of protective cork cells. These two new meristems and the tissues that result from them represent secondary growth; the wood and the bark constitute the secondary plant body.

The internal structure of leaves (Fig. 1.7) is rather similar to that of the primary stem axis: the surface is covered by epidermis, and a hypodermis may occur in many leaves. All of the leaf tissues interior to the epidermis are collectively called mesophyll, and this consists of photosynthetic parenchyma (usually but not always palisade parenchyma and spongy mesophyll) and vascular bundles, each of which, as in shoots and roots, contains xylem and phloem. In addition, in many leaves, the vascular bundles have large masses of sclerenchymatous fibers associated with them.

This very brief outline of plant structure is meant only as a review, not as an introduction to new material, and all of the terms and concepts are probably at least somewhat familiar to

you already. They, and many more, will be discussed in the remainder of this book. If you think about some of the plants that you come into contact with every day, either as houseplants, ornaments, or food, you will probably encounter at least several that do not seem to fit some of the details of the outline just presented. This is no problem, because, as mentioned before, we must expect and appreciate diversity. We must not view unusual structures as exceptions but rather as valuable, instructive, alternative solutions to the problems of function and survival.

As you read and think about the material presented in this book and as you study plants on your own, it is important always to keep two questions in mind. First, what are the **alternatives** to a particular structure? Do other structures exist that perform the same function? If so, why is one structure employed sometimes and the other structure at other times? If no alternative exists, why not? Is an alternative theoretically impossible? Second, what are the **consequences** of a particular structure? Each aspect of anatomy is not only the result of metabolism, but also the framework within which metabolism happens; thus, each structure and each variation may have significant ramifications.

For example, all of the higher plants are multicellular. An alternative exists: unicellularity. Why is it that none of the higher plants is unicellular? Because a consequence of multicellularity is that it allows a division of labor at a cellular level, some cells of the body can become adapted for photosynthesis, some for mineral and water absorption, some for reproduction. Because each of these processes must occur in the same cell of a unicellular organism, that cell cannot specialize for any one function. But each cell of a higher plant can specialize, and the plant as a whole is much more efficient and sophisticated. A further consequence of multicellularity is that each cell then becomes dependent on all other cells, and damage to one part of the plant may result in death of the whole plant, even of the cells not originally damaged. In unicellular organisms, a cell will die only if it is damaged directly. These two questions about alternatives and consequences are powerful analytical tools and, if applied to the topics that follow, will make plant anatomy a much more dynamic and interesting subject.

PART I

Subcellular Anatomy

It is important to begin an analysis of plant anatomy with a consideration of the subcellular organization, because the structure that is visible by light microscopy or even by the naked eye is actually the pattern formed by the functioning and the arrangement of the subcellular parts. The anatomy of a plant is the physical manifestation of that plant's metabolism, of the physiological interaction of the plant's DNA, its cytoplasm, and its environment. Although many aspects of a plant's metabolism must be considered as processes that occur at the tissue level (such as water movement through the xylem or water retention by the epidermis), many aspects occur as purely cellular and subcellular phenomena. The size and shape of cells are important, but without understanding cellular and subcellular aspects of wall deposition, softening, encrusting, and resorption, it would not be possible to understand either cell size or shape; it would be possible only to describe and catalog these two fundamental parameters. Similarly, a knowledge of subcellular anatomy and physiology is necessary for understanding many other aspects of anatomy and physiology at the tissue, organ, or whole-plant level.

An interesting aspect of the subcellular organization is that at this level it is really impossible to distinguish between anatomy and physiology, between structure and function: by analyzing the structure of a ribosome as it mediates protein synthesis, one automatically analyzes the biochemistry of the process, and vice versa. It is easy to see that the sequence of amino acids in a protein is the basis of both its physiology and its structure; in reality, anatomy and physiology are just the names we apply to describe the two methodological approaches to studying a process. As our knowledge and understanding of subcellular biology improves, we will be better able to understand cellular and tissue-level biology, and the distinctions between anatomy and physiology will continue to fade away.

The Cell

Concepts

Plants have multiple levels of organization. At the most fundamental level are the individual macromolecules such as proteins, nucleic acids, and carbohydrates; the synthesis of these is controlled in such a way that those macromolecules having the proper structures are produced at the correct time and place. These molecules in turn are assembled into the **organelles**, the microscopic components of **cells**. The cells of a plant do not occur at random but are carefully organized into **tissues**, and various types of tissues are grouped into **organs**. Finally, at the highest level, the organs together constitute the **plant**.

Cells are universally regarded as the primary units of life, because all organisms are composed of cells and because almost all cells are capable of living independently (either naturally as in many algae and protozoans, or artificially in laboratory culture), whereas their parts cannot. Furthermore, in many cases cells seem to be the primary units of differentiation and morphogenesis: an individual cell or layer of cells may have characteristics that make it quite distinct from its neighbors. In some of the simplest multicellular algae, this individuality predominates, and it is often more beneficial to consider the alga not as a single plant but rather as a colony of distinct individuals. In the vascular plants, although cells may differentiate individually as mentioned above, they are integrated with the surrounding cells to form unified, functioning tissues. The differentiation of a particular cell may be controlled by other cells and in turn may influence the development of still others; morphogenesis can then be seen to be acting at the tissue, organ, and whole-body levels as well as at the cellular level. Because one cell can differ from another only if one or more of its organelles or macromolecules differ from those of the other cells, morphogenesis acts at these levels, too.

Plant anatomy has been concerned historically with the association of cells into tissues and organs and has not been as concerned with the organelles, except the cell wall. This probably is because most organelles, although detectable with light microscopy, are not easily distinguishable and frequently require special preparation that may introduce artifacts. Further, because the functions of many organelles were not understood until recently, the labor spent in studying them did not yield much useful information. With the advent of the electron microscope, the study of cell and organelle structure was revolutionized, and the new discipline of **ultrastructure** was born. This field has tended to remain rather distinct from classical plant anatomy. Our knowledge of both anatomy and ultrastructure is now so extensive that there is no reason for this separation to continue. Anyone studying plant anatomy, even if the studies use solely light microscopy, must be aware of and concerned with the ultrastructural basis of the differences visible at the light-microscope level. Similarly,

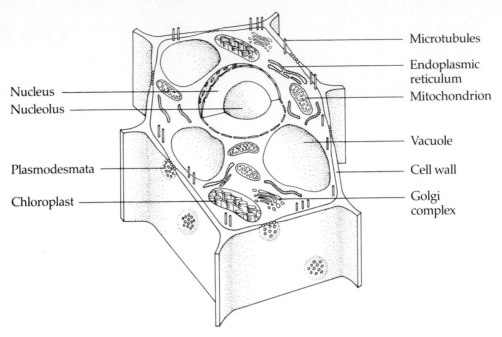

Nucleus

Nucleolus

Plasmodesmata

Chloroplast

Microtubules

Endoplasmic reticulum

Mitochondrion

Vacuole

Cell wall

Golgi complex

Figure 2.1
This is a diagrammatic representation of a plant cell and the organelles that might be found in it.

electron microscopists must thoroughly understand that there is no such thing as a "typical" plant cell or organelle. All are specialized, and the ultrastructure can have no meaning other than in the context of the tissue and the tissue-based metabolism in which it occurs.

The Cell

It is difficult to give a precise definition of just what a cell is. Similarly, it is impossible to describe an average plant cell—none exists. As with so many concepts, structures, and phenomena in biology, there is such diversity that a full understanding, achieved through the study of various examples, is much more important than a terse definition. Plant cells almost always have a **cell wall** (Figs. 2.1, 2.2); only some reproductive cells are known to lack them. The dead, water-conducting cells of xy-

lem consist only of cell walls; although it would be reasonable to argue that these are actually just the remnants of cells whose other parts have been digested away, they function only when dead and empty, and normally they are described as cells. Living cells always contain **protoplasm**, simply because it is defined as the contents of living cells, excluding the wall. The protoplasm of a single cell is a **protoplast**, and these are always bounded by a membrane called the **plasma membrane**, or **plasmalemma** (pl.: either plasmalemmae or plasmalemmas). Protoplasts typically contain one **nucleus** (pl.: nuclei), which contains the genetic material DNA and which controls the cell metabolism. Certain cells in the phloem have protoplasts lacking nuclei when mature, and other cells, especially in certain flower parts, may have several (then the cells are bi-, tri-, or multinucleate). Plant protoplasts apparently always have one or several water-filled **vacuoles**; in some cells,

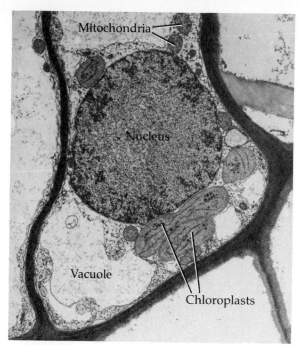

Mitochondria

Nucleus

Vacuole

Chloroplasts

Figure 2.2
This is a parenchyma cell of *Echinocereus engel-mannii;* the nucleus dominates this section of the photograph. Several small chloroplasts are visible, as are vacuoles and mitochondria. The cell walls are prominent, and plasmodesmata can be seen. × 1000.

vacuoles constitute almost the entire cell volume. **Mitochondria** (sing.: mitochondrion), responsible for aerobic respiration, are always present, as are **plastids**. The plastids constitute an extremely versatile, multifunctional type of organelle: in green tissues, plastids have developed as **chloroplasts**; in storage tissues, they might develop as **amyloplasts** (which store starch) or as **elaioplasts** (which store oil). In many petals and fruits, plastids accumulate pigments and become **chromoplasts**.

Protoplasts also universally contain **endoplasmic reticulum** (a system of tubules involved in intracellular transport), **ribosomes** (particles which mediate protein synthesis), **microtubules** and **microfilaments** (both skele-

tal elements, providing shape and strength to the protoplast), and usually **dictyosomes** (involved in packaging material for export from the cell). Plant cells are premier synthetic factories, and many of them produce quantities of specialized chemical compounds which are stored as droplets, crystals, or particles, all generally known as **ergastic substances**. All of the protoplasm, excluding the nucleus and vacuoles, is collectively called the **cytoplasm**; all of the cytoplasm, excluding the structured organelles mentioned above, is the **hyaloplasm** (that is, the hyaloplasm is the ground substance, the liquid between all of the other organelles).

Every cell of a particular plant is connected to its neighboring cells by fine strands of cytoplasm that run through minute holes in the walls; the cytoplasm and the hole together are termed a **plasmodesma** (pl.: plasmodesmata). Therefore, all of the protoplasm of any plant is in reality a continuous phase called the **symplasm**; all the protoplasm of one plant is the **symplast**. But plants are not made up solely of cells. Because most cells do not abut tightly against their neighbors, especially at their corners, **intercellular spaces** occur (Fig. 2.3). In many tissues they are small and isolated, but in leaves intercellular spaces can constitute over half of the tissue volume and are important in the circulation of carbon dioxide and oxygen within the leaf (Fig. 2.4). The intercellular spaces and the space occupied by the wall are called the **apoplast**. The symplast plus the apoplast constitute the entire plant.

Cells, like entire plants, have life cycles. In meristematic and expanding regions, cells are actively growing and dividing, while in other parts of the plant the cells are growing and differentiating or are full-sized and fully matured. In these latter regions, such as leaves, wood, or bark, the cells, being mature and not dividing, are obvious, discrete entities and may persist as such either living or dead for hundreds or thousands of years. Living parenchyma cells of *Kingia* may be 400 years old (Lamont, 1980);

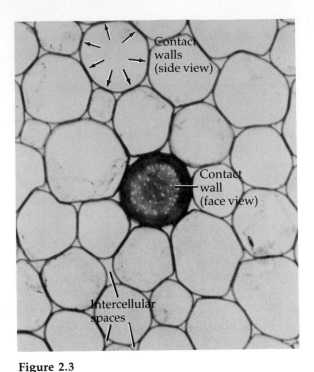

Figure 2.3
These are cells in the stem of *Ranunculus;* they appear empty because the vacuoles constitute almost all of the protoplasm. The stem was cut in such thin sections that both the front wall and the back wall were removed from every cell. The dark circular plate in the center of the photo is a wall between two cells that was not cut away. It appears to have holes in it, but those are thin areas (primary pit fields). × 100.

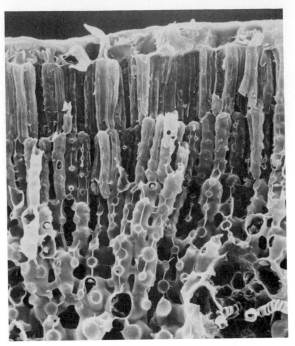

Figure 2.4
This scanning electron micrograph shows the interior of a leaf of *Eucryphia cordifolia.* Large intercellular spaces enhance the circulation of carbon dioxide to the chloroplasts that are located in the columnar cells. The small projections hold the cells apart, providing circulation space and a large surface area for absorption, while the round contact areas permit the cells to maintain cytoplasmic contact with each other. × 200. Micrograph generously provided by G. Montenegro.

dead but functional wood cells of bristle cone pines can be 7000 years old. In meristematic regions, cells can be a little more problematical, at least philosophically: after two sister cells are "born" by the division of their mother cell, the mother cell ceases to exist, and two new cells take its place—but all of the protoplasm and most of the wall are just the same as before, merely being packaged as two units instead of one. Furthermore, cell division may occur for only a brief time and be accompanied by little growth: a packet of four cells may be very obviously produced by the division of one cell, then the subsequent division of the two

daughters. These packets can then act as more of a unit than the individual cells within them. More dramatically, the large cells of the apical meristems of many ferns divide asymmetrically to produce one very large daughter and one small one. The large daughter cell is virtually identical to the mother cell; discussions of such an apical cell usually treat it as though it were a single, persistent cell rather than a series of new cells produced with each division.

To stretch our concept of the cell even further, size should be considered. Meristematic cells are often the smallest, frequently being isodiametric with dimensions of about 10 μm on each side (Mauseth, 1982b, 1984a). One μm (micrometer) = 1 millionth of a meter. Parenchyma cells are often larger, with dimensions of 20 μm $\times$ 20 μm $\times$ 100 μm being common. Very long cells (up to 11,000 μm) occur in the wood of pines and other conifers, and certain types of latex-producing cells become truly gigantic, stretching from roots into shoots and ramifying extensively, for example *Euphorbia supina* (Rosowski, 1968). Cells of this size (millions of cubic micrometers in volume) are multinucleate (Mahlberg and Subharwal, 1967) and are often termed **coenocytes**. A reasonably impressive example of this is a coconut: when full of "coconut milk," the entire cavity is one single cell, and the liquid is the protoplasm, which contains thousands of nuclei (Cutter et al., 1955).

Our concept of cells as structurally diverse entities has expanded with the advent of a new analytical method, **stereology** (Fagerberg, 1980; Steer, 1981; Toth, 1982; Weibel, 1979). This is a means for quantifying the volumes and surface areas of organelles within a cell (or of cells within a tissue, and so on). Stereological studies carried out to date indicate that cells that might at first appear very similar actually contain different quantities of a particular organelle (Table 6.1; Mauseth, 1982b) or that the organelles are distributed asymmetrically in the cell rather than uniformly (Moore, 1983).

This discussion of variations in cell structure is meant to be an introduction to cellular diversity and to prepare you for the more detailed treatments that follow. As you read and think about cellular diversity, it is important to bear in mind continually the functions of the structures being discussed. Cells are constructed remarkably logically, as are their individual modifications and specializations.

Membranes

The cells of both plants and animals are literally filled with membranes; the plasmalemma surrounds the entire protoplast, and most organelles are constructed of membranes (Figs.

2.1, 2.2). One of the most important properties of biological membranes is that they are **semipermeable**; that is, certain ions and molecules can pass readily through them while others cannot (Alberts et al., 1983). This property allows the membranes to act as barriers, so that the material exterior to a membranous vesicle or organelle cannot mix freely with the material inside of it. In addition, most membranes contain molecular pumps that are able actually to force certain metabolites across the membrane. This is important in all cells, but it is easily appreciated in the root's absorption of minerals and in the phloem's ability to move sugars to very specific locations. Another critical feature of most membranes is that they actually contain many important enzymes responsible for metabolism. This allows enzymes to be positioned precisely within the cell; the cell can then become highly ordered and differentiated. It also helps isolate certain products: reactants can come to the enzyme from one side of the membrane, and products are released to the other side. If this happens in a small vesicle, the vesicle fills with highly concentrated product which is isolated from the rest of the cell.

At present, the most widely accepted model of membranes is the **fluid mosaic model**, which postulates that membranes are composed of two layers of lipids. The individual lipid molecules, which are rather linear, associate weakly side by side, so individual molecules can move within the plane of their layer but cannot leave it. Embedded within this lipid bilayer are proteins of various types, some being enzymes, some being molecular pumps, some stabilizing the membrane and maintaining its structure. Some proteins are so large as not to fit within the lipid bilayer, so one or both ends project into the liquid of the cell. Some of the proteins may attract others by the positive and negative charges on their lateral surfaces, and they associate into clusters or large particles (Fig. 2.5). Because the bilayer acts as a two-dimensional solution, proteins can diffuse throughout the membrane, moving within its plane.

Figure 2.5
This cell (from the plasmalemma of *Pteris tremula*) was frozen in liquid nitrogen, then fractured; the break passes through the middle of membranes, revealing the protein particles embedded in them. The large ridges are areas where the membrane is pressed against cellulose fibrils in the wall below. × 10,000. Micrograph generously provided by A. Koller and E. Roberts.

Currently we believe that probably all cell membranes are constructed in this manner, and with electron microscopy they all look very similar (membranes are not visible by light microscopy). However, we know that the membranes of each organelle are distinct in their protein content: mitochondrial membranes contain enzymes for aerobic respiration, but the membranes of other organelles do not. The lipid compositions of each are also different, and because of this the organelles remain as individual structures; they do not readily fuse with each other (Thorpe, 1984; Tolbert, 1980a).

The Nucleus

The nucleus is a fundamental organelle, being the primary repository of the genetic information necessary to maintain and control the cell structure and metabolism and, by extension, to control the structure and metabolism of the entire plant (Figs. 2.2, 2.6). Chemically the nucleus is composed of **deoxyribonucleic acid (DNA)**, **ribonucleic acid (RNA)**, proteins, and water. The DNA is complexed with one class of proteins, the **histones**, to establish a stable, functional structure for the DNA, allowing it both to replicate itself prior to nuclear division and to direct the synthesis of RNA that is essential to protein synthesis (Cook and Laskey, 1984). Because DNA is accurately replicated, each daughter cell at division receives a daughter nucleus that is identical to the original mother nucleus. Therefore, all cells of any plant have nuclei that contain the same genetic information (aside from rare errors in replication); thus they all have the same genes. The various types of cells are able to develop their own distinctive structures and metabolisms because each cell controls which genes are expressed at any time and which are not (Alberts et al., 1983). Regions of DNA that contain quiescent genes, genes not actively directing the synthesis of RNA, complex with histones in a way that causes them to stain intensely with the histological stains safranin and lead citrate, making them appear very dark in either light or electron micrographs. These regions are called **heterochromatin** (Fig. 2.6). Regions of DNA with active genes (**euchromatin**) complex with histones such that a light stain results; this difference is readily visible in either light or electron microscopy.

The nucleus is the primary repository of information in the cell, but it is not the only archive. Plastids and mitochondria also have their own sets of genes (Hoober, 1984; Schiff, 1980), which occur as small circles of naked DNA (DNA not complexed with histones). But even more than this, it must be emphasized

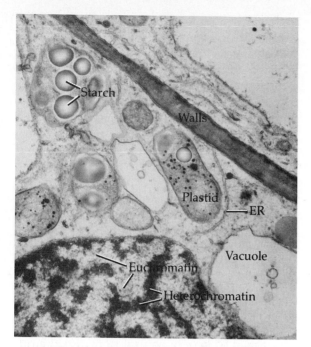

Figure 2.6
It is extremely common for the nucleus to be located near a wall like this, although they can sometimes be found suspended in the center of a cell. This nucleus contains patches of euchromatin (light color) and heterochromatin (dark color). The two membranes of the nuclear envelope are also visible, but the magnification is too low to show nuclear pores. Four plastids are present, and they are so undeveloped that they would be considered proplastids; three contain starch deposits. Mitochondria and endoplasmic reticulum are also visible. The walls between the two cells are extremely thin. × 750.

that the very components of an existing cell or organelle are themselves vital sources of information because they act as templates or frameworks for the synthesis of more of the same component. As a protein is synthesized, it can fold into the proper shape by itself. If it is to be part of a multienzyme complex, it can also associate with the other members of the complex automatically, guided by electrostatic charges

on its surface or by hydrophobic bonding; this is **self-assembly**. But for structures larger than this, pure self-assembly appears to be inadequate: there is no evidence that a fully competent nucleus, if placed in a small amount of hyaloplasm, would be capable of directing the synthesis of a totally new plasmalemma. It can definitely direct the insertion of new material into a pre-existing plasmalemma, but to build one without a starting template is probably impossible.

Nuclei were discovered by Robert Brown in 1831. As they have been studied with various techniques since then, various terms have been employed to describe them. Heterochromatin and euchromatin have already been mentioned. The liquid phase surrounding these two types of chromatin is variously called the **nucleoplasm** (most widely used), **karyolymph**, **karyoplasm**, or **nuclear sap**. During preparation for nuclear division (**karyokinesis**), the DNA-histone complex begins to coil in a complex manner, such that the chromatin seems to "condense" into visible **chromosomes** (Alberts et al., 1983; Cook and Laskey, 1984). By the middle of division, the chromosomes have condensed so much that they are fairly easy to study by light microscopy, and individual types of chromosomes can be identified by their length, shape, and size. At this point, two especially important observations can be made: (1) each chromosome has two identical parts (the **chromatids**) which indicates that indeed the DNA had duplicated prior to karyokinesis, and (2) each chromosome has an identical mate (a **homolog**) in the same nucleus; that is, the chromosomes occur in **homologous pairs**. The cell is **diploid** because it actually has two complete, homologous sets of chromosomes. With the completion of karyokinesis, each chromosome breaks into its two component chromatids, and the chromatids of each chromosome go to opposite sides of the cell and then form two new daughter nuclei. Each chromosome now contains just one chromatid, but each daughter nucleus has two of each type of chromosome (there are still homologous pairs), so each gene is present as two copies in each nucleus.

Normally two copies of each gene are sufficient to produce adequate amounts of RNA for directing protein synthesis. But in plant cells that are very active metabolically, having only two copies appears to be insufficient. Cells are able to overcome this in several ways; the simplest is to have the nucleus replicate its DNA again, such that there are now four copies of each gene. This can be repeated indefinitely until the nucleus becomes enormous and there are hundreds of copies; such a nucleus is said to be **polyploid** rather than diploid. This process is rather rare in animals, but Jordan et al. (1980) state that it is the rule rather than the exception for plant cells, and possibly up to 80% of the cells of a plant are polyploid rather than diploid. Once a nucleus begins to replicate its DNA and becomes polyploid, the process can continue until it contains 8192 copies of the DNA, as in *Phaseolus*, or even 24,576 copies, as in *Arum maculatum* (Barlow, 1985; Jordan et al., 1980).

In a second method, the DNA is replicated, then the nucleus divides but the cell does not, so both daughter nuclei remain in one cell, again with the result that four copies of each gene are now present. Such binucleate cells are not really uncommon, but for the process to repeat itself, producing tetranucleate or multinucleate cells is fairly rare, happening almost exclusively in some secretory cells (Mahlberg and Sabharwal, 1967; see Chapter 9), certain xylem cells (List, 1963), and in some types of endosperm (Cutter et al., 1955; see Chapter 20). The opposite extreme, the loss of the nucleus from a living cell, occurs only in the conducting elements of the phloem (Evert, 1977); in this instance, the protoplast is maintained in healthy condition by means of metabolic control provided by the nucleus of an adjacent cell. We still do not know why nuclear degeneration should be necessary for the proper differentiation or functioning of these cells.

Structurally, the nucleus is composed of the chromatin and nucleoplasm and is bounded by a **nuclear envelope** (Fig. 2.6) which is not visible by light microscopy. Electron microscopy, however, shows that the nuclear envelope is actually a set of two membranes, the **inner envelope** and the **outer envelope**, the two being separated by the **perinuclear space**. This perinuclear space is usually narrow (Roberts and Northcote, 1971), being only about 20 nm wide. One nm (nanometer) = 1/1000 micrometer. In some species, however (*Isoetes*, *Ochromonas*), the space can accumulate and store material (Kruatrachue and Evert, 1978; Severs, 1976), becoming quite large. This space can also swell during the formation of sperm cells (Bell, 1978) and during the nuclear fusion of sexual reproduction. Because nuclear genes control the synthesis of messenger RNA (mRNA) necessary for the protein synthesis that occurs in the cytoplasm, it is only natural to expect the nuclear envelope to have passageways to facilitate the transport of mRNA out to the cytoplasm. These are the **nuclear pores**, which are 70 nm in diameter and which cover the nuclear surface in densities of between 6 to 25 pores per square micrometer (Fabbri and Bonzi, 1975; Gunning and Steer, 1975), the density of pores being correlated with nuclear activity (Jordan et al., 1980). The pores may be distributed at random, but frequently they have special patterns: in spores of *Equisetum* the pores form a girdle around the nucleus. In other cases, they occur in rows, spirals, or clusters.

Nuclei are variable in volume, not only from species to species, but also changing as the DNA is replicated and new histones are synthesized during the cell cycle, or as the nucleus becomes polyploid. Also, extremely active nuclei, such as those of apical meristems or certain glands, are much larger than the less active nuclei of mature tissues within the same plant, probably because they contain more RNA and synthetic enzymes. In meristematic cells, nuclei are the dominant organelle, constituting as much as 36% of the protoplast volume in *Echinocereus engelmannii* (Table 6.1), but as cells expand and mature, nuclei do not grow an equivalent amount, so they constitute a smaller fraction of cell volume (0.23%; Mauseth, 1981b).

Nuclear shape is also variable; many are spherical, but highly lobed nuclei are not uncommon (Gunning and Steer, 1975; Yeoman et al., 1970). A nucleus with a lobed surface has a greater surface-to-volume ratio than does a round nucleus with an equal volume; this permits a more rapid transfer of mRNA. In mature cells the nucleus is often flattened against the side of the protoplast, pressed against the wall. Many fiber cells and other elongate cells have such a narrow diameter that a spherical nucleus is impossible; it must instead be a slender, tapered spindle in order to fit inside the cell.

Although nuclei can be found in any part of a cell, it is probably safe to suggest that the position of the nucleus is rarely if ever random. In highly asymmetric cells such as those of the epidermis, the nuclei are usually precisely positioned near the inner wall in some species, as in *Tristerix aphyllus* (Mauseth et al., 1985), near the outer wall in others (Fig. 10.10; *Phthirusa, Aloe verrucosa*). There are numerous examples of complicated but orderly movements of nuclei during differentiation: many cells undergo an asymmetric division to produce a large cell and a small cell (during the formation of root hairs, guard cells, companion cells), the division being preceded by the movement of the nucleus to the proper site. It has been found by careful fixation and analysis that the nuclei and other organelles of certain root cap cells are also precisely oriented (Moore, 1983; Moore and Pasieniuk, 1984).

Within the nucleus are one or more small bodies, the **nucleoli** (sing.: nucleolus). These are areas rich in the precursors for ribosomal RNA (rRNA), the structural RNA that is part of the ribosomes. Extensive light microscopy studies have shown nucleoli to be dynamic, capable of swelling and shrinking rapidly, capable of changing shape and of forming, then losing "vacuoles" (Barlow, 1970; Johnson and Jones, 1967). However, electron micrographs show no nucleolar membranes at all, and virtually nothing that is an obvious structural basis for such active shape changes. Because the nucleolus is the source of precursors for ribosomes, its size and activity are well correlated with the cell's level of protein synthesis.

The Vacuole

Vacuoles are almost always the most voluminous organelles of mature cells, frequently constituting more than 90% of the total protoplast volume and leaving the rest of the protoplasm as just a thin layer pressed against the wall (Figs. 2.2, 2.3, 2.10). The structure is simple: just a vacuolar membrane, the **tonoplast**, and the liquid that it encloses, the **vacuolar sap**. In meristematic cells and other small cells, rather than one large vacuole, there are many small ones, and the term **vacuome** is used to describe collectively all of these vacuoles of one cell. As cells grow, the numerous small vacuoles fuse, forming the one single large vacuole characteristic of mature cells. Quiescent vascular cambium cells, dormant during the winter, are some of the few large cells that have a vacuome of many small vacuoles instead of a large solitary one, as *Pinus strobus* (Srivastava and O'Brien, 1966). The origin of vacuoles is still not established with certainty, but much evidence suggests that vesicles derived from the endoplasmic reticulum may swell to become small vacuoles (Marty et al., 1980). Cells of the tapetum in anthers are the only cells reported to completely lack vacuoles (Gunning and Steer, 1975).

Previously, vacuoles had been considered to be rather simple, static structures with relatively little metabolism or function, but this idea is being changed rapidly. If the vacuole absorbs water and presses outward, but the wall is strong, then pressure (**turgor pressure**) builds up inside the cells. This is just as useful mechanically as the pressure used to inflate tires or the hulls of submarines. Although fibers are frequently given great credit as mechanical components of plants, think about what happens to leaves or whole plants when water is insufficient: they wilt. Normal shape is restored with adequate watering, demonstrating that the turgor pressure of the vacuole is the primary support of living organs.

Vacuoles are critically important in cell growth: protoplasts can secrete protons (H^+) that weaken the wall, and simultaneously the vacuole is able to absorb water rapidly, swelling and thereby pressing the tonoplast and cytoplasm against the wall, exerting sufficient turgor pressure on it to stretch it to a larger size. Using this mechanism, plants achieve an "inexpensive," rapid growth, involving little more than water absorption, whereas animal growth, which does not involve vacuoles, always requires massive amounts of synthesis of proteins, lipids, and the whole fabric of protoplasm.

Plants undergo many rapid movements: the opening and closing of guard cells, leaf movements in sensitive plant (*Mimosa pudica*) and Venus flytrap (*Dionea*), and flower opening. All are based on the tonoplast's ability to absorb or lose water very quickly.

Vacuoles also act as storage reservoirs (Grob and Matile, 1979). It had been thought that the main storage was one of waste products, because plants have nothing equivalent to a kidney system, and most wastes must be sequestered inside the vacuole. An interesting newer discovery is that vacuoles may act as active, temporary, controllable repositories for useful materials, especially calcium, absorbing it at certain times to maintain low cytoplasmic levels, then releasing it at other times to raise cytoplasmic concentrations and thereby activate certain enzymes that are sensitive to calcium (Marty et al., 1980).

Although vacuoles frequently appear empty when viewed by either light or electron microscopy, it is not unusual to observe crystals, fibrils, particles, or droplets. It may be difficult to identify their chemical nature, but their presence is often used as an important character for taxonomic comparisons. This identification must be done carefully and should involve living material, because the vacuolar contents can precipitate as the tissues are dehydrated in preparation for microtomy, and crystals may occur in the dead preparation even though they were not present in the living tissues. Alternatively, the acids present in many fixatives can dissolve certain types of crystals.

Vacuoles also act as **lysosomes** (digestive organelles), actually capable of breaking down and recycling the components of old, unneeded organelles (Marty 1978; Marty et al., 1980; Prat et al., 1977). Small vacuoles fuse with old organelles and, by means of their lytic enzymes, digest them (Matile and Moor, 1968).

There is also now great appreciation for the dynamic nature of the shape of vacuoles in mature cells. Although at maturity a cell's vacuole usually remains stable, if the plant is wounded, cells adjacent to the wound become mitotically active. This activity is accompanied by changes in the tonoplast that result in the vacuole's being subdivided and reduced in volume; in *Sedum*, from 86% of cell volume to 41% (Moore, 1982). Wounding also causes the formation of a cytoplasmic bridge across the vacuole, through which cell division will occur (Barckhausen, 1978). Likewise, vascular cambium cells are able to change their vacuome from one large vacuole in the spring growing season to many small vacuoles in the winter dormant season, then change it back again (Robards and Kidwai, 1969).

The Plastids

Plastids are extremely dynamic organelles, capable of dividing, growing, and differentiating into a variety of forms, each with a specialized metabolism and structure (Kirk and Tilney-Bassett, 1978; Possingham, 1980). They are perhaps most familiar as the green **chloroplasts** of photosynthetic tissues (Fig. 2.7), the starch-containing **amyloplasts** of storage tissues (Figs. 2.8, 2.9), and the richly colored **chromoplasts** of many flower petals and fruits. These are all considered to be variations of the same organelle because they are interconvertible and because they develop from the same precursor (**proplastids**, Fig. 2.6; Gunning and Steer, 1975). Plastids are now widely regarded as having evolved from highly adapted **endosymbionts** (Gray and Doolittle, 1982). They probably originated as photosynthetic procaryotes, perhaps similar to the blue-green alga *Prochloron*, which were engulfed but not digested by a primitive eucaryotic cell (Olson, 1981). The relationship between these two cells was presumably mutually beneficial, and through millions of years of evolution the symbiosis has been perfected. There is abundant evidence consistent with this theory (Hoober, 1984): the plastids contain their own DNA (**ctDNA**) which, like procaryotic DNA, is circular, is not complexed with histones, and is about 45 μm long (enough to code for about 150 average-size proteins). Plastids contain their own ribosomes, and these too are similar to those of procaryotes in their size, structure, and sensitivity to antibiotics. For really successful symbiosis, the metabolism of the two organisms must become highly integrated and coordinated, such that the endosymbiont (the plastid) becomes controlled by the nuclear genes of the "host" (the plant cell). Thus many of the proteins in the plastids are now coded by nuclear genes, not by plastid genes, and many of them are produced on cytoplasmic ribosomes (Hoober, 1984).

The fact that plastids have their own DNA means that they can arise only from pre-existing plastids, not from vesicles budded off of other organelles, as had been proposed several times (reviewed by Possingham, 1980). This further implies that some tissues which have no obvious need for plastids might contain them primarily to pass them on to their daugh-

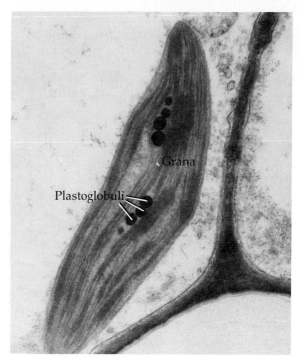

Figure 2.7
This is a large, well-developed chloroplast; it has abundant internal membranes, and grana stacks are visible. The dark spheres are osmiophilic bodies (plastoglobuli). $\times$ 1000.

ter cells, which will need them. It has been shown (Mauseth, 1982b) that shoot apical meristem cells of *Trichocereus pachanoi* dedicate 4 to 6% of their cytoplasm to plastids, which is sufficient to ensure a large number (about 35) per cell. This is enough that each daughter will receive at least one during mitosis, even though there is no special mechanism to guarantee their distribution. In certain species, some cells do contain so few plastids that it is not uncommon for one daughter cell to receive none, so it and the cells derived from it can never produce chloroplasts. When this occurs in developing leaves, the leaves have patches lacking chloroplasts, thus appearing white, pink, or purple, and the leaf is said to be **variegated**.

Certain types of plastids (chloroplasts, amyloplasts, chromoplasts) are easily visible with even a primitive light microscope, whereas other types (proplastids, **etioplasts**—plastids that have developed in the dark and that lack chlorophyll) are so small and simple that they are difficult to distinguish. This, coupled with the fact that all types are interconvertible, led to some confusion, both in understanding and in terminology. With the aid of electron microscopy, our understanding has increased greatly, but there is still some confusion of terminology. Proplastids are the small precursor organelles found in meristematic regions and in some internal tissues. Chloroplasts are the large, highly structured, chlorophyll-containing, photosynthetic plastids. Chromoplasts, containing large amounts of carotenoid and xanthophyll pigments, are the nonphotosynthetic plastids. If plastids store large amounts of starch on a relatively permanent basis (more than a few hours), they are amyloplasts; if they store oils, they are elaioplasts. The term **leucoplast** is not specific but rather has been used to refer to colorless plastids in general.

Proplastids

Proplastids, as just mentioned, are the precursor structures and therefore are small (1 to 2 μm in diameter), spherical in shape or somewhat cylindrical (Fig. 2.6; Hoober, 1984). As is true of all stages, they have ctDNA, ribosomes, and a double membrane system. The outer membrane is smooth, and the inner membrane is also simple, with only occasional folds projecting into the inner liquid (the **stroma**). Proplastids are capable of division by a method that resembles bacterial binary fission, but little is known about the process (Possingham, 1980; Possingham and Lawrence, 1983; Possingham and Saurer, 1969; Whatley, 1980).

Chloroplasts

Chloroplasts are much more complex than proplastids (Fig. 2.7; Hoober, 1984). They also have ctDNA, ribosomes, and the two membrane systems, but the inner membrane projects into the stroma as an extensive, interconnected system of flattened, broad vesicles called **thylakoids**. These are arranged as parallel sheets that in certain areas touch each other, forming discrete stacks. The stacks are **grana** (sing.: granum), and the thylakoids that run between grana are **frets**. This arrangement seems to be universal in the land plants, except that in the bundle sheath cells of C_4 plants, the grana are infrequent and small and have occasionally been overlooked, leading to reports of **agranal** chloroplasts in some of these species (Chapter 12; Jensen, 1980).

The rate of chloroplast photosynthesis usually exceeds the rate of transport of carbohydrates out of the chloroplast, so much of the newly synthesized material is stored temporarily in the stroma as small grains of **assimilatory starch** (Fig. 2.6). At night, this material is depolymerized and transported out. Some of the pigments involved in electron transport, especially plastoquinone and tocophorylquinone, can accumulate as droplets; these are called either **plastoglobuli** (sing.: plastoglobulus) or **osmiophilic bodies** because they absorb the stain osmium tetroxide (Fig. 2.7).

The size and number of the chloroplasts, as you might expect, are extremely variable, but they tend to be oval or lenticular, approximately 5 to 10 μm long, and there are from 5 to 500 per cell (Gunning and Steer, 1975; Hoober, 1984).

Amyloplasts

Amyloplasts are plastids that are specialized for long-term starch storage (Figs. 2.8, 2.9). Whereas the assimilatory starch that accumulates temporarily in chloroplasts is the result of photosynthesis in situ, the **storage starch** of amyloplasts is the result of transport into the organelle. Each amyloplast may contain one to several starch grains, each constructed by the layered deposition of starch around a nucleation center called the **hilum** (pl.: hila; Sterling,

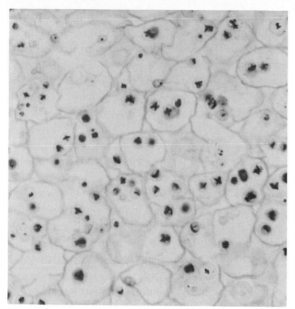

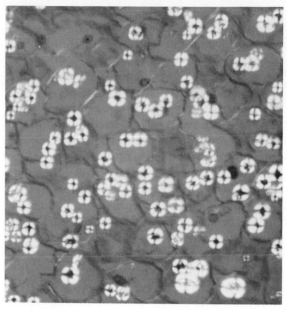

Figure 2.8
These cells are filled with starch granules. Using only light miscroscopy such as this, it is not obvious that these are really a type of plastid (amyloplasts). Several cells appear to have very few amyloplasts; it is possible that they really had fewer than the others, but it is more likely that some fell out when the material was cut and stained. × 100.

Figure 2.9
One of the easiest tests to be certain that particles such as those in Figure 2.8 are really starch is to view them with polarized light. Starch will always appear bright with a dark cross. Notice that these are the same cells and amyloplasts as in Figure 2.8. × 100.

1968). In some plants, experiments suggest that the layering of the grain is related to the daily cycle of photosynthesis (Buttrose, 1962), but in others, this relationship does not seem to exist (Innamorati, 1966). The deposition of starch around the hilum is frequently asymmetric; also, two grains can grow together, forming a compound grain. Surprisingly, starch grains are not as simple as one might expect— instead they have specific sizes and shapes, such that they are frequently an extremely valuable means of identifying plants (Esau, 1965a). Ease of identification is especially important when the plants have been ground or milled into flour or animal feed. Starch-grain identification may be the primary means for verifying that a purchaser has received (and is

eating) the type of grain that he or she has paid for. In some groups of plants, the starch grains are complex enough that they have been used as a means of studying the evolutionary relationships among various species (Mahlberg, 1975, 1982).

One type of amyloplast deserves special mention: the **statoliths**, which are large, dense amyloplasts found in the root cap and in the nodes of many shoots (Fig. 6.34). They are always found settled to the lowest side of a cell and are probably involved in the perception of gravity (Juniper, 1976), either by interacting with the plasmalemma on the lower side of the cell or by causing the other organelles to be excluded from this region and concentrated on the upper side of the cell (Hensel and Sievers,

1980; Juniper and French, 1973; Moore and Pasieniuk, 1984).

Elaioplasts

Elaioplasts are plastids that store large amounts of oil. Care must be used with this term; many plastids, especially chloroplasts and chromoplasts, normally always contain at least a small amount of oily, osmiophilic material (Fig. 2.7). Also, oils can be stored by other means (single-membrane bounded spherosomes) that can greatly resemble elaioplasts, especially with light microscopy (Fig. 2.11). Unless special care has been taken to ascertain whether a structure is truly an elaioplast or a spherosome, caution should be used.

Chromoplasts

Nonangiosperms are a remarkably green group of plants. Of all the evolutionary advances that occurred as angiosperms came into being, the most beautiful was the ability of plastids to greatly increase their normal production of carotenoids, accumulating large droplets of these bright yellow, orange, and red pigments, thereby becoming chromoplasts (Burgess, 1985). Chromoplasts, being plastids, have ctDNA, ribosomes, and the double-membrane system. The inner membrane may fold into the stroma to produce large undulating sheets that do not associate into grana (Gunning and Steer, 1975; Laval-Martin, 1974). The pigments are diverse in their composition and therefore also diverse in color and in the means of deposition; over 30 different types of pigments are found in the chromoplasts of *Capsicum* (Burgess, 1985). Droplets are frequent (*Ranunculus, Tulipa*). In carrot (Ben-Shaul, Treffry, and Klein, 1968) and tomato (Ben-Shaul and Naftali, 1969) the carotenoids are present as large crystals that are initiated in special thylakoids, then expand, deforming the surrounding membranes. The pigments can be associated with filaments in fruits of *Capsicum annuum* (Spurr and Har-

ris, 1968) and associated with bundles of narrow tubules in *Cucumis sativus* (Smith and Butler, 1971). Although all types of plastids are interconvertible, chloroplasts and chromoplasts are especially so, the chloroplasts routinely changing to chromoplasts as fruits ripen.

Etioplasts

If plant parts that typically occur in lighted conditions (stems and leaves) are grown in the dark, they become deformed (**etiolated**) and contain etioplasts rather than chloroplasts. Light is necessary for the conversion of the precursor protochlorophyll (which is colorless) into chlorophyll and for the proper formation and stacking of thylakoid membranes (Hoober, 1984; Thorpe, 1984). In this etiolated condition, the etioplasts develop a **prolamellar body**, a paracrystalline association of vesicles derived from the inner membrane. When an etiolated plant is placed in light, the prolamellar body is broken down and produces the thylakoid system of a normal chloroplast.

The cell wall and the organelles just described are large, prominent structures that can be fairly easily and reliably studied by light microscopy and that have played a major role in plant anatomy. The following organelles, except for the wall, are difficult to study in plants using only light microscopy, and consequently most of our knowledge of them has been derived from electron microscopy. There are fewer observations, but because it is clear that they are also important in the differentiation and morphogenesis of plants and their cells, much attention is being given to them at present.

Mitochondria

The mitochondria are responsible for aerobic respiration (Douce, 1985; Thorpe, 1984) and are universally present in plant cells (Fig. 2.2). Like plastids, they are now believed to be another class of endosymbiont and have their own circular, naked DNA (**mtDNA**), 70s ribosomes, and a double-membrane system (Hanson and Day, 1980; Schiff, 1980). The inner membrane is always highly folded, projecting into the stroma as sheetlike or tubular **cristae** (sing.: crista; Öpik, 1974). Unlike the plastids, which can alter their metabolism and internal structure, the mitochondria are quite constant internally. In material cut into thin sections, mitochondria appear as round or oval organelles, occasionally being long and tubular. By using special techniques for observing whole, unsectioned cells or by reconstructing a cell from serial sections, it is frequently found that mitochondria may be long, branched cylinders (Atkinson et al., 1974). Their shape is dynamic, and they can fuse with each other or break into smaller mitochondria (Duckett and Toth, 1977). In several plants (*Micromonas* and *Chlorella*, both algae) it is known that all the mitochondria of a cell fuse into a single giant netlike mitochondrion just before cell division; it then breaks up into individual mitochondria afterward. Because of their ability to fuse and divide, the number of mitochondria per cell, usually given as between 200 to 3,000 (Clowes and Juniper, 1968), may not be especially instructive. Stereological analyses have found that they constitute from 1 to 8% of the protoplasm volume (Berlin et al., 1982; Mauseth, 1981b, 1984a; Moore, 1983; Moore and McClelen, 1983a).

Just as with plastids, because mitochondria contain their own DNA, they can arise only from pre-existing mitochondria. Whereas a cell can survive without plastids, mitochondria must be present in every cell.

Endoplasmic Reticulum

As its name implies, the endoplasmic reticulum (**ER**; pl.: endoplasmic reticula) is a network (reticulum) located in the interior of the cell, the endoplasm (Fig. 2.6; Chrispeels, 1980). This is one of the most dynamic and variable of the organelles: it may be a network of tubules, cisternae (flat vesicles), or fenestrated sheets, or any combination of these (Gunning and Steer, 1975). It is connected to and continuous with the outer nuclear envelope, and it may form a cup around plastids and mitochondria (Steer, 1974).

The structural diversity of the ER reflects its functional diversity. One of its primary roles is in protein synthesis and transport. Many ribosomes, while synthesizing proteins, become attached to the ER, giving it a rough appearance and a new name: the **rough ER** or **RER** (**granular ER** is occasionally used). The proteins produced in this manner may pass through the membrane of the RER and accumulate in its lumen. The network of the ER can then channel this protein to specific parts of the cell, or the ER can bleb off vesicles that migrate through the cell, carrying the protein inside themselves.

Some regions of the ER do not associate with ribosomes and are known as **smooth ER** or **SER** (Chrispeels, 1980; Thorpe, 1984). These regions differ significantly from the RER and are involved in synthesis of lipids and fats. As we all know from personal experience, we animals are very proficient at synthesizing and accumulating fats; plants, however, make only small amounts of lipid, and SER tends to be rather rare, the exception being in glands, as in *Arctium lappa* (Schnepf, 1969a) and *Lonicera* (Fahn and Rachmilevitz, 1970); pollen (Cresti et al., 1985); and seeds such as *Sinapis alba* (Bergfeld et al., 1978) that are oily. Measurements of ER in plant cells are rare (Briarty, 1980; Briarty et al., 1979; Buckhout et al., 1981), but they show that the ER increases greatly as seeds store material during development and again as they break dormancy and germinate.

Because the ER can be involved in both lipid and protein synthesis, it is ideally suited for the assembly of membrane, and it does appear to have a central role in membrane flow. As just mentioned, the ER can form vesicles that move through the cell. If these fuse with the membranes of another organelle, the vesicle membrane becomes part of that organelle's membrane. Obviously, if the ER is to be an efficient routing mechanism, its vesicles cannot fuse indiscriminately with all other organelles. The ER is somehow able to produce vesicles of various classes, each with a particular type of contents and membrane. The basis of this is unknown.

The ER is involved in establishing and maintaining protoplast heterogeneity in another way. Although the ER often appears as a random system of tubes, cisternae, and sheets, if it is mapped in three dimensions, it is shown to be highly ordered and exquisite in its distribution and association with other organelles (Chrispeels, 1980). It often has a special association with the wall and is capable of either directing materials to it (Northcote, 1969) at specific locations or doing just the opposite, covering it and preventing the deposition of material into the wall—pollen (Heslop-Harrison, 1971b) and possibly xylem (Pickett-Heaps, 1968). During cell division the ER appears to contribute to a framework in which the cell plate (the new, growing cell wall) is constructed (Hepler, 1982).

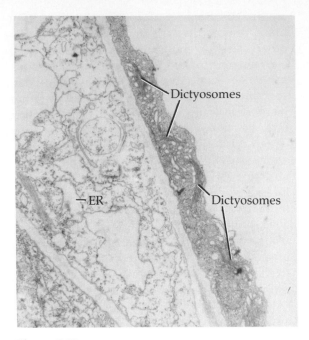

Figure 2.10
These three cells were beginning to differentiate as fiber cells when they were fixed; the predominant activities would have been elongation of the cell and the deposition of extremely thick walls. In one cell, four prominent dictyosomes are visible; the vesicles that they were producing were carrying material for wall construction. In the middle cell, there are numerous irregular vesicles interconnected by short tubes; these are all part of the endoplasmic reticulum (ER). × 1000.

Dictyosomes

Dictyosomes are small stacks of flattened vesicles whose margins are not smooth, but rather are broken up into a lacework of interconnected tubules (Figs. 2.1, 2.10; Mollenhauer and Morré, 1980). Dictyosomes function in the processing of material as it is moved through or out of the cell. Some of the vesicles that are derived from the ER accumulate on one side of the dictyosome (the **forming face**), then fuse together to form a new dictyosome vesicle. As the material in this vesicle is being processed,

typically by having sugars added onto the proteins (Alberts et al., 1983; Thorpe, 1984), another vesicle will form to the side of it. Simultaneously, an enlarged, mature vesicle will be released from the opposite side (the **maturing face**) and will move away, carrying its load of processed material. A vesicle is in the dictyosome only temporarily as it moves from forming face to maturing face, and it is somewhat difficult to think of the dictyosome as being a separate entity; it is more like a flow pattern of vesicles moving from ER to other organelles (Robinson and Kristen, 1982).

Alternately, the material that enters a dictyosome may be moved to the interconnected tubules on the periphery. These swell into large vesicles that then detach and move away.

The primary function of the dictyosome-processed vesicle is the secretion of carbohydrates—nectar, cell wall material, mucilages (Figs. 9.10, 13.5)—or sugar-protein complexes—glycoproteins of certain mucilages and some walls (Trachtenberg and Mayer, 1981; Mollenhauer and Morré, 1980). Plants secrete much less than animals do; consequently the dictyosomes are usually fewer and smaller in plant cells, frequently appearing inactive (that is, with just small vesicles, not large ones). They are functional most obviously in gland cells (Fahn, 1979), mucilage cells (Mauseth, 1980a; Trachtenberg and Fahn, 1981) and wall material in fibers (Fig. 2.10; Mauseth, 1977).

In plants the individual dictyosomes of a cell usually do not associate or interact with each other and are at most only sparsely interconnected, whereas in many animal cells they are closely aggregated and interconnected, forming a large complex called the **golgi apparatus** or **golgi body**. An exceptional aggregation occurs during pollen tube growth, and as many as 25,000 dictyosomes can be associated in a Golgi apparatus (Rosen, 1968). It is very common to see an individual dictyosome referred to as a golgi body, so care must be used when this term is seen in plant literature (Mollenhauer and Morré, 1966, 1980).

Microbodies

With the aid of the electron microscope, small, rather nondescript bodies were discovered and given the noncommittal name of **microbodies**, because their function was unknown (Beevers, 1979; Tolbert, 1980b). Now, on the basis of histochemical tests we know that there are at least two classes of microbodies: (1) **peroxisomes**, which are associated with chloroplasts and are the sites of the photorespiration of glycolic

acid, and (2) **glyoxysomes**, which are involved in the utilization of the acetyl-CoA produced during the mobilization of lipid reserves, especially during the germination of oily seeds (Beevers, 1979; Cooper and Beevers, 1969; Thorpe, 1984).

In all the preceding organelles, membranes are a major component, but the following are not membranous in nature.

Ribosomes

Ribosomes are the small particles (17 to 20 nm in diameter) that are responsible for protein synthesis (Fig. 2.10; Alberts et al., 1983; Davies and Larkins, 1980). They consist of two subparticles, each composed of **ribosomal RNA (rRNA)** and protein. The two halves are synthesized in the nucleolus, migrate out of the nucleus to the cytoplasm, and then combine with each other and with **messenger RNA (mRNA)** to translate it into protein. Because the mRNA is large enough for several ribosomes to read it and translate it simultaneously, the ribosomes are frequently found as clusters called **polysomes** or **polyribosomes**. Polysomes are found either floating freely in the hyaloplasm or attached to the ER, converting it to RER. Ribosomes occur in all cells, but they are most abundant in those with the highest rates of protein synthesis, such as the protein-storing cells of some seeds (Fig. 2.15). Although too small to be visible by light microscopy, if ribosomes are especially abundant, their high content of RNA and protein causes the cell to stain intensely with many dyes. Except for proteinaceous seeds, plants do not usually produce large amounts of protein and do not have high densities of ribosomes as might be found in the pancreas, for instance (Lin and Gifford, 1976).

As mentioned previously, plastids and mitochondria contain their own 70s ribosomes which are more procaryotic in nature. The ribosomes that are synthesized in the nucleus and that function in the cytoplasm are slightly different in size, shape, and sensitivity to antibiotics; these are called **80s ribosomes** (Thorpe, 1984).

Microtubules

Microtubules are narrow (23 to 27 nm in diameter), straight protein tubules that are below the limit of resolution of a light microscope; they have been studied primarily by rather laborious electron microscopy techniques (Gunning and Hardham, 1982; Gunning and Steer, 1975; Lloyd, 1982; Thorpe, 1984). Fortunately, new fluorescently labelled antibodies are now available for marking the tubulin monomers of the microtubules, and elegant light microscope studies of the distribution of microtubules in whole cells are now being made (Jeffree et al., 1982). This is important because microtubules are critically involved in establishing and maintaining cell shape and differential chemistry. They are frequently described as a "skeletal" system, but this is mostly true of animals; in plants it would be more accurate to analogize them to a template or scaffolding: they temporarily establish a shape which is then permanently maintained by the wall (Marchant, 1982). Arrays of microtubules can catch and guide vesicles to a site of synthesis or can exclude them from such a site (Goosen-De Roo, 1973a, b; Hepler and Fosket, 1971; Hepler and Newcomb, 1964; Pickett-Heaps, 1967a). There is accumulating evidence that they direct the orientation of the cellulose microfibrils as walls are being deposited (Heath and Seagull, 1982; Herth, 1985; Robinson and Quader, 1982).

During karyokinesis (nuclear division), microtubules form the spindle that orients and moves the chromosomes (Forer, 1982). Subsequently, during cytokinesis (cell division), the microtubules form the phragmoplast that controls the deposition, placement, and orientation of the new cell wall (Gunning, 1982).

Microfilaments

Microfilaments are extremely slender (5 to 7 nm) protein structures found in both plants and animals (Lloyd, 1982). Only recently has it become established that plant microfilaments are similar to those of animals, being composed of actin and being involved in movements within the cell (Thorpe, 1984). Plant protoplasm is extremely dynamic: currents of hyaloplasm flow swiftly (**cyclosis** or **cytoplasmic streaming**), carrying organelles with them. Individual organelles can move independently: chloroplasts can move to positions of optimal light in the cell, dictyosome vesicles can migrate to regions of growth or secretion, and mitochondria and nuclei can also move. At present it is not possible to state exactly which organelles can be moved by microfilaments and under which conditions this movement occurs. The drug cytochalasin B inhibits the action of microfilaments, and treatment with this drug stops cyclosis (Bradley, 1973), inhibits the movement of chloroplasts (Schönbohm, 1973; Wagner et al., 1972), and interrupts the differential, site-specific growth that dictyosomes are involved in (Franke et al., 1972). But these studies must be interpreted carefully at present, as the mechanism of movement is not yet established, and cytochalasin B may have several effects on the cell.

Hyaloplasm

In the past, hyaloplasm, also called **cytosol** (Kelly and Latzko, 1980), was often not rigorously defined; sometimes it was considered to include the ER and dictyosomes, and other times not. Now that stereological techniques allow precise measurement, precise definitions are important. For stereological purposes hyaloplasm is designated as the nonmembranous portion of the cytoplasm, excluding the lumens of all organelles (Mauseth, 1980a). In other words, it is the liquid phase of the cytoplasm in which the other organelles are located and which contains dissolved in itself enzymes, metabolic precursors, and products. Although many types of specialized metabolisms occur in the nuclei, plastids, mitochondria, and the like, a major amount of the basic cellular physiology occurs in the hyaloplasm (Kelly and Latzko, 1980). It is the most abundant portion of the cytoplasm, constituting 91% of it in certain cells in *Sinapis* (Havelange et al., 1974; see Table 6.1). Cells that are very active metabolically appear to be densely cytoplasmic by both light and electron microscopy; this is due to a decrease in the volume of the vacuole and an increase in the relative proportion of the hyaloplasm (Mauseth, 1984a).

Ergastic Substances

Plants are capable of synthesizing and accumulating a large number of compounds (Goodwin and Mercer, 1983). At one time, it was believed that these were either waste products or nutritional reserve compounds. At present, there is a tendency to avoid dismissing compounds as being mere wastes with no value. Instead, it is thought that many of these substances are selectively advantageous in making the plants either less nutritious or bitter and unpalatable, thereby deterring animals from eating them (Harborne, 1982). In addition, many crystals are effective in inhibiting insects from being able either to chew on the plants or lay eggs in them. For many of the substances that can be seen in most histological preparations, we do not know either their composition or the selective advantage that they confer. With new techniques that are capable of analyzing microscopically small amounts of material, advances are being made.

Starch

Starch was described earlier in this chapter with regard to the amyloplast form of plastids (Figs. 2.8, 2.9). It is mentioned again here because it is the most common ergastic substance, being found in all land plants except possibly the parasite *Thonningia* (Mangenot, 1968). It is, of course, most abundant and easily observed in the starchy fruits and vegetables such as potatoes and wheat, but it can occur in almost all cells. In the green photosynthetic tissues it may be present (as assimilatory starch) only during the day because it is transported to other parts of the plant at night.

The detection of starch by light microscopy is simple and reliable: when the tissues are treated with potassium iodide (Johansen, 1940), the starch grains immediately turn blue-black. When viewed with a polarizing microscope, the starch grains shine brilliantly except for a dark cross centered on the hilum (Figs. 2.8, 2.9).

Lipids

Lipids (other than those in the membranes) are also frequently present in many materials, certainly those that are oily, such as peanuts and avocados (Fig. 2.11; the terms *lipids*, *fats*, and *oils* are generally used interchangeably). These may occur either in the elaioplasts or chloroplasts, or they may accumulate in spherosomes instead. Their distribution in plants is more restricted than that of starch, typically being confined to seeds, fruits, and flower petals, but they can occasionally be seen in other tissues.

Lipids are easily detected because they occur as spherical droplets in the cell, and they readily absorb the stains Sudan III and Sudan IV. Only rarely is lipid in the crystalline state encountered, an example being the needleshaped fat crystals in the endosperm of the palm *Elaeis* (Küster, 1956).

Tannins

These, like lipids, are not a single specific compound but rather are a family of related substances (Fig. 3.2; Goodwin and Mercer, 1983).

The tannins are phenols, and as such are capable of complexing with proteins; this is the basis of the leather tanning industry, which relies on tannins extracted from plants. It is thought that tannins can deter insect feeding by interacting with and damaging the proteins of the gut. Tannins occur in the vacuole, the cytoplasm, or even in the wall, and, when present in quantity, the vacuole or whole cell appears yellow, red, or brown after staining with safranin. The color may be either smoothly uniform or finely or coarsely granular. It is most typically found in hard, tough material such as the leaves of oaks and conifers, in bark and seed coats, and in immature fruits, giving them their astringency (Ting, 1982). Tannins can occur in soft tissues, being located in most of the cells of the tissue or only in individual, isolated cells—**tannin idioblasts** (Zobel, 1985).

Crystals

Crystals are common in many plants and can occur in many shapes and sizes (Fig. 2.12; Gunning and Steer, 1975; Metcalfe, 1983c). In higher plants, calcium oxalate is the substance that occurs most commonly (Arnott, 1976; Arnott and Pautard, 1965; Francheschi and Horner, 1980; Pentecost, 1980); calcium carbonate is rather rare, as is calcium malate. The crystals of calcium oxalate can take on the following forms.

DRUSES.   **Druses** are crystals that have numerous faces and whose points are acute and sharp (Fig. 2.13). They tend to be about 5 to 10 μm in diameter, and typically there is only one druse per cell (Francheschi and Horner, 1980; Sunell and Healy, 1979).

RAPHIDES.   **Raphides** are long, thin, sharply pointed crystals that always occur in large numbers packed together in a bundle (Fig. 2.14;

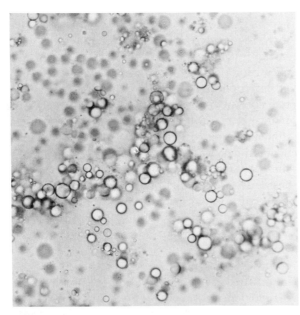

Figure 2.11
These are the fat bodies or elaioplasts from an avocado fruit. Notice that they are round, which is often a good indication that the particle is soft and is held together by a membrane; solid particles are frequently angular. × 200.

Figure 2.12
These are crystals from the wood of *Crinodendron patagense*. The shape and location of crystals are important characters for both taxonomists and physiologists. Although each crystal appears to be free, each had formed within a cell. × 1000. Micrograph generously provided by G. Montenegro.

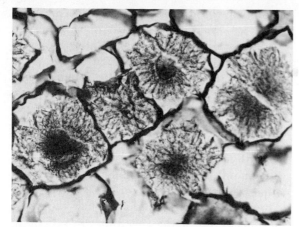

Figure 2.13
Star-shaped crystals such as these are druses, and they are extremely common. They can be abundant in some tissues, occurring in virtually every cell. × 500. Material prepared by J. Menard.

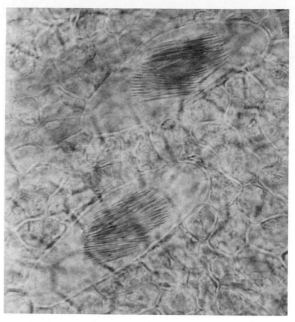

Figure 2.14
These two cells contain packets of raphide crystals in the leaves of taro (*Colocasia esculenta*); each crystal is needle-shaped. Raphides always occur in bundles like this. × 400. Micrograph generously provided by L. A. Sunell and P. L. Healey.

Horner et al., 1981; Kausch and Horner, 1983; Sunell and Healey, 1985; Tilton and Horner, 1980a). During the preparation of tissues, some of the cells that contain raphides are broken open, releasing the crystals, so it may appear as though they can occur individually. The raphides in *Dieffenbachia* are under pressure in the cell, and if fresh tissue is lightly minced with a razor and then viewed as a fresh mount in water, it is possible to watch the individual crystals shoot out. If leaves of the plant are eaten (not recommended) the crystals cause temporary paralysis of the throat muscles, giving the plant its common name, dumb cane.

Cells that contain raphides often have characteristic distributions within a plant (Wheeler, 1979) and can be useful in taxonomy (Dahlgren and Clifford, 1982; Metcalfe, 1983c).

STYLOIDS.   **Styloids** have the same basic shape as raphides, being long, narrow, and pointed, but they occur singly or in pairs, not in bundles, and are very large. They usually continue growing to such an extent that the cell itself is deformed and finally has the same shape as the crystal. Styloids are rare, occurring in the Iridaceae, Agavaceae, and Liliaceae and some others (Cronquist, 1981; Metcalfe, 1983c; Ter Welle and Mennega, 1977).

The growth and development of crystals is difficult to study, because they cannot be cut (they shatter) and, when they are present in the tissue, the microtome knife tends to push them, which shreds the surrounding cells. After fixation, the material must be treated to dissolve the crystal before microtoming, so that only the outlines can be studied. A crystal often becomes so large that the cell is deformed and ultimately has the same shape as the crystal; with light microscopy, it then appears as though the crystal is in an intercellular space. However, electron microscopy has shown not only that the crystals always occur within cells, but also that they develop within the central vacuole or within a vacuole of their own (Francheschi and Horner, 1980). Older reports of crystals occurring in intercellular spaces or within walls should be reinvestigated.

The presence, absence, and shape of crystals are extremely important characters for studying the evolutionary relationships of plant species. Another factor is the distribution of the crystals within the plant body. These cells never occur at random but are always located in specific regions, such as the hypodermis, near vascular bundles, in rows, or in clusters (Metcalfe, 1983c).

Calcium carbonate is present in a small number of plants (Moraceae, Acanthaceae, Cucurbitatceae, Urticaceae) as **cystoliths** ("cell rocks"; Fig. 10.39): these are crystals that form on a narrow ingrowth of the cell wall that projects deeply into the cell (the cell is a **lithocyst**, or "rock cell"). The calcium carbonate is deposited onto this unusual wall, forming a large, often irregularly shaped mass. A single cell may contain either one or several cystoliths, and the shape of the crystal can be important taxonomically. Cystoliths typically develop in epidermal cells, and when this occurs the epidermal cell usually becomes quite large and projects deep into the leaf. The leaves of *Pipturus* can have up to 140 cystoliths per square millimeter of epidermis (Nicharat and Gillett, 1970).

Calcium carbonate can also be deposited in the walls of cells that are not lithocysts (Metcalfe, 1983c; Pobeguin, 1951, 1954). Such encrustations have been found in walls of vessels in wood, as in some Aceraceae, Lauraceae, and Urticaceae, among others; the surfaces of leaves, as in some Frankeniaceae, Saxifragaceae, and Tamaricaceae (Metcalfe, 1983c), and some trichomes, as in Boraginaceae and Hydrangeaceae.

Silica Bodies and Stegmata

Silica bodies and **stegmata** (sing.: stegma) are deposits of silicon dioxide, and they are much more common in the monocots than the dicots. They are never as angular as calcium oxalate crystals, but they have very precise shapes, the shape often being characteristic of a particular family or genus (Ter Welle, 1976a, b). In the palms, they are hat-shaped, rectangular in the Heliconiaceae, troughlike in the Musaceae, sandlike in Zingiberaceae, conical in the Cyperaceae, and amorphous in the grasses (Metcalfe, 1963). The term *silica body* is used for the silica deposits in the grasses and Cyperaceae; *stegmata* is used for the other families. Like calcium oxalate crystals, they do not occur at random in the plant but rather are frequently restricted to just the epidermis or to cells immediately adjacent to vascular tissues. In the very large monocots, such as palms, where an entire plant cannot be placed in an herbarium for study, much of the taxonomy is based on microscopic characters of the leaves and pieces of the stem. In these cases, the stegmata and silica bodies are especially valuable diagnostic characters.

Silica can also be deposited directly in the walls of cells (Hodson et al., 1985; Lewin and Reimann, 1969; Postek, 1981; Sangster, 1985; Ter Welle, 1976a, b). Silica is often difficult to detect by light microscopy, but new methods have been published recently (Dayanandan et al., 1983).

Protein Bodies

Protein bodies occur mostly as reserve material in seeds and fruits (Fig. 2.15). The protein may occur as amorphous bodies or as distinctive shapes (Gunning and Steer, 1975; Lott, 1980).

Mucilages

Mucilages are common in many desert plants, and it is thought that they are capable of binding water, thereby preventing the plant from desiccating in the dry environment. There is no reliable histochemical test for mucilage, nor is it obvious in sectioned material, but if mucilage is present, it is easily detected because of

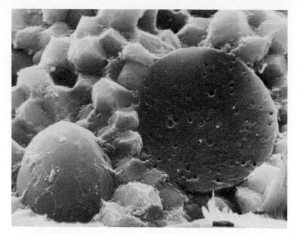

Figure 2.15
Protein bodies from the cotyledons of a sunflower seed. × 1000. Micrograph generously provided by R. D. Allen.

its slimy nature. Mucilage is treated in more detail in Chapter 9 (Figs. 9.6, 13.5).

The Cell Wall

The cell wall is special to plant anatomy for several reasons. All plant cells (except the sperm and some egg cells) have walls, and the walls are easy to see, often even with just the naked eye or a handlens. Walls are persistent, being preserved very well during specimen preparation or even in death. In trees, most of the wood and bark is composed only of cell walls, the actual protoplasts having died and degenerated. In woody plants, the walls constitute the majority of the plant body. Walls also are one of the primary reasons that plants and animals differ as much as they do. Because each cell is encased in a rather rigid, inflexible wall, cell movements are not possible, and the development of muscle is precluded, as is motility and the elaborate sensory and nervous systems that motility requires.

The cell wall has historically been treated as a nonliving secretion that was not truly part of the cell. As more studies demonstrate the dynamic nature of the wall and reveal that a large number of chemical and physical actions occur in it (Colvin, 1981; Darvill et al., 1980; Fincher and Stone, 1981; Kolattukudy, 1981; Labavitch, 1981), the wall is becoming accepted as an organelle and as an integral part of the cell's metabolism. Great advances are being made in the study of the ultrastructure, chemistry, and physics of walls and the bases of their diversity. Modifications and specializations of the protoplasts are almost always accompanied by some distinguishing modification in the walls, and we should not expect all discoveries to be universally applicable to all walls. Fortunately we do not have to fear that wall biology will be without any general principles, because the wall seems to be a two-part system—a basic, standard **primary wall** present in all cells and a **secondary wall** present in only certain cells. The secondary wall serves as the structure to which most modifications are made. In this section, only the primary wall will be considered; in subsequent chapters, the modifications and specializations of the secondary wall will be discussed where relevant.

Chemical Structure of the Wall

The four major components of walls are cellulose, hemicelluloses, proteins, and pectic substances (Goodwin and Mercer, 1983; Thorpe, 1984). **Cellulose** is a precisely defined polymer composed purely of glucose molecules linked to each other by β-1,4 bonds. The molecules contain from 8000 to 15,000 glucose monomers and are 0.25 to 5.0 μm long. This type of bonding causes the molecules to be flat and ribbon-like, and it allows the formation of internal hydrogen bonds which stabilize the molecule. The molecules can lie parallel to each other and form more hydrogen bonds between themselves, crystallizing and producing aggregates called **microfibrils**. Each microfibril contains 40 to 70 chains, all lying side by side, and these can be seen in electron micrographs (Thorpe, 1984). The exact nature of the arrangement of

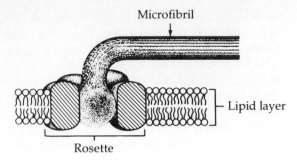

Microfibril

Lipid layer

Rosette

Figure 2.16
This diagram represents a membrane with a cellulose-synthesizing rosette embedded in it. The rosette extends completely through the membrane, so precursors can be absorbed on the cytoplasm side and the cellulose polymer (the microfibril) extruded on the other side. Diagram generously provided by T. H. Giddings, Jr. Reproduced from *The Journal of Cell Biology,* 1987, vol. 84:327, by copyright permission of The Rockefeller University Press.

the glucan chains within the microfibrils is still a point of controversy. The mode of synthesis is now thought to be the basis of the crystallization: complexes of cellulose-synthesizing enzymes are embedded in the plasmalemma in the form of **rosettes** (Figs. 2.16, 2.17; Giddings et al., 1980; Herth, 1984, 1985; Herth and Weber, 1984). The enzymes are believed to receive activated glucose from the cytoplasm side and add it to growing molecules of cellulose which extend out of the other side of the plasmalemma (the side facing the wall). Because the cellulose synthases are aggregated into a rosette, the growing molecules are automatically aligned, and they crystallize immediately. In such a system, it is reasonable to expect that the growing microfibrils will interact with pre-existing wall material and will be anchored rather solidly.

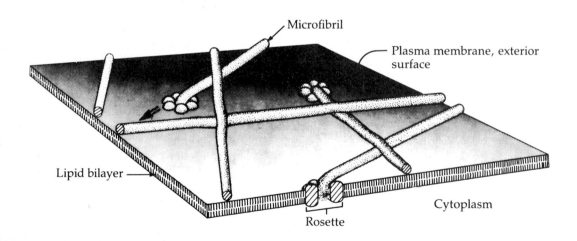

Microfibril

Plasma membrane, exterior surface

Lipid bilayer

Rosette

Cytoplasm

Figure 2.17
This diagram depicts the view of a plasma membrane as seen from the outside. As each rosette produces a microfibril, the microfibril becomes attached to the pre-existing wall (not shown here); as more cellulose is added to the end of the microfibril, the rosette is pushed forward within the membrane. Diagram generously provided by T. H. Giddings, Jr. Reproduced from *The Journal of Cell Biology,* 1987, vol. 84:327, by copyright permission of The Rockefeller University Press.

Thus as growth of the molecules continues, the synthase rosettes or aggregates must float forward in the plane of the plasmalemma, which should be fairly easy according to the fluid mosaic model of membrane structure. If there is no mechanism to orient the movements of the rosettes, the microfibrils should be deposited at random, which indeed does occur in most primary walls.

The rosettes, which are aggregates of cellulose-synthesizing enzymes, can also aggregate into very large arrays that have as many as 16 rows of rosettes (Fig. 2.18; Giddings et al., 1980). The entire array moves as a unit through the membrane, depositing parallel rows of cellulose crystals. The array moves such that rows of rosettes are aligned, and the microfibrils of one rosette can associate with the microfibrils of the other rosettes in the same row, forming a much larger structure termed a **fibril**. The greater the number of rosettes in the row, the greater the number of microfibrils in the fibril and the larger its diameter. These fibrils are more typical of the secondary wall, which is deposited interior to the primary wall after the cell has stopped growing.

In this method of deposition, new microfibrils or fibrils are deposited interior to the existing wall elements, producing the layered nature of the wall; this process of adding new material only to the interior face of the wall is called **apposition**. An alternative theory (deposition by **intussusception**) sought to explain the occasional interweaving of microfibrils by postulating that the growing end of the molecule was not embedded in the plasmalemma, but rather in the wall, and that the microfibril thus grew as a root does, weaving around and between obstacles, the existing microfibrils. This theory has little support and, as Figure 2.17 shows, it is unnecessary; microfibrils growing by plasmalemma-associated enzyme complexes can interdigitate.

Much attention has been given to the mechanism that could be responsible for the orientation of the cellulose microfibrils and fibrils. Re-

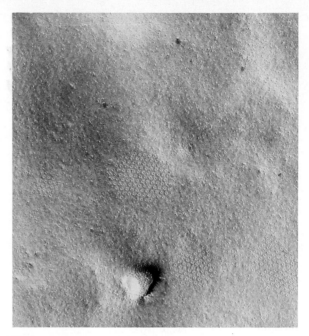

Figure 2.18
This micrograph shows several hexagonal arrays of rosettes. The entire array moves forward as a unit, and, as each rosette produces a microfibril, that microfibril reacts and crystalizes with the adjacent, aligned microfibrils. × 10,000. Micrograph generously provided by T. H. Giddings, Jr. Reproduced from *The Journal of Cell Biology*, 1987, vol. 84:327, by copyright permission of The Rockefeller University Press.

cently, studies have centered on microtubules as possibly forming a template just inside of the plasmalemma (Herth, 1985; Robinson and Quader, 1982). The rosettes, which extend completely through the membrane, could interact with the microtubules and could be guided by them. Numerous studies have shown that the microtubules and microfibrils are aligned.

In addition to the cellulose microfibrils and fibrils, the wall contains large amounts of hemicelluloses (Darvill et al., 1980; Goodwin and Mercer, 1983). These are not precisely defined groups of chemicals, as cellulose is, but rather a mixture of polymers of varying composition and which are highly branched. Two important features occur in the hemicelluloses: (1) they have a flat backbone with β-1,4 bonds from which short side chains emerge, and (2) because the side chains can interact with cellulose, the hemicelluloses co-crystallize with the cellulose, coating it and effectively gluing microfibrils together.

Also present in varying amounts are structural proteins that have very unusual properties. They are rich in the amino acid serine and can contain as much as 25% hydroxyproline; the sequence ser-hyp-hyp-hyp-hyp occurs frequently (Thorpe, 1984). The backbone protein may have arabinose tetrasaccharides linked to every hydroxyproline, but the hydroxyprolines have been reported to carry glucose and galactose molecules also. The function of these wall proteins is unknown, but because they tend to act as rigid, extended rods, it may be that they can crosslink other components and act as structural elements.

Gross Structure of the Wall

As a cell finishes karyokinesis and begins cytokinesis, a complex of microtubules and ER (the **phragmoplast**) forms between the two new daughter nuclei (Gunning, 1982; Gunning and Steer, 1975). The microtubules appear to be capable of trapping dictyosome vesicles that then fuse into one large flat cisterna (Hepler, 1982; Jones and Payne, 1978). The carbohydrates that the vesicles had contained are synthesized into the two new primary walls and the middle lamella that binds them together. Unfortunately the details of this process are unknown; it may be that the vesicles contain only the pectic substances necessary for the middle lamella while their membranes carry the cellulose-synthesizing rosettes. As the new walls form inside the cisterna, the membrane becomes transformed into plasmalemma. The new structure—middle lamella, two primary walls, and plasmalemmas—is termed the **cell plate** (Fig. 2.19), and it continues to grow at its margins as the phragmoplast expands outward, trapping more vesicles that fuse with the cell plate. The

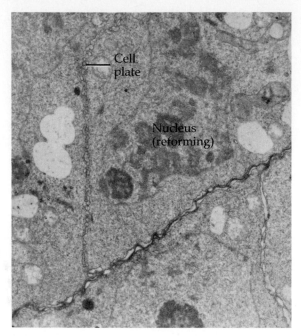

Figure 2.19
This cell has just finished nuclear division (karyokinesis) and is in the process of cytoplasmic division (cytokinesis). The cell plate is forming as vesicles begin to fuse in the center of the cell. If you look carefully at some of the vessels, you can see a line in the center, showing that the very first layers of the two walls are being formed. × 1000.

phragmoplast and cell plate are most often found at or near the center of the cell, but in some cases they are placed asymmetrically and result in the formation of one large daughter cell and a small one. Rarely, a curved cell plate occurs—in the vascular cambium (Figs. 14.8, 14.13) and in pollen grains (Fig. 19.24). This positioning within the cell is extremely precise, but the mechanism responsible for it is unknown. In many cells, even before karyokinesis begins, a **preprophase band** of microtubules appears in the position where the future cell plate will attach to the existing cell wall (Gunning, 1982; Pickett-Heaps and Northcote, 1966a, b), but the preprophase band is not a universal feature of dividing cells, and in those

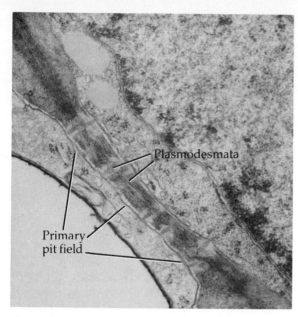

Figure 2.20
This high-magnification micrograph shows several plasmodesmata in longitudinal section. Notice that the wall is especially thin in this area, and much thicker at the edges of the micrograph. The thin area with the plasmodesmata is a primary pit field (see Figure 2.3, clear areas in the face view of the contact wall). × 5000.

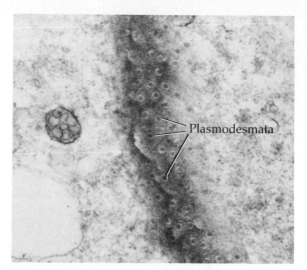

Figure 2.21
This section passed through a wall at a shallow angle, so the plasmodesmata can be seen in face view. Notice how abundant they are. × 5000. Micrograph generously provided by R. Fulginiti.

where it does occur, it disappears long before the phragmoplast forms. The phragmoplast disassembles as it approaches the margins of the cell, and the plasmalemma of the cell plate fuses with the existing plasmalemma of the mother cell; the two primary walls meet the existing wall as does the new middle lamella (Cronshaw and Esau, 1968). The wall at this stage is extremely thin; deposition may continue over the entire surface of the protoplast, and the old wall as well as the new one is augmented, especially if the cell is expanding.

Plasmodesmata

The formation of the new wall inside the phragmoplast is interrupted at sites where ER and microtubules of the mitotic spindle still occur (Hepler, 1982). In these areas, the cell plate has a hole through it, and as the wall thickens these holes remain as links from one protoplast to the other. They are lined with the new plasmalemma that is forming from the dictyosome vesicles, and the whole structure is termed a **plasmodesma** (pl.: plasmodesmata); see Figs. 2.20, 2.21, 2.22 and Gunning and Robards, 1976. Initially, the plasmodesma is just a small hole in the new wall (approximately 60 nm in diameter), but in most, a central structure called a **desmotubule** forms. It was at first thought that the desmotubule was a section of ER that had been trapped as the cell plate formed, but there is little support for that interpretation now (Jones, 1976; Robards, 1976), although the exact nature of the desmotubule is unknown.

Some plasmodesmata appear to be formed at random, but most occur clustered together (Fig. 2.22; Gunning, 1978). In the clusters their

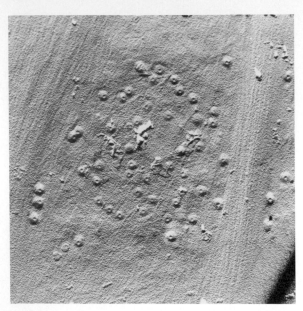

Figure 2.22
This freeze-fracture has exposed a primary pit field
and its plasmodesmata in *Ophioglossum petiolatum.*
Notice that the primary pit field appears to be in
a depression, because the wall is especially thin
there. Also, there are plasmodesmata outside of the
primary pit field. × 5000. Micrograph generously
provided by E. Roberts and A. Koller.

density can be as high as 60 plasmodesmata
per square micrometer of wall surface (Jones,
1976). As the new wall continues to thicken,
cellulose microfibrils are deposited between
plasmodesmata but not over them, so they become tunnellike. There is even more wall deposition in regions that lack plasmodesmata, so
the areas of clusters are relatively thin and are
known as **primary pit fields** (Figs. 2.3, 2.22).
With light microscopy, the primary pit fields
appear as slight depressions in the wall when
viewed from the side and as lighter, more
transparent regions in face view.

There is still debate as to whether the plasmodesmata act as channels for the movement
of material from one cell to another (Gunning and Robards, 1976). However, as Robards
(1976) pointed out, a cell with a typical size
and density of plasmodesmata has between
1000 and 10,000 connections with its neighboring cells. Furthermore, even if the wall is thin
such that the plasmodesmata are short, between one-third and one-half of the plasmalemma of the cell will be located inside of plasmodesmata. They cannot be an inconsequential
feature of the cell. Their possible role in the
transport of material from cell to cell is considered in Chapter 8 (sugar and mineral transport) and Chapter 9 (secretion).

PART II

Simple Tissues

Although almost any microscopical study of plants will reveal what appears to be an amazing variety of types of plant cells, actually only three fundamental types are considered to exist: parenchyma, collenchyma, and sclerenchyma. Each is distinguished from the other two solely on the basis of the properties of the wall, as will be described in the following chapters: parenchyma has thin walls, collenchyma has unevenly thickened walls, and sclerenchyma has thick, strong walls. In addition to being cell types, these are also tissue types. Any region that is composed purely of parenchyma cells is parenchyma tissue, and similarly for collenchyma and sclerenchyma cells.

The conversion from a typology basis to a functional basis in the field of plant anatomy is well illustrated here. Although we still use the terms parenchyma, collenchyma, and sclerenchyma, we now recognize that some cells have the properties of two types and that other cells, as they mature, can be converted from one type to another, especially from parenchyma to sclerenchyma. This should not be surprising, because the differences in wall properties between the three types is related to function. If a cell or tissue needs little skeletal support, a thin wall is sufficient (parenchyma), but if much support or rigidity is needed, then extra wall material must be deposited (sclerenchyma). In many situations the amount of wall strength needed by a tissue changes as the tissue ages, matures, or changes function.

There are many ways in which cells and tissues could be classified; photosynthetic tissues, secretory tissues, reproductive tissues, and absorptive tissues are among the possibilities that are frequently used. These are valuable, and you will notice immediately that the basis of classification is function. If a person is interested in storage, then a classification based on the nature of the storage product (such as starch, oil, protein, and tannins) or duration of storage (short-term, long-term) might be more significant than one based on the nature of the wall and that results in parenchyma, collenchyma, and sclerenchyma. The reason that this latter classification is used here is that most other classifications usually can be considered subdivisions of parenchyma. All reproduction, all absorption, and all secretion (other than of the wall itself) occur in parenchyma cells, and almost all photosynthesis and storage occur in parenchyma. There is no reason to be dogmatic, however; whichever classification system is most clearly useful and the best aid for a particular study is the one that should be used.

3

Parenchyma

Concepts

A fundamental classification of plant cells according to the nature of their walls is as follows: if there is only a thin primary wall, the cell is a **parenchyma** cell (Fig. 3.1); if the primary wall is rather thick and somewhat malleable, the cell is a **collenchyma** cell; **sclerenchyma** cells are those that have both a primary and a secondary wall. This classification is especially useful for collenchyma and sclerenchyma because each of these two types contains little variation; collenchyma and sclerenchyma cells are always easily recognizable.

Within the classification "parenchyma," however, there is tremendous diversity with regard to the cell's size, shape, metabolism, and functional role; these various types of parenchyma have in common only the thin primary wall and lack of a secondary wall. In herbaceous plants, almost all of the plant body is one type of parenchyma or another; only a small proportion of the plant consists of collenchyma or sclerenchyma. Even in shrubs and trees, large parts of the wood and bark may consist of parenchyma.

Parenchyma cells are the oldest type of eucaryotic cells, having been the first to evolve (Stewart, 1983), and even today the entire body of any alga consists solely of parenchyma cells (Bold and Wynne, 1978). After plants moved onto land and became more upright, they re-

quired special support, and then collenchyma and sclerenchyma evolved as modifications of parenchyma. The parenchyma of those first land plants probably was rather simple, like the parenchyma of the mosses, liverworts, and the lower vascular plants today (Scagel et al., 1982). In these groups almost all of the parenchyma is involved in photosynthesis. But as land plants advanced, division of labor became more sophisticated, and discrete regions of the plant body could become specialized. Some of the original cells could change from generalists to specialists, and whole new types of metabolism arose. Resin canals, lined with secretory parenchyma cells, became common in the gymnosperms, as did tannin-containing cells. With the advent of angiosperms, even more diverse types of parenchyma developed: mucilage cells, gum cells, laticifers, food bodies for ants, cells capable of rapid movement in many plants, and the highly pigmented cells of flowers and fruits.

Despite all this diversity, parenchyma cells and parenchymatous tissues are actually rather easy to understand; the key is to consider their functions and to think about which structures might be specially adapted to facilitate those functions. It is possible to subdivide parenchyma into at least five classes based on function: synthetic parenchyma, structural parenchyma, boundary parenchyma, transport parenchyma, and storage parenchyma.

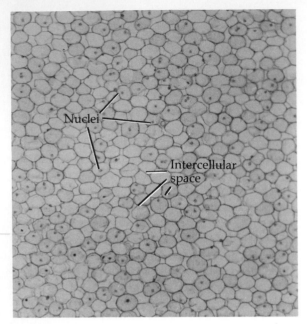

Nuclei

Intercellular space

Figure 3.1
This is a mass of pure parenchyma from the body of *Homalocephala texensis;* it is probably the type of tissue that most people envision when they think of parenchyma: the cells are large, have thin walls, and do not appear specialized. × 100.

Examples

Synthetic Parenchyma

PHOTOSYNTHETIC PARENCHYMA. In the chlorophyll-rich photosynthetic parenchyma (**chlorenchyma**), the objective is to convert light energy into chemical energy and store that energy in the bonds of carbohydrates (Fig. 2.4). To do this, light must be intercepted, and carbon dioxide must be absorbed into the cell. Because light is diffuse and because carbon dioxide dissolves into the cell's water only slowly, an optimal anatomy is small cells with large surface areas, such as cylinders. The cells should touch each other as little as possible, freeing the maximum amount of surface for the absorption of carbon dioxide. A large central vacuole would press the chloroplasts into a uniform layer next to the wall, preventing self-shading within the cell and placing the chloroplasts exactly at the site of carbon dioxide absorption. Chlorenchyma with exactly these characters is found in most leaves and stems (Fig. 2.4; Meyer, 1962; Montenegro, 1984).

The large surface area of the chlorenchyma tissues allows water to evaporate out of the cells into the intercellular spaces and then be lost through the stomata. Where water is plentiful, this is no disadvantage, but in deserts the water loss may be unacceptable. In desert plants, the chlorenchyma cells abut each other as much as possible and minimize the exposed surfaces (Crassulaceae, Mesembryanthemaceae). Although the cell's ability to absorb carbon dioxide is thus reduced, water conservation is apparently more important.

MERISTEMATIC PARENCHYMA. All meristematic cells are parenchyma cells, and these are probably the most elaborate of all synthetic cells: they are able to absorb sugars, water, and inorganic nutrients and use them to synthesize whole cells. The newly created protoplasm must be divided into new cells by the processes of nuclear division (karyokinesis) and cytoplasmic division (cytokinesis). An optimal arrangement for a cell with such a function is a small size with small numbers of each organelle, allowing rapid duplication and division. Also, a small, isodiametric cell is able to enlarge to any size and shape, however irregular, whereas a large meristematic cell could not shrink to a small, unusual shape.

One exception to this arrangement is quite instructive. Some of the meristematic cells that produce wood cells are extremely long (Figs. 14.6, 14.7). They produce wood cells that are also very elongated (up to 8 cm), and it is thought that perhaps it is more advantageous to have a long meristematic cell than to require extensive elongation of the daughter cells as they mature into wood cells (see Chapters 14 and 15 for a more complete discussion). There seems to be a trade-off, however, between ease of differentiation and ease of duplication. The increased length of the meristematic cells may permit easier differentiation, but these cells divide more slowly than do the small cells of other meristems in the same plant (Bannan, 1962).

SECRETORY PARENCHYMA.   One of the striking features of angiosperms is the number of types of parenchyma cells that secrete large quantities of substances that can be moved to the exterior of the plant or to a cavity or duct inside it (Fig. 3.2). The secreted material can pass from the cell to the site of deposition molecule by molecule, or by vesicles that fuse with the plasmalemma, or by breakdown of the entire cell. In all cases, the cell walls in the area of material transfer are usually thin and extremely permeable, permitting the movement of the secretion out of the cell. If the secretory cells are on the surface of the plant, then merely allowing movement across the wall may be adequate for secretion. But in the many cases when the secretory cells are internal, the secreted material should accumulate in a cavity or duct rather than diffuse throughout the plant by moving through the wall or intercellular spaces. The deposition is localized and prevented from flowing laterally through the walls by layers of hydrophobic material embedded in the cell walls in certain areas. The outer walls, closest to the rest of the plant, are the sites through which material enters the secretory cells; these walls typically have a high density of plasmodesmata.

Structural Parenchyma

AERENCHYMA.   In many cases, the cells themselves are not really so important as the intercellular spaces around them. Because diffusion through a volume of gas is much more rapid than through an equal volume of liquid (Salisbury and Ross, 1985), bulky organs must develop tissues in which the intercellular spaces are large and interconnected, forming a con-

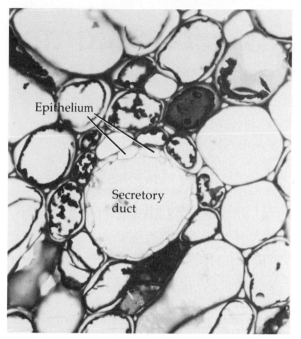

Figure 3.2
This is a secretory duct in the root of *Clusia*; it has a lining (epithelium) of parenchyma cells. Such parenchymatic linings often have the dual function of producing the secretion and of keeping it isolated from the rest of the plant. The surrounding cells are other parenchyma cells of the cortex; some have produced tannins. × 200.

tinuous gas phase that ramifies throughout the tissue (Fig. 3.3; Parkhurst, 1982; Turrell, 1936). This is especially well developed in the spongy mesophyll of leaves, where rapid circulation of carbon dioxide is important, and in the submerged stems and petioles of aquatic angiosperms, where the rhizomes and roots must be supplied with oxygen (Fig.3.4; Hulbary, 1944; Stant, 1964, 1967; Teal and Kanwisher, 1966), such as in water lilies (Laing, 1940), rice (Barber et al., 1962) and mangroves (Scholander et al., 1955). Aerenchyma also produces a tissue that is relatively strong despite its light weight (Fig. 3.5; Williams and Barber, 1961).

If the intercellular spaces are to occupy more than about one-half of the volume of the tis-

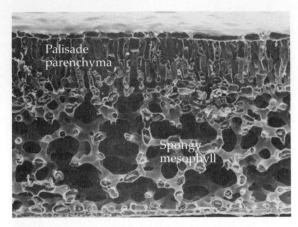

Figure 3.3
Cross section through a leaf of *Laurelia philippiana*. The uppermost layer (the epidermis) has no intercellular spaces; the next two layers (of palisade parenchyma) have small intercellular spaces. But in the lowest layers, the intercellular spaces constitute over half the volume. Imagine how freely the carbon dioxide can circulate once it enters through a stoma in the lower epidermis. × 100. Micrograph generously provided by G. Montenegro.

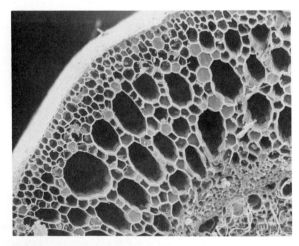

Figure 3.4
Stem of *Ludwigia terrestre:* the cortex is almost pure aerenchyma, formed when large regions of one layer of cells separated from the next layer. These channels provide an excellent means for gas circulation and permit the construction of a large stem using almost no building material. × 100. Micrograph generously provided by G. Montenegro.

Figure 3.5
This is the aerenchyma from the interior of a leaf of banana (*Musa*). This construction permits the leaf to be flexible, an important quality for leaves as large as these (two or three meters long). × 100.

sue, then the cells may develop lobes or arms, actually becoming star-shaped, as in *Juncus*, *Musa*, and *Scirpus* (Geesteranus, 1941). If the air spaces are even larger than this, then very often the tissue consists of sheets of cells that surround huge chambers as in *Eichhornia*, water lilies, and *Typha*.

Boundary Parenchyma

Many of the interfaces between two regions of a plant or between a plant and its environment are composed of parenchyma cells.

EPIDERMIS. The epidermis constitutes the boundary between the plant and its environment. Because both can be so diverse, epidermises are similarly diverse, but some generalizations can be made. On the shoots, an important function of the epidermal cells is the retention of water inside of the plant. For this, special modifications must be made to the outer wall to make it less permeable to water: hydrophobic, waxy chemicals are deposited in it (see Chapter 10 for details). Notice that the epidermis can be considered to be composed of secretory parenchyma also, because it must secrete the waterproofing substances. The large air spaces found in aerenchyma and chlorenchyma would be disadvantageous, so they are not formed (Figs. 3.3, 3.4). The intercellular spaces that do exist (stomata) are adjustable, being opened in the morning, which allows the absorption of carbon dioxide, and closed in the evening, which prevents water loss. The epidermis of the roots must not be impermeable to water, because water and water-soluble salts must be absorbed through it.

Because the epidermal parenchyma is the outermost layer, exposed to wind and animals, it must be resistant to tearing and puncture. The epidermal cells are attached to each other especially firmly; in many plants, they are so strongly interconnected that the epidermis can be pulled off as a large sheet (Figs. 10.6, 10.9). Epidermal cells could, therefore, be classified as structural parenchyma as well.

ENDODERMIS.   In roots, the conducting tissues are isolated from the cortex parenchyma by a boundary of cells that constitute the endodermis. The endodermal cells form a cylinder around the vascular tissues, and like the epidermal cells they abut each other tightly with no intercellular spaces, not even stomata. This arrangement prevents substances from diffusing into the vascular tissues by simply passing through the intercellular spaces. The radial walls between adjacent endodermal cells are impregnated with a hydrophobic substance, so material cannot pass through these walls either. Endodermises occur in a few stems and rarely in leaves (Fig. 3.6); a detailed description is given in Chapters 11 and 13.

Transport Parenchyma

TRANSFER CELLS.   In many areas in a plant, material must be transferred rapidly and in large quantity over a short distance—into a gland, into and out of sieve elements, into a developing embryo. In these cases, **transfer**

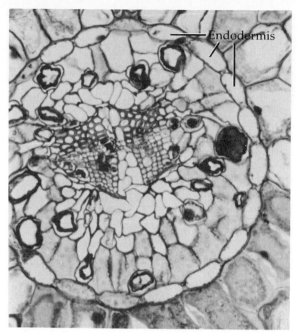

Figure 3.6
This is a cross section of the center of a needle-leaf of spruce (*Picea*). The outermost cells are chlorenchyma; the innermost ones are the vascular tissues. This is an unusual leaf in that it has an endodermis that acts as a boundary between the conducting region and the rest of the leaf. × 100.

cells are often present (Fig. 3.7; Gunning et al., 1968; Gunning and Pate, 1974). These are cells that have one or several walls that are highly modified, having many ridges and papillae on their inner face, forming a labyrinth (they are called either **labyrinthine walls** or **transfer walls**). Because the plasmalemma remains pressed against all parts of the wall, conforming to the larynth, it has a much greater surface area than if the wall were simply flat, and this presumably allows the plasmalemma to have many more molecular pumps that can actively transport substances. Where transfer cells move material from one cell to another, they have close contacts with both cells, leaving few, small intercellular spaces. But transfer

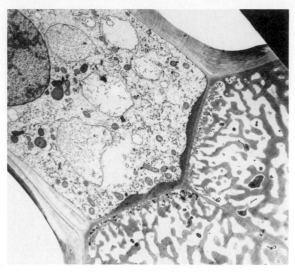

Figure 3.8
A plant of *Frithia pulchra;* it has been grown in cultivation, with the leaves exposed. In nature the plant would be embedded in the soil with only the leaf tips visible. × 0.5. Photograph generously provided by L. Simpson and R. Moore.

Figure 3.7
Labyrinthine walls in two root cells of the orchid *Sobralia macrantha.* × 4000. Micrograph generously provided by D. H. Benzing, D. W. Ott, and W. E. Friedman.

cells can also collect material from the intercellular spaces, and then pass it to another cell. In this situation, they have much of their surface exposed, allowing efficient absorption.

SIEVE ELEMENTS. In the phloem, sugars are conducted over long distances by specialized parenchyma cells called sieve elements. They have a very functional structure in which the cells are elongate and have rather large holes to allow the easy passage of phloem sap. These are discussed in detail in Chapter 8.

LIGHT TRANSMISSION. An unusual type of "transfer parenchyma" occurs in *Frithia, Lithops,* and some other members of the family Aizoaceae, the "window leaf plants" of the African deserts. The whole plant is subterranean, with only the leaf tips reaching the soil surface, so photosynthesis must take place below ground (Fig. 3.8). To accomplish this, a leaf with a transparent channel acts as an optical fi-

ber to carry light from the surface to the photosynthetic tissues below (Fig. 3.9). The plant remains in the cooler, more moist subterranean environment, so water storage and conservation are maximized (Marloth, 1909; Vogel, 1955). The optimal anatomy for light transmissibility is large cells (therefore fewer opaque walls), thin walls, and no air spaces (air-water interfaces can act to scatter light back out of the leaf). This is exactly the structure that they have (Eller and Nipkow, 1983; Simpson and Moore, 1984).

Storage Parenchyma

Most of the plant parts that are used for food—seeds, fruits, tubers—are composed primarily of storage parenchyma. In these organs, the parenchyma cells usually store starches (cereal grains, potatoes—Fig. 2.8), proteins (beans, peas), or oils (avocados, safflower—Fig. 2.11). Cells of this type are often so filled with their

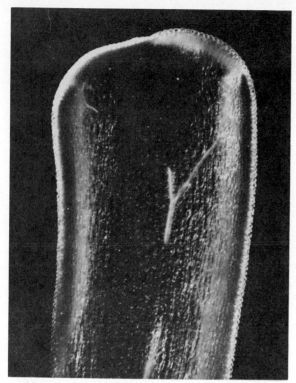

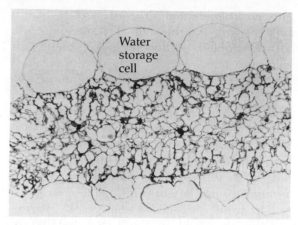

Figure 3.10
Cross section of a leaf of *Carpobrotus* (ice plant); although some of the epidermal cells are small (as in Figure 3.3), many become enormous, filled with water. These plants are leaf succulents of semi-arid regions. × 50.

Figure 3.9
This is a longitudinal section through a leaf of *Frithia pulchra*. Notice that the center is translucent and the sides chlorophyllous. Try to imagine this buried up to its upper surface, the central parenchyma conducting light and the peripheral chlorenchyma carrying out subterranean photosynthesis. × 8. Photograph generously provided by L. Simpson and R. Moore.

storage product that the vacuole is almost completely absent and the other organelles are obscured. The cells are usually of an intermediate size, perhaps to retain an optimal nucleus-to-cytoplasm ratio during the processes of product accumulation and release, which require active metabolism and nuclear control.

Succulent plants, such as cacti and euphorbias, are composed in large part of parenchyma cells that store water. Such cells are large, and the bulk of their volume consists of a greatly expanded vacuole surrounded by a thin

layer of cytoplasm (Fig. 3.10). The cells are much bigger than those that store starch or protein, and the nucleus-to-cytoplasm ratios are smaller, but the process of water uptake and release is probably simple and requires less metabolism and less nuclear control than does the storage of either protein or starch.

A less common means of storage is by increasing the amount of hemicelluloses in the wall. These can be deposited to such an extent that the walls become remarkably thick. Later, the hemicelluloses are digested out of the walls, the resulting sugars are used by the plant, and the walls become quite thin. This mechanism occurs in the seeds of *Coffea*, *Diospyros* (Fig. 5.4), and *Phoenix*.

Cell Shape

Because the walls of parenchyma cells are thin, they have often been considered to be weak, especially when compared to the thick walls of fiber cells. This view led many investigators to suspect that the shape of a parenchyma cell was a result of surface tension. If a cell were to grow in isolation, it would be spherical; if two cells of the same size were side by side, the wall between them would be a flat plate (Fig. 3.1). A casual examination of cells in callus culture or in the pith of many plants appears to support that view, and many studies were carried out on the shapes of parenchyma cells (Higinbotham, 1942; Hulbary, 1948; Marvin, 1944).

Soap bubbles were considered to be ideal models of parenchyma cells (Matzke, 1945, 1946) as were spheres of soft material (usually lead shot), which were pressed together to simulate a tissue (Matzke, 1939). The result of these studies was the conclusion that parenchyma cells have 14 faces; a geometrically regular polyhedron with eight hexagonal and six quadrilateral faces (an **orthic tetrakaidecahedron**) is frequently mentioned as the ideal. Dormer (1980), however, points out that tissue composed of such polyhedra can exist only if subject to uniform stress and only if not formed by cell division—two criteria that all plants tissues fail. More importantly, the attempt to treat cell shape as purely a phenomenon of surface tension suggests that cells have no control over their shape. But numerous examples of cells with flattened, elongate, curved, and even stellate shapes are known (Figs. 3.3, 3.5). Dormer (1980) states that the more beneficial use of mathematical analyses of cell shape (and also tissue and organ shape) is (1) to identify those elements that are mathematical necessities and (2) to determine which aspects of shape are not governed by geometric requirements. The latter aspects are controlled by the organism and are suitable for study.

Development

Phylogenetic Development

Tissues that are made up purely or predominantly of parenchyma cells are frequently re-

ferred to as parenchyma, and this is correct for the nonalgal plants. But the green algae, which are composed solely of parenchyma cells, may be unicellular or exist as loose aggregations of cells (colonies) or filaments, so they really do not consist of tissues.

In several groups of green algae (Chlorosarcinales, Ulvales, Chaetophorales, and Charophyceae), the algae have a rather solid body that superficially resembles the parenchymatous tissues of higher plants. However, in almost all of these, the "tissues" are actually composed of numerous filaments that are associated side by side. Such a "tissue" is called **pseudoparenchyma**.

Graham (1982) has traced the evolution of "true" parenchyma (in which each cell is connected to all of its neighbors by plasmodesmata) within the family Charophyceae (Fig. 3.11). The genus *Coleochaete* contains members that have a small discoid body that is attached to the surface on which it grows. In *C. irregularis* and *C. soluta*, the disc expands as filaments grow radially outward, with the cells occasionally dividing longitudinally to produce a new branch filament. In *C. soluta*, the longitudinal division may occur in the terminal, outermost cell, allowing the branch filament and the main filament to be equal. In *C. scutata* and *C. orbicularis*, longitudinal division within the terminal cell produces two adjacent, interconnected daughter cells that give rise to two files of interconnected cells, not to two filaments. Even if the terminal cells undergo only transverse divisions after this, the cells in these two files will be interconnected by plasmodesmata, even though they are not sister cells, and a true parenchyma tissue results. At present, there is great interest in the Charophyceae as possible ancestors to the land plants (Stewart and Mattox, 1975). The earliest plant fossils that are clearly terrestrial are truly parenchymatous.

Ontogenetic Development

All meristematic cells are parenchyma cells, as is the zygote itself, so parenchyma cells come

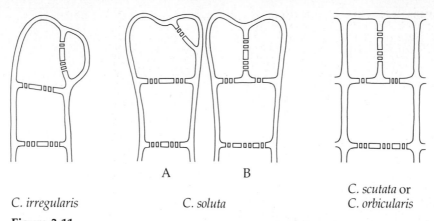

C. irregularis   C. soluta   C. scutata or
                                                              C. orbicularis

Figure 3.11
The cells of *Coleochaete irregularis* divide only transversely, so the branch that
is forming may branch again, but no cell can ever be completely surrounded
by other cells. *C. soluta* normally branches as in (A), but sometimes it divides as in (B); this results in an interface of three cells, all connected by
plasmodesmata (represented by broken lines). If those two terminal cells
then divided transversely, a true three-dimensional tissue would be formed,
as in *C. scutata*. Diagram based on information provided by L. Graham.

from pre-existing parenchyma cells. However,
as they mature, great variations can occur in
their metabolism, size, shape, and associations. It is impossible to discuss all variations
here; much of the rest of this book is dedicated
to parenchyma cells and tissues, and the development of each type will be described.

4

Collenchyma

Concepts

support
protection

Collenchyma is one of two specialized mechanical tissues, the other being sclerenchyma. In those plants that grow slowly, the regions actually undergoing growth are small, and turgor pressure in the parenchyma alone provides sufficient support. But most plants grow rapidly, and the elongating regions are often long and slender. They need a reinforcing tissue that can be functional even while growing, and it must be composed of cells that can themselves expand.

Growing regions also contain the tissues that insects and other herbivores seek most often, and they are susceptible to fungal attack and wind damage. They must be capable of undergoing wound healing and tissue regeneration more often than mature areas do. The cells of this region, including the support cells, should not be terminally differentiated if possible; instead they should be capable of dedifferentiation and renewed cell division.

The elongating young tissues of most plants are photosynthetic, so the reinforcing tissues should be at least transparent and, if possible, should also contain chloroplasts.

Collenchyma is a tissue that has all of these characteristics (it is also a cell type: as with parenchyma, there are both collenchyma cells and collenchyma tissue). The most striking and most important feature of collenchyma cells is their wall (Figs. 4.1, 4.2). It is considered to be a primary wall, although it is much thicker than the typical primary walls found in other cells, and its composition is unusual. The wall is not thickened uniformly on all surfaces, but rather only in specific sites: thin-walled areas alternate with regions that are very much thickened.

Types of Collenchyma

Collenchyma cells can be short and isodiametric or long, tapering fiberlike cells, and intermediate forms also exist. The pattern of thickening is used as the basis for classifying different types of collenchyma (Duchaigne, 1955; Müller, 1890).

Angular Collenchyma

In **angular collenchyma**, the extra wall material is deposited in the angles where two or more walls meet (Figs. 4.1, 4.2). The deposition may be restricted to just the corners, or it may extend along the face of the wall, becoming thinner farther from the angle. Examples of plants with angular collenchyma are *Begonia*, *Beta*, *Cannabis*, *Ficus*, and *Vitis*.

Lamellar Collenchyma

Lamellar collenchyma is very much like angular collenchyma, but because the cells are

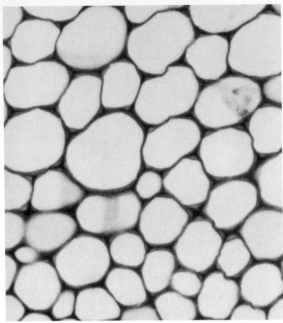

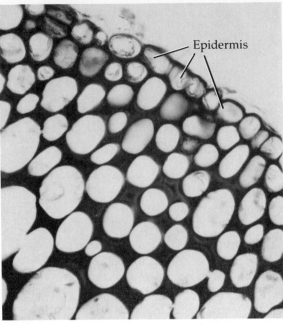

Figure 4.1
These collenchyma cells in *Nerium* have only rather
thin walls, indicating that they were fixed while
young. The original shape of the cells is still visible
because the middle lamella is dark, and the collenchymatous wall layers are lighter. Note that the
corners have much more material than do the
middle regions of the walls. × 100.

Figure 4.2
The collenchyma cells of this water lily petiole are
more fully developed than the cells of Figure 4.1:
their walls are much thicker, and the cell lumen appears round. × 90.

regularly arranged, the thickenings at one angle
of one cell tend to be placed against the thickenings of the three adjacent cells (Fig. 4.3).
This results in a tissue-level pattern of tangential lamellae, or layers. The effect is further emphasized because the thickenings extend farther along the tangential walls than along the
radial ones, which may be thin over almost all
of their surface. Examples are *Artemisia afra*,
Eupatorium, *Rhamnus*, and *Sambucus*.

Lacunar Collenchyma

If collenchyma forms around a cavity or small
chamber inside the plant, the thickenings may
occur along the walls facing the cavity lumen,
and **lacunar collenchyma** results. Examples
are *Althaea*, *Asclepias*, *Malva*, and *Salvia*.

Annular Collenchyma

Duchaigne (1955) designated **annular collenchyma** as a fourth type, distinguished by
having cell lumens that are round in cross section (Figs. 4.4, 4.5). But as Esau (1936) and
Fahn (1982) have pointed out, in the other
three types the walls may become massive, and
the thickened areas may cover virtually all of
the faces evenly; the lumen in which the protoplast is located becomes small and round.

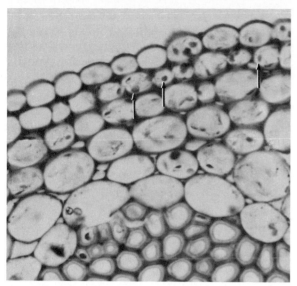

Figure 4.3
This stem has lamellar collenchyma. The inner wall
of each epidermis cell is collenchymatous, as are
the tangential walls of the second, third, and
fourth layers. Chloroplasts can be seen in most of
the collenchyma cells (arrows). × 75.

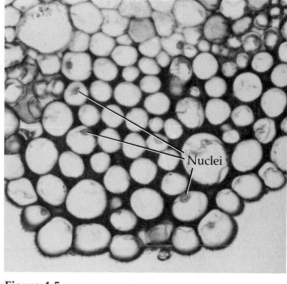

Figure 4.5
This micrograph of the midrib of a leaf of *Sambucus
canadensis* shows how localized collenchyma tissue
can be: it is restricted to the area just below the
midrib and does not extend out into the thin blade
region. Nuclei are visible in several cells. × 75.

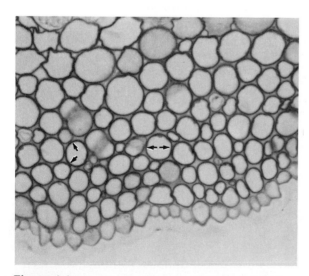

Figure 4.4
These collenchyma cells in the midrib of a *Nerium*
leaf have a thin, dark layer of lignin on their inner
surface (arrows). × 75.

Thus annular collenchyma is not so much a
fourth type as it is an advanced stage in the development of the other three types.

Other types might be expected to occur in
individual species or genera; for example, in
the collenchyma in *Mammillaria magnimamma*
(Fig. 4.6), the radial walls are thicker than the
tangential walls and the corners. This unusual
pattern may be an adaptation that permits
more sunlight to penetrate into the chlorenchyma below.

Nature of the Wall

Chemically and structurally the thickened portions of the collenchyma cell wall are distinct
from the thin primary walls of parenchyma
cells. They contain large amounts of pectic

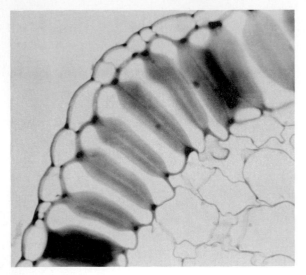

Figure 4.6
The radial walls of these collenchyma cells in *Mammillaria magnimamma* are the ones that are the thickest, whereas the tangential walls are thin. Notice that only one single cell layer is collenchymatous: the next layer inward has no sign of wall thickening, and the inner wall of the epidermis has only a trace of extra wall material. × 100.

substances and water (as much as 60% of their fresh weight may be water). By electron microscopy, the wall is seen to be layered; 4 to 7 layers in *Heracleum sphondylium* (Majumdar and Preston, 1941); 20 or more in celery (Beer and Setterfield, 1958), with the orientation of the cellulose microfibrils changing with each layer. In alternate layers, the cellulose microfibrils are oriented longitudinally, while in the intervening layers they have a transverse disposition. In Chapter 2, it was mentioned that cytoplasmic microtubules are believed to be responsible for guiding the deposition of microfibrils in the wall. Chafe and Wardrop (1970) have shown that when the orientation of microfibril synthesis changes from one layer to the next, the microtubule orientation is changed also, and the two remain parallel.

Not all transverse layers have the same orientation: the innermost layers (which are the ones most recently deposited and youngest) are more nearly transverse, whereas the outermost layers are more nearly longitudinal. Chafe (1970), Wardrop (1969), and Preston (1982) suggested that these layers are initially deposited with a transverse orientation and that as the cell elongates the outermost, oldest ones are stretched into a more longitudinal direction parallel to the axis of elongation. The inner layers, which have not undergone so much stretching, retain more of their transverse nature.

The layers vary also in composition, with layers rich in cellulose and poor in pectic compounds alternating with layers that are poor in cellulose but rich in pectins. The pectin-rich layers are those in which the cellulose microfibrils are oriented longitudinally. These layers are also the ones that are most responsible for the unevenness of the wall's width. In the areas that are thickened, the pectin-rich layers with longitudinally arranged microfibrils are most prominent, but in the thin areas of the wall these same layers are thinner than the other layers that have the transversely oriented microfibrils (Chafe, 1970). It is important to point out that the lamellar nature of the collenchymatous wall is sometimes not visible by light microscopy but is always evident by electron microscopy.

There is still some controversy about the nature of these alternating wall layers. In the angular collenchyma of celery, Beer and Setterfield (1958) reported that many of the lamellae were restricted to just the thickened wall areas located at the corners of the cells, but Chafe (1970), after studying all types of collenchyma in ten species, reported that all lamellae were continuous around the cells, that none of the layers was discontinuous. Although the micrographs in Beer and Setterfield's paper appear quite convincing, Chafe suggested that their use of methacrylate as an embedding medium (which is known to swell) may have caused disruption of the regions in which the lamellae are very thin and that the absence of layers in the thin regions may thus be an artifact.

The deposition of the thickened areas begins before the cell is completely elongated (Esau, 1936; Roland, 1967), since collenchyma occurs in growing tissues and organs. At this time the cells are short, about 20 to 40 μm in celery petioles (Beer and Setterfield, 1958), but they elongate as the organ grows—up to 500 to 2000 μm in celery, up to 547 μm in *Eryngium* (Wardrop, 1969). Physical agitation, such as shaking, can stimulate the cells to deposit walls that are even thicker than normal (Walker, 1957, 1960). As already mentioned, it is thought that the transverse cellulose microfibrils become progressively reoriented toward a longitudinal arrangement, but no other details of wall rearrangement are established.

Several investigators have studied the mechanical properties of the collenchymatous wall (Jaccard and Pilet, 1975, 1977, 1979; Jarvis et al., 1984; Pilet and Roland, 1974). The **plastic** nature of the wall is probably its most important feature. Having plasticity means that the material, even though strong, can be stretched or pushed into a new shape, without cracking or breaking, and retain its new shape when the tension or pressure is released. It was found (Ambronn, 1881) that collenchyma is capable of supporting 10 to 12 kg per mm^2, but it begins to stretch plastically when subjected to as little as 1.5 to 2 kg per mm^2. Plasticity is an essential property because collenchyma occurs in organs that are constantly altering their shapes as they grow, adjusting them toward the mature forms. If collenchyma were not deformable in a plastic way, normal growth would be impossible.

After the organ has reached maturity, the collenchyma can retain its character or be altered by the deposition of lignin. Briefly, lignin is a strong, waterproof polymer that can be deposited between the cellulose microfibrils; this deposit strengthens the wall and makes it water-resistant. Deposition of lignin is called

both lignification and, because it is characteristic of sclerenchyma, sclerification. The process seems to occur most often when the organ will persist for a long time (see the following chapter for a more detailed discussion).

When the organ has reached its mature form, collenchymatous plasticity is no longer advantageous; instead a sclerenchymatous **elasticity** is more functional. Unlike a plastic wall, an elastic one will return to its original size and shape after it has been stretched and then released. This is critically important in mature tissues, because the cells, tissues, and organs have attained their proper shapes and any stretching is not due to normal growth but to being deformed by wind, water, or animals.

To sclerify a collenchyma cell, a new layer of wall material is deposited interior to the innermost layer of the collenchymatous wall (Fig. 4.4). This layer constitutes a true secondary wall. It has a high cellulose content and is irregular at first because it follows the contours of the primary wall (Calvin and Null, 1977; Wardrop, 1969), but as deposition continues the inner surface becomes rounded in cross section. Lignin is deposited in the wall in a centripetal manner. This lignin is different chemically from that which is in the walls of xylem elements: it contains a preponderance of syringyl units, whereas the lignin of xylem has guaiacyl units. Concurrently with the deposition of this lignified secondary wall in the collenchyma, cell expansion ceases (Chafe, 1970; Wardrop, 1969).

During the deposition and lignification of the secondary wall, the collenchymatous primary wall remains in place, it is not digested nor resorbed. Reports of its becoming thinner are probably due to a slight dehydration: upon drying, the collenchymatous wall can shrink to as little as 1.0% of its original thickness. In some species, the primary wall can also become lignified, and the cell becomes indistinguishable from a sclerenchyma cell (Duchaigne, 1955). It must be emphasized, however, that the majority of sclerenchyma cells do not pass through a collenchymatous stage during their development. Examples of plants with sclerified collenchyma cells are carrot, *Eryngium*, *Medicago sativa*, and members of the families Bignoniaceae, Piperaceae, and Polemoniaceae (Metcalfe, 1979c).

The dynamic nature of the wall is further illustrated if the collenchyma is cut and forms a wound-sealing periderm. The collenchyma cells then digest away the thickened regions, convert to cork cambium cells, and begin cell division (Chlyah and Tran Thanh Van, 1984). In those plants that retain the cortex for many years, allowing it to expand while secondary tissues develop interior to it, the collenchyma cells also become modified. They can become larger and their walls thinner, but this thinning might just be due to dehydration and concomitant shrinkage due to age.

Position in the Plant

Because collenchyma is the mechanical tissue of growing regions, it is typically found only in young organs. But if the organ is persistent, such as a leaf or the cortex, then the collenchyma is, of course, likewise persistent, although it may become sclerified. In stems the collenchyma usually is peripheral, being either directly below the the epidermis (the epidermis itself can be composed of collenchyma cells) or located a few layers below the epidermis and separated from it by parenchyma. The collenchyma may occur as a continuous cylinder (*Sambucus*) or as individual bundles (*Cucurbita, Pastinaca*). It is not unusual to find collenchymatous cells associated with the phloem of the vascular bundles.

In leaves, collenchyma occurs primarily in the petiole and the main veins of the blade. It occurs as a prominent cap under the phloem in those bundles that are large enough to protrude as ribs on the underside of the leaf (Fig. 4.5). Bundles of collenchyma may also run along the margins of the leaf, forming a tough border that is resistant to tearing, thus helping to maintain the integrity of the blade. Many monocots, which form sclerenchyma early, do not contain collenchyma. The presence or absence of collenchyma and its abundance and distribution are often important characters in taxonomic studies (Joubert et al., 1984; Leitão, 1984).

Subterranean roots only rarely contain collenchyma, examples being *Diapensia* and *Vitis* (Metcalfe, 1979c); this is logical because most roots would not need the type of strength that collenchyma offers. The rind of fruits that are soft and edible often are collenchymatous (Considine, 1982; Rao and Dave, 1980).

In some species, leaves are able to undergo movement in response to touch or light. The portions of the petiole that act as joints or points of movement must not be too strong or rigid. Fuhrman and Koukkari (1981) found that in *Abutilon theophrasti* all of the petiole has a prominent layer of collenchyma just below the epidermis, except at the joint where movement occurs. In the joint, the collenchyma is located in the very center of the petiole, where it can supply strength without having either to swell or shrink as the joint flexes. Satter et al. (1982) found that in *Samanea* the collenchyma is also centrally located in the region of flexure. Furthermore, they discovered that the thick walls of the collenchyma cells are rich in potassium and chloride ions. They postulated that the walls might act as reservoirs to store these ions when they are pumped out of the protoplasts during the cell contraction that is the basis of leaf movement.

Similarly, Leach et al. (1982) found that the walls of collenchyma cells can serve as storage sites for defensive antibacterial compounds. The compounds are agglutinins, and they are thought to immobilize the bacteria when they try to penetrate the cell wall. On the other hand, whereas many cell wall components are rather resistant to enzymatic attack, the pectins are easily digested. Pathogenic fungi can use the pectin-rich layers of collenchyma as a good source of nutrients (Antonova and Golovin, 1983).

The collenchyma that is associated with the phloem may stain differently from the more exterior collenchyma, and it may respond distinctly to a deficiency of boron (Spurr, 1957). For this reason, Esau (1965a) has suggested that only the cells in a peripheral location in the plant should be called *collenchyma*, while those associated with the vascular bundles should be referred to as *collenchymatous*. This seems a reasonable precaution, but few people have adopted it.

5

Sclerenchyma

Concepts

Sclerenchyma is the second main type of mechanical tissue that occurs in plants, the first type being the collenchyma of the last chapter. Although both collenchyma and sclerenchyma are reinforcing tissues, the means they use to achieve their properties, and these properties themselves, are quite different. Whereas collenchyma is strong by means of its thick plastic primary wall, sclerenchyma is elastic because of its secondary wall, an additional wall layer deposited interior to the primary wall.

An elastic wall is one that can be deformed by either tension or pressure (just like a plastic wall) but which resumes its original size and shape when the deforming force is removed. Sclerenchyma is an ideal tissue for reinforcing an organ that has developed its mature shape and size: because the organ already has its proper form, almost anything that deforms it is a detrimental, abnormal perturbation. If the mature organ were reinforced by plastic tissues, then forces such as wind, passing animals, heavy snow, or large fruits would cause deformations that would be permanent. Because of the elasticity of sclerenchyma, after the wind stops or the snow melts or the fruit is released, the leaves and branches return to their proper shapes and positions.

As mentioned earlier with regard to collenchyma, another important aspect of reinforcement is resiliency—the ability to deform without breaking; this is also a feature of sclerenchyma. Elasticity and resiliency are properties of both sclerenchyma cells themselves and of the tissues they form.

In some cases, the strength and elasticity of the secondary walls are not used to support parts of the plant but rather to provide resistance to the mouth parts, claws, and ovipositors of animals. The characteristics of the secondary walls make sclerenchyma cells ideal as a protective layer around stems, seeds, and unripe fruits, such that insects and larger animals cannot bite through them, digest them, or lay eggs in them. In this regard, another property of secondary walls is also important: they are indigestible. There is no animal that has enzymes to break down these walls; even termites succeed only by the action of actinomycetes in their digestive tracts. Because secondary walls are not nutritious, most animals do not seek sclerenchyma as a food source. This alone is an enormous benefit to the plant.

Nature of the Wall

As mentioned, sclerenchyma cells achieve their properties through the secondary walls (Figs. 5.1, 5.2, 5.3). It had been thought that these walls are deposited only after the cell has attained its final size and shape (Kerr and Bailey, 1934), but Juniper et al. (1981) have shown that deposition can begin even while the cell is still elongating. The secondary wall is formed by apposition from cellulose syn-

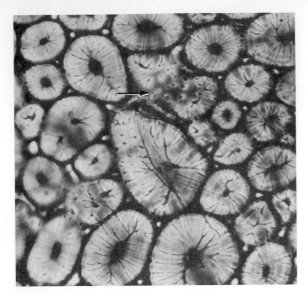

Figure 5.1
These sclereids in the shell of a coconut (*Cocos nucifera*) show how thick the secondary wall can become; the small black oval in the center of each is the lumen of the cell. The wall is much more voluminous than the protoplast. In several cells, the walls show numerous striations, indicating the many fine layers that contribute to the wall. The pits (the dark lines radiating away from each lumen) are elongate and most are branched. At the arrow, the surface of a cell is visible, and the pits are seen to be rather circular in cross section. × 100.

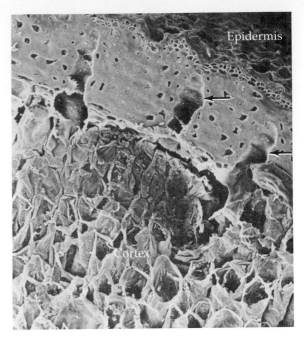

Figure 5.2
This scanning electron micrograph shows the surface of a large cactus, *Trichocereus chilensis*. The epidermis has thin walls, but the next six or seven layers are sclereids with extremely thick walls, which protect the plant from animals and fungi and help conserve water. The three canallike regions (arrows) are located below stomata and permit carbon dioxide to pass through the sclerenchyma layer. × 75.

thase complexes in the plasmalemma, as described in Chapter 2. Herth (1985) confirmed this specifically for the secondary wall. The orientation of the microfibrils is extremely orderly, and usually it changes twice during deposition such that the secondary wall has three layers: the outer layer (**S1**), the central layer (**S2**), and the inner layer (**S3**). It is S1 that Juniper et al. (1981) found to be present during elongation; S2 and S3 were deposited after cell extension had ceased. In S1 the cellulose microfibrils are deposited transversely to the long axis of the cell, as is also true of S3. Both of

these layers are typically rather thin. S2 is the predominant, thickest layer, and its microfibrils are oriented almost parallel to the long axis of the wall; the size and nature of S2 make it the principal mechanical layer. There may be more layers than these three, but a "warty layer" that was occasionally described is now known to be degenerated cytoplasm, not wall material (Core et al., 1979).

Like the primary wall, the secondary wall contains cellulose, hemicelluloses, and pectic substances, and in addition as much as 18 to 35% of the secondary wall may be composed of

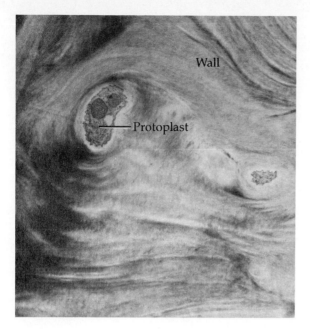

Figure 5.3
This is an electron micrograph of a sclerenchyma cell, showing the very thick, prominent secondary wall. × 1000. Micrograph generously provided by W. M. Harris.

lignin. Lignin is not a precisely defined chemical such as cellulose, whose composition is always β-1,4-linked glucoses. Instead, lignin is an amorphous, heterogeneous plastic formed by the free-radical polymerization of various alcohols, especially *p*-coumaryl alcohol, coniferyl alcohol, and synapyl alcohol. The actual alcohols and their relative abundances vary from species to species (Pearl, 1967). If the lignin contains mostly coniferyl alcohol, it is termed **guaiacyl lignin**; this is found in most gymnosperms (Vance et al., 1980). In the dicots there is an abundance of synapyl alcohol, and the polymer is called **guaiacyl-syringyl lignin**; in many monocots (and some dicots), additional aromatic acids are included (Nakamura and Higuchi, 1976; Vance et al., 1980).

The precursor alcohols are synthesized in the protoplast by the conversion of the amino acid phenylalanine into cinnamic acid, which is then used in the production of the other alcohols (Goodwin and Mercer, 1983). These then cross the plasmalemma and permeate the cell wall where they encounter enzymes (laccases). The enzymes dehydrogenate some of the alcohols to yield monomeric free radicals, which then polymerize at random, forming an extensive, three-dimensional network that surrounds and encases the cellulose microfibrils. There is increasing evidence that the lignin is actually bound covalently to the cellulose, but the nature of this linkage is still unknown (Goodwin and Mercer, 1983; Vance et al., 1980).

Lignification begins at the corners of the cells, in the region of the middle lamella and the primary wall; then it progresses between the adjacent cells by passing along the middle lamella and primary wall (Wardrop, 1971). These two regions are the most heavily lignified; the secondary wall becomes lignified last, and it contains a slightly lower proportion of lignin. Occasionally, S3 is not lignified at all. Studies with radioactive tracers indicate that, once the lignin is deposited, it is stable and does not turn over. With few exceptions (Alexandrov and Djaparidze, 1927), it remains in the wall until the organ decays (Goodwin and Mercer, 1983).

The presence of lignin dramatically alters the nature of the secondary wall in at least three ways: (1) by forming an extensive, cross-linked network, it anchors the cellulose microfibrils and provides much more strength and rigidity to the wall; (2) by being very inert itself, it provides a stable, resistant, protective coating around the other wall components, shielding them from chemical, physical, and biological attack (Millet et al., 1975; Ride and Pearce, 1979); and (3) by providing a waterproof barrier around the microfibrils, lignin regulates the hydration of the cellulose and thus the elasticity of the wall. Consequently, the lignified walls of sclerenchyma are stronger, more hydrophobic and waterproof than an equivalent amount of primary wall. Although wood-rotting fungi do have enzymes capable of degrading lignin—see Pearl (1967) and Vance et al. (1980)—heavily lignified tissues are quite resistant to decay (for example, bald cypress). In addition, parenchyma cells (which typically are not lignified) can deposit a protective layer of lignin in their walls when they are attacked by fungi (Aist, 1976; Ride and Pearce, 1979).

The presence of lignin is closely associated with the vascular plants: it occurs in all plants that have vascular tissues, but it is not found in algae. Some of the mosses have a ligninlike material, but there is considerable doubt that this is true lignin (Goodwin and Mercer, 1983).

Whereas water and most dissolved substances move easily through the primary wall, passage through a lignified secondary wall is extremely slow. For the protoplast to continue nutrient and gas exchange, it is essential that some sites on the primary wall remain uncovered and act as channels for the entry and exit of materials. These areas where the primary wall is not covered by secondary wall are called **pits** (Figs. 5.1, 5.2, 5.4, 5.5). Sclerenchyma cells typically occur in bundles, as in fibrous stems and bark (Fig. 5.6) or very large masses, as in the shells of nuts and the "stones" of certain fruits such as peaches and cherries (Fig. 5.1); therefore, the cells adjacent to a sclerenchyma cell are usually sclerenchyma cells themselves. Thus they must exchange material with other cells that also have thick secondary walls, and it is obvious that the pits of one must match the pits of the neighboring cells. The functional unit is the **pit-pair**, consisting of the two pits (Fig. 5.5). One pit is separated from the other pit of the pair by the **pit membrane**, which is actually the two primary walls and the middle lamella. There is no correlation between the primary pit fields of the primary wall and the pits in the secondary wall. When the sclerenchyma cell begins deposition of S1, it covers most primary pit fields and plasmo-

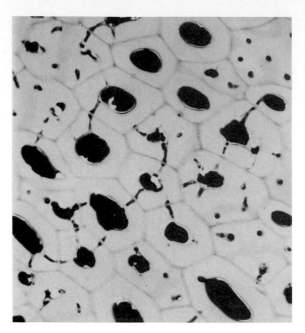

Figure 5.4
These sclereids of *Diospyros* (persimmon) endosperm have fewer but wider pits than do those of Figure 5.1, and it is possible to see that a pit of one cell meets a pit in neighboring cells. The primary walls and middle lamellas are visible as a dark line; where they cross the pit pairs they are known as a pit membrane. × 100.

desmata (Lawton et al., 1979); the pit membrane is not especially rich in primary pit fields.

In sclerenchyma that is specialized for conduction (the tracheary elements of the xylem), the pits must be wide and elaborate in order to allow the rapid movement of large amounts of water and solutes into and out of the individual cells (the structure of those pit-pairs is described in Chapter 7). In the sclerenchyma that is not concerned with rapid conduction, the pits can be rather small and not at all complicated structurally. If the secondary wall remains rather thin, then the pit is just a small, round hole in the secondary wall where the primary wall is exposed. This is a **simple pit**

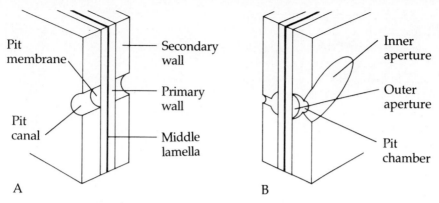

Figure 5.5
These diagrams show the structure of pits; in (A) is a pair of simple pits. These have been drawn with an intermediate depth, but they could have been deeper or shallower, narrower or wider. The significant feature is that the sides are more or less parallel to each other. The pit is merely the absence of secondary wall: initially the cell has only a primary wall, and pits are areas where no secondary wall is deposited. A slightly more complex pit is shown in (B): as the wall is deposited the aperture does not remain round but instead gradually shifts to oval.

(Fig. 5.5). But sclerenchyma cells usually have very thick walls. As more and more secondary wall is deposited, especially in S2, the pit must become deeper. It will typically remain a narrow, parallel-sided tunnel, and its **inner aperture** (the rim that faces the protoplast) is just a small, round hole about the same size as the **outer aperture** (the rim facing the pit membrane); the tunnellike portion of the pit is called the **pit cavity**, or **pit canal**.

The number and spacing of simple pits are quite variable. If the wall contains relatively few and if they are widely scattered, then each is unbranched and runs straight through the wall, being oriented perpendicular to the surface of the wall. But if many occur close together in a wall that is very curved, then as the wall becomes thickened individual pits may intersect (Fig. 5.1). As wall deposition continues, only one pit canal is formed instead of two or three, and a **branched pit** results (it would have been more accurate to have named it an

anastomosing pit from a developmental point of view). In a special type of xylem cell (a **fiber-tracheid**—Fig. 5.7), the deposition of the secondary wall is such that the inner aperture is modified; instead of remaining round, with each successive layer of microfibrils the inner aperture becomes more and more oval and finally elongate. If the wall becomes very thick, then the inner aperture may become a long, narrow slit (Figs. 5.5, 5.7). This does not happen at random; instead the orientation of slit-like inner apertures of all the pits of one wall are parallel to each other and perpendicular to the orientation of the other half of the pit-pair. Occasionally the two inner apertures of the pit-pair are aligned parallel to each other.

Nature of the Protoplast

For a long time, it was widely believed that sclerenchyma cells always died after the com-

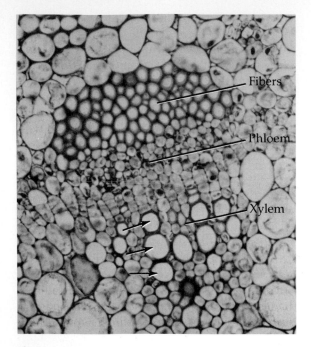

Figure 5.6
These fibers are to the exterior side of a vascular bundle, where they provide maximum protection and strength. The arrows indicate several trachery elements—that is, conducting sclerenchyma. × 100.

pletion of the deposition and lignification of the secondary wall, and mature, living sclerenchyma cells were thought to be very rare (Kallen, 1882; Spackman and Swamy, 1949). Undoubtedly, the majority of sclerenchyma cells do die as the final stage of differentiation. But recently, many instances of mature, long-lived sclerenchyma, especially fibers, have been reported, both in the dicots and the monocots—Czaninski, 1964; Dumbroff and Elmore, 1977; Fahn and Arnon, 1963; Fahn and Leshem, 1963 (70% of the 60 species examined had living fibers); Gregory, 1978; Wolkinger, 1969, 1970a, b. The sclerenchyma of *Tamarix aphylla* can live for 16 to 21 years (Fahn and Arnon, 1963). It is really not surprising that some fibers should be found to be living at maturity, because many of them in young wood and

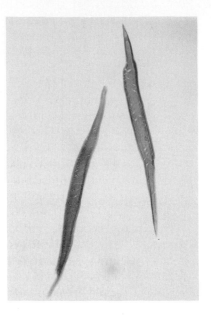

Figure 5.7
These fiber-tracheids of *Weingartia westii* are rather short for fibers, but they have pits with elongate inner apertures. The centers of several pits are bright and rather circular, indicating that there are small chambers, making these fiber-tracheids. Note that in each cell the inner apertures of some pits are elongate in one direction and that the apertures of others are perpendicular. This is because both the front and back wall are visible, and the two sets of pits appear superimposed. These cells were prepared by maceration: they were treated with acid until the middle lamella dissolved. × 250.

bark serve a function in nutrient storage (see Chapters 15 and 16), and it is necessary for them to retain living protoplasts to facilitate the processes of nutrient loading and unloading.

Types of Sclerenchyma

Sclerenchyma cells are classified primarily according to shape: short cells are **sclereids** (Figs. 5.1, 5.8), and long ones are **fibers** (Figs. 5.7, 5.9). This is not an especially good basis for classification, because short sclereids and long

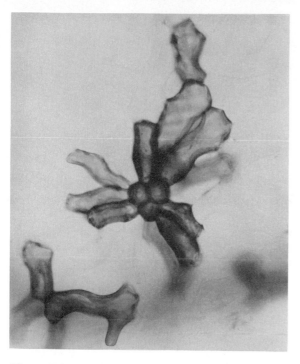

Figure 5.8
These are sclereids in the leaf of *Hedyosmum bonplandianum*. It is possible that the presence of such cells makes the leaf inedible (see Fig. 5.12), or possibly they act as water reservoirs. Two types of sclereids are present: in the center of the cluster are isodiametric brachysclereids; the outer ones are macrosclereids. These cells are part of a leaf clearing; the leaf was treated with bleach to make it transparent, then the lignin was stained with the dye basic fuchsin. Material generously provided by C. Todzia. × 150.

Figure 5.9
This is a maceration of bamboo. The long, narrow cells are extraxylary fibers; the shorter, wider ones are parenchyma cells. × 90.

fibers merely represent the extremes of a continuum of sizes; there are also long sclereids and short fibers. Furthermore, this classification ignores the fact that all of the conducting elements of xylem are actually sclerenchyma but cannot be considered to be either fibers or sclereids. A better classification would be to divide sclerenchyma into **conducting sclerenchyma** and **nonconducting sclerenchyma** and then subdivide the latter into sclereids and fibers. Because the conducting sclerenchyma of xylem is so specialized, it is treated separately in Chapter 7, which deals with the xylem; here only nonconducting sclerenchyma is discussed.

Sclereids

Sclerenchyma cells that are very short are termed sclereids, but there is a continuous range in size of sclerenchyma cells from isodiametric to long fibers, and the distinction is at least partially artificial. The difference in shape of fibers and sclereids is nevertheless important: it causes them to have different properties. In a mass of long, tapered fibers, all of the walls are really lateral walls oriented parallel to each other, and there are no true end walls, so the tissue tends to be somewhat flexible (Fig. 5.10). But in an equivalent mass of more-or-less isodiametric sclereids, which do have end walls, there are thickened walls ori-

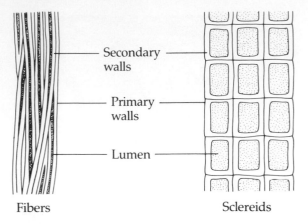

Fibers   Sclereids

Figure 5.10
A mass of fibers can flex and twist, but it resists
stretching. A mass of sclereids, because of the
cross walls, resists motion in all directions.

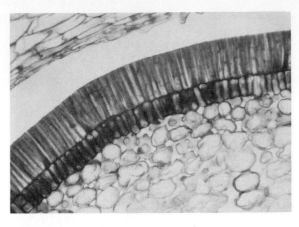

Figure 5.11
This is the seed coat of bean, consisting of an outer
layer of macrosclereids and an inner layer of developing osteosclereids.   × 90.

ented in all three planes, so the tissue is quite
rigid.

Sclereids can occur in any part of the plant
body, being associated with other types of cells
or forming sheets or bands of pure sclereids
wherever a plant needs strong, resistant, inflexible protection, such as around seeds, as
the "stone" of many fruits (Figs. 5.1, 5.11) or
as coverings for stems (Fig. 5.2), bud scales, or
immature fruits. When sclereids occur individually as idioblasts, they do not form a hard
layer, but rather they cause the tissue to be
coarse and gritty. Such sclereids can be found
in leaves of *Trochodendron* (Foster, 1945a, b),
Mouriria (Foster, 1946), *Hedyosmum* (Figs. 5.8,
5.12), and many others (Rao and Das, 1979). It
is possible that the presence of sclereids makes
the tissues inedible. Sometimes sclereids are
apparently formed only because a more appropriate type of cell would be difficult. As the
phloem expands and its bundles of fibers are
pushed apart, for example, the intervening
areas may become filled with parenchyma cells
that are then converted to sclereids. The production of fibers might possibly be more ad-

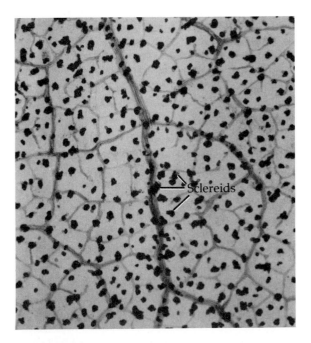

Figure 5.12
This leaf clearing of *Hedyosmum* shows how abundant sclereids can be. The continuous pale lines are
the vascular bundles of the leaf. Material generously provided by C. Todzia. × 50.

vantageous, but their formation would require extensive and difficult intrusive growth.

Many types of sclereids have been described, based on shape (Bloch, 1946; Tschirch, 1889):

1. **Brachysclereids** are isodiametric (Fig. 5.8). These are also called **stone cells** and are probably most familiar as the "grit" in pear fruits where they occur in small clusters. Brachysclereids are perhaps most common of all types because they are frequent in wood, bark, and the shells and husks of many fruits.

2. **Macrosclereids** are rod-shaped (Fig. 5.11) and are frequently deposited in layers only one or two cells thick. These are often found as an external boundary layer in seed coats and the outer covering of fleshy fruits.

3. **Astrosclereids** are not really star-shaped but rather are highly branched (Fig. 5.13), often with long arms. They are found in some leaves, but they are not especially common (some Loranthaceae, Menyanthaceae, Myristicaceae).

4. **Osteosclereids** are, as their name implies, bone-shaped; their ends are enlarged or even branched (Fig. 5.14).

5. **Trichosclereids** are long and hairlike; they differ from fibers in that they branch. Like astrosclereids, trichosclereids are most often found in leaves (some Plumbaginaceae; *Monstera*).

Other types, also based on shape, occur occasionally. Rao (1951a–c) classified sclereids on the basis of the tissue in which they originate: epidermis, palisade tissue, or spongy mesophyll. De Roon (1967), on the basis of studies of the Marcgraviaceae, established a very elaborate set of terms based on shape and position. Unfortunately, the construction of complex classifications with the concomitant naming of many types usually leads to the assumption that all the classes are equally important. Actu-

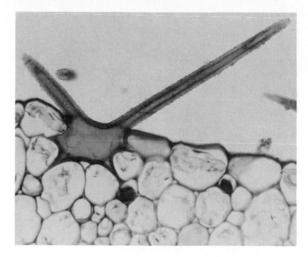

Figure 5.13.
This is only a small part of an astrosclereid, showing part of two arms that project into a large intercellular space in a water lily petiole. × 100.

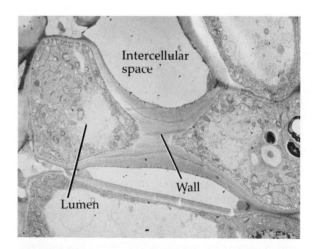

Figure 5.14
An early stage in the development of an osteosclereid. The cell lumen was originally cylindrical, but then wall deposition in the midregions prevented further enlargement. At the time the specimen was fixed, turgor pressure was causing the ends to swell, establishing the bonelike shape. Micrograph generously provided by W. M. Harris. × 1000.

ally these classifications probably do not represent any fundamental quality of the sclereids; just as parenchyma cells can acquire many shapes, so can sclereids, and little importance should be placed on minor shape variations. Instead, emphasis should be given to understanding the function, the biology, and the consequences of the sclereids. I agree with Metcalfe (1979c) that the classification of Tschirch (presented earlier) should be retained. An extensive list of the distribution of the different types of sclereids within tissues and families is given by Metcalfe (1979c).

Development of Sclereids

Sclereids are initially small, cytoplasmic cells when young; if they are to become brachysclereids, they are indistinguishable from parenchyma cells until secondary wall deposition begins. Although it is sometimes stated that they can be distinguished due to having very large nuclei, this is also true of many types of parenchyma cells. The other types of sclereids become recognizable as such as they begin to elongate, branch, or form their lobed ends (Fig. 5.14). In *Camellia* (Boyd et al., 1982), *Memecylon* (Rao, 1957), *Nymphaea odorata* (Gaudet, 1960), *Olea* (Arzee, 1953), and *Trochodendron* (Foster, 1945a), the lobes of the sclereids are able to invade free intercellular spaces and perhaps even to force their way between neighboring cells. In sclereids with rather regular shapes, secondary wall deposition and lignification begin after the mature shape is attained. Wall deposition may be uniform over the whole cell, or it may occur mostly on one end of a macrosclereid while the other end remains thin, as with *Cassia* and *Desmodium* (Arnholt, 1969). In the osteosclereids of *Pisum* seeds (Harris, 1984), the secondary wall is laid down in the midregions of the cell while the cell is still columnar; the center then resists expansion, but the ends (which are still composed only of primary wall) continue to swell, establishing the bonelike shape. In the astrosclereids of *Camellia* (Boyd et al., 1982) the ini-

tial deposition of the secondary wall is also in the cell center, and the branched ends remain thin and capable of continued intrusive growth.

Although sclereids may terminate secondary wall synthesis while the wall is rather thin, typically they continue adding more wall until only a tiny lumen remains with just a small amount of dense cytoplasm and a nucleus. During the deposition of the secondary wall, the cytoplasm contains many active dictyosomes (Boyd et al., 1982; Harris, 1983, 1984). Sclereids typically have many prominent simple pits (Figs. 5.1, 5.2, 5.4), both branched and unbranched; the pits are narrow and circular in cross section, not elongate or slit-shaped. A large proportion of the protoplasm occurs as long, narrow arms filling the pit canals.

The initiation of sclereids can occur at any time in the ontogeny of an organ. In *Camellia*, Boyd et al. (1982) found that all developing sclereids were of a similar age, indicating that they had all begun development simultaneously while the leaf was only 5 to 6 cm long, whereas in *Trochodendron aralioides* (Foster, 1945a) they arise over a long period. Furthermore, it is not unusual for mature parenchyma (Haberlandt, 1918) or collenchyma (Beer and Setterfield, 1958; Wardrop, 1969) to deposit a secondary wall and become converted to sclereids (see Chapter 4).

It had been thought that both sclereids and fibers died immediately after completing the formation of the secondary wall, but this opinion is changing rapidly as sclerenchyma is examined more carefully. The numerous large pits are adequate for maintaining sclerenchyma protoplasts in a living, healthy condition; by retaining living protoplasts, the capacity for modification is maintained. Alexandrov and Djaparidze (1927) reported that in the ripening of quince fruit the brachysclereids could delignify their walls. This agrees with and helps reinforce our current concept of the wall (both primary and secondary) as a living, dynamic organelle, rather than just an inert secretion. On the other hand, in many cases there is no obvious reason to maintain the sclereid protoplast. In seed coats and the shell of nuts, only the embryo and endosperm need to remain alive. The sclerenchyma can function while dead, so a capacity to modify itself would be unnecessary.

Fibers

Sclerenchyma cells that are long are, by definition, fibers. They can be short enough to intergrade with sclereids, or they can be as long as 6 cm, as in *Cannabis sativa* (hemp) and *Linum usitatissimum* (flax), or even 55 cm, as in *Boehmeria nivea* (ramie). Lengths of 1000 to 10,000 μm are not unusual. Tables of lengths and widths are given in Catling and Grayson (1982) and Kirby (1963). Whereas sclereids only occasionally occur as a part of the vascular tissues, fibers frequently do, and they are classified on this basis as being either **xylary** or **extraxylary fibers**. This is probably the most fundamental classification for nonconducting sclerenchyma; it should possibly replace the classification of fibers versus sclereids. The xylary fibers are believed to actually be modified tracheids—that is, to have evolved from tracheids (Fig. 5.15). A tracheid that is specialized for support can be produced by increasing the amount of secondary wall that it develops and by its producing simplified pits. With maximum modification, the cell is a **libriform fiber** with very thick walls and simple pits (Fig. 5.16). With less modification, the cell is a structure that is intermediate between a true tracheid and a libriform fiber: a **fiber-tracheid** (Fig. 5.7). In species with complex xylem (such as oaks), tracheids, fiber-tracheids, and libriform fibers all occur together. Some cells have intermediate characters, so an entire range of cell types exists. It can be extremely difficult to distinguish between cell types, even for experts. Basically, libriform fibers have the thickest secondary wall, and this wall is almost complete, being interrupted only by simple, narrow pits. The pits may be circular in cross section (as in scle-

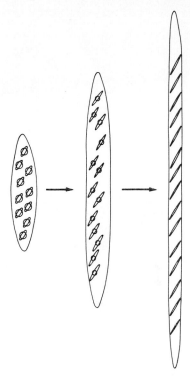

Figure 5.15
This diagram represents the presumed evolution of tracheids into fiber-tracheids, then into libriform fibers. The cells increased in length, the walls became thicker, and the pits became narrower and simpler.

reids) or their inner aperture (the one farther from the primary wall) can be slit-shaped. The outer aperture (the one against the primary wall) is small. In fiber-tracheids, the secondary wall is somewhat thinner and the pits somewhat wider; usually the inner aperture is oval, not round. The outer aperture of fiber-tracheids is slightly expanded, forming a small **pit chamber**. As will be explained in greater detail in Chapter 7, tracheids have thinner walls, the inner aperture of their pit is round, and their pit chamber is much larger.

Thus fiber-tracheids and libriform fibers occur only in the xylem and appear to be obviously descended from tracheids. Extraxylary

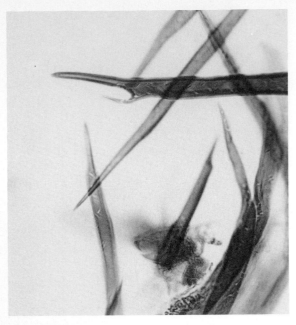

Figure 5.16
This libriform fiber from *Leptocereus* is unusual in having a forked end. This is identifiable as a libriform fiber (rather than a fiber-tracheid) because the pits are small and without a pit chamber. × 300.

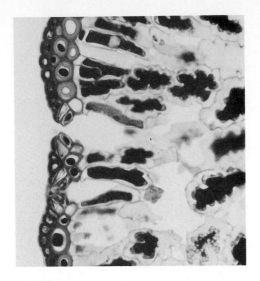

Figure 5.17
This cross section of a needle-leaf from *Cedrus deodora* shows why most conifer leaves are so tough and leathery: both the epidermal cells and the hypodermal ones are sclerenchymatous. It is not possible to tell in a cross section like this, but the epidermal cells are short sclereids and the hypodermal cells are long fibers. Notice the lobed chlorenchyma cells in the interior of the leaf and the prominent stoma and guard cells. × 100.

fibers occur in tissues other than the xylem and do not show characteristics of tracheids (Figs. 5.9, 5.17, 5.18). Their secondary walls are thick (frequently almost filling the lumen of the cell), and their pitting is always simple, never like that of fiber-tracheids. They do not have pit chambers, and the inner aperture is circular, not oval. Their walls may be heavily lignified, especially in the monocots, but in other plants (flax, hemp) there is little or no lignin. Instead, the secondary wall can consist of as much as 75 to 90% cellulose (M. Harris, 1954). They resemble elongate sclereids more than they resemble xylem elements. Although they do not occur in the xylem, they can be associated with it as a cap (Fig. 5.6) or sheath (Fig. 11.29) to the vascular bundles, especially in the monocots. Phloem frequently contains large numbers of fibers (Fig. 11.21), either mixed with the conducting cells or as a layer on the outer edge of the vascular bundle. Extraxylary fibers are also frequently found just beneath the epidermis, especially along the margins of leaves (Fig. 12.3). The spines of cacti (Mauseth, 1977) and succulents are composed of extraxylary fibers.

Development of Fibers

PHYLOGENETIC DEVELOPMENT. The distinctions between xylary and extraxylary sclerenchyma are satisfyingly clearcut and obvious, but we should consider them further. The ability to deposit secondary cell walls and to lignify them had evolved by the Mid-Silurian Period, 420 million years ago: *Cooksonia* had

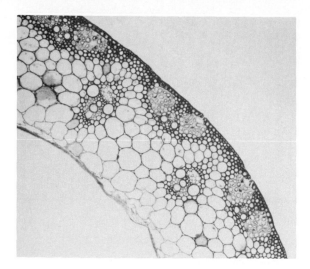

Figure 5.18
This is a cross section of a wheat stem, and the outer layers of fibers are clearly visible. Just as in Figure 5.6, you can see that the exterior position of the sclerenchyma offers maximum protection. × 100.

tracheids, one of the types of conducting sclerenchyma (Stewart, 1983). Once the morphogenetic mechanism for controlling this had been established for the xylem, it was then available in all cells of the plant, of course. Any mutation that allowed this suite of genes to become active in the outer cortex would have resulted in a plant that was stronger and more resistant to herbivore attack, giving it a strong selective advantage. There is no need to postulate a separate origin for extraxylary sclerenchyma but rather just a separate evolutionary sequence for its perfection.

ONTOGENETIC DEVELOPMENT. The fiber initials can be derived from a variety of sources: the procambium produces cells that can mature into primary xylem or phloem fibers; the vascular cambium produces cells for the secondary xylem and phloem fibers; and cells in the ground tissues of the cortex or mesophyll of leaves can also develop into fibers. Even epidermal cells in some grasses and Cyperaceae

differentiate into fibers (Thielke, 1957). In the monocots, which often have long, tough, leathery leaves, the vascular bundles are enclosed in a sheath of fibers, and large bundles that contain only fibers may occur also.

If the fiber initial is produced in primary tissues that are still elongating, such as the procambium, the cortex of a young internode, or the mesophyll of a long, monocot leaf, then the initial itself will be elongate and may continue to grow for a considerable period before beginning secondary wall deposition (Juniper et al., 1981). Such fibers can become extremely long: in *Boehmeria nivea*, the fiber initials are about 20 μm long and must grow for months before reaching their final length of 550,000 μm (Aldaba, 1927). If the fiber initial is derived from a cell of a nonelongating tissue, such as the vascular cambium or the mesophyll of many dicot leaves, then the initial will probably be rather short and begin maturing at once, but it may undergo intrusive growth. These cells never become very long, even if they do grow intrusively (Maiti, 1980).

The deposition of the secondary wall has been reported to begin even while elongation is occurring. The midregions of the cell, being attached to the surrounding cells, do not elongate, so deposition can begin there. The secondary wall then extends toward the ends of the fiber while the ends continue to grow (Aldaba, 1927). During this phase, S1 will exist as a long cylinder running almost to the ends of the fiber, S2 will be a somewhat shorter cylinder, and S3 will be even shorter. When cell elongation stops, then the wall layers can catch up with the ends, and all three layers are completed. It is not unusual, however, for wall synthesis to cease before the secondary wall is completed; the cell tips then remain with only primary wall.

The organelles that appear primarily responsible for the deposition of the massive secondary wall are the dictyosomes and the endoplasmic reticulum (Fig. 2.10; Boyd et al., 1982; Gunning and Steer, 1975; Juniper et al., 1981;

Lawton et al., 1979; Mauseth, ·1977; Pizzolato and Heimsch, 1975). The cytoplasm typically stains very heavily, and the vacuoles are small. Nuclei are often spindle-shaped, perhaps because the lumen of the cell is so narrow that a round nucleus could not fit in it. In those regions where the secondary wall is being deposited, the number of rosettes in the plasmalemma may be as high as 191 per square micrometer (Herth, 1985).

As is true for sclereids, we now believe that, although most fibers are dead at maturity, many do remain alive for years after their differentiation. Although the pits are often narrower and less frequent than those of many sclereids, they would be sufficient for the maintenance of the protoplasts.

Unusual Types of Fibers

GELATINOUS FIBERS. In **gelatinous fibers** (also called **mucilaginous fibers**), the secondary wall, especially the innermost layer (the **G-layer**) may have an unusual chemistry, being very poor in lignin and hemicelluloses but rich in cellulose. The G-layer may occur to the inside of S3, or it may replace either S3 or both S2 and S3 (Core et al., 1979). It is thought that, due to the lack of lignin and hemicelluloses, the G-layer has looser connections between its cellulose microfibrils. Thus, like the walls of collenchyma cells, it is able to absorb large amounts of water and swell, even to the point of filling the entire lumen of the cell. This layer shrinks irreversibly when dried, making it difficult to study histologically (Dadswell and Wardrop, 1955).

SEPTATE FIBERS. After a fiber has become elongated, it may resume mitosis and cytokinesis, resulting in a chain of cells within the original fiber cell wall (Fig. 5.19; *Coleus*: Pizzolato and Heimsch, 1975; *Ribes*: Parameswaran and Liese, 1969; *Vitis*: Spackman and Swamy, 1949). The new primary walls separating each new daughter cell may remain thin, but there may

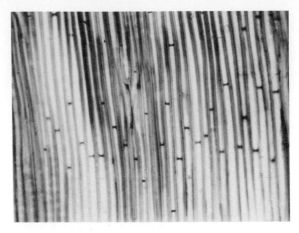

Figure 5.19
These are septate libriform fibers in the wood of *Triomma malaccensis*. Note how dark the septa are but that the dark staining does not extend into or along the secondary walls of the original mother cell. In several cells you can see at least two septa, and many more would be visible if entire cells were shown. × 75.

also be deposition of secondary wall over them. These cross walls are called **septa** (sing.: septum). This process can occur in fiber-tracheids, libriform fibers, and extraxylary fibers. Pizzolato and Heimsch (1975) compared the development of xylary and phloic septate fibers in *Coleus*. In the primary phloem, the fibers elongate greatly (1.2 to 2.8 mm), whereas those in the secondary xylem grow only to 0.5 to 0.8 mm. In the phloic fibers, the nucleus divides several times, and about three septa are formed, producing four uninucleate or binucleate cells. This occurs after the deposition of the secondary wall has begun, so the primary walls of the septa cannot attach to the primary wall of the fiber (Fig. 5.20). Even the septa are then covered by secondary wall layers as synthesis continues in all regions of the fiber. In the xylary fibers, however, there is a single mitosis and cytokinesis, resulting in just two uninucleate cells; this happens as deposition of the secondary wall ceases, so the septa remain thin primary walls only.

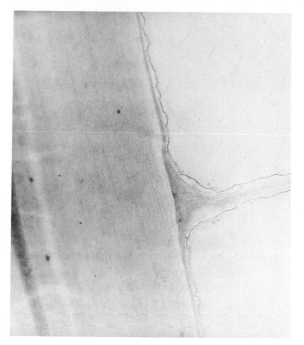

Figure 5.20
This electron micrograph shows the junction of the thin, new septum with the thick, old pre-existing secondary wall in *Coleus*. Micrograph generously provided by T. D. Pizzolato and C. Heimsch. × 1000.

Septate fibers usually remain alive with an active metabolism and play a role in the storage of material, the most common products being starch, oil, and crystals. Septate fibers are common in both the secondary xylem (many dicots) or as extraxylary fibers in some monocots, such as bamboos and palms (Tomlinson, 1961).

SUBSTITUTE FIBERS.   The secondary xylem may sometimes contain elongate cells with thin secondary walls; these have been called **substitute fibers** by Sanio. But based on studies by Fahn

and Arnon (1963) and Fahn and Leshem (1963), it has been suggested (Fahn, 1982) that they should actually be considered xylem parenchyma.

Fiber Identification

Natural fibers are still extremely important commercially, being used to make paper, rope, and other substances. In many cases it is most convenient to process a tree into pulp and ship the wood as a watery mixture. Naturally, the purchaser needs a method to verify the quality and identity of the pulp that has been received. Consequently, the fibers of many plants have been studied carefully to identify characters that are of taxonomic importance; among these are length, width, the ratio of the wall thickness to the diameter of the lumen, cross markings left by surrounding cells, and the presence of crystals or silica deposits (Carpenter and Leney, 1952; Catling and Grayson, 1982; Maiti, 1980; Strelis and Kennedy, 1967).

Terminology

The terms just presented and used are those currently most widely accepted. However, as the concept of sclerenchyma developed, a variety of terms were introduced that are now rarely used. Mettenius (1865) used the word **prosenchyma** to describe fibers, and Haberlandt introduced **stereide** as a term for the individual cells of sclerenchyma, regardless of their length. **Sterome** is still occasionally used to refer collectively to all of the sclerenchymatous tissues of an organ or a plant. Long, branched sclereids have been called both "internal hairs" and "ground tissue hairs," especially in translations from German.

PART III

Complex Tissues

Complex tissues are those composed of more than one type of cell. Xylem and phloem were recognized as complex tissues very early, xylem because it contains both sclerenchymatous conducting tracheary elements and parenchymatous storage cells and phloem because of parenchymatous conducting sieve elements and sclerenchymatous fibers. However, as we learn more about plant physiology, the concept of complex tissues becomes even more valid. Previously it was thought that the sieve elements were the sole functional units of phloem transport, but now we know that in many plants the companion cells are critically involved in this because they mediate sieve element loading and unloading.

Similarly, apical meristems had been considered small groups of homogeneous, juvenile cells, but recent research has shown them to be composed of distinct populations, each with its own properties.

Nectaries and secretory ducts are often complex tissues. In studying secretory tissues it is easy to focus attention on just the secretory cells themselves, but in many cases other cells are also involved, to accumulate the secretion, to form a storage space to keep it isolated from other tissues, and to transport in precursor materials.

The epidermis is almost always complex, containing a combination of regular cells, guard cells, trichomes, storage cells, secretory cells, cork cells, silica cells, and more.

It is important to point out that the complex tissues are mixtures of several types of the simple tissues: parenchyma and sclerenchyma in xylem and phloem or two or more types of parenchyma in secretory tissues and the epidermis. The complex tissues may be sharply delimited spatially from simple tissues (the vascular tissues may be separated from the parenchyma of the cortex by an endodermis), or the two types may grade into each other (the outermost layers of secretory tissues around many ducts and cavities are almost indistinguishable from the adjacent cortical parenchyma).

6

Apical Meristems

Concepts

There are two means by which an organism can organize its growth: plants use **localized growth** in which discrete regions (the **meristems**) are responsible for cell division and growth (Figs. 6.1, 6.2), whereas animals use **diffuse growth** in which the entire organism grows in all parts simultaneously. This difference has extensive ramifications: a plant, except as a seedling, is a mixture of (1) the constantly juvenile meristematic cells, (2) differentiating, maturing derivatives of these cells, and (3) fully mature adult cells, which also have been derived from the meristem. Localized growth thus allows a plant to have fully functional, fully developed organs and tissues even while it is still growing. The meristems, being sources of immature, undifferentiated cells, are sources of developmental plasticity and allow plants to respond selectively should the environment change.

The cells of meristems are considered to be **totipotent**; that is, they are totally competent to develop into any type of cell that occurs in the plant's life cycle. This certainly is true for the cells of shoot apical meristems: their derivatives develop into complete shoots that bear flowers and fruits and which either naturally or artificially can be induced to form roots. In plants that produce new, adventitious shoot buds on their roots, the same totipotency is true of the root apical meristem cells. The basal meristems of grass leaves, the vascular cambium, and the cork cambium are also all meristems, yet they are usually not included in discussions of totipotency. The concept of totipotency appears to have lost importance as a result of recent developments in our concepts of the biochemistry of differentiation (Halperin, 1986).

We now believe that all cells have a complete **genome** (set of genes) and that differentiation occurs by differential activation and repression of certain specific genes (Alberts et al., 1983). We no longer believe that development and morphogenesis are due to the elimination of certain genes from certain cells. Therefore, all cells should be regarded as totipotent, and at present, in fact, we can use tissue culture techniques to cause parenchyma, collenchyma, procambium, vascular cambium, epidermis, and microspore mother cells to revert to calluses from which whole plants can be produced (Henke et al., 1985; Vasil, 1984). We expect that in the future we will be able to do this with all living cells that have nuclei; therefore all will be shown to be totipotent. Related to the concept of totipotency is the concept that other types of cells are differentiated but that meristematic cells are not.

Another consequence of growth by meristems is **open**, or **indeterminate** growth; that is, plants tend not to have a fixed size. As long as conditions are favorable, the plant can continue to grow indefinitely. Although many plants do have a characteristic size, this is mostly an environmental and statistical phenomenon. They cannot survive the harsh

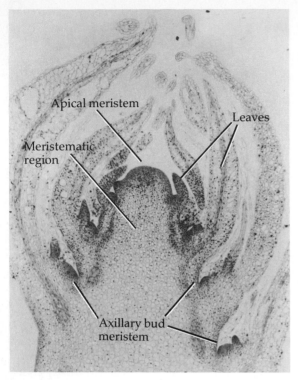

Figure 6.1
This longitudinal section of *Opuntia polyacantha* shows the apical meristem, the meristematic region, leaves in various stages of development, and also axillary bud meristems. Although the leaves appear to be arranged in two rows, in fact there are many more sets of leaves that were cut away during preparation. × 40. See Figure 6.2 for a three-dimensional view.

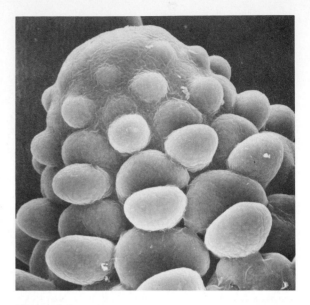

Figure 6.2
This SEM of *Anemopsis californica* shows the arrangement of primordia on an apex: the uppermost dome is the apical meristem itself. It stops, by definition, at the level of the youngest leaf primordia. The small protrusions at the apex are leaf primordia, and these are progressively larger and older toward the base of the apex. The uppermost are solitary, but the lower ones have axillary buds (flower buds) just above them. × 20. Micrograph generously provided by S. C. Tucker.

winter or summer conditions (many annuals are really perennials if protected in a greenhouse); or their susceptibility to disease, wind, drought, or unusually severe winters kills them; or they reach the point at which they can no longer properly control the long-distance transport of materials between very long roots and shoots. Probably few plants die of old age (except for **monocarpic** plants, those that automatically die after they flower). Thus, under exceptionally favorable conditions, exceptionally large individuals result, which is not the case with animals.

With the concentration of growth in the meristematic regions, the plant becomes especially dependent upon the meristems and thus is sensitive to damage to these small regions. Loss of the meristem could mean no further growth. This concentration also means, however, that only a small part of the plant is especially vulnerable and in need of special protection. In many cases, particularly in the shoot and root meristems, the plant is able to produce many "reserve meristems," and these are maintained dormant, in a quiescent state, ready to be utilized if the active apical meristem is damaged.

Recently, our concepts about meristems have changed in two areas: it had been thought that meristems were composed of undifferentiated, unspecialized cells, and that they were responsible for growth. We now consider meristematic cells to be highly differentiated, just like any other cell type (Mauseth, 1982c). The interpretation of meristem cells as undifferentiated was related to the concept of totipotency: it had been thought that perhaps the meristem cells were the only ones with complete genomes and that differentiation was accompanied by differential gene loss. Now that we know that this is not true, we should abandon both concepts, totipotency and unspecialized meristem cells. The structure, ultrastructure, and physiology of meristem cells are all highly different from those of all other cells and are specialized for the role of cell division *in an orderly fashion*. I emphasize "in an orderly fashion" because it is common to see callus cells in laboratory culture being described as meristematic, while in fact these cells have little in common with plant meristems. All they do is divide at random. One of the greatest problems for genetic engineers today is to switch these cells over from unorganized callus to organized, controlled, specialized meristem cells that produce shoots or roots. Even though many types of meristem cells resemble each other superficially at both the light microscope and electron microscope levels, there are enough distinctions that light microscopists have been able to recognize many classes of meristems and even to subdivide individual meristems into specialized, distinct regions. With the use of histochemical techniques, this differentiation is even more pronounced. Currently, stereological techniques of quantitative ultrastructure are being applied to meristems, again with the result that we discern many types of meristem cells, each specialized to its function.

The second area where our ideas are changing is that of growth. The interpretation that meristems are responsible for growth results partly from semantic problems of terminology. The terms "meristem" and "meristematic region" are often used interchangeably by nonspecialists. However, **meristematic regions** are regions of cell division and growth, and they tend to be difficult to delimit exactly (Maksymowych et al., 1985). Good examples are shoot and root tips. The meristematic region extends from the very apex back down the stem or root as far as cell division and expansion are occurring reasonably rapidly (Fig. 6.1). "Reasonably rapidly" is impossible to define: we do not want to extend the concept of meristematic region so far down the shoot or root as to include all cell division or growth—that might encompass most of the plant. Even though the concept of a meristematic region sounds vague and ill-defined (it is), the concept is useful, because there really is a region of rapid cell division and growth. That it grades imperceptibly into mature regions should not be any great difficulty for plant biologists who are accustomed to thinking in terms of gradients.

Every meristematic region contains a meristem that is usually much more discretely localized than the rest (Figs. 6.1, 6.3, 6.8). In the examples of shoots and roots, each has an apical meristem at its extreme apex that extends back not for millimeters, but usually at most 200 μm. In the meristems, cell division and expansion occur, but typically at rates much lower than for the rest of the meristematic region. This is especially well recognized in roots, where the subapical region is termed the "zone of elongation." We do not use that term for shoots, but it is firmly established that young internodes are the most rapidly elongating parts of the shoot.

One of the main activities of a meristem, as distinct from a meristematic region, is to act in establishing patterns and to act as a source of cells that are "mitotically young" and genetically sound (Klekowski, 1988). Beginning in the 1920s, plant physiologists and anatomists have carried out many experiments that clearly establish that if the shoot or root apical meristems are removed, growth of the remaining meristematic region continues, but that only those tissue patterns that had already been es-

A

Figure 6.3
(A) This is an apex of wheat (*Triticum aestivum*) just after it converted from vegetative growth to reproductive growth; the floral bracts (leaf primordia) are just beginning to emerge. Whereas the leaf primordia in Figure 6.2 are hemispherical, those in wheat are thin, wide flaps of tissue. This apex is much taller and narrower than those of Figures 6.1 and 6.2; the upward growth, compared to either radial growth or to the growth of primordia, is much greater here than in the others. In other types of apices, the surrounding tissue can grow more rapidly and leave the apical meristem as the low point of a depression, just the opposite of this wheat apex. × 120. (B) This is the same type of apex, somewhat older, showing that the floral apices have become organized and are growing outward, producing their own primordia for stamens and carpels. × 120. Micrograph generously provided by J. S. Gardner, W. M. Hess, and E. J. Trione.

B

tablished are able to develop (Steeves and Sussex, 1972). Similarly, the vascular cambium is reponsible for the patterns that occur in the secondary xylem and phloem, even though the cambium is the region least active mitotically (Chapter 14). Finally, as the blade begins to develop on a young leaf primordium, a marginal meristem arises and establishes the

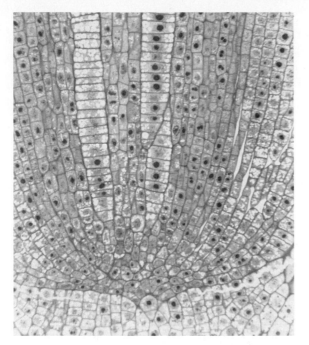

Figure 6.4
Root apical meristem of wheat. The files of cells can be traced almost but not quite to the center of the meristem: this is an open root meristem. The two rows of giant cells will develop into vessels; each cell is an immature vessel element. × 150. Micrograph generously provided by R. Moore.

proper number of cell layers for the blade, then a broad meristematic region causes the rapid expansion and differentiation that fulfills that pattern (Chapter 12).

Related to this role of pattern establishment is the role of being a reservoir of cells that are genetically sound and have gone through very few mitotic cycles. As a cell passes through the S phase (DNA synthesis phase) of its cell cycle, it exposes its DNA to extraordinarily dangerous situations. As it opens for replication, it is more sensitive to thermal damage and also to mispairing of bases. Data for plants are not available, but in humans thermal disruption causes the loss of 5000 purine bases

(adenine and guanine) per day per nucleus, and spontaneous deaminations of cytosine to uracil in the DNA occur at the rate of 100 per genome per day (Alberts et al., 1983). Although the replicases that synthesize the new strands of DNA are extremely accurate, they are faced with the problem that the DNA bases can exist in rare tautomeric form in the ratios of one part to 10,000 or 100,000, and these can mispair with the open DNA. Obviously, if all of these errors went uncorrected, the genome would be useless within two or three rounds of cell division. Cells have elaborate and efficient "proofreading" and error correction mechanisms; yet mutations still accumulate at the rate of one base-pair change in about one billion base-pairs per replication. In the diffuse type of growth found in animals, the adult body size can be achieved without any cell line (one cell and its descendents) actually undergoing very many replication-division cycles. Even so, the progenitors of the sex cells are set aside very early, while the embryo contains only a few hundred cells, thereby reducing the possibility of errors caused by replication. If plants were to grow truly by their meristems, then each cell of these tiny meristems would have to undergo millions or billions of replications to produce the enormous number of cells that constitute a plant; long before adult size were attained, the genome of the meristem cells would be filled with errors.

One adaptation that helps avoid this outcome is referred to as the multistep meristem: the meristem, through occasional divisions, provides cells to the rest of the meristematic region. Whereas the cells of the meristem must function for the entire life of the plant, those passed to the meristematic region must persist for only a short time, usually one growing season or less. The cells of the young shoot internodes function meristematically only for a few days, until the internode is grown. Cells that pass from the vascular cambium into the xylem mother cell region function only for several weeks until the new xylem is formed. The cells of the meristematic region need to produce only a limited number of cells each of which are destined to differentiate into specific types of cells. Any errors that happen in them will be somatic mutations, perhaps occurring in genes unnecessary for the cell type to be formed. Thus by not growing and dividing rapidly, the meristems of plants ensure that the meristematic regions will constantly be supplied with new, genetically healthy cells, and the plants will be able to grow indefinitely (Klekowski and Kazarinova-Fukshansky, 1984a, b)—at least for 11,000 years so far in bristlecone pines.

Types of Meristems

There are various ways to classify meristems, each reflecting a particular aspect of meristem biology.

Position of the Meristem

One of the most useful classifications is based on the position of the meristem relative to the derivatives it produces:

1. **Apical meristems** are located at the apex of the organ (shoot, root, trichome, gland, etc.) that they produce (Figs. 6.1, 6.2, 6.3, 6.4).

2. **Basal meristems** are located at the base of the organ (Fig. 6.5).

3. **Intercalary meristems** occur between (are intercalated between) their derivatives, some occurring above the meristem and others below it, and the meristem contributes cells to both sides (Fig. 6.6).

4. **Lateral meristems** are those meristems located along the periphery of an organ, the two main ones being the vascular cambium (Figs. 14.2, 14.4) and the cork cambium (Figs. 17.1, 17.8). Because the vascular cambium produces wood to one side of itself and bark to the other, it is also an intercalary meristem.

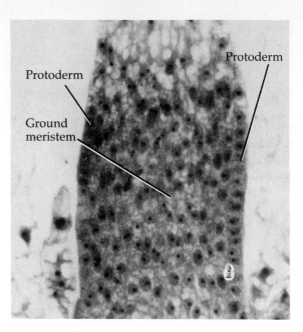

Figure 6.5
This spine of *Opuntia polyacantha* (prickly pear cactus) grows by a basal meristem; the basal region contains only meristematic cells and can be divided into a protoderm and a ground meristem. Cells flow upward out of the meristem, elongate, and then differentiate into fibers. × 500.

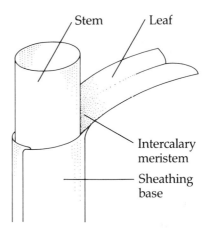

Figure 6.6
Internodes and the leaves of many monocots grow by means of meristems located within the internode or leaf; these are intercalary meristems. Very often, intercalary meristems are not so compact nor so distinct as are basal or apical meristems.

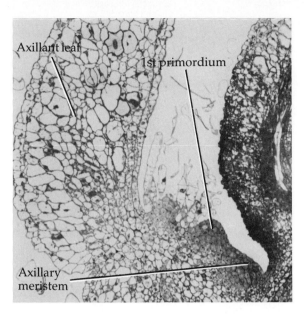

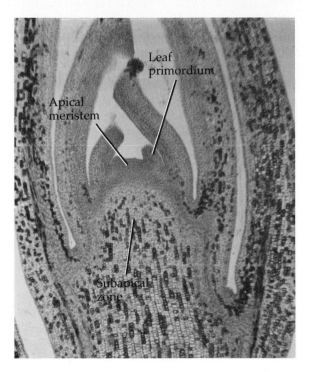

Figure 6.7
Just after a leaf primordium forms, an axillary meristem becomes organized above it. In this longitudinal section of *Opuntia polyacantha*, you can see the axillary meristem, the first primordium it has produced, the large axillant leaf to the left, and a younger leaf to the right. The primordium produced by the axillary meristem would have developed into a spine rather than a leaf like the two in the micrograph. × 100.

Figure 6.8
Although the apical meristem itself is small, the subapical region is extensive and is the zone in which most cell division and elongation occur. × 200.

5. **Axillary meristems** are the apical meristems of buds that are themselves located in the axils of leaves (Fig. 6.7).

Types of Derivatives Produced by the Meristem

A second useful classification of meristems differentiates between them on the basis of the type of derivatives produced.

1. A **protoderm** gives rise to cells that differentiate into epidermis cells (Figs. 6.5, 6.8).

2. A **procambium** (pl.: procambia) produces the primary vascular tissues.

3. The **ground meristem** is one that produces rather large amounts of more or less homogeneous tissues—for example, the pith or cortex, or the masses of sclerenchyma fibers within a spine (Figs. 6.5, 6.8).

4. **Promeristems** are meristems that give rise directly to other meristems or to other, distinct parts of the same meristem (Figs. 6.11, 6.12).

Sequence of Formation of the Meristems

The third classification is based on the sequence of formation of tissues: the shoot and root apical meristems that were present originally within the seed or embryo are **primary**

meristems, and **primary tissues** are those that are produced by these meristems. Any meristems that differentiate within these primary tissues are **secondary meristems**, and they produce **secondary tissues** (examples are the vascular cambium and cork cambium that arise in primary parenchyma but are never present in seeds; see Chapters 19 and 20). This classification is not really useful scientifically. The meristems of branches in many species arise as groups of cells that redifferentiate from young shoot tissues, so they should strictly be called secondary meristems (Figs. 6.1, 6.7), but they are identical in structure and function with the primary shoot apical meristems. Further, the axillary buds that arise on these branches would have to be called tertiary meristems. Today, we still speak of primary tissues as those just defined, but the term primary meristem is rarely used. Secondary tissues are those derived from the vascular cambium and cork cambium, but no other use or significance is customarily attached to this terminology.

The Vegetative Shoot Apical Meristem

In Vascular Cryptogams

In the vascular cryptogams (the Psilotophyta, Microphyllophyta, Arthrophyta, and Pterophyta) the shoot apical meristems are very simple structurally (Gifford, 1983). In the first three divisions listed above and in some of the ferns, there is a very large, unmistakable **apical cell** (also called an **apical initial**; see Figs. 6.9, 6.10). It is shaped like an inverted pyramid with either three or four faces pointing downward (the **cutting faces**) and a single triangular or square face directed upward, forming part of the outer surface of the plant. This apical cell divides in an orderly fashion, cutting off a narrow, flat cell by means of an asymmetric division in which the phragmoplast and cell plate occur close to one of the cutting faces. The next division of the apical cell will be a similar

Figure 6.9
The single apical cell of this *Davallia solida* (a fern) is prominent and easy to detect. Look at the cells on either side and notice how flat they are; you can see how they were cut off from the apical cells. Notice also, however, that farther from the apical cell, the derivatives have subdivided, so an entire meristematic region is present and important, not just one single cell. × 200. Micrograph generously provided by J. G. Croxdale.

asymmetric division along an adjacent cutting face. Division continues until all faces have produced small, thin daughter cells; then the process is repeated.

It is frequently stated that all cells of the shoot in these plants can be traced back to this apical cell, and, although that is technically

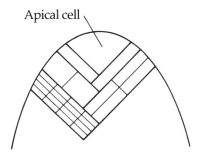

Figure 6.10
The derivatives of an apical cell divide and subdivide, forming packets of cells.

true, it must be emphasized that they cannot be traced back to it directly. That is, the daughter cells produced by the apical cell also divide rapidly and produce large packets of cells that develop into segments of the shoot (Fig. 6.10). Thus, it would be safe to interpret these apical meristems not as consisting of a single apical cell but rather as consisting of two zones: the the apical cell and a subjacent zone of rapid division (Croxdale, 1976; McAlpin and White, 1974). These meristems conform to the multistep organization just described, and the apical cell really acts as a slowly dividing reserve of one genetically sound cell.

In many ferns, rather than just a single apical cell, there may be a small cluster or a short row of prominent apical cells. They function just like the single apical cell, but they avoid cutting off daughter cells along faces that are in contact with other initials.

In Seed Plants

The shoot apical meristems of both gymnosperms and angiosperms do not contain any cells that are as distinctive as the apical cell of the vascular cryptogams, although in the last century (when studies of the ferns were dominating morphological thought) several were reported. After it was accepted that the apical meristems of seed plants are not organized as in ferns, it was initially thought that they were uniform groups of homogeneous cells. The first real step toward our current concept of apical meristems was formulated by Schmidt in 1924, when he postulated that shoot apical meristems consist of two distinct zones, the outer **tunica** and the inner **corpus** (Figs. 6.11, 6.12). These two are distinguished one from the other on the basis of the orientation of cell divisions: the cells of the tunica divide only with **anticlinal walls**, walls perpendicular to the surface of the tunica (Fig. 6.13). The two daughter cells both lie within the plane of the tunica; because they do, the tunica can grow as a sheet but cannot increase in thickness. To be-

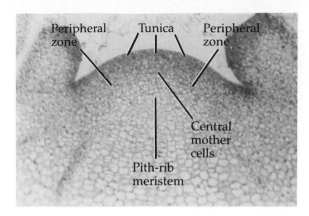

Figure 6.11
This apical meristem of *Mammillaria elongata* has complete zonation. First examine the outermost layer carefully; it is uniform and distinct across the apical meristem. This is the tunica. Now study the next two to five layers interior to the tunica; they are orderly in the peripheral regions but rather irregular at the center. The orderly regions are part of the peripheral zones; the irregular parts are the central mother cells. The pith-rib meristem is the lower central region, about even with or a little lower than the two leaf axils. It is rather easy to see the zonation (the central mother cells may be difficult), but trying to identify the exact boundaries between zones is usually impossible. Only the tunica is truly distinct. × 160.

come thicker, a tunica cell would have to divide with a **periclinal wall** (a wall parallel to the surface of the tunica), which does not occur. The tunica can, however, consist of several distinct layers. One single layer (**monostratose**) is probably the most common structure, followed by two to four (**multistratose**), but as many as 18 layers occur in *Xanthorrhoea media* (Staff, 1968). The corpus is covered by the tunica and, at the time of Schmidt, was thought to be composed of cells that could divide in any direction and could, therefore, grow in three dimensions, not just two like the tunica.

Work with the monocots and especially with gymnosperms revealed that periclinal divisions could occur in the tunica layers; in the

Figure 6.12
The zonation present in Figure 6.11 may be easier to see after you study this diagram.

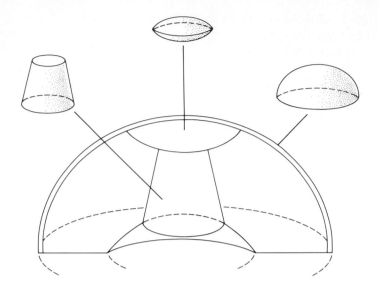

gymnosperms these divisions can be frequent enough that the individuality of these layers is lost (Fig. 6.14; Klekowski, 1988; Ruth et al., 1985). In the monocots, the periclinal divisions are most common in the inner layers of the tunica if it is multistratose, but the outermost layers tend to have only anticlinal divisions. It is now common to use the terms **mantle** and **core** instead of tunica and corpus, respectively,

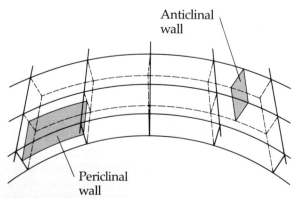

Figure 6.13
An anticlinal wall is perpendicular to the surface of a tissue or organ. Periclinal walls are parallel to the surface, and they increase the tissue's potential thickness.

whenever periclinal divisions occur in the outermost layer (Popham and Chan, 1950). This is not just a semantic distinction; an important corollary to the lack of periclinal divisions in the tunica is that the tunica, because it cannot grow in thickness, cannot contribute to the corpus. Therefore, the two zones are absolutely distinct, separate populations of cells, which could possibly differ in other fundamental ways.

The next step in our understanding of meristems was the discovery that the corpus itself is usually not homogeneous but contains several zones (Fig. 6.12; Boke, 1941; Foster, 1938, 1941). A variety of different types of zonation has been reported (Popham, 1952, 1960), but the one that is most common and has been studied most thoroughly is the **usual angiosperm type**. In this type of meristem a tunica of one or more layers overlies a corpus, and the corpus itself is composed of three zones: an uppermost zone of **central mother cells**, below this the **pith-rib meristem**, and surrounding both the **peripheral zone** (also called the **flank meristem**), which is shaped like a truncated hollow cone (Mauseth and Niklas, 1979). These zones can be recognized on the basis of several features:

In Figure 6.13, the labels read:

Anticlinal wall

Periclinal wall

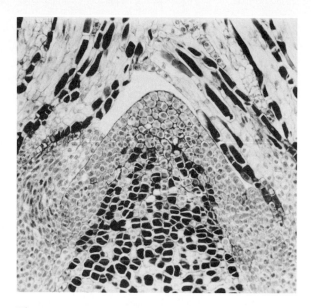

Figure 6.14
This is an apical meristem of *Pinus* (pine), and it has all of the zones that are shown in Figures 6.11 and 6.12, with the central mother cells being rather easy to see. Look at the outermost layer of the apex very carefully and you will see three or four clear examples of periclinal divisions; this is a mantle/core, not a tunica/corpus. × 160.

1. The sizes and shapes of the component cells. The central mother cells are usually much larger than those of the other zones and tend to be cuboidal, whereas the cells of the peripheral zone and the pith-rib meristem are smaller and tabular (brick-shaped—Niklas and Mauseth, 1981).

2. The density of the protoplasm when stained with various histological or histochemical dyes. The peripheral zone cells stain most densely, the pith-rib meristem cells less so, and the central mother cells least. The differences are a function both of the concentration of proteins and ribosomes in the cytoplasm and of the amount of vacuoles in each cell (Table 6.1; Lyndon and Robertson, 1976; Havelange and Bernier, 1974; Havelange et al., 1974; Mauseth, 1981a, 1982b, 1984a).

3. The planes of cell division. The cells in every zone are able to divide in any plane, but it is only in the central mother cell zone that divisions seem to occur more or less equally in all planes. In the pith-rib meristem, division is predominantly such that the new cells are aligned vertically, and rows of cells result (these rows are called **ribs** in older terminology, thus the term *pith-rib* meristem). Longitudinal divisions also occur, increasing the number of rows (Figs. 6.11, 6.14; Boke, 1951). The peripheral zone cells also divide predominantly anticlinally, giving rise to rows of cells, just as in the pith-rib meristem. Occasional periclinal divisions increase the number of rows, and the thickness of the peripheral zone increases toward its base.

4. The ultrastructure of the component cells. At first glance, most apical meristem cells appear remarkably similar ultrastructurally (Chandra Sekhar and Sawhney, 1985; Cragg and Willison, 1980; Cecich, 1977; Sawhney et al., 1981), but quantitative techniques have shown that each zone is composed of cells that have characteristic architectures (Table 6.1; Lyndon and Robertson, 1976; Havelange and Bernier, 1974; Havelange et al., 1974; Mauseth, 1981a, 1982b, c, 1984a; Orr, 1981, 1985).

All of the foregoing supports an interpretation of the shoot apical meristem as a collection of several distinct meristems that function together to establish the pattern of the shoot being produced (Gahan, 1981; Mauseth, 1982c). Hanstein in 1868 had proposed a similar idea, dividing the meristem into distinct groups of initials: the **dermatogen** (which would produce the dermal tissues), the **periblem** (which would produce the cortex), and the **plerome** (which would produce the central tissues of the shoot). The present-day concept, although somewhat similar, is different in important ways. Hanstein thought of his zones as **histo-**

gens, groups of initials that differ fundamentally from one another and are capable of producing only certain types of cells. Notice that he proposed this in 1868, long before the concept of genes or DNA-based differentiation would become established. In the morphogenic concepts of those times, the various histogens were thought to differ from one another in fundamental ways as much or more than one species of plant or animal varied from others. Thus there could be no flow of cells from one histogen to another. Our concepts today are based on the premises that (1) all cells have equal potential and (2) as cells flow from the central mother cell zone into the peripheral zone or the pith-rib meristem, their metabolism is adjusted (as witnessed by the changes in size, shape, and ultrastructure) and that perhaps even some genes are activated and others repressed. The peripheral zone produces the cortical cells not because the peripheral zone cells are fundamentally unique (as the periblem was conceived to be) but because its metabolism has been so channeled. If somehow those same cells were located in the pith-rib meristem, their metabolism would be altered, and pith cells would be produced by them. It has been suggested that at least part of the reason that the cells of each zone have unique metabolisms is that the differentiation of their derivative cells has already begun even while the cells are still in the zones (Gahan, 1981; Mauseth, 1982c).

A variety of other types of corpus zonation has been described, but only the **Opuntia type** seems to be at all common. It is more or less identical to the usual angiosperm type except that a **cambium-like transition zone** is located between the central mother cells and the pith-rib meristem. This zone seems to be a temporary phenomenon, varying during the plastochron and even temporarily disappearing entirely in certain species. Of the other types of zonation, very little is known aside from the original reports.

General Features of Vegetative Shoot Apical Meristems

The actual size and shape of shoot apical meristems in angiosperms are extremely variable. They can be as small as 80 μm in diameter or as large as 1500 μm—this range occurs even within a single family, the Cactaceae (Boke, 1953, 1954; Mauseth, 1978b). Meristems as large as 3,000 μm wide occur in *Cycas revoluta* (Foster, 1940). In the only family studied (Cactaceae again), seedlings have meristems that are much smaller than those of the adults of the same species (Fig. 6.15). In some (*Pereskia, Rhipsalis*) the apices enlarge relatively little with age; in others the increase is tremendous: in *Homalocephala* and *Echinocactus* the adult apex is three thousand times as large as that of the seedling (Mauseth, 1978b). Meristems tend to be rather hemispherical in shape (Figs. 6.1, 6.2, 6.11; Gardner et al., 1985; Tucker, 1985), but they can be quite elongate like a parabola

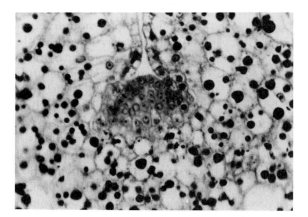

Figure 6.15
Whereas all the apical meristems in the previous illustrations have been medium-sized, this one of a seedling of *Melocactus fortalezensis* is extremely small, being only six cells in diameter. It has only a tunica/corpus organization, but older plants of this species have larger apices and complete zonation. This apex had already produced two leaf primordia (not visible). × 400.

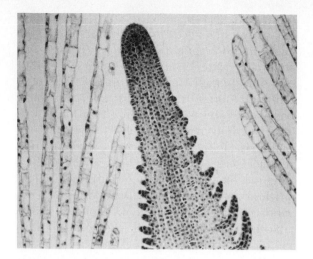

Figure 6.16
An apex of *Elodea* is extremely elongate; its height above the youngest leaf primordia is almost twice its diameter at the level of the primordia. × 160.

(Figs. 6.14, 6.16), or flat, and even concave in some seedlings (*Mammillaria brandegeei*). The shape changes as meristems go through their cycle of leaf production. (The time between the initiation of one leaf and the initiation of the next is called a **plastochron.**) The height of a

meristem is measured from the axil of the youngest leaf primordium, because the occurrence of a leaf primordium at a site indicates that a dramatic event has occurred and that there is more than just shoot apical meristem metabolism occurring there (Figs. 6.17, 6.18). As soon as a new leaf primordium is recognizable, our reference point for measuring apical height and diameter is changed, and the shoot apical meristem, by definition, becomes shorter and narrower (Fig. 6.17). After the initiation of a leaf, several hours or days pass before the initiation of the next. During this time, the apical meristem is growing, but our point of reference stays the same (the youngest leaf axil), so the apex becomes taller and wider. When the next primordium is recognizable, the reference point shifts to it, and suddenly the apex is short and narrow again. We say that the shoot apical meristem goes through "maximal" and "minimal" phases during each plastochron, but the distinction is really artificial: if the regions between leaf primordia are examined, they merely grow uniformly and do not cycle like this. However, the concept of maximal and minimal phases is a useful method for measuring how far into the plastochron a shoot apical meristem has progressed.

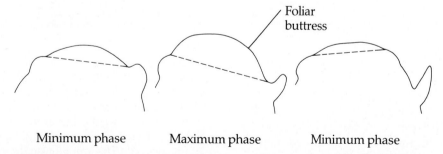

Foliar
buttress

Minimum phase   Maximum phase   Minimum phase

Figure 6.17
Most leaf primordia become visible as a swollen area on the apex, an area called the foliar buttress. In measuring the diameter and height of the apex, the swollen area causes the apex to be large if it is considered part of the apex, but it causes the apex to be small if it is considered to be part of the leaf, thus redefining the boundaries of the apex.

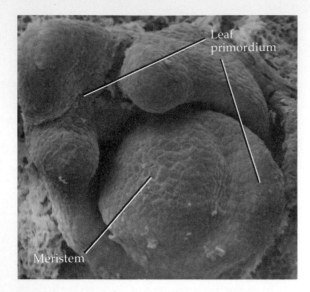

Figure 6.18
This SEM shows an apex of *Ranunculus repens*
(buttercup) just after a foliar buttress has become
so distinct that it is obviously a leaf primordium, so
the apex is in its minimum phase. In the time required for the formation of this primordium,
the next older leaf primordium has developed
lobes and become rather large. × 150. Micrograph
generously provided by R. D. Meicenheimer.

In some species, leaf primordia tend to be
large (Fig 6.18), while in others they are rather
small. Leaf primordia must achieve a certain
shape and distinctness from the meristem before we are able to identify them and call them
primordia; when a large primordium is formed
on a small apex (which is not rare), the growth
of the primordium before it becomes truly distinct causes the meristem to bulge. Not yet
willing to call this a primordium, we use the
term **foliar buttress**, and the apex is said to
be in its maximum phase, as noted earlier
(Fig. 6.17).

So far, the shoot apical meristem has been
treated as a mass of cells that grows uniformly,
aside from the production of leaf primordia. If
all cells of a meristem grow and divide at absolutely uniform rates, then a straight shoot will
result that grows upward; whichever cell or

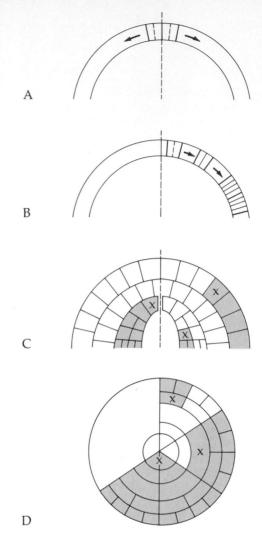

Figure 6.19
(A) If a cell can remain at the center of the apex,
it will contribute cells continuously to the axis.
(B) But if it is on the side or interior of the apex, it
will be "pushed out" (actually the apex will grow
away from it); once it is out of the meristematic region, it will stop dividing and will have produced a
finite patch. (C) This can be visualized in a chimera; a periclinal chimera is analyzed with respect
to the number of layers it affects. (D) A sectoral
chimera is analyzed with regard to the circumferential portion it affects. X marks the position
in which a mutation arises, forming the chimera.
The patch of cells derived from the mutant cell is
shaded.

cells are at the absolute apex and are perfectly aligned with the axis of uniform growth will remain permanently in that position (Fig. 6.19). As any of them divide, one daughter cell will not be aligned with the axis so that as the other, inner daughter cell (which is still aligned) expands, it pushes the nonaligned one farther out. The inner, aligned cells act as apical initials. But if one of these inner cells (or one of its recent derivatives) begins to grow more rapidly than the others, the center of most active growth will shift toward these cells, and the cells that are growing more slowly will not be on the axis. They will begin to flow out of the meristem, and whichever cells are aligned around the new axis will be a new set of initials. They can retain this role only until they slow down or a new center of more rapid growth develops. Thus, even in large meristems, if the central mother cells are mitotically active, there should be one or several apical initials that can be replaced as the growth center shifts.

Such a theory fits perfectly with experimental data on **polyploid chimeras** (Dermen, 1945, 1947; Klekowski, 1988; Stewart and Dermen, 1979; Satina et al., 1940). If shoot apical meristems are briefly exposed to colchicine to temporarily block mitosis, those cells that are in metaphase while the colchicine is present cannot complete chromosome movement, because the drug disrupts the spindle microtubules; but the other events of mitosis do continue. New nuclei are formed, but because the chromosomes have not moved from the metaphase plate, they are all incorporated into one nucleus that is twice as large as it should be; this polyploid nucleus is easily recognizable by its size (Fig. 6.19). If the colchicine is then washed out and the meristem is allowed to grow for several days or weeks, the cells, the polyploid one included, will grow and divide normally. The descendents of the polyploid cells will be recognizable as a patch of cells with large nuclei.

When many individuals are treated this way

and then analyzed, many types of patches result. If the affected cell had been in the subapical region and had almost finished all of its divisions at the time of colchicine treatment, then it would give rise to only a small patch of cells. If the affected cell had been higher in the meristem and still had much growth and division ahead of it, then the resulting patch would be larger. The largest patches ever found included either half or one-third of the circumference of the shoot, indicating that the plant had had two or three initials at the time of treatment (Dermen, 1945). In other cases, the largest patches involved only one-fourth or less of the circumference, indicating that there had been four or more initials. Some of the patches extended into the apical meristem, indicating that the affected cell was still acting as an initial. In other cases, the patch did not reach all the way to the apex: the growth center had shifted, causing the initial to lose its central position, allowing it to be pushed out by the cells of the new growth center.

In plants that have quiescent central mother cells (see next section), such large patches should not be possible; instead, all of the uppermost cells of the peripheral tunica, peripheral zone, and pith-rib meristem that border on the central mother cells would act as the active initials. There would be a cup-shaped layer with dozens of cells. The experiments needed to verify this have not yet been performed.

When the chimera is analyzed with regard to the amount of circumference it affects, it is considered a **sectoral** or **mericlinal chimera** (Fig. 6.19). If it is analyzed with regard to the number of layers it affects, it is a **periclinal chimera**. On the basis of studies of periclinal chimeras in several dicots and numerous monocots (Stewart and Dermen, 1979), it has been concluded that the apical meristem contains three layers of initials: LI, LII, and LIII. LI is the outermost and produces the epidermis; LII is immediately below it and produces the one to several layers just below the epidermis; LIII, the innermost, produces the bulk of the

shoot axis tissues. Such an organization probably occurs only in plants that have small meristems with active central mother cells. As
mentioned earlier with regard to sectoral
chimeras, plants with large or inactive central
mother cell zones should not have such an organization, but the experiments to determine
this have not yet been performed.

Activity of the Shoot Apical Meristem

Perhaps because of their dynamic nature as
much as the importance of their product,
shoot apical meristems have been the subject
of a great many studies and several comprehensive reviews (Cutter, 1965; Gifford, 1954,
1983; Green, 1980; Halperin, 1978; Hicks, 1980;
Steeves and Sussex, 1972). It is now firmly established that in some species, such as *Zea
mays* (Clowes, 1961, 1978a), the entire shoot
apical meristem is mitotically active and cells
"flow" from the central mother cells into the
peripheral zone and pith-rib meristem or from
the central tunica into the peripheral tunica.
From these, they continue to "flow" into the
developing cortex, pith, or epidermis, respectively. In other species, such as *Helianthus annuus* (Steeves et al., 1969), the central mother
cells are not active mitotically; tritiated thymidine is not incorporated into DNA in these
cells. The active meristem consists only of the
peripheral tunica, peripheral zone, and pith-
rib meristem. In at least one species, both
types occur in the same plant. In *Opuntia polyacantha* the long shoot apical meristems have all
zones active, but the short shoot apices, with
zonation identical to that of the long shoots,
have the central mother cells and pith mitotically inactive (Mauseth, 1976, 1984a; Mauseth
and Halperin, 1975). The idea of a quiescent
center in the shoot apical meristems was advo-
cated by Buvat (1955) and Nougarède (1965),
who divided the shoot apical meristem into a
méristème d'attente (**waiting meristem**) and an
anneau initial (**initiating ring**). The concept

was a more extreme version of the multistep
meristem idea, in that all vegetative growth
was postulated as being due to activity in
the initiating ring, whereas all reproductive
growth would be the responsibility of the waiting meristem. The latter maintained its genetic
material in safe condition by being completely
inactive mitotically and as inactive as possible
physiologically. It "waited" until it was the
proper time to flower; then the waiting meristem swelled and was transformed into the floral apex. This is perfectly reasonable and completely analogous to the situation in animals
(Klekowski, 1988): human spermatogonia are
held as inactive as possible until puberty,
about 12 to 14 years after birth, and primary
oocytes (which are the precursors of eggs) can
remain dormant for up to fifty years.

Unfortunately, the merits of this idea seem
to have been overlooked, perhaps because the
méristème d'attente concept became linked to
an earlier idea of control of leaf initiation, in
which theoretical leaf-forming stimuli were
postulated to move upward in spirals emanat-
ing from pre-existing leaves termed multiple
foliar helices (Plantefol, 1947). This hypothesis
was not accepted then and still has no experimental evidence in its favor. Reaction to this
theory was strongly negative, and regrettably,
Nougarède's very good idea was thrown out
with Plantefol's hypothesis.

During the dispute over the active vs. inactive central mother cells, much experimental
work was done that has great importance aside
from its direct results on shoot apical meristems. The experiments showed that simply
counting mitotic figures was not a reliable
guide to whether or not one region was more
or less mitotically active than another: with
similar activity, a region with larger cells will
automatically have fewer nuclei, therefore
fewer mitotic figures (Clowes, 1961). To replace simple counts, the idea of **mitotic index**
was developed. In this method, the number of
mitotic figures is divided by the total number
of nuclei present, eliminating differences in

cell size. This too was shown to be only partially reliable: it will work only if the two groups of cells that are being compared spend equivalent percentages of their cell cycles in mitosis. But if one group requires, for instance, 10% of the cell cycle time to complete mitosis while the other population requires 30%, the second group will have a mitotic index three times higher than the first, even if the same percentage of cells are dividing and total cell cycle times are equal. Such differences are entirely possible and not at all unusual. The most reliable method for evaluating mitotic activity is to follow rather laborious double-labeling techniques with radioactive tracers (Mitchison, 1971). These techniques have been followed with shoot apical meristems and have shown what was mentioned earlier—some do have active central mother cells, but in others these cells are either totally quiescent or cycle very slowly.

Branching

Dichotomous Branching

In the vascular cryptogams, branching results from a dichotomous division of the apical cell. If only a single prominent apical cell is normally present, it first undergoes a median division to produce two daughter cells of similar size. Then, as these both cut off daughters on all sides, the two initials move away from each other, each organizing a shoot below itself. Alternatively, in some plants (*Lycopodium complanatum*), the central initials stop dividing, and two new growth centers arise on the sides of the apex; they organize new apical cells and then begin growth as two new apices.

Dichotomous branching is extremely rare in the seed plants, being confirmed in only the following: (1) palms—*Hyphaene* (Hallé and Oldeman, 1970), *Nypa fruticans* (Tomlinson, 1971), and *Chamaedorea* (Fisher, 1974); (2) Flagellariaceae—*Flagellaria indica* (Tomlinson and Posluszney, 1977); (3) Asclepiadaceae—*As-*

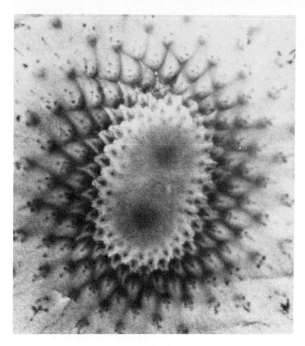

Figure 6.20
This is a cross section through an apex of *Mammillaria perbella* just as the apex is dividing dichotomously. At this point the apex has become elongate, and two groups of central mother cells are visible. A careful examination of the spirals of leaves will show that this is one apex dividing, not an apex and an axillary bud. × 17. Micrograph generously provided by N. H. Boke.

clepias syriaca (Nolan, 1969); and (4) Cactaceae—*Echinocereus reichenbachii*, *Mammillaria parkinsonii*, and *M. perbella* (Boke, 1976; Boke and Ross, 1978). In the mammillarias the apices, which normally are large, expand even more. The central mother cell zone and the pith-rib meristem become extremely wide; then pith-rib activity at the geometric center of the apex decreases (Fig. 6.20). Simultaneously, a depression forms between the two daughter apices as these two regions begin to grow more rapidly than the center. Finally, the region of the original central mother cell zone is converted to two peripheral zones, and the dichotomy is complete (Boke and Ross, 1978).

Axillary Buds

The typical method of branching in seed plants is by the formation of axillary meristems just above each leaf primordium (Figs. 6.1, 6.7). Two somewhat different methods have been reported. In the first type, after a leaf primordium has been established, the apical meristem grows upward, leaving it behind, and the cells around it begin differentiating into stem tissues (Fink, 1984). In most cases, this involves an increase in the relative volume of the vacuoles, but this does not happen to the cells immediately above and adaxial to the leaf primordium. These remain as a small cluster of meristematic cells, often with just a tunica-corpus organization. In many species, they become demarcated from the shoot by the formation of a **shell zone**, a concave cup of flat, tabular cells that are temporarily meristematic (Shah and Patel, 1972). The shell zone persists briefly; then its cells convert to bud tissues such as procambium, ground meristem, or pith. This allows vascular tissues to differentiate through the zone to the axillary meristem (Larson and Pizzoloto, 1977). The meristem can remain simple or can organize a fully zonate corpus and begin initiating its own leaf primordia, which usually develop into protective bud scales (Fig. 6.21).

In other species, the axillary meristems are established as all the tissues around the original leaf primordium begin to become more vacuolate, but then the axillary cells stop enlarging their vacuoles. They reduce the size of their vacuome and become more cytoplasmic again, forming the axillary meristem. This method for the formation of an axillary meristem requires a redifferentiation of the partially mature parenchyma cells back to the meristematic condition. After redifferentiation, the processes of development are as already described.

If there is strong apical dominance, these buds will become quiescent until the active shoot apical meristem on the main shoot dies or grows too far away to be able to exert its in-

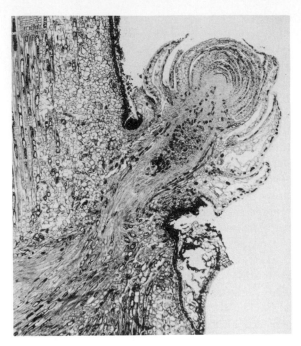

Figure 6.21
Although an axillary meristem begins as just a small group of cytoplasmic cells, they usually develop into a large apical meristem, especially if they are induced to grow out as a lateral branch. This bud of *Taxus baccata* (English yew) has several leaf primordia and a strong connection to the vascular tissues of the main shoot. × 30. Micrograph generously provided by S. Fink.

fluence over the bud, which can then resume activity and grow into either a vegetative or a reproductive branch (Steeves and Sussex, 1972). Such interrupted development is termed **prolepsis** (Hallé et al., 1978), and such buds are **proleptic buds**. In many plants, especially those in warm tropical areas—for example, *Myrsine floridana* (Wheat, 1980)—the axillary meristems do not undergo a dormant period but rather develop continually (**syllepsis**) and are **sylleptic buds**.

Adventitious Buds

Adventitious buds occur in many plants; they are defined as any buds that do not arise in the axil of a leaf. They are commonly produced on roots, as in *Salix, Clusia rosea, Desmaria, Notanthera* (Kuijt, 1982) and *Stichianthus minutiflorus*, where even adventitious flower buds form (Winkler, 1931); in leaves, as in *Kalanchoe*; in invasive tissues of parasites as in *Tristerix aphyllus* (Mauseth et al., 1984) and *Pilostyles* (Rutherford, 1970); and in bark, as in *Tilia platyphyllos* (Fig. 6.22) and *Fraxinus excelsior* (Fink, 1983)—especially the bark of tropical trees that bear their flowers on the mature trunk. These last are the **cauliflorous plants**: *Artocarpus integrifolia, Swartzia schomburgkii,* and *Couroupita guianensis* (Fink, 1983).

Adventitious buds can be extremely important in the architecture of the trees in which they occur. They may be either **exogenous**, arising from superficial layers of cells as do the axillary buds, or **endogenous**, arising deep within the tissues. If an exogenous bud, either axillary or adventitious, remains inactive for several years, it may become overgrown by the bark and appear to be endogenous (Fink, 1983, 1984).

Development of the Shoot Apical Meristem

Phylogenetic Development

It is at least theoretically possible to trace the origins of shoot apical meristems all the way back to the green algal ancestors of the land plants. In the green algae, many types of meristem occur, apical as well as basal and intercalary. With the transition to land, basal and intercalary meristems are at a disadvantage. It seems likely that the first land plants were rhizomatous; that is, the main shoot was prostrate on or in the mud of lake and river banks. With this condition, the horizontal shoot is held rather firmly by the substrate, so growth by means of a basal or intercalary meristem would be difficult. An apical meristem, which is the only moving part, would permit easier growth. Even on upright shoots, basal and in-

Figure 6.22
This bud of *Tilia platyphyllos* (large-leaved lime) is buried in the bark because it is not an axillary bud but an adventitious bud. It was produced when bark parenchyma began to proliferate, then became organized and formed a small shoot apex. × 45. Micrograph generously provided by S. Fink.

tercalary meristems suffer the disadvantage of having to support the weight and flexion of the organ that they produce. Also, all water and nutrient transport between shoot and root system would have to pass through the meristematic zone, which would be an interruption of the vascular tissues; this would not be advantageous.

Apical meristems suffer from none of these disadvantages; the primary problem that they face today is that, in being elevated, they are more exposed to insect and herbivore predation—a problem that was clearly not quite so pressing 420 million years ago, 80 million years before the advent of land animals. Thus shoot apical meristems were established very early, the cryptogamic type with a single apical cell evolving first. We are fortunate enough to have available several fossil sphenopsids that have the shoot apical meristems beautifully preserved. These have single pyramidal cells with three cutting faces in *Sphenophyllum* (Good and Taylor, 1972) and three or four cutting faces in two species of *Calamites*. Unfortunately, we do not have fossils of the trimerophyte or progymnosperm apices, which might be the groups where the transition to the seed plant type occurred (Stewart, 1983).

Ontogenetic Development

In the cryptogams, the first division of the zygote frequently produces a suspensor cell and an embryogenic cell (Foster and Gifford, 1974). The first divisions of the embryogenic cell result in a tetrad of four cells. In many cases, especially for certain ferns, it is believed that the apical initial is set aside as one of these four cells. In other cases, only after the embryo has become slightly enlarged is the apical initial recognizable.

In the seed plants, the first division of the zygote also typically produces a suspensor cell and an embryogenic cell (Chapter 19). Further development is variable, but often the embryogenic cell divides twice to reach the four-cell stage, then passes to the eight-cell stage by means of curved phragmoplasts that divide each cell into an inner member and an outer one (Figs. 20.9, 20.10). This division establishes the precursors of the outermost tunica. Between this stage and germination of the seed, little is known. From a survey of germination and seedlings in the Cactaceae (Mauseth, 1978a), it was discovered that, although all adult members of this family have either the usual angiosperm type of shoot apical meristem or the Opuntia type, at germination the embryos could have various degrees of zonation. All had at least the tunica-corpus level of zonation (Fig. 6.15); some had fully zonate corpuses; in others the central mother cells were distinct in the corpus but the peripheral zone and pith-rib meristem were not; and many had the corpus completely homogeneous. Those

that lacked complete zonation developed it within several weeks after germination. An analogous situation was found in *Picea abies* (Gregory and Romberger, 1972) and *Pinus* (Cecich and Horner, 1977; Riding and Gifford, 1973). Thus zonation can develop in the embryo, during germination, or in the growing seedling. It is interesting that normal shoot and leaf initiation and development occur with zonation, without it, or while it is partially developed.

The Reproductive Shoot Apical Meristem

Within the angiosperms, a large number of types of flowers and inflorescences have evolved. Therefore, it might be expected that there is also diversity in the apical meristems involved in their production. In considering reproductive shoot apices, it must be remembered that some plants can have at least four types of shoot apical meristems: juvenile vegetative apices, adult vegetative apices, inflorescence apices, and floral apices.

Very few angiosperms are capable of flowering immediately after germination; they must go through a **juvenile period** that lasts from several weeks to many years before they become **adult** and capable of flowering. The anatomy and general habit of juvenile plants are often more or less identical to those of the adult, but in some the two phases are noticeably different (Figs. 6.23, 6.24, 6.25; Dobbins et al., 1983). Citrus and ivy are frequently given as examples (Frydman and Wareing, 1974; Stein and Fosket, 1969), but plants that form **cephalia** are even more dramatic: the cephalium is the adult phase of the plant. Examples include *Melocactus* (Fig. 6.26), *Backebergia*, *Cephalocereus* (Gibson, 1978b; Rauh, 1979a). A study of the shoot apical meristems during the transition from juvenile to adult in *Melocactus matanzanus* revealed that the apex remained remarkably constant: zonation, shape, and cell shape

Figure 6.23
The shoots of *Marcgravia rectifolia* are dimorphic: the upper shoot is the juvenile stage, the lower shoot is the adult stage. × 0.5. Photograph generously provided by D. R. Dobbins, H. Alden, and D. Marvel.

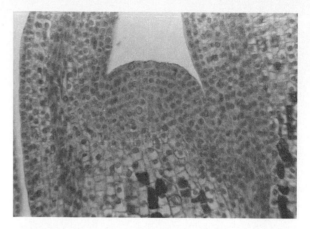

Figure 6.25
This is the shoot apical meristem of the adult phase of *M. rectifolia*; it is more rounded than the juvenile phase apex, but in general it is quite similar at the light microscope level. × 272. Micrograph generously provided by D. R. Dobbins, H. Alden, and D. Marvel.

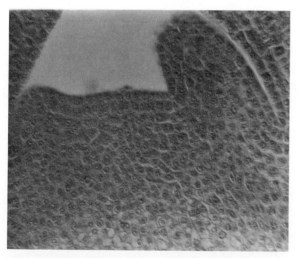

Figure 6.24
This is the shoot apical meristem of the juvenile phase of *Marcgravia rectifolia;* compare it with Figure 6.25. × 272. Micrograph generously provided by D. R. Dobbins, H. Alden, and D. Marvel.

Figure 6.26
This is *Melocactus matanzanus,* which also is dimorphic like *Marcgravia* (Fig. 6.23). The lower portion with the large spines is the juvenile phase; when it became reproductively mature, it began its adult growth, the cephalium. The two parts are formed by the exact same meristem, which undergoes only minor changes during the juvenile/adult transition. × 2.0.

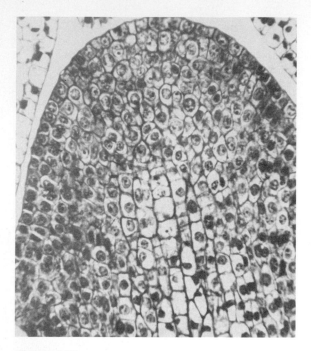

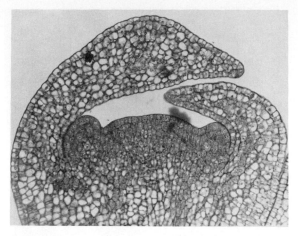

Figure 6.28
The floral apex of *Citrus sinensis* (sweet orange) produces just a few flower parts, and it loses the zonation of the corpus very quickly. At an earlier stage, it had full zonation; at a later stage, there would be even less. × 112. Micrograph generously provided by E. M. Lord and K. J. Eckard.

Figure 6.27
This is a flower apex of *Michelia fuscata*; it has a tunica/corpus organization, and the corpus also has a peripheral zone, central mother cells, and a pith-rib meristem. It seems possible that this full zonation is related to the fact that this apex will produce a large number of flower parts (many stamens and carpels), not just a few as in *Citrus* (Fig. 6.28). × 410. Micrograph generously provided by S. C. Tucker.

within each zone did not change (Niklas and Mauseth, 1981). The shoot apical meristem was able to produce two totally different types of shoot without any detectable changes at the light microscope level. In *Marcgravia rectifolia* also, the two types of apices are remarkably similar (Figs. 6.24, 6.25; Dobbins et al., 1983).

In some plants the transition from the vegetative condition to the reproductive one can be reliably controlled by light treatments. These have been the most favorable plants for study, because exact time sequences can be obtained. Many studies are therefore centered on the

conversion of an actively growing vegetative apex into an active reproductive one (either a floral apex or an inflorescence apex). In these plants, a tentative generalization is that, if the reproductive apex must produce many primordia—many bract primordia if it is an inflorescence apex, many flower parts (sepals, petals, stamens, and carpels) if a floral apex—then it retains the full zonation of the vegetative apex (Fig. 6.27; Tucker, 1960). If it must function only briefly, producing just a small number of flowers or flower parts, zonation in the corpus is lost (Fig. 6.28; Lord and Eckard, 1985). The apex enlarges, often dramatically, and most cells in the center lose the pith-rib nature and have more of a central mother cell appearance or actually look like young parenchyma. The outer layers continue to be very cytoplasmic, resembling a multistratose tunica; this resemblance is even more pronounced as the interior cells become more vacuolate

(Corson, 1969; DeMaggio and Wilson, 1986; Havelange, 1980; Havelange and Bernier, 1974; Havelange et al., 1974; Orr, 1981, 1985; Tucker, 1960).

Many physiological and cytohistological studies have been made of the transformation of the vegetative apex to a floral one. Details differ from species to species, but in general the changes are what would be expected considering that the reproductive apex produces its primordia much more rapidly than the leaf primordia are produced by the vegetative apex. There is a rise in the mitotic index, a related increase in DNA synthesis, and an increase in nucleolar diameter, presumably as new ribosomes are prepared that will be necessary for the increased rates of protein synthesis in the metabolically more active cells (Havelange, 1980; Havelange and Bernier, 1974; Havelange et al., 1974; Orr, 1981; Marc and Palmer, 1982).

It should be emphasized, however, that many, perhaps most, floral apices never pass through a vegetative phase: they are initiated as floral apices. One needs only think about the florets (miniature flowers) of a dandelion or sunflower or any of the numerous plants that produce flowers (not flowering branches) in leaf axils. These are more difficult to study, but it seems safe to draw a generalization that these have only a tunica-corpus organization, that the corpus itself is not zonate.

In sum, however, despite the critical importance of flowers to the biology and even the concept of the angiosperms, very little is known about reproductive apices, and the generalizations drawn must be regarded as somewhat tentative. Exceptions must always be expected, especially when they would be logical.

Death of the Shoot Apical Meristem

In a variety of species, the shoot apical meristem is **determinate**; that is, after producing a certain number of leaves, it dies. Although unusual, this death is a normal part of its mor-

phogenesis. In *Gleditsia* (Blaser, 1956), *Ulex europaeus* (Bieniek and Millington, 1967), and *Euphorbia aggregata*, the axillary buds begin growth in a normal fashion but soon begin to produce only small, scale-like leaves. Subsequently, a transformation of the apex begins, and leaf production ceases. Mitotic activity rises throughout the apex, then stops completely, and all cells of the meristem begin to elongate and sclerify, especially the cells of the tunica and central portions of the corpus. With this transformation of the meristem cells into fibers, the meristem itself is converted to an elongated, needlelike thorn. In *Carissa grandiflora* (Cohen and Arzee, 1980), which similarly produces spines, the apical meristem loses all zonation in the corpus, then dies; neither differential activity nor conversion to fiber cells was reported. In *Salix pentandra* (Juntilla, 1976) and *Syringa vulgaris* (Garrison and Wetmore, 1961), the apical meristem is not modified, but rather becomes cut off from the shoot below it by the formation of a corky layer. This isolates the meristem, causing it to die and be sloughed off. In several species, the apical meristem either becomes disorganized and parenchymatous, as in *Alstonia* (Figs. 6.29, 6.30; Mueller, 1985) and *Opuntia*, or it shrinks and becomes sclerified, as in *Pinus* (Fig. 6.31; Ewers and Schmidt, 1985) after it finishes leaf production.

The Root Apical Meristem

In Vascular Cryptogams

The root apical meristems in vascular cryptogams resemble the shoot apical meristems in that there is a single prominent, readily identifiable apical cell (Fig. 6.32; Eastman and Peterson, 1985; Gifford, 1983; Kurth, 1981). The cell may have three or four cutting faces that are directed upward and produce cells for the body of the root. In addition, the one face of the apical cell that is directed outward, away

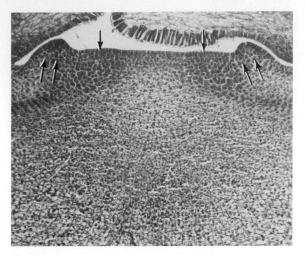

Figure 6.29
Alstonia scholaris has determinate shoots that stop growing after a certain number of leaves have been formed. Then the two uppermost axillary buds enlarge as shown here (single arrows). The two axillant leaf primordia are also visible (double arrows). It would be easy to mistake this for a dichotomy (as in Fig. 6.20) if a careful study were not performed. See next figure for a later stage. × 100. Micrograph generously provided by R. J. Mueller.

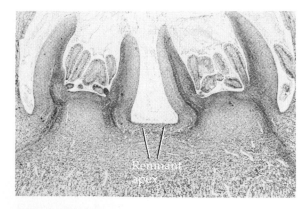

Figure 6.30
This is a later stage of development of *Alstonia scholaris*, showing that the two uppermost axillary buds have already established their own shoots. Notice the remnant of the original shoot apex, located between the bases of the two buds. × 26. Micrograph generously provided by R. J. Mueller.

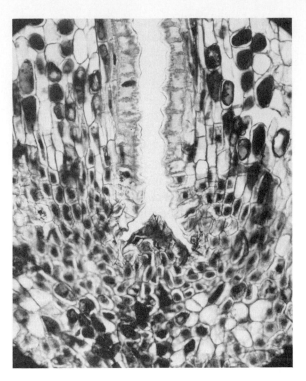

Figure 6.31
The dwarf shoots of *Pinus longaeva* (bristlecone pine) are also determinate, like those of *Alstonia* (Fig. 6.29), but these shrink and sclerify after they finish leaf production. × 165. Micrograph generously provided by F. W. Ewers and R. Schmid.

from the root, is also a cutting face, unlike the corresponding face in shoot apical cells. It is necessary to have cell division along this face in order to produce the precursor cells for the root cap.

As is true of the shoot apex, although all cells of the root can be traced back to this single apical initial, this one cell is not the entire meristem. The cells that it produces are also mitotically active and constitute a multicellular meristem. In 1954, Buvat and Roger-Laird extended the idea of the quiescent méristème d'attente to the root tip of *Equisetum arvense*, and later the idea of an inactive root apical initial was applied to many vascular cryptogams. Recent studies have investigated this carefully

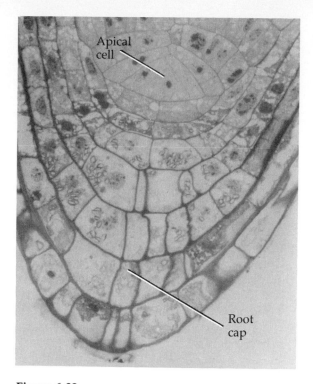

Figure 6.32
The root apex of vascular cryptogams, just like the shoot apex, has a prominent apical cell. Around it are packets of cells that have been derived from it. An important difference is that in the root apical cell, all sides of the cell are cutting faces, the most distal side producing cells that will become part of the root cap. × 670. Micrograph generously provided by A. Eastman and R. L. Peterson.

and have established not only the fact that the entire root apex is mitotically active, but also the cycling time of the apical cell itself: 15 to 37 hours for *Equisetum scirpoides* (Gifford and Kurth, 1982); 28 hours for *Azolla filiculoides* (Gifford and Polito, 1981); and 12 to 25 hours for *Marsilea vestita* (Kurth, 1981).

In Seed Plants

Root apical meristems in the seed plants are similar to the shoot apical meristems in being large, multicellular structures having distinct

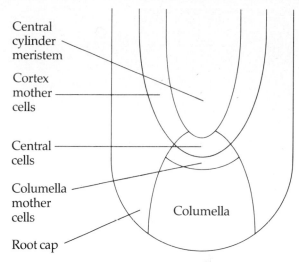

Figure 6.33
Root apical meristems have a zonation that is almost identical to that of shoots (Figs. 6.11 and 6.12), but the names of the zones differ. Also, roots typically possess root caps and consequently have an extra zone, the columella mother cells. Figures 6.11 and 6.12 illustrate the zonation of shoot apical meristems.

zonation. The terminology for the zones is somewhat variable, but four zones are consistently recognized (Fig. 6.33): (1) the **central cylinder meristem** (or **central cylinder mother cells**), (2) the **cortical initials** (or **cortex mother cells**), (3) the **central cells** (or **permanent initials**), and (4) **columella mother cells** (or **calyptrogen**). As the names imply, the central cylinder meristem gives rise to the innermost tissues of the root, generally all of the vascular tissues; the cortical initials produce the cortex; the columella mother cells produce the **columella**, the central region of the root cap (Fig. 6.34); and the central cells act as a promeristem and produce cells that move to the surrounding zones.

An understanding of the functioning of the root apex has been somewhat difficult because two distinct but interconvertible types of root exist and because the activity of the meristem can vary. The simpler of the meristems to

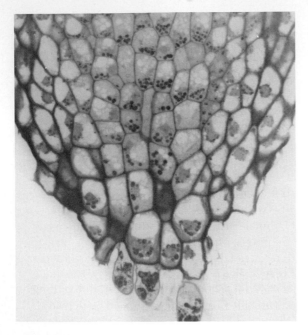

Figure 6.34
This is the root cap of *Phaseolus vulgaris* (kidney bean), showing the central columella cells with their prominent starch grains or statoliths. The starch grains are heavy enough to be always located on the bottom of the cell, so they are commonly assumed to be involved in the perception of gravity. However, although it cannot be seen in this photograph, most of the other organelles are also stratified, so they could just as easily be involved. × 460. Micrograph generously provided by J. S. Ransom and R. Moore.

understand are the **closed meristems**; these are meristems in which the root cap is totally distinct, structurally and developmentally, from the root proper (Clowes, 1981b). If the root epidermis is followed from older regions toward the apex, it can be identified easily even where it reaches the center of the meristem. The root cap and its meristematic cells are located exterior to the protoderm, and the cortical initials are located interior to it. If the root cap is ignored, the root itself grows exactly as does the shoot apex. In an **open meristem**, the cortical initials and the columella

mother cells are not so distinctly delimited, and, at the interface between the two, the cells may at one time produce cortex, at another time produce root cap. There is no separate protoderm, but rather the epidermis develops at times in one layer, at times in another, and it is not unusual for the epidermis and part of the outer cortex to be derived from the root cap cells.

Clowes (1981b) has indicated that the nature of open meristems can be understood on the basis of the **quiescent center** in the root (Fig. 6.35). Whereas a quiescent central mother cell zone is unusual in shoots, a quiescent center is perhaps universal in roots. In closed meristems, the quiescent center includes either just the central cells or the central cells and part of the surrounding zones; it does not include any columella mother cells. The strict boundary of the quiescent center maintains the strict boundary of the apical zones, and the root cap remains distinct from the root.

In open meristems, the quiescent center can fluctuate with time, sometimes extending ahead of the central cells, sometimes not. If, as a root is growing, the quiescent center is small, the cells ahead of the central cells all act like columella mother cells, and the cortical initials are located to the sides of the central cells. If the quiescent center then expands forward, some of the columella mother cells become quiescent. Because these cells now do not divide and flow rapidly forward as they had been doing, cell flow dynamics are altered; the cells on the lateral edges of this extended quiescent center now act like cortical initials and protoderm (Armstrong and Heimsch, 1976; Clowes, 1981b). The cells at the end of the quiescent center, which had been columella cells, are now the columella mother cells. If the quiescent center shrinks, the boundaries will shift again. This flexibility, which allows a cell to temporarily become an initial, was a major argument against the histogen concept as formulated by Hanstein.

The quiescent center cannot be detected

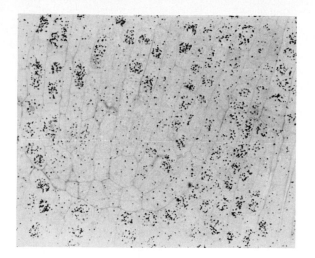

Figure 6.35

The quiescent center is not obvious in ordinary preparations of root meristems, but this is an autoradiogram of *Ephedra chilensis*. The roots were given radioactive thymidine, and those that were synthesizing DNA absorbed it; thus all the nuclei except those of the quiescent center became radioactive. After the material was fixed, microtomed, and placed on slides, it was coated, in the dark, with photographic emulsion and left for several days. The radioactivity "exposed" the emulsion, the slides were developed just like a photograph, and the result is clusters of silver grains over the radioactive nuclei and nothing over the nonradioactive quiescent center nuclei. × 340. Micrograph generously provided by R. L. Peterson and J. Vermeer.

without special techniques. Counts of mitotic figures and mitotic indices can give a hint of the location of the quiescent center, but it is best to supply tritiated thymidine and make autoradiographs (Fig. 6.35; Peterson and Vermeer, 1980; Torrey and Feldman, 1977). The quiescent center is then visible as a central region in which few or none of the nuclei have incorporated any of the thymidine, indicating that the cells are either completely quiescent or, more commonly, have very long cell cycle times.

If these cells were permanently quiescent, they would be useless. They are, like the central mother cells of the shoot, a reserve of ge-

netically healthy cells. Because the root apex is pushed through the soil, it is in constant danger of being damaged. The root cap is an effective but imperfect means of protection. If a large part of the root apex or root cap is damaged, the quiescent center disappears immediately as all cells become mitotically active. This results in the formation of a whole new apex derived from the central cells. Once the damage has been eliminated and a new, functional apex established, the central cells immediately become inactive, and a new quiescent center is formed. No matter how often the apex is damaged, as long as the quiescent center is not destroyed, a new apical meristem can be regenerated (Ball, 1956; Barlow, 1974; Feldman and Torrey, 1976).

Ultrastructure of Meristematic Cells

Apical meristems were among the first tissues ever studied by electron microscopy and have received attention ever since. It was concluded that meristematic cells resemble each other strongly because only rarely were distinct differences observed between the several zones of any meristem, as the quiescent center of roots (Clowes and Juniper, 1964) and the central mother cells of *Helianthus* (Sawhney et al., 1981) and *Pinus* (Cecich, 1977). However, quantitative studies of root apical meristems (Moore and McClelen, 1983a; Moore and Pasieniuk, 1984), shoot apical meristems (Havelange, 1980; Havelange and Bernier, 1974; Havelange et al., 1974; Mauseth, 1980c, 1981a, 1982a–c, 1984a), as well as leaf and spine primordia (Mauseth, 1982a) have shown that every population of meristematic cells has its own characteristic, recognizable ultrastructure (Table 6.1).

Because the differences between individual meristematic zones are so subtle as to require stereology to detect them, it is possible to give a general description that is reasonably valid for most meristematic cells. The cells tend to be isodiametric and small, between 10 to 20 μm

Table 6.1. The mean relative volumes and standard deviations of cell components in the shoot apical meristems of *Echinocereus engelmannii*. In the top line you can see that the nucleus constitutes only 22.5% of the volume of a cell in the central tunica, but it is 36.2% of the volume of a peripheral zone cell. In the lower set of figures, the volumes are expressed as percentages of the cytoplasm—that is, all of the protoplasm excluding the nucleus and vacuole. Adapted from Mauseth, 1981a.

| No. of samples | Tunica | | Corpus | | |
	Central tunica 9	Peripheral tunica 12	PZ 13	CMC 11	PRM 9
Expressed as a percentage of the protoplast volume:					
Nucleus	22.5 ± 5.3	34.3 ± 7.8	36.2 ± 5.0	28.3 ± 4.3	27.1 ± 4.7
Nucleolus	0.96 ± 0.47	1.63 ± 1.32	1.41 ± 0.64	0.82 ± 0.49	0.90 ± 0.44
Vacuole	35.8 ± 12.2	10.4 ± 5.3	9.68 ± 5.66	17.9 ± 6.7	33.0 ± 8.6
Mitochondria	5.14 ± 1.46	7.81 ± 1.81	6.91 ± 0.78	7.52 ± 1.61	4.58 ± 1.05
Chloroplasts	5.99 ± 2.04	4.48 ± 1.55	6.30 ± 1.88	7.57 ± 2.21	4.55 ± 1.10
Dictyosomes	0.17 ± 0.18	0.58 ± 0.31	0.35 ± 0.13	0.35 ± 0.17	0.36 ± 0.30
Hyaloplasm	30.3 ± 8.1	42.3 ± 5.2	40.4 ± 4.6	38.3 ± 5.0	30.8 ± 5.2
Expressed as a percentage of the cytoplasm volume:					
Nucleus	55.5 ± 14.4	63.6 ± 19.6	67.6 ± 12.3	54.0 ± 13.9	60.4 ± 22.7
Vacuole	95.1 ± 50.1	19.1 ± 9.7	18.4 ± 11.8	35.1 ± 16.4	86.6 ± 33.8
Mitochondria	12.5 ± 0.97	14.1 ± 2.9	12.8 ± 1.7	13.9 ± 1.6	11.3 ± 1.1
Chloroplasts	14.9 ± 5.8	8.20 ± 2.92	11.7 ± 3.6	14.0 ± 3.3	11.3 ± 2.5
Dictyosomes	0.41 ± 0.4	1.07 ± 0.6	0.65 ± 0.24	0.65 ± 0.29	0.90 ± 0.71
Hyaloplasm	72.1 ± 6.7	76.5 ± 4.8	74.8 ± 4.1	71.4 ± 3.26	76.4 ± 2.0

on each side; cells with unusual shapes (fiber-like or stellate) are never found in apical meristems, although certain cells in the vascular cambium are elongate and narrow. The walls are thin and weak, characteristics that allow the growth and expansion of the cells (Figs. 6.36, 6.37). The nucleus is always a dominant part of the cell, frequently constituting as much as 50% of the volume. In many cases, it is the only organelle that is easily detected by light microscopy, because all of the other organelles are so small and few. The vacuome consists of numerous small vacuoles, and the plastids are present as very simple proplastids. In some plants, these may have a few grana, but typically they do not; they also have little chlorophyll, giving the apex a pale, light-green color. Mitochondria are present and constitute about 12% of the cytoplasm. Dictyosomes and ER are usually rather sparse, and only rarely are they observed producing vesicles.

Because the meristems of axillary buds become dormant soon after they are formed, the ultrastructure of their cells differs significantly from that of the active meristems. The dormant cells, which have mostly been studied during winter conditions, have a high content of lipids—up to 4.5% of the cytoplasm volume (Mauseth, 1984a)—and other storage material, and inactive dictyosomes. In *Tilia europaea*, the ER is organized near the periphery of the cells (Cragg and Willison, 1980). Also in *T. europaea*, the nuclear pores are extremely large (175 nm in diameter) in the quiescent bud. The chromatin may become very fibrillar, as in *Tradescantia paludosa* (Booker and Dwivedi, 1973). The dormant cells also are slightly more vacuolate than when they are activated in the spring. Similar ultrastructure occurs in the dormant apex of the main shoot (Lynch and Rivera, 1981).

The ultrastructure of meristematic cells, com-

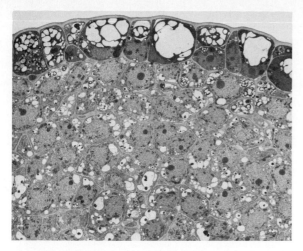

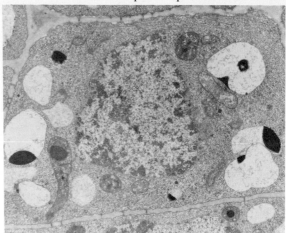

Figure 6.36
This transmission electron micrograph of a tomato shoot apex shows the general features of meristem cells. The tunica is distinct from the corpus, but the central mother cells (center) are not obviously different from the peripheral zone cells. All are cytoplasmic, with large nuclei, small vacuoles, and very thin walls (primary walls only). × 4200. Micrograph generously provided by K. N. Chandra Sekhar and V. K. Sawhney.

Figure 6.37
This is a magnified view of a meristem cell from tomato; note that the plastids are proplastids, all vacuoles are small, and the cytoplasm is filled with ribosomes. × 15,700. Micrograph generously provided by K. N. Chandra Sekhar and V. K. Sawhney.

pared to that of some of the other cell types to be discussed, is relatively simple. But this is somewhat deceptive; it must be remembered that the plant transports minerals, water, and sugars into apical meristems and that it "transports" cells and protoplasm back out in the form of stems, leaves, and roots. These very simple cells are involved in very sophisticated transformations of raw material into plant material. These involve not just the synthesis of

proteins, lipids, nucleic acids, and carbohydrates, but also their assembly into organelles and cells. Although electron microscopy shows a static view of simple proplastids and nondescript mitochondria, in actuality these organelles are rapidly growing, dividing, producing ATP, hormones, and other metabolites. The apical meristems and their cells are extraordinarily dynamic parts of the plant.

Concepts

Xylem is a complex tissue that has many functions in a plant, the three most obvious being (1) conduction of water and solutes, (2) mechanical support of the whole plant and of its parts, and (3) storage of water or nutrients. Because xylem has so many functions, it contains a variety of rather diverse types of cells. The sclerenchyma cells involved in the mechanical support are discussed in Chapter 5, and the parenchyma cells needed for storage are treated in Chapter 3. Here, only the conduction aspects of the xylem will be analyzed. The integration of conduction with storage and mechanical support is discussed in Chapters 11 and 15.

Conduction in the xylem is performed by two types of sclerenchyma cells called **tracheids** and **vessel elements** (Figs. 7.1, 7.2, 7.3), which together are generally called **tracheary elements**. The vessel elements are connected end to end to form a **vessel** (Fig. 7.4). Movement of water through the tracheary elements is affected by two main problems: (1) the ability of water to enter and to exit the element and (2) the ability of the element to avoid collapse. For water to enter or leave the cell, the walls should be as thin as possible, or should even have large holes; but to resist collapse these same walls should be as thick and rigid as possible, with few or no holes. The resulting element must be a compromise between these two conflicting demands: a cell with a thin primary wall that is reinforced in places by a layer of secondary wall. The elements either have no large holes in the primary wall (which makes them tracheids) or they have large holes at each end (then they are vessel elements). At maturity and while functioning, these cells consist solely of walls. The protoplasm has died and completely degenerated; it could only interfere with water flow if it were to persist.

Tracheary elements show great variability in size, shape, and type of secondary wall. These factors differ from species to species, and even from one part of a plant to another (Baas et al., 1976; Butterfield and Meylan, 1980; Gibson, 1977, 1978a, b; Mauseth and Ross, 1988; Zimmermann et al., 1982). There have been many attempts to explain this variability, but the most successful, I believe, are those of Carlquist (1975; 1980b) and Zimmermann (1983). They applied the concepts of water relations that have been developed by plant physiologists and used them to analyze which types of xylem anatomy could optimize water transport and minimize damage during water deficit (drought).

The basics of water relations are rather simple. The capacity of water or of any substance to do work is called its **chemical potential**; for water, we usually substitute the name **water potential** (Nobel, 1983; Salisbury and Ross, 1985). Pure water at one bar of pressure (one atmosphere of pressure) has, by definition, a water potential of zero. Water potential has three simple components, pressure potential, osmotic potential, and matric potential:

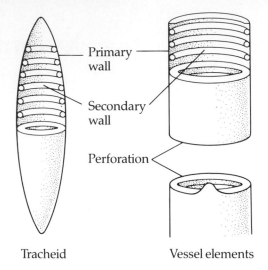

Figure 7.1
Tracheids tend to be long and spindle-shaped; the one shown here is really rather short and could easily have been over 50 times longer in relation to the vessel element that is shown. The important features to notice are that the secondary wall is deposited interior to the primary wall, the secondary wall does not completely cover the primary wall, and the primary wall of the tracheid is complete, whereas that of the vessel element has two perforations.

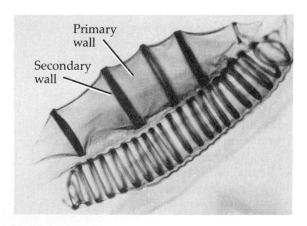

Figure 7.2
Two tracheids from *Notocactus schumannianus*. The primary wall is the thin, gray "envelope"; the secondary wall is the set of rings in one, the helix in the other. × 1000.

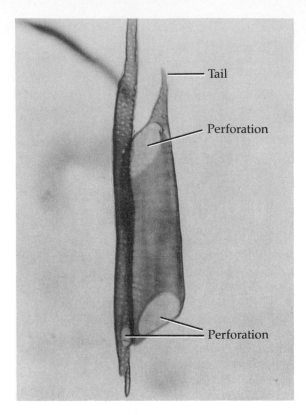

Figure 7.3
Vessel elements in *Prunus,* as in almost all other angiosperms, can be of various sizes. These two are actually rather similar; the important feature is that they both have perforations. Obviously, the wider vessel element can conduct more water with less friction than can the narrower vessel element. Unlike the tracheids of Figure 7.2, it is not possible here to see the primary wall; it is almost completely covered by secondary wall, except for the pits. × 500.

WP = PP + OP + MP. If pressure is applied to water by compressing it or by pumping it, the water has more capacity to do work, so the water potential becomes more positive (becomes larger) because the pressure potential component is becoming more positive. Because water is so sticky, it is possible to pull on it (you can

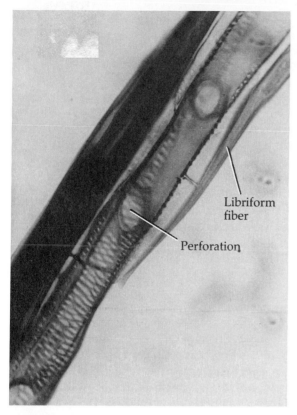

Libriform
fiber

Perforation

Figure 7.4
These vessel elements remained together as part of
a vessel when the tissue was macerated. Three
simple perforations are visible as is the pitting on
the lateral walls. × 500.

restricted in their ability to move freely. Their
capacity is always less than that of molecules
in pure water; the water potential becomes
more negative because the osmotic potential
component is negative (this component can
never be positive). Note carefully that, when
an acid is dissolved in water, the solution
might be more powerful than the original
water, but that the water molecules themselves
within the solution are less active. Water can
also hydrate the surfaces of solid objects such
as soil particles and cell walls, also decreasing
its capacity. This effect on water is the matric
potential; it too is always negative, because it
always restricts the water's capacity for work.

Four other characteristics of water are important with regard to transport:

1. Water always tends to move, if possible,
   from a region where the water potential is
   more positive (where it can move more
   freely) to a region where it is more negative
   (regions where its movement is restricted).

2. It is **adhesive**; it tends to interact with and
   stick to many substances.

3. It is **cohesive**. The water molecules can hydrogen bond to each other very strongly,
   and the group of molecules tends to act as a
   unit.

4. Water is heavy.

As a leaf or any other aerial part of the plant
is exposed to the atmosphere, it tends to lose
water to the air, both through the cuticle
(**transcuticular transpiration**) and through the
stomata (**transstomatal transpiration**). The leaf
cells dry slightly, and the water potential becomes more negative. This establishes a gradient of water potential (a gradient of dryness)
within the leaf, and water gradually moves out
of the xylem and passes cell by cell to the intercellular spaces or the epidermis where it is lost.
But because of their cohesive nature, the water
molecules cannot leave the water column in
the xylem freely—those water molecules that

hold and stretch a drop between your fingers)
and create a tension (negative pressure) in the
water. When this happens, the pressure potential and therefore the water potential become more negative, and the water, being held
this way, has less capacity.

The osmotic potential measures the extent to
which solutes reduce the capacity of water to
do work. When substances are dissolved in
water, the water molecules interact with the
solute molecules, hydrating them, and the
water molecules tend to be slowed down and

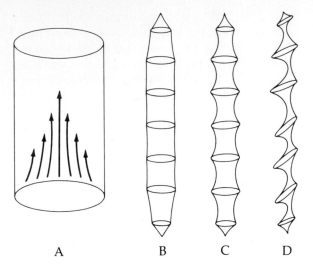

Figure 7.5
(A) The flow of water in a tracheary element is not uniform. The water closest to the walls is hydrogen bonded to the walls and moves slowly. The water in the center is not so restricted by this friction and moves faster. (B) As the water is pulled upward, it tends to pull the side of the cell inward because of its cohesive and adhesive properties; this causes the cell to buckle (C) and even collapse (D).

leave, being hydrogen bonded to the others, tend to pull them along. Because the entire mass of water is hydrogen bonded throughout, the departure of these molecules drags the water column up through the roots and stem. This is similar to pulling on the top of an icicle: even though you touch only the uppermost water molecules, the whole icicle is lifted because all of the water molecules are interconnected.

Three forces will tend to counteract this upward movement: (1) the water in the soil tends to stay there because it is bound to the soil particles, (2) water is a heavy liquid, so lifting it is difficult, and (3) because the water adheres to the walls of the tracheary elements, it tends to stick to the sides of the cells (Fig. 7.5A). This resistance is especially important; air frequently has such a strong capacity to absorb

water (it usually has a very negative water potential) that it can pull the water upward against almost any resistance. The water column will be under tension and will become narrower, and its adhesion to the walls of the tracheary elements will tend to cause these elements to collapse inward (Fig. 7.5B, C, D). The walls of the tracheary elements must be strong to prevent their buckling; a secondary wall must be present to prevent collapse. This wall restricts the capacity for water to enter or leave the element, but there is no alternative.

The friction caused by the adhesion to the wall can be minimized by increasing the radius of the element. There are fewer micrometers of circumference for each square micrometer of cross-sectional area available for transport as the radius increases (or just consider that the adhesion affects the edges of the water column much more than it affects the center; Fig. 7.6). The ability of capillary tubes like tracheary elements to permit the movement of water is called **conductance**, and it is related to the fourth power of the tracheary element radius by the Hagen-Poiseuille law (Nobel, 1983; Zimmermann, 1983). Imagine three tracheary elements with diameters of 10, 20, and 40 μm. Their cross-sectional areas are then 78 μm^2, 314 μm^2, and 1256 μm^2 (CSA = πr^2), proportionally 1, 4, and 16 due to the squaring of radii in the formula. But because conductance is related to the fourth power of the radius, the relative flow rates will be 1, 16, and 256. If the three tracheary elements were present together in a vascular bundle, the largest would carry 93.8% of the water; the 20-μm-diameter one would carry only 5.9%; and the narrowest would carry almost nothing—0.4% (Kuo et al., 1972; Kuo et al., 1974).

From this consideration, it is obvious that even a slight increase in the diameter of a tracheary element greatly increases the conductance, and one might expect to find only elements with large diameters. But narrow ones still exist (Fig. 7.7). Why? The standard answer used to be that the plant (or the xylem) must be

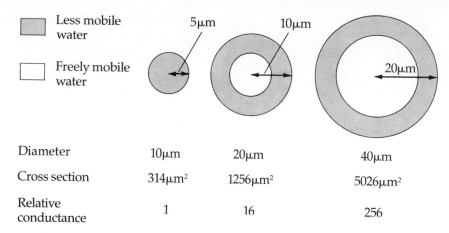

Diameter	10μm	20μm	40μm
Cross section	314μm²	1256μm²	5026μm²
Relative conductance	1	16	256

Figure 7.6
The layer of water that is next to the wall is supported by the wall, and it acts as though it were stronger than ordinary liquid water. If the tracheary element is narrow, this boundary effect reinforces all of the water column; in larger elements, the center molecules are too far from the wall to be reinforced by it. They can flow easily, but they can also cavitate with less tension on the water.

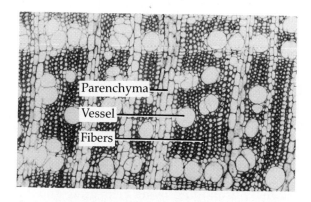

Figure 7.7
This wood (secondary xylem) of *Didierea procera* contains vessel elements with a range of sizes. When water is plentiful in the soil and is moving easily, the large-diameter vessels are responsible for virtually all conduction. As the soil becomes dry and tension increases, these will cavitate and fill with embolisms, while the water columns in the narrow ones are still intact from the strength provided by adhering to the walls. As tension mounts, these will begin to flow faster and carry a greater percentage of the water. × 100.

primitive. However, the fact is that under certain circumstances the adhesion is actually advantageous in that the wall, being hydrogen bonded to the water, reinforces and strengthens the water column (Fig. 7.6). A similar water column in a glass tube (with minimal adhesion) would **cavitate** (break) much more easily. It may be that the tallest trees can reach their heights only because of the reinforcing that the walls provide to the water. If the tension on the water column becomes sufficiently severe, the cohesive and adhesive forces will be overcome and cavitation will occur: an embolism will form in the element. An embolism is a partial vacuum that contains some water vapor; it is frequently called an "air bubble." This embolism will expand until it meets solid membranes (Slatyer, 1967). If it occurs in a tracheid, the functioning of the tracheid is stopped because the hydrogen bonding of the water has been broken (Fig. 7.8). If it happens in a vessel element, the embolism spreads from element to element through the perforations; the entire

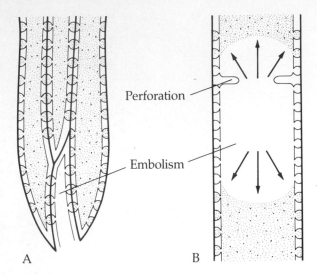

Figure 7.8
(A) When the water column in a tracheid cavitates, the embolism expands rapidly, because the transpiration tension pulls the upper part of the broken water column upward, while gravity pulls the lower part downward. When the embolism reaches the ends of the tracheid, it stops expanding, because it would require an enormous pull to break the water's surface tension, which would be necessary for the air-water interface to pass through the pit membranes. (B) A simple perforation offers no resistance at all to the expansion of an embolism, and the entire vessel will become filled with gas. A reticulate or scalariform perforation plate offers some resistance, and, if the tension on the water is not too great, it may stop the air-water interface.

vessel becomes useless. Under conditions of possible cavitation, it may be more advantageous to have narrow vessel elements so that the walls can reinforce the water column; if the water tension is severe, it may be advantageous to have tracheids (which are short) rather than vessels (which are long) to minimize the damage. The friction of the walls and the pit membranes is much less of a problem than is cavitation.

Walls of Tracheary Elements

Patterns of Secondary Wall Deposition

A number of types of patterns have evolved for the structure of the reinforcing secondary wall of tracheary elements. It is possible to establish which appeared first in the fossil record, so we know which are more primitive and which are more advanced. However, each has certain advantages and disadvantages, and it is not appropriate to equate "primitive" with "nonadaptive" and "advanced" with "better."

ANNULAR.   The annular type is the simplest of the secondary wall patterns that are possible (Figs. 7.1, 7.2, 7.9A). The wall is deposited not over the entire surface of the primary wall as was described for fibers and sclereids in Chapter 5 but is laid down as narrow rings inside the primary wall, each ring being distinct from the others of the cell. This arrangement provides considerable strength against collapse without using much wall material, and it also leaves large amounts of the primary wall uncovered and available for the water to enter and exit the cell. As in most fibers and sclereids, the secondary wall is lignified and waterproof (Wardrop, 1981).

The annular type of secondary wall pattern is not really very strong and will collapse fairly easily, but it has the advantage of allowing the tracheary element to be extensible, since the rings are independent of each other. That is, as the surrounding living cells grow and expand, they can cause the tracheary element, even though it is dead, to elongate as well, because they are attached to it by their middle lamellas. This type of tracheary element can differentiate in growing tissues, then function and elongate simultaneously, being used to supply water to the youngest parts of the plant. Since it can be stretched, it will neither hinder the surrounding expansion nor be torn away from the adjacent cells. As the cell is elongated, the rings become separated from each

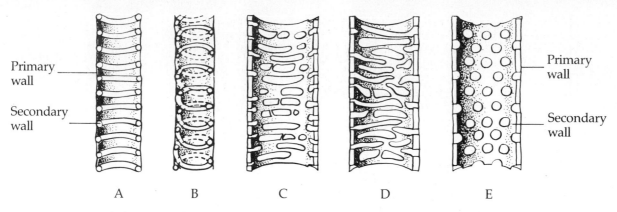

Primary wall

Secondary wall

Primary wall

Secondary wall

A   B   C   D   E

Figure 7.9
These are the types of secondary wall that may be found in both tracheids and vessel elements: (A) annular, (B) helical, (C) scalariform, (D) reticulate, and (E) circular bordered pits.

other. Obviously, it cannot do this indefinitely, and it may be torn apart ultimately.

HELICAL.   This pattern of deposition is similar to the annular type, but the secondary wall occurs as one or two helices (Figs. 7.2, 7.9B). This has the same advantages and disadvantages as the annular type. While young, the helix of secondary wall material is tightly coiled, but as the cell is stretched by the growth of the surrounding cells, the helix is also stretched out. This type of cell can also be torn apart by expansion in the adjacent cells.

SCALARIFORM.   In the scalariform or "ladder" type, the secondary wall is much more extensive than in the annular or helical types, and usually at least half of the primary wall is covered (Fig. 7.9C). In stained preparations, the secondary wall looks dark and appears like the rungs of a ladder, and the noncovered spaces are quite restricted, appearing like flat, broad ovals. An important feature is that the secondary wall occurs not just as transverse bands, but as vertical ones as well, so this element is reinforced in all directions. Not only will it resist collapse inward, but it will also resist being extended by the growth of surrounding cells.

It would either tear away from them or it would inhibit their growth. This cannot be used for those tracheary elements that occur in organs that are still growing and elongating but only those where primary elongation has ceased.

RETICULATE.   The secondary wall thickenings in this type are not as regular as in the scalariform type but rather form an irregular net shape (Fig. 7.9D). There is even more vertical component, and, as with the scalariform type, it is impossible for the reticulate type to expand.

CIRCULAR BORDERED PITS.   This type involves the maximum coverage of the primary wall by the secondary wall. Almost the entire surface is covered, the exceptions being small areas called pits, as was true for sclereids and fibers (Figs. 7.9E, 7.10, 7.11). This offers the maximum strength against collapse but leaves the minimum area free for water entrance and exit.

The pitting deserves special attention: if the cell leaves just a narrow, straight-sided channel, the pit is a simple pit (Fig. 7.10). But this has both minimum strength (since it is a hole) and minimum conductivity (because it is so narrow). It can be greatly improved with slight

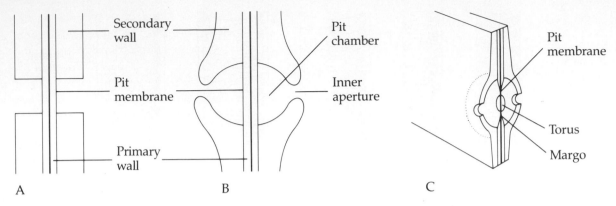

Figure 7.10
Pitting in secondary walls. A simple pit (A) has parallel sides and no reinforcement around the inner aperture. A bordered pit (B) is reinforced by
extra wall material. Side view (C) shows a pit membrane that has developed
as a torus and margo.

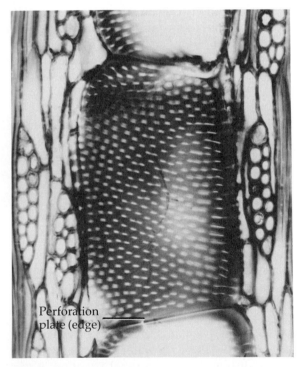

Figure 7.11
The lateral-wall pitting of vessel elements provides
a means by which water moves from vessel to vessel, into tracheids, or even to the surrounding cortex or mesophyll. This vessel element of *Machaerium
laterifolium* has alternate circular bordered
pits. × 500.

modification. It can be made larger without
being weaker if the inner aperture is reinforced
with extra wall material, thereby forming a
border (Figs. 7.10, 7.12, 7.14). But the outer aperture is the area of greatest friction and slow-
est water movement, because it is there that
the water traverses the pit membrane (the two
primary walls and the middle lamella; see
Chapter 5). An optimum structure would be to
have this aperture maximally wide and the
inner aperture somewhat narrower, only wide
enough to admit or receive water as rapidly as
it can pass across the pit membrane (Figs. 7.12,
7.13). Considering the whole cell, not just the
pits, maximum strength with maximum conductivity can be attained if the pits are circular
and packed as close together as possible so
that the border of one reinforces the border of
the others around it. The best close packing is
an **alternate** arrangement, in which the pits are
arranged in diagonal rows and tend to be hexagonal instead of truly circular (Fig. 7.11), but
opposite close packing (arranged in horizontal
rows) is also successful and occurs frequently.
Circular bordered pitting seems to have been
somewhat difficult to achieve, because the
ferns and lower vascular plants do not have it.
They do have borders on the scalariform secondary wall (**scalariform bordered pitting**).

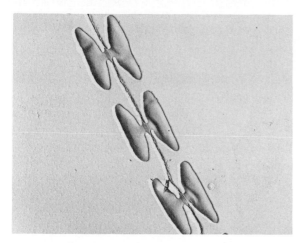

Figure 7.12
Longitudinal sections of bordered pits in *Liquidambar styraciflua* (sweet-gum). The inner apertures are narrow, and the outer ones are wide, providing maximum area for water passage through the pit membrane. × 10,000. Micrograph generously provided by E. Wheeler.

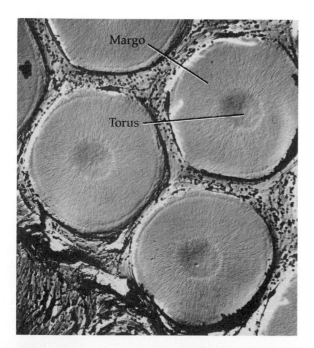

Figure 7.13
These pit membranes are differentiated into a torus and a margo: *Celtis laevigata* (hackberry). × 30,000. Micrograph generously provided by E. Wheeler.

In the conifers and several of the more primitive angiosperms, the pit membrane has been specially modified: the peripheral portions are digested such that only a loose network of fibrils remains, and water can pass with a minimum of friction (Bailey, 1957a, b; O'Brien and Thimann, 1967). This is the **margo** (Figs. 7.13, 7.14). The central region of the membrane, however, is thickened with secondary wall material and lignified and becomes the **torus**. Apparently, as a tracheary element stops functioning in water conduction and its water column breaks, the water is pulled rapidly out of the cell by the tension on it. This rapid movement causes the torus to be displaced and pressed tightly up against the border, thereby sealing the pit against any further water movement (Harris, 1954). It is thought that this may aid in keeping the water of the younger regions, which are still conducting, from moving into the older regions; it may also aid in sealing cavitated tracheids from uncavitated ones.

In the mechanism just described, the movement of the pit membrane is advantageous, but in many cases such movement would not be, especially if there is no torus to act as a seal. In fact, the rapid flexing of the pit membrane that would occur as water rushes through it when cavitation occurs might possibly even rupture the pit membrane, allowing the embolism to pass through the pit. An intact pit membrane is an invaluably important protection against the spread of embolisms; the spaces between its microfibrils are so small that the air-water interface at the edge of the embolism cannot pass through unless the water is under extraordinary tension. The bubble cannot cross intact pits and is therefore confined to individual tracheids or vessels (Slatyer, 1967). In some plants, there are elaborate ingrowths from the inner edge of the pit border, and the pit is said to be **vestured** (Fig. 7.15; Cassens, 1980; Ohtani and Ishida, 1973). These do not occur at random, but actually form a brushlike surface that might act to support the pit membrane and prevent it from

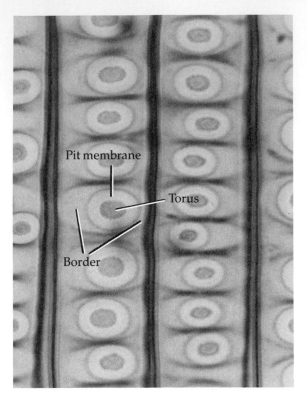

Figure 7.14
Although the SEM views of circular bordered pits provide high resolution and depth of field, most people will examine such pits by light microscopy and will see images like this. The borders are prominent; the pit membrane is distinguishable, as is the torus. × 400.

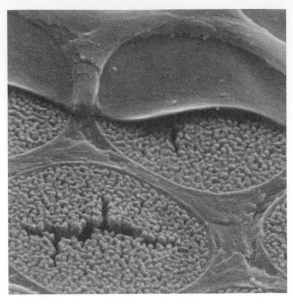

Figure 7.15
Vestured pits in vessel elements of *Lagerstroemia subcostata* (crape myrtle). Part of a pit membrane is visible. × 1000. Micrograph generously provided by J. Ohtani and S. Ishida.

flexing or stretching so much that it would break (Carlquist, 1982b; Zweypfenning, 1978). Vestured pits are difficult to distinguish by light microscopy, but with the aid of electron microscopy they have been found in 30 families (Carlquist, 1982b).

Because xylem is a complex tissue rather than a simple one, the tracheary elements may be located adjacent to various types of cells. If the contact cell is another tracheary element, then the pits are members of pit-pairs and are usually abundant; this is termed **intervascular pitting**. If the neighboring cell is a fiber, then the pits are few and very small, or there may

be none at all. When a parenchyma cell is the partner, the common wall may contain **half-bordered pits**; that is, they are bordered pits on the tracheary element side but simple indentations of the primary wall on the parenchyma cell side. It is not uncommon for a single tracheid or vessel element to be in contact with all three types of cell, and each side of the tracheary element will then have the appropriate type of pitting.

All five of these types of secondary wall deposition can occur in both tracheids and vessel elements. Any single plant and even individual vascular bundles might contain all of these types of tracheids and vessel elements. Cells with the annular and helical secondary wall will form while the tissues are still expanding (O'Brien, 1981), and usually they will function only briefly, although in some plants they are the only type ever formed. After elongation has stopped, most plants will then begin to produce one or several of the other three types, scalariform being quite common in the ferns, and circular bordered pits being most abundant in the seed plants. As can be imagined, intermediate types can occur, especially on long tracheids, where one portion of a cell may be scalariform, another portion reticulate. Enormous variation is found (Bierhorst, 1960).

Perforations

The maximum modifications of the pits still leave the pit membrane in place; regardless of how thin it might be, the membrane gives more friction and resistance than if it were removed and completely absent. In vessel elements, areas on the end walls are modified to eliminate the primary walls. When that is complete, there is a large hole between the two elements: the **perforation** (Figs. 7.1, 7.3, 7.4, 7.16, 7.17; Meylan and Butterfield, 1981). Obviously, the only way that a perforation can be effective is for it to be aligned with a perforation on an adjacent vessel element. The perforations may rarely be located on the side walls of a vessel element (Fig. 7.17), and there may also be more than two perforations on an element, but by far the most common structure is for the vessel element to have rather transverse, distinct end walls, each with one perforation. This end wall is called the **perforation plate**. If it contains only one perforation, it is a **simple perforation plate**; if it has two or more, it is a **compound perforation plate**. The single perforation of a simple perforation plate frequently occupies almost the entire surface of the plate; all that is left is a small rim at the very edge of the vessel element.

Four types of perforation have been reported. **Simple perforations** are large, circular, or oval. The rim of the perforation plate may be thickened and appear like a border, but this perforation is not related to circular bordered pits; it is larger and forms by a different mechanism. Water can flow easily through simple

Figure 7.16
This is a scalariform perforation plate in the wood of *Sphenostemon platycladum*, and it has the most common form: a long oval with numerous thin bars. × 500.

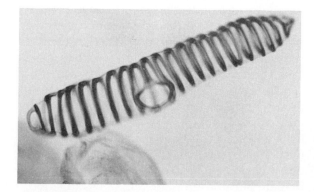

Figure 7.17
This vessel element of *Notocactus* shows an unusual feature: a perforation located on the side of the cell, not just on the end walls. × 300.

Figure 7.18
An ephedroid perforation plate contains several small, round perforations, rather than just one large one.

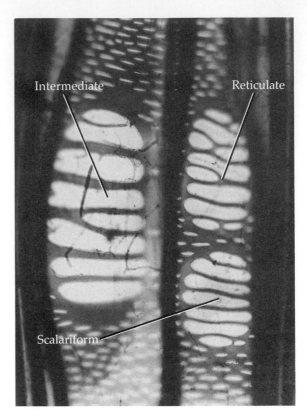

Figure 7.19
Of these three perforation plates in *Gilibertia smithiana*, one is reticulate, one is scalariform, and one is intermediate. × 300.

perforations, and a transverse perforation plate provides sufficient area for water movement. Because of its large size, the value of r^4 is great, and there is little resistance to water flow through a simple perforation. This is the most common type. **Ephedroid perforations** (also called **foraminate perforations**) are groups of small, circular perforations (Fig. 7.18). Ephedroid perforation plates are rare, occurring in only a few families such as the Begoniaceae, Dilleniaceae, Ephedraceae, Ericaceae, and Myrothamnaceae (Chalk, 1983a). If the perforations resemble a net, then they constitute a **reticulate perforation plate** (Fig. 7.19); this, too, is a rather rare form. The **scalariform** type resembles scalariform pitting and is related to it: a scalariform perforation is created when en-

zymes digest the unprotected primary wall out of regions that look very much like scalariform pits (Figs. 7.17, 7.20). The bars that remain provide reinforcement that prevents the perforation plate from collapsing, but they also cause considerable resistance to water flow because the r^4 value of each narrow perforation is small. Because of this impedance, many perforations are necessary; a slanted end wall provides more area for a large number of perforations, up to hundreds in a single compound perforation plate (Fig. 7.21). By light microscopy, these cells are difficult to distinguish from tracheids, because it can be hard to deter-

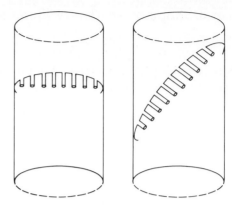

Figure 7.21
If a vessel element has a radius of x, then the cross-sectional area of a transverse perforation must be less than πx^2, and, if scalariform, the bars will occupy a significant portion, creating an impediment to water flow. If the perforation plate is slanted, then it is oval in outline, and has a larger area through which water can pass.

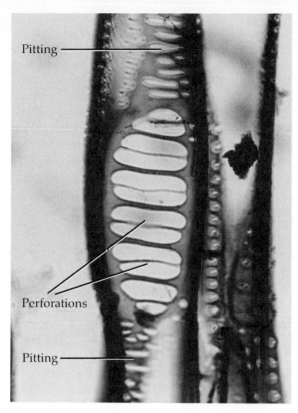

Pitting

Perforations

Pitting

Figure 7.20
These scalariform perforation plates of *Meryta macrocarpa* are unusual in having two types of bars, thick ones alternating with thin ones. × 300.

mine if the pit membrane is present or digested away (Fahn, 1953). Scalariform perforation plates occur in many angiosperm families, especially in the ones that are considered primitive.

In order for perforations to be functional, the perforation of one vessel element must be aligned with the perforation of the contiguous element. "Blind perforations," like blind pits, would be useless. The alignment of the two halves of a perforation is usually close to perfect, but in long scalariform perforation plates, the ends may be slightly misaligned, such that the last several bars of one cell face only a

pitted wall in the other cell (Meylan and Butterfield, 1981). Even when the alignment of the perforations is correct, mismatching of the types of perforations formed results in dimorphic perforation plates (Butterfield and Meylan, 1975; Chalk, 1933; Meylan and Butterfield, 1973, 1975; Ohtani and Ishida, 1973; Parameswaran and Liese, 1973; Thompson, 1923). Apparently it is not terribly unusual for the simple perforation of one vessel element to face a scalariform perforation on the adjacent vessel element. Simple perforations facing reticulate ones have also been reported (Chalk, 1933; Gottwald and Parameswaran, 1964; Muller-Stoll and Suss, 1969; Parameswaran and Liese, 1973), but Meylan and Butterfield (1981) concluded that scalariform/reticulate dimorphic perforation plates are rare. They also illustrated a perforation plate in which both halves are scalariform but in which the bars of one were oriented at an angle to those of the other (their Fig. 3.7).

Size and Shape of Tracheary Elements

Tracheids

Because tracheids have the friction of pit membranes hindering the entrance and exit of water, maximum conduction occurs if the cells are as long as possible so that each water molecule traverses as few membranes as possible. In a plant that is 10 m tall with tracheids 500 μm long, each water molecule must pass through at least 20,000 pit membranes, but if the tracheids are 1 mm long, there are only 10,000 membranes to cross. Also because water must enter and leave this way, the end wall should be as long as possible, so that the greatest number of pits can be placed on it; to make a very long end wall without having a very wide cell, the end wall must be slanted. In tracheids it slants so much that it intergrades with, and is usually indistinguishable from, a side wall: the tracheids are long, tapered, spindle-shaped cells.

Tracheids can be quite wide, especially in gymnosperms where they are the only conducting element. In angiosperms, where most conducting can be done by the vessel elements, tracheids are narrower, stronger, and more resistant to collapse; they function in the dry periods when the water is being transported by great tension and the vessels have cavitated. A similar situation occurs in many conifers in which tracheids with large diameters are produced in the spring (the moist season when conduction is easy) and tracheids with narrow diameters develop in summer (the dry season when conduction is difficult; Fig. 7.7).

Vessel Elements and Vessels

With the evolution of vessel elements, angiosperms obtained the potential for an exquisite two-phase type of conduction that is partially developed with the large-diameter and narrow-diameter tracheids of conifers. Vessel elements are very good at conduction when water is readily available and moves in high volume. Under these conditions, a wide vessel element is advantageous, with an end wall (perforation plate) that is transverse or almost so, such that the rim of it is as small as possible (circular instead of a long oval as it would be on a tapered end wall, as in Figure 7.11). The length of the element is not so important for conduction, since there is no need to minimize the number of pit membranes traversed; but, probably due to the logistics of trying to produce a very wide cell, it is better for the cell to be short. A cell that is both very wide and very long would be so large that growth and differentiation might be slow and difficult, which may also explain why tracheids, which are long, tend to be narrow.

Another factor that may influence the length of vessel elements is that, if the vessel requires extra reinforcement over and above that provided by the secondary wall, the perforation plate rim or the scalariform bars act as an extra means of strengthening the cell. If the vessel is composed of shorter elements, then these extra reinforcements will be closer together.

The tips of vessel elements are narrow and pointed, projecting a short distance beyond the perforation plate. These are termed **tails** (Figs. 7.1, 7.3; Chalk and Chattaway, 1934), and they typically are pitted.

Vessel elements are often described as being round in cross section, while tracheids are said to be angular; this difference is an automatic consequence of the geometry of the packing of cells. In a mass of cells that are all about the same size, all cells are angular, tending to become hexagonal. This is true both of tracheids and of vessel elements, but only if the vessel elements occur packed with other vessel elements or any other cells that are as large as they are. In a mass of cells where most are small and a few isolated ones are large, then the large ones appear round; they have so many small cells around them that the large cells have numerous contact faces, creating the overall effect of roundness.

Although vessels (not vessel elements) are long, they do not extend from root tip to leaf tip—except in shrubs (Zimmermann and Jeje, 1981)—so they must have ends (Handley, 1936). The terminal vessel elements are unusual in having only one perforation; the other end is not perforated. It must be emphasized here that vessel elements have both perforations and any of the five types of secondary wall mentioned above. The presence of perforations on their end walls facilitates longitudinal conduction through the vessel, and the pitting on their side walls allows water to move from vessel to vessel or from the vessel into the adjacent tracheids, parenchyma, or other tissues.

Development

Phylogenetic Development

The evolution of conducting elements occurred soon after plants began a terrestrial existence. In *Cooksonia*, there are beautiful tracheids with annular thickenings. The oldest fossils that actually contain these tracheids come from Upper Silurian deposits, but the sporangia of *Cooksonia* have been found as early as Mid-Silurian, at about 420,000,000 years before present (Stewart, 1983).

Cooksonia is a very simple plant, consisting of a horizontal rhizome and upright stems that had no leaves. But it already had several effective adaptations to terrestrial living besides vascular tissue: there were stomata with guard cells, epidermis with cuticle, and rhizoids (but no roots). Another very early vascular plant was *Baragwanathia*; it was more complex than *Cooksonia* in that it had "leaves," but the xylem contained only annularly thickened tracheids like those of *Cooksonia*.

Vessel elements evolved from tracheids and are **polyphyletic** in origin; that is, the ability to form perforations has evolved several times. Vessel elements occur in *Selaginella* (Duerdan, 1934), *Equisetum* (Bierhorst, 1958), four ferns—

Actiniopteris, Pteridium, Regnellidium, and *Marsilea* (Mehra and Soni, 1971; Singh et al., 1978; Tewari, 1975), three very unusual gymnosperms—*Ephedra, Gnetum,* and *Welwitschia* (Martens, 1971; Muhammad and Sattler, 1982), and in most angiosperms. It is interesting that, aside from the angiosperms, the vessels in the other groups are anomalies—none of these genera is large, and the occurrence of the vessels in them is rather unexpected. Further, the presence of vessels has not led to any great evolutionary or ecological success on the part of the plants that contain them aside from the angiosperms. The conifers, which completely lack vessel elements, are extremely successful, much more so than the angiosperms in several environments.

EVOLUTION OF THE PERFORATION PLATE.   Except for the sporadic occurrence of vessels in the nonflowering plants, the evolution of vessel elements and perforation plates is linked firmly to the evolution of the Anthophyta. Within the angiosperms, vessel elements apparently did not evolve until after the dicots were well established (Bailey, 1944, 1957, 1966; Cheadle, 1953). Today there are still five families believed to have never had vessels. They are referred to as **primitively vesselless**: Amborellaceae (sometimes listed as the Monimiaceae), Chloranthaceae (the genus *Sarcandra*), Tetracentraceae, Trochodendraceae, and Winteraceae (Chalk, 1983a). In the other families, the vessel elements are well developed. Young (1981) believed that the ancestors of the angiosperms had vessels and that these five families are secondarily vesselless (that is, that they lost the ability to produce vessels). He argued that, if the primitive angiosperms were vesselless, then vessels would have had to have arisen ten distinct times within the angiosperms in order to have the distribution which occurs today. Carlquist (1983) disagreed with this position and summarized the evidence for the more widely accepted idea that the original angiosperms had only tracheids.

We have almost no fossil record of the early angiosperms that could help us understand how vessel elements arose. In an attempt to elucidate how this evolution might have occurred, the xylem of existing plants has been studied. It is assumed that those vessel elements that are most like tracheids (long, narrow, spindle-shaped, with long, slanted perforation plates) are relictually primitive (Bailey, 1944; Carlquist, 1980b, 1983). These features are often associated with scalariform perforations, so it was assumed that vessel elements arose from scalariformly reinforced tracheids, then advanced toward a more ideal shape of short and wide with transverse, simple perforation plates (Bailey, 1953; Cheadle, 1943, 1944, 1953; Fahn, 1982). Based on this primitive-versus-advanced concept, a scheme of evolution was suggested. In the dicots, vessel elements first arose in the woody plants in the secondary xylem and later the primary xylem; in the monocots, they first appeared in the roots and then later in the stems and leaves. In the monocots the plant was considered to have a variety of different types of vessel elements, each at a different stage of evolution and each evolving independently.

Such a suggestion is difficult to accept today; all cells of a plant share the same set of genes, and all cells within that plant that begin to differentiate into a vessel element have the same potential. The presence of dimorphic perforation plates proves this. It is more logical that the reason that different parts of a plant have different types of vessel elements and perforation plates is that they have distinct requirements with regard to the amount of water that must be transported, the tension it might be under, and the dangers of cavitation. Young plants with only primary xylem are usually shaded by larger plants, so their leaves do not heat up and transpire rapidly. Their water columns are short, and the active absorption of water by the roots might be sufficient to move water even without transpiration. When that same plant begins to produce secondary xylem, it is larger, its leaves are probably in a more stressful environment, its water columns are subject to more tension, and the mechanical support of the trunk and branches is important. Thus it is not surprising that the type of cells produced in the primary xylem and secondary xylem are different.

These concepts are relatively new to plant anatomy, and the earlier investigators did not have them to work with when they were developing their ideas. Instead, they thought in terms of maximum conductance: anything that is wider and with a larger perforation must be more "ideal." Because conductance was seen as the only consideration, then there should be one and only one optimum structure. Any vessel element that did not have that structure was thought to be relictually primitive, thus the distinct "levels of evolution" within a single plant. The primary xylem was even described as a refugium of primitive characters (Bailey, 1944) rather than being a tissue that is highly adapted to the needs of seedlings, leaves, flowers, fruits, and the tips of roots and shoots.

Undoubtedly the vessel elements did arise from tracheids, so the first elements may have been long and narrow, but we cannot automatically assume that all such vessel elements are relictual. This shape can be very adaptive in environments of severe water stress. Although the dominant trend has been for vessel elements to become shorter during evolution (Carlquist, 1980b), there is no obvious reason why they could not become longer again if that were adaptive. A second trend may be irreversible: simple perforations are thought to have evolved by the loss of the bars in scalariform perforation plates. Perhaps once this evolution becomes complete, the genes for bar formation are gone. A simple perforation might not be able to give rise to a scalariform one. But the xylem of many plants is heterogeneous in that some parts have simple perforations and other parts have scalariform ones. If any part of the plant (protoxylem, roots, flowers) retains scalariform perforation plates, then that genome is present in all cells and could be expressed in the xylem of other parts of the plant if the proper mutations occurred. It is not so uncommon for characteristics that are associated with the xylem of juvenile parts to appear in the xylem of more mature parts, if the environment changes and they become selectively advantageous. Such an expression of "juvenile" characters in "adult" tissues is called **paedomorphosis** (Carlquist, 1962, 1980b, 1985). The concept of reversibility is discussed by Baas (1973) and Carlquist (1969).

An important aspect that is completely unknown is the mechanism that determines the length of the vessels (not vessel elements). These can be as short as a few centimeters or as long as several meters (Zimmermann, 1983), and the size is not random. Certain organs have only long vessels, others have only short ones, but we know nothing about how length is controlled.

Ontogenetic Development

Both tracheids and vessel elements begin as parenchyma cells derived either from the procambium or the vascular cambium, and initially they have dense protoplasm, small vacuoles, and thin primary walls (O'Brien, 1981; Barnett, 1981). The cells enlarge, becoming more vacuolate and retaining just the primary wall until they are almost full sized. The cells may become multinucleate, but much more commonly they remain uninucleate but with large, endopolyploid nuclei (Bailey, 1920; Barlow, 1985; Lai and Srivastava, 1976; List, 1963). However, as O'Brien (1981) pointed out, it is very common for somatic plant cells to be multinucleate or endopolyploid, so the occurrence of these phenomena during the differentiation of tracheary elements may not be especially significant. As the deposition of the secondary wall begins, arrays of microtubules become aligned over those portions of the primary wall where the secondary wall will form (Esau and

Charvat, 1978; Hardham and Gunning, 1979; Pickett-Heaps, 1968; Shininger, 1979). The pattern of cytoplasmic streaming has also been reported to be identical to the pattern that will develop in the secondary wall (Sinnott and Bloch, 1945). It is thought that the microtubules are instrumental in directing the deposition of new wall material either by catching the dictyosome-derived vesicles and guiding them to the plasmalemma in this region, or by the microtubules' attaching to the plasmalemma and pulling it away from the existing primary wall, thereby creating a space where wall material can accumulate. However correct these hypotheses may be, they are not by themselves sufficient: a mechanism is needed to explain the intercellular communication required for the secondary wall of one cell to match that of the adjacent cell (Hepler and Fosket, 1971; O'Brien, 1981).

Plasmodesmata seem not to be involved, because the deposition is not oriented with regard to the primary pit fields. These may be left uncovered as part of the pit membrane, but they are just as likely to be covered by secondary wall (Barnett, 1981; Barnett and Harris, 1975).

In developing vessel elements, the end walls are distinct from the other walls at an early stage. Meylan and Butterfield (1981) suggested that this area be called the **perforation partition**, because it is not always an end wall. The primary wall at the site of the future perforations becomes unusually thick (Esau and Hewitt, 1940; Meylan and Butterfield, 1972), but no secondary wall is deposited on it. This wall has almost no cellulose (Benayoun et al., 1981) but rather is mostly noncellulosic polysaccharides. The plasmalemma in the region of the perforation partition forms many vesicles (Czaninski, 1968) which may be removing material from the wall. As the cytoplasm begins autolysis (Wodzicki and Brown, 1973; Wodzicki and Humphreys, 1972), the wall at the site of the perforation becomes granular, then numerous small holes appear, then it breaks down and disappears (Meylan and Butterfield, 1972; Murmanis, 1978). The actual removal of the perforation plate apparently results when the transpiration stream begins to move rapidly, tearing it away.

The enzymes that attack the perforation partition also attack other areas of unprotected primary wall: the pit membranes become at least partially degraded, resulting in the primary wall's being just a mat of fibrils (O'Brien, 1981). Any region that is covered by lignified secondary wall is protected from degradation.

Structure in Relation to Function in Water Conduction

All of the aspects of xylem structure and water relations are quite simple, but they can certainly be overwhelming when presented all at once, as has just been done. In order to put them in perspective and see how they are interrelated, it is best to stop and think about individual plants. Consider several types of situations that might exist:

1. Soil always moist, soil water always available. Under conditions such as this, water is not a limiting factor for the plant, and because any water that is lost by transpiration can be replaced, it is advantageous not to use the material and energy resources of the plant to construct a thick cuticle, wax layers, or hypodermis. It is easier and less expensive just to replace the water that is readily lost. There would be a voluminous transpiration stream, so the xylem might be expected to have many vessels, each of large diameter; such an arrangement minimizes the water's contact with the vessel element walls (a wide vessel element has low surface/volume ratio). This arrangement decreases the water's adhesive interaction with the cellulose, thus decreasing the friction, and the water flows easily. Vessels are more advantageous than tracheids, because water does not have to cross pit membranes except when passing from vessel to vessel. Although each pit membrane is thin and offers

little resistance, if the plant is 10 m tall and has tracheids that are each 1 mm long, then each water molecule would have to be pulled through 10,000 pit membranes, and the friction would be considerable. If the plant had vessels, each 1 m long, then each water molecule faces only 10 pit membranes, and flow would be greatly improved. Because each vessel is 1000 times as long as each tracheid, every cavitation would cause 1000 times more damage, but under conditions of abundant water such as these cavitation will probably be very rare. Because the water moves easily, the tracheary elements do not have to be especially strong, and less wall thickening will also enhance water entry and exit from each element, because more of the primary wall remains uncovered. The vessel elements would be expected to have simple perforations; these offer the least resistance to water flow and, although they are not so strong as scalariform perforation plates with their bars, this is not a problem since water moves easily and with little tension.

2. Soil frequently dry, soil water frequently limiting. In conditions such as this, as might be found in deserts or on the rainshadow side of mountains, water is typically a valuable resource that is carefully conserved. Because water that is lost to transpiration cannot be replaced immediately, it is selectively advantageous for a plant to waterproof its surfaces with a thick cuticle and waxes, produce a thick-walled epidermis, and perhaps even reduce the transpiration surface by having needle leaves (as in the conifers) or other very reduced leaves (stem succulents). If water movement in such plants is slow at all times (the environment is rather dry year round), then the large conducting capacity described in the last section would be unnecessary. Futhermore, because the soil is frequently dry, the plant may not be able to pull water out of it at times, even though it will still be losing at least some water to the atmosphere despite whatever waterproofing adaptation might exist in the stem. The water column will be under extreme tension, and cavitation will be an important risk. The tracheary elements should be narrower to maximize the wall's ability to reinforce the water column. Vessels might be shorter to minimize the damage that occurs if cavitation does happen, and under extreme conditions tracheids or vascular tracheids might be employed instead of vessels to further minimize cavitation damage.

3. *Soil alternates between dry and moist, soil water sometimes available, sometimes limiting.* This is probably the most common situation, and we find a variety of successful adaptive strategies. If the dry periods are not too long and severe, typical leaves or leaves with some xeromorphic characteristics might be capable of surviving; if the dry season is more adverse, then leaves might be more ephemeral, being expanded rapidly during the rainier, more moist season and being shed in the dry season. With regard to conduction capacity, if the plants have some capacity to store water, then they may need to accumulate it rapidly when it is available. Thus there may be large vessels that function only in the moist season and that are abandoned in the dry one, being allowed to cavitate because they have already functioned to transport what water had been available; like the leaves, they are expendable. If the dry season is severe and protracted, then there may be no other xylem production until the following year. But, if during the dry season there is some soil water available, then a second type of tracheary element may be formed to allow limited conduction of the little water that is available. This element might consist purely of tracheids or it might be tracheids mixed with specialized vessel elements that are narrow and short. These vessel elements would be likely to have scalariform perforation plates, because the scalariform bars act as reinforcement to prevent collapse and the rate of water movement is so small that the friction they offer is inconsequential.

Other scenarios of water availability can be imagined, and for each it is possible to postulate a type of xylem architecture that would fa-

cilitate conduction and minimize damage. An extremely important principle to remember is that water does not merely move directly from root tip to leaf veinlet. Each part of a plant may have its own special, distinct problems and needs with regard to water availability, movement, and loss. Consider a stem succulent such as a barrel cactus. It grows in a habitat where soil moisture is completely unavailable except for a few days each year. It has a large barrel-shaped body with a broad system of very long, cablelike roots that lie only millimeters below the soil surface. When rains occur, they are torrential and tend to be so brief and intense that most of the water runs rapidly over the surface and is lost in flash floods. Very little penetration occurs, and the soil dries within several days. The roots, being so shallow, can absorb this temporary abundance, but they must conduct it rapidly into the shoot before the soil and the roots dry out. Therefore, the roots need a high-volume, temporary conducting capacity. But once in the base of the body, the water needs to move only a meter or so as it is distributed to all parts of the cortical parenchyma, and much of this conduction is accomplished by transfer directly between parenchyma cells. The water will later be moved slowly and in small quantities to the growing regions of the plant, so the shoot needs a low-volume, long-term conducting capacity whose structure will be very different from that of the root. Finally, when the plant flowers, the petals and stamens are such water-loss liabilities that they are extremely ephemeral and wilt on the same day that the flower opens. They need rapid, temporary conduction with the minimum expenditure of wall material, but the ovaries, which develop into a large fleshy fruit, need moderate-volume, long-term capacity. There is no reason whatsoever to assume uniformity of xylem structure throughout the plant when each part has such distinct needs. In other plants in less extreme environments it must also be expected that each organ produces its own type of xylem, specialized for the functions of that particular organ.

One thing must always be remembered about the environment, its variability—especially with respect to the expected lifetime of the plants growing in it. First, many environments and seasons are not as most people envision them. People from north temperate regions (which would include northern North America, northern Europe, and the U.S.S.R.) think of winters as being harsh, cold, and stressful, with spring and summer being the optimum growing seasons. But nearer to the equator, in the mediterranean climate regions, the situation is just the opposite. Summers are too hot and dry for many plants to survive other than as seeds, but winters are cool and moist; autumn is the season that brings relief from stress. This misconception is important because the north temperate regions just mentioned are the sites where most botanists and plant anatomists have lived and worked, and this has affected ideas and concepts of stress and adaptation. Had botanical studies been centered almost exclusively in the tropics for the last two hundred years, many concepts of "normal" and "stressful" would be quite different.

A second important aspect of the environment that must be considered is that perennial plants must be adapted to the heat and water stress patterns that might occur during their expected lifetimes, not just during the "average year." For plants that tend to live 100 or more years, the average climatic conditions are less important than the worst possible conditions that might be expected in that lifetime,

and especially in the time between germination and the beginning of reproductive age. Because many trees do not flower for the first time until they are ten or more years old, this is an important consideration.

Third, it is important to note the microhabitat of the plant: two plants growing side by side can be in totally different environments if one is growing at the edge of a stream or pond while the other is just a few meters back, or if one is tall and extends to the sunlight while the other is short and is an understory plant. Similarly, two shrubs will have different water conducting requirements if one has perennial leaves and the other is deciduous during droughts (Baas and Zweypfenning, 1978; Carlquist, 1966; Chowdhury, 1941).

Finally, the concept of stressful conditions varies from evironment to environment and from species to species. Weather that might be considered quite pleasant in the Chihuahuan desert would kill most plants in the Olympic rainforest and vice versa. This is perhaps especially true in the tropics, which are too frequently considered absolute paradises for plants. It is not unusual to hear them described as being constantly warm and moist, but this is not true. Certainly they are not as cold or dry as many places, but, for many of the plants that grow there, one week without rain or mist might be fatal, as might one day of temperatures lower than 40°. Stress cannot be evaluated in general principles, but only with regard to the plants and animals that are native to the regions being considered.

Concepts

In the previous chapter we saw that, as primitive land plants began to grow upright, the tissues responsible for gathering water and inorganic nutrients were separated from those responsible for photosynthesis. This effectively began the differentiation of the plant body into two distinct portions: one aerial, photosynthetic, and autotrophic; the other lower, non-photosynthetic, heterotrophic, and able to survive only due to the sugars supplied by the upper portion. Xylem arose as a mechanism for the long-distance transport of water and nutrients upward into the aerial portions of the plant. Likewise, phloem evolved as a means to translocate organic nutrients, especially the sugars produced by photosynthesis, downward into the lower parts.

The evolution of phloem had other important consequences in addition to keeping the lower, heterotrophic portions of the plant alive. As mentioned earlier, land plants must have localized meristems to achieve large size; diffuse growth is not functional. If there is no long-distance transport of nutrients into a meristem, then the meristem can grow only very slowly. Thus the evolution of phloem not only permits a type of forced feeding of meristems and primordia, but also permits much more rapid growth and differentiation than would otherwise occur. Furthermore, the presence of phloem facilitates greater control over which

parts of a plant grow most rapidly: at different times in a year, sugars can be transported predominantly to meristems and leaf primordia, or to reproductive structures, or to storage tissues.

It is interesting that two such different systems of long-distance transport evolved. The conducting elements of xylem are dead while functioning, and water movement is controlled abiologically by gradients of water potential. The plant really has very little ability to control water movement. Phloem, by contrast, must be living in order to function. Rather than having the nutrients merely pulled passively through it, the sugars, amines, and other nutrients are loaded into it by molecular pumps in the plasmalemmas (Evert, 1982; Giaquinta, 1980; Kursanov, 1984); these pumps force certain molecules into the phloem in some areas (**sources**) and extract them in other areas (**sinks**). Thus, movement in the phloem is very dynamic, and every aspect of it is under the active control of the plant. Because of this, nutrients can be directed to special regions in order to support growth, and they can be kept away from other regions to suppress it. Furthermore, whereas tracheary elements function for months or years, the conducting elements of phloem often function for only days or weeks. The phloem elements must be constantly replaced. The plant might be able to take advantage of this fact to further its control: the elements are replaced if growth is to continue but not replaced if growth must be halted.

Rather surprisingly, the mechanism by which material is translocated through the phloem is still unknown (Ting, 1982). At present, the Münch Pressure Flow hypothesis is receiving much attention and little criticism. This hypothesis proposes that sugars and other metabolites are loaded into the conducting elements at sources, that this causes the osmotic potential and water potential to become more negative, and that water thus flows into the cell. The absorption of water would normally cause a cell to become very turgid, but in the conducting elements the increased pressure merely pushes the sugar solution out through large, open pores into the next element. The entire tubular network of the phloem becomes pressurized. At sinks, the sugars are actively unloaded from the conducting elements, causing the osmotic potential and water potential to become less negative. Water moves out of the cells, and a flow is established from source to sink. The osmotic and water potentials of the source (photosynthetic cells or storage parenchyma cells in the spring when they are releasing their nutrients) are not affected osmotically by their loss of sugar, because they are creating sugar (from carbon dioxide or from starch) as rapidly as they are loading it into the phloem. Similarly, the potentials in the sinks (growing regions or storage tissues in the summer when they are accumulating material) are not affected, because they are either respiring the sugar for energy, converting it to other compounds, or polymerizing it to starch as rapidly as the phloem delivers it.

Phloem can translocate large amounts of material quickly: phloem sap contains between 50 and 300 mg of dry matter per milliliter, and 80 to 90% of this is sugar. Amino acids occur at concentrations of 20 to 80 mg/ml, and other components of phloem sap are sugar alcohols, sugar phosphates, organic acids, organic phosphates, growth regulators, nucleic acids, vitamins, and inorganic substances (Crafts and Crisp, 1971). The velocity of translocation ranges between 10 and 100 cm/hr (up to 300 cm/hr in some cucurbits), and the rate varies from 0.5 to 5.0 g of dry matter per square centimeter of phloem per hour. Some simple calculations provide instructive results: the "average" conducting element is 50 μm long; at a velocity of 100 cm/hr (which equals 1,000,000 μm/hr), each solute molecule passes through 1,000,000/50 = 20,000 conducting elements in one hour. Schumacher (1948) calculated that a plant of *Bryonia dioica* (a vine 7.4 m long) contained 80 million conducting cells in its phloem but that these same cells had a combined volume of only 4 cm^3.

Structure of Phloem

Phloem is a complex tissue that can consist of solely parenchyma or of both parenchyma and sclerenchyma. Any sclerenchyma present will be in the form of fibers, or very rarely sclereids, as have been described in Chapter 5. Nonconducting parenchyma is present as storage cells (these are described in Chapter 3) or secretory cells (Chapter 9). Within a vascular bundle, very often less than half of the volume of the phloem is composed of conducting cells (Lawton and Canny, 1970). The subject of this chapter is the conducting parenchyma and the associated cells (Fig. 8.1). The integration of conduction, storage, and mechanical strength is presented in Chapters 11 and 16.

Conducting Cells

The conducting elements of the phloem are the **sieve elements** (Hartig, 1837; Nägeli, 1858), which are of two types: **sieve cells** in almost all nonangiosperms and at least one of the most primitive angiosperms, *Austrobaileya scandens* (Srivastava, 1970), and **sieve tube members** in the majority of the flowering plants (Fig. 8.2). The sieve elements, being parenchyma cells, have a primary wall and no secondary wall.

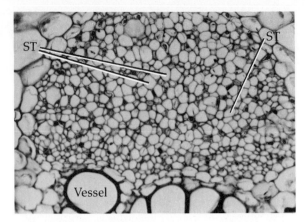

Figure 8.1
This is a cross section of a vascular bundle showing the phloem and several vessels. The phloem is made up of larger, lighter cells (sieve tube members) and smaller, darker cells (companion cells). The companion cells look like they have been cut out of one of the adjacent sieve tube members. × 500.

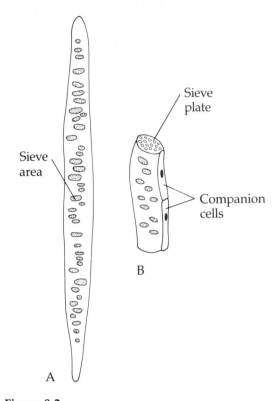

Figure 8.2
Both sieve cells and sieve tube members are parenchyma cells with modified primary pit fields (sieve areas) and plasmodesmata (sieve pores). In sieve cells (A), the sieve areas are either uniformly distributed on the surface of the cell or are slightly aggregated near the ends. They taper and often do not have a distinct end wall. The sieve tube members (B) differ in that they are wider, having a distinct transverse end wall with a prominent sieve area (the sieve plate).

The walls of sieve elements are virtually never lignified. A rare exception occurs in some cereal grasses (Kuo and O'Brien, 1974), but Cartwright et al. (1977) have shown that in those cases the lignified sieve tube members neither conduct nor store material. The primary wall is unusual in that the plasmodesmata of the primary pit fields have become extremely large, often more than 1 μm in diameter (Fig. 8.3): *Larix* and *Pinus*, 1–2 μm, and *Ailanthus*, 14 μm (Hepton and Preston, 1960; Murmanis and Evert, 1966). Because these plasmodesmata are so dilated and specialized, their name is changed to **sieve pores;** the primary pit fields where the sieve pores occur also have a new name: **sieve areas** (Figs. 8.3, 8.4).

In the simplest type of differentiation, all plasmodesmata and primary pit fields of the cell become equally enlarged, and all parts of the cell surface are rather similar; such an element is a sieve cell (Fig. 8.2). But this is not an optimal arrangement; because phloem transports mostly longitudinally, it would be best to have those sieve areas on the two ends be especially large with wide sieve pores, whereas the sieve areas on the side walls could be rather small, not much different from regular primary pit fields and plasmodesmata. An element like this is a sieve tube member, and the end walls are called **sieve plates** (Figs. 8.2, 8.5, 8.6). If a sieve plate has only a single sieve area, it is **simple**; with more than one, it is a **compound**

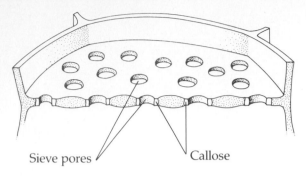

Sieve pores   Callose

Figure 8.3
A sieve pore is a modified plasmodesma and is lined by the plasmalemma. There may be a thin layer of callose between the plasmalemma and the wall; by the time the sieve pores have become fully differentiated, the cytoplasm has become mictoplasm.

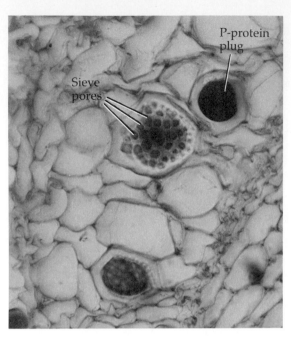

P-protein plug

Sieve pores

Figure 8.5
This section of *Cucurbita* (squash) shows two plates and the P-protein plugs of one other. The sieve pores are large and appear black with a gray outline. The black is due to the P-protein that is precipitated into the pore, and the gray is callose, which also would not be present while the sieve tube member was functioning. × 300.

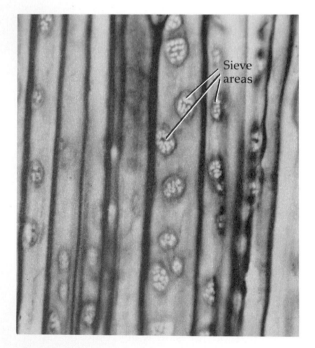

Sieve areas

Figure 8.4
These are the sieve cells in *Pinus strobus,* showing prominent sieve areas on the side walls. The pores in these cells are small in comparison to those of sieve plates (Fig. 8.5), but actually these sieve areas are larger and more easily seen than most. × 400.

Figure 8.6
Sieve plate of *Cucurbita pepo* (pumpkin), showing two sieve pores on the left. Small aggregations of endoplasmic reticulum are visible. × 20,000. Micrograph generously provided by R. Warmbrodt.

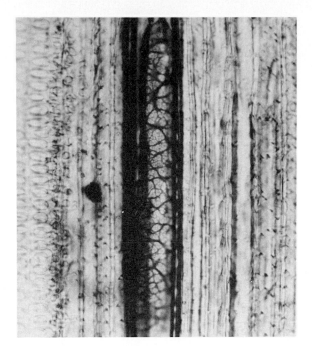

Figure 8.7
A spectacular compound sieve plate from the phloem in the air roots of *Monstera*. The sieve tube members are large and are involved in long-distance transport. The dark webbing is the ordinary primary wall of the sieve plate, and the white polygons are the masses of callose. The pores appear small and are difficult to distinguish at this magnification, but that is primarily due to the plugging by the callose. × 300.

sieve plate (Fig. 8.7). The pores in sieve plates can have a diameter of up to 10 μm, as in *Cucurbita* or even 14 μm, as in *Ailanthus altissima*. The average sieve pore diameter, based on a survey of 126 dicots, was 1.18 μm (Esau and Cheadle, 1959). Sieve plates can occur on side walls occasionally.

An entire column of sieve tube members constitutes a **sieve tube** (Figs. 8.8, 8.9); each internal sieve tube member has two sieve plates, but the two members that terminate a sieve tube each have just one. The distal wall does not have sieve pores (Pizzolato, 1983). Unfortunately we have no measurements of the length of individual tubes, and the theoretical con-

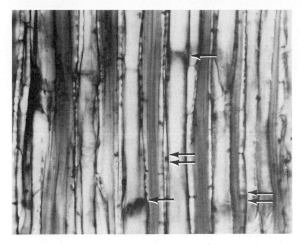

Figure 8.8
Longitudinal section of phloem; several sieve tube members and sieve plates (arrows) are easily distinguishable. These are unusually wide, and it is rare to obtain such good preservation without large masses of callose and P-protein. The long, narrow cells with nuclei are probably companion cells (double arrows), but it is difficult to be certain from only a longitudinal section. × 300.

Figure 8.9
The sieve plates in this *Sechium edule* (also a cucurbit) are not so perfectly transverse as the sieve plates in Figure 8.8, and one is located on the side of the cell. Although these P-protein plugs are artifacts caused by less-than-perfect fixation, they actually make it much easier to locate and study the sieve plates. × 300.

sequences of having sieve tubes of different lengths have not been treated. Whereas an embolism in a vessel member will incapacitate the entire vessel, damage to a sieve tube member will be restricted by efficient plugging mechanisms. Furthermore, movement of material into and out of sieve tube members is very dynamic, so the blocked region probably can be circumvented easily. The conductance of the phloem is related to the fourth power of the radius just as is true of the xylem conductance. But in the phloem it is the diameter of the sieve pores rather than the diameter of the whole cells that is important, because sieve tube members never have the equivalent of one large, simple, low-friction perforation. Even the largest of the sieve pores is much smaller than the diameter of the cell (Fig. 8.5).

This difference in the size and distribution of sieve areas requires that sieve cells and sieve tube members have somewhat different shapes. Because the sieve cells have only small sieve areas with small sieve pores (range: 0.05 to 2.0 μm) distributed over the entire surface, it is best if their endwalls are extremely slanted, creating a large area of overlap with the sieve cell above or below it. This large overlap area provides room for many sieve areas, and flow can occur between vertically adjacent cells. The cells themselves tend to be long, often over 1 mm—2750 μm in spruce and 3500 μm in *Pinus* (Kursanov, 1984). Because sieve tube members have such large areas and pores on their endwalls, it is sufficient for them to have a transverse endwall. To increase the surface area of the sieve plate and to maximize the area available for the sieve pores, the sieve tube member should be as wide as possible, up to about 40 μm. Sieve tube members are short, only about 50 to 150 μm in length.

The functional significance of the length of sieve elements must be analyzed carefully, keeping in mind that these cells are derived from the same initials as are the tracheary elements: procambium or vascular cambium. As described in Chapter 7, the length of the tra-cheary elements has important consequences, and it has been suggested (Carlquist, 1975) that the sizes of the cambial cells are related to the needs of the xylem cells. The sizes of the sieve elements are probably influenced by the cambium cells and thus are indirectly related to the needs of the xylem. However, there is suggestive evidence that the length of sieve tube members is functionally important: in some members of the Winteraceae, the cells that will mature into sieve tube members undergo cross divisions during development, and the members are much shorter than the cambial cells that produce them (Esau and Cheadle, 1984). If the sieve plate is involved in pumping material across it as has been proposed by some (Ting, 1982), then a sieve tube composed of short members would have more pumps than one composed of long members. It was suggested (Chapter 7) that vessel elements are short merely to facilitate their tremendous growth in width; such an explanation does not seem appropriate for sieve tube members, because they are always rather narrow, only rarely being as wide as 20 μm.

Companion Cells and Albuminous Cells

Closely associated with the sieve elements are specialized cells called **companion cells** in the angiosperms (Figs. 8.1, 8.10) and **albuminous cells** in the nonangiosperms (Sauter, 1974). Until recently, it was thought that these cells were concerned solely with controlling the metabolism of the adjacent conducting elements. This view was based on the fact that the sieve elements lose their nucleus during differentiation, so that close association with the nucleated albuminous cells or companion cells would allow the maintenance of the sieve cell or sieve tube member cytoplasm. Unfortunately there is no experimental information about the amount of control that albuminous cells or companion cells may exert. When the concept of control was being formulated, the means of sieve-element loading and unloading was un-

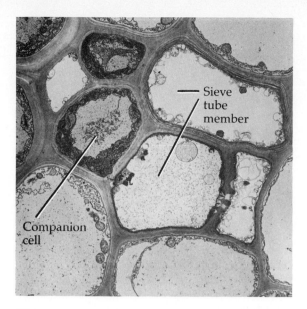

Figure 8.10
These two companion cells are densely cytoplasmic, whereas the adjacent sieve tube members appear almost empty of cytoplasm. Several nonconducting parenchyma cells are also visible.
× 5000. Micrograph generously provided by R. Warmbrodt.

known. Now, with the aid of radioactive tracers, it has been established that, in at least some species, companion cells are the primary means for accomplishing these tasks (Evert, 1977; Kursanov, 1984).

Phloem loading has been studied most extensively in the minor veins of leaves where newly synthesized sugars are loaded for export. In the minor veins, the sieve tube members are very small, and the associated parenchyma cells are much larger. These cells, both companion and noncompanion, are called **intermediary cells** because they mediate the accumulation and loading of material. In some plants (sugar beet) the intermediary cells have smooth walls, but in others the cells are transfer cells with labryinthine ingrowths. These greatly increase the surface area of the plasmalemma, thereby increasing the number of molecular pumps that it can contain. Two types of intermediary cells can be recognized. **A-type intermediary cells** are companion cells, and they have labyrinthine ingrowths most well developed on all walls except those that contact the sieve tube member and that contain the numerous plasmodesmatal connections (Evert, 1977). **B-type intermediary cells** are not companion cells, and the labyrinthine ingrowths occur on all faces and are most prominent on the sieve tube member contact face. Material can be absorbed into intermediary cells from the apoplast by means of labyrinthine walls that face intercellular spaces. The path for movement of nutrients from intermediary cells into sieve elements appears to be either across the plasmalemma or through the plasmodesmata (Gunning et al., 1968).

Because intermediary cells can accumulate material from the apoplast (van Bel and Koops, 1985), it has been proposed that they are also responsible for retrieving materials that have leaked from the cells and are moving apoplastically as a result of transpiration. By being retrieved, these substances, especially the salts, can be loaded into the phloem and recycled through the plant.

At present, there seem to be no reports of labyrinthine walls in the phloem of gymnosperms. Because many of the intermediary cells of angiosperms have smooth walls, however, it is reasonable to assume that the albuminous cells and other phloem parenchyma in the gymnosperms are also intermediary cells (Sauter, 1980).

In most plants, the walls between a sieve cell and its albuminous cell or between a sieve tube member and its companion cell have a great number of connections (Fig. 8.10). These are rather large plasmodesmata on the albuminous cell or companion cell side, but sieve areas on the sieve element side (Esau, 1969; Gunning and Robards, 1976; Kursanov, 1984). These provide a substantial cytoplasmic continuity between the two types of cells. The attachment is such that, when tissues are macerated (treated with acids to dissolve the middle lamella, allowing the cells to separate from each other), the albuminous cells or companion cells remain attached to their associated sieve elements.

It had previously been thought that albuminous cells differed from companion cells in that any parenchyma cell adjacent to a sieve cell could differentiate into an albuminous cell, whereas companion cells were thought to be only sister cells to their sieve tube members. It is now established that the latter idea is not always true: following the cell division that establishes the sieve tube member, the sister cell may either differentiate directly into a companion cell or undergo repeated divisions (Shah and Jacob, 1969; Sokolova, 1968; Srivastava and Bailey, 1962). Part or all of these derivatives may differentiate as companion cells (Fig. 8.11). Thus sieve tube members as well as sieve cells may have one or more associated cells which may or may not be sister cells. Furthermore, as Evert (1977) pointed out, companion cells and other phloem parenchyma cells intergrade with each other and are not always unequivocally distinguishable from each other. With the discovery of the role of B-type intermediary cells, the companion cells are now considered to be less critically important.

Development

Phylogenetic Development

Sieve elements are some of the most labile cells in any plant; they function briefly and then collapse, and they degenerate rapidly after the plant dies. Only a very few species have sieve areas and sieve pores large enough to be studied easily with light microscopy; in almost all plants they are so tiny as to be invisible after standard fixation and staining. Phloem can be reliably studied only under the best of conditions and after special precautions have been taken; therefore, few anatomists have bothered with it. In the fossils, even those that show ex-

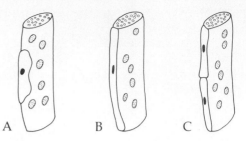

Figure 8.11
The relationship of companion cells or albuminous cells to the associated sieve element can take different forms. In some cases, especially in conifers, the albuminous cell is small, and much of the sieve cell has no contact with it (A). In many angiosperms, the companion cell is as long as the sieve tube member (B), but in others it will subdivide into a row of two (C) or more cells. If it subdivides, some of the daughters may develop into ordinary parenchyma, not into companion cells.

cellent preservation of other tissues typically offer no useful details of the phloem structure, rare exceptions being specimens of *Palmoxylon* (Bonde and Biradar, 1981) and *Rhynia* (Satterthwait and Schopf, 1972). The phylogeny of phloem must be inferred from extant plants.

Rather surprisingly, the phloem of vascular cryptogams has been almost completely ignored until recently (Esau, 1969; Lamoureux, 1961; Warmbrodt, 1980; Warmbrodt and Evert, 1974, 1978, 1979a, b). The studies of Warmbrodt and Evert have shown that in most of these plants the sieve cells are long, tapered and anucleate. They usually have membrane-bounded protein bodies called **refractive spherules** (Evert and Eichhorn, 1974; Fisher and Evert, 1979). The sieve areas are poorly developed, being only slightly modified primary pit fields. In many species, the presence of refractive spherules was the most reliable means of identifying a cell as a sieve cell. The sieve cells have numerous plasmodesmatal connections with the surrounding mesophyll cells. In addition, the adjacent vascular parenchyma cells and mesophyll cells can have extensive wall ingrowths.

Sieve tube members that resemble sieve cells are believed to be relictually primitive: the sieve areas on the sieve plates are not strikingly different from those on the lateral walls. Specialization involved increasing the diameter of the cell, making the end wall more horizontal, and having it bear only one sieve area that has very large pores (Cheadle, 1948; Cheadle and Whitford, 1941; Zahur, 1959). As with the xylem, different parts of a plant have individual needs for translocation. Considering the rapid growth of very large squashes, pumpkins, and watermelons, it should certainly not be surprising that the fruit stalks of these plants have some of the largest sieve plates and pores of all the angiosperms. However, the roots and leaves of these same plants do not have exceptional requirements and do not have exceptional sieve tube members. It would be erroneous to believe that a single plant had several levels of phloem evolution within itself. The evolution of the sieve plate appears to have occured several times; in addition to the angiosperms, sieve elements with sieve plates have been reported for four species of *Equisteum*— *E. arvense*, *E. giganteum* (Agashe, 1968), *E. hyemale* (Lamoureux, 1961; Dute and Evert, 1978), and *E. telmateia* (Lamoureux, 1961)— and for one species of fern—*Cyathea gigantea* (Shah and Fotedar, 1974).

Ontogenetic Development

Sieve element initials are derived either from the procambium or the vascular cambium. They are initially very cytoplasmic cells, as would be expected (Esau and Thorsch, 1985). As the cell matures, a new layer is deposited on the primary wall; this is called the **nacreous layer** because in fresh sections it is lustrous (Behnke, 1971; Esau, 1969; Evert, 1977; Kuo, 1983). The nacreous layer is often about 1 μm thick, but may be so extensive that it almost fills the lumen (Fig. 16.10), as in *Laurus*, *Magnolia*,

Persea, and *Rhamnus* (Fahn, 1982). With age, it usually shrinks and becomes much thinner. It is reported to be completely absent in some of the aquatic angiosperms, such as Hydrocharitaceae, Cymodoceaceae, and Posidoniaceae (Kuo, 1983). In both *Cucurbita* and *Pinus*, the nacreous layer is composed of many layers of cellulose microfibrils that are densely packed (Deshpande, 1976). Esau (1969) considered the nacreous layer to be a type of secondary wall; however, there is no evidence that the nacreous layer of sieve elements is homologous with the secondary wall of sclerenchyma. The function of the nacreous layer is still unknown, but Kursanov (1984) speculated that it might be a means of increasing the volume of the apoplast, which would facilitate radial transport of nutrients.

As the cell matures, **callose** is deposited within the matrix of the primary wall, on the borders of the plasmodesmata (Esau et al., 1962; Evert et al., 1971; Kursanov, 1984). Callose is a polymer of glucose units linked together by β-1,3 bonds. The universal presence of callose in sieve elements is an important aid to the study of phloem: callose can be stained with aniline blue, which then fluoresces a brilliant lemon yellow when viewed with a microscope using ultraviolet light (Esau, 1948). Initially the callose occurs as a collar or cone around the plasmodesma, just exterior to the plasmalemma, and encrusts the primary wall (Esau et al., 1962). The collars at each end of the plasmodesma do not contact each other but are separated in the region of the middle lamella. The plasmodesmata begin to increase in diameter beginning at the middle lamella. It is not known if this is due to a digestion or a compaction of the wall material on the edge of the plasmodesmata, although Esau and Thorsch (1984) refer to it as lysis in *Echium*. This increase in diameter causes the cytoplasmic channel to become larger, forming a **connecting strand**. The original endoplasmic reticulum tubule (desmotubule) disappears. The mature connecting strand is an open channel lined by the plasmalemma and containing cytoplasm (Evert et al., 1973).

The nucleus degenerates chromatolytically (it becomes less stainable) in the angiosperms but pycnotically (it becomes shrunken and very dark) in the gymnosperms (Evert, 1977). It is now becoming apparent that nuclear breakdown is not universal: nuclei do not degenerate at all in the Taxaceae (Evert et al., 1970), oat (O'Brien and Thimann, 1967), *Neptunia* (Shah and James, 1968) and others (Evert and Deshpande, 1970). In other taxa, the breakdown is only partial (Kollmann and Schumacher, 1961; Walsh and Evert, 1975). Kursanov (1984) concluded that the sieve cells of conifers have a more pronounced tendency to retain their nuclei than do the sieve tube members of angiosperms.

During differentiation, the endoplasmic reticulum becomes modified such that it forms parallel sheets along the edges and in the corners of the cell (Esau and Gill, 1971). The tonoplast breaks down, allowing the vacuolar contents to mix with the hyaloplasm (Dute, 1983); this extremely watery mixture is occasionally called **mictoplasm**. The few organelles that remain (plastids, mitochondria, and ER) are located immediately adjacent to the plasmalemma (Fig. 8.10), and the mictoplasm occupies all of the center of the cell and is continuous from cell to cell through the sieve areas. The plastids may store either protein (**P-type plastids**) or starch (**S-type**). The method of storage varies, and numerous subtypes can be identified by electron microscopy. The ultrastructure of these phloem plastids has become an extremely important taxonomic character, especially in the classification of the dicot Order Caryophyllales (Behnke, 1975, 1976).

The cell also synthesizes large amounts of **P-protein** (Esau and Cronshaw, 1967). The nature of the P-protein varies from species to species, being fibrillar, tubular, granular, or crystalline (Alosi and Alfieri, 1972; Cronshaw and Esau, 1967; Evert, 1982; Hoefert, 1979, 1980; Parthasarathy and Mühlethaler, 1969). It

is distributed along the sides of the cell, usually not as a smooth layer but as clumps and small aggregations. After the tonoplast degenerates, the P-protein occurs as filaments in the thin, parietal layer of residual cytoplasm. The P-protein constitutes a mechanism for sealing the sieve elements whenever they are ruptured. The contents of the sieve elements are under pressure; if the phloem is cut, the mictoplasm can surge out. The rapid drop in pressure causes the P-protein to be carried to and accumulate at the sieve areas as long filaments (Fig. 8.12). They form a cap called the **P-protein plug** or **slime plug** (Fig. 8.9); P-protein was formerly called **slime**. In addition, starch and protein granules from broken plastids can contribute to the plug (Kallarackal and Milburn, 1983; Walsh and Melaragno, 1981).

It is occasionally proposed that the P-protein is involved in translocation, but Evert (1982) argued convincingly against that, pointing out that P-protein does not occur in many monocots, even in the very large palms where long-distance transport is important, and that it does not occur in gymnosperm sieve cells. Also, P-protein is frequently found in companion cells.

The intact sieve element contains large amounts of suspended callose which is also pressure sensitive. When the element is damaged and surging begins, the callose precipitates and contributes to the plugging action (Fig. 8.12). The callose penetrates many areas and actually fills the sieve pores. Because it is deposited in response to injury, it is referred to as **wound callose**. Callose and P-protein together quickly seal the damaged region; without them, the mictoplasm of many cells would be pushed out before the pressure in the sieve elements dropped sufficiently.

The callose and P-protein are actually so sensitive to conditions in the phloem that almost any disturbance can cause them to precipitate. This can result in numerous artifacts during fixation and processing of tissues for examination, and it had been thought that

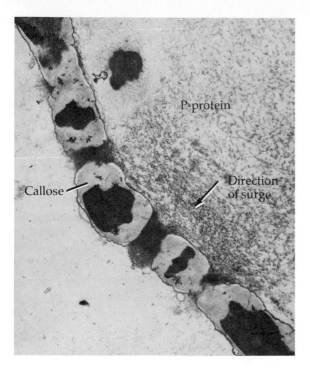

Figure 8.12
Callose and P-protein coagulated at a sieve area. In this area, the callose is abundant but would not have been sufficient to have plugged the plate. The P-protein is present on the upper side of the plate, indicating that the phloem had been damaged below and the phloem sap was surging downward. × 1000.

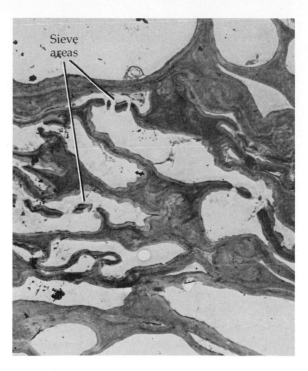

Figure 8.13
When phloem stops transporting material, it is said to collapse; this micrograph shows that this is true. The cells not only die, but also do not remain cylindrical; the walls buckle, and there is extensive folding. See Figure 8.5 for a light microscopic view of this. × 600.

even in normal, functioning sieve elements the sieve pores were lined with callose so heavily that they were completely occluded. The P-protein was frequently seen to be precipitated throughout the cell, further blocking it. Such a structure would have been incompatible with the bulk movement of material through the sieve elements, and it caused tremendous difficulty in all attempts to explain how phloem functioned. By using very careful techniques to prepare the phloem, being especially careful to fix whole plants or whole parts still attached to a living plant so that the vascular bundles were not cut until after the tissues had been

thoroughly fixed, it has been possible to ascertain that in functioning phloem there is no P-protein plug and at most only a thin layer of callose near the sieve pores (Eschrich, 1963; Evert and Derr, 1964). Evert (1977) discusses this in detail and gives four criteria that can be used to evaluate whether the fixation has resulted in artifacts.

A sieve element is usually active in transport for less than four or five months (Derr and Evert, 1967; Ghouse and Hashmi, 1979; Tucker and Evert, 1969). When it has completed its functioning, the cell collapses (Fig. 8.13), and the callose precipitates naturally and seals the sieve area completely. Under these conditions, the callose is called **definitive callose** (the term **callus** may be seen in older books, but it is no longer accepted). In certain dicots (*Polygonatum, Smilax, Suaeda, Tilia*, and *Vitis*) the phloem functions for two or three years (Fahn, 1982; Fahn and Shchori, 1967). In the autumn, definitive callose is formed in *Vitis*, but in the spring it is resorbed and transport resumes (Esau, 1948; Bernstein and Fahn, 1960). In palms and other long-lived monocots that do not have secondary growth, the primary phloem must function for many years, even hundreds of years. Structurally it is indistinguishable from the phloem that functions for only a few weeks (Parthasarathy and Tomlinson, 1967).

Protophloem

The preceding description is based on the metaphloem and secondary phloem. Protophloem is formed as an organ is elongating. Being anu-cleate, the protophloem cannot grow with the surrounding tissues. It is soon stretched to the point of breaking, as was described for protoxylem. Protophloem elements may actually function for only one or two days, perhaps less in some plants or organs. Sieve cells and sieve tube members occur in protophloem, but companion cells and albuminous cells do not. The sieve areas on both sieve cells and sieve tube members are so small as to be extremely difficult or impossible to detect even with the best methods of light microscopy. In many cases protophloem is identifiable only because its cells are anucleate and their walls are thickened. After the elongation of the organ destroys the protophloem sieve elements, the elements may collapse completely and be undetectable in older tissues. Nonconducting parenchyma that occurred in the protophloem may persist and may continue as parenchyma, or it may become collenchyma or even develop into phloem fibers.

Secretory Cells and Tissues

Concepts

All plant cells are secretory. The cell wall is a form of secretion, and many compounds are actively transported out of the cell: sugars, hormones, nitrogenous and sulfur-containing compounds, and inorganic salts. Cell wall components such as the cuticle, waxes, lignin, and suberin are secretory products. Furthermore, the ER and dictyosomes, which are the most important organelles for most types of secretion, are present in all cells.

Using this fundamental metabolism as a basis, an amazing variety of secretory products, structures, and mechanisms have evolved. In order to study the anatomy of secretory structures and to analyze the biology of secretion, it is necessary to analyze the function or selective advantage that secretion confers upon a plant. The advantage will vary depending on the nature of the secreted material, but it is possible to break secretions down into two very rough categories that are not necessarily mutually exclusive. The first category contains those secretory systems that are involved with the plant's own fundamental metabolic needs, and the second category contains those that mostly facilitate the plant's interaction with other organisms. In the first type are classified glands that remove excess water or salts from the plant, or areas that accumulate and isolate waste products so that they do not damage the living cytoplasm. Even the most advanced and sophisticated of animals, which can carefully select their food, must have systems capable of eliminating the indigestible components and excreting their own metabolic wastes. Plants have the same problems and typically take in excessive amounts of certain salts or water, which must then be sequestered (internal secretion) or dumped outside of the plant body (external secretion). Metabolic wastes and by-products must also be handled properly or they poison the protoplasm of living cells.

In the second category are such structures as the scent-producing glands and the nectaries that attract pollinators to flowers. Beyond the very fundamental level of secretion biology of the first category, secretory tissues have evolved into extremely elegant mechanisms by which plants can interact with other living organisms in their environment. Scent glands and nectaries have already been mentioned; these can have such precise anatomy and chemistry that the plant attracts only a single species of pollinator. On the other hand, many secretory systems function to keep animals away—stinging hairs and poison cells. Finally, the digestive glands of the carnivorous plants are a means by which the plants can interact particularly definitively with animals.

There are many other types of secretion and many other roles for the secreted material. Numerous types of secretory tissues and glands have evolved, and a large percentage of the basic types have originated several or even many times. It is extremely difficult to make broad generalizations, and it is important always to keep the function of the secretory structure in mind while the anatomy is being studied. It is critical to remember that many secretory structures and glands evolved from structures that did not originally have a secretory function. Additionally, the function of a gland can change with evolution, such that the current role is not the same as that of the original gland. In these cases, the secretory structure that exists and functions today might be far from ideal; its capacity and efficiency may be severely limited by its evolutionary history. Secretory biology is one of the areas in plant anatomy in which it is often easy to see how history has affected present-day biology.

Methods of Classifying Secretory Systems

The basis for classifying and naming anatomical structures should be the same as that for classifying and naming species. It should be a natural system that brings together structures that are homologous due to their being in one evolutionary sequence, and it should separate those that are merely analogous because they have separate evolutionary origins. Anatomists have been rather successful with this method of classifying, but certain areas are very problematic, secretory structures being one of them. Secretory structures have evolved so many times that an enormous number of terms would be required if each were to have a unique name and terminology. Furthermore, there have been so many convergences that very similar structures would have different names. The pragmatic approach has been taken: secretory structures are classified on artificial bases such as the type of secretory product or the type of structure. The natural, evolutionary approach has been used only in specific studies of certain genera and families (Mahlberg and Pleszczynska, 1983; Mahlberg et al., 1983).

It is most important in using an artificial system to remember that it is artificial—that it brings together structures and mechanisms that are not necessarily related and that therefore may share only the most superficial of characteristics. It is counterproductive to attempt to find similarities or generalizations that have no reason to exist, as has been done all too frequently, usually by definition. It is similarly unwise to be dogmatic about which structures or mechanisms do or do not fit into certain categories, because the categories, being artificial, depend solely upon the opinions of the person who created them. Because we do not yet know everything about plant anatomy, all extant categories must be based on incomplete information. They are, or should be, designed only to help our understanding. It is in the area of plant secretion that the typology method of anatomy has persisted longest, and until recently the most comprehensive and authoritative reviews of secretion were written as though the various categories of glands, ducts, and cavities were natural types. Fortunately, the newest review (Fahn, 1979) discards that method and emphasizes function and biology, giving this area a modern treatment.

The following artificial classifications are in use. It will be obvious that any particular secretory cell, duct, or gland can be classified by all methods, and the choice of classification will depend on whether the structure is the object of a physiological, ecological, ultrastructural, or some other type of study.

Nature of the Secretory Product

The most common basis for classification is the nature of the secreted material.

Nectaries secrete sugar solutions called **nectar** (Bentley and Elias, 1983). By far the most common sugars are sucrose, glucose, and fructose, but other simple sugars and short polysaccharides occur in at least trace quantities: maltose, melobiose, and raffinose. These can contain traces of amino acids and other acids, as well as numerous other organic compounds.

Figure 9.1
Portion of *Ancistrocactus scheeri* showing the numerous spines and one extrafloral nectary. It is not immediately obvious that the nectary actually is a modified spine. The extrafloral nectaries of other closely related cacti are distinctly spinelike. If only this picture or this one species were available for study, anatomical analysis could be extremely difficult. × 2.

Figure 9.2
This leaf of *Rudbeckia laciniata* has hydathodes in the notches of the leaf margin, and each hydathode has a large drop of water. × 0.5. Photograph generously provided by N. R. Lersten and J. D. Curtis.

The function of nectar is to attract pollinators. Nectaries are subdivided on the basis of their position on the plant into **floral nectaries** and **extrafloral nectaries** (Fig. 9.1). There is an extensive literature on nectaries and nectar composition; the works of Baker and Baker (1973a, b, 1975), Fahn (1979) and Zimmermann (1932) are especially important.

Hydathodes are regions of leaf that secrete water (Fig. 9.2; Haberlandt, 1894). The water is almost pure, having very little material dissolved in it.

Salt glands, as their name implies, secrete inorganic salts. These generally occur on plants that grow in brackish coastal areas or in desert regions that have saline soils. In some cases these appear to serve as a means of removing salts that the roots are unable to exclude. In other species, it may be that the accumulation of salts on the surfaces of the plants makes them unpalatable to herbivores.

Osmophores secrete the odors and perfumes that guide pollinators to flowers (Fig. 9.3). These are mostly oily substances, often volatile small terpenes (Kisser, 1958; Vogel, 1962). In the aroids, which have the repugnant, stinking odor of rotting flesh, the osmophores also produce amines and ammonia (Smith and Meeuse, 1966).

Digestive glands are located on insectivorous plants; they secrete the enzymes necessary to digest the victims (Heslop-Harrison, 1976).

Adhesive cells cover the surface of the attachment organ of parasites (Fig. 9.4). The cells secrete the adhesive material that allows the attachment organ to cling firmly to the host as penetration occurs (Mauseth et al., 1985).

The preceding secretions are typically found outside the plant body, whereas the following are more often located inside the body.

Resin ducts are elongated cavities that contain the sticky resin, or "pitch," so commonly seen in trees, especially in conifers (Fig. 9.5). The resin is a complex mixture, but the pre-

Figure 9.3
The inner surface of the upper sepal and all surfaces of the petals are osmophores in this flower of *Restrepia hemsleyana.* × 1.0. Photograph generously provided by A. M. Pridgeon and W. L. Stern.

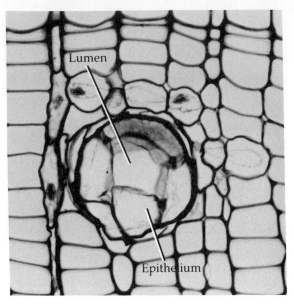

Figure 9.5
Cross section of pine wood, showing a resin duct. It has a small lumen, surrounded by large parenchymatous epithelial cells. Almost all of the adjacent cells are tracheids. × 200.

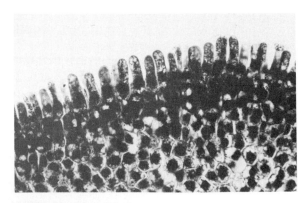

Figure 9.4
These are the hairs at the end of a haustorium of the mistletoe *Tristerix aphyllus* seedling. When they contact the surface of a host plant, they secrete adhesive material that attaches the parasite firmly to the host. × 150.

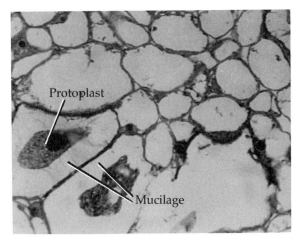

Figure 9.6
In this small area of a leaf of *Opuntia polyacantha*, three mucilage cells are visible. All were active at the time they were preserved, and mucilage was accumulating between the protoplast and the cell wall. In two, nuclei are visible. × 200.

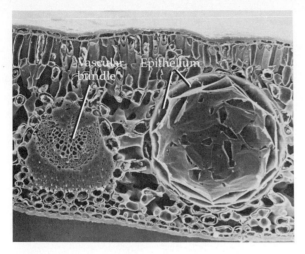

Figure 9.7
This leaf cross section of *Tepualia stipularis* shows a large oil chamber. The outer area is made up of thin, flat epithelial cells; the inner wall of each is extremely thin. All of the oil has drained out during specimen preparation, so a face view of the inside of the chamber is available. × 100. Micrograph generously provided by G. Montenegro.

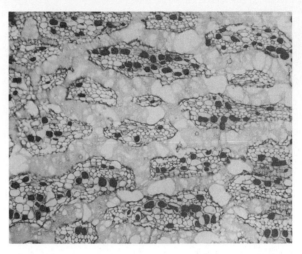

Figure 9.8
These gum ducts in *Lannea coromandelica* have a complex development: much of the gum is formed by lysigeny of the protoplast, but a significant portion results from the alteration of cell walls. × 80. Micrograph generously provided by K. Venkaiah and J. J. Shah.

dominant compounds are terpenes (Benayoun, 1977, cited in Fahn, 1979).

Mucilages are secretions that contain carbohydrate; they are slimy and have an extremely high water content (Fig. 9.6; Trachtenberg and Mayer, 1981, 1982).

Oils are frequently deposited into large cavities, but they can also be moved to the outside of the plant where they are aromatic (odiferous; Fig. 9.7). Oils are an especially diverse group of compounds, and they vary from those that have such a low molecular weight that they are volatile to those that have such a high molecular weight that they are waxy (Goodwin and Mercer, 1983).

Gums are located inside of the wood of certain tree species. The gums result from the modifications of the cell wall (Fig. 9.8).

Myrosin cells contain the enzyme myrosinase in their central vacuoles. This, by itself, is a rather harmless protein, but when the cell is broken, as by a chewing insect or other animal, the enzyme mixes with its substrate compounds—thioglucosides—which occur in surrounding cells. The resultant reaction produces toxic mustard oil (isothiocyanate).

Laticifers secrete **latex** (Fig. 9.11). It is almost impossible to define *latex*. For our purposes the intuitive understanding of latex as the "milk" of milkweeds, euphorbias, poppies, and so on will be sufficiently accurate. Laticifers are an excellent example of an artificial classification. Esau (1965a) pointed out that laticifers occur in as many as 12,500 species in 900 genera of dicots and monocots, and one of fern (*Regnellidium*; Labouriau, 1952), and they have

Figure 9.9
This diagram illustrates three mechanisms of secretion: in (A) is eccrine secretion, showing molecular pumps in a membrane. Each pump forces a molecule of secretion across the membrane. Granulocrine secretion is illustrated in (B); the secretion is first accumulated into vesicles that then migrate to the plasmalemma and fuse with it, releasing the secretion. The vesicle membrane is incorporated into the plasma membrane; extra membrane is removed when coated vesicles pinch off from the membrane. In (C) is holocrine secretion in which the entire cell breaks down, simultaneously liberating the secretion and becoming a part of it.

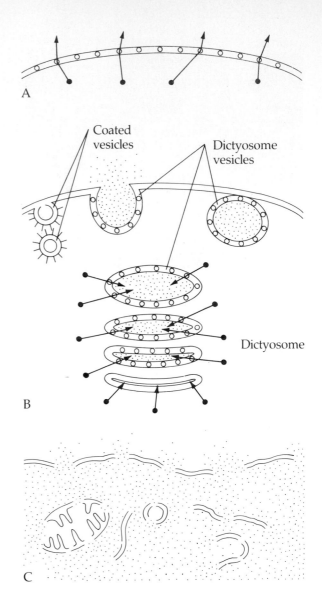

had many separate evolutionary origins. The nature of latex is diverse: it may contain carbohydrates, acids, salts, alkaloids (Fairbairn and Kapoor, 1960), sterols, lipids, tannins, mucilages, terpenes, camphors, rubber (Bonner and Galston, 1947), protein, vitamins (Urschler, 1956), crystals, starch grains, and even living protozoans. The one characteristic usually considered to unite all these secretions under the term *latex* is that they are typically creamy, white, and thick (*latex* comes from the Latin *lac* for milk). Latex, however, can be clear, yellow, or orange as well. Despite this diversity, the terms *laticifer* and *latex* are quite useful as long as they are used descriptively and as long as their artificiality is kept in mind.

Gases fill the intercellular spaces. Although these spaces are frequently referred to as air spaces, they never really contain air, not even in leaves. The gases contained in them are rich in CO_2 if the tissue is heterotrophic and rich in O_2 if it is photosynthetic. They may also contain the gaseous hormone ethylene. Such spaces and their "secretions" have not been considered glandular, but this may really have been just an artifact of the difficulty of studying such small gas pockets. Now, with the aid of gas chromatography, this difficulty may perhaps be resolved, and these may soon be considered another form of secretion and glandular tissue.

These "glands" are critically important flotation devices for most aquatic plants.

Mechanism of Secretion

There are several mechanisms by which material can be secreted, and these too serve as a basis for classification.

Eccrine secretion is the mechanism whereby the secretory product is transported out of the cytoplasm as individual molecules (Fig. 9.9A; Fahn, 1979). The driving mechanism is usually membrane-bound molecular pumps that recognize the molecules to be secreted, bind to them, then force them across the membrane (either the plasmalemma or the tonoplast), with ATP supplying the energy. This process allows the accumulation of secretory product against a concentration gradient. In some cases the secretory product is more concentrated in the cell so that it can be secreted passively, by osmosis alone. Water secretion of hydathodes is due to the presence of pressure in the conducting elements that run to the hydathode.

In **granulocrine secretion**, the secretory product is first accumulated inside of vesicles, through the action of the endoplasmic reticulum, the dictyosomes, or both (Figs. 9.9B, 9.10; Fahn, 1979). The accumulation may be caused by actual synthesis of the product in the vesicle, by synthetases located in the vesicle's membrane, or by synthesis in the hyaloplasm followed by transport across the vesicle membrane by ATP-driven molecular pumps. Granulocrine secretion is similar to the eccrine type of secretion, but with the very important difference that the pumps are located in membranes other than the plasmalemma or tonoplast. Therefore the product is accumulated in numerous sites inside the cytoplasm, not in one large central vacuole or outside the protoplast.

After accumulation within the vesicles, the product is secreted by the movement of the vesicles to the plasmalemma, followed by membrane fusion. As a result the vesicle membrane becomes part of the plasmalemma, and the interior of the vesicle becomes part of the exterior of the cell (Gunning and Steer, 1975). If secretion is happening rapidly, there is not only a movement of much secretory product, but also a large scale "flow" of membrane. To prevent the growth of the plasmalemma, membrane must be removed from it. This is done by means of **coated vesicles** that have a **clathrin** coating

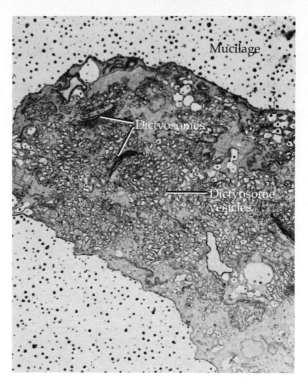

Figure 9.10
Portion of a mucilage cell in *Echinocereus engelmannii*, showing the granulocrine method of secretion. The mucilage (which can be recognized by the presence of dark particles) is first accumulated into dictyosome vesicles. These then migrate to the plasmalemma, fuse with it, and release the mucilage to the exterior of the protoplast. × 1000.

(Alberts et al., 1983). The coated vesicles migrate back into the cytoplasm where the membrane components can be reused. Thus granulocrine secretion is accompanied by a considerable amount of membrane recycling and is fundamentally different from eccrine secretion.

Both the eccrine and the granulocrine methods of secretion are grouped together as subtypes of **merocrine secretion** (Fahn, 1979). This is defined by the fact that the secretory cell remains alive during secretion. It may actually continue living indefinitely and secrete actively for a long period of time, because the secretory process is not necessarily detrimental.

In contrast to merocrine secretion, in **holocrine secretion** the product is liberated by the complete breakdown and disintegration of the cell (Fig. 9.9C). In this method the entire contents of the cell form part of the secretion, and cell lysis may be accompanied by extensive alteration in the organelles. In some cases, the organelles appear to maintain rather normal ultrastructure, and it may be possible that some metabolism occurs in the secretion. A secretion produced by this method obviously must be extremely complex and probably variable, unlike the simpler, purer secretions that can be produced by either of the two merocrine methods.

In at least several cases, the "secretion" is never located in the protoplast but always outside of it, so transport out of the protoplast does not occur. This external formation of the secretion occurs only in those structures where the product is modified wall material; examples are gum canals (Fig. 9.8)—*Acacia senegal, Commiphora mukul* (Setia et al., 1977), *Lannea coromandelica* (Venkaiah and Shah, 1984), *Rhus glabra* (Fahn and Evert, 1974)—and kino veins such as *Eucalyptus*.

Purpose of the Secretion

Another method of classification is based on the purpose of the product (Frey-Wyssling, 1935, 1972). In this system, materials that would be considered waste products would be called **excretions** (Wiesner and Linsbauer, 1920), those that still have a function (such as nectars) would be **secretions**, and those that are more or less simply passed through the plant without really participating in the metabolism (water of hydathodes, salt of salt glands) would be **recretions**. As Schnepf (1969b) and Fahn (1979) have pointed out, it is difficult to determine the function of most secretory products, and it is rarely safe to categorize any substance as purely a waste product with no function. Materials that are definitely waste products are often unpalatable and also not nutritious, so they may deter feeding. Similarly, the salts of

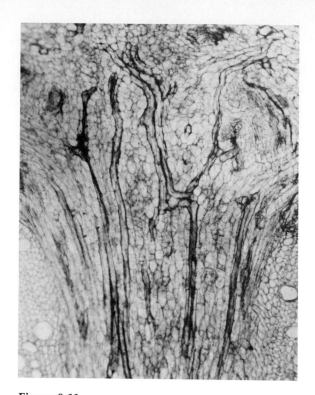

Figure 9.11
These long cells are nonarticulated laticifers. This is an unusually thick section: laticifers are often so irregular that in a section of normal thickness (about 7 μm) only small, glancing pieces of the laticifers are visible. In a thick section (about 15 μm) more of the laticifer is present within the section. Even so, this picture does not show all of any of the laticifers; what appear to be ends are just the points where the elongate cells turn and run out of the section. These are branched laticifers. × 100.

salt glands make the plant less useful as food for animals; plants with these glands occur in deserts and coastal areas where fresh water is scarce and animals must therefore avoid consuming excessive salt. It is not possible to state whether this is primarily a recretion (salts just passing through the plants) or a secretion (with the function of defense). Like so many things in plant biology, it has both aspects simultaneously. Fahn (1979) suggested that the terms *excretion* and *recretion* not be used, and

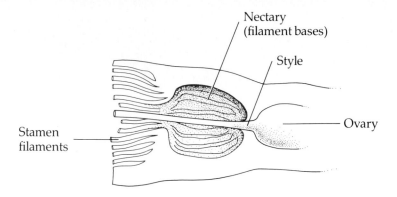

Nectary
(filament bases)

Style

Stamen
filaments

Ovary

Figure 9.12
The nectar of *Neobuxbaumia* is produced by the bases of the stamens; the stamen filaments bend together and form a dike that retains the secretion. The nectar is completely enclosed until a bat forces it open. Is the secretion internal or external?

that *secretion* is adequate for all. Secretion will be the only term used in this book.

Accumulation Space

An important consideration for any type of secretion is the space where it accumulates. One basic distinction is whether that space is inside the plant body (**endogenous secretion**) or outside it (**exogenous secretion**; Kisser, 1958). The osmophores that produce the fragrances of flowers liberate their product to the exterior of the plant, and there is no accumulation at all, whereas the contents of laticifers (Fig. 9.11) and resin ducts (Fig. 9.5) are always internal unless the plant is damaged. However, the distinction between internal and external may not be so simple as it may seem at first. Many nectaries secrete their nectar around the base of the ovaries; it accumulates there because the sepals, petals, and stamen filaments form an open cup that contains it. But in many species, such as *Matucana*, *Neobuxbaumia* (Fig. 9.12), and *Oroya* (Buxbaum, 1950), the nectar chamber is effectively sealed because the filaments fold over the secretory area. In *Psychotria bacteriophylla*, the young leaves enclose the apical meristem, producing a completely sealed chamber into which they secrete a mucilage that nourishes bacteria. Although an apical meristem is, by morphological definition, an external plant part, here it is functionally internal (Horner

and Lersten, 1968). Finally, during the maturation of the fruits of Lardizabalaceae, the surface forms grooves that infold, forming sealed cavities that are completely lined by epidermis (Zheng-Hai, 1963); secretion across the "outer" wall of these epidermal cells could arguably be considered either external or internal.

With regard to internal secretions, important distinctions can be made as to whether the secretion is localized within individual cells or extracellularly. If extracellular, then the shape of the accumulation space—whether it is a **cavity** (also called a **chamber**) or a long **duct** (the same as a **canal**)—will provide some information as to the mechanism of development of the secretory structure. A cavity usually begins at one point, often a single cell, and then develops centrifugally for a short distance (Fig. 9.23). The cavity ceases to expand when it reaches the proper size. Ducts, however, typically develop for a long period after their initiation, as differentiation continues at one or both ends. This growth is one of **recruitment** in which parenchyma cells in the vicinity of the growing duct are converted into secretory cells (Mauseth, 1980a; Wittler and Mauseth, 1984a). The conversion of these parenchyma cells at the ends of the ducts is usually similar to the processes involved in the original initiation of the duct, so here there is a process of continual initiation, as opposed to the discrete initiation of cavities. Another morphogenic aspect of ducts is the control of the direction of growth.

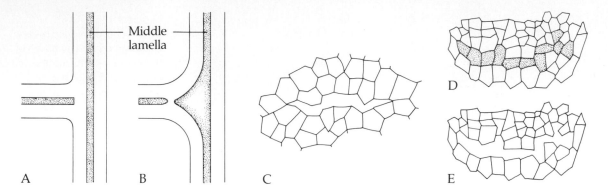

Figure 9.13

These diagrams represent two methods for the formation of an intercellular space: in (A) and (B), a schizogenous space forms because the middle lamella of a new cell plate does not connect with the middle lamella of the existing wall, and the junction is a point of weakness. In (C), a large schizogenous space forms as an entire region of middle lamella breaks down; the surface on one side of the space can fit against the surface on the other side. (D) and (E) represent a lysigenous formation; when the stippled cells in (D) lyse, an irregular cavity results (E). In contrast to a schizogenous cavity, the surface on one side will not necessarily fit against the surface on the other.

In some, the shape of the mature duct is so irregular that probably no control exists, and the direction of growth is random. But in others the ducts run in a straight line or branch in a regular manner, and a precise control mechanism must be operating to guide recruitment.

Another important distinction is whether the space forms **schizogenously** (the separation of cells from each other; Fig. 9.13), **lysigenously** (the actual breakdown of whole cells; Fig. 9.13), or by a process including both (**schizolysigenously**). If the cavity is at least partially lysigenous, then the secretion is holocrine, and it may be a complex mixture of many components, some of which are modified cell wall and protoplast, as mentioned above. However, if the cavity is entirely schizogenous, then the secretion will be merocrine and will consist of materials elaborated by healthy organelles.

Schizogeny is thought to be related to the action of the phragmoplast and cell plate during cell division (Martens, 1937, 1938; Sifton, 1945,

1957; see Chapter 2). When the new cell plate contacts the existing walls, the new middle lamella stops at the first, innermost cellulosic layer and does not penetrate further to contact the existing middle lamella located on the other side (Fig. 9.13A). With continued deposition of wall material, the cell plate is continuous with new wall layers deposited over the whole surface of the new daughter cell. But as turgor pressure builds, the cell will tend to swell and become rounded; this causes the points of weakness at the junction between the cell plate and cell wall to tear and form an intercellular space. In almost all parenchyma tissues this happens to a small degree at least, and tiny, isolated intercellular spaces are common. With more expansion, the intercellular spaces enlarge as the junctions tear apart. It is not known if this is facilitated by the secretion of pectinases or not. With sufficient expansion, the intercellular spaces of one corner will meet those of other corners, and an extensive, ramifying space is formed that can be filled with either gases or liquid secretions.

In the lysigenous development of an intercellular space, there is actual breakdown (**lysis**) of the material that occupies the space (Fig. 9.13D, E). This may be a lysis of the protoplasm only or of the entire cell, including even the cell wall (Wittler and Mauseth, 1984a, b).

These two processes can occur sequentially in the formation of a space; when this happens, the initial cavity is formed schizogenously, then is expanded as the cells bordering the cavity lyse, thereby contributing to the cavity and to the secretory material.

If a cavity accumulates material by active transport of secretory product into the cavity, pressure will build up. This can cause the surrounding cells to become flattened or concave, thereby forming layers that have the appearance of being a special lining, which may be termed an **epithelium** (Figs. 9.7, 9.20, 9.24). In some cases, this appearance is augmented by periclinal divisions of these cells, and the epithelium becomes multilayered. It should be emphasized that an epithelium is defined only on the basis of shape; in some species, it may be very active metabolically and may contribute to the secretion by eccrine, holocrine, or granulocrine processes. For example, in *Mammillaria*, epithelial cells produce latex within themselves and then lyse; the latex duct becomes wider (Fig. 9.23; Mauseth, 1978a, b). But the epithelium does not become thinner because adjacent cortical cells are converted to epithelial status. The epithelium "migrates" through the cortex, and the duct becomes progressively wider. In other species, the epithelial cells do not appear active, and they may represent a type of inert lining for the cavity or duct (Fig. 9.5).

The fact that plants are a combination of two distinct phases, the symplast and the apoplast, causes some problems for secretory structures. The secretion is produced by the symplast (protoplasm), and then it is moved to the apoplast (intercellular space or outer wall of

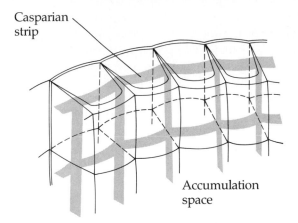

Casparian strip

Accumulation space

Figure 9.14
This diagram represents a small part of the surface of an accumulation space. The cells have ordinary walls, but the radial walls are impregnated with suberin, which makes them impermeable to most hydrophilic substances. Once material is secreted into the accumulation space, it cannot escape by seeping through the walls.

the epidermis) unless it is a type that is stored within the protoplasm. The apoplast is usually so permeable that if no special precautions are taken the secretion could actually permeate much of the plant, rather than remaining localized and isolated at a specific site. Even in glands that project from the epidermis, it would be possible for the secretion to be trapped under the cuticle and be forced to seep through the gland wall back into the plant. This is especially true of watery secretions such as those produced by nectaries, hydathodes, and salt glands. In these cases the apoplast of the secretory area must be isolated by the deposition of waterproofing materials (usually suberin and lignin) in the form of bands called **Casparian strips** (Fig. 9.14). If a gland has a supporting stalk and a secretory head, the Casparian strips are found in the stalk (Fig. 10.38) or they surround the basal cells. Similarly, impermeable strips may be found in the radial walls of epithelial cells, thereby isolating internal cavities or ducts. Such obvious permeability barriers are not at all universal; where they are absent, isolation apparently is maintained by the viscosity of the secretory product, by the compactness of the wall, or by the absence of intercellular spaces. Casparian strips and suberin are more often studied in endodermal cells; a detailed explanation of them is provided in Chapter 11 in the section "Endodermis").

Examples

Thousands of types of secretory structures exist, but only a few examples can be presented and analyzed. The ones that are discussed below were chosen as examples that illustrate and demonstrate the principles mentioned above. They are arranged in order of increasing complexity so that only a few aspects are needed to understand the first, a few more to understand the next, and so on. The arrangement is not meant to suggest an evolutionary relationship.

Secretion by a Single Internal Cell

MYROSIN CELLS.   Myrosin cells represent only a simple level of organization of secretory cells, but they are excellent for an introduction into the principles of secretion. First, the adaptive advantage of myrosin cells is apparently the defense of the plant (Fahn, 1979). As mentioned earlier, the secretion product is the enzyme myrosinase, which can catalyze a reaction that produces toxic mustard oil from thioglucosides (Björkman, 1976). The secretion is compartmentalized inside the central vacuole, within individual, living cells (Werker and Vaughan, 1974, 1976). The secretion is released and the lethal reaction facilitated when animals chew on the plant and disrupt the vacuoles. The plant itself does not control the release.

The distribution of the secretory system within an individual plant is important. For myrosin cells to be an effective deterrent, they must occur in all parts of the plant that otherwise might be edible. The myrosin cells are located in the tissues of fruits and seeds, as well as in leaves, petioles, and stems (Heinricher, 1884; Sharma, 1971; Werker and Vaughan, 1974, 1976). The cells may occur as individual, small idioblasts, or they can be arranged in rows. Each cell may be rather isodiametric, or they can grow intrusively between adjacent parenchyma cells, becoming several millimeters long in *Moricandia* and *Diplotaxis*. In *Sinapis alba* they are extensively branched (Werker and Vaughan, 1976).

Ontogenetically, as myrosin cells develop from young parenchyma cells, the endoplasmic reticulum (ER) becomes extensive and is covered heavily with ribosomes (Werker and Vaughan, 1974, 1976). As the myrosinase accumulates within the RER, swellings form, and these later pinch off to become individual vesicles. The vesicles fuse with each other until the cell finally has just one large central vacuole that contains the myrosinase.

Phylogenetically, myrosin cells appear to have had two separate origins, because they occur only within a few families of two separate orders of flowering plants: Order Geraniales (Families Limnanthaceae and Tropaeolaceae) and Order Capparales (Families Capparidaceae, Cruciferae, Moringaceae, and Resedaceae; Fahn, 1979). Because the myrosin cells have had two distinct evolutionary origins, there is no reason to expect similarities between the two types except where function requires it. A secretion like this must be sequestered away from the substrate and it must be located in areas where insects are likely to chew, as opposed to being located deep inside of woody tissues. To be strictly correct and to avoid any risk of confusion, it might be best to give the myrosin cells of the Capparales a name that is different from the one used for them in the Geraniales. This naming would increase the amount of terminology, but it would emphasize that what we now call myrosin cells represent two types of convergent, analogous structures, not one uniform group of homologous ones.

MUCILAGE CELLS.   Mucilage (Figs. 9.6, 9.10, and 9.24) is defined by its slimy nature, which is due to its polysaccharide content. The nature of the polysaccharide can vary, and among the monomers that it can contain are arabinose, fucose, galactose, galacturonic acid, and xylose (Juniper and Pask, 1973; Trachtenberg and Mayer, 1981). The mucilage carbohydrates have a tremendous capacity to absorb and hold water, and tissues of *Opuntia* will remain hydrated even when immersed in 75% alcohol. Most of the functions of mucilage-secreting tissues are believed to be related to this water-retaining capacity. Desert succulents almost always contain mucilage cells, apparently as a means of water retention. Many seeds have a layer of mucilage cells on their surface (Fahn and Werker, 1972; Garwood, 1985; Hyde, 1970; Jordan et al., 1985); these absorb water and swell, forming a moist envelope around the seed that permits imbibition and germination (Harper and Benton, 1966). Cells of root caps secrete mucilage into the soil ahead of the growing root (Fig. 13.5). This lubricates the soil particles and makes passage of the root much easier. Many carnivorous plants secrete mucilage that traps insects. It has been suggested that mucilage is used as a food reserve; it is not yet certain if the plants themselves can digest the mucilage carbohydrates, but soil bacteria proliferate in the root cap mucilage, and *Psychotria bacteriophylla* is able to maintain an active bacterial culture in its leaf primordia by supplying the bacteria with mucilage (Horner and Lersten, 1967, 1968; Lersten, 1974a, b).

Despite the numerous functions for mucilage, the details of its synthesis and transport seem to be remarkably similar in all species studied. *Opuntia* will be used as an example here. The mucilage cells occur as idioblasts, each cell being separated from other mucilage cells by chlorenchyma cells (Figs. 9.6, 9.10;

Mauseth, 1977, 1980b; Trachtenberg and Fahn, 1981). As the mucilage cell initial becomes recognizable by its large nucleus and dense cytoplasm, the dictyosomes begin to increase in volume and activity. About 0.5% of the volume of a chlorenchyma cell is dictyosome, but up to 3.5% of the mucilage cell volume is composed of dictyosomes (Mauseth, 1980b). The endoplasmic reticulum (ER) does not seem to be involved in secretion, because it is never especially prominent. Mucilage-containing vesicles are budded off from the periphery of the dictyosomes; these vesicles move to the plasmalemma and fuse with it, depositing the mucilage to the exterior of the protoplast (granulocrine secretion; Fig. 9.9B). As the mucilage accumulates between the plasmalemma and the cell wall, the protoplast shrinks. Ultimately, all of the organelles break down, and the entire cell degenerates, leaving only mucilage (Mauseth, 1980b; Trachtenberg and Fahn, 1981). As would be expected of a cell that is responsible for water conservation, mucilage cells are located throughout all parenchymatous tissues in *Opuntia*.

Mucilage cells also appear to be polyphyletic, existing in many families, some of which are not closely related to the others: Bombacaceae, Cactaceae, Didiereaceae, Dipterocarpaceae, Malvaceae, Sterculiaceae, and 46 others (Metcalfe and Chalk, 1983). However, in all cases studied, the secretion mechanism is similar to that just described for *Opuntia*: *Hibiscus esculentus* (Mollenhauer, 1967; Scott and Bystrom, 1970), *Althaea rosea* (Bouchet and Deysson, 1971; Mollenhauer, 1967), *Rumex* and *Rheum* (Schnepf, 1968).

In root tips and in carnivorous plants, the mucilage must be moved entirely out of the cell to be an effective lubricant or insect trap. The secretion in roots has been studied intensively (Whaley and Leech, 1961; Northcote and Pickett-Heaps, 1966; Juniper and Pask, 1973; Moore and McClelen, 1983a, b; Paull and Jones, 1976). In roots, the middle lamella begins to weaken as the cells become secretory, and a large apoplastic space develops around the mucilage cells of the root cap. Many of the cells separate completely from the root and are sloughed off into the soil (Fig. 13.5). In either case, the cell is no longer closely surrounded by other cells in a compact tissue; the mucilage can permeate the walls, so it does not accumulate and does not result in the shrinkage of the protoplast. Large quantities of mucilage can be secreted, and roots that are carefully grown in a humid chamber may develop a sizable mucilage drop. Considering the fact that the mucilage is so easily released in root caps, it must be that in internal, idioblastic mucilage cells, the mucilage is trapped by the turgor of neighboring parenchyma cells, not by any impermeability of the wall of the mucilage cell.

NONARTICULATED LATICIFERS.   Laticifers are an extremely heterogeneous group, not only metabolically, but also developmentally and structurally. They are typically classified into two fundamental types: (1) **articulated laticifers** are those made up of many cells, and (2) **nonarticulated laticifers** are those that consist of a single cell.

In a small number of species, such as *Cryptostegia, Jatropha* (Dehgan and Craig, 1978), and *Parthenium argentatum* (Metcalfe, 1967) the nonarticulated laticifers are small, rather isodiametric idioblasts, somewhat resembling myrosin cells or mucilage cells. But typically the nonarticulated laticifers are extraordinarily long cells, often extending from the root up into the stem and leaves (Fig. 9.11). In some species they are unbranched: *Cannabis* (Moraceae); *Cyclanthus* (Cyclanthaceae; Wilder and Harris, 1982); *Urtica* (Urticaceae); and *Vinca* (Apocynaceae). In others, they branch frequently, forming an even more extensive network: *Asclepias, Ceropegia, Cryptostegia* (Asclepiadaceae); *Broussonetia, Ficus, Maclura* (Moraceae); *Nerium* (Apocynaceae); and *Euphorbia, Jatropha* (Euphorbiaceae). The tip of the nonarticulated laticifer may project into the margins of the shoot and root meristems, but,

as the meristem grows away, the laticifer continuously invades the newly formed stem and root tissue (Blaser, 1945; Mahlberg, 1959b). The growth of the laticifer, at the very tip, is intrusive; Wilson et al. (1976) have found that the laticifers of *Asclepias syriaca* contain abundant pectinase. They believe that the laticifer secretes the pectinase ahead of itself, thereby softening the middle lamella and loosening the tissues that will be invaded. Pectinases have also been found in *Nerium oleander* (Allen and Nessler, 1984). However, once it has invaded young tissues, the laticifer must expand along with the surrounding cells; this must be symplastic growth. If a laticifer produces a branch in a mature portion of the root or shoot, then it grows only intrusively, because the surrounding tissues, being mature, are not growing.

Nonarticulated laticifers can occur in any part of the plant, most often in the softest regions such as the pith and cortex. But they can invade leaves and leaf gaps as well as wood and phloem. When a nonarticulated laticifer becomes extremely large, it usually is multinucleate; it is probably necessary to have hundreds or even thousands of nuclei to control the metabolism of such an elongate cell (Dehgan and Craig, 1978; Mahlberg, 1959a; Mahlberg and Sabharwal, 1966). They maintain an active metabolism as long as the surrounding tissues remain alive. The first initials of nonarticulated laticifers can be detected in the embryos of some species: *Cryptostegia* (Blaser, 1945), *Nerium oleander* (Mahlberg, 1961), and *Euphorbia* (Mahlberg and Sabharwal, 1968; Rosowski, 1968; Schaffstein, 1932). These then grow and branch, invading new tissues as the plant grows. In some species, new initials are formed in new tissue, and an older plant will have more laticifers than a younger plant.

The ultrastructure has been studied in very few nonarticulated laticifers, and probably the organellar structure is more closely related to the chemical nature of the latex than to whether the laticifer is articulated or not. The intrusive growth of nonarticulated laticifers is a phenomenon that offers the possibility of studying cell-cell relationships, but this area seems to have not yet been studied.

Secretion by a Single External Cell

STINGING HAIRS.   Stinging hairs are produced by four families of dicots: Euphorbiaceae, Hydrophyllaceae, Loasaceae, and Urticaceae (Fahn, 1979). Like myrosin cells, stinging hairs are very simple (Fig. 9.15), but they offer good lessons about secretion. They are defensive mechanisms, and they store toxins within the central vacuole (Haberlandt, 1886; Thurston and Lersten, 1969). The irritating material may contain acetylcholine, histamine, and 5-hydroxytryptamine, as well as other compounds (MacFarlane, 1963; Saxena et al., 1965). The stinging hairs are distributed on the plant so as to maximize defense. They are epidermal cells (subepidermal in *Dalechampia* and *Tragia*) that project outward, so animals will encounter the hairs before they reach the rest of the plant tissues. As with myrosin cells, the plant itself has no control over the release of the secretion from the hair; that happens whenever animals brush against the stinging cell. The plant does, however, facilitate the release, because the hair is sharply pointed and the wall is impregnated with silica (Thurston, 1974). This makes the tip glasslike; it breaks easily, resulting in a sharp edge that can penetrate animal tissues and inject the toxin much like a hypodermic needle. In *Dalechampia* and *Tragia* the tip of the hair contains a sharp crystal of calcium oxalate that punctures the animal's skin (Thurston, 1976).

The toxins appear to be produced by RER that forms vesicles (Marty, 1968; Thurston, 1974). The vesicles fuse to form one large vacuole at the hair apex. Marty (1968) reported that the dictyosomes appear active and may participate in the secretion, but Thurston (1974) presented evidence that the dictyosomes are involved in depositing silica in the wall instead.

The four families in which stinging hairs are known to occur are placed in four separate

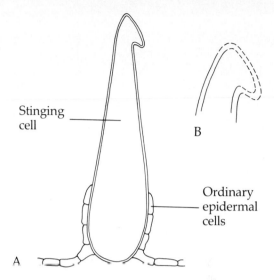

Figure 9.15
The stinging hairs of *Urtica dioica* (stinging nettle) are large cells filled with toxin; the tip is pointed and glassy, which makes it brittle and easily broken. The small lip also allows it to catch in an animal's skin and break off (B), thus releasing the poison. Based on data presented by E. L. Thurston (1974).

orders: Urticaceae: Urticales; Euphorbiaceae: Euphorbiales; Hydrophyllaceae: Solanales; and Loasaceae: Violales (Cronquist, 1981). Thus the families are unrelated, and the stinging trichomes represent four separate analogous glands that are the result of extreme convergent evolution.

NECTARY OF *LONICERA*.   Nectaries of *Lonicera* are extremely simple, consisting of just a single epidermal cell that protrudes above the surrounding epidermal cells (Feldhofen, 1933; Fahn and Rachmilevitz, 1970, 1975; Frei, 1955). The lower portion is rather narrow and stalk-like; the upper portion is expanded and is responsible for the secretion. When secretion is about to start, the uppermost region of wall develops labyrinthine ingrowths, becoming a transfer wall. Simultaneously, the cuticle is re-

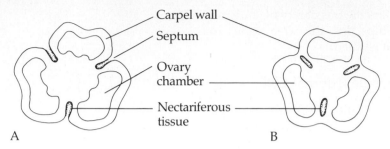

Carpel wall

Septum

Ovary
chamber

Nectariferous
tissue

A   B

Figure 9.16
A common type of septal nectary that might be found in many
types of monocots. (A) is a cross section near the top of the structure, where the three carpels are only partially fused and deep
grooves are present in the septa. The epidermis in the grooves is
secretory. (B) is a cross section lower in the structure, where the
carpels are more completely fused; because the outer edge is fused,
the only way the nectar can escape is by migrating upward to the
region of (A).

leased from the wall, creating an accumulation
space between the wall and the stretchable
cuticle. The RER and mitochondria increase
greatly. At the time of anthesis, when secretion occurs, the RER increases even more, and
all the lamellae are arranged parallel to each
other. Vesicles are pinched off from the edges;
they migrate to the plasmalemma and fuse
with it, releasing their contents.

This type of nectary, consisting of a single
glandular hair, is extremely common (Fahn,
1979; Uphof, 1962), and it has evolved repeatedly. The structure is so simple and functions
so well that numerous types of secretion can be
mediated by it. Because it is so common, it is
difficult to know whether two similar hairs are
homologous, both having evolved from a common ancestor, or whether they are analogous,
the products of two independent lines of convergent evolution.

SEPTAL NECTARIES.   Septal nectaries represent
part of the problem in trying to categorize secretory structures: the actual secretion is carried out by epidermal hairs as in *Lonicera*, but
the secretory epidermises are located on cavities that are functionally internal (Fig. 9.16;

Fahn, 1979). During the formation of the
flower, the carpel primordia swell and press
against each other; the contact faces between
the primordia are the septa. Only the outer
and inner edges of the septa fuse; the majority
of the contact surfaces retain their individual
epidermal nature, even though pressed together. These regions develop the secretory
hairs (Fahn, 1979).

The hairs undergo division to become multicellular just before secretion begins (Fahn and
Benouaiche, 1979). The nectar appears to be
synthesized in both the RER and the dictyosomes. The nectar is released as the vesicles
fuse with the plasmalemma; the nectar then dif-
fuses through the intact cell wall and accum-
mulates in the septal chamber, gradually filling
it. At the top of the nectary, the carpels are not
fused together (Fig. 9.16A), so the nectar can
leak out of the open top of the chamber when it
becomes completely full. This type of nectary
is found only in monocots, and 59% of the
monocots studied have them (Daumann, 1970).
The ones that have been studied resemble each
other strongly, and the families that contain
them are closely related (Dahlgren and Clifford,
1982). It seems possible that these structures

are homologous with each other—that is, that they all are derived from the same ancestral structure—whereas the stinging hairs are only analogous with each other. Examples of plants with septal nectaries are *Allium*, *Gasteria*, *Muscari*, *Eichhornia*, *Pontederia*, and *Musa* (Daumann, 1970; Fahn and Benouaiche, 1979; Grassman, 1884).

Secretion by Internal Complexes of Cells

HYDATHODES. In the previous examples, it was emphasized that it is important to know the function of a secretory system in order to understand its anatomy. Hydathodes represent the opposite situation: understanding the anatomy first has been the means for discovering the function. When the ground is moist and the air is cool and humid, the leaves of many plants—over 350 genera in 115 families (Frey-Wyssling, 1941)—especially grasses, bear small droplets of water (Fig. 9.2). Although often mistaken for dew, it is actually a secretion of almost pure water. This particular secretory process is called **guttation**. Some tropical plants (*Colocasia*, for example) can guttate up to 100 milliliters of water in one night (Kramer, 1956).

Hydathodes were first studied by Haberlandt in 1894 and are basically a small amount of very loose parenchyma located at the end of one or several small veins in a leaf. These cells constitute a tissue called the **epithem** (Fig. 9.17; De Bary, 1877). Surrounding the parenchyma in most hydathodes is a layer of tight-fitting cells called the **sheath** (Gardiner, 1883; Kurt, 1930; Johnson, 1937). The cells of the sheath abut each other tightly, and often their adjacent walls are cutinized (Johnson, 1937), forming them into an endodermis-like covering around the hydathode; nothing can enter or leave the hydathode by diffusing through the apoplast. The outer surface of the hydathode is the leaf epidermis. There is always at least one stoma over the hydathode; there are often more, as many as ten in *Crassula argentata*

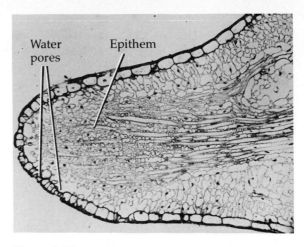

Figure 9.17
This section of a leaf of *Rudbeckia laciniata* shows the structure of the hydathodes presented in Figure 9.2; you can see several stomata, the epithem, and abundant xylem. × 100. Micrograph generously provided by N. R. Lersten and J. D. Curtis.

(Rost, 1969; Roth and Clausnitzer, 1972). The stomata are called **water pores** (De Bary, 1877), and they often differ from regular stomata in that they are larger and their guard cells usually cannot close the stomatal pore. However, Steinberger-Hart (1922) reported that the water pores of *Achillea*, *Impatiens*, and *Tropaeolum* could be opened and closed by movements of the guard cells.

Until recently, there has not been a convincing explanation of why a plant would exude water, when obtaining and retaining water are usually of prime importance. One explanation was that roots could absorb large amounts of water and force it upward into the stem; if the water were excessive, it might waterlog and damage the shoot tissues. The hydathodes were postulated as being vents to permit the release of this superfluous water. This may be the primary function in some species, but three observations suggest that this is not the complete explanation, at least for many plants (Höhn, 1950, 1951; Klepper and Kaufmann, 1966).

First, it was discovered that many hydathodes contain transfer cells, those cells that have special wall ingrowths (Renaudin and Capdepon, 1977; Renaudin and Garrigues, 1966). Perrin (1971) found them in the hydathodes of *Taraxacum*, *Cichorium*, and *Papaver*. Transfer cells, in this case in the sheath, are able to transfer forcefully large amounts of material across their plasma membrane. Because the water that is secreted is almost pure, they must be pumping ions out of the hydathode, not into it.

The second important observation is that aquatic angiosperms have hydathodes, especially on their submerged leaves (Fahn, 1979). In these plants, there cannot be any water movement due to transpiration like that in terrestrial plants. The only means of moving water through the plant is by having the roots pump ions into the xylem cells, causing the water potential to become more negative, and water to flow inward. Obviously, aquatic plants do not need to move water through their xylem just to distribute water; any cell can absorb water directly from the environment. Instead, the water is circulated as a means of transporting mineral nutrients that have been absorbed by the roots from the mud. The transfer cells of the sheath actively move the minerals from the vein endings into the adjacent mesophyll. The remaining water, which had been used only as a transport medium, is discharged through the water pores.

The third observation is that hydathodes are most frequent on very young leaves and that their water pores develop before normal stomata do. Immature, expanding leaves need a very high supply of minerals, but they have only poorly developed vascular tissues, and they are too small to transpire very much. Water flow that is powered by normal transpiration would not be sufficient to provide an adequate supply of minerals (Höhn, 1950). But hydathodes are able to unload the minerals actively, causing the water to flow faster and to bring in even more minerals.

Hydathodes are more complex than the other types of secretory systems described so far. They must have a vascular supply because rather large amounts of material are processed. They must have a more elaborate means (the sheath) of isolating the secretory cells from surrounding cells. The means by which the secretion moves to the exterior of the plant (water pores) is more elaborate, and in some species can be regulated by the plant itself.

ARTICULATED LATICIFERS.   Articulated laticifers are fundamentally different from nonarticulated laticifers in their development and structure (Figs. 9.18, 9.19). Each is actually a row or file of individual laticiferous cells. In *Allium* (Huang and Sterling, 1970), each cell is connected to the two adjacent laticiferous cells only by plasmodesmata, but in others there is usually some form of perforation of the common wall. In *Musa*, a small flap of wall breaks open, allowing the latex of one cell to merge with that of other cells, and allowing all of the latex to flow out if the laticifer is cut (Skutch, 1932; Tomlinson, 1959). It is not unusual for the entire transverse wall to be digested away and for the mature articulated laticifer to look somewhat like a xylem vessel element. (When the end wall is completely digested, the mature articulated laticifer can be mistaken for a nonarticulated laticifer, and several have been so described.)

As the plant grows, the articulated laticifers differentiate upward in the stem (and downward in the root). This growth is by the continuous recruitment of the nearby parenchyma cells, which are converted to laticifer cells. As these new cells differentiate adjacent to older, existing laticifer cells, the common wall becomes perforated, and the new cells are added to the laticifer, much as new vessel elements are added in the xylem. In some species the laticifers are said to be **non-anastomosing**, because any single laticifer (which is a row of cells) never merges with another laticifer (Fig. 9.19): *Achras* (Sapotaceae), *Allium* (Liliaceae),

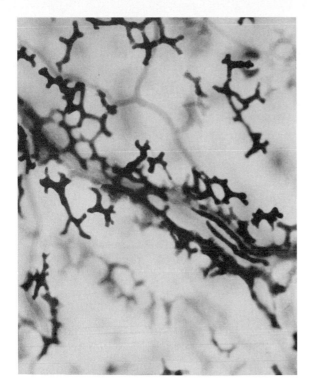

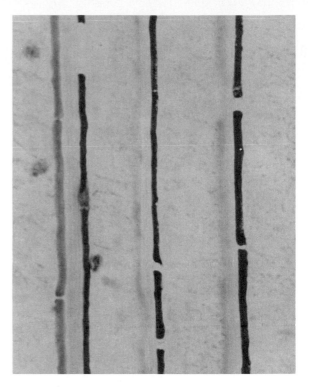

Figure 9.18
Leaf clearing of *Jatropha;* treatment with sodium hydroxide has bleached out all chlorophyll and made most cells transparent. Simultaneous treatment with the stain basic fuchsin has colored the latex in the laticifers. Notice that each is a short, highly branched cell that connects with other laticifers; the cells constitute an articulated laticifer. Although all cells appear distinct, a great deal of latex oozes out if the leaf is cut, so the cells must be interconnected and latex able to flow from one to another. × 50.

Figure 9.19
Four parallel vascular bundles in a leaf clearing of banana (*Musa*); the vascular tissue is pale gray. Next to each bundle is an articulated laticifer (the bundle on the left is accompanied by two laticifers). The treatment has caused the latex to shrink slightly, so the individual cells can be distinguished. × 80.

Ipomoea (Convolvulaceae), and *Musa* (Musaceae). In other species they are **articulated anastomosing laticifers**, because one laticifer can fuse with others, forming an extensive three-dimensional network that permeates the entire plant (Fig. 9.18): *Argemone, Papaver* (Papaveraceae); *Carica* (Caricaceae); *Cichorium, Lactuca, Taraxacum, Tragopogon* (Compositae); *Hevea, Manihot* (Euphorbiaceae). In order to do this, the laticifers must be able to branch. It

must be kept in mind that the cells of the laticifer do not grow out to form the branch; instead, regular parenchyma cells that lie between two existing laticifers are induced to differentiate into latex-bearing cells (Blaser, 1945; Vertrees and Mahlberg, 1978). Once this is complete, a bridge has been formed between the two original laticifers, and the whole complex acts as one single laticifer.

As with nonarticulated laticifers, the articulated laticifers can occur in all parts of the plant body, either in young tissues or older ones, vegetative or floral. In *Taraxacum kok-saghyz,* they are present in the seedling, at germination. In *Papaver somniferum* (the opium poppy), laticifers are present in the seedling and young plant, but they are largest and most abundant in the seed capsule. In the tree from which commercial rubber is extracted (*Hevea brasiliensis*), the most important laticifers are in the bark; these are the ones tapped for the production of rubber.

Because *latex* is a term that covers dozens of types of secretions, it should not be surprising that the ultrastructure is extremely diverse. Indeed it is at the level of electron microscopy that it becomes obvious that most laticifers are not at all related to most of the others. In the nonarticulated laticifers, the cells must remain alive and healthy in order to be able to grow intrusively into new tissues. The latex is either held in vacuoles in the cytoplasm or deposited in the central vacuole. In many articulated laticifers, the cells lyse and degenerate at the final stage of development (Heinrich, 1967; Neumann and Müller, 1972); secretion is holocrine. In *Hevea*, the rubber is present as globules called **lutoids**. The lutoids are complex, about 1–5 μm in diameter and surrounded by a membrane. Inside the lutoid are smaller particles (Southorn, 1960, 1964). As much as 30% of the latex is rubber.

Because the laticifers of *Papaver somniferum* are the source of opium and heroin, their ultrastructure has received considerable study (Dickenson and Fairbairn, 1975; Nessler and Mahlberg, 1977, 1978; Thureson-Klein, 1970). The cytoplasm contains numerous ER vesicles that contain morphine (Nessler and Mahlberg, 1977).

In several families (Asclepiadaceae, Euphorbiaceae), both articulated and nonarticulated laticifers have been reported, and in a few instances the two types occur together in the same plant: *Jatropha* (Dehgan and Craig, 1978); *Stapelia bella, Trichocaulon* (Schaffstein, 1932). You will have noticed that the discussions dealing with both nonarticulated and articulated laticifers have been devoid of any functional

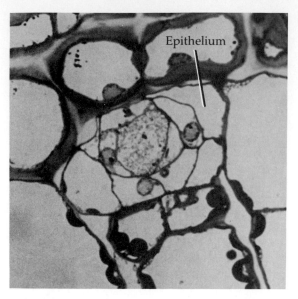

Figure 9.20.
Cross section of a latex duct of the "semi-milky" type in *Mammillaria heyderi*. At its present state, it is small, with a narrow lumen and a well-defined epithelium, each cell of which has a prominent nucleus. × 500. Micrograph generously provided by G. Wittler.

analysis. This is because their role and adaptive significance in all species is still completely unknown (Metcalfe and Chalk, 1983).

LATEX DUCTS IN *MAMMILLARIA*. A phylogenetic, developmental study of this genus has revealed an interesting set of stages in the evolution of a complex secretory system (Hunt, 1971; Wittler and Mauseth, 1984a, b). This is a large (about 200 species) genus of cacti, and the basic phylogenetic relationships of the subgenera have been outlined by Hunt (1971). He suggested that the subgenus Hydrochylus ("watery" mammillarias, that is, with no latex) contained primitive members and that from plants similar to these arose two lines of evolution: (1) the subgenus Subhydrochylus ("semi-

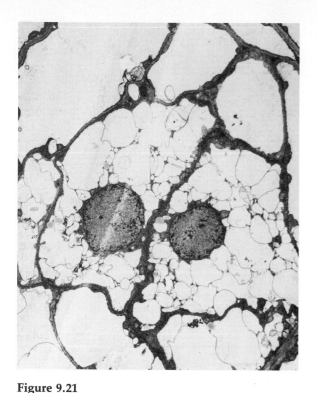

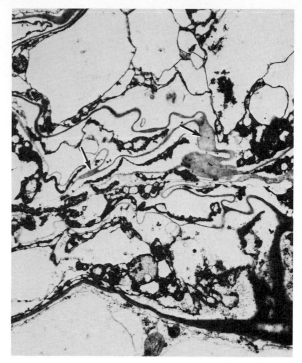

Figure 9.21
Cross section of a latex duct in *Mammillaria heyderi* at a younger stage than in Figure 9.20. The two large central cells will degenerate, forming the lumen of the duct; the protoplasm had already begun to fill with vesicles. This is not simply lysigeny, however, because the walls undergo a remarkable conversion to create a schizogenous accumulation space (see Fig. 9.22). × 1000. Micrograph generously provided by G. Wittler.

Figure 9.22
In *Mammillaria heyderi*, a schizogenous space is formed, but not by the method illustrated in Figure 9.13 (A) and (B). Instead, layers within the wall begin to swell, forming bulbous pockets (arrows) which act like cleavage planes, and one layer separates from another as the swelling spreads through the wall. The swollen layers dilate more and become the accumulation space as the remaining wall lamellae are torn apart. Although it at first appears as though there are five or six walls separating six or seven cells, there is just one (delaminated) wall separating two cells. × 500. Micrograph generously provided by G. Wittler.

milky" mammillarias) and (2) subgenus Mammillaria ("milky" mammillarias). Certain species of Hydrochylus (*M. lasiacantha* and *M. plumosa*) contain cortical regions of tissue whose cells are extremely watery and contain few chloroplasts and whose walls are unusually thin. These areas are in the same position as the latex ducts of subgenus Subhydrochylus (Figs. 9.20, 9.21; Mauseth, 1978b; Wittler and Mauseth, 1984b). In this latter group the cells in these subepidermal regions of the outer cortex begin differentiation by complex modifi-

cations of their walls: these begin swelling at certain sites, forming **bulbous pockets** that expand throughout the wall (Fig. 9.22). Simultaneously, other flat regions arise in the wall and stain darkly. The walls are converted to large regions that are noncompact and whose loose microfibrils alternate with dense, dark regions. By light microscopy these walls appear to be

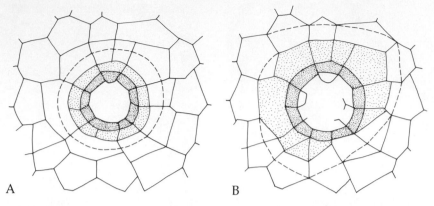

A   B

Figure 9.23
Recruitment growth in diameter. In (A), the lumen is surrounded by epithelium cells. The inner layer (heavily stippled) has almost completed its development and is ready to undergo lysis. In (B), that layer has degenerated,
and the lumen is now larger, but the epithelium is not narrow, because
nearby parenchyma cells undergo cell division and produce new layers of
epithelium.

empty epithelial cells (Mauseth, 1978b). The
wall ultimately breaks down to form the lumen
and the first "secretion." Simultaneously, the
protoplasts become modified by filling with derived vesicles (Fig. 9.21); the protoplast lyses,
mixing with the wall material and adding to
both the lumen and the secretion (Wittler and
Mauseth, 1984b). The adjacent cells may undergo the same transformation. This duct is
therefore a schizolysigenous one, and the secretion is holocrine.

In subgenus Mammillaria, there is no schizogeny, and the walls are not modified as just
described for subgenus Subhydrochylus (Wittler and Mauseth, 1984a). Instead, the first step
of differentiation is the production of numerous vesicles of diverse types and from diverse
sources: chloroplasts, ER, dictyosomes, and
plasmalemma. The wall becomes thinner as material is apparently removed from it, thereby
forming more vesicles. Finally, the walls rupture as digestion continues; by this time, the
protoplast has been converted to latex. Adjacent cells undergo division and create a
smooth, well-defined epithelium (Mauseth,

1978a). These ducts, like those of Subhydrochylus, increase in size with age as the innermost epithelial cells lyse and nearby cortical
cells are converted to epithelial cells (Fig. 9.23).

The latex ducts in *Mammillaria* have some
features in common with the articulated laticifers described above, but differ in that the entire wall breaks down, rather than just the end
walls. Also, although the final secretion is a
mass of vesicles and is extremely heterogeneous, with diverse types of vesicles derived
from many different organelles, not just the ER
and dictyosomes.

MUCILAGE CANALS.   Mucilage canals or ducts
are formed in a small number of families: Cactaceae (Fig. 9.24; Mauseth, 1980a), Sterculiaceae, and Tiliaceae (Metcalf and Chalk, 1983).
In *Sterculia bidwilli* (Bouchet and Deysson,
1974), the ducts develop as the cells in long
files divide, forming concentric rings of cells.
After the cells fill with mucilage, the walls as
well as the protoplasts degenerate, and the
duct lumen is filled with mucilage and cell debris. In *Nopalea* (Mauseth, 1980a), mucilage

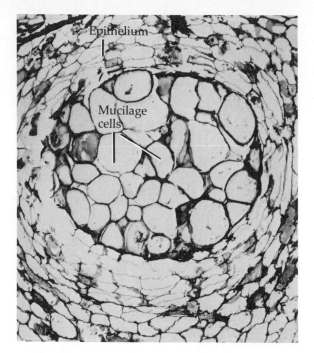

Epithelium

Mucilage
cells

Figure 9.24
Mucilage canal of *Nopalea;* the central cells are
filled with mucilage, but many remain alive. The
middle lamellas between the mucilage cells break
down, and the cells are free of each other; they
float in the secreted mucilage. × 100.

ducts are formed as cells, located just exterior
to the phloem, fill with mucilage. At the same
time, the middle lamella breaks down but the
walls do not. A liquid is secreted into the space,
and the duct becomes filled with whole mucilage cells that float in this liquid. When the
plant is cut, the liquid and the mucilage cells
flow out. Similar ducts are formed in *Uebelmannia gumifera* and in *Pereskia.*

Considering that mucilage cells are polyphyletic, it is interesting that the subcellular
metabolism is so similar (dictyosome-mediated
granulocrine secretion), whereas the organization above the cellular level is diverse: idioblasts, ducts, root cap, insect traps, and seed
coat epidermis. The similarity of secretion may
indicate that this mechanism is a modification
of the ubiquitous physiology used to build
walls: the deposition of carbohydrate for the
growing cell plate is mediated by dictyosomes
and their vesicles. Those vesicles also carry
polysaccharides, and it may not be too difficult
to modify this wall-construction mechanism
into one that secretes other polysaccharides
that have different functions. The diversity of
organization above the cellular level illustrates
how anatomy must be adapted for function:
mucilage that conserves water inside the plant
and mucilage that lubricates root tips can be
identical chemically, but they must be secreted
in different ways and in different places in
order to fulfill their roles.

Secretion by External Complexes of Cells

EXTRAFLORAL NECTARIES OF *ANCISTROCACTUS.*
Extrafloral nectaries of *Ancistrocactus* are glands
that secrete a sugary nectar that attracts ants to
the plant (Figs. 9.1, 9.25). Although it is frequently suggested that these ants protect the
plant from other insects and weeds, such defense has been proven only for some *Acacia*
species and for some tropical ant-plants. The
function in other plants is not yet definitively
established.

The extrafloral nectaries in *Ancistrocactus* are
modified spines, and as such, they have a
simple anatomy (Mauseth, 1982d). Most cactus
spines have only an epidermis and a core of fibers (Boke, 1980; Mauseth, 1977). Both types of
cells are sclerified and dead at maturity. In the
nectaries of this genus, however, the spine remains short, and all the cells remain alive. Instead of being closely packed together as in
spines, the "fiber" cells are rather loosely arranged, especially near the top of the gland.
The intercellular spaces are interconnected to
form a large apoplastic accumulation space.
The nectar is secreted into this space and then
moves to the top of the gland, where it is released as the cuticle is ruptured. Each gland

Figure 9.25
This longitudinal light micrograph shows the internal structure of an extrafloral nectary of *Ancistrocactus scheeri* (whole nectary is in Figure 9.1). This had been secreting when it was preserved; the entire epidermis is lifted up and broken off. In this gland, only the upper cells are dense and secretory, but if it had not been collected the next lower cells would have become active, then the cells below them. The water and sugars for the secretion are brought in by a set of vascular bundles at the base. × 50.

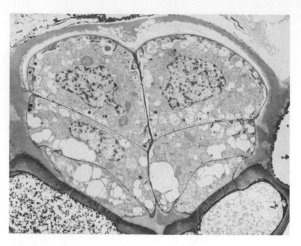

Figure 9.26
This electron micrograph of a salt gland of *Tamarix aphylla* (tamarisk tree) shows its simple but elegant structure. The basal cells have numerous plasmodesmatal connections with the surrounding cells, the uppermost cells have transfer walls, and the lateral walls of the basal cells are waterproof and prevent the salt solution from re-entering the plant by migrating through the wall under the cuticle. × 5000. Micrograph generously provided by A. M. Bosabalidis and W. W. Thomson.

functions only briefly, and when secretion ceases the interior cells collapse and die.

These nectaries strongly resemble the nectaries of *Vinca* when they are mature, but the two have evolved from very different ancestral structures: spines for *Ancistrocactus*, carpels for *Vinca* (Bonnier, 1879; Rachmilevitz and Fahn, 1973). The tremendous evolutionary convergence shows how closely structure is related to function.

SALT GLANDS.   Many of the plants that grow in saline habitats (coastal marshes and swamps or poorly drained desert valleys) have the special problem of avoiding the accumulation of excessive amounts of salt in their bodies. The roots are apparently only partly successful in preventing the entry of harmful quantities of salt, and it is necessary for the plants to have salt glands to secrete actively the ions that are unavoidably transported up to the leaves. Salt glands occur on many of the flowering plants and even some ferns that occur in such habitats, but this type of secretion is especially characteristic of the families Avicenniaceae, Acanthaceae, Chenopodiaceae, Frankeniaceae, Plumbaginaceae and Tamaricaceae (Fig. 9.26; Fahn, 1979). Numerous types of ions have been detected in the secretions—Na^+, K^+, Mg^{++}, Cu^{++}, Cl^-, SO_4^{--}, NO_3^-, PO_4^{-3}, and HCO_3^- (Thomson, 1975)—and the process requires metabolic energy.

Salt glands are polyphyletic, and numerous types exist. In some, the gland is basically a small hair and functions much like the stinging hairs, sequestering the salt in its central vacuole. Such glands occur in *Atriplex* (Volkens, 1887; Black, 1954; Osmond et al., 1969; Smaoui, 1971).

Multicellular salt glands are more prominent and offer good examples of a more elaborate type of gland. In *Avicennia* (a coastal mangrove plant), the glands occur on both surfaces of the leaves. Each consists of two or four **collecting cells**, one **stalk cell** and eight **secretory cells** (Chapman, 1944; Shimony et al., 1973; Walter and Steiner, 1937). The collecting cells are located at the level of the surrounding epidermal cells, and they absorb salts from the epidermal cells and from the mesophyll cells. The salts are passed to the secretory cells. It is still not established how the salt movement occurs (Thomson, 1975), but the transverse walls of the stalk cell contain numerous plasmodesmata.

In the secretory cells, the vacuole is small, the nucleus is large, and there are many mitochondria. The plastids are degenerated. Many small vesicles occur near the plasmalemma, and multivesiculate bodies are common between the plasmalemma and the cell wall. It has been suggested that the vesicles carry the salt in a granulocrine manner (Shimony and Fahn, 1968; Shimony et al., 1973). Once secreted, the salts pass through the cell wall; they then pass to the surface of the gland by means of small channels in the cuticle.

It is necessary to prevent the salts from seeping along the wall and moving back into the plant. This is accomplished by having the longitudinal side walls heavily cutinized, with the cutin permeating all wall layers. The stalk cell acts like an apoplastic barrier.

In *Tamarix*, the glands are similar, consisting of two basal collecting cells and six secretory cells (Fig. 9.26; Bosabalidis and Thomson, 1984; Brunner, 1909; Campbell and Strong, 1964; Thomson and Liu, 1967). A cuticle covers all of the gland, and a special Casparian strip region

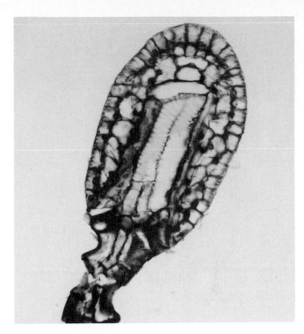

Figure 9.27
This is the secretory head of a digestive gland of sundew (*Drosera*). × 80.

prevents back flow of secreted salt. These glands are more elaborate than those of *Avicennia*, however, because the contact walls between the collecting cells and the secretory cells are transfusion walls.

Salt glands can be quite elaborate, and many of them show all the features that can be associated with a sophisticated secretory system: There is (1) a means of absorbing (over a large surface area) the material to be processed, (2) a means of transporting it within the gland (transfer walls or fields of plasmodesmata), (3) a discrete group of cells specialized for the actual secretion, and (4) a means of isolating the secretion and preventing it from leaking back into the symplasm.

DIGESTIVE GLANDS.   The digestive glands of *Drosera* (sundew) represent an increased level of complexity for the structure of a gland (Figs. 9.27, 9.28). In general, the digestive glands are

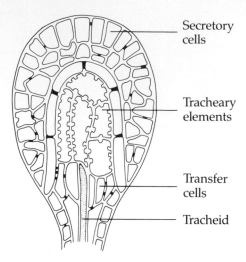

Secretory cells

Tracheary elements

Transfer cells

Tracheid

Figure 9.28
This diagram interprets the structure of a sundew gland (Fig. 9.27) and is based on research by Gilchrist and Juniper (1974). The outermost layers are secretory cells; the next layer consists of endo-dermis-like cells; the interior cells are short, wide tracheary elements. Between the head and the stalk is a layer of transfer cells, which are not easily de-tectable in Figure 9.27.

very similar to the salt gland of *Avicennia* and *Tamarix*. However, they are larger and have a small supply of tracheary elements.

The stalk contains two layers of parenchyma cells: an outer epidermis of small cells and an inner layer of elongate cells that have numerous plasmodesmata in their end walls (Ragetli et al., 1972). At the center of the stalk is a slender row of tracheary elements that extends to the top of the gland. In the base of the head, the tracheids are surrounded by a layer of transfer cells, and both the transfer cells and the tra-cheid mass are enclosed by a cap of endoder-mislike cells (Gilchrist and Juniper, 1974; Lloyd, 1942; Schnepf, 1961; Williams and Pickard,

1974). The endodermis is connected to the epi-dermis of the stalk, completely enclosing the transfer cells. Above the transfer cells are two layers of secretory cells that have dense cyto-plasm. The secretory cells also have transfer walls, apparently to aid the movement of mate-rials. The entire gland is covered by a cuticle, but there are many pores in the cuticle, which lies over the secretory cells. In the upper half of the gland, the pores account for 12% of the sur-face area (Williams and Pickard, 1974).

These cells are functionally quite complex; in addition to secreting digestive enzymes, they must also absorb the digested nutrients (Juniper and Gilchrist, 1976). Thus there is both an upward-outward secretion as well as an inward-downward absorption. Not all di-gestive glands are this complex: in *Dionaea* (Venus' flytrap) they are short, stalked glands that do not have a vascular supply, much like the salt glands (Lloyd, 1942). In *Pinguicula*, the gland has just a single basal cell, one stalk cell, and two to eight secretory cells in the head; the secretion is prevented from returning to the plant apoplastically by the cutinization of the lateral walls of the stalk cell (Heslop-Harrison and Knox, 1971; Schnepf, 1961, 1963; Vogel, 1960).

Other types of glands can be as complex as the digestive glands of *Drosera*. The mucilage-secreting glands of *Drosophyllum* have an al-most identical structure, with xylem in the stalk and a well-defined, caplike endodermis to isolate the secretion (Lloyd, 1942). These glands secrete a sticky mucilage that traps in-sects for later digestion. The secretion is gran-ulocrine as in the other structures which se-crete mucilage, but the organization of the gland is different from that of water-storage tissues or roots. This is to be expected, because the function is different.

Concepts

The epidermis is the outermost layer of the primary plant body (Figs. 10.1, 10.2). Developmentally it is derived from the tunica in the shoots (or from the outermost layer if the tunica is multistratose). As mentioned in Chapter 6, in the roots there may not be a unique protoderm layer, but still the root epidermis is easily recognized by its position and the presence of root hairs. It is the epidermis that is the interface between the plant and its environment, and therefore the epidermis is in a unique situation. On one side are the living tissues that are regulated as carefully as the plant can, but on the other side is the exterior environment over which the plant exerts little or no control. A large number of forces, both biological and nonbiological, may impinge upon the epidermis, and the number of different types of contingencies that any single cell may face is very large. It should be no surprise that the epidermis can be an extremely complicated tissue and that it has a plasticity in its development that is greater than that of most other tissues. Any cell derived from the protoderm has a wide range of potential fates. Furthermore, the epidermal cells are typically long-lived and are commonly able to alter their anatomy and physiology if it is necessary. Even the simplest dermal system will have at least two cell types (ordinary epidermal cells and guard cells if in the shoot, ordinary epidermal cells and root hairs if in the root), and it is not un-usual to find ten or more distinct epidermal cell types on a single plant. As was true of all the tissues discussed in earlier chapters, these cell types do not occur at random. There is no need to try to memorize numerous rules or generalizations, for epidermal cells are elegantly suited to their functions and are understandable in a functional context. When considering the following functions of the epidermis, it is helpful to remember that, as most plants age, the epidermis is replaced by bark (Chapter 17), which has most of the same functions but a very different structure.

Functions of the Epidermis

Water Regulation

The primary function of most epidermises (the plural *epidermi* is obsolete and rarely seen) is to control the movement of water into and out of the plant. The roots of most plants are embedded in the soil or some similar substrate and are able to extract water from the soil because the protoplasm of root cells is "drier" than the soil (their water potential is more negative). Air is almost always dry enough to absorb water from the shoot tissues, so the aerial epidermis must be resistant to water movement (loss), in contrast to the root epidermis, which must be designed to facilitate movement (absorption). Two examples can show how efficient epidermises can be. If a

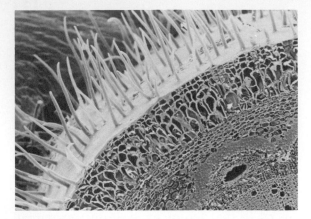

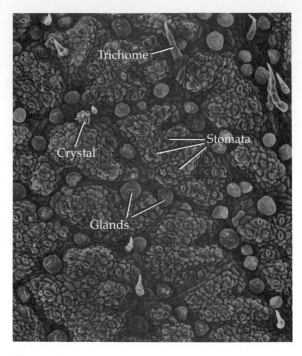

Figure 10.1
This scanning electron micrograph of *Lithraea caustica* gives a good perspective of the epidermis and its relationship to the rest of the plant. Most of the epidermis forms a smooth, tight sheet over the body. Numerous hairs project outward and will partially shade the underlying tissues, slow down air movement near the surface, and hinder the movements of at least small insects. × 300. Micrograph generously provided by G. Montenegro.

Figure 10.2
The epidermis of *Haplopappus glutinosus* has only a few small trichomes, but there are numerous glands. The epidermis is rough, being low over vascular bundles but bulging upward between them. Stomata are abundant. × 100. Micrograph generously provided by G. Montenegro.

healthy mesophytic plant is uprooted and its roots are kept away from water, the plant will begin to wilt within hours; it had been losing water by transpiration, but the roots were replacing the water so efficiently that the plant had remained turgid until the roots were removed from the soil. An uprooted xerophytic succulent, however, has such an efficient shoot epidermis that there is very little transpiration, and the plant remains turgid and alive for a long time (as much as nine years for giant cacti); but if the shoot epidermis is removed, even without uprooting, the water loss will be so rapid that the plant will die within hours. Carlquist (1975) has pointed out that plants rarely wilt in their natural habitats; thus the root and shoot epidermises typically work together in an excellent balance.

In some species, there are modifications such that the roots are less permeable and the leaves are absorptive (Benzing, 1980; Benzing and Pridgeon, 1983). This is completely the reverse of what we would expect, but the plants with this arrangement live in environments where the air is more moist than the soil: tropical cloud forests and rainforests. The "tank" bromeliads live epiphytically, and their roots, serving primarily to attach the plant to a tree or vine, have little or no absorptive function. Instead, the leaves form a tight rosette that creates a basin that catches rainwater and nutrient-containing dust and debris (Fig. 10.3). The submerged leaf bases are the absorptive system for these plants. The exposed portions of the leaves, however, are extremely waterproof. The leaves of carnivorous plants must also be absorptive (Joel and Juniper, 1982).

The leaves of most rainforest epiphytes, ranging from cacti to orchids to philodendrons, are

Figure 10.3
Aechmea pineliana is one of the numerous tank bromeliads. The leaf bases abut so tightly that they form a tank that catches rain, dust, and debris. The exposed parts of the leaves are waterproof, but the submerged bases absorb both water and nutrients.

highly waterproof. It might be expected that in such environments, where the humidity is always high, the leaves and stems would have very permeable epidermises to allow them to absorb moisture from the air, thus reducing the amount of xylary conduction and the dependence on roots. But for these plants the most severe stress is mineral deficiency: their roots, being attached only to the bark of a tree or to the bare rock of a cliff face, are on the worst of all possible "soils." There are almost no available, dissolved minerals, and whatever is absorbed and taken into the plant must be protected. Rainwater, if it seeped into the leaves and stems through the epidermis and then leaked out again, would carry away valuable minerals from the protoplasm and the apoplast (Kramer and Kozlowski, 1979; Patterson, 1975; Tukey, 1970). In one year, one acre of apple trees can lose up to 30 kg of potassium from their leaves, from leaching by rainwater. Therefore, leaves must keep out all water that does not arrive through the xylem, and they must assure that water exits only as a gas by transpiration, not as a liquid that can carry salts with it.

Protection Against Sunlight

A second important function of the epidermis is to protect the plant against excessive solar radiation (Martin and Juniper, 1970). Although all autotrophic plants need visible light for photosynthesis, in some occasions the light (especially the ultraviolet and the infrared wavelengths) can be too intense. It can cause damage to the plant by overheating the protoplasm and by bleaching the chlorophyll. Deserts, seashores, and high alpine areas, with their clear skies, are obvious examples of environments that can have an excess insolation, but even more moderate environments, such as the canopies of forests, can be damagingly bright. The same is true of cloud and rainforests during occasional clear days; the radiation level may not be especially intense in an absolute sense, but it may be too bright for the rather delicate plants that live there. The epidermis, because of its reflectivity, can prevent a significant fraction of the light from entering the leaf (McClendon, 1984).

Protection Against Other Organisms

The epidermis is the first line of defense against many biological pests; these can be extremely diverse (Levin, 1973; Mabry et al., 1977; Rodriguez et al., 1984). Obvious pathogens such as bacteria and fungi (Fig. 10.4) are ubiquitous in all environments (Martin and Juniper, 1970). **Epifoliar** lichens and algae in the tropics grow on the perennial leaves of trees and vines, intercepting the light that their hosts receive and thereby interfering with photosynthesis. A smooth epidermis can allow the spores of these lichens and algae to fall off before they germinate and become established. Most parasitic angiosperms (such as mistletoes and dodders) attack their host's primary tissues, so the epidermis must be resistant to these (Fig. 10.5; Kuijt, 1969; Mauseth et al., 1985). Insects, of course, are a constant menace to plants, and there are numerous epidermal

Figure 10.4
This is the epidermis of a leaf of *Luzunaga radicans* from central Chile. The leaf is covered with fungal hyphae and spores, as well as dust and debris. × 80. Micrograph generously provided by G. Montenegro.

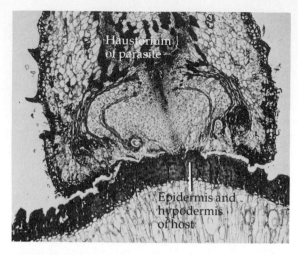

Figure 10.5
This photograph shows the haustorium of the mistletoe *Tristerix aphyllus* invading the cactus host *Trichocereus chilensis*. The epidermis and hypodermis of the host not only constitute a waterproof barrier but also coincidentally protect it from the attacks of normal mistletoes. Unfortunately for the cactus, within this particular species of mistletoe, a new mechanism of invasion has evolved, allowing it to pass through the stomata and bypass the regular protective mechanisms. × 100.

adaptations to prevent insects from being able to chew, suck, lay eggs on, or even land on the plant. In many cases, epidermal glands contain substances so toxic or noxious that even large animals do not eat them (stinging nettles, for example; Chapter 9). Of course this sometimes backfires on the plant: the active substance of marijuana (THC) is produced by the epidermis, so the epidermis is responsible for the plant's being harvested and burned.

Protection Against Nonbiological Agents

Wind damage is a constant threat to most plants, primarily through the violent torsions and flexions that may occur as leaves and young shoots flutter in even gentle breezes. Without a flexible, tear-resistant covering, especially to protect the edges of leaves (deeply lobed leaves in particular) they would become torn and shredded very quickly. Banana leaves provide an excellent example of wind damage. The

leaves are produced with an entire lamina, not a compound one, but thin regions of non-reinforced tissues are deposited at certain sites. After expanding, the leaves are rapidly torn to their "compound" state by air movements.

In arid, sandy regions, wind-carried dust is an extreme threat. Dust particles are frequently the predominant erosive force in such environments; the plants must be capable of withstanding intense sandstorms.

Epidermal Function in Reproduction

A more positive, less defensive role of the epidermis is in reproduction, at least in angiosperms. In the flowering plants, a number of mechanisms have evolved to encourage or ensure cross pollination. One of the simplest mechanisms is that in which the epidermis and subepidermal tissues of the anther break open early, releasing pollen grains while the epidermis of the stigma is still in a nonreceptive condition (Figs. 19.9, 19.25). When pollen is brought to a receptive stigma, there may be an elaborate set of interactions whereby the stigmatic epidermis "tests" the pollen and "rejects" it if it is not of the proper type (Fig. 19.26; Heslop-Harrison, 1978; Heslop-Harrison and Heslop-Harrison, 1982). If it is a compatible type, then its growth can be encouraged by the epidermis that lines the inside of the stylar canal. Epidermises may even be involved in getting the pollen to the stigma: the color, texture, and reflectivity of the petals and other flower parts, along with scent, are important cues that allow pollinators to identify and visit the proper species (Meeuse and Morris, 1984). All of these characters are dependent upon the nature of the epidermis. Similarly, many flowers have floral or extrafloral nectaries in which the epidermis is either secretory or important in accumulating or releasing the nectar for the pollinators (Fahn, 1982).

Epidermal Function in Secretion

Finally, the epidermis can be extremely active. It may contain cells (especially in grasses) which by their absorption or loss of water can cause leaves to open when moisture is available and close when it is not. And epidermal trichomes can be involved in attracting, guiding, trapping, and digesting insects. These functions were covered in Chapter 9.

Types of Epidermal Cells

There are certainly other functions for epidermal cells than those just mentioned, some of which are general, others that are found only in one or two species. Considering this diversity of function, it is somewhat surprising that there seem to be only four basic types of ma-

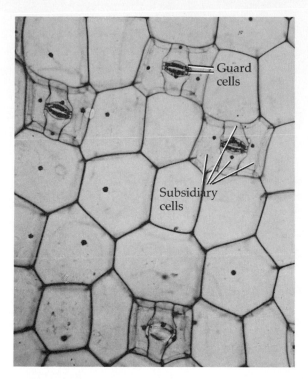

Figure 10.6
This is a leaf peel of *Zebrina* (wandering Jew). The epidermis is grasped with a pair of fine jeweler's forceps and then peeled slowly off the plant. The large polygonal cells are the ordinary epidermal cells. Stomata and guard cells are visible; each stoma is surrounded by a complex of four subsidiary cells. Except for the stomatal pores, there are no intercellular spaces. × 150.

ture epidermal cells: ordinary epidermal cells, guard cells, trichomes, and root hairs.

Ordinary Epidermal Cells

Ordinary epidermal cells are defined by default: they include all that are not of the other types to be described (Figs. 10.6, 10.7). Ordinary epidermal cells are basically the cells that lie between the more specialized cells of the epidermis, and they are typically the most numerous and cover the greatest proportion of

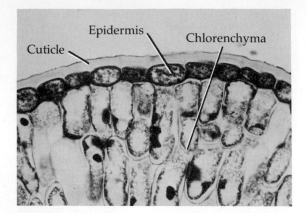

Figure 10.7
This is a cross section through a leaf of the conifer *Taxus* (yew); the ordinary epidermal cells are small and densely cytoplasmic (very different from Figure 10.6). All of their walls are extremely thin, but the outer wall has a thick layer of cutin on it. This epidermis is quite homogeneous—no trichomes or glands occur, and, although stomata are present, none is visible in this section. × 200.

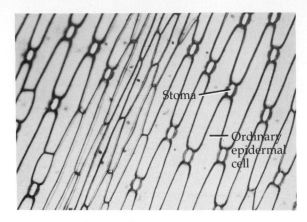

Figure 10.8
This epidermal peel of *Iris* leaf shows that the epidermal cells are elongated parallel to the long axis of the leaf. The stomata are also aligned with leaf elongation. The narrow band of smaller cells that lacks stomata was located above a group of fibers. × 150.

the plant body (Montenegro, 1984). They can have almost any shape, but they are most often tabular. Tall or columnar epidermal cells are rare, being mentioned in only 18 families by Metcalfe and Chalk (1979), such as Chrysobalanaceae, Ericaceae, and Thymelaeaceae. Tall cells are most common in seed coats and bud scales, where they seem to be correlated with the need for extra protection.

When seen in face view, the shape of the ordinary epidermal cells frequently bears some relation to the organ that they cover, or to the tissues immediately beneath them. In elongated nodes and petioles or the long, strap-shaped leaves of many monocots, the epidermal cells are elongated parallel to the organ (Fig. 10.8). The epidermal cells that lie over the veins of leaves are usually elongated parallel to the vein (Fig. 10.2). In some palms, such as *Bactris*, and cycads, such as *Encephalartos* and *Ceratozamia* (Linsbauer, 1930), the ordinary epidermal cells are rhomboidal. The anticlinal

walls of epidermal cells in dicots are often sinuous (Fig. 10.9), but those in monocots are either straight or sinuous (Ayensu, 1972; Cutler, 1969; Metcalfe, 1971; Tomlinson, 1961). Neither the causes nor the selective advantage (if any) is known for the undulations when they occur in anticlinal walls. The environmental conditions that exist during leaf development are known to affect them (Watson, 1942). In leaves, the size and shape of epidermal cells are usually characteristic of the position of the cells: those above veins or sclerenchyma bundles tend to be small and flat; those on the margins are more fiberlike; the cells of the upper surface often differ from those of the lower surface (compare Figure 10.7 with 10.15). An almost universal feature of ordinary epidermal cells is that they are firmly attached to each other along their sides and attached less firmly to the tissues beneath. In many plants (especially monocots with flat leaves and some succulent dicots with soft mesophyll) the epidermis can be peeled as a sheet away from the underlying cells (Figs. 10.6, 10.8, 10.9). Inter-

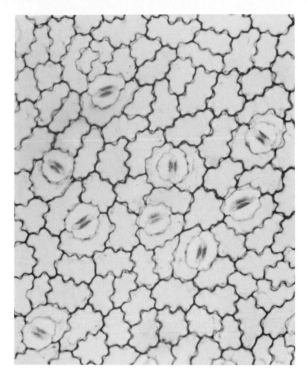

Figure 10.9
The ordinary epidermal cells of *Coffea arabica* (coffee) have undulate, sinuous anticlinal walls, which is rather typical of dicots. × 100.

cellular spaces do not occur between ordinary epidermal cells; the only spaces present are the stomatal pores.

The ordinary epidermal cells may be required to undergo a great deal of growth (both cell division and expansion) during the development of the organ that they cover. Whereas many twigs have a small diameter that requires little epidermal growth after the cell leaves the apical meristem, succulent stems, tubers, and rhizomes may reach a diameter of 50 cm or more, requiring tremendous growth. The same is true of even modest-sized leaves that undergo great increases in surface area as the leaf primordium matures. The leaves of the palm *Raphia* may be gigantic, up to 16 meters long (Tomlinson, 1961). This great expansion of the

ordinary epidermal cells may not occur until late: the ovaries of most angiosperms may be quite small, but after pollination and fertilization they may grow rapidly and extensively into very large fruits (watermelons and pumpkins, for example), requiring renewed expansion in the epidermis (Roth, 1977).

Ordinary epidermal cells usually remain alive for the lifetime of annual or biennial plants but are often replaced by bark in perennial plants. This replacement may happen within the first year, or the epidermis can remain alive and functional for as long as 20 years, as in *Acer striatum* (De Bary, 1884) or even 40 years, as in *Firmiana*. In many cases the epidermal cells themselves undergo the cell divisions necessary to produce the first bark (Figs. 17.1, 17.2).

Ordinary epidermal cells are usually unremarkable cytologically; they have all of the typical organelles, but the plastids are present as proplastids or leucoplasts. Chloroplasts are rare (Martin and Juniper, 1970), except in aquatic angiosperms, such as *Myriophyllum*, *Ranunculus*, *Potamogeton*, and *Sagittaria* (Linsbauer, 1930) and in ferns (Meyer, 1962; Wylie, 1948). There is usually very little storage in ordinary epidermal cells, the exception being the accumulation of anthocyanin pigments in the petals of many flowers (Fig. 19.6; Hess, 1983). Other brightly colored parts of some species, such as the red-purple leaves of *Zebrina pendula* and red cabbage, are also due to epidermal anthocyanins. An interesting aspect of the protoplasts of ordinary epidermal cells in many species is the asymmetric position of the vacuole, nucleus, and cytoplasm, especially if the cell is at all columnar (Figs. 10.10, 10.11).

Most of the functions mentioned in the introduction to this chapter are based on modifications to the walls of the ordinary epidermal cells. Epidermal cells, particularly long-lived ones, may have extremely thickened walls (Fig. 10.11). These are usually interpreted as primary walls, although Esau (1965) suggests that the very thick walls might in fact be secondary walls. True secondary walls are deposited in

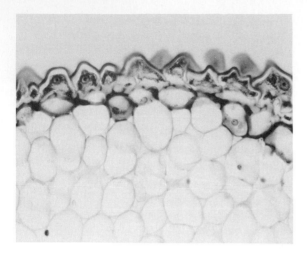

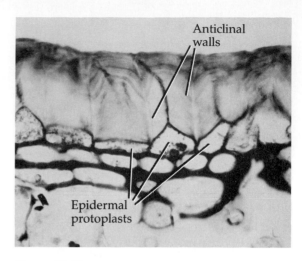

Figure 10.10
The ordinary epidermal cells of *Ariocarpus fissuratus* (star cactus) are papillose, with areas projecting outward. The protoplasts are also asymmetric: the vacuole is always at the base of the cell, and the nucleus is always at the top of the papilla. × 150.

Figure 10.11
This is the epidermis of a desert xerophyte, *Echinocereus enneacanthus*, which has remarkably thickened radial and outer tangential walls. The basal portion of the radial walls do not thicken at all, nor does the inner tangential wall. × 150.

the epidermis of many hardened seed coats. Lignin is rare, but it does occur in the epidermal walls of the leaves of ferns, cycads, and many conifers, and also in the epidermal cells that lie above bundles of sclerenchyma in the leaves of grasses, rushes, sedges, and some other plants, such as *Eucalyptus*, *Laurus nobilis*, *Nerium oleander* (Linsbauer, 1930).

When epidermal cells have thickened walls, the thickenings are almost always asymmetrical (Fig. 10.11); they are rarely uniform as would occur in fibers or sclereids. Instead, the outer periclinal wall—*Ammocharis*, *Cocculus laurifolius*, *Hakea leucoptera*—or inner periclinal wall—certain species of Epacridaceae and Euphorbiaceae (Solereder, 1908)—or both (Fig. 10.13), is thickened, but the anticlinal walls are not. Or the outer wall and the anticlinal walls are thickened, but not the inner periclinal wall (*Echinocactus grusonii*). The outer periclinal wall of *Acer pennsylvanicum* and *Camphora officinarum* are over 30 μm thick, and in extreme xerophytes such as *Crinum* and *Brunswigia*, it has been re-

ported as 100–200 μm thick (Linsbauer, 1930). It is not uncommon for the thickening on a particular wall to be uneven, with either warts, protrusions, or ridges occurring (*Erodium cicutarium*, *Lycopodium*, *Opuntia*, *Potentilla anerina*). In other cases, the thickened walls, especially anticlinal ones, may contain depressions, thin areas, or large windowlike pits called **fenestrae** (sing.: fenestrum). These can vary from being just irregular depressions to being canallike and even resembling bordered pits in *Candollea* (Solereder, 1908). True plasmodesmata do occur, and primary pit fields are present; as the primary wall thickens, these take on the appearance of simple pits. Plasmodesmata have been reported (Schnepf, 1959; Sievers, 1959) on the outer periclinal wall; these are termed **ectodesmata** (sing.: ectodesma). Franke (1971b) has suggested that ectodesmata are not related to plasmodesmata and that they be called **teichodes** to emphasize that. He believes that instead they are responsible for moving material to and from the plant surface (Franke,

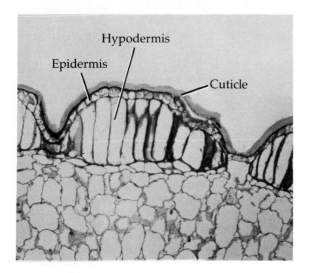

Figure 10.12
The cuticle of this *Uebelmannia pectinifera* stained very differently from the wall and also began to separate from it. The layer of cutin is almost as thick as the entire epidermal cell protoplast. This is an unusual plant in that it has a warty surface, caused by uneven expansion of hypodermal cells, not epidermal cells. × 150.

1971a; Scott et al., 1958). Very little information is available about the formation of the ectodesmata; this is especially unfortunate because Metcalfe (1979a) mentions that they are more common than is generally realized.

CUTIN AND THE CUTICLE.   Because one of the main functions of the shoot epidermis is water retention, it must have waterproof walls. This is accomplished by depositing a layer of the hydrophobic material **cutin** on the outer epidermal wall (Figs. 10.7, 10.12). The cutin is a complex, high-molecular-weight lipid polyester that results from the polymerization of certain fatty acids. In the angiosperms, several types of cutin exist, and they can be separated into three broad classes based on the nature of the fatty acid monomers that predominate in them: (1) cutins that contain mostly fatty acids that are 16 carbons long (C_{16}), (2) those with mostly

C_{18} fatty acids, and (3) cutins that contain more or less equal amounts of both (Holloway, 1982b). The type of cutin can vary from one part of a plant to another, leaves and fruits often having very distinct types, for example (Baker and Holloway, 1970; Baker and Procopiou, 1975). The cutins of the gymnosperms and the cryptogams seem to lack the C_{18} monomers. The fatty acids are produced by the epidermal protoplasts, apparently in the endoplasmic reticulum. The precursors are transported to the outer wall of the epidermal cell by an unknown mechanism. Granulocrine transport by small vesicles has been reported (Frey-Wyssling and Mühlethaler, 1959; Heide-Jørgensen, 1978), but many suspect that the process is really eccrine, involving individual molecules. Polymerization was considered to be possibly an automatic event, mediated by oxygen, but enzymatic control of polymerization is now considered to play an important role (Holloway, 1982b).

As the fatty acids migrate outward, they can begin to polymerize while still in the wall, and the cutin will form a matrix surrounding the cellulose microfibrils. With continued secretion, a layer of pure cutin builds up on the outer surface of the wall proper (Priestley, 1943; Holloway, 1982a; Frey-Wyssling, 1959). The mixture of cutin plus wall material is called the **cuticular layer**, while the cap of pure cutin is known as the **cuticle** (for clarity, the **cuticle proper**, because the usage of the term *cuticle* has been variable). With new studies involving electron microscopy and histochemistry, more complex layering is frequently reported, and it is apparent that there may be variability from species to species (Holloway, 1982a). The cuticular layer is usually thin (less than 0.1 μm) but can be as much as 0.5 μm thick in xerophytes, as in *Clivia minimata* (Holloway, 1982a); the cuticle proper can frequently be 5 μm or more in thickness. The process of deposition of cutin in the wall to form the cuticular layer is called **cutinization** (Fritz, 1935, 1937; Roelofsen, 1959), and the deposition of pure cutin on the outside of the wall to form the cuticle proper is called **cuticularization** (Esau, 1965), although the two processes may really be identical. In many plants, the cuticle proper is separated from the cellulose wall by a **pectin layer** (Frey-Wyssling, 1976), which is rich in pectic substances and possibly may be equivalent to the middle lamella of the inner surfaces of the cell. The pectin layer is reported to be continuous with the middle lamellas of the anticlinal walls.

Because cutin is so impervious, it is extremely difficult to prepare it for electron microscopy; despite this problem, many excellent studies have been performed (these are summarized in Holloway, 1982a; Martin and Juniper, 1970). The cuticular layer tends to have a fibrillar organization, due at least in part to the presence of cellulose fibrils. The cuticle proper may be homogeneous and amorphous in some species, such as *Eucalyptus perriniana* (Hallam, 1970) and *Prunus persica* (Schneider and Dargent, 1977), but in others it may be lamellate, as in *Agave americana* (Wattendorff and Holloway, 1980) and *Pseudotsuga menziessii* (Sargent, 1976), or reticulate, as in *Citrus sinensis* (Thomson and Platt-Aloia, 1976). A classification and lists of species are given by Holloway (1982a). Platelets of wax often occur embedded in the cuticle (Baker, 1982; Sargent and Gay, 1977; Roelofsen, 1952).

The cuticle is responsible for giving the epidermis many of its properties; it is waterproof and is the primary means of water retention. Because it is shiny and reflective, it is capable of deflecting some of the excess solar radiation (Martin and Juniper, 1970). It reflects ultraviolet light, thereby protecting the DNA below from the mutagenic effects of sunlight. Cutin is completely indigestible; nothing is known that can metabolize it. Thus it is an excellent protection against fungi and bacteria, as they have no enzymes capable of digesting holes in the cuticle to facilitate invasion. Also, the cuticle can be so smooth that spores fall off or are shaken off in the wind.

Although the cutin is typically deposited on

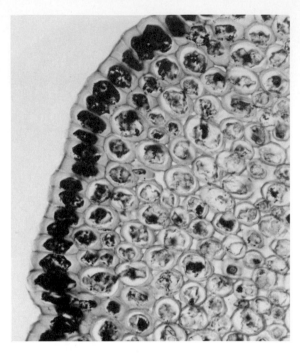

Figure 10.13
Cutin has been deposited in all walls of the epidermal cells of this cactus, *Homalocephala texensis*. The outer and inner walls are most heavily cutinized; the protoplasts are still alive, presumably being maintained by plasmodesmatal connections to the subjacent tissue. × 150.

the outer wall of the epidermal cells, in some plants, especially of arid regions, the cutin may be deposited on the anticlinal walls and even on the inner periclinal wall (Fig. 10.13), although usually in lesser amounts (Lyshede, 1982). The deposition of cutin seems to be at least partially influenced by the environment: more is deposited when conditions are drier. In this regard, even the mesophyll cells near stomata, where the internal "air" can be rather dry, may produce a small amount of cutin. The only epidermal cells that lack it are those near the root apex and those with root hairs (Bonnett and Newcomb, 1966). Even the epidermal cells in completely protected positions, such as on the inner side of carpels, have a cuticle.

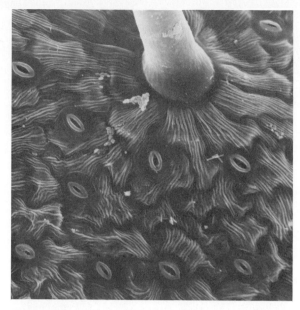

Figure 10.14
The cuticle of *Ludwigia terrestre* has characteristic striations that radiate from the stomata and from the base of the trichome. The guard cells do not have cuticular wrinkles. × 150. Micrograph generously provided by G. Montenegro.

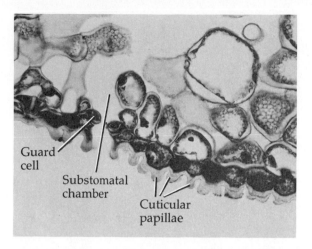

Figure 10.15
The lower epidermis of *Taxus* (yew) and the epidermal cells are markedly papillose. The papillae consist of thick areas of cuticle overlying thick areas of wall. Compare this lower epidermis with the upper epidermis (Fig. 10.7) of the same leaf (same section). On the upper epidermis the cuticle is smooth, without any papillae. × 150.

In some species, the cuticle has complex patterns of striations (Fig. 10.14), bumps, and wrinkles (Figs. 10.15, 10.16; Stace, 1965; van Staveren and Baas, 1973; Wilkinson, 1971). A list of variations with examples is given by Wilkinson (1979). In some cases these are sufficiently characteristic to be useful in taxonomy (Carr et al., 1971; Ihlenfeldt and Hartmann, 1982).

WAX.   Wax is a universal adjunct to the outer wall of epidermal cells (Figs. 10.17, 10.18, 10.19; Baker, 1982; Eglinton and Hamilton, 1967; Johnson and Riding, 1981). Like cutin, wax is not a specific compound but rather an extremely heterogeneous polymer that results from the interaction of very long-chain fatty acids (up to 34 carbons), aliphatic alcohols, and alkanes in the presence of oxygen. Many of the precursors are branched-chain or cyclic com-

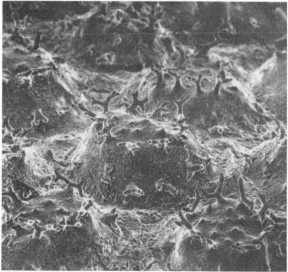

Figure 10.16
These epidermal cells of *Anisocalyx vaginatus* show antlerlike protuberances that are both cuticular protrusions and wall material. × 100. Micrograph generously provided by H. D. Ihlenfeldt.

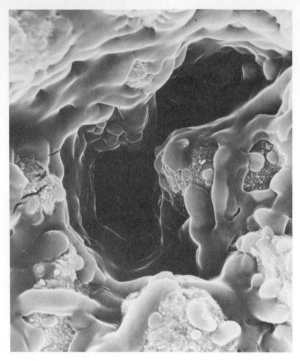

Figure 10.17
This is a stomatal cavity of *Trichocereus chilensis*. The guard cells are at the base of the crypt and are not visible in this photograph. The smooth areas are regions covered by an oily wax. × 400. Micrograph generously provided by G. Montenegro.

Figure 10.18
One of the most common forms of wax is small plates as shown here on *Eucalyptus alba*. Because the wax is not a solid block, there can be some air circulation between the plates. × 200. Micrograph generously provided by N. D. Hallam.

pounds. As is true for cutin, the actual means by which the wax precursors arrive at their destination has not been settled yet. There are reports of channels or canals that run through the outer wall and cuticle that could act as a means for carrying the precursors to the surface (Hall, 1967a, b; Lyshede, 1978; Miller, 1985; Scott et al., 1957), but channels such as those proposed could easily be artifacts produced during specimen dehydration or microtomy. Several investigators have found no evidence of any canals, and it seems possible that the precursors migrate through the wall and cuticle individually, then polymerize spontaneously when they reach the surface and the presence of oxygen (Baker, 1982; Heide-Jørgensen, 1978).

With careful analysis, it has become obvious that there are two distinct classes of wax: (1) **epicuticular wax** on the surface of the cuticle proper and (2) **intracuticular wax**, which occurs as particles within the cutin matrix. The intracuticular waxes are composed mostly of short chain (C_{18}) monomers, rather than long chain ones.

The amount of epicuticular wax can vary enormously, from a thin layer (5 to 10 μg per cm^2) to a cap that is over 5 mm thick in *Ceroxylon andicola* (Martin and Juniper, 1970) or containing 410 μg per cm^2 in *Malus pumila* (Holloway and Baker, 1970). Probably more important than the amount is the nature of the wax. In some plants, such as jojoba (*Simmondsia*), it is more oily than waxy (Fig. 10.17)

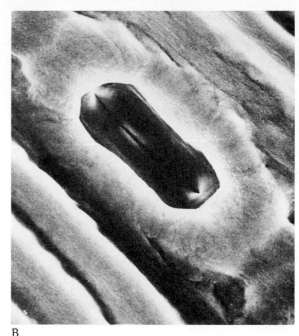

A   B

Figure 10.19
(A) shows rods or threads, another common form of wax deposition. Notice
that they are longest on the subsidiary cells that surround the guard cells.
(B) shows the same surface treated with chloroform to dissolve the wax. In
the dewaxed state, any breeze would immediately sweep away water molecules that had escaped out of the stomatal pore. White pine (*Pinus strobus*).
× 500. Micrograph generously provided by R. W. Johnson and R. T. Riding.

and forms a uniform layer over the whole
surface. In other species, it polymerizes into
plates (Fig. 10.18; *Citrus, Vitis vinifera*), rods
(Fig. 10.19; *Chelidonium majus, Ginkgo, Picea*),
granules (*Eucalyptus, Rhododendron*), or other
forms (Baker, 1982; Schieferstein and Loomis,
1956). These can have a variety of effects. They
may cause the surface to be extremely hydrophobic and nonwettable, an important consideration in the application of exogenous substances such as herbicides or growth regulators
(Price, 1982). On the other hand, they can
create capillary surfaces that increase wet-
ability and allow dew to be distributed into
a cooling layer over a plant, as in *Copiapoa
haseltoniana* (Mooney et al., 1977). In the form of
rods or platelets, the wax can create a dead air

space around stomata, and any water molecules that bounce out of the stoma, because
they are not immediately blown away, may
bounce back in by chance (Figs. 10.19, 10.23).
The irregular arrangement of rods and platelets
also makes them effective sunscreens. Most
light striking their surfaces will be reflected
away from the tissues below; any that strikes
the wax with the proper angle will penetrate
that bit of wax but will probably be reflected
by another particle beneath. The wax layer
of *Echeveria bracteosa* reflects 25% of the inci-
dent light. This reduces the damage to pigments and reduces heating (McClendon, 1984;
Mooney et al., 1977). Waxes with this arrangement cause the grey color of plums, grapes,
and apples.

Waxes are also effective against insects; with the correct composition, the wax will be gummy and sticky and can jam an insect's mouth parts; simultaneously it can stick to the claws on the insect's feet such that they cannot grip the plant effectively. This is the mechanism used by carnivorous pitcher plants to catch flies. The rim and sides of the pitcher are covered with fragile, loose wax plates; once the insect's feet become covered with these, it looses its grip, slips, falls into the trap, and drowns (Lloyd, 1976).

Certain air pollutants and acid rain can interfere with the development of the cuticle and wax and cause significant damage to the epidermis and interior tissues (Percy, 1985; Percy and Riding, 1978; Riding and Percy, 1985; Swiecki et al., 1982).

Guard Cells and Stomata

The epidermis of shoot systems must not be adapted to be absolutely impermeable, even in the driest of environments, because the cutins and waxes that restrict the movement of water also block the entry of carbon dioxide into the photosynthetic tissues. It would have been ideal if a polymer had evolved that was impermeable to water but allowed the free movement of carbon dioxide and oxygen, but such a compound does not occur in either plants or animals (consider the tremendous amounts of water that our lungs lose every time we exhale). Instead, a mechanism has evolved that allows the epidermis to adjust its permeability as required. The epidermis contains holes (**stomatal pores**) whose size can be increased or decreased by the swelling of the adjacent **guard cells** (Figs. 10.8, 10.9, 10.14, 10.15, 10.20). The term **stoma** (pl.: stomata; physiologists usually use *stomate*) is sometimes used to mean the pore only, sometimes the pore and the guard cells. For clarity, I recommend the terminology of Esau (1965) in which *stoma* refers to both the pore and the guard cells, and *stomatal pore* indicates the intercellular space only.

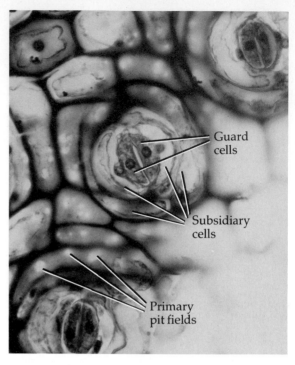

Figure 10.20
This is a stomatal complex of *Leuchtenbergia principis* in face view. The stomatal pore is visible, as are the two guard cells (and their nuclei). There are two sets of subsidiary cells parallel to the guard cells. × 400.

Stomata are found on virtually all green parts of a plant, especially the leaves and stems. On the leaves, they are typically more abundant on the abaxial surface, with the upper surface having fewer or even none. The adaxial surfaces of leaves typically have about 100 stomata per mm^2, and in the leaves of many deciduous trees the density can be ten times as high (Ting, 1982). Densities as great as 2230 per mm^2 occur in *Miconia pycnoneura* (Howard, 1969), whereas in *Macropanax*, *Schefflera*, and *Ozoroa* the entire epidermis consists of virtually only guard cells and immediately neighboring cells (Solereder, 1908; Wilkinson, 1979). In the floating leaves of water lily (*Nymphaea*), abaxial stomata would be submerged and useless; only

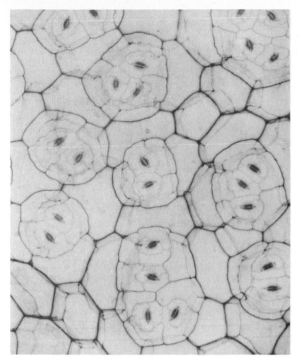

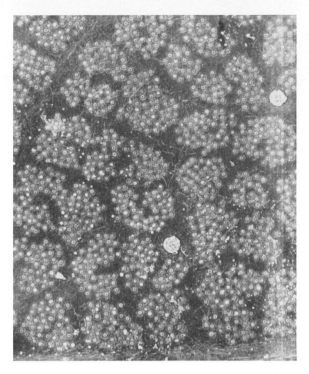

Figure 10.21
The stomata of this *Begonia semperflorens* occur in clusters of two to four each. Such clustering must affect the diffusion of both carbon dioxide and water. × 200.

Figure 10.22
The clusters of stomata in this *Nothofagus nitida* are much larger than in the *Begonia* of Figure 10.21, and the intervening areas that lack stomata are also larger. Notice that the density of stomata within each cluster is very high. This photo also shows a considerable amount of epifoliar fungi. × 80. Micrograph generously provided by G. Montenegro.

the adaxial surface has stomata. Similarly, the aquatic angiosperms, in which the entire plant is submerged and incapable of transpiration, have been reported by Solereder (1908) to lack stomata totally: Ceratophyllaceae, Nymphaeaceae, Podostemonaceae, and some Ranunculaceae. In *Callitriche heterophylla*, some leaves are submerged and have few stomata; other leaves are emergent and have numerous stomata (Deschamp and Cooke, 1985). Very low densities occur in certain cloud forest plants: 22 stomata per mm^2 in *Peperomia emarginella* (Howard, 1969).

Within a leaf, the stomata can be clustered and restricted to certain regions, as in *Begonia semperflorens* (Fig. 10.21), *Saxifraga sarmentosa*, and many others (Fig. 10.22; van Cotthem, 1971; Wilkinson, 1979). The clustering of stomata is an anatomical feature that deserves more study. If stomata are close to each other, the water that they lose coalesces into a small, humid zone that reduces further transpiration (Fig. 10.23E, F; Ting, 1982). Stomata do not often occur over solid masses of cells such as the fiber bundles of veins or the sclerenchyma of the leaf margins. Such tissues do not have intercellular spaces, and stomata located above them would be useless for absorbing carbon dioxide. Wilkinson (1979) described several examples of leaves with unusual distributions of

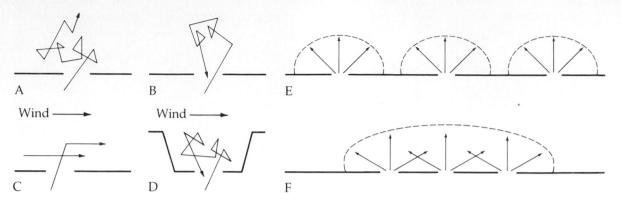

Figure 10.23
This diagram attempts to depict the movement of water molecules as they exit
a stoma. (A) With calm air, the water molecule bounces against air molecules, repeatedly changing direction, but gradually leaving. (B) With calm
air, there is a small possibility that the water molecule will bounce back into
the stoma. (C) With a wind, the water molecule is swept away too quickly to
reenter the stoma. (D) If the stoma is sunken (as in Figures 10.15, 10.17, or
10.19), then the wind cannot carry water away so quickly. (E) If the stomata
are widely separated from each other, then each builds up a humid zone
around it, and the surface of this zone loses water to the drier air. (F) If the
stomata are clustered, the humid zones overlap, and a water molecule might
bounce back into any of the stomata, not just the one from which it came.

stomata. In certain species of *Saxifraga*, the stomata only occur near the leaf tip, while in other
species they are all near the leaf margin. In
Daphne petraea, (Hill, 1931) they occur only in
two bands immediately adjacent to the midrib,
but not over the margin. Such distributions
must affect the absorption and internal diffusion of carbon dioxide; it may be that such patterns, rather than being adaptive, are simply
not so disadvantageous as to be a liability to
the plant. On the other hand, a little more information about the biology of these species
might allow us to postulate a useful role for
such a distribution. In *Mimosa cruenta*, the stomata are also distributed in an unusual pattern. They are located throughout the upper
surface, but on the lower surface they are restricted to just one of the two longitudinal
halves of the leaflet. This must cause poorer
absorption of carbon dioxide on the other half

of the leaflet, but the key to understanding the
pattern may be in the fact that this is a species
that has "sleep movements": at night the leaflets fold against each other so that the portion
with stomata is covered by the next leaflet.
When fully folded, no stomata are exposed; all
are fully protected (Wilkinson, 1971). It seems
reasonable that this stomatal pattern may be an
effective means of minimizing water loss at
night and thus offsets the presumed disadvantage of poorer distribution of carbon dioxide in
the leaflet.

Some nongreen portions of the plant also
have stomata, most notably many petals
(Maercker, 1965), stamens (Kenda, 1952), fruits
and seeds (Jernstedt and Clark, 1979; Rugenstein and Lersten, 1981); but often these are
nonfunctional. Roots never have stomata, with
one known exception. Lefebvre (1985) discovered them in roots of pea; he did not know if

they were capable of opening, but he suggested that they might allow oxygen into the root tissues. Because most roots do not have stomata, it appears that simple diffusion through the epidermal cells is sufficient. Many parasites, such as mistletoes (Loranthaceae) are hemiparasites: they draw water and minerals from the host, but they still are green and carry out photosynthesis. These plants have stomata. The holoparasites (Balanophoraceae, Monotropaceae, Orobanchaceae, Rafflesiaceae) obtain all of their resources from their hosts; they lack chlorophyll and are unable to photosynthesize. These are reported not to have stomata (Wilkinson, 1979).

If the ordinary cells of the epidermis show a strikingly regular orientation, then the stomata will usually be aligned with that pattern (Figs. 10.8, 10.24). Otherwise it is much more typical to have the stomata showing no obvious regularity in their orientation.

Two basic types of guard cells occur: dumbbell-shaped ones in the grass family and Cyperaceae (sedges), and crescent-shaped ones in the rest. In the grasses and sedges, the dumbbell shape results from the fact that the elongated guard cells are thin-walled on the ends but have thick walls along the middle (Fig. 10.25A). As the cells absorb water, the ends swell, but the middle remains narrow; thus the midregions of the two guard cells are pushed apart by the enlarged ends, opening the stomatal pore. When the guard cells lose water, the ends shrink, midregions move together, and the pore is closed. The dumbbell shape is present only when the guard cells are turgid. In grasses, it has been found that, when the guard mother cell undergoes cytokinesis to form the two guard cells, the division is incomplete, and the two protoplasts are continuous at their ends (Brown and Johnson, 1962; Pickett-Heaps, 1967; Ziegler et al., 1974).

In the other type of guard cell, the walls are also asymmetrically thickened, but in a different manner (Fig. 10.25B, 10.26, 10.27). The wall adjacent to the pore (the **ventral wall**) is thicker

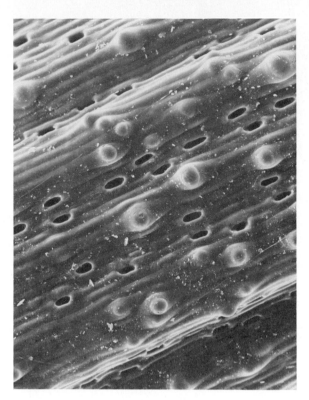

Figure 10.24
This is *Retanilla ephedra* in the Rhamnaceae; it bears only microscopic leaves, and the stems (shown here) carry out all photosynthesis. The holes are actually flanked by subsidiary cells, with the guard cells recessed below them. × 100. Micrograph generously provided by G. Montenegro.

than the opposite wall (the **dorsal** or **back wall**). The classical hypothesis for the opening of these stomata was based on the idea that the dorsal wall, being extremely thin, was more extensible and thus would stretch, while the ventral wall would not. This would cause the cell to arch into a crescent shape, opening the pore. Recently, an alternative hypothesis has been proposed that emphasizes the additional importance of the radial orientation of the microfibrils in the thickened regions: as the dorsal wall swells outward, the microfibrils allow it to pull the ventral wall with it (Aylor

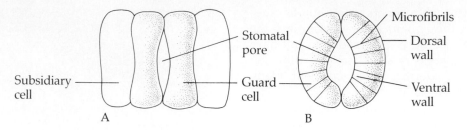

Figure 10.25
These drawings represent the two types of guard cells. In (A) are the
dumbbell-shaped guard cells of grasses and sedges. As the ends swell, the
central regions are pulled apart, opening the stomatal pore. There is no
similar opening on the other side, because the walls of the adjacent subsidiary cells are thin, flexible, and firmly attached to the guard cell. In (B) the
crescent-shaped guard cells, which are by far the most common, open because the microfibrils are radially arranged and the walls are differentially
thickened. The lower rigidity and higher flexibility of the back walls allow
them to swell outward, and the radial microfibrils pull the inner walls along.

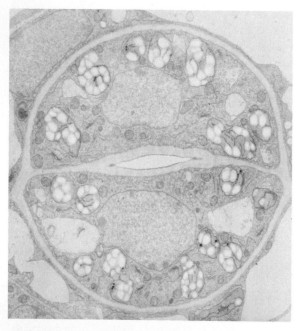

Figure 10.26
This electron micrograph of guard cells of *Vigna
sinensis* (cowpea) shows the central nuclei and
other organelles. It hints at the asymmetric thickening of the wall, but that is shown better in Figure
10.27. × 900. Micrograph generously provided by
B. Galatis and K. Mitrakos.

et al., 1973; Palevitz and Hepler, 1976; Raschke,
1979). If the microfibrils were not so arranged,
the guard cells would swell into a semicircular
shape, and the pore would remain closed.

In both types of guard cells, the critical
changes in shape result from the changes in
pressure and volume that occur as water is absorbed from or released to the surrounding
cells (Edwards and Meidner, 1979). The functional unit is, therefore, not just the guard cells
but the **stomatal complex**: the guard cells together with the adjacent epidermal cells. If
these adjacent cells are distinct in size, shape,
or cell contents, then they are termed the **subsidiary cells** (Figs. 10.6, 10.9, 10.19). The
movement of water between the guard cells
and adjacent cells is controlled by changes in
the water potential of the cells as potassium
ions (and possibly also hydrogen ions and sugars) are transported between these two cell
types (Salisbury and Ross, 1985; Ting, 1982). It
is important that K^+ not be able to diffuse
freely between guard cells and adjacent cells.
This is possibly the reason that plasmodesmata
are broken during the differentiation of guard
cells (Carr, 1976; see "Ontogenetic Development" later in this chapter). Electron micro-

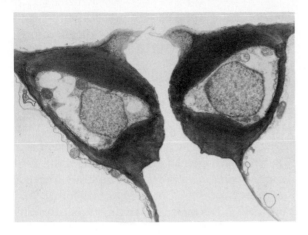

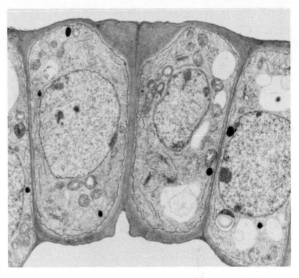

Figure 10.27
This cross section of the guard cells of *Vigna sinensis* shows the wall asymmetry; notice how thin the back wall is. Although a small region of the ventral wall is also thin, the inner regions of the periclinal walls are very thick. These guard cells also have prominent outer ledges. × 700. Micrograph generously provided by B. Galatis and K. Mitrakos.

Figure 10.28
These are young, immature guard cells of *Vigna sinensis* with the middle lamella just breaking down to form the stomatal pore. × 500. Micrograph generously provided by B. Galatis and K. Mitrakos.

scopy has shown the absence of complete plasmodesmata in the walls between mature guard cells and adjacent cells (Figs. 10.26, 10.28), and aborted ones have been seen in *Phaseolus vulgaris* (Peterson and Hambleton, 1978; Willmer and Sexton, 1979). Complete ones are present during the early stages of epidermal differentiation, but they are ruptured before maturation is complete.

In addition to the special radial arrangements of microfibrils responsible for the opening and closing of stomata, the walls of the guard cells frequently have the same modifications as ordinary epidermal cells. They always have a cuticular layer and a cuticle proper at least on the outer periclinal wall and often also on the inner one (Wilkinson, 1979). If waxes are present on the epidermis, they will usually be present on or near the guard cells and adjacent cells. This aspect is variable; in some cases the guard cells have either more wax or a different type of wax than that of the ordinary

epidermal cells, but in other cases the guard cells have little or no wax. The wax of subsidiary cells may be copious and project over the stomatal pore or even form a cup around it (Fig. 10.19).

Such accumulations of wax near the stoma are just one means of reducing the water loss that inevitably occurs when the stoma is open. Another method is to have the guard cells recessed into depressions in the epidermis termed **stomatal cavities**, as in *Franklandia fucifolia*, *Bredemeyera colletioides* (Napp-Zinn, 1974; Wilkinson, 1979), and *Trichocereus chilensis* (Fig. 10.17; Mauseth et al., 1984, 1985). A third method is to have the adjacent cells arch up and over the stoma (Fig. 10.15, 10.29; *Corpuscularia lehmanni*, *Odontophorus marlathii*). In some cases the guard cells are so deeply recessed that they do not appear at first to be part of the epidermis, although they invariably are (Fig. 10.15). The result of the recession of the stoma into a stomatal cavity is the lengthening

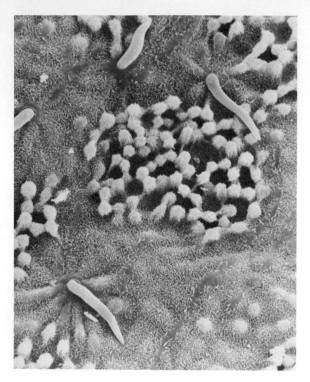

Figure 10.29
The epidermis of *Crinodendron patagense* is extremely complex. The stomata are clustered and surrounded by elevated subsidiary cells; the epidermal cells have abundant wax except at the base of the trichomes. The cells at the trichome base are arranged radially. Notice also that trichomes do not occur within the stomatal clusters. × 200. Micrograph generously provided by G. Montenegro.

of the diffusion pathway (Fig. 10.23). If the length is doubled, the resistance to diffusion is also doubled (Ting, 1982), and water will be conserved. An important modification of the guard cells themselves is the formation of one or two ridges of cuticle-covered wall material on the wall facing the stoma (Fig. 10.27). If these **ledges** protrude toward the pore, then, as a stoma closes, the ledges of one guard cell contact the corresponding ledges of the other guard cell, forming an especially tight seal.

Ledges are not universally present, and when they do occur, they are usually on the outer, exterior side of the guard cell. Interior ledges also exist. In some species, the ledges are either too short to meet, or they project outward, parallel to each other; these may be important in lengthening the diffusion path of the stoma, rather than sealing it.

Recently, a great deal of attention has been given to the shape and arrangement of the subsidiary cells of stomatal complexes (Fryns-Claessens and van Cotthem, 1973; Pant, 1965; Patel, 1978; Stevens and Martin, 1978a; Tomlinson, 1974; Wilkinson, 1979). This has come about primarily due to the rapid development of the very important field of taxonomic anatomy (Carlquist, 1961). Plant taxonomy is utilizing more and more characters for the analysis of the evolution and natural groupings of species. Anatomy offers a much larger number of characters than does flower morphology, which has always been and continues to be the primary basis of classification systems. Furthermore, in many cases complete plants are not available for study, not only in the case of fragmentary fossils but also when the plant has extremely large organs—few herbaria contain complete leaves of palms or philodendrons. Fortunately, epidermal anatomy has numerous characters that are often good indicators of the family and genus of the plant, and sometimes even of the species (Metcalfe and Chalk, 1979; Tomlinson, 1961, 1974). With increasing attention to leaves and epidermises, greater numbers of arrangements of guard cells and subsidiary cells have been recognized (Payne, 1979; Rasmussen, 1981; Stebbins and Khush, 1961; van Cotthem, 1970). Some of the current systems are so comprehensive that a rather large number of types and names have been recognized. Many of these are so rare that they are useful primarily only for specialized studies of the families and genera in which they occur. The presently available pioneering studies probably will be followed soon by ontogenetic-phylogenetic studies that will re-

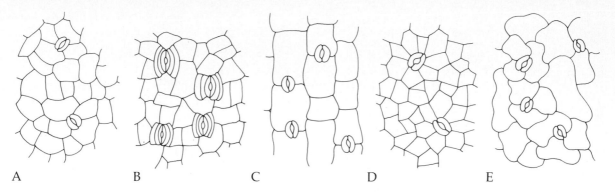

A   B   C   D   E

Figure 10.30
Types of stomatal complexes. (A) anomocytic; (B) paracytic; (C) diacytic;
(D) actinocytic; (E) anisocytic.

duce this current abundance of types and names to a more manageable system. Here only the five most common types are described:

1. The **anomocytic type** includes those epidermises in which there are no obvious subsidiary cells. The guard cells appear to be embedded in ordinary epidermal cells (Fig. 10.30A). It is important to point out here that stomatal complexes are based on light microscopy only; it may be that in these plants the surrounding cells are functionally subsidiary cells, lacking plasmodesmatal connections with the guard cells. Anomocytic stomatal complexes can be found in Capparidaceae, Cucurbitaceae, Geraniaceae, Malvaceae, Papaveraceae, Ranunculaceae, Scrophulariaceae, and Tamaricaceae. It must be mentioned that any given family may contain several stomatal types (Metcalfe and Chalk, 1950) and that even one leaf may have more than one type (Pant and Banerji, 1965; Pant and Kidwai, 1967). The anomocytic type was formerly known as the **ranunculaceous type** in a system established by Vesque (1889).

2. The **paracytic type** is easily recognizable: each guard cell is accompanied by one or more subsidiary cells that are aligned parallel with it (Fig. 10.30B). These occur in the Convolvulaceae, Leguminosae, Magnoliaceae, and Rubia-

ceae, among others. This was formerly known as the **rubiaceous type**.

3. In the **diacytic type** (formerly the **caryophyllaceous type**), there are two large subsidiary cells that completely surround the guard cells and are aligned perpendicular to them (Fig. 10.30C). These are common in the Acanthaceae, Caryophyllaceae, and others.

4. In the **actinocytic type** the guard cells are surrounded by many subsidiary cells aligned radially around them (Fig. 10.30D). This type did not occur in the classification of Vesque.

5. **Anisocytic type** (formerly **cruciferous type**) stomatal complexes can be somewhat difficult to recognize (Fig. 10.30E). They consist of three unequally sized subsidiary cells, which may not be very distinct from ordinary epidermal cells. This type can be found in the Cruciferae, Solanaceae, and others.

In conjunction with these five types and the others that have been proposed, it is important to know something of the development of the stomatal complex. It is possible for two mature stomatal complexes to have similar structure but have differing developmental patterns (Patel, 1978; Payne, 1970; Tomlinson, 1970). In **mesogenous stomatal complexes** the subsidiary cells and the guard cells are all derived from the same mother cell, while **perigenous**

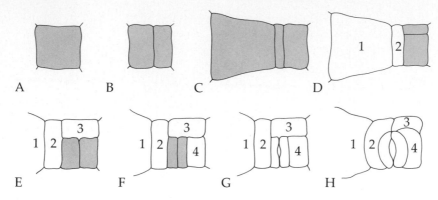

Figure 10.31
This is an example of a mesoperigenous stomatal complex. Subsidiary cell 4 is a sister cell to the guard mother cell, but the other subsidiary cells are less closely related; 1 is really no more closely related to the guard cells than are the ordinary epidermal cells that are not shown. Adapted from Patel (1978).

stomatal complexes are those in which the subsidiary cells are not immediately related to the guard cells. In the **mesoperigenous type** (Fig. 10.31), some of the subsidiary cells are related to the guard cells, and others are not (Pant, 1965).

These classifications emphasize the shape and arrangements of subsidiary cells; however, subsidiary cells can often also be recognized by their contents (Wilkinson, 1979). In *Clusia* they contain a yellow-brown pigment (Howard, 1969); in *Tabebuia* they are translucent, and in *Microsemma salicifolia* (Metcalfe and Chalk, 1950) and some Cyclanthaceae (Wilder, 1985) they contain a granular material. The subsidiary cells can be obvious because they are more papillose, as in *Melanochyla auriculata* (Wilkinson, 1979), or less papillose, as in *Cratoxylum* (Baas, 1970), than ordinary epidermal cells.

As the classifications have become more complex, certain problems have become apparent. First, many of the studies in which stomatal complexes are to be used as taxonomic or evo-lutionary evidence must use mature, dead material. It is not possible to study the developmental patterns, and as mentioned earlier, two different types of cell division patterns can result in the same type of stomatal complex. A related problem is that a rather minor change in the pattern of cell division or expansion can result in stomatal complexes that look dramatically different but which are, in fact, closely related (Tomlinson, 1974). There is no widespread agreement as to how, exactly, subsidiary cells should be defined. Some investigators feel that they must be related developmentally to the guard cells; others state that that is not necessary. Furthermore, although subsidiary cells are easy to recognize in simple types such as paracytic or diacytic complexes, in more complicated types the outer cells intergrade with ordinary epidermal cells and investigators can often disagree as to how many cells should be considered subsidiary cells. Eggli (1983) has carefully shown that very large patterns can also exist around stomata, some encompassing as many as 87 cells.

Another significant problem is that all classification schemes are based on incomplete and insufficient information. Tomlinson (1974) described how a single study of four species was used as a generalization for all orchids (20,000 species) and how this was then incorporated into numerous methods of stomatal taxonomy. The palms were similarly classified, although no actual study of their stomata could be found. Although the types of stomatal complexes may be constant in certain taxa, in many they are variable in either developmental or mature form. The most significant problem with the classification of stomatal complexes is that there is no theory of the role, function, or adaptive advantage of the various types. As far as we know, the guard cells and immediately adjacent cells work together in exchanging potassium and water, so that the stomatal pore can be opened and closed. There seems to be no evidence that any of the various arrangements affect this either positively or adversely, and the abundance of families with anomocytic stomatal complexes demonstrates that visibly distinct subsidiary cells are not necessary for stomatal functioning. Certain characteristics of subsidiary cells, such as the presence of waxes, can affect pore diameter and the length of the diffusion pathway, but this is related to the thickness of the wax, not to the arrangement of the subsidiary cells. I must be careful to emphasize that just because we do not know of a function for the various types of stomatal complexes, it does not mean that no function exists. However, it seems that it would be more productive to study the biology of the types rather than just their classifications. The hypodermal cells should also be studied, because in many species they have a more extensive contact with the guard cells than do the subsidiary cells (Fig. 10.15).

Trichomes

The term *trichome* (or hair) is applied to an artificial grouping of all cells that project

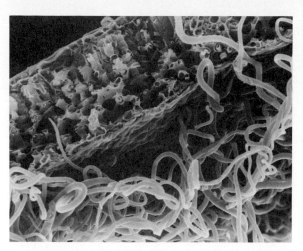

Figure 10.32
The surface of *Proustia pyrifolia* is densely covered by a mass of long, tangled trichomes. The layer of hairs is thicker than all the rest of the leaf. × 100. Micrograph generously provided by G. Montenegro.

markedly out of the plane of the epidermis (Figs. 10.14, 10.29, 10.32; Theobold et al., 1979; Uphof, 1962). By artificial, I mean that trichomes have evolved many times, so that the structures described by this term are not all homologous with each other. As was discussed in regard to the secretory tissues, from a purely scientific standpoint it would be best to be able to identify each independent evolutionary line of homologous structures and give them each their own name. At present our knowledge is insufficient to do this, and actually it might result in such a multitude of terms as to be more of a hinderance than an aid. For us, the optimal solution is to continue using the classifications based on structure or function, recognizing that they are artificial, and not to search for generalizations or dogmas that have no reason to exist.

In the past there have been attempts to differentiate rigidly between concepts such as trichome, spine, thorn, prickle, hair, wart, papilla, and emergence (Ramayya, 1964; Theobald et al., 1979). The hopelessness of such efforts is now accepted, and all of these are "officially" recognized under the name of trichome as long as they are primarily epidermal in origin. (Do not expect a definition of "primarily.") Structures that are modified leaves, stipules, or branches are not considered trichomes. A case can be made for designating the term *emergence* for any trichomelike structure that is not a modified leaf, stipule, or branch but that does contain some nonepidermal tissues, but so far there is no complete agreement on this.

Before we describe the trichome types, several comments are in order. A single plant may contain thousands or millions of individual trichomes; collectively these constitute the **indumentum** of the plant. The extensive terminology of trichomes is described by Payne (1978). Even if the trichome is simple in structure, ranges of variation in size, shape, contents, and structure must be expected. This is especially true if the trichomes occur on many different surfaces such as the upper and lower epidermises of leaves and on petioles, nodes, and internodes. All of these can dramatically affect the nature of a trichome, and distinctly different types may be found on different organs. Carlquist (1961) has referred to the suite of trichome types of a species as its **trichome complement**; this is a valuable concept when the trichomes are being examined as part of a taxonomic or ecological study.

Trichomes may also change with age: they frequently have deciduous parts, being complex (even glandular) while young (Rodriguez et al., 1984) but simple when older, because many parts have abscised (Fig. 10.33; Hammond and Mahlberg, 1977). It is not uncommon, especially in palms (*Lantania verschaffeltii*, for example), for the entire trichome to break off, so that a leaf will appear to have no trichomes when mature even though it was covered while still developing. Probably no plant completely lacks trichomes. Many, possibly most, plants have more than one type, and these may be extremely different from each other. Numerous classifications have been proposed (Dickison, 1974; Foster, 1950; Metcalfe and Chalk, 1950; Solereder, 1908; Theobald et al., 1979; Uphof, 1962).

The functions of trichomes are wonderfully diverse (Johnson, 1975; Levin, 1973). The glandular trichomes can secrete water, salt, nectar, mucilage, terpenes, adhesives, digestive enzymes and irritants that sting (Chapter 9). Other trichomes absorb water and salts (Benzing and Pridgeon, 1983; Pridgeon, 1981). The nonglandular trichomes can, along with the cuticle and waxes, protect against excessive sunlight. As they die and dehydrate, their walls become more refractile and scatter light. Such a structure is also a deterrent against insects, because the hairs can tangle the feet or impale the insect (Fig. 10.32); being dead and empty, the trichomes are of little nutritive value to the insect. Plants sometimes modify this and have short, curved hairs that all point in the same direction. Such an arrangement makes it easy for an insect to walk only in that direction. This easy little walk usually brings the insect to a digestive gland (Hooker, 1875; Lloyd, 1976). The list of functions and modifications of trichomes is almost endless, and these are some of the most beautiful and fascinating structures in plant anatomy. A brief introduction to them follows.

NONGLANDULAR TRICHOMES.   **Unicellular trichomes** are extremely common. They consist of a single cell that projects above the surrounding surface (Fig. 10.1). If only a small part of the outer wall projects, these may be called warts if short or papillae (sing.: papilla) if long (Fig. 10.34). These are extremely common and especially useful to us: cotton fibers are the long (up to 6 cm) unicellular trichomes of the seed coat (Fig. 10.35; Stewart, 1975). Extensive lists of families that have each type of trichome are given by Metcalfe and Chalk (1979); diagrams and examples are given by Theobald et al. (1979) in the same book.

Multicellular trichomes may be either uni-

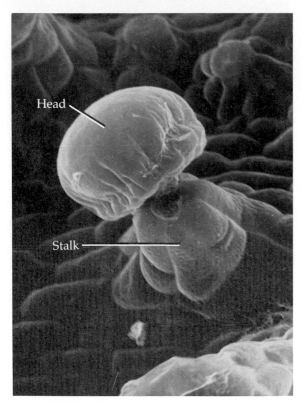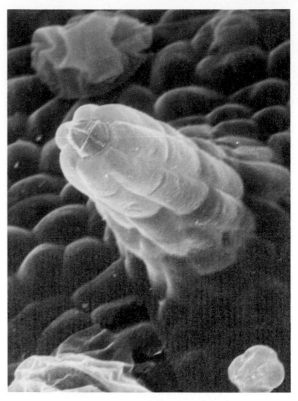

Figure 10.33
On the left is a glandular trichome of marijuana (*Cannabis sativa*); the secretory head is in the process of abscising. On the right is the stalk that remains. If only mature leaves were examined, it would be easy to conclude erroneously that the hairs were nonglandular. × 200. Micrograph generously provided by C. Hammond.

seriate, if they consist of one row of cells, **biseriate**, or **multiseriate**. These may be long or short, with few or many cells, and they may be branched or unbranched. If branched (Fig. 10.36), the arrangement of the branch cells can be important taxonomically. **Stellate hairs** (*Lesquerella engelmanni, Shepherdia canadensis, Olearia astrotricha*) and **candelabriform hairs** (*Alternanthera stellata, Lavandula vera*) occur as do some that look like complete trees. If all the branch cells interconnect along their sides, they form a disk or shield-shaped structure. If

these have a stalk, they are called **peltate hairs** (*Lesquerella schaffneri, Olearia albida*), but if the stalk is absent or extremely short they are **scales** or **squamiform hairs** (Fig. 10.37; *Croton argyranthemus, Olea*, Bromeliaceae). In the Bromeliaceae, the peltate hairs are especially complex and are critically important for taxonomy; plants can be identified to subfamily, occasionally even to genus, on the basis of a single trichome (Benzing, 1980). Peltate trichomes can sometimes be difficult to recognize if the ordinary epidermal cells beneath them are ex-

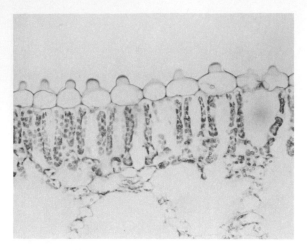

Figure 10.34
The epidermal cells of this *Colocasia* (elephant's ear) leaf have short projections: are these ordinary epidermal cells with papillae or are they short trichomes? × 150.

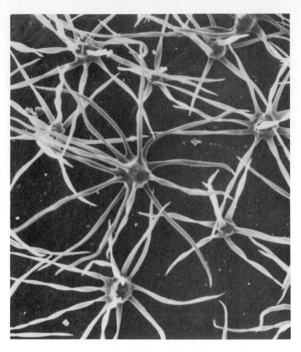

Figure 10.36
These are multicellular branched hairs in *Aetoxicon puncteatum.* × 100. Micrograph generously provided by G. Montenegro.

Figure 10.35
This scanning electron micrograph of cotton seeds (*Gossypium hirsutum*) shows long, unicellular trichomes. These are harvested and are the "fibers" of cotton cloth. × 100. Micrograph generously provided by J. M. Stewart.

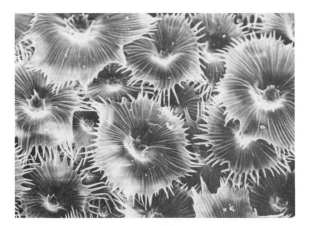

Figure 10.37
These squamiform hairs form a dense covering over the ordinary epidermal cells. There is a short stalk, then all of the upper cells radiate outward, being laterally attached except at their tips. × 100. Micrograph generously provided by G. Montenegro.

tremely thin and the stalk is short; under these conditions the trichome appears to be just ordinary epidermal cells.

GLANDULAR TRICHOMES.   Glandular trichomes can have virtually all the forms just described for nonglandular trichomes, but of course there may be special modifications due to their secretory nature. It is not uncommon for all of the cells of a glandular trichome to be secretory, but in the more elaborate ones there is a marked differentiation of regions within each trichome. There may be a glandular **head** elevated by the **stalk** or **neck** and attached to the epidermis by a **foot cell** or **basal cell** (Figs. 10.33, 10.38; Chapter 9). Rarely, the surrounding epidermal cells are modified into collecting cells. As was mentioned in Chapter 9, the secretory region may be isolated by Casparian strips so that the secretory product cannot flow back into the plant apoplastically. Glandular trichomes, like all shoot epidermal cells, are covered with cuticle (Cutler et al., 1982). In many cases, the secretory product accumulates below the cuticle, which is lifted away from the cellulosic portion of the wall. In other cases, the cuticle tears, and the secretion is liberated immediately. If the secretory trichome is very large, the stalk may actually contain vascular tissue with prominent tracheary elements (*Drosera, Prunus amygdalus*). In several studies, the transverse walls of the stalk cells have been found to contain a very high number of plasmodesmata, which presumably facilitate the flow of material to the head (Robards, 1976). Similarly, the foot cell and the collecting cells may be transfer cells with labyrinthine walls adjacent to ordinary epidermal cells or mesophyll cells.

Root Hairs

Root hairs are often considered just another type of trichome, but they are given special mention here because they are almost universal in occurrence and are critical for the ab-

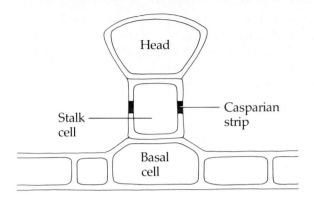

Figure 10.38
This diagram shows the three parts that are frequently present in a glandular trichome: basal cell, stalk cell, and head (or secretory cell). All are shown as being single cells, but within a trichome there may be several of each.

sorption of water and nutrients (Russell, 1977). Root hairs are almost always unicellular (Fig. 13.6). One example of multicellular root hairs is known: *Kalanchoe fedtschenkoi* (Popham and Henry, 1955).They can be very long—from 80 to 1500 μm, but a length of about 200 to 300 μm is more typical (Dittmer, 1949). Their diameter varies from 5 to 17 μm. In some plants, every root epidermal cell grows out as a root hair, but in others the cells undergo an unequal division, and the short cell (the **trichoblast** or **piliferous cell**) grows out as the root hair (Dosier and Riopel, 1978). Although a few species have crooked, angular root hairs, such as *Amaranthus torreyi* and *Agropyron elongatum* (Dittmer, 1949), most are straight. The elongation of so many cells greatly increases the surface area available for absorption. In a single rye plant, there are about 14 billion root hairs, with a surface area of almost 400 m^2 (Ting, 1982). Like almost all other types of trichome, root hairs live only briefly, dying within several days of their formation. They are frequently sloughed off the root, but, if they remain, their walls are thickened and may become lignified and suberized (Artschwager, 1925; Cormack, 1949). The functioning of the root hair is affected by the soil and by the interior tissues of the root. For this reason, the full discussion of root hairs is given in Chapter 13 rather than here.

Unusual Epidermal Cells

Although there are only four basic types of epidermal cells, as just described, certain genera or families do have other types of epidermal cells that are often very unusual and characteristic. These occur in so few groups that they are valuable taxonomic characters.

LITHOCYSTS. **Lithocysts** (stone cells) are large epidermal cells that protrude into the underlying tissues and that contain a large crystal of calcium carbonate (Fig. 10.39; Metcalfe, 1979b; Pireyre, 1961; Pobeguin, 1951, 1954). A papilla of wall material grows into the cell lumen and then becomes a nucleation site for the calcium carbonate. Pectins and silica are also deposited (Davis, 1984; Hiltz, 1950); otherwise the calcium carbonate would not be stable. The resulting complex crystal is a **cystolith** (cell stone). These are characteristic of Acanthaceae, Cucurbitaceae, Moraceae, and Urticaceae, but they also occur in a few species of ten other families (Metcalfe, 1979b). Usually each cell contains a single cystolith and exists as an isolated idioblast, but groups of lithocysts are present in the families Opiliaceae and Boraginaceae. Both the crystal and the cell are so large that they protrude from the epidermis, but they project inward into the mesophyll, not outward like a trichome. The interior portion of the lithocyst becomes so swollen that it appears at first glance to be a mesophyll cell, but they are almost always epidermal. Rarely, pith parenchyma cells can develop as lithocysts (Metcalfe, 1979b).

SILICA CELLS AND CORK CELLS. When seen in face view, grass epidermises contain **long cells** and **short cells**, usually grouped together as

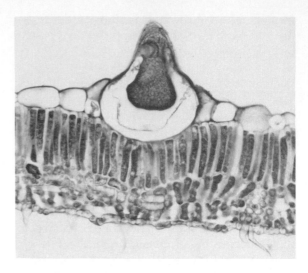

Figure 10.39
This lithocyst (the cell) is an epidermal cell; the cystolith (the crystal) is an encrustation on a short ingrowth of the cell wall, visible near the apex on the left. × 500.

pairs. The long cells are rather ordinary, but the short cells are modified into either **silica cells** or **cork cells**. The silica cells contain silica bodies with various shapes (round, elliptic, dumbbell, or saddle-shaped). These are extremely useful taxonomically and may also occur in other monocot families. The silica is not abundant until the cells begin to senesce (Lawton, 1980). The cork cells have walls encrusted with suberin, and their lumens are often filled with ergastic substances.

BULLIFORM CELLS. The epidermis of many monocots, especially the grasses and sedges, may contain a specialized type of cell called **bulliform cells** (Fig. 10.40). These are very large, thin-walled epidermal cells that are arranged in long bands parallel to the length of the leaf; rarely, they constitute the entire upper surface of the leaf (Esau, 1965). They seem to act as points of flexure: if they are turgid and

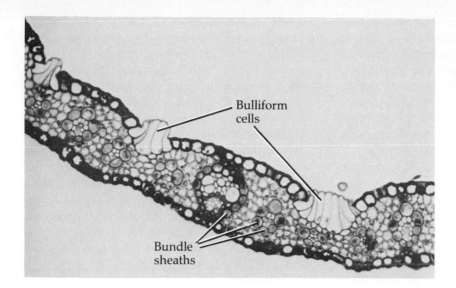

Figure 10.40
This portion of a leaf cross
section of *Tripsacum dac-
tyloides* shows three sets of
bulliform cells in the upper
epidermis. Although they
are epidermal cells, they are
obviously much larger and
less thick-walled than other
epidermal cells. This grass
also has very prominent
sheaths around the vascular
bundles. × 100. Material
prepared by W. V. Brown.

swollen, the leaf is open and flat, but if they
lose water and become flaccid the leaf folds,
minimizing its exposed surface area. Shields
(1951) reported that such movements may involve other cells as well as the bulliform cells. It
may be that the primary function of bulliform
cells is the opening of the leaf as it expands
from the bud; most young monocot leaves are
tightly rolled as they undergo the first stages of
growth, and they must later unroll to expose
the adaxial surface. Ellis (1976) discussed in detail the literature on bulliform cells, and he
suggested caution in assigning them a role in
leaf movement. He pointed out several instances in which they did not seem involved in
leaf flexure: in some species they fill with silica
and become rigid although the leaf remained
flexible, while in other cases the flexure is in
response to bright or dim light, and the bulliform cells are turgid under both conditions.

Multiple Epidermis and Hypodermis

The epidermis frequently is completely different in many properties from the tissues immediately interior to it, but in many instances, the
layer or layers adjacent to the epidermis are
functionally related to it. The most frequent
example is when a tough, resistant dermal system is needed and the surface is covered by
two to many layers of thick-walled cells. Re-
sponding to other selection pressures, some
species that have the opposite characteristics
have evolved. The leaf epidermis and subjacent
tissues of many *Piper* and *Peperomia* species
store water, and the dermal system, consisting
of large thin-walled cells, may constitute over
half of the volume of the leaf.

Such multistratose primary dermal systems
can come about in either of two ways. First, if
the cells that are derived from the outermost
layer of the tunica undergo periclinal divisions,
then a structure with multiple layers is produced, and all layers have the same developmental origin; this is known as a **multiple
epidermis** (Linsbauer, 1930). In a multiple epidermis, the outermost layer looks like a typical
uniseriate epidermis, containing guard cells,
trichomes, cuticle, and waxes. Alternatively, if
the cells of the outermost layer of the tunica do
not divide periclinally, but the cells derived
from inner layers of tunica or from the corpus
develop special characteristics distinct from

those of the inner cortex, then these layers constitute a **hypodermis** (Figs. 10.5, 10.12; De Bary, 1884). In some plants, both a multiple epidermis and a hypodermis occur together. If only mature tissues are examined, it can be extremely difficult or impossible to distinguish a multiple epidermis from an epidermis plus hypodermis; developmental stages are necessary for unambiguous determination.

Hypodermises are common, but multiple epidermises are rare, occurring in some species of Begoniaceae, Chenopodiaceae, Moraceae, Piperaceae and the roots of many epiphytic orchids and arums. These roots are often aerial and exposed, running along a branch or rock, or they may dangle in the air for many meters. The entire surface except for the extreme tip is covered with a multiple epidermis called the **velamen**. The cells are large, thick-walled, and dead at maturity; it has been suggested that the velamen is capable of absorbing and storing water and minerals and that the root cortex later gradually absorbs the water from the velamen (Benzing et al., 1982; Capesius and Barthlott, 1975).

Multiseriate epidermises often develop over only a portion of a surface, such as just above a vascular bundle. If a multiseriate epidermis occurs over a surface that must have stomata, these develop while the epidermis is still uniseriate (Fahn and Dembo, 1964).

A hypodermis often plays a mechanical role, being either collenchymatous or sclerenchymatous, but hypodermal cells of almost any nature can occur. In the stems of *Uebelmannia* (Fig. 10.12; Mauseth, 1984) and in the seeds of *Maurandya* (Elisens, 1985), the irregular warty or winged surfaces of the organs are the result of the elongation of hypodermal cells rather than the epidermal ones.

Root Epidermis

In many species, the root epidermis does not develop from one distinct layer in the manner of the shoot epidermis, which is always derived from the outermost layer of the tunica (Chapter 6). Furthermore, the root epidermis and the shoot epidermis each develop from their own distinct meristems, and these meristems have their own origins in different parts of the embryo. Because of this, under a strict ontogenetic interpretation, the root epidermis would have to be considered to be a totally distinct tissue, one that is only functionally similar to the shoot epidermis. Linsbauer (1930) and Guttenberg (1940) have considered the two to be separate, and they call the root epidermis a **rhizodermis**. Similarly, the root cortex is called an **epiblem**. However, at present the terms epiblem and rhizodermis are rarely used.

Structurally, the root epidermis is simple, containing just ordinary epidermal cells and root hairs; guard cells do not occur (except in pea as noted earlier). Also, apart from root hairs, none of the numerous types of trichomes occur, neither the glandular nor the nonglandular varieties. A cuticle is present, but it is very thin. Further details of the root epidermis are given in Chapter 13 in connection with the functioning of the whole root.

Development of the Epidermis

Ontogenetic Development

As described in Chapter 6, the epidermis of the shoot is derived from the outermost layer of the tunica. The cells are initially small and rather isodiametric, having all the cytological features of apical meristem cells (Mauseth, 1982a, b). As the cells flow out of the meristem, they divide less frequently and begin to enlarge, usually elongating most in the same direction as the organ is. The walls and cuticle remain thin until the organ and the epidermis are almost full sized. Even in those species which have very thick cuticles or wax layers, there are no especially obvious changes in the ER or dictyosomes at the time of secretion (Hallam, 1970).

Stomata begin to develop early, while the organ is still small. The guard mother cell becomes recognizable as the smaller of two cells produced by an asymmetric cell division (Bünning and Biegert, 1953; Bonnett, 1961; Pickett-Heaps and Northcote, 1966; Stebbins and Jain, 1960). The guard mother cell is connected by plasmodesmata to all surrounding cells (Carr, 1976; Galatis and Mitrakos, 1980; Kaufman et al., 1970; Landré, 1972; Ziegler et al., 1974). In some species, the cell wall appears to thicken slightly, and the plasmodesmata are broken. In other species, the plasmodesmata are lost after the guard mother cell divides to form the two guard cells. Ziegler et al. (1974) reported that, as the mother cell divides in *Zea mays*, the new wall does not form any plasmodesmata. In pea (Singh and Srivastava, 1973), however, this wall does contain plasmodesmata, but they are broken as the pore is formed. The two young guard cells are small and have no stomatal pore between them (Fig. 10.28). However, as they enlarge, the walls develop their asymmetric thickenings, and the middle lamella between them swells. Microtubules appear along the cell plate where the pore will form, but they are absent at the ends where the two developing guard cells will remain in contact (Galatis and Mitrakos, 1980; Kaufman et al, 1970; Peterson and Hambleton, 1978; Srivastava and Singh, 1972). The cell plate begins to thicken, especially where it contacts the outer periclinal wall; this establishes the outer ledges (Fig. 10.27). In paradermal sections, the cell plate becomes lens-shaped. Next, the middle lamella, which holds the two ventral walls together, becomes **electron-translucent** (it does not absorb stains for electron microscopy). This may be due to the digestion of the middle lamella (Peterson and Hambleton, 1978; Stevens and Martin, 1978b). The pore begins to open as the two ventral walls start to break apart, beginning where they meet the inner periclinal walls (Fig. 10.26; Galatis and Mitrakos, 1980; Ziegenspeck, 1944). The hypodermal cells immediately below the guard mother cell also separate from each other and from the developing guard cells (Johnson and Riding, 1981). If the guard cells are to be elevated or sunken, that happens next, as the surrounding cells expand differentially. As the underlying cells enlarge, the ordinary epidermal cells continue to divide, and new guard mother cells arise between existing stomata.

Uniseriate trichomes also often begin development as the smaller of two sister cells that are produced by an unequal division. This small trichoblast may enlarge, or it may push outward while still small. The root hairs are perhaps the most extensively studied type, and they will be described here. The root hair trichoblast begins to grow outward only after the underlying region of root is no longer being pushed forward through the soil. The nucleus and cytoplasm move into the protuberance as it elongates outward. If careful measurements are made, it can be shown that only the very tip of the root hair is growing, all other parts being mature (Sievers, 1963a, b). This is called **tip growth**, and it also occurs in other cylindrical cells, such as pollen tubes and fungal hyphae. The cytoplasm is aggregated into the tip, and there is a high concentration of dictyosomes. It is not unusual for root hair nuclei to be polyploid (Cutter and Feldman, 1970a, b). As you can imagine, the great diversity in the types of trichomes that occur on shoots is accompanied by an equal diversity of developmental pathways, and generalizations are difficult to find.

Phylogenetic Development

The epidermis and all of its associated structures are features only of thallophytes; they occur in none of the algae (Edwards et al., 1982). Indeed, it is precisely by the presence of cuticle and stomata that paleobotanists determine if a fossil was derived from an early land plant or not. The earliest known vascular plants (*Cooksonia* and *Rhynia*) both had a true epidermis with cuticle and stomata (Stewart, 1983).

Because these are both advanced enough to be vascular, and because vascular tissues would have developed only in land plants, not algae, it is reasonable to assume that the epidermis was already present and well developed in the ancestors to *Cooksonia* and *Rhynia*. The earliest stomata discovered (*Asteroxylon, Rhynia*) are anomocytic, and paracytic stomata first appear in Mesozoic fossils (Mersky, 1973). Comprehensive descriptions of the cuticles of the earliest land plants are given by Edwards et al. (1982) and by Thomas and Masarati (1982).

A detailed discussion of current thought on stomatal evolution is given by Wilkinson (1979). As mentioned earlier in this chapter, the stomata and stomatal complexes have been used for phylogenetic-taxonomic purposes in many groups. When these are based on small taxa or single families, they provide useful data, but as Tomlinson (1974) described, there are still too few studies to attempt to use them for a single scheme of angiosperm evolution. The same is true for trichomes (except that trichomes can be so much more complex than stomata that convergent evolution is probably less of a problem), and trichomes are one of the most important characters for taxonomy at the level of species and genus. The studies are too numerous to list; only a few are mentioned here: Carlquist (1958), Carolin (1971), Cowan (1950), Ellis (1976), Heitzelman and Howard (1948), Metcalfe (1960, 1961, 1971), Metcalfe and Chalk (1950, 1979), and Tomlinson (1969). Even more are listed by Theobold et al. (1979).

PART IV

The Primary Vegetative Body of the Plant

In the previous ten chapters, the various types of cells and tissues were examined and discussed. Merely knowing the individual cells is insufficient for understanding how the tissues function. Similarly, we must go beyond the individual tissues themselves and consider how they are arranged within a plant, how they are interassembled, and how they interact with each other. None of the tissues functions alone, and, perhaps just as importantly, each differentiates within the environment of the body created by the tissue itself along with the other tissues. The plant must be considered in its totality.

Plants go through several phases of growth. Immediately after germination, a seedling begins to develop into a small, herbaceous plant that consists of a stem, leaves, and roots. These organs are soft, not woody, and they constitute the primary vegetative body of the plant. Later the plant will form reproductive organs, and many species will form wood and bark; they may become massive. The wood and bark, being secondary tissues, constitute the secondary body of the plant. Secondary tissues are always formed slowly within primary tissues. A plant, no matter how old or large, will never be without primary tissues: the tips of twigs and roots are always new increments of primary body.

The primary body is usually small, but it is somewhat complicated, because the root-shoot axis produces and integrates leaves, branches, and lateral roots. The vascular connections between all of these must be efficient and orderly. It is not sufficient simply to have the proper cells and tissues; they must be arranged in the correct pattern.

11
Stem

Concepts

As always, before embarking on the study of a new subject, it is important to think about what one already knows of that subject. Shoots are composed of stems and leaves, the stems being the axis of the shoot. Stems, of course, are extremely familiar to all of us already because we see them every day in gardens and parks, along streets, on campus, in flower pots and cooking pots. Stems tend to be long and narrow with leaves attached, except that some (rosette shoots such as lettuce and cabbage) are more leafy than others (asparagus); some (vines) are longer than others (bulbs); and some (stolons and runners) are narrower than others (tubers). Actually, stems are extraordinarily variable.

The variations, however, are not difficult to understand; having long internodes with leaves spaced far apart is an efficient mechanism for a plant that is otherwise immobile to move to a different locality. If the old, original parts of the plant die and the newer parts form adventitious roots, then the plants actually do migrate through their habitat. For a plant that is in an optimal location, long internodes are not necessary, and short ones may be advantageous. If conditions are very good, a wide stem provides parenchyma for storage of carbohydrates and other nutrients.

These modifications and the ones to follow are, of course, all related to the functions of the stems. These functions include (1) the production and (2) elevation of leaves and reproductive tissues, as well as (3) the long distance transport of water and nutrients. These three functions are common to most stems, but they are not the only functions; most stems are also involved in (4) storage, either short-term—several weeks for annual plants—or long-term—many years in long-lived monocarpic perennial plants such as century plant (*Agave*) and fishtail palms (*Caryota*). The storage may be of either nutrients or water. Other stems are parts of shoots that serve primarily as (5) perennating organs: it is not uncommon for underground rhizomes, tubers, corms, and bulbs to survive while the above-ground parts of the plant die during the winter or summer stress. Examples include lilies, irises, and potatoes (Montenegro et al., 1983). Stems are sometimes designed to fall apart easily (**disarticulate**); then each part can be carried by water, as with *Eichhornia, Lemna*, and *Pistia* (van der Pijl, 1982), or by animals, as with *Opuntia* and *Tillandsia*, to another area where the stem puts out new roots, thereby being (6) a means of dispersal. A small but significant number of stems owe much of their form to the fact that they are modified (7) to house symbiotic organisms. These stems, such as *Cecropia, Polypodium carnosum*, and *Solanopteris* (Huxley, 1980), can become quite elaborately modified to accommodate the ants or other organisms that live inside them.

From everyday experience alone it quickly becomes obvious that stems can be a rather diverse group of entities, but we can extend this even further by considering some less common plants. Whereas most individual twigs produce only about five to ten new leaves per year, some (*Ariocarpus*) produce only one or two, and others (*Cleistocactus*) can initiate more than one thousand from each apical meristem every year (Rauh, 1979b). Most conifers, many monocots—especially the palms, yuccas, and the Australian grass trees (Xanthorrhoeaceae)—and a few dicots, such as *Magnolia*, retain their leaves for many years, so the stems are very leafy. Some dicots never have more than two mature leaves at any time; "living stones," *Argyroderma*, and *Conophytum* for example, which typically occur in drier environments (Rauh, 1979a).

The longevity of the apical meristem is similarly variable. In most plants, the apex becomes dormant during times of stress and is reactivated in the subsequent growing season; the shoot axis that is produced is a unitary, continuous product of one apical meristem. These shoots are said to be **monopodial** (Foster and Gifford, 1974; Zimmermann and Brown, 1974). Palms and conifers are examples of the many plants with monopodial growth. But in other plants (*Betula, Gleditsia, Salix, Ulmus*), each apex produces only a small number of leaves (perhaps only two or three); then it aborts (Fig. 6.31) or becomes either a floral apex or a dormant bud that looks much like an axillary bud. One of the uppermost true axillary buds becomes active and replaces it in forming the shoot system, only to cease also after a few leaves have been produced. The shoot axis here is **sympodial** and is really not one structure but a series of branches on branches. This will be obvious if it exhibits a zigzag nature, but frequently the axis appears perfectly straight, and only very careful analysis of the activity of the apex reveals that it is not monopodial (Foster and Gifford, 1974; Hallé et al., 1978; Troll, 1935, 1937; Zimmermann and Brown, 1974).

A diversity of stem forms can occur within a single plant, not solely between different species. The flowers of angiosperms and the cones of conifers are examples of highly modified stem systems, but right now let us consider only vegetative stems. In all plants, the very first leaves produced by a seedling are at least slightly different, in size, shape, or arrangement on the axis, from those of an older plant. The tissues of the stem may also differ; if the difference is really dramatic, the plant is said to be **heteroblastic** (Allsopp, 1965; Cutter, 1965; Dobbins et al., 1983; Kaplan, 1973). Usually the transition from the "juvenile" type of leaf and stem to the "adult" or "mature" type is gradual, but in some cases it is abrupt and dramatic (Fig. 6.23). Ivy and citrus are the most widely known examples (Stein and Fosket, 1969), but much more extreme transitions occur in some cacti that form a cephalium (*Melocactus, Backebergia*) as mentioned in Chapter 6 (Fig. 6.26). Most people, including most botanists, would assume that it is a separate species or genus artificially grafted onto the juvenile shoot (Rauh, 1979a). In the transition from juvenile to adult in these, virtually all aspects of the stems and leaves change dramatically (Gibson, 1978b; Niklas and Mauseth, 1981). This is reasonably easy to understand: the two types of shoots have totally different functions. The juvenile is responsible for photosynthesis and energy acquisition, and the adult is responsible for reproduction; its flowers and fruits must be more carefully protected.

Many stems have become remarkably adept at imitating leaves: they are planar and more or less determinate, and they frequently have other aspects of leaf appearance (Figures 11.1 and 11.2; *Epiphyllum, Muehlenbeckia, Opuntia, Phyllanthus, Ruscus aculeatus*; Bruck and Kaplan, 1980; Metcalfe, 1983a). These **cladodes** (equally frequently called **phylloclades**; also called **cladophylls**) may augment the photosynthesis of the leaves, or they may be the primary photosynthetic organs of plants whose true leaves are extremely reduced. Interest-

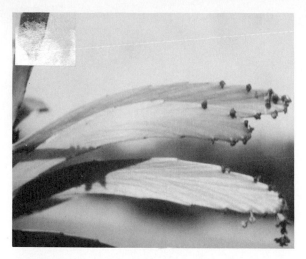

Figure 11.1
These appear to be the leaves of *Phyllanthus*, but actually they are the lateral shoots. Because they resemble leaves so closely, they are called cladodes. You can tell that these are not leaves because they have flowers on them; see Figure 11.2. × 0.3.

Figure 11.2
If you were able to tell that the structures in Figure 11.1 were not really leaves, try your skill on these structures of *Phyllonoma*. The presence of fruit shows that they are not leaves, either. × 0.1.

ingly, cladodes can mimic not only leaves but also their diversity by imitating the modified leaves known as bud scales; the "scales" of asparagus are really cladodes.

A plant may produce two different types of stems simultaneously; the branches of a plant almost always differ from the trunk at least in that they grow less vertically than the trunk does. In many plants the branches differ from the main stem in their anatomy and the types of leaves they bear; their internodes may be extremely short, and the plant is then said to consist of **long shoots** (the main shoot) and **short shoots** (the lateral shoots). Examples are pines, *Larix*, *Ginkgo* (Critchfield, 1970), cacti (Leinfellner, 1937; Mauseth, 1977), and many more (Hallé et al., 1978).

Finally, one of the most instructive modifications of stems is their more or less complete absence in certain dicots. Several parasitic flowering plants (*Tristerix aphyllus*, *Rafflesia arnoldii*, and *Pilostyles thurberi*) exist vegetatively solely

as an endophytic mass of filaments embedded within their hosts. These endophytes lack stems and leaves as well as roots (Mauseth et al., 1984, 1985; Rutherford, 1970). Only at the time of flowering is a short section of rather normal stem produced. This lack of a vegetative stem system is possible because the functions normally associated with a stem (production of leaves and flowers, support, and conduction) are not necessary for this growth phase of these remarkable plants. Similarly, *Campylocentrum*, *Harrisella*, and *Taeniophyllum* consist of primarily a photosynthetic root system that produces flower buds adventitiously or from a diminutive residual bit of shoot (Benzing et al., 1983; Dressler, 1981). A large vegetative shoot system never occurs in these species, either. See Benzing et al. (1983) for a

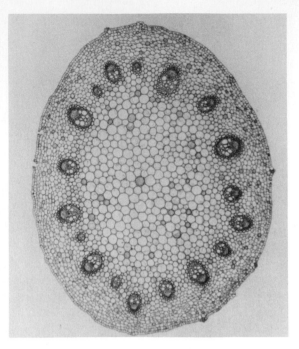

Figure 11.3
This cross section of a stem of *Ranunculus* (butter-cup) shows the arrangement of the tissues. From outside to inside: epidermis, cortex, vascular bundles, pith. The parenchyma between the vascular bundles is called a medullary ray. × 40.

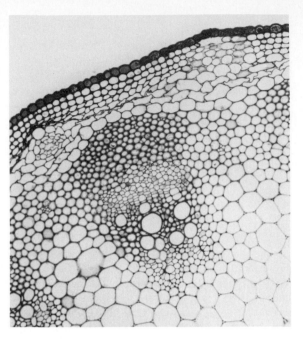

Figure 11.4
This section of *Helianthus annuus* (sunflower) shows the arrangement of tissues within a bundle. On the outer edge is a thick layer of phloem fibers, then the conducting tissues of the phloem are interior to the fibers. In the xylem several large vessels are obvious. × 160.

discussion of the advantages of shootlessness in certain environments.

Considering the tremendous number of possibilities for the functions and structures of stems, it is truly remarkable that there is only one single basic type in all of the vascular plants. In cross section, there is an outermost epidermis that overlies the cortex; the cortex in turn surrounds the vascular tissues, or **stele** (Figs. 1.3, 1.4, 11.3; van Tieghem and Douliot, 1886). Among all of the plants, the only variation allowed is in the disposition of the xylem and phloem within the stele. Considering only the seed plants, the stele always consists of **vascular bundles** (Fig. 11.4), each containing primary xylem and primary phloem; the bundles are arranged in a ring (gymnosperms and di-

cots) or in a complex network (monocots). All of the modified stems described earlier have this arrangement.

Arrangement of Tissues

Epidermis

The epidermis is always the outermost layer of the stem; there is never the equivalent of a root cap (Figs. 11.4, 11.5, 11.21). The structure of the epidermis was described in Chapter 10. Aquatic angiosperms may not have stomata in the epidermis of their stems; it has been claimed that these plants do not have a true epidermis (Linsbauer, 1930), but that is not generally accepted.

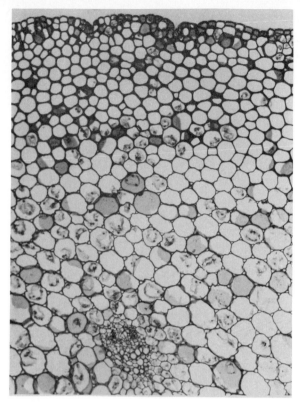

Figure 11.5
Unlike the thin cortex of *Helianthus* (Fig. 11.4), the cortex of *Peperomia* is quite thick; also, it has its own network of bundles (cortical bundles), one of which is visible here. × 160.

Cortex

The cortex is the region between the epidermis and the outermost cells of the vascular cylinder (Figs. 11.3, 11.4, 11.5, 11.21). The cortex is frequently composed purely of rather uniform parenchyma, and it always contains at least some of this type of cell. When other cell types are present, they are usually in the form of a collenchymatous or sclerenchymatous hypodermis or as a sclerenchyma sheath just exterior to the vascular bundles (Fig. 11.6). In many cases the cortex is quite simple, consisting mostly of chlorenchyma and mechanical tis-

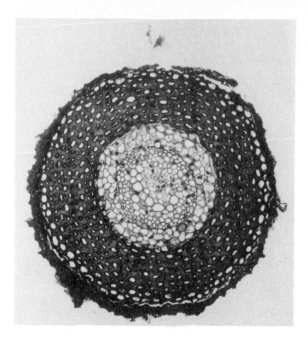

Figure 11.6
The cortex may contain some sclerenchyma in many plants, but in this *Hymenophyllum australae* (a fern), the entire cortex consists of heavy-walled fibers. Only the stele contains some parenchyma. × 160.

sues, but, if internal secretory tissues occur in any vegetative portion of the plant, they will almost certainly occur at least in the cortex. The same is true of idioblasts.

The cortex is often narrow, being a few millimeters thick, but in the herbaceous stems of many monocots the cortex is very thick (rhizomes of *Iris*, *Musa*; corms of *Gladiolus*; bulbs of *Allium*, *Tulipa*). Those of dicots are almost always very narrow, with a few exceptions, such as some rosette plants like *Apium* (celery), cacti, *Geum*, *Plantago*, *Sempervivum*, *Taraxacum* (Troll and Rauh, 1950; Rauh and Rappert, 1954). The cycads also have this shape (Stevenson, 1980). A feature that is inversely correlated with stem width is stem length; these stems are also much shorter than those of the more typical dicots. Studies of the monocots (Ball, 1941; Boke, 1944, 1951; Clowes, 1961; DeMa-

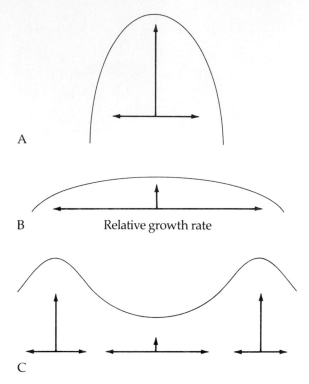

A

B   Relative growth rate

C

Figure 11.7
In the subapical region, the stem grows in both
length and width. In (A), the growth in length predominates, and a long, narrow shoot results. (B) If
radial growth is greater, the stem is shorter and
may even grow as a flat disk. (C) The area of maximal upward elongation does not have to be directly
below the apex, but can be to the side of it.

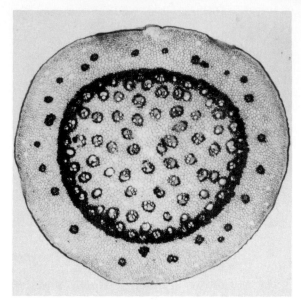

Figure 11.8
Cross section of a stem of the monocot *Anthurium
polystichum*. Rather than a single ring, there are numerous bundles. Although no pattern is obvious
from this single section, in three dimensions the
bundles are part of an orderly, predictable network.
× 18. Micrograph provided by J. French.

son, 1979a, b, 1980, 1983; Skutch, 1932; Stevenson and Fisher, 1980; Zimmermann and Tomlinson, 1967, 1968, 1969) have shown that their
primary growth is not significantly different
from that of the narrower dicots but is based on
a greater amount of transverse growth versus
longitudinal growth in the tissues immediately
below the shoot apical meristem (Fig. 11.7). In
the central ground meristem of dicots, growth
is almost exclusively longitudinal, and the surrounding provascular tissues form a cylinder
around the developing pith. The cortical and
pith cells undergo very little radial expansion.
But in these monocots, the ground meristem,

which is producing the numerous bundles and
intervening parenchymatic matrix (the matrix
is called **conjunctive tissue** in monocots), expands mostly in a radial direction (Figs. 11.8,
11.19, 11.20). As a result, the provascular tissues are pushed primarily outward as a disk
instead of primarily upward as a cylinder
(Fig. 11.7). The provascular tissues are very active mitotically, producing large numbers of
bundles and conjunctive tissue.

Based on histological studies, DeMason
(1979b, 1980, 1983) concluded that there was a
broad region of mitotic activity, called the **primary thickening meristem** (Figs. 11.9, 11.10).
This encompasses the provascular tissue, the
outer regions of the conjunctive tissue, and the
inner regions of the cortex. Zimmermann and
Tomlinson (1968) had proposed that the term
primary thickening meristem be abandoned be-

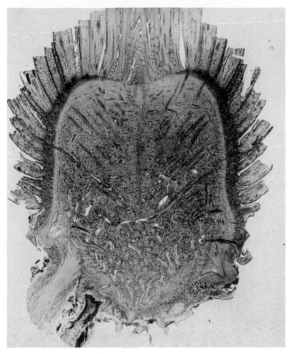

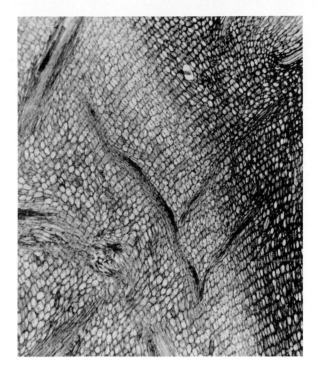

Figure 11.9
This longitudinal section of *Yucca whipplei* illustrates the type of growth diagramed in Figure 11.7(B): radial growth is very much accentuated. The lateral regions are somewhat higher than the center as in Figure 11.7(C), due to the presence of the primary thickening meristem (shown magnified in Figure 11.10). The numerous axial bundles and leaf traces produce a complex pattern, but it is important to realize that there is an underlying pattern (Fig. 11.18). × 5.0. Micrograph generously provided by P. K. Diggle and D. A. DeMason.

Figure 11.10
The primary thickening meristem of this *Yucca whipplei* is a broad zone of cell division; within the zone are strands of procambium that develop into the vascular bundles. The mitotic activity of the primary thickening meristem contributes to the conjunctive tissue, the vascular tissue, and the cortex. × 50. Micrograph generously provided by P. K. Diggle and D. A. DeMason.

cause it had been used with various meanings and because they felt that cell division in the center of the stem was more important. DeMason's autoradiographic evidence, however, shows that mitotic activity is not located centrally, but rather as a thick disk; she proposed (1983) that the term *primary thickening meristem* be retained. There have been no studies in the dicots that are as detailed as those of DeMason's, but the work of Bailey (1963), Boke (1944, 1951), and Yarrow and Popham (1981) shows that a roughly similar process can occur in

some dicots, although the expansion of the pith seems to be much more diffuse and less well organized than it is in monocots. DeMason (1983) felt that the term *primary thickening meristem* should be restricted to the monocots, but she did not suggest a name for the process in dicots or cycads. The term is so commonly used for all plants with large, fleshy columnar stems that it will continue to be employed until a thorough study is made in the dicots and alternative terminology is proposed.

An exceptional case of cortical outgrowth in

the dicots occurs in the cacti; they develop a thick cortex, as just mentioned, but, in addition, the regions immediately interior to the leaves grow outward even more (Boke, 1944, 1953, 1961; Buxbaum, 1950). This results in a conical structure called a tubercle; if all the leaves are aligned in vertical rows, the regions of outgrowth may merge and form ribs. These can project out several centimeters, and in *Echinocactus, Trichocereus* and other large cacti the entire cortex can be up to 20 or 30 cm thick (Mauseth et al., 1985).

In plants such as these that form thick, fleshy, columnar stems, radial growth is dominant while the plant is a seedling, and the stem becomes wider without becoming taller. This phase is called **establishment growth** (Fisher and Tomlinson, 1972; Tomlinson and Esler, 1973; Zimmermann and Tomlinson, 1970). After the proper width is achieved, the ratio of growth is shifted into a balance with longitudinal expansion, and the plant becomes taller.

In some palms, there is a swelling at about mid-height in the stem (*Actinophloeus, Jubaea chilensis, Roystonea*). This results from the continued slow division and expansion of tissues in this region. The cells mainly involved are the outermost cells of the bundle sheaths and the conjunctive tissues. Tomlinson (1961) has named this **diffuse secondary thickening**. The basalmost portions of many palms are quite swollen; this is due to the production of deeply seated adventitious roots that grow downward through the stem tissues, pushing them outward.

CORTICAL BUNDLES. The cortex of a small number of species contains vascular bundles other than leaf traces; these are called **cortical bundles**, and their function is unknown (Figs. 11.5, 11.11). Their presence does not seem to be correlated with the distribution of water and nutrients throughout a large, wide, fleshy cortex, because many genera that have cortical bundles are not succulent—*Bladhia* (Ogura, 1937); *Calycanthus* (Fahn and Bailey, 1957);

Figure 11.11
The cortex of this *Epiphyllum* (orchid cactus) is extremely expanded; the innermost vascular bundles run longitudinally up the stem—these are the axial bundles. Most of the vascular network that is visible consists of cortical bundles. × 25.

Chimonanthus (Balfour and Philipson, 1962); *Idiospermum* (Blake, 1972), *Nyctanthes* (Kundu and De, 1968)—while many plants that do have a large cortex do not have such bundles. In the Cactaceae, some species have a cortical system of bundles, while other species that are just as succulent do not (Boke, 1961; Gibson and Nobel, 1986; Mauseth, 1987a,b). In the cacti where they do occur, they may be responsible for carrying sugars from the photosynthetic outer cortex to the rest of the plant; the cortex is persistent and photosynthesizes for many years. The phloem of the old cortical bundles appears quite healthy, and in *Melocactus* new phloem is produced year after year (Mauseth, 1988). In a review of cortical bundles, Howard

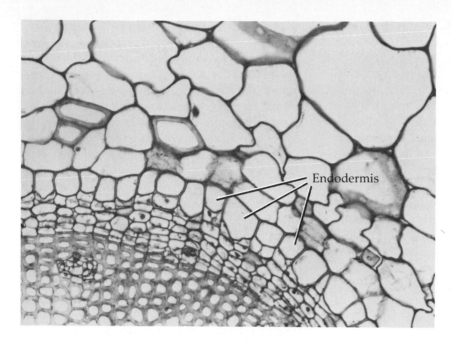

Endodermis

Figure 11.12
There is a prominent endodermis in this stem of *Salicornia;* it is the single layer of
cells with prominent walls.
Although it is not easy to see
in a black-and-white photograph, the radial walls of the
endodermal cells stain heavily, due to the presence of
Casparian strips. × 400.

(1979) concluded that in some plants the cortical bundles could run the length of the stem
without connecting to the stelar axial bundles
at all, or they could connect with them at the
node and also enter the leaf as extra traces. In
Chimonanthus, the cortical bundles arise after
the stelar bundles have been formed (Balfour
and Philipson, 1962).

ENDODERMIS.   The innermost layer of the cortex of the stem in some plants is the **endodermis** (Clarkson and Robards, 1975). This is a
unistratose cylinder of cells that line the inner
boundary of the cortex and the outer boundary
of the stele (Figs. 11.12, 13.8, 13.9). There are
no intercellular spaces between endodermal
cells (Gambles and Dengler, 1982b), and the
cells of the endodermis are characterized by
the presence of hydrophobic **suberin** encrusted
into their primary walls, rather than adcrusted
onto them (Robards et al., 1973; van Fleet,
1942). Suberin is closely related to cutin chemically (Kolattukudy, 1980a, b), differing from
it primarily in the relative quantities of the
monomers present: suberin contains a higher

proportion of dicarboxylic acids, phenolic compounds, and very long chain (C_{20} to C_{26}) acids
and alcohols. As is true of cutin and wax, very
little is known about the ways that these monomers are transported or crosslinked to form the
resulting polymer suberin. However, the result is a dense, hydrophobic matrix that thoroughly permeates the wall, resulting in an impervious region that hinders the diffusion of
most molecules. In the first stages of differentiation, the suberin is deposited in strips (**Casparian strips**; Figs. 9.14, 11.12, 13.9) within all
four radial walls of the endodermal cells, the
inner and outer faces remaining free of suberin
(Gunning and Steer, 1975; Haas and Carothers,
1975; Warmbrodt and Eschrich, 1985). It has
been reported that the wall regions covered by
the Casparian strips contain no plasmodesmata
(Bonnett, 1968; Hajibagheri et al., 1985) or only
very few (Haas and Carothers, 1975). The plasmalemma of the cell somehow becomes attached to the wall at the Casparian strips: even
in strongly plasmolysed cells the plasmalemma
does not pull away from these areas. This has
the important consequence of dividing the apo-

plastic space of the shoot into two sections, one exterior to the endodermis, the other interior to it.

In some species, the endodermis can continue to develop beyond this stage; in the secondary stage, a thin layer of suberin may be deposited over all walls, such that the cell is dependent upon its plasmodesmata for material exchange (Clark and Harris, 1981; Haas and Carothers, 1975). In the tertiary stage, a new layer of cellulose is deposited interior to the existing wall. This layer is lignified and is laid down unevenly such that the inner periclinal wall is thickest, the radial walls are slightly less thickened, and the outer periclinal wall has only a thin layer (Clark and Harris, 1981; Clarkson et al., 1971; Haas and Carothers, 1975). Such elaborations are more characteristic of endodermises in roots than in stems; a more complete discussion is given in Chapter 13.

The importance of the endodermis in those few stems that have one appears to lie in the ability of the endodermis to channelize the water of the stele and keep it from leaking into the stem cortical tissues (Zimmermann, 1983). Water usually is pulled through the xylem up to the transpiration surfaces of the leaves; on the way, small amounts move out into the stem cortex and pith. In addition, the roots are able to force water weakly upward into the stem, because they generate root pressure. If the leaves suddenly are lost (due to an early frost, insect attack, or some form of dieback) while the roots remain healthly, then the root pressure could cause damage by forcing water out of the stem xylem into the cortex in such quantities that it would fill the intercellular spaces. The cortex would become waterlogged, and oxygen and carbon dioxide could not diffuse through the tissues. This would be especially harmful in the aerenchyma tissues of submerged stems and in the large, fleshy storage tissues of tubers and rhizomes, especially in the autumn when the foliage dies while the soil is still warm enough for the roots to be active. Root pressure is not strong enough to move wa-

ter very high, so only the lower, basalmost tissues of the stem are in danger; the root pressure is too weak to push water outward in the higher portions of the stem. But in low-growing, rhizomatous, tuberous, stoloniferous, or bulbous stems, root pressure could cause real damage. An endodermis in the stem (and even in the lower leaves) can overcome this. The Casparian strips can keep the water from leaving the stele apoplastically, and, if the endodermal cells adjust their water potentials, water cannot exit through them symplastically either (Zimmermann, 1983). This danger does not arise very often; stem endodermises are rare, occurring mostly in rootlike stems such as rhizomes (Fig. 11.20).

Pith

The pith is almost always purely parenchyma (Fig. 11.3, 11.4), frequently lacking any pigmentation, although brightly colored pith cells do occur in many members of the Caryophyllales. It usually has no chlorophyll because it is too deep within the stem to receive adequate light for photosynthesis, although in *Theligonum* chlorophyll is more abundant in the pith than it is in the cortex (Metcalfe, 1979d). Sclerenchyma is only rarely present, but when it does occur, it may be composed of either sclereids or, even more rarely, fibers located on the edge of the pith, close to the vascular bundles, as in the cycads, Araliaceae, Malvaceae, and Polygonaceae (Solereder, 1908). The pith of *Myzodendron* is almost wholly sclerenchymatous. It is not unusual for the cells on the perimeter of the pith to differ from those in the center. This outer region is sometimes referred to as the **perimedullary zone**, and it is frequently useful in systematics (Doyle and Doyle, 1948; Metcalfe, 1979d). The pith was formerly called the *medulla*; this term is never used now, but it does survive in the adjectival form, as in *perimedullary*.

As parenchyma, the pith can have a role in storage, although it is often too small to be very significant, except in some tubers such as potatoes, where it constitutes almost the entire stem (Artschwager, 1924; Reed, 1910). In many plants, the pith cells mature early, long before the surrounding tissues have completed their expansion. Thus the pith is torn apart and becomes hollow as the adjacent vascular tissues continue their elongation. It may do this irregularly, or individual layers may separate from adjacent layers, forming the **septa** (sing.: septum) of a **chambered pith**, as in *Phytolacca americana* (Mikesell and Schroeder, 1980), *Juglans* (Holm, 1921), and *Pterocarya* (for a complete list, see Metcalfe, 1979d). The septa may be composed of brachysclereids in Annonaceae, Magnoliaceae, and Theaceae (Solereder, 1908). Such pith chambers may become occupied by ants (*Ceropegia, Solanopteris brunei*; Gómez, 1974; Wagner, 1972) or can fill with gas and be advantageous as flotation devices. But most of the genera and species just listed are not aquatic, nor do they form ant-associations, and there is no immediately obvious function for the hollow pith. It is possible that the true function of pith in many plants occurs immediately after its formation, in the subapical meristem. It may participate in establishing whatever fields or gradients of hormones necessary for the formation of the vascular tissues in their proper places. That is, the young pith may act as a matrix around which the vascular tissues form.

Between the vascular bundles, the pith parenchyma is in contact with the parenchyma of the cortex. These regions are called **medullary rays** or **interfascicular regions** (Figs. 11.3, 11.4). In the medullary rays it is not possible to distinguish pith cell from cortex cells.

MEDULLARY BUNDLES.   In several families vascular bundles, called **medullary bundles** (Fig. 11.13), run through the pith—Amaranthaceae (Wilson, 1924), Cactaceae (Boke, 1961; Gibson and Nobel, 1986; Mauseth, 1988), Chenopodiaceae, Melastomataceae (Lignier, 1887), Nyctaginaceae, Piperaceae, and Polygonaceae

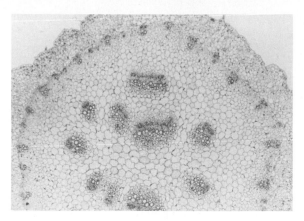

Figure 11.13
This cross section of a stem of *Boerhaavia* shows a rare phenomenon: the presence of medullary bundles. These are even more unusual because they are so much larger and more prominent than the ordinary vascular bundles of the ring. × 100.

(Davis, 1961; Howard, 1979; Pant and Mehra, 1964). These may be complete bundles with both xylem and phloem, or they may consist of phloem only; fibers are usually absent. The bundles may divide and anastomose, forming a complex network. In the Melastomataceae and Piperaceae, the medullary bundles may turn outward and contribute to the leaf vasculature. In the cacti, the extensive reticulum of medullary bundles is connected to the network of cortical bundles by means of bundles that run horizontally through the medullary rays (Boke, 1961). The role or function of the medullary bundles is completely unknown. As with the cortical bundles, stems with a rather narrow pith may have medullary bundles, while stems with a much more massive pith may lack them. In *Melocactus*, new xylem and phloem are added to the medullary bundles each year.

Primary Vascular Tissues

ARRANGEMENT OF THE BUNDLES.   The stele of stems in all seed plants is composed of a set of

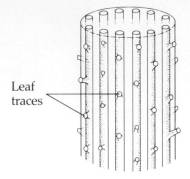

Figure 11.14
The vascular cylinder of *Carnegiea* (saguaro) consists of bundles that rarely or never merge with each other; they run up the stem, giving off leaf traces.

vascular bundles, **axial bundles**, also called **cauline bundles** (Dormer, 1972). They are located in one ring in all gymnosperms and most dicots (Figs. 11.3, 11.13), a series of concentric rings in a very small number of dicots (some Compositae and Piperaceae), and distributed throughout a ground parenchyma in the monocots (Fig. 11.8; often described as being "scattered," but this suggests random distribution and irregularity that is incorrect). When viewed in three-dimensional reconstructions or in cleared material, a much more meaningful pattern emerges. The simplest arrangement (which is perhaps also the rarest) is one in which the axial bundles remain independent of each other throughout their length, and the stem really has a vascular system composed of individual rods (Fig. 11.14). This occurs in *Carnegiea*. In probably all other dicots and gymnosperms, the bundles that appear as individuals in cross section are actually portions of a cylindrical reticulum (Gibson, 1976; Goffinet and Larson, 1982a; Larson, 1975). Each bundle fuses, to some degree, with adjacent bundles at points higher and lower in the stem, and each bundle is capable of branching to produce either more bundles for the stem, to supply bundles to leaves (**leaf traces**), or to branches (**branch traces**; Figs. 11.15, 11.16, 11.17). The pattern of branching and anastomosis can be

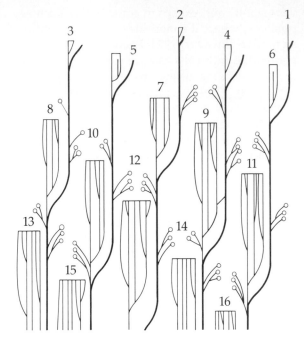

Figure 11.15
This is a diagrammatic representation of the vascular system of the shoot of *Pereskia humboldtii*. This type of diagram conveys a great deal of information rapidly and is extremely useful, but you need to look at it systematically in order to understand it. First look at a heavy vertical line, such as 2; this is one of the axial bundles or sympodia. Notice that none of the sympodia has connections with the other sympodia. This is an open stele. However, each gives off a trace (2 produces 7), which then branches. The abrupt termination by a horizontal line is the method used to show that it is a leaf trace. In this case, each leaf receives vascular tissue from only one sympodium. The sympodia also give off short lines that end in circles. These are the traces to the axillary bud; notice that each bud receives a vascular supply from two different sympodia. Diagram generously provided by A. C. Gibson.

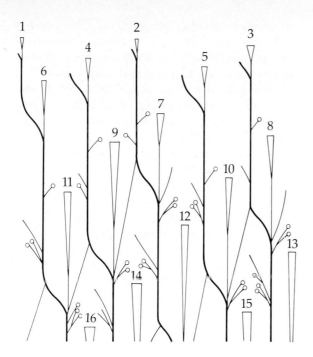

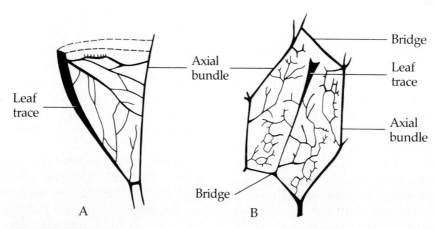

Figure 11.16
This is the vascular diagram of *Pereskia aculeata*. Notice that this is a closed stele because each sympodium is connected to the adjacent ones by bridges. The leaf traces and branch traces do not divide so much as in *Pereskia humboldtii* (Fig. 11.15), but there is a short trace that vascularizes the parenchyma of the node. Diagram generously provided by A. C. Gibson.

Leaf trace

Axial bundle

Bridge

Leaf trace

Axial bundle

Bridge

A

B

Figure 11.17
This is a diagram of a node of *Opuntia phaeacantha;* although only a part of the system is shown, you can see that the vascular bundles form a diamond. The bridging bundles are as large as the sympodial bundles, and it can be difficult to decide which is which. In this species, as in many others, the axial bundles form a net. Diagram generously provided by A. C. Gibson.

either rather simple or enormously complicated, and a large number of patterns exist. In some, the axial bundles remain rather distinct for several internodes; each may interconnect with others by smaller bundles, but the individuality is obvious (Fig. 11.15). Each axial bundle and its series of leaf traces is referred to as a **sympodium** (pl.: sympodia), and the stele is said to be **open** (Dormer, 1945). In other plants, the interconnecting bridges between bundles are as large as the bundles themselves, so in reality there is a true reticulum, and sympodia cannot be identified. Such steles are **closed** (Figs. 11.16, 11.17). In an open stele, damage to any part of a sympodium will adversely affect all of the leaves that are supplied by that sympodium. But in a closed stele, there is such extensive interconnection that there is a unified conducting system for the plant, rather than a set of separate systems. As a consequence any portion of the shoot can receive water from any or all roots and is not dependent on just one. Any damage remains localized, and leaves a short distance above or below the damage remain healthy. The same is true for roots receiving carbohydrates from the leaves. Thus, damage to any one bundle can be compensated both above and below the point of damage.

In the majority of the monocots, the bundles occur throughout the stem, not just in a single ring as in the dicots and gymnosperms. Most botanists believe that the angiosperms are monophyletic, having had only one origin. Therefore, the vascular system of the monocots must have had the same origin as that of the dicots, even though at present it appears to be quite distinct structurally. In reality, there may not really be as many differences as there appear to be at first. Recently, several investigators have made careful studies of monocot steles, and orderly patterns have been revealed (Fig. 11.18; Bell, 1980; French and Tomlinson, 1981a–d; LaFrankie, 1985; Zimmermann, 1983; Zimmermann and Tomlinson, 1965, 1968, 1969, 1970, 1974; Zimmermann et al., 1974, 1982). The small palm *Rhapis excelsa* has been studied

in greatest detail (Zimmermann, 1983) and is described here. The stele consists of numerous sets of axial bundles; if we track them upward in the stem, we can see that they "move" inward toward the center of the stem (Fig. 11.18). At their innermost point, they turn abruptly outward, and give off leaf traces that enter the leaves. The axial bundle also sends out numerous bridges, then turns upward again. The bridges connect this one bundle with the adjacent axial bundles. If the leaves were widely separated, and if every bundle gave off leaf traces at every node, and if each leaf received just a small number of traces, then the monocot stem would greatly resemble that of a dicot or conifer, and would appear to be made up of a single ring of bundles. But instead, an axial bundle moves upward and inward for several nodes, and the traces for all of these nodes overlap it. Also, each leaf receives many traces, with the result that there appear to be numerous bundles "scattered" through the ground tissue (the conjunctive tissue; Figs. 11.19, 11.20). The appearance is even more complicated, because the leaf traces may move upward or circumferentially a great distance, and while they do they superficially resemble axial bundles. A similar situation occurs in some dicots that have a thick cortex: the leaf traces appear to be "extra" bundles of the stele. These can be confused with cortical bundles; they can be found in Araliaceae, Calycanthaceae, Melastomataceae, and Proteaceae. These "cortical bundles" (really leaf traces) give the stems of these plants, which are all dicots, the appearance of being monocots.

PRIMARY VASCULAR TISSUES WITHIN THE BUNDLES.   The axial bundles are usually **collateral**; that is, they contain both primary xylem, on the inner side of the bundle, and primary phloem, on the outer side (Figs. 11.4, 11.21, 11.22, 11.23, 11.24). But **bicollateral bundles** exist; these have a second strand of primary phloem (**internal phloem**) interior to the xylem (Fig. 11.25). When care is taken to follow the paths of bicollateral bundles, it is usually found

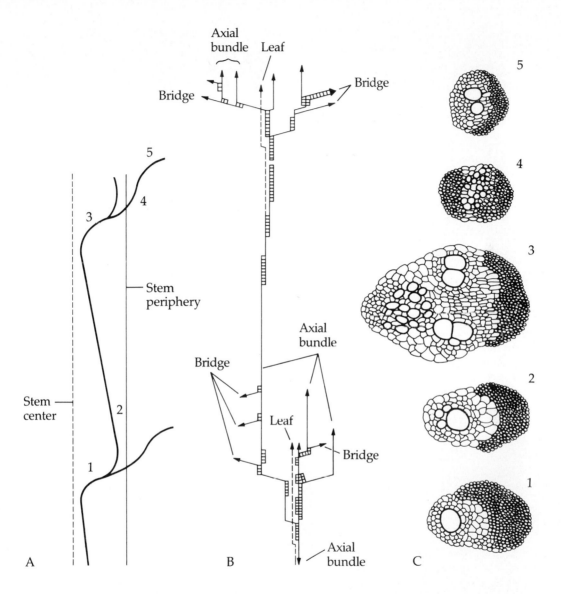

Figure 11.18
Although the vascular system of most monocots appears formidable at first glance, actually the pattern is simple. As shown in (A), a major bundle moves inward gradually, then outward rapidly, simultaneously giving off leaf traces and connecting to other axial bundles by bridges. In (B), the length of the vessels at each level is shown, and the crossbars indicate intervessel pitting. The vessels are very short in the leaf traces and bridges, but long in the axial region. The diagrams in (C) are cross sections located at the levels indicated by the corresponding numbers in (A). Where there are long vessels, they are also wide, but at the nodes the vessels are both short and narrow, and the passage of embolisms would be restricted. Drawn based on information published by Zimmermann (1983) and Zimmermann and Sperry (1983). *AB* = axial bundles. *Br* = bridge.

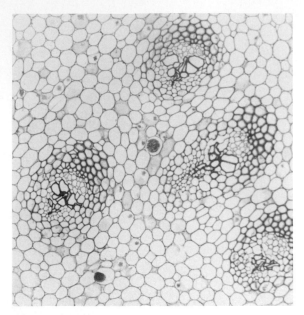

Figure 11.19
This is the conjunctive tissue in the stem of *Dracaena fragrans;* as in most monocots, it consists of rather thick-walled parenchyma that has numerous intercellular spaces. In this species the boundary between the conjunctive tissue and the outermost cells of the bundles is difficult to detect. × 100.

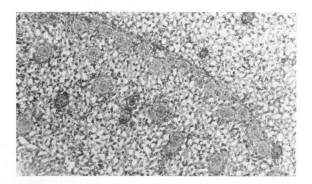

Figure 11.20
The conjunctive tissue of an *Acorus* rhizome is aerenchyma, as is the cortex. *Acorus* grows in bogs, and the rhizome is always under water. The stele consists of a ring of bundles on the edge of the conjunctive tissue, then numerous bundles within the ring. If you look closely, you can see a fine line between the conjunctive tissue and the cortex; that is an endodermis. A higher magnification of these bundles is presented in Figure 11.30. × 40.

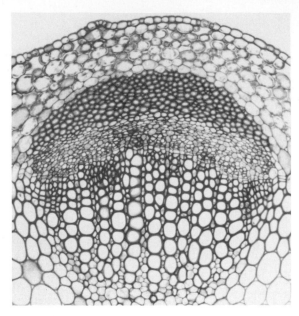

Figure 11.21
This bundle of *Trifolium* (clover) is large with abundant conducting elements. The xylem is arranged in radial rows, the vessels of one row being separated from those of adjacent rows by parenchyma. The phloem has a thick cap of phloem fibers; the conducting phloem is quite abundant. High magnification micrographs are shown in 11.26 and 11.27. × 160.

that the internal phloem is really rather independent of the rest of the bundle, branching and anastomosing at different sites than where the other tissues do, especially in the families Apocynaceae, Asclepiadaceae, Compositae, Conolvulaceae, and Solanaceae. The bundles are truly bicollateral only in small areas. In Cucurbitaceae and Myrtaceae, the internal phloem is more uniformly associated with the xylem.

The external phloem can also be somewhat independent of the xylem. As early as 1884, Koch reported that the external phloem of *Ecballium elaterium* could anastomose with the phloem of adjacent bundles even when the xylem did not form a similar anatomosis. This occurs in a few other species and families

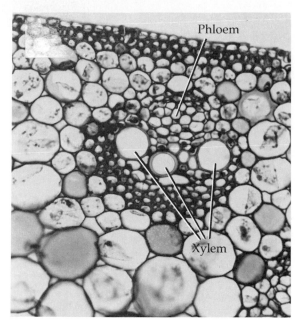

Figure 11.22
Vascular bundle from a corn stem. The sieve tube members and the companion cells form a regular pattern. × 100.

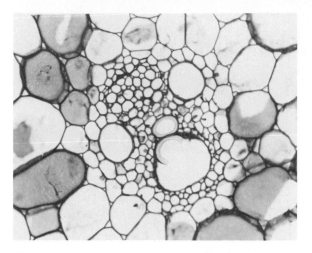

Figure 11.23
As the protoxylem is stretched by the growth of the surrounding cells, its primary walls may be broken, as has happened to one protoxylem element. From cross sections such as this, it is not possible to be certain if the protoxylem elements are tracheids or vessels; longitudinal sections would be necessary. *Zea mays* (corn). × 400.

(Aloni and Sachs, 1973; Zimmermann, 1922), but is really quite rare. Most bundles are truly collateral.

Within the xylem of any bundle, there will be a gradation in the diameter of the tracheary elements. In all extant plants, the tracheary elements with the smallest diameters (**protoxylem**) are located on the inner side of the mass of xylem, and those with larger diameters (**metaxylem**) are located on the outer side (Figs. 11.21, 11.23, 11.24, 11.25, 11.26). The reasons for the differences in size and names becomes obvious from a developmental study. Just below the shoot apical meristem, all cells are small, cytoplasmic, and very active mitotically. Some of those that will differentiate into tracheary elements stop dividing and begin to grow. Almost immediately, those on the inner side begin to differentiate to form the first xylem elements (hence the name protoxylem).

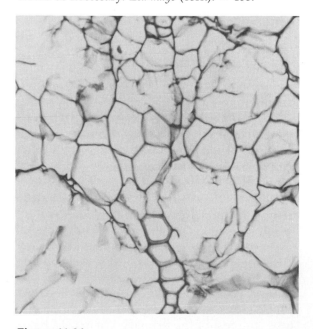

Figure 11.24
In contrast to the large bundle of Figure 11.21, some bundles are delicate with few conducting elements of any kind. This is a bundle from *Euphorbia resinifera*, a small succulent from Africa. × 400.

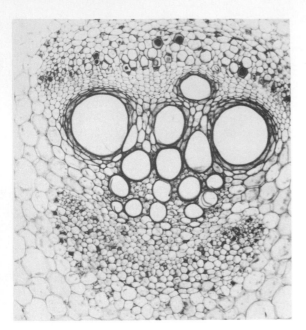

Figure 11.25
This is a bicollateral bundle from pumpkin; notice that there is phloem to the exterior and interior of the xylem. × 160.

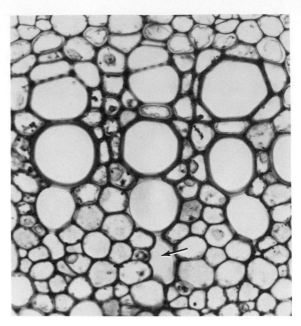

Figure 11.26
This high magnification of the xylem of *Helianthus* (sunflower) shows the small protoxylem elements that have been torn (arrow). The metaxylem elements are all much larger, and they are *never* torn, because they form after the stem stops elongating in that region. There is abundant xylem parenchyma, especially in the protoxylem. × 635.

Because they have little time to grow, they will be short and narrow. The other cells continue to enlarge, but soon the cells just outside the first also begin to differentiate. Because these have had more time to grow, they are larger. This pattern of growth continues until all the protoxylem has formed. Because these elements are being produced in growing tissues, it is important that they be extensible longitudinally; they must have only annular or helical thickenings on their walls. This allows them to be weakly plastic, and even though they are dead, they are pulled into a longer shape because their middle lamellas glue them to the adjacent cells, which are still living and expanding. By the time the expansion of the stem stops, the rest of the primary xylem cells have become quite large, and they differentiate to be the metaxylem cells. Because they differentiate in the nonexpanding portion of the stem, they

may have any type of secondary wall thickening. Much of the protoxylem is damaged by tissue expansion; the primary wall can even be broken (Figs. 11.23, 11.26). Consequently the metaxylem will be the main pathway for conduction until secondary xylem is formed. In plants that lack secondary growth, metaxylem is the main means of conduction during the entire life of the plant. In annuals and biennials this is one or two years, but in palms and many other long-lived monocots, as well as in ferns (which never develop wood), the metaxylem must function for many years.

In many plants there is an abrupt transition from protoxylem to metaxylem, at least on the basis of cell diameter (Fig. 11.23). Because of this, there was considerable emphasis placed on these two types of primary xylem, and many attempts were made to identify distinguishing characteristics (Bierhost and Zamora, 1965; Bugnon, 1925; Esau, 1943c). However, these attempts have never been uniformly successful, and in many plants the transition from first-formed protoxylem to last-formed metaxylem is so gradual that it is impossible to identify a last protoxylem element and a first metaxylem one (Figs. 11.21, 11.24, 11.25).

The protoxylem often contains more parenchyma than tracheary elements; either tracheids or vessel elements may be present, but fibers typically are not. The metaxylem is usually more complex, having more conducting cells and less parenchyma; fibers may also be present. The metaxylem may contain either tracheids or vessels. Although annular and helical secondary walls may be present in metaxylem tracheary elements, scalariform, reticulate, and pitted patterns are much more common.

An analogous situation occurs in the phloem of the bundles: the outermost cells stop growing early and differentiate as **protophloem**; the inner cells grow for a longer time, so they are larger when they finally differentiate into **metaphloem** (Figs. 11.21, 11.22, 11.27). As was mentioned in Chapter 8, the protophloem may differ significantly from the metaphloem, often lacking companion cells and possibly even not having recognizable sieve cells in the gymnosperms (Esau, 1965a; Smith, 1958). The sieve elements that are present collapse quickly, and the residual parenchyma cells may then expand and become either collenchymatous or sclerenchymatous (Blyth, 1958), changing the appearance and physical properties of the protophloem.

The metaphloem, like the metaxylem, is the main pathway for conduction until secondary tissues are produced, if they ever are. In long-lived, nonwoody plants such as monocots and ferns, the metaphloem must function for the lifetime of the plant. Also, vines can elongate so rapidly that extremely long stems are

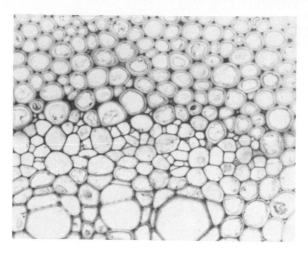

Figure 11.27
The phloem of *Helianthus* shows typical characters: the cells are angular, and the small, dark companion cells are easy to identify. Unlike the protoxylem, the protophloem is not obvious; it collapses quickly and rather completely. Dark spaces at the edge of the metaphloem are always a good first guess, but longitudinal sections may be necessary to be able to identify the protophloem with certainty. This metaphloem contains nonconducting parenchyma mixed in with the sieve tubes and the companion cells; the lack of any obvious pattern is a good clue that this is not a monocot bundle. × 635.

produced before secondary growth starts, so the metaphloem is responsible for very long-distance transport. If secondary growth does occur, the metaphloem sieve elements collapse when this begins. The sieve elements of the metaphloem are larger than those of the protophloem, and their sieve areas are much more distinct (Esau, 1969). In the dicots, both nonconducting parenchyma and companion cells are present, interspersed with the sieve tubes, although fibers are lacking (Esau, 1965a). In the monocots, nonconducting parenchyma is typically absent, and the metaphloem consists of just sieve tube members and companion cells in a regular array (Figs. 8.1, 11.22). The monocot bundle very often has a prominent sheath

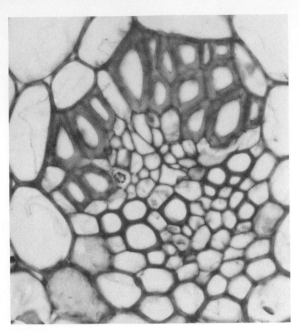

Figure 11.28
This bundle from the stem of *Aloe* contains xylem, phloem, and a fibrous cap; compare it with other bundles that occur in the same stem (Fig. 11.29). × 400.

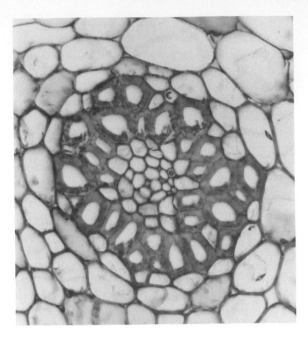

Figure 11.29
This bundle is from the same section as that in Figure 11.28. This bundle, however, lacks xylem conducting elements and contains only phloem and fibers. × 400.

of fibers around it (Figs. 11.22, 11.28, 11.29; Cheadle and Uhl, 1948); fibers may be present mixed in with the conducting cells, but this is less common. Esau (1965a) concluded that the protophloem of monocots is often quite recognizably distinct from the metaphloem, but in the dicots the two types of tissue intergrade gradually and can be difficult to delimit.

Probably of great importance are the arrangement of tracheids and vessels within each bundle and the arrangement of sieve tubes and sieve cells. Because the metaxylem tracheary elements (either tracheids or vessel elements) are so much wider than those of the protoxylem, they have a much greater radius and so greater conductivity and less resistance, which are related to r^4 (see Chapter 7). In the lower part of a stem with only primary tissues, water will move upward through the outer metaxy-

lem elements. But when it encounters the last, uppermost functional elements in any given vessel or file of tracheids, it must move centripetally to the elements of a file that extends farther up the stem. At the very top, the water will have moved all the way inward to the only elements that are differentiated: the newest protoxylem elements.

In the angiosperms, the interconnections between the individual vessels of the bundles are important, but there has been little study of this (Zimmermann, 1983). It is common among the dicots to see the vessels arranged radially, with all the vessels of any row in full contact with each other along their tangential faces. Such an arrangement would facilitate the centripetal transfer of water just described. At any given section, the vessels of any file often appear to be separated by parenchyma cells, fibers, or tracheids, from the vessels of the nearby files of the same bundle. Such an arrangement would provide a good "insulation" against cavitation: as the tension on the water columns increased, they would tend to pull water in from adjacent parenchyma cells. If they were instead surrounded by other vessels, one of which was cavitated and air-filled, they might pull in an air bubble from it, and that would cause an embolism. At least occasionally there are contacts between the vessels of two adjacent files, and this may be sufficient to maintain the three-dimensional interconnection of the water columns.

In the monocots, metaxylem vessels tend to be very large and to be isolated from each other. It is not unusual for a bundle to have only a few vessels (one to five) at any cross-sectional level, and each vessel is surrounded by parenchyma (French et al., 1983; Zimmermann, 1983). *Rhapis excelsa* (Zimmermann, 1983) has only one main metaxylem vessel per axial bundle; these average 25 cm in length, but can be up to 50 cm long (Fig. 11.18). In larger palms, there may be more vessels, but it is not known how they interconnect. In *R. excelsa*, just below the point where the axial bundle produces leaf traces and bridges that communicate with the adjacent axial bundles, many vessels only about 5 cm long appear in the bundle and interface with the axial vessel (note that these are vessels, not vessel elements). By having numerous short vessels in this region, and by having direct intervessel pitting, water can move up the stem, into leaves, or from bundle to bundle, following the paths of greatest transpiration pull and least resistance. Most important, however, is the fact that the presence of short vessels in this region of numerous junctures causes the air embolisms that are formed when leaves abscise to be confined to those short vessels. They cannot cause damage over a long distance, certainly not into the large metaxylem vessels of the axial bundles.

It is certain that leaf trace complexes are also important for other monocots as well as the di-

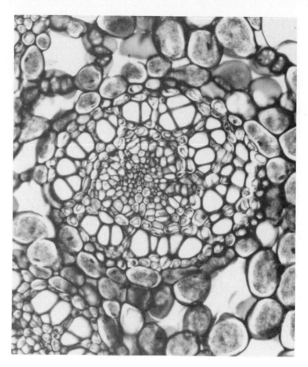

Figure 11.30
The bundle in the stem of this monocot (*Acorus*) is amphivasal, with the xylem completely surrounding the phloem. The phloem is not especially obvious as phloem; companion cells and sieve tube members are difficult to identify. × 400.

cots, but none has been studied (Zimmermann, 1983). In this regard, the arrangement of xylem and phloem might be important. As mentioned earlier, bundles are usually collateral or bicollateral, but in many monocots, the xylem completely surrounds the phloem (an **amphivasal bundle**; Fig. 11.30). They may occur as the bundles of the primary body (*Acorus, Convallaria majalis*, some Xanthorrhoeaceae), but they also constitute the secondary bundles of the arborescent, "woody" monocots (*Aloe arborescens, Dracaena*; Fig. 18.13, 18.14). In some angiosperms and ferns the phloem surrounds the xylem (an **amphicribral bundle**; Fig. 11.31). Amphicribral bundles are interesting because

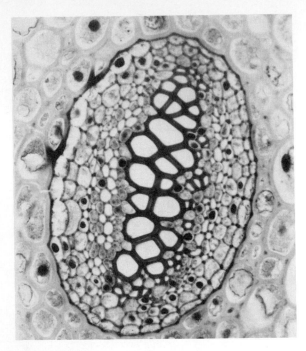

Figure 11.31
In an amphiphloic bundle, the phloem surrounds the xylem. This is from the fern *Polypodium*, so companion cells and sieve tube members are not present; the sieve cells can be identified as the ones that appear to be empty, surrounded by phloem parenchyma cells that contain starch. × 400.

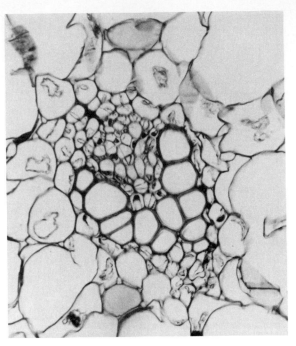

Figure 11.32
V-shaped bundles are not common; this one was located in the stem of *Smilacina* (false Solomon's-seal). The protoxylem is at the bottom (toward the center of the stem); the metaxylem is near the top. × 400.

they occur in unusual situations: in the dicots, as medullary bundles in *Begonia*, *Mesembryanthemum*, *Rheum*, and *Rumex*. Having the xylem completely surrounded by phloem obviously presents some problems for unloading of water out of tracheary elements. And similarly, the loading of phloem in amphivasal bundles would seem at first glance to be difficult. In other cases, mostly in monocots, the bundles may have the xylem arranged in a U shape almost completely surrounding the phloem (*Asparagus aphyllus*, *Kingia australis*), or in a V shape with the phloem positioned at the two free ends of the V (Fig. 11.32). These arrangements are restricted to a small number of genera in just a few families, so it may

be that they represent anomalous conditions that are neither selectively advantageous nor disadvantageous.

NODES. The stems of all seed plants and ferns bear lateral appendages (either leaves or reproductive structures); there are no exceptions. The points at which the appendages are attached to the stem are called **nodes**; because vascular tissues of the appendages must connect with those of the stem, the nodes are complex anatomically. Between the nodes are regions (**internodes**) where there is only stem tissue. In the internodes, the vascular bundles tend to run directly up the stem, branching or fusing rarely, and they are, in general, rather

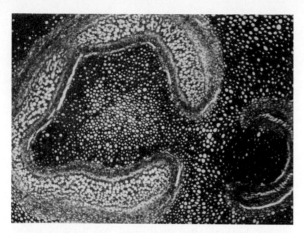

Figure 11.33
This is a solenostele in the fern *Adiantum* (maidenhair fern); the vascular tissues form a complete cylinder with phloem on both the exterior and the
interior. The small arc on the right is the large leaf
trace; the portion of the stele next to the leaf trace
has no vascular tissue; it is just parenchyma. This is
the leaf gap. × 160.

easy to study. If an internode is cut in cross
section, then the vascular tissues are also cut
in cross section, and all tissues are easy to recognize. Similarly, a longitudinal section of an
internode contains the tissues arrayed in orderly, longitudinal patterns.

At the nodes, bundles move out into leaves
and buds, and bridging bundles unite adjacent
cauline bundles. It can be difficult to obtain
good cross sections or longitudinal sections,
and it is often not easy to recognize cell-cell relationships. Perhaps for this reason, very few
studies of tissue differentiation or function
have been performed on nodes. Instead, nodal
anatomy has been studied almost exclusively
for taxonomy and phylogeny.

From a functional standpoint, the node is
usually the site where the axial vasculature
connects to the lateral vasculature (Figs. 11.15,
11.16, 11.17, 11.33). In many plants, the leaf
traces branch off from the axial bundle at the
level of the node, and then they turn and extend almost horizontally to the leaf (Howard,
1979). Such a junction often seems to consist of
short, irregular tracheary elements. In other
species, the leaf trace separates early but extends upward like an axial bundle; at the node,
the entire bundle turns outward and enters the
leaf. In *Serjania*, the leaf trace runs in the cortex
for 17 nodes before entering its leaf (Johnson
and Truscott, 1956).

Three-trace nodes are extremely common in
the dicots; the median trace often arises lower
in the stem than do the two lateral traces.
Once the traces have separated from the axial
bundles, they may fuse with each other, or
they may split into extra traces; this is especially characteristic of the lateral leaf traces.
When the laterals split, the two resulting traces
usually enter the same leaf (Howard, 1970);
in some Asteraceae that have whorled leaves
crowded close together, one of the resulting
traces enters one leaf and the other enters an
adjacent leaf (Carlquist, 1957). If the leaf has
stipules, these may receive vascular tissues
from the leaf traces, or they may be supplied
with their own independent traces (Ozenda,
1949). The palms, aroids, and many other
monocots with large sheathing leaf bases receive numerous leaf traces that arise from many
axial bundles.

In plants that have leaves that die before
the plant does (deciduous leaves or persistent
dead leaves) the node must be a site that is capable of stopping the passage of embolisms in
the vessels. As the leaf dies, all of its vessels
will ultimately cavitate. If it had large, high
conductance vessels connecting it to the axial
bundles, then the embolisms would be pulled
into the stem, and the metaxylem would be in-
capacitated (Zimmermann, 1983). As early as
1928 Meyer recognized this and concluded
from a study of the literature that vessels are
not continuous from stem to leaf. This may be
true of many plants. It certainly is in the palm
Rhapis excelsa; the wide axial metaxylem vessels

must transfer water to the wide leaf trace meta-xylem vessels by means of an intervening set of tracheids (Fig. 11.18; Zimmermann and Sperry, 1983). In *Populus deltoides* another means of blocking may have been found (Isebrands and Larson, 1977a, b); just below the point of leaf abscission is a "constricted zone" where there are vessels but where they are especially narrow. It seems that a tracheid bridge would be more effective than narrow vessels alone in stopping embolisms, but it may be that the axial vessels below the constricted zone are distinct from the leaf trace vessels above it. Larson (1976) found that the constricted zone was the point where vessel differentiation begins. The procambium was continuous from stem to leaf, but vessels actually first appeared in the constricted zone, then extended basipetally and acropetally. It may be that the ones that extend basipetally are different from the acropetal ones; if no vessel extends in both directions, the constriction zone would be an area of vessel-vessel overlap without direct vessel continuity. It is typical for xylem development to begin like this at the leaf base (Esau, 1965b), and it may be that this ensures vessel discontinuity just below the point of leaf abscission. Unfortunately there are almost no studies of the actual vessel arrangement within the leaf trace.

If the main function of a constricted zone or region of vessel discontinuity is to block the passage of embolisms from dying leaves into the stem, then plants that do not have vessels should not have constricted zones. Ferns have only tracheids, and in one study of the holly fern *Cyrtomium falcatum* (Gibson et al., 1984), it was found that indeed they do not have a constricted zone. Because ferns have neither vessels nor secondary growth, they have a simple vascular system ideal for experimental study (Nobel, 1978; Woodhouse and Nobel, 1982).

The parenchyma in the node has usually been treated as a continuation of either the pith or the cortex, and typically no attention is given it. However, it is now known that it can contain transfer cells (Gunning et al., 1970;

Kirby and Rymer, 1975; Zee and O'Brien, 1970), and these are assumed to affect the solutes carried by the vascular tissues. Fisher and Larson (1983) found the nodal parenchyma to have thick walls and numerous plasmodesmata; they speculated that these might be involved in transferring material between the axial and the lateral systems. The nodal parenchyma may also contain resin canals, laticifers, and other secretory tissues (Artschwager, 1943; Howard, 1979).

Although the leaf traces naturally attract our attention at nodes, the axial bundles themselves may have a structure within the node that is different from their structure within the internode. In *Clematis*, the nodal regions of the axial bundles developed nonextensible xylary elements earlier than did the internodal regions; nodal metaxylem elements were continuous with internodal protoxylem elements (Esau, 1965b). In *Coleus*, Bruck and Paolillo (1984) found that xylem formation began in the nodes and then extended both upward and downward into the internodal portions of the bundles. Whereas many species have bridges between adjacent bundles in the node, in a few species there is either an extensive network at the node or a complete transverse plate of vascular tissue (Dormer, 1972; Hitch and Sharman, 1971). In *Ricinus*, each internode contains 60 or more bundles, and each leaf may receive as many as 23 traces. At each node the pith is traversed by a dense network of bundles (Reynolds, 1942). In *Peperomia*, the internodal bundles lose their identity in a nodal plate; from the edges emerge leaf traces and from the top come "new" axial bundles (Dormer, 1972). The internodal vessels in Dioscoreaceae end in a nodal *xylem glomerulus*: a mass of short, irregular tracheids that resembles a jigsaw puzzle (Ayensu, 1972; Brouwer, 1953). From the top of the xylem glomerulus emerges the set of vessels for the next internode. Similarly, each node contains a *phloem glomerulus*, a mass of short, thin-walled cells that have small sieve areas, visible only by electron microscopy (Behnke, 1965).

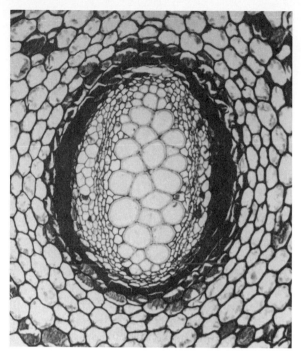

Figure 11.34
This is a haplostele of *Osmunda*. It is a protostele with a solid mass of xylem that is more or less circular. × 160.

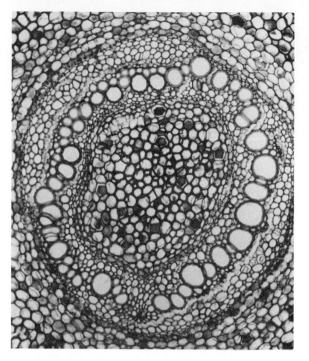

Figure 11.35
An amphiphloic siphonostele. It is amphiphloic because there is phloem on both sides of the xylem, and a siphonostele because there is a pith in the center. Because there are no leaf gaps visible, they must be scarce, so this is a solenostele. *Marsilea.* × 160.

Development of the Stele

Phylogenetic Development

The arrangements of vascular bundles in a ring or distributed throughout a ground parenchyma are not the only ones possible. The patterns of xylem and phloem are considered to form two basic types (Foster and Gifford, 1959; Smith, 1955). The first is a **protostele** (Fig. 11.34), defined as being a stele in which the xylem is located as a solid mass in the center, with no pith. The phloem surrounds this core of xylem. The second basic type of stele is the **siphonostele** (Fig. 11.35), in which the xylem, although still located interior to the phloem, is not a solid mass but rather has parenchyma in its center. The protostele is definitely older than the siphonostele phylogenetically. It was

the pattern that occurred in the first vascular plants (Stewart, 1983), and it still is the predominant pattern found in the shoots of the lower vascular cryptogams as well as the roots of almost all plants (Bold et al., 1980; Foster and Gifford, 1974). Siphonosteles occur regularly only in the shoots of ferns and seed plants and in roots that have a broad stele, mostly roots of monocots (von Guttenberg, 1968).

Ferns and fern allies were extremely popular subjects for study in the last century, and much importance was attached to the nature of their steles. It was thought that they represented a primitive stage in the evolution of the seed plants. It is important to realize that at that time the term *primitive* carried with it the concept of *inferior* and *imperfect*. Rather than thinking of the vascular tissues of the ferns as being adapted to the habit and habitats of the ferns, it was considered that these were just intermediate stages in the evolution of the more "perfect" vascular system of the angiosperms. It was thought that they might serve as indicators of evolutionary relationships, so every variation and subtlety was cataloged and named, and an extensive terminology exists. In many instances the terminology was meant also to reflect the supposed phylogeny. At present, the terminology still remains, but our ideas of the phylogeny have changed greatly; the ferns are now believed to be a separate line of evolution that diverged very early. Their origins probably were in the trimerophytes (Fig. 1.1), and they are only remotely related to the seed plants. We still accept the idea that the protostele, being the first to appear in the fossil record, was the first type of stele. Out of it arose variations of the basic type of protostele, the most important variation being the siphonostele. It is now rather well accepted that siphonosteles arose at least twice, once in the line of evolution that produced the present ferns, and once in the progymnospermopsida–seed plant line. Although our terminology was developed with the idea that the ferns and other vascular cryptogams were the basic groups of tracheophyte evolution, it still is a rather good terminology because almost all of the variations and diverse types are restricted to the nonseed plants.

PROTOSTELES.   The simplest type of protostele (and earliest in the fossil record), is the **haplostele** (Fig. 11.34), in which the xylem mass is circular in cross section (*Rhynia, Selaginella*). If the xylem margin is not smooth, but rather undulates, it is an **actinostele** (Fig. 11.36; *Lycopodium, Psilotum*). If it occurs not as one solid mass but as a series of plates and small cylinders, it is a **plectostele** (Fig. 11.37; *Lycopodium*).

SIPHONOSTELES.   Siphonosteles consist of two basic subtypes: in **amphiphloic siphonosteles**

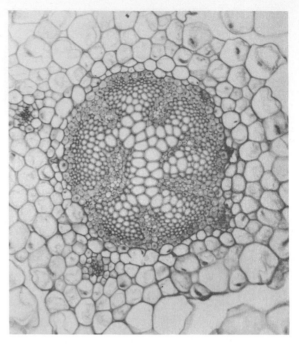

Figure 11.36
Because this protostele (no pith) of *Lycopodium* (club moss) has a star-shaped mass of xylem, it is an actinostele. Notice that this stele, as well as those in Figures 11.34 and 11.37, has the protoxylem on the exterior of the xylem mass and the metaxylem located in the center. × 160.

phloem occurs both to the inside and the outside of the xylem (Fig. 11.35), and in **ectophloic siphonosteles** the phloem is only exterior (Figs. 11.3, 11.4). Only the cryptogam lines of evolution have amphiphloic siphonosteles.

Because the protostele occurs before the siphonstele in the fossil record, it is obvious that the two types of siphonostele have evolved from it. The means by which this occurred has been much debated (see Metcalfe, 1979d, for a detailed discussion). One theory of the origin of siphonosteles postulates that the nature of the stele is a consequence of the diameter of the axis and of the stele itself. The idea is that the morphogenic control mechanisms that are responsible for controlling xylem differentiation are not able to function over large ex-

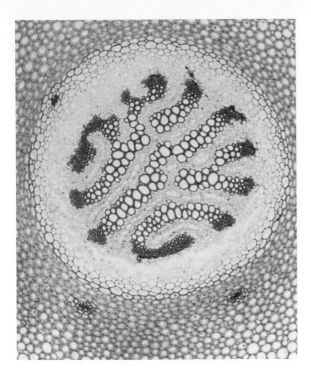

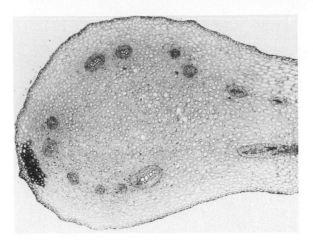

Figure 11.38
Each bundle here in this *Polypodium* is amphiphloic;
the spaces between the bundles are leaf gaps (several leaf traces and a petiole are visible on the
right). Because the gaps are so abundant, this is a
dictyostele. × 40.

Figure 11.37
The tissue between the plates of xylem is phloem,
not pith, so this is a protostele; because the xylem
consists of plates, it is a plectostele. Look carefully
at the phloem and you can see several large sieve
cells. They are not usually so easily visible. *Lycopodium annotinum.* × 100.

panses; thus, as the protostele becomes wider
in a more robust plant, it reaches a critical diameter. With any further increase, the mechanism is unable to control the center of the mass,
which therefore matures as parenchyma instead of sclerified tracheary elements (Sporne,
1962). Such a mechanism seems entirely reasonable in many ferns, because any single
shoot may have both protostelic and siphonostelic regions, depending on the vigor and diameter of the shoot when those regions were
being formed (Bierhorst, 1971). In many individuals, this switching of stele types may occur
frequently. However, switching is not the case
in seed plants: regardless of how small or nonvigorous they may be, the shoots are never
protostelic.

A second theory takes into account the fact
that, when leaf traces leave the axial protostele
and move toward the leaf, often fewer tracheary elements and more parenchyma are found
immediately above the point of departure from
the stele. Bailey (1956) was of the opinion that
these regions, if they were to become larger
and protrude more deeply into the stele, would
form isolated patches of parenchyma within
the stele. If they become even larger and if the
leaf traces occurred close to each other, the parenchymatous regions would merge, forming a
pith and a siphonostele. In this hypothesis, the
pith is really an extension of the cortex.

SIPHONOSTELES IN FERNS. In the amphiphloic
siphonostele of ferns, as a leaf trace departs
from the stele toward the leaf, a **leaf gap** (a
region of parenchyma) occurs in the stele
just above the point of departure (Figs. 11.33,
11.38, 11.39). If the leaves are not close together, then neither are the leaf traces nor the
leaf gaps. In a cross section, the siphonstele
will appear as either a complete cylinder or
as an almost complete cylinder with one leaf

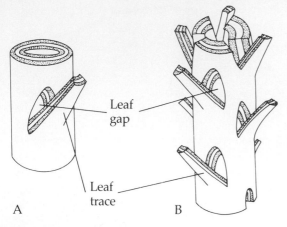

Figure 11.39
If leaf gaps are rare (A), the stele is mostly a cylinder, but as leaf gaps become more common, the stele begins to appear to consist of interconnected bundles (B).

gap. This type of rather complete amphiphloic siphonostele is called a **solenostele** (Fig. 11.33). In other cryptogams, the leaves, and therefore also the leaf traces and gaps, are so close together that a cross section reveals mostly leaf gaps, and the stele appears as a series of bundles; this is a **dictyostele** (Figs. 11.38, 11.39; Bierhorst, 1971; Foster and Gifford, 1974). These bundles are amphicribral (phloem surrounding xylem; Fig. 11.31), and sometimes each bundle is referred to as a **meristele**. The terminology to this point can be applied only to the vascular cryptogam line of evolution; the angiosperms that have bicollateral bundles are sometimes said to have dictyosteles, but these plants had a totally distinct origin for their bundles, so extending this term to them would make the classification artificial and misleading.

In the ferns, the dictyostele evolved as the leaf gaps became more numerous and larger, and they became more closely spaced. At some point, the gaps became the most striking feature, and the siphonostele appeared to be a set of bundles rather than a dissected cylinder.

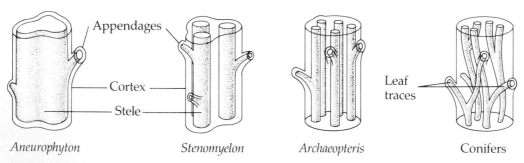

Aneurophyton   *Stenomyelon*   *Archaeopteris*   Conifers

Figure 11.40
This diagram represents a hypothesis about the origin of the siphonostele in seed plants: the original condition was a haplostele; as leaf traces came off, the margins became undulate (an actinostele); with deeper undulations, the protostele edges became separated, forming a siphonostele. Based on the information and ideas presented by Namboodiri and Beck (1968a, b).

SIPHONOSTELES IN SEED PLANTS. Ectophloic siphonosteles occur only in the seed plant line of evolution. An ectophloic siphonostele always appears to be composed of separate bundles (much like a dictyostele) and is termed a **eustele** (Fig. 11.3). The monocot stele has been named an **atactostele** (Fig. 11.8).

The origin of the eustele in the seed plant line of evolution has been studied by Namboodiri and Beck (1968a, b) and Slade (1971), and it is very different from the origin of the dictyostele. The original protostele was a haplostele (Fig. 11.40). Later, in plants such as *Aneurophyton*, the xylem protruded slightly along the radius that produced a leaf trace and indented in the areas that did not produce traces; this resulted in a xylem mass with undulating margins (an actinostele). As the margins became more undulate, the protruding regions became isolated as the indented areas merged, resulting in a stele composed of interconnected rods and plates with parenchyma surrounding the xylem. This parenchyma is not a "new" tissue, but just cortical parenchyma that extends into the regions between the xylem rods and plates. This stage is represented by the fossil *Stenomyelon*. In *Archaeopteris*, there was a greater number of individual rods, and these were disposed in a rather regular ring around a well-defined central region of parenchyma, easily distinct enough to be called a pith.

The revelation that the seed plant eustele is not a dissected siphonostele has had several consequences. First, it has made the interpretation of certain developmental-physiological studies easier. There had been some difficulty in correlating studies of vascular differentiation in ferns and seed plants, and there had also been trouble in deciding whether to analyze the seed plant eustele as an axial system that gave off leaf traces or as a series of leaf traces whose bases merged to form a network that looked like an axial system (Meyer, 1928). The problem is now resolved in favor of the former.

Also, the anatomy of the node is more complex than that of the internode, because axial bundles branch to produce leaf traces, and there may be a fusion of axial bundles. On the basis of fern anatomy, seed plant nodes were interpreted as consisting of leaf gaps, axial bundles, and leaf traces. It is now established that leaf gaps do not exist in seed plants as they do in ferns.

The study of nodal anatomy is an important source of information for phylogenetic and systematic studies, and an extensive literature exists. It is possible that the term *leaf gap* will be retained and used with the understanding that in the seed plants it represents something different from the leaf gap of cryptogams. The first classifications of nodal structure emphasized the number of leaf traces that are associated with each leaf (Hasselberg, 1937; Pierre, 1896); one-trace, two-trace, three-trace, and multitrace nodes were listed. Sinnot (1914) emphasized instead the leaf gaps: unilacunar, trilacunar, and multilacunar. The two aspects were combined by Marsden and Bailey (1955): unilacunar–three trace, and so forth. The extensive literature that deals with the taxonomic-phylogenetic aspects of nodal structure has been reviewed in detail by Howard (1979a).

Unusual Steles

In some ferns, such as *Marattia*, *Matonia*, and *Pteridium*, there are two or more concentric vascular cylinders, termed **polycyclic steles**, not just one (Fig. 11.41; Bierhorst, 1971; Fahn, 1982; Farmer and Hill, 1902; West, 1917). The outer vascular cylinder forms in the young plant, and the inner cylinder forms in the upper portions of the stem as the plant becomes larger. They are interconnected at the base of the inner stele, and apparently water and nutrients can move from one to the other.

In some roots and shoots, several steles run parallel to each other, forming **polysteles** (Scott, 1891). In cross section, the steles may be organized into a ring, or they may be scattered. The individual steles can anastomose with each other to form a network. Polysteles have been reported in Acanthaceae, Gunneraceae, Nymphaeaceae, Palmae, Parnassiaceae (Arber, 1925; Metcalfe, 1979d; Tomlinson, 1961). The functional and adaptive significance of polystely is in need of study.

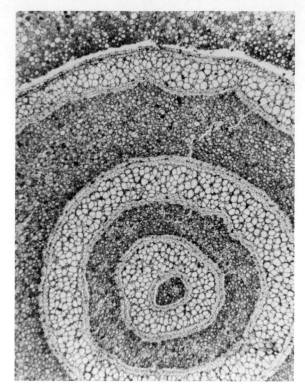

Figure 11.41
Matonia pectinata stems have concentric steles; in this micrograph, three steles are visible. The xylem is easy to identify; the phloem consists of narrow bands along the surfaces of the xylem cylinders. Between the outer phloem of one stele and the inner phloem of the next is a band of dark parenchyma. × 40.

Concepts

Leaves constitute one of the three basic organs of higher plants, the other two being stems and roots. Leaves are very familiar to all of us, and their vital role in photosynthesis makes them an abundant and conspicuous feature of our biotic surroundings. But photosynthesis is only one of the functions fulfilled by leaves; there are numerous others. Of the three organs, leaves have been the most plastic evolutionarily, and numerous types have evolved in response to various types of selection pressures. The need for photosynthesis has resulted in **foliage leaves** (rarely also called **trophophylls**), but leaves also function to

1. Protect developing vegetative buds: **bud scales** or **cataphylls**.

2. Protect developing inflorescences: **floral bracts** or **hypsophylls**.

3. Protect individual flowers: **sepals**.

4. Store nutrients for the embryo: **cotyledons**.

These five types of leaf are almost universally present in the flowering plants, and the gymnosperms, which lack flowers, have leaves in their cones that are similar to hypsophylls and sepals in function. The collective term for all the leaves of a plant is **phyllome** (Arber, 1950). In certain individual families or genera of angiosperms, the phyllome contains a variety of leaf types, because the leaves have been modified into spines, tendrils, glands, insect traps, and food storage organs (bulbs). In many of these, the structure of the leaf has been so extensively altered evolutionarily that it can be quite difficult to determine that the structure is indeed a leaf. On the other hand, for many desert succulents, the large surface area of foliage leaves would present too great a danger of water loss; in these species, selection favored the suppression of foliage leaf development. Although these plants are often described as being "leafless," this is seriously misleading: the shoot apical meristems of all seed plants and ferns produce leaf primordia, and these undergo at least a small amount of development. These leaves may stop growing and be mature while still microscopic, but none of the ferns or seed plants is ever actually truly leafless.

Structure of Foliage Leaves

The dominant function of foliage leaves is, by definition, photosynthesis. In order to carry out photosynthesis, the leaf must absorb carbon dioxide from the air, and it must absorb sunlight. Light is gathered efficiently by chlorophyll, and a leaf should contain a band of chlorenchyma that is not thicker than the penetrability of light (Figs. 12.1, 12.6); otherwise the lower cells would be shaded by the upper ones. Consequently, leaves should be thin. Because light comes from one direction, leaves

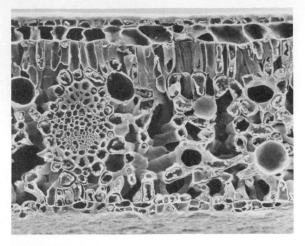

Figure 12.1
This SEM of a leaf of *Laurelia sempervivens* shows the three-dimensional arrangement and relationships of leaf tissues. From the top, you see a thin epidermis with a thick cuticle, a hypodermis with large cells, a palisade parenchyma only one cell thick, spongy mesophyll with large intercellular spaces, and the lower epidermis. A vascular bundle is visible, with xylem in the upper half; phloem fibers are also present. × 400. Scanning electron micrograph generously provided by G. Montenegro.

Figure 12.2
This is a leaf clearing of *Coleus blumei;* the leaf was treated to bleach out all pigments, then a dye specific for lignin is applied, allowing the xylem to be studied. Notice the petiole, the midrib, and the secondary veins that come off the midrib. Tertiary veins come off the secondaries and also can come directly off the midrib. × 1.0. Photograph generously provided by W. A. Russin and R. Evert.

should be flat; to prevent one leaf from shading those below, the lamina is projected away from the stem by a narrow petiole (Fig. 12.2). In tropical climates that never freeze and never have prolonged periods of drought, leaves can persist for many years and can be very large. To support their weight, a great deal of sclerenchyma may be present; and, to avoid being torn by wind, the leaves are frequently compound, allowing the individual leaflets to flutter on the rachis without the entire leaf's acting like a single sail. A compound leaf is also able to avoid overheating better than a simple leaf with the same area. When the wind blows over a large smooth surface, a thick undisturbed boundary layer of air serves as a thermal insulator; if that surface is divided into many

small pieces, the wind becomes turbulent when it flows by, and the turbulence can carry away heat (Gates, 1980; Nobel, 1983). Large perennial leaves do not function well in cold or dry climates, and only the evolution of small deciduous leaves allowed plants to migrate successfully into numerous temperate areas. Once leaves became deciduous, a very different structure became advantageous. Because a plant will abscise a leaf after six to eight months, it is necessary for the plant to quickly recover the investment of energy, carbon, and nitrogen it spent while making the leaf; therefore, large amounts of sclerenchyma should be avoided, so the leaf must be smaller. With reduced size, being compound is less important. The correlation between investment in a leaf and leaf longevity is further exemplified by xeromorphic leaves: these are typically very hard (**sclerophyllous**) to prevent animals from eating them. To recover the investment, these leaves must be, and are, perennial, although they are small and occur in temperate areas. Examples are conifers (leaves up to 12 years old), *Buxus*, *Hakea* (5 years), *Ilex*, and *Rhododendron* (3 years; Howard, 1979a). There are numerous ecological and mechanical constraints acting on leaves, and the structure of present-day foliage leaves is an efficient, elegant adaptation to those constraints.

Epidermis

The epidermis on leaves has the same general properties as were described in Chapter 10, but of course the need for efficient photosynthesis will influence the presence or absence of particular epidermal characters. Because the leaf is dorsiventral, the epidermis on the upper (**adaxial**) surface is in an environment that is different from the epidermis on the lower (**abaxial**) surface, and it has different functions. During the day, most leaves are at least as warm as the surrounding air, so convection currents rise from them, and even a mild wind would carry air away from the surface (Nobel, 1983). Such air movement would carry any transpired water vapor away from the leaf. Consequently, the adaxial epidermis typically has few stomata or even none (species of Aceraceae, Begoniaceae, Epacridaceae, and Portulacaceae), whereas the abaxial epidermis has many. The water that is lost from these cannot be carried away by convection currents; furthermore, the abaxial epidermis frequently is covered with trichomes that further prevent air movements from sweeping water away. If the leaf floats on the surface of a pond or stream, then abaxial stomata are useless, and all will be on the adaxial side of the

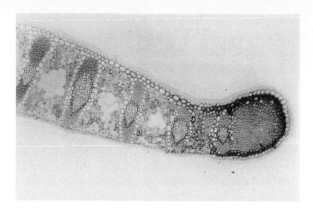

Figure 12.3
This leaf of *Phormium tenax* (New Zealand flax) has a thick mass of fibers along the margin of the leaf; with this reinforcement it is resistant to tearing. The rest of the surface has little sclerenchyma, but each bundle has a sheath and fibrous bundle sheath extensions. × 160.

leaf (Figs. 12.4, 12.8; Nelumbonaceae, Nymphaeaceae, Ranunculaceae). Leaves of submerged plants (Nymphaeaceae, Podostemonaceae) are reported not to have stomata at all; neither do the leaves of many nonphotosynthetic, achlorophyllous parasites (Balanophoraceae, Monotropaceae, Rafflesiaceae; Metcalfe and Chalk, 1979).

The upper epidermis characteristically has a thicker cuticle and more waxes than the lower epidermis, probably both to prevent transcuticular transpiration and to reflect away excessive sunlight. Epidermal cells of both surfaces tend to be flat and tabular, only rarely having warts or papillae. All walls are usually thin, except on leaf margins where it is not uncommon to find thick-walled epidermal cells overlying a sclerenchymatous hypodermis (Figs. 12.3, 12.7). This arrangement reinforces the edge of the leaf and prevents wind-induced tearing.

A hypodermis is frequently present on the margin of a leaf, as just mentioned, but over the majority of the leaf surface one may or may

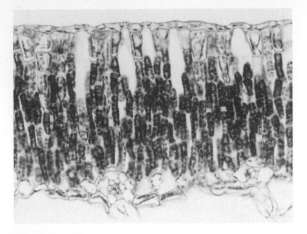

Figure 12.4
This leaf of water lily (*Nymphaea alba*) has a tremendously thick zone of palisade parenchyma, being as much as seven layers deep in some places. The cells of all layers are filled with chloroplasts. The adaxial (upper) epidermis is visible and has numerous stomata. × 400.

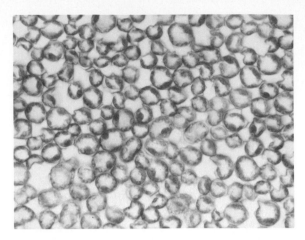

Figure 12.5
This is a paradermal section (cut parallel to the surface) of a leaf of *Ligustrum* (privet). Compare it with the SEM view (Fig. 12.1) and the cross-sectional view (Fig. 12.4). You can see how close the cells are, but they have little contact; there is a very large surface-to-volume ratio, which favors the absorption of carbon dioxide. × 635.

not be present. It is present in Chloranthaceae, and Monimiaceae, Piperaceae; for a complete list of families which have this and other special leaf characters, see Metcalfe and Chalk (1979). A hypodermis occurs more often in leaves that are tough and sclerophyllous; thinner, softer leaves generally do not have a hypodermis. The hypodermis will often be most prominent over the largest leaf veins, especially the midrib.

Palisade Parenchyma

All the leaf tissues other than epidermis and veins are collectively known as the **mesophyll**. Although the mesophyll resembles the cortex and is continuous with it by means of the petiole, the leaf tissues are never called cortex; that term is restricted to stems and roots.

Immediately below the upper epidermis on most dorsiventral foliage leaves is the **palisade parenchyma**, a layer of columnar chlorenchyma cells (Figs. 3.3, 12.1, 12.4, 12.5). This tissue is responsible for most photosynthesis, and its structure is remarkably well suited for it. The long, narrow cylindrical cells are rich in chloroplasts; these are pressed firmly against the plasmalemma and wall by a large central vacuole. Carbon dioxide dissolves only slowly into water or hyaloplasm, so a large absorptive surface area is needed, which is provided by having the cells separate from each other. They almost touch, being separated by only a few micrometers (Figs. 12.1, 12.5), but this is enough so that almost the entire surface of each cell is exposed and carbon dioxide can circulate freely around and between the cells, dissolving into the whole surface. Because the chloroplasts are pressed near to the wall, the carbon dioxide is immediately pulled out of the hyaloplasm, so no matter how much moves across the wall, the hyaloplasm is never saturated, and a steep diffusion gradient exists. In some species, small projections, as in Xanthorrhoea-ceae (Fahn, 1954) and *Eucryphia cordifolia* (Montenegro, 1984), or lobes, as in *Lilium*, ensure that the cells cannot maintain extensive contact faces. It is important that the palisade cells not be separated from each other by too great a space, because apparently the contact region between these cells and the upper epidermis forms a capillary system that functions to distribute water to the cells from the veins (Wylie, 1943). The palisade cells contact the epidermis in the same way that the bristles contact the back of a brush. With the cells or the bristles widely separated, nothing happens if a drop of water is inserted. But with a close spacing, the water droplet will immediately spread across the entire epidermis and come into contact with all of the cells. Because most palisade cells do not actually touch a vein, this distribution is important.

The palisade parenchyma may be arranged in two rows, three rows, or even five (Figs. 12.4, 12.6; *Berberis trifoliolatus*, *Picconia excelsa*). This occurs in plants exposed to strong sunlight, and the lower layers will receive enough light to photosynthesize efficiently. If several layers are present, the innermost is typically rather irregular and contains cells that are shorter than those of the outermost layer. Also, because these plants are receiving such strong sunlight, they are usually also xerophytes and have other unusual characters. In a small number of plants, again mostly xerophytes (*Atriplex portulacoides*, *Artemesia*, *Myoporum*, *Sonneratia alba*; common in the Crossomataceae, Myrtaceae, and Platanaceae), the palisade parenchyma is found on both the upper and lower side of the leaf (Fig. 12.7). The presence of very penetrating sunlight probably also makes this possible. The structure of the leaf can be completely reversed if necessary: some species of *Frankenia* grow as cushion plants in the severe, cold climates high in the Andes (Espinosa, 1932; Hauri, 1917; Qùézel, 1966, 1967). The tiny leaves have their adaxial surface permanently pressed against the opaque stem: in

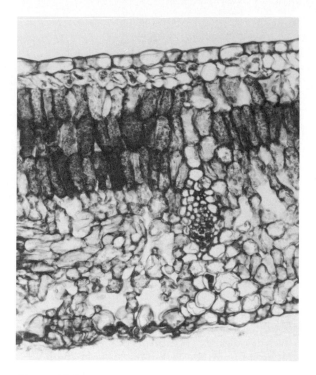

Figure 12.6
This leaf of *Berberis* (barberry) shows a three-layered zone of palisade parenchyma that has almost no intercellular spaces; it also has a spongy mesophyll that is quite compact. These two features are usually good indicators of a xerophyte. There is much less surface area in this leaf than there is in *Ligustrum* (Fig. 12.5), so water will be lost from the cells much more slowly. × 400.

these leaves, the palisade parenchyma must be purely abaxial (Metcalfe, 1979).

Leaves in which the upper and lower surfaces are similar are said to be **isolateral** or **isobilateral**, whereas leaves in which the two sides are different are **bifacial** or **dorsiventral**. The term *bifacial* is much more accurate because it unambiguously indicates that the two surfaces differ, whereas *dorsiventral* has gradually come to indicate simply that a leaf (or any other structure) is flattened top-to-bottom as opposed to being radially symmetrical or laterally flattened. Isolateral leaves often occur on

Figure 12.7
This leaf of *Eucalyptus* is iso-bilateral; although there is an adaxial and an abaxial side, both have a palisade parenchyma, and it is difficult to tell which side is which without looking closely. The clue is in the vascular bundles: the xylem is on the adaxial side. × 160.

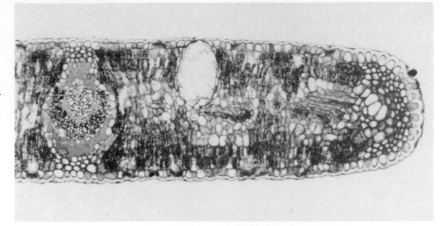

Figure 12.8
This leaf is just the opposite of the *Berberis* (Fig. 12.6). Instead of a compact, dense tissue, we see gigantic air spaces in the spongy mesophyll. This is not from a desert plant; these are the floating leaves of water lily (*Nymphaea alba*). At this magnification it is hard to see, but the lower epidermis has no stomata; they are all in the upper epidermis (Fig. 12.4). × 100.

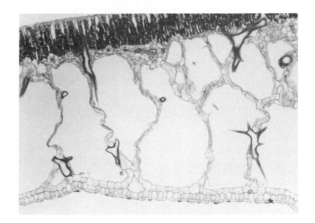

submerged plants, and the mesophyll is homogeneous, without being differentiated into palisade and spongy mesophyll. Many grasses in temperate regions also lack palisade parenchyma, with the mesophyll cells being arranged radially around the vascular bundles. (See the section "Kranz Anatomy" later in this chapter.)

Spongy Mesophyll

Between the palisade parenchyma and the abaxial epidermis is the **spongy mesophyll**, a region of mesophyll in which the cells are very widely separated from each other, such that the apoplast is typically larger than the symplast (Fig. 12.1). The leaves of several species have

been studied, and it has been found that the apoplast of the whole leaf can range between 4.1% and 52.3% of the leaf volume (Figs. 12.8, 12.11; Byott, 1976; Fagerberg and Culpepper, 1984; Fisher, 1985; Veres and Williams, 1985). A consequence of the presence of spongy mesophyll is that the abaxial epidermis is held away from the palisade parenchyma, creating an extremely open aerenchyma and greatly enhancing the circulation of carbon dioxide. If there were no spongy mesophyll, the palisade parenchyma would contact the abaxial epidermis, and the inner aperture of the stomata would be densely surrounded by palisade cells (Fig. 12.9). Molecules of carbon dioxide that entered a stoma would be confined by these cells and could not diffuse away. If the carbon

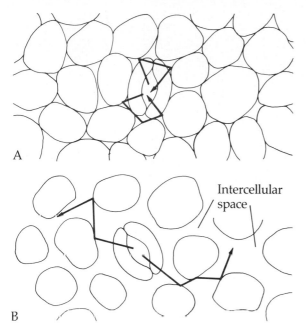

Figure 12.9
(A) If the stomata were located against a palisade parenchyma, the entry of carbon dioxide would be inhibited, because the molecules would bounce off the walls of the cells and might bounce back out of the stomatal pore. (B) With a spongy mesophyll around the stoma, the molecules of carbon dioxide can penetrate deep into the leaf quickly.

dioxide could dissolve into the cell instantaneously, this would be no problem, but, due to the low solubility of carbon dioxide, when a molecule strikes a cell it usually bounces off. If it were restricted to the vicinity of the stoma, it probably would bounce back out of the leaf. With the spongy mesophyll present, once a molecule of carbon dioxide enters the stoma, over half the volume around it is intercellular space, and the molecule will soon be too far away from the stoma to be lost.

The spongy mesophyll also allows the leaf to be flexible; as it flutters in the wind, it bends and flexes slightly, causing the apoplast to be mildly compressed and expanded. This may cause the gases inside the leaf to be circulated more rapidly and thoroughly than could be achieved by diffusion alone.

In isobilateral leaves the spongy mesophyll will occur as a band through the center of the leaf and will hold the two sets of palisade parenchyma apart (Fig. 12.7). In many isobilateral leaves, a typical spongy mesophyll is not present; instead, there is just a layer or two of small, noncolumnar cells. Other types of mesophyll cell arrangements have been reviewed by Meyer (1962).

Because the intercellular spaces of leaves are so critical to photosynthetic efficiency, they have been studied extensively. In all except a few aquatic leaves, the spaces form schizogenously as the epidermis and vascular bundles continue to expand after the spongy mesophyll and palisade parenchyma cells stop. As soon as the spongy mesophyll cells finish expanding, the growth of the other tissues tears them from each other; shortly thereafter the palisade parenchyma cells stop expanding and are similarly separated. Total leaf expansion ends shortly, and the palisade parenchyma cells remain close together.

Surface-to-volume ratios are important in all processes in which material that is to be metabolized must be transported into or out of the site of metabolism. In leaves, two such ratios are important (Nobel, 1983). First, the **external surface-to-volume ratio** is a measure of the ability of the leaf to absorb carbon dioxide and to lose water, related to the amount of photosynthetic tissue present. Under mild conditions in which water is generally available, this ratio will be large, because the leaves are big, thin, and planar (Fig. 12.2). But in habitats with water stress, the leaves are smaller and thicker and tend to be cylindrical, all of which reduce the transpirational surface and thus reduce this ratio. Second, the **internal surface-to-volume ratio** measures the ability of the leaf to dissolve carbon dioxide in relation to the amount of photosynthetic tissue available. This ratio is almost always much larger than the first. The palisade parenchyma has a much higher amount of surface per unit volume than does the spongy mesophyll, from 1.6 to 3.5 times more (Turrell, 1936). But this large internal surface area, in addition to facilitating carbon dioxide uptake, also allows water to leave the cells rapidly (Nobel, 1983; Stålfelt, 1956). However quickly water is lost out of the stomata, a large internal surface area causes the internal air to remain humid, and more water can be lost. Xerophytic plants tend to have a low internal surface-to-volume ratio, because there is increased contact between mesophyll cells, or there may be no intercellular spaces at all (Fig. 12.10). This greatly reduces the ability to absorb carbon dioxide, but the benefit of water conservation apparently offsets this. Stereological methods especially designed for studying such leaf parameters have recently been developed (Fig. 12.11; Parkhurst, 1982).

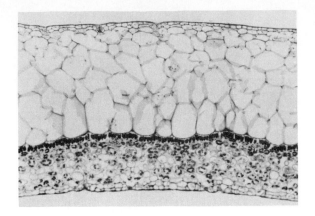

Figure 12.10
Although most leaves do not store material, the leaves of *Peperomia* have a thick layer of colorless, water-filled tissue just above the palisade parenchyma. These plants are xerophytic epiphytes that live in dry sites within rainforests. × 320.

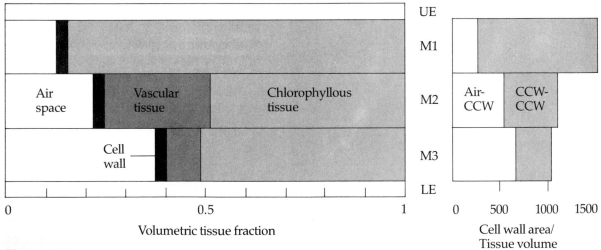

Figure 12.11
These diagrams represent the volumes and surface areas of the tissues in the various parts of a leaf of *Coreopsis*. UE and LE are the upper and lower epidermis and obviously have none of the types of tissues measured. M1 is the upper mesophyll layer: it has a low fraction of its volume as air, most as chlorophyllous tissue, and very little as vascular tissue or cell wall. Most of the wall area occurs as contact faces with other walls. In the middle layer, which contains the vascular bundles, there is an increase in vascular tissue (of course) and also an increase of air space; there is a corresponding decrease in the volume of the chlorophyllous tissue. To the right are the surface areas per tissue volume in cm^2/cm^3. Over half of the wall area in M2 and M3 faces intercellular space rather than another wall. In the paper from which this diagram was taken (*Amer. J. Bot.* 69:31–39), there is a similar analysis of *Heliopsis*, and this method of presentation allows the two leaves to be compared quickly. Diagram generously provided by D. F. Parkhurst.

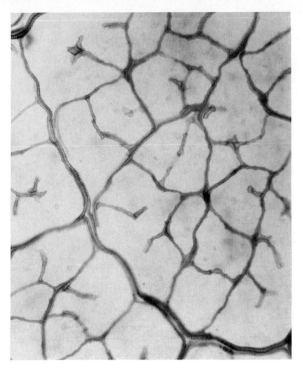

Figure 12.12
This is a leaf clearing of *Buddleja* (butterfly bush).
The areoles can be seen, as well as the vein endings and the xylem of the minor veins. The veins
may end blindly or they may branch. Notice that
no part of the mesophyll is far from a minor vein.
× 160.

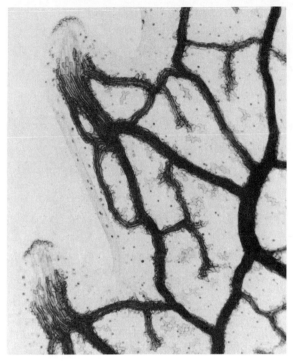

Figure 12.13
The veins of *Hedyosmum arborescens* have a sheath of
very irregular sclereids. The sclereids extend all the
way to the end of the bundles and cover even the
tip. The teeth on the margin of the leaf are hydathodes. × 40. Material prepared by C. Todzia.

Vascular System of the Dicot Foliage Leaf

The vascular system of the foliage leaf functions in several ways. It brings water into the
leaf from the stem, and it distributes the water
throughout the leaf mesophyll. The phloem
component absorbs the carbohydrates that
have been produced by photosynthesis and
carries them out to the stem. In order to accomplish these two types of distribution and
absorption, the vascular tissues of the leaf
must have intimate contact with the parenchyma cells. The relatively large veins that enter through the petiole branch into finer and
finer bundles. In dicots, there is usually one
large central vein (the **midrib**) from which
branch **secondary veins** (or **first order laterals**);
from these come **tertiary veins**, and so on
(Fig. 12.2). The small portions of leaf that are
bounded by the finest veins are called **areoles**
(Figs. 12.12, 12.13). The midrib typically runs
down the center of the leaf, but in many genera it branches near the base, and no single
midrib is present. The actual number of variations is enormous, and an extensive terminology has been established (Hickey, 1973, 1979).
As with the classification of types of stomatal
complexes, it is not necessary to learn all of the
variations. With leaf venation, the branching
pattern of the veins is probably the result of

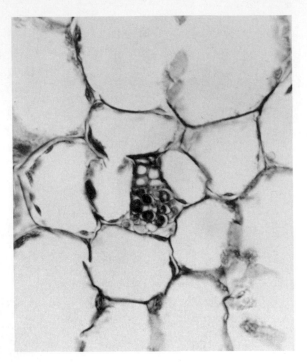

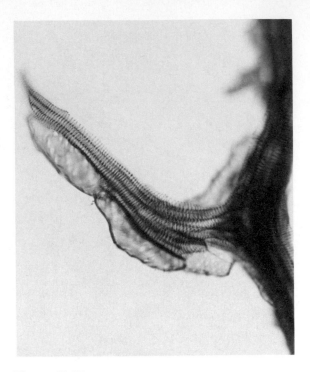

Figure 12.14
The minor veins can be extremely delicate, as the leaf clearings (Fig. 12.12) have shown; here you can see one in cross section in *Alluaudia procera*. Notice the large parenchyma cells that have full contact with the vein. × 635.

Figure 12.15
This is a vein ending in *Hedyosmum bonplandianum*; there is a relatively large number of long, spirally thickened conducting tracheids accompanied by three unusual cells. They have secondary, lignified walls with irregular pits. These may be just another type of sclereid (Fig. 12.13), but they have such a large, open lumen that the possibility exists that they function in water storage. × 400. Material prepared by C. Todzia.

growth and expansion patterns in the developing lamina. It is more important to learn the principles that govern the structure and organization of the veins. The venation pattern is extremely useful in taxonomy, because the patterns formed by the major and minor veins are constant and often very characteristic. They are especially valuable when trying to identify a fragment of a fossil leaf or a piece of milled plant material to be used for food or drug manufacture.

Loading and unloading of the veins happen primarily in the finest of veins, called the **minor bundles** (Figs. 12.12, 12.13, 12.14, 12.15, 12.17; Evert, 1978; Evert et al., 1977; Fisher and Evert, 1982a, b; Franceschi and Giaquinta,

1983a, b; Gunning et al., 1974; Russin and Evert, 1985). Consequently they are the longest component of all; in *Amaranthus retroflexus*, they constitute 95% of the total vein length, providing an extensive surface for loading (Fisher and Evert, 1982b). These veins may end blindly in the leaf mesophyll, or they may run from a larger vein on one end to a larger vein on the other. These minor veins always contain primary xylem, which is usually in the form of annularly or spirally thickened tracheids (Fig. 12.15; Strain, 1933). This is necessary because

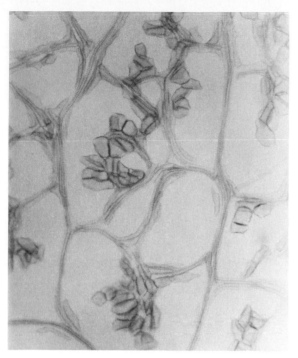

Figure 12.16
The vein endings of this leaf of *Euphorbia millii* (crown of thorns) have terminal tracheids: they are short and irregular and have the appearance of a parenchyma cell that redifferentiated into a tracheid. × 160.

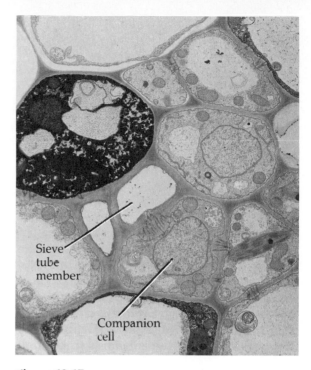

Sieve tube member

Companion cell

Figure 12.17
This is a transmission electron micrograph of a minor vein of *Populus deltoides* (cottonwood); especially important to examine are the several cells of vascular parenchyma. They are situated all around the sieve tube members and companion cells. Several prominent areas of plasmodesmatal connections are visible. × 9160. Micrograph generously provided by W. A. Russin and R. F. Evert.

they are established during the final stages of the expansion of the lamina. Larger, pitted tracheids and vessel elements usually are not found in the minor veins, nor is there ever any secondary growth. In veins that end blindly, the last several tracheids are often unusual (Figs. 12.15, 12.16; Strain, 1933; Tucker, 1964). They may be quite large and irregularly shaped, and frequently they do not actually contact each other, but instead lie free in the mesophyll parenchyma. These are termed **terminal tracheids** (Fahn and Arzee, 1959; Foster, 1956), and there is some thought now that they serve to store water for the leaf, rather than distribute it (Pirwitz, 1931; Zimmermann, 1983).

Primary phloem may or may not be present

in the minor veins (Esau, 1967). When present, it is in the form of small sieve tube elements (Figs. 12.14, 12.17); companion cells may be either present or absent. Interest in these has increased tremendously with the discovery that many of the parenchyma cells of the minor veins are actually transfer cells (Gunning et al., 1968; Gunning and Pate, 1969, 1974; Pate and Gunning, 1969, 1972). The transfer cells have large contact surfaces with surrounding chlorenchyma cells and with the apoplastic space. This allows them to absorb sugar over a large area and then pump it forcefully into the sieve

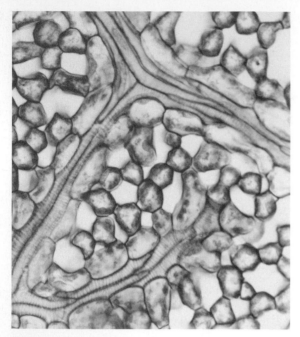

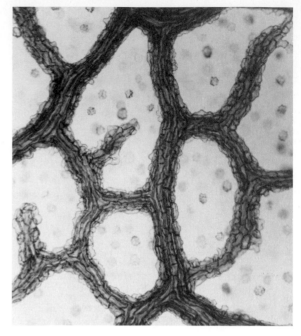

Figure 12.18
The relationship of the bundle sheath to the vein and mesophyll is very clear in this paradermal section of *Ligustrum* (same leaf as Fig. 12.5). The tracheids and sheath cells fit perfectly together; then the sheath cells interface with the mesophyll. × 635.

Figure 12.19
The minor veins of *Hedyosmum toxicum* are massive; even the vein endings are very wide with several rows of tracheids, and the bundle sheaths contain large cells. The mesophyll contains many druses. × 100. Material prepared by C. Todzia.

tube members. This active loading also initiates transport, according to the Münch Pressure Flow hypothesis. In some species, this loading occurs in the next larger set of veins, because the very finest veins contain a phloem composed solely of nonconducting parenchyma; sieve tube members and sieve areas do not occur in the minor veins of these species.

The actual conducting elements of the xylem and phloem seem never to occur suspended in a large intercellular space; the veins always have a sheath of tight-fitting parenchyma cells that keep the tracheids and sieve tube members from being free in the spaces between mesophyll cells (Figs. 12.1, 12.18, 12.19). This **bundle sheath** is often quite conspicuous on larger bundles but can be easily overlooked on the

fine bundles. The sheath cells of minor bundles that are transfer cells are classified into two types. **A-type sheath cells** are actually the companion cells of the sieve tube members, and they can be distinguished by having wall ingrowths on all sides (Evert, 1977). **B-type sheath cells** are not companion cells, and they have labyrinthine ingrowths most extensively developed on the wall that contacts the sieve tube members (Chapter 8). This discovery is quite recent, and the number of plants that have been carefully examined is not great. Other arrangements of minor vein sheaths should be expected.

The fine veins are attached to larger veins of higher orders. These contain a greater amount of both xylem and phloem, and the individual

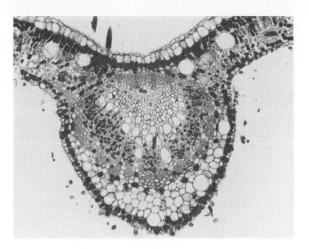

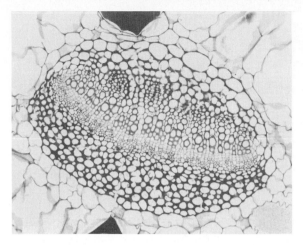

Figure 12.20
The midrib is always much more massive than the
minor veins: it must be able to conduct as rapidly
as all the minor veins combined if it is to keep them
supplied with water and if it is to be able to export
sugar efficiently. This midrib of a species in the Annonaceae has an extremely common form. Above
the vascular tissue is a mass of fibers; below that is
xylem, first protoxylem and then metaxylem; next is
a substantial amount of phloem, with a lower zone
of phloem fibers. Notice that the midrib bulges outward below the lamina; this is almost always true
of the midrib and larger veins, that they do not
project above the lamina but below it. × 160.

Figure 12.21
This midrib of *Carpobrotus* differs from that in Figure 12.20 by having large amounts of collenchyma,
which is also extremely common. Notice how
orderly the rows of xylem cells are and that a cambiumlike region is located between the xylem and
the phloem. However, procambium can have the
exact same appearance (see Fig. 12.22). × 160.

conducting elements are usually larger. The
bundle sheath parenchyma is also more prominent. Vein loading and unloading occur from
these as well.

In very large leaves, numerous higher orders
of veins may exist, but in simple, small leaves,
the midrib may be next (Figs. 12.20, 12.21,
12.22, 12.23). With each higher order, the veins
become larger and more complex, and vessels
may be found in the xylem. In a few plants
with perennial leaves, a true cambium arises in
the midrib and produces secondary tissues,
but the amount is always very small (Samantarai and Kabi, 1953). The amount of primary
vascular tissue can be extensive, however, and

it is in the large leaf veins that the similarity of
primary growth and secondary growth can
most easily be seen (Fig. 12.22). Large amounts
of metaxylem can occur, and most of it is arranged in perfect rows, just as if it had been
produced from a vascular cambium. The midrib may contain more than one vascular bundle
(Fig. 12.23), and if so the several bundles can
be arranged in a flat row, an arc (*Ambrosia*), or
as a ring of bundles, as in *Liriodendron* and
Vitis (Plymale and Wyle, 1944).

In addition to the larger amounts of conducting tissues, greater amounts of other cells
are found, including xylem parenchyma and a
much more substantial bundle sheath (Fig.
12.21). The bundle sheath of larger veins is
composed of parenchyma most frequently, but
sclereids and fibers can also occur. An important feature of the larger bundles in some species is the presence of **bundle sheath extensions**: these are wings composed of cells located

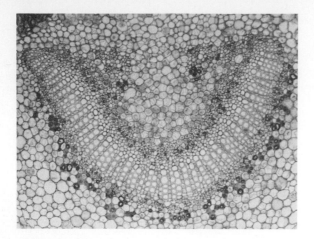

Figure 12.22
In this midrib of *Nerium oleander* (oleander), the xylem looks as though it had been formed by a vascular cambium. The primary xylem of cauline bundles is usually not so regular, but it frequently is in leaf petioles and midribs. You can tell that this is not secondary growth by looking at the phloem: phloem cells are not aligned with xylem cells. There is also phloem on the upper side of the xylem and in the mesophyll between the upper ends of the xylem. × 160.

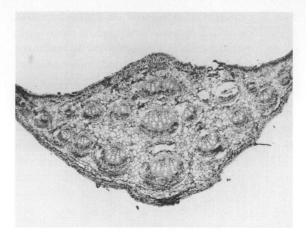

Figure 12.23
If you look closely, you might notice that this midrib of *Couroupita guianensis* (cannon-ball tree) is a bit complex. The leaves themselves do not appear especially complicated. × 40.

between the bundles and one or both of the epidermises (Fig. 12.7). The bundle sheath extensions may consist of either parenchyma cells or fibers, and they are thought to play a major role in the distribution of water away from the xylem out to the mesophyll (Pizzolato et al., 1976; Wylie, 1943, 1947, 1949, 1951). Movement from the veins to the epidermis may be by apoplastic capillary action, although the actual mechanism by which water moves through the mesophyll and epidermis is still not known for certain (Meidner and Sheriff, 1976; Sheriff, 1982). As a rule, all of the largest veins of a leaf have bundle sheath extensions while none of the finest veins do, but the variation is great. In *Quercus calliprinos* 94% of the total veins length have bundle sheath exten-

sions, whereas in *Q. boissieri* only 71% of the length is so accompanied, and in *Pistacia palaestina* just 50% (Fahn, 1982). In *Olea* and *Pistacia lentiscus*, extensions do not occur at all.

Whenever the bundle sheath or bundle sheath extensions contain significant amounts of sclerenchyma, then a mechanical role in support or protection must be considered. Even though many leaves do have sclerified veins, most are able to wilt and droop, indicating that support is supplied almost fully by turgor pressure. The sclerenchyma that accompanies the bundles may be more important in deterring insects from sucking or chewing the vascular tissues.

Petiole and Node of Dicot Leaves

The lamina of most dicot leaves is attached to the stem by means of a petiole (Fig. 12.2), although **sessile leaves** lack a petiole and their lamina is attached directly to the stem. As mentioned above, one of the main functions of the petiole is to hold the lamina away from the stem so that the leaves fill a greater space without overlapping and shading each other (Chazdon, 1985; Horn, 1971; Niklas and Kerchner, 1984). The petiole also permits a rather free movement of the leaf in wind (e. g., quaking aspen); such movement not only may help cool the leaf (Nobel, 1983) but also may deter insects from landing on it. The abscission zone, which is responsible for the removal of the leaf, is located in the petiole; because the petiole is narrow, the broken, vulnerable area that results from abscission is quite small.

As described in Chapter 11, the vascular tissues for the leaf separate from the stele at the node. The leaf trace or traces then turn outward and upward toward the petiole (Howard, 1979a, b). If the cortex is narrow (which is the typical case), the leaf traces have only a short amount of cortex to traverse. The cortex in this nodal region often is slightly swollen, being somewhat wider than the cortex of the internode. In some cases, it is very prominently swollen and may be referred to as a **leaf base** or (rarely) as a **podarium** (pl.: podaria). As the leaf traces traverse the cortex, they may fuse with each other or may remain independent; it is also possible for them to branch. Once the leaf traces have actually entered the petiole, fusions, branchings, and other types of arrangements are common. The anatomy of the petiole is an excellent taxonomic character, but it is important to examine the anatomy at the base, middle, and apex of the petiole because of the extensive arrangements that occur (Dehgan, 1982; Howard, 1979b). The bundles, at any particular level, may be arranged in a closed or open circle (*Aquilegia, Geranium, Ricinus*), an arc (*Olea, Nicotiana*), a plate (Fig. 12.24; *Betula*), scattered (*Rumex*), or any combination of these. The bundles themselves may be collateral, bicollateral, or concentric. The arrangement frequently changes as the bundles traverse the petiole; there is not a constant relationship between xylem and phloem. A classification of petiolar anatomy has been established (Howard, 1962) and recently updated (Howard, 1979b).

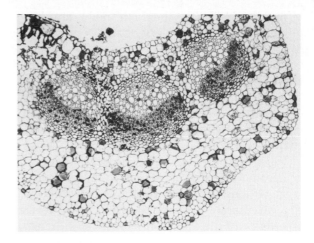

Figure 12.24
This midrib of *Saururus cernuus* (lizard's tail) has relatively abundant phloem. The xylem consists of a mixture of predominantly metaxylem vessels and parenchyma. × 100.

The nonvascular, nonepidermal portions of the petiole are simply called "ground tissue." Although it is continuous with the cortex, it is never referred to as cortex. Large amounts of collenchyma or fibers may be present, and laticifers and other secretory structures may be well developed. Secretory ducts, such as resin canals and laticifers, will often be continuous from the cortex into the petiole, but the number may decrease at the base of the petiole. Howard (1979a) stated that the number of resin canals at the abscission zone of the petiole is usually equal to the number of leaf traces present. An endodermis may (rarely) be present; if so, it may encircle the entire set of bundles, or each bundle may have its own endodermis. If there is a single dominant midrib, then all of the petiolar bundles must come together and form it, of course. But if the lamina has several large veins that arise at its base, then each of these may come from a separate petiolar bundle.

Howard (1979a, b) has stressed the importance of considering the stem, node, and leaf as a continuum, which is certainly very true. A great deal of emphasis has been given to the course of vascular bundles in the stem, but there has been very little study of how the various leaf traces are distributed to the parts of the leaf (Fisher, 1985). Just as the stelar reticulum and intervessel pitting allow points of damage to be circumvented, it seems reasonable to assume that the same safeguards exist in leaf traces and petioles, especially if the leaf is at all large or long-lived. Similarly, any given point of a vascular bundle must be able to conduct as well as all of the lower order veins that branch off it (Fisher, 1970). Another important consideration in this continuum is the extreme danger of cavitation that leaves represent. Leaves are at the top of the transpiration stream, and the water columns are subjected to the greatest tension in leaves. The leaves are favored food objects for numerous animals, and any time a vessel is bitten it will automatically cavitate. Furthermore, leaf abscission in the autumn is perfect for causing air-seeding of vessels. Obviously there must be a mechanism for preventing the embolisms that arise in the leaves from passing into the stem. At the abscission zone, the xylem has often been found to contain only tracheids (Meyer, 1928). If it does have vessels, the vessel elements in that region are especially narrow or discontinuous (*Populus deltoides*; Larson and Isebrands, 1978; Chapter 11). By either mechanism, an embolism is blocked from passing through the petiole and into the stem. An elegant feature of some leaves is the presence of exceptionally wide vessel elements at the distal ends of the leaf vessels. With such an arrangement, when cavitation finally does take place, it will most probably occur in these wide terminal elements; as the embolism expands, the water in the column will be pulled back into the stem. If the embolism started in the petiole instead, the entire column on the distal side of the embolism would be trapped in the leaf and would be lost.

An **abscission zone** occurs in almost all dicot leaves; even evergreen dicots abscise their pe-

rennial leaves. It is usually visible as a groove or constriction around the petiole, and its color may differ from that of the rest of the petiole. In sessile leaves, the abscission zone may be the only remnant of the petiole. The abscission zone consists of two parts, a **separation layer**, or **abscission layer**, and a **protective layer**. The separation layer functions to ensure the abscission of the leaf and to define the point of rupture. All leaves eventually die; if no separation layer were present, they would remain indefinitely on the stem and be a point of infection and decay. If they did break off without a separation layer, extensive tearing would result, just as usually happens when nonsenescent leaves are torn off before the separation layer has formed. The stem epidermis is ripped and is an excellent entry site for fungal spores and bacteria. The separation layer, however, is a plate of small, thin-walled cells that is oriented transversally across the petiole. The vascular bundles contain fewer fibers in this region, and usually vessels do not cross it, but are instead linked by tracheids, which prevents the spread of embolisms.

The separation layer may form only just before leaf fall or may differentiate early and then remain quiescent, possibly depending on the predictability of the environment. If the summers are mild enough that leaves never need to be abscised early, the precocious formation of a separation layer would not seem necessary. But if adverse conditions can arise at any time such that the leaves must be dropped suddenly, then the full abscission zone should be preformed and waiting. When the leaf is to be abscissed, numerous metabolic reactions occur in the abscission zone, the final one being the degradation of the middle lamella (Facey, 1950) and the hydrolysis of the cellulose in the walls (Lieberman et al., 1982; Moline and Bostrack, 1972).

The protective layer is located just proximal to the separation layer and consists of a group of cells that are able to undergo cell division to form a sealing layer. They can force their way into ducts and laticifers to seal them off, and they can push tyloses into vessels. Once formed the walls of these cells become suberized or filled with gum to form an impermeable barrier. All cells that are exterior to the protective layer die.

The leaves of numerous species undergo a slow photonastic closure and drooping at night ("sleep movements"), and others, such as the sensitive plant (*Mimosa pudica*) can close rapidly in response to touch. This motion is due to a loss of turgor in cells of part of the petiole called the **pulvinus** (pl.: pulvini). The pulvini are large swollen regions at the base of the petiole, and they are composed predominantly of large parenchyma cells. When stimulated, these cells lose water and collapse, causing the petiole to droop. During the recovery period, water moves back into the pulvinus cells, causing them to swell and lift the leaf.

Vascular System of Monocot Leaves

Many monocots have leaves that are long and strap-shaped and that grow from a basal meristem. The new tissues are generated and are pushed upward by cell division and expansion at the basal meristem, and this results in the entire leaf having a linear aspect. The majority of the vascular bundles run parallel to the long axis of the leaf and to each other (Fig. 12.25). There is usually still a hierarchy of veins, with large veins alternating with small ones. In some species (especially grasses), there is a prominent midrib. The long, parallel veins are not independent of each other, but are interconnected by very fine **commissural bundles** that run perpendicularly to the large veins, linking them (Fig. 12.25). The large veins have a structure similar to those of dicots, consisting of xylem, phloem, bundle sheaths, and bundle sheath extensions (Fig. 12.26). The extensions in monocots are much more fibrous and are more important in support than is turgor pressure. Many of the monocot leaves become long, and maintaining them upright or horizontal would be too difficult. Instead, the sclerenchymatous bundle sheath extensions, combined

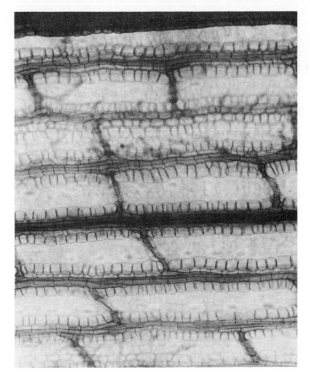

Figure 12.25
You can tell immediately that this leaf clearing is different from all the previous ones in this chapter: the veins are all parallel to each other. This is an almost infallible indication that the plant is a monocot, in this case corn (*Zea mays*). The bundles have a prominent sheath of parenchyma cells; the sheath cells appear to lie on only two sides of the bundle, but in fact they are located on all sides. The fine cross bundles are commissural bundles. × 160.

with sclerified leaf margins, make the leaf flexible and tear-resistant.

Because of their basal meristems, monocot leaves are technically indeterminate and could, theoretically, become enormously elongated. In fact, however, environmental conditions impose a rather strictly defined, predetermined length for them, because many monocots with this type of leaf are either annual or suffrutescent. The leaf or the whole plant dies after one growing season, and the leaves do not have

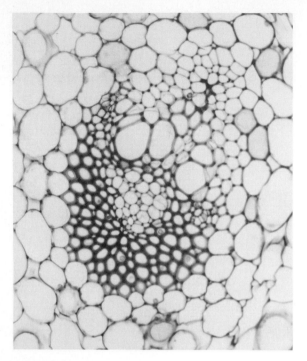

Figure 12.26
This is a vascular bundle of *Dracaena*, shown in cross section. It has a structure that is basically similar to that of a dicot but rather easily distinguishable as being from a monocot. The metaxylem vessels are quite wide; the phloem consists only of large sieve tube members and companion cells in an orderly array. There is a large amount of narrow fibers, in this case just near the phloem. × 250.

time to become extremely long. New leaves are produced in the next season, either on a new part of the shoot or on a wholly new plant. An important principle is involved in this process: such leaves must not grow indefinitely because they do not function well if too elongate. As a strap-shaped leaf becomes longer, the total volume of leaf that is maintained by each bundle becomes greater. If the leaf doubles in length, then the amount of water and sugar that must be transported in each bundle also doubles. But these plants do not have secondary growth, and the primary tissues are limited (as described for monocot stems in Chapter 11), so their total capacity is strictly determined. Be-

cause the leaves live only one year, they rarely grow long enough to exceed the conducting capacity of their bundles. However, in many common ornamentals of this type (irises and gladiolus, for example) the tip of the leaves die long before the rest of the leaf, which may indicate that water conducting capacity has been exceeded: there is only enough xylem to maintain a specific length of leaf.

Some leaves of this type are perennial; examples are agaves and yuccas. Their leaves become wider as they age, and more veins are initiated to maintain a specific maximum spacing between adjacent veins. Each individual vein still has the problem of having an increasing load as the length increases. This problem has been solved by allowing the leaves to become thicker as well as wider, and there are veins in the interior tissues (Fig. 12.27). This may not at first seem like a good solution, because in addition to extra length there is now extra volume to be maintained, and each bundle still is getting longer. But consider how conduction needs are related to leaf size. The water conduction needs are related to the surface area, and therefore to the amount of stomatal and transcuticular transpiration. By increasing the thickness, the surface area increases very little or not at all, and the surface-to-volume ratio actually declines. The phloem capacity that is needed is related to the amount of chlorenchyma that is present, and this too is a function of surface area. Thick leaves have no more chlorenchyma than thin ones; they just have more colorless aerenchyma filler.

As in the dicots, there is a correlation between the amount of energy and material invested in a leaf and the length of time it persists on the plant. Annual leaves tend to be smaller, softer, and thinner, whereas perennial leaves are larger and much more sclerenchymatous. In the more xerophytic groups, especially the epiphytes such as most orchids and bromeliads, the perennial leaves are extremely tough and leathery, although they do not become especially large. The extra reinforcing in these leaves occurs in two ways. First, many of

Figure 12.27
This is a cross section through a small part of a leaf of *Yucca*. First, notice that not all the bundles are vascular bundles: some contain just fibers (fiber bundles); some are fibrovascular bundles. The relative amounts of tracheary elements, sieve elements, and fibers vary continuously. Next, look at all the vascular bundles carefully; you will see that in all of them the xylem is in the adaxial side, the phloem in the abaxial side. × 100.

Figure 12.28
The venation of this *Calathea* is easy to see without clearing; the bundle sheaths are so dark that they are visible against the lighter mesophyll. Although the leaf is broad like that of a dicot, the venation appears to be all parallel except for the midrib. This is from the family Marantaceae, a monocot. × 0.5.

the bundles contain fibers only; they carry no vascular tissues at all. Such **fiber bundles** are more frequent in the outermost portions of the leaf, where their strength will provide the greatest support and defense (Fig. 12.27). Second, the sheaths around bundles are thicker and more consistently fibrous than those of annual leaves. Instead of just one layer of fibers, the sheath may consist of several layers. Not only is the proportion of vascular tissue less in these bundles, but the absolute amount is often much smaller. These are **fibrovascular bundles**, and they occur throughout the leaf. The large amounts of sclerenchyma would

seem to create an obvious problem with regard to loading and unloading of the vascular tissues. Water can probably still be carried out of the xylem by capillary movement through the apoplast. The loading of the phloem must also occur through the sheath. It is known that apoplastic movement of sugars can occur, but the heavy sheath does not seem optimal.

Some monocots have broad leaves, not strap-shaped ones, and they at first appear to be just like the broad leaves of dicots. However, most of them do not have reticulate venation; instead, all of the secondary veins that come from the midrib run to the margins parallel to each other (Fig. 12.28). From the secondaries come perpendicular tertiaries that cross the short space to the adjacent secondary. The tertiaries look just like commissural bundles. These leaves do not have reticulate venation like the dicots. Examples of these are the palms, some Araceae (*Dieffenbachia, Philodendron*), the

Zingiberaceae (gingers), and *Musa* (banana). In others (Dioscoreaceae, some Araceae, and Trilliaceae), the broad leaf does have a venation that is reticulate and does not have any of the parallel orientation that is often considered to be universal in monocots; these leaves also have true petioles.

In all three types of monocot leaves, it appears that the venation pattern is related to the growth pattern of the leaf: in strap-shaped, grasslike leaves, there is a basal meristem, and the bundles differentiate toward it. In broad leaves with lateral parallel venation, the majority of the growth of the lamina must take place outward from the midrib. In the monocots with true reticulate venation, growth probably occurs in length and width simultaneously, as it does in dicot leaves.

Truly compound leaves occur in some Dioscoreales and Arales; these are leaves in which the leaflets develop from leaflet primordia during growth. **Pseudocompound leaves** are those that grow as a single, simple lamina, which then tears apart into an apparent compound condition, as occurs in the palms, Cyclanthaceae, and bananas. Although these leaves are pseudocompound because they grow as a single unit, they develop predetermined lines of weakness that facilitate the tearing, which is thus a normal developmental process, not wind damage.

Stem-Leaf Junction of Monocots

Petiolate leaves with a broad lamina and reticulate venation are very rare in the monocots, being restricted primarily to the dioscorealean families, some Araceae, and Smilacaceae. In these there are a small number of leaf traces that depart from the atactostele and enter the petiole.

In all other monocots, the leaves have been extensively modified, and have been considered to be **phyllodes**, actually being flattened and broadened petioles (Arber, 1918; Kaplan, 1970). After extensive work on several mono-

cots and dicots, Kaplan (1975) concluded that the long strap-shaped and sword-shaped leaves of monocots are indeed leaves, not petioles. Instead of forming a slightly thickened midrib and the two wings of the blade, they produce an extremely enlarged midrib and no blade. The entire leaf is just petiole and midrib. The ability to produce a distinct petiole and broadened lamina was lost, and now most plants simply produce a thin, flat structure from a basal meristem. In palms and many aroids, the leaf has evolved to have not only a more complicated structure, but actually one that resembles the structure that was lost. They have a very prominent and well-defined **pseudopetiole** and a broad lamina, which may even be pseudocompound. The prefix *pseudo* is used to indicate that the morphogenic developments that produce the pseudopetiole and the pseudocompounding are not directly related to the genetic programs that were present in the ancestral leaves.

These monocot leaves typically have at least two parts, a leaf base and a lamina. Many also have a third part, the pseudopetiole. The leaf base always has a broad attachment to the stem, usually completely encircling it and forming a **sheath** (Fig. 12.29). A very few dicots, in the families Polygonaceae and Umbelliferae, also have sheathing leaf bases. If the two margins almost meet or overlap on the far side of the stem, the base is said to be an **open sheath**, but, if they actually fuse together, it is **closed**. The veins in the sheath are parallel to each other and to the stem. All of them enter the stem, providing an extensive vascular connection. The smallest veins do not penetrate deeply into the stem, but instead turn downward and remain in the periphery (Zimmermann, 1983). Often they end blindly in the cortex. The largest bundles penetrate to the center of the stem, then turn downward and connect with axial bundles of the atactostele. As they traverse the stem, they form bridges with adjacent bundles, as described in Chapter 11. Veins of intermediate size penetrate into the stem far

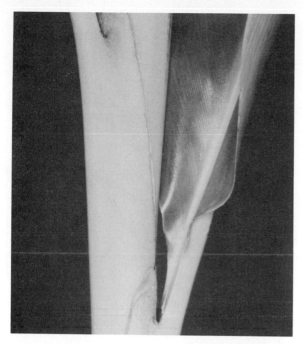

Figure 12.29
In this photograph of *Alpinia*, you can see the important parts present in many monocot leaves: the sheathing leaf base, ligule, a pseudopetiole, and a lamina. × 0.5.

may be a small flap of tissue from the adaxial epidermis, just from the point where the sheath attaches to the blade; this structure is a **ligule** (Fig. 12.29; grasses and some Pontederiaceae). Ligules may be veined or membranous (Chaffey, 1982, 1985; Tran, 1963), and it is assumed that, because they are pressed against the stem, they function to prevent water and debris from falling between the leaf sheath and the stem (Chaffey, 1985). If there are two vertical projections from either side of the attachment point, they are **stipules** (some Hydrocharitaceae, Potamogetonaceae, and Pontederiaceae). If the flaps project from the upper corners of the sheath, they are **auricles** (Poaceae). The selective advantages of each of these have not been clearly established.

The pseudopetiole, if one exists, has numerous fibrovascular bundles, and if large, as in palms and Cyclanthaceae, there are also many fiber bundles. It may be round in cross section, but frequently there is an adaxial groove (Tomlinson, 1961). In the feather palms (those with pinnately compound leaves), the pseudopetiole continues into the rachis, but in the fan palms (palmately compound ones) it stops at the point where all the leaflets are attached. At the distal end of the pseudopetiole in fan palms is a ligulelike structure called the **hastula**.

If there is no pseudopetiole, then the lamina is attached directly to the leaf base, and the vascular bundles and fiber bundles are continuous from one into the other. The basal meristem of the leaf is located just above the junction with the leaf base, and the bundles must pass through it. If the leaf is growing rapidly, the conducting elements exist as immature cells in this region, and the veins are discontinuous through the meristem (see the section "Ontogenetic Development" later in this chapter). This is obviously a problem in water movement and may be part of the reason that leaves like this never become large. In very slow-growing leaves with basal meristems, such as agaves and yuccas, the actual movement of cells out of the meristematic zone is slow

enough to make immediate contact with the bridges and outermost axial bundles. Because monocots lack secondary growth, the irreplaceable vessels of the stem must be even more carefully protected from embolisms originating in dying leaves. The large metaxylem vessels of the stem are connected to the large vessels of the leaf by a bridge of extremely narrow tracheids in the leaf trace (Zimmermann and Sperry, 1983), or the stem-leaf junction consists of numerous short overlapping vessels (Sharman, 1942). This is, unfortunately, an impediment to the easy flow of water to the leaf, but it is necessary for the protection of the water columns in the stem (Zimmermann, 1983).

The leaf base may consist of just the sheath, but several modifications have evolved. There

enough that vascular differentiation can proceed across the zone, and the bundles are complete. New tracheary elements and sieve elements must be formed continuously, but differentiation can keep up with cell flow.

It is not unusual to find several abscission zones within a perennial monocot leaf. The lamina and much of the pseudopetiole are initially abscised in the palms, but the leaf base remains at least temporarily. This too is finally cut off by a second abscission zone that occurs very close to the point where the sheath actually joins the stem.

Structure of Fern Leaves

In ferns, the organization of the leaf is remarkably similar to that of dicots (Ogura, 1972). They can range from being thick and leathery (*Cyrtomium*) to being "filmy" (Hymenophyllaceae) with the lamina being only one cell thick except at the veins (Bierhorst, 1971). There is always a petiole; the lamina is compound (rarely simple as in *Asplenium nidus*); and the internal organization is like that of dicots (Warmbrodt and Evert, 1979), except that the epidermis contains chloroplasts and that vessels never occur in the xylem. The phloem is often quite simple, consisting of sieve cells that may not be well differentiated. The vascular bundles are always collateral, and, in most species, all end freely either at the leaf margins or in areoles. The finest veins usually contain only tracheids, but in two species of *Salvinia* the vein ends consist of only sieve elements (Warmbrodt and Evert, 1978). In the heterosporous ferns, the leaf veins are concentric, amphicribral, and sheathed by an endodermis. They contain numerous intercellular spaces in *Marsilea macropoda* and *Regnellidium diphyllum*.

The primary differences between mature fern leaves and those of dicots are that most ferns have an open venation that is usually dichotomously branched (Fig. 12.30), and that only a small number of ferns have reticulate venation (Fig. 12.31; Wagner, 1979; Warmbrodt

Figure 12.30
You can tell immediately that this is not a dicot or a monocot because of the dichotomous venation. Such bifurcations can exist in the leaves of seed plants, but they are not so closely spaced. This is the fern *Adiantum*. The black object is a dense collection of sporangia. × 40. Material prepared by D. Lemke.

and Evert, 1979). After a vascular bundle enters a region of the lamina, it may bifurcate and run to the edge without ever fusing with another bundle. In certain species there may be some anastomosis, but the open nature is still obvious. The spacing between veins is greater in ferns than it is in angiosperms.

Structure of Gymnosperm Leaves

The leaves of cycads are also like those of dicots, being large, dorsiventral, and having a

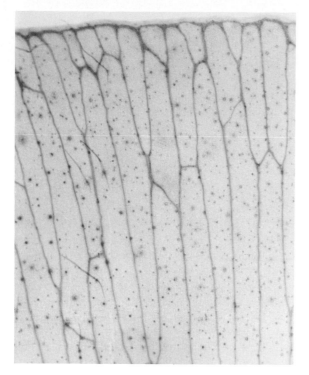

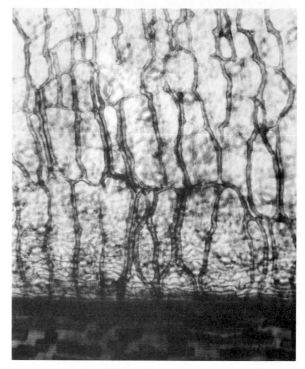

Figure 12.31
Very few ferns have reticulate venation; this is one that does, *Marsilea*. There is an underlying pattern of dichotomies, but numerous cross bridges occur. As in the seed plants, the veins are so close together that no part of the mesophyll is far from a vein. × 40.

Figure 12.32
Clearing of a leaf of a cycad, *Cycas rumphii*. The heavy dark region is the vascular bundle that runs through the leaflet; from it come the accessory transfusion tracheids. These do not form bundles, not even minor veins; instead they are single files of elongate, irregular tracheids. × 100.

petiole; they all have a compound lamina. The venation is open and dichotomous, and a small amount of secondary xylem develops in the midrib. The leaves of cycads differ from those of angiosperms, however, in having **transfusion tissue** (Fig. 12.32); this is a set of special **transfusion tracheids** located radially around the regular vascular tissues of the bundles (Pant, 1973). The transfusion tracheids are very short and wide, often almost isodiametric, but with pitted walls. The sheath of transfusion tissue can be three or four times the width of the regular vascular tissues of the bundle, and it extends to an endodermis. Just exterior to

the endodermis, the palisade parenchyma and spongy mesophyll of the leaflet begin. These latter two tissues do not abut each other as in angiosperms, but instead are separated by a layer of **accessory transfusion tissue** that is made up of horizontally elongate transfusion tracheids (Fig. 12.32; Lederer, 1955). The transfusion tissue and the accessory transfusion tissue contact each other and are responsible for moving water out to the mesophyll. This is a much more elaborate system than that of the capillary action used by the angiosperms. It probably is related to the fact that the vascular bundles are widely separated in the cycads, and large regions of mesophyll are not close to minor veins. In the angiosperms, the fine veins are much more closely spaced, and transfusion tissue is not necessary (or, because transfusion tissue is not present, the veins must be closely spaced?). One study of six dicots found that their leaves have an average of 102 cm of veins in each square centimeter of leaf blade (Plymale and Wylie, 1944), and the average distance between veins is only 103 μm (Wylie, 1939). Transfusion tissue is common in the gymnosperms, being present in the conifers also, so the ancestors of the angiosperms probably had it. It is interesting that angiosperms changed to a method of closely spaced bundles instead of transfusion tissue. It may be that the needs for efficient sugar accumulation and phloem loading were the important factors.

Conifer leaves are simple, not compound (Feustel, 1921; Florin, 1931; Laubenfels, 1953), and, except in *Larix* and *Taxodium*, they are perennial (up to 33 years in *Pinus longaeva*; Ewers, 1982). Like the leaves of cycads, they have xeromorphic characters, being tough, leathery and heavily sclerified, even the epidermal cells being very thick-walled (Gambles and Dengler, 1982b). The leaves are scalelike in several genera—large scales in the Araucariaceae, but small, almost microscopic scales in the Cupressaceae (Fig. 12.33). In the large family Pinaceae (*Abies, Cedrus, Larix, Picea, Pinus, Pseudotsuga, Tsuga*), the leaves are elongated needles. In scale leaves, there may be several vascular bundles, but in the needle leaves just one or two central, unbranched bundles occur (Fig. 12.34). The bundles of many conifer genera have a vascular cambium that produces secondary phloem and little or no secondary xylem (Fig. 12.35; Elliot, 1937). In *Pinus longaeva*, one to two new cell layers of secondary phloem are produced per year (Ewers, 1982). The vascular tissues are surrounded by a sheath of transfusion tissue, then an endodermis (Figs. 12.34, 12.36; Gambles and Dengler, 1982a; Ghouse, 1974; Ghouse and Yunus, 1974, 1975; Lederer, 1955; Sutherland, 1933). The transfu-

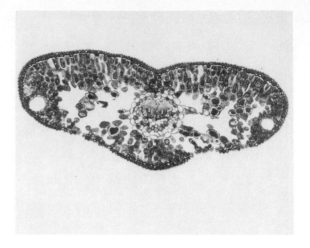

Figure 12.33
This is a cross section of a gymnosperm leaf, *Pseudotsuga* (Douglas fir). It has a palisade parenchyma and a spongy mesophyll, which is rather unusual for conifers. The epidermis and hypodermis are thick-walled; there is only one vein, and it is surrounded by a prominent endodermis. Resin canals are characteristic of conifer leaves, and two are present here. $\times$ 100.

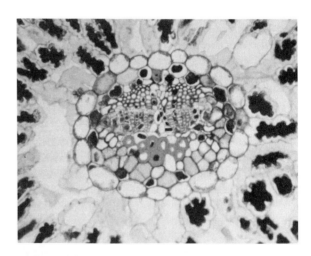

Figure 12.34
This is a magnification of the bundle of *Cedrus deodora* (deodar cedar), showing the xylem, phloem, and the transfusion tissue. Also around the outside you can see the lobed chlorenchyma cells that are characteristic of some conifers. $\times$ 250.

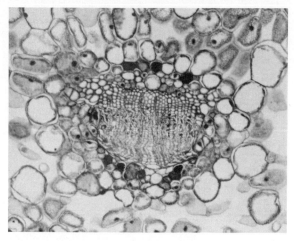

Figure 12.35
This is a long-lived leaf of *Taxus* (yew); it has developed a cambium that is producing secondary phloem. Secondary xylem is not formed. × 350.

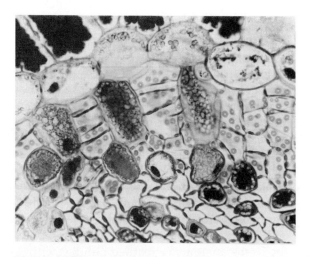

Figure 12.36
This is a magnification of the transfusion tracheids present in a leaf of *Pinus monophylla;* these appear to be short in this cross section, but a longitudinal section would reveal that they are columnar. They have prominent, circular-bordered pits on all walls. They occur only to the inside of the endodermis, not to the outside. Compare with the accessory transfusion tracheids of Fig. 12.32. × 400.

sion tissue consists of living transfusion parenchyma and dead transfusion tracheids. Ultrastructural studies of pine needles showed that the endodermal cells do not have true Casparian strips in their walls (Carde, 1978; Walles et al., 1973). Between the endodermis and the epidermis is a compact chlorenchymatous mesophyll. In pine and several other genera, the mesophyll consists of just two or three layers of lobed cells (Fig. 12.34), but in *Abies, Araucaria, Cunninghamia, Podocarpus, Sequoia,* and *Taxus* there is a palisade parenchyma and spongy mesophyll (Fig. 12.33). The conifers all have resin canals in their leaves; these are lined at least by a thin-walled parenchymatous epithelium and in some species by a fibrous sheath with lignified walls. The resin canals may be either elongated chambers, as in *Picea* (Marco, 1939) or complete canals that are connected to those of the stem, as in *Cryptomeria* and *Cunninghamia* (Cross, 1941, 1942). The leaves are so narrow that accessory transfusion tissue is not necessary. Resin canals occur in the mesophyll. In conifers with long, needle-shaped leaves, such as *Pinus taeda*, the problem of conduction is similar to that in the strap-shaped leaves of monocots: the vascular bundles at the base of each needle must have a greater conductance than at the tip, and the total capacity may be limited.

Structure of Modified Leaves

Kranz Anatomy and C_4 Photosynthesis

One of the most critically important steps in photosynthesis is the actual attachment of a molecule of carbon dioxide to a molecule of the sugar ribulose-1,5-bisphosphate (RuBP); this results in two molecules of 3-phosphoglyceric acid and the first step of carbon assimilation. The enzyme that catalyzes this reaction is RuBP carboxylase, and it is obviously an important enzyme. Without it, photosynthesis and therefore life would be impossible. Unfortunately, RuBP carboxylase has some serious drawbacks, one of which is that it has only a low affinity for carbon dioxide. It can also accidentally put oxygen instead of carbon dioxide onto RuBP, and the resulting molecules are then photorespired away with a significant loss of energy for the plant. There must be little possibility of improving the enzyme, because its amino acid sequence is one of the most highly conserved ones known; virtually all mutations affecting RuBP carboxylase are selectively disadvantageous.

However, several diverse groups of angiosperms have encountered a partial solution: **C$_4$ photosynthesis** and **Kranz anatomy**; the alternative type of photosynthesis is C$_3$ (Brown, 1975). In C$_4$ photosynthesis the mesophyll cells use a different enzyme to pick up carbon dioxide very efficiently and use it to make oxaloacetic acid (OAA, which has four carbon atoms, hence the name C$_4$). The OAA can be converted to other acids, but their important feature is that they are all transported to the bundle sheath cells, where they decompose, releasing the carbon dioxide again. Because the carbon dioxide is collected throughout the leaf and then released in just one layer of cells, it achieves a very high concentration in the bundle sheath. The chloroplasts of the bundle sheath cells can pick up the carbon dioxide because it is in high concentration, and RuBP carboxylase can work efficiently. The bundle sheath cells in C$_4$ plants are large and prominent, with a large number of chloroplasts. Whereas it is possible to overlook the sheath in many other plants, the sheaths in C$_4$ plants cannot be missed (Fig. 12.37), and it was originally thought that they occured only in C$_4$ plants. The actual photosynthetic fixation of carbon dioxide (C$_3$ photosynthesis) does not happen in the mesophyll parenchyma, and a palisade parenchyma does not occur. Further, for efficient transport between mesophyll and the bundle sheath, a radial arrangement of chlorenchyma cells around the bundles is better. These radially aligned mesophyll cells were given the German name Kranz (wreath; Haber-

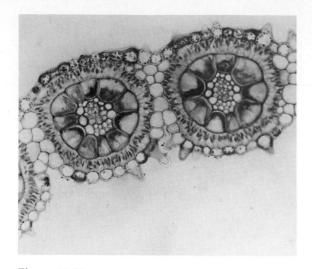

Figure 12.37
This is a leaf of a C$_4$ grass (*Schedonnardus*), and the Kranz anatomy is not particularly difficult to see: the bundle sheath is double, with the outer sheath consisting of very large cells that have their chloroplasts close to the bundle. The inner sheath is a mestome sheath of small endodermal cells. Around the sheath is the chlorophyllous mesophyll of narrow cells that radiate away from the sheath. Originally, the term *Kranz* was applied only to this radial mesophyll, but now it is occasionally used to refer to the large outer sheath also. Notice that there is a single layer of clear mesophyll cells between the two bundles. × 400.

landt, 1914), and we still use this term even in English, saying that the leaf has *Kranz anatomy*. A spongy mesophyll is usually lacking; this may be due to the efficient uptake of carbon dioxide, but that is not certain. In cross section, the Kranz anatomy (wreathed bundles with radial mesophyll and no palisade parenchyma) is very distinctive (Fig. 12.37). As would be expected, the differentiation of these leaves differs slightly from that of nonKranz leaves (Dengler et al., 1985). The bundle sheath may consist of just one layer of cells (typical of the panicoid grasses) or of two layers (most often found in the festucoid grasses). If only one layer is present, it is typically composed of thin-walled parenchyma cells that contain numerous small chloroplasts. It is not uncommon for this sheath to contain a large amount of starch, and it is occasionally referred to as a *starch sheath* (a term also applied to the developing endodermis in roots). In double-layered sheaths, the outer layer is like that just described, but the inner one, called the **mestome sheath** (Brown, 1958) is an endodermis, having suberized layers on the inner surface of its walls (Carolin et al., 1973; O'Brien and Carr, 1970). Grasses with a double sheath may have either C_4 or C_3 metabolism, but those with a single bundle sheath on their major leaf veins are all C_4 (Hattersley and Watson, 1976).

The C_4 metabolism is based on very common, ubiquitous enzymes and intermediates that occur in all plants. In order to make these work to form the C_4 pathway, the main things needed are intercellular transport systems and a place of accumulation, both of which are also universally present in plants. It is not really surprising that C_4 metabolism and Kranz anatomy have evolved repeatedly in at least ten families: Aizoaceae, Amaranthaceae, Compositae, Chenopodiaceae, Euphorbiaceae, Poaceae, Portulacaceae, and others (Carolin et al., 1975; Laetsch, 1974). It is interesting that the Kranz anatomy always evolved simultaneously with the C_4 metabolism.

Xeromorphic Foliage Leaves

The term *xeromorphic* is used to indicate plants that have an anatomy and growth form that are especially adapted to desert conditions, particularly conditions in which water is frequently scarce. In using the term, and especially in trying to understand xeromorphic characters, it is important to realize that desert environments differ from mesic ones in many ways other than just water availability. Desert soils are typically alkaline and very rich in minerals, rich to the point of being saline and toxic. They are poor in the organic material that helps hold water in the soil. Temperatures vary between extremely hot and severely cold on both a daily and an annual basis. Because the sky is clear and dry, insolation is high. Many streams and ponds exist only immediately after a rainstorm, and plants are often the only source of moisture for animals, so protection against predation is necessary.

There are various ways to meet these challenges: ephemeral plants are able to germinate, grow, set seed, and die within a matter of weeks. Their whole life cycle can occur in the spring or autumn or after a heavy rain. Their adaptation to xeric conditions is the ability to complete a life cycle while the desert is temporarily mesic. Other plants grow on the banks of streams or ponds that always contain water or that dry out so briefly that the soil remains humid. Although these plants are situated in desert regions, they live in mesic microhabitats. Insolation, predation, and temperature fluctuations are their main problems, and they rarely show xeromorphic characters.

Plants that occur in xeric microhabitats and that are perennials are the ones that show well-developed xeromorphic characters; such plants are called **xerophytes** (Maxinov, 1931). These characters include a thick-walled epidermis and hypodermis, covered by a dense, waxy cuticle. Trichomes are typically abundant (Oppenheimer, 1960). These features are selectively advantageous for many reasons. All are effective in blocking excess sunlight (Hartmann, 1979; Heide-Jørgensen, 1980). They are also good deterrents against insect feeding and egg laying (Levin, 1973). Salt glands may be present in many, and the coating of salt may also protect against feeding and sunlight, as well as keeping salts from accumulating in the protoplasm. At least some of these characters are directly induced by the environment: plants of *Prosopis velutina* grown outdoors have a cuticle ten times thicker than that of plants grown indoors (Bleckmann et al., 1980).

The exterior surfaces of xeromorphic leaves are reduced, which decreases the transpirational surface area, but the density of stomata increases, perhaps to allow more rapid gas exchange in the brief periods when humidity permits stomatal opening. Leaves may be smaller, more cylindrical, and more succulent. They are frequently expendable, being abscised even though they might be only a few weeks old and quite healthy (Didieriaceae, *Fouquieria*, *Pereskia*), being too much of a transpirational liability if a sudden drought should occur. However, these species are able to send out a new set of leaves within days of a heavy rainfall. To further reduce water loss, the stomata may be sunken, or they may be restricted to well-protected crypts (*Ficus*, *Nerium*, *Ouratia*).

Inside the leaf, the surface area is also reduced, as cells are packed together and are also smaller. The spongy mesophyll may be lost altogether, with only palisade parenchyma remaining. As the volume and surface area of the leaf apoplast diminish, less water can be lost from each cell to the apoplast. The vascular bundle sheaths are frequently sclerenchymatous. Water storage cells are often present; they may occur either as large mucilage cells (cacti) or as regions of thin-walled, colorless parenchyma cells (*Salsola kali*). In *Salicornia* (Fahn and Arzee, 1959) and *Pogonophora schomburgkiana* (Foster, 1956), the water storage cells are **tracheoid idioblasts**: short, tracheidlike cells that occur dispersed within the palisade parenchyma or the rest of the mesophyll. It is thought that they hold water by means of hydrogen bonding to their walls (Haberlandt, 1914; Zimmermann, 1983).

Submerged Foliage Leaves

A small number of ferns and angiosperms, but no gymnosperms, have migrated back into aquatic habitats. There are various degrees of this, of course, ranging from growing in marshy, swampy areas, to growing as floating, unanchored plants, to growing completely immersed (Stodola, 1967). In the first two types (marsh and floating plants) the leaves remain in the air and have an anatomy that is not especially unusual. But when a plant is anchored to the bottom of a stream, pond, or ocean and has its leaves fully submerged, then major changes occur, and the plants are called **hydrophytes** (examples: *Ceratophyllum*, *Eleocharis*, *Myriophyllum*, *Najas*, *Sagittaria*, *Utricularia*, *Vallisneria*).

In previous chapters, several functions were ascribed to the cuticle, such as protection from excess sunlight and deterrence of insects as well as water conservation. Aquatic plants do not need to conserve water and protect their tissues from drying out. The cuticle is very reduced, and the cell walls tend to be thin. By coincidence, immersed leaves are protected from bright sunlight by the water itself, and aquatic angiosperms suffer little predation by aquatic animals.

It is frequently stated that the thinness of the cuticle allows the leaves to take up nutrients directly from the water, but, if this is true, it is probably of minimal importance. All aquatic angiosperms are delicate and small when compared to the "average" land plant, but they are massive and bulky compared to most algae and many invertebrates. Minerals must still be taken up in the roots and transported to the leaves; because there is no transpiration, xylem flow occurs by root pressure. As mentioned in Chapter 9, hydathodes are common on submerged leaves, and these may help move water by pulling minerals out of the xylem stream at its apex (Lüttge and Krapf, 1969).

Because most plants are buoyant in water, submerged plants can use it for support. However, plant walls and protoplasm are slightly more dense than water, and they sink (wood floats only as long as air is trapped inside). In order to use the water to remain upright, the plants must be full of **gas chambers**. These are typically called *air chambers*, but they are filled with gases generated by the plant, not with air. The gas chambers are abundant in leaves, which is not surprising because all leaves except certain xeromophic ones have a large spongy mesophyll gas chamber. But in submerged plants the gas chambers are present in the stem and roots as well. These are not necessary for buoyancy, because only the uppermost portion of a plant needs to float in order to support the whole plant. Their function in the stems and roots is probably to allow the diffusion of oxygen from the upper chlorenchyma down to the lower, nonchlorophyllous parts. Almost any land plant will be killed by keeping its roots constantly flooded and thereby cutting off the supply of oxygen. Soils must periodically drain or dry slightly to maintain the soil oxygen content. Since the soil at the bottom of streams and oceans does not dry, the plants must have an internal ventilation system.

In this regard, it is interesting that most aquatic angiosperms have an endodermis in the stem and leaves. This seems to be involved in channelizing the flow of water and keeping it confined to the xylem. In transpiring land plants, water is pulled upward as cells lose water to the air; inside the plants, water automatically goes wherever it is most needed. But in submerged plants, it is pushed up by root pressure produced as the root endodermal cells pump ions into the stele. If the stems and leaves had no endodermis, then the pressurized water could squeeze out of the xylem anywhere and fill up the gas chambers, causing the plant to drown. But with the endodermis, the flow is channelized, and in any region water and minerals are allowed out only in quantities adequate for that region.

With support provided by the buoyancy of the plant, it might be expected that fibers would be competely absent. However, fibers also provide for elastic flexibility, and they protect against tearing. Plants exposed to wave action and tidal fluctuations need such protection, and indeed fiber bands can still be found on leaf margins and other places where tearing might otherwise happen.

For land plants, the lamina of a leaf presents a problem because it acts like a sail and catches wind. This trait can put a lot of torque on a branch (Nobel, 1983), especially in high winds; trees frequently blow over in hurricanes or wind storms. Larger leaves are almost always compound, allowing the leaflets to flex individually. Because water is much denser and more massive than air, leaves with a flat, broad lamina would be extremely disadvantageous; even a mild current or tidal action would put tremendous pressure on the plant. The solution is either to have finely dissected compound leaves or to allow the leaves to be very thin and cylindrical. Such shapes offer little resistance, and water moves easily around them. These shapes also have a high external surface-to-volume ratio, but because transpirational water loss is not a factor, this is not a problem.

Cataphylls

Cataphylls (more commonly called bud scales) function as a means of protecting the apical or axillary shoot tips during times of dormancy. Annual plants do not have terminal resting buds and consequently lack bud scales. Many tropical plants grow continuously or have brief quiescent periods (two to four weeks) and similarly lack buds and bud scales, while in others (*Hevea braziliensis, Licania macrophylla, Mangifera indica*) there are resting buds with cataphylls (Foster, 1928). Temperate, perennial species especially must have a means of protecting their apical regions from damage. Several dangers confront the meristems and leaf primordia within a bud, but the most important are probably desiccation and insect predation. Winter winds are capable of absorbing a great deal of moisture, and the immature vascular tissues are not able to conduct water rapidly during winter. Furthermore, the young stems and leaves are the richest sources of proteins in the nonfloral parts of the plant, and are a favored source of food for insects.

To combat these dangers, the bud scales remain small and are either sessile or have only an extremely short petiole (Fig. 12.38; Cross,

Figure 12.38
This bud scale of *Homalocephala texensis* has a solid mesophyll and very little vascular tissue. The cuticle is extremely thickened. × 100.

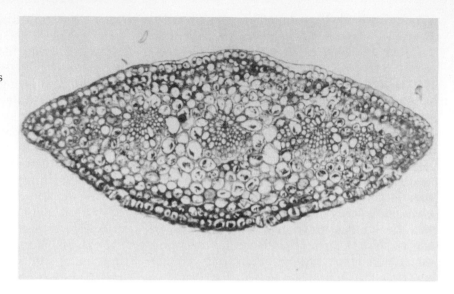

1936, 1938; Doak, 1935; Foster, 1935a, b, 1937; Sacher, 1955; Schneider, 1913). Their longitudinal growth stops much earlier than it would in a foliage leaf of the same plant, and the marginal growth starts sooner. In plants with large leaves, the marginal growth of the bud scales is less than that of the foliage leaf, but in the conifers with needle leaves, the cataphylls are broader than the needles. The epidermis may be thick-walled, but more commonly it forms a cork cambium and produces a thin, protective, corky bark (see Chapter 17). The cork is a good water retention device, and it is also completely indigestible by insects. Resins, oils, and waxes may also be secreted (Curtis and Lersten, 1974; Dell and McComb, 1978). Secretion and cork formation often occur mostly in just the outermost, exposed portions of the bud scales. Stomata are rare (Sacher, 1955) and may be lost when the cork is formed. The vascular supply remains simple, since very little conduction is needed, either into the scale or out of it. The mesophyll is usually a homogeneous spongy mesophyll, but a palisade parenchyma may be present in green bud scales. Secretory cells and cavities may occur, especially ones that produce pungent oils. It must be emphasized that bud scales are not just stunted foliage leaves that stop developing early, as was once thought; see Foster (1928) for an extensive, detailed review. Instead, they have their own special role, and natural selection favors mutations that make them more specialized and efficient. Goffinet and Larson (1982b) found that, when *Populus deltoides* was artificially induced to become dormant, the first signs of bud scale morphogenesis occurred only in primordia that were formed after induction. Any leaf primordia that were present before induction, even if extremely immature at the time of induction, continued to develop as foliage leaves. A stunted foliage leaf would be a liability because of its poorly developed epidermis and its lack of any other means of protection.

Hypsophylls

Hypsophylls (floral bracts), protect the inflorescences as they develop. They function in a fashion that is similar to that of cataphylls, but they more typically are formed at the beginning of the growing season and function as the plant is active and producing inflorescences.

The environmental conditions that they shield against are milder than those faced by cataphylls. Consequently, hypsophylls tend not to be as resistant as cataphylls, and they more often are green and photosynthetic. It might be expected that anti-insect mechanisms would be more often present in hypsophylls, because they function when insects are more abundant. In reality, there have been very few studies of either cataphylls or hypsophylls; foliage leaves have received much more attention.

The anatomy of sepals and cotyledons will be considered in Chapters 19 and 20.

Development

Ontogenetic Development

ONTOGENY OF BIFACIAL LEAVES. Leaf primordia are initiated by the shoot apical meristem, as discussed in Chapter 6. The use of periclinal chimeras (Dermen, 1947, 1951; Satina and Blakeslee, 1941) has shown that several of the outermost cell layers of the stem participate in leaf formation; the protoderm gives rise to the leaf epidermis, and the next layer or two produce the mesophyll and vascular tissues.

In the cryptogams, especially the ferns, an apical meristem is formed in the tip of the leaf primordium and can function for an extended period (Croxdale, 1976; Hirsch and Kaplan, 1974). Some ferns (*Lygodium*) have indeterminate leaves that at first glance appear to be stems (Mueller, 1982). In the seed plants, apical meristems have been claimed to occur in the leaves, but convincing evidence is scarce. The majority of the leaf growth in length occurs by cell division throughout the young primordium. Simultaneously, the primordium becomes thicker as the cells divide in an orderly manner with walls that are parallel with the surface of the nearby stem. These divisions can be localized enough that they are considered to constitute an **adaxial meristem** or **ventral meristem** (Fig. 12.39; Foster, 1936; Troll, 1939).

At first the leaf primordium is just a small protrusion or short cone, but its length increases more rapidly than its width or thickness, and it becomes cylindrical. Two longitudinal ridges begin to protrude from the two sides of the leaf primordium, or both are located on the adaxial face. These are the **marginal meristems**, and they represent the initiation of the lamina (Franck, 1979; Postek and Tucker, 1982). As development continues, the two ridges become taller and slightly platelike. A close examination of the edges shows that the marginal meristems are composed of **marginal initials** and **submarginal initials** (Figs. 12.39, 12.40; Maksymowich and Erickson, 1960). It is important to consider some of the theoretical aspects of meristems at this point. All leaves have a limited, determinate width, so the leaf marginal meristems do not need to function indefinitely. Even very broad leaves are formed in less than a year, and all plants have so many leaves that no single one is absolutely essential. Therefore, the marginal meristem cells do not have such an archival function, and any somatic mutations that occur in their cells will affect just a small, temporary part of the plant (Klekowski, 1988).

A second aspect of meristems is that it is not efficient to produce a large number of derivative cells from a small number of meristem cells; in shoots and roots, the large subapical meristems are responsible for most of the growth and volume increase. In leaves, this is even more important; although most leaves are narrower than a stem is long, most are much wider than an internode. Thus the amount of width growth of even small leaves is considerable and in palms it is truly fantastic. "Subapical" meristems are essential for leaves, and they do exist; they are called **plate meristems** and consist basically of the entire lamina. While it is young, all regions of the blade undergo growth in length and width, just as young internodes undergo diffuse growth in length and diameter. As the cells of the lamina divide, the new walls are perpendicular to the plane of the

Figure 12.39
This cross section of a *Plumeria* (frangipani) leaf shows the activity of the adaxial meristem and the growth of the lamina just after formation of the marginal initials. × 400.

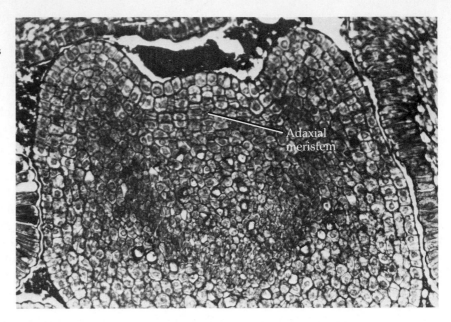

layer, so the layer grows in area but not in thickness. As a result, the layers remain distinct, and their development is easy to study (Figs. 12.39, 12.40).

One of the primary functions of apical meristems in shoots and roots is to establish patterns, and the same is true in leaf marginal meristems. Notice that the marginal meristems in Figures 12.39 and 12.40 resemble shoot meristems: the marginal initials are equivalent to a protoderm, and the submarginal initials are equivalent to a corpus or ground meristem. Cell division patterns support this analogy; the marginal initials generally divide only with anticlinal walls, whereas the submarginal initials can divide in various planes to establish several layers of cells. Careful studies of cell division patterns (Foster, 1936; Kaufman, 1959) indicate that the various layers of the lamina are initiated at the margins, then are expanded by the plate meristems.

In lobed leaves, the marginal meristems are initiated as described above, but the level of mitotic activity does not remain uniform. In the regions that will grow out as lobes, the rate of cell division is high, but, where sinuses will develop, there is less division and expansion (Fuchs, 1975).

If a leaf is to be compound, then the marginal meristems do not develop. Instead, two rows of small projections are formed; these are the leaflet primordia (Foster, 1932, 1935). Typically they form in a basipetal direction as the leaf primordium grows outward, but they can be formed acropetally. As the leaflet primordia become larger, they form either marginal meristems (if the leaf is once compound) or more leaflet primordia (if twice compound). Once the marginal meristems are formed on the leaflets, development is as described above.

As the lamina expands, provascular tissue is first discernible in the midrib; later, it differentiates outward from the midrib (Franck, 1979; Postek and Tucker, 1982). The minor veins differentiate simultaneously after the other veins have formed.

Studies of leaf development in the monocots have centered almost exclusively on the grasses, and their leaf development differs from the method that occurs in dicots. It begins as a leaf primordium is formed, just as in the dicots; but, after the primordium has become a

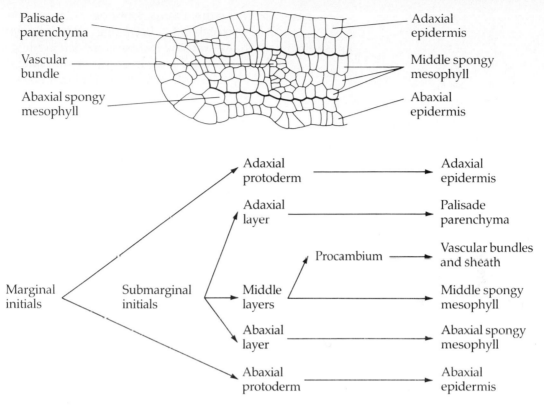

Figure 12.40
This diagram shows the cell lineages involved in leaf development: the marginal initials produce only epidermis; the submarginal initials produce the mesophyll. The derivatives of the submarginal initials can divide with periclinal walls to produce more layers. Anticlinal division within each layer then causes the lamina to expand.

small cone, the regions of the shoot apex just at the two sides of the primordium also become active as primordial tissue (Esau, 1943b; Kaufman, 1959; Sharman, 1942). The cells next to these become active, and the entire process continues until the primordium appears to be seated on a short sheath that completely encircles the base of the shoot apical meristem (Fig. 12.41). If there is to be a sheathing leaf base, then all of this short sheath becomes mitotically active and grows upward, simultaneously with the enclosed internode or slightly more rapidly than it. The apical portion on one side becomes more active and grows outward to

form the lamina. The young lamina grows briefly by diffuse mitotic activity, establishing a larger, parenchymatic primordium. The procambium of the midvein and the two or four largest lateral veins differentiate acropetally from the leaf base toward the leaf tip. Next, protophloem appears at the base and also differentiates acropetally within the new procambial strands (Esau, 1943b; Forde, 1965; Patrick, 1972; Sharman, 1942). Finally, protoxylem is initiated at the leaf base, and it too develops toward the tip of the leaf primordium. The protophloem contains sieve tube members without companion cells, and the protoxylem contains

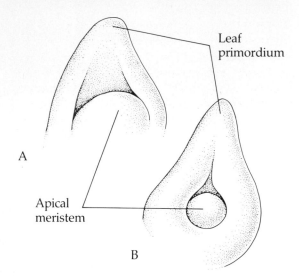

Figure 12.41
As the leaf primordium grows laterally, it becomes
hood-shaped and surrounds the shoot apex.

annularly or helically thickened vessels. Then a
basal meristem is formed as the apical cells begin to mature and stop cell division.

As the leaf elongates upward by means of
this basal meristem, the longitudinal veins
grow at their base, and all differentiation of
procambium, protophloem, and protoxylem is
now basipetal. Forde (1965) carefully studied
the differentiation of protophloem during this
phase in *Lolium perenne* and found that protophloem sieve tube members could differentiate
within the basal meristem but that the continued growth of the meristem would then de-
stroy them, so they had to be replaced constantly. The leaf grew 3 cm per day, and enough
sieve tube members could mature each day to
keep this region supplied. Thus there is a
phloem connection across the meristem that is
repeatedly broken and reformed. However, it
should be pointed out that, during such stages
of growth, the basal meristem is probably the
most important sink for nutrients, not the leaf
blade itself. Also, while the leaf is this small,
it retains its photosynthate for its own
development.

As the lamina continues to grow, the shoot
also expands, becoming wider circumferentially, and the leaf basal meristem must widen
to keep up with it. New, smaller lateral veins
appear, either de novo (Soper and Mitchell,
1956) or by the splitting of the bases of existing
veins (Forde, 1965). From these small intermediate veins, the commissural bundles extend transversely toward the adjacent bundles
(Esau, 1943b). At this point, all differentiation
is basipetal. Whereas the protophloem of the
major veins may differentiate within the basal
meristem, that of the smaller, intermediate
veins only extends to the vicinity of the meristem but not into it, so it does not contribute to
vascular continuity. This process was found to
continue in corn until the leaf tip emerged
from the sheath of the next older leaf (Sharman, 1942). At that time a wave of final vacuolation and maturation began at the tip and progressed down the leaf, causing it to elongate
greatly and rapidly. At this point stomata appeared in the epidermis, chlorophyll was synthesized in the mesophyll, and the metaphloem
and metaxylem began to differentiate. The
basipetal wave of maturation can make it appear as if the metaxylem vessels are mature at
one end while differentiating at the other, but
Sharman was careful to point out that several
chains of short, overlapping vessels are actually produced, not one single long one, and
each vessel differentiates all at once, while the
series differentiates basipetally. The amount of
overlap is extensive, about 1000 μm long at
each contact area. Simultaneously, the basal
meristem becomes disorganized, and full vascular continuity is established between the leaf
and the axial bundles of the stem. When mature, the larger bundles contain several large
metaxylem vessels, xylem parenchyma, a large
xylem lacuna, metaphloem consisting of an
orderly array of sieve tube members and companion cells, and crushed protophloem (Esau,
1943a, b).

The grasses that have been studied have determinate leaves. Corn leaves reach a predetermined length, then the basal meristem disor-

ganizes, and growth stops. Other monocots such as Iridaceae and Liliaceae have leaves that grow for a much longer time; the distal portions of these leaves are mature and very active photosynthetically, while the base is still growing and immature. It seems possible that in these the problem of conduction across the basal meristem is more important, but these plants apparently have not yet been studied in this regard.

ONTOGENY OF UNIFACIAL LEAVES   All the leaves discussed to this point have been bifacial leaves; that is, they have an adaxial and an abaxial side. They might also be isobilateral, in which the two sides look alike but are two sides nonetheless. **Unifacial leaves** have only one side from a developmental viewpoint (Goebel, 1928), and they occur in many monocots, especially ones with cylindrical leaves or leaves that are **ensiform** (Figs. 12.42, 12.43; sword-shaped, flattened from side to side, not top to bottom). Examples are onion, irises, *Acorus calamus*, and *Sansevieria*. Their development is easy to understand, but you will find it much easier to follow if you have some green onions handy (the ones with the small white bulb and long green leaves). This description is based on Kaplan's study of *Acorus calamus* (1970). The leaf primordium forms as just described for bifacial monocots leaves, and the sheath forms a complete covering around the stem. In onions, it forms the large white fleshy leaves of the bulb, the part that is cut into onion rings; this is homologous to the leaf sheath in any other monocot. The vascular bundles here have normal orientation, with the xylem nearer the adaxial surface and the phloem closer to the abaxial side. Notice that the onion rings are complete circles, so the sheath is a closed one. At the apex, in the region of the original leaf primordium, the adaxial meristem is formed. This becomes much more active than ever occurs in bifacial leaves (Fig. 12.44); furthermore, marginal meristems are not formed (Kaplan, 1970, 1975). Theoretically, the lamina consists of just a very thickened midrib. This is an important

point, because an alternative theory suggested that the original leaf apex arched over the shoot apex and stopped growing, while a new meristem was envisioned as arising on the abaxial surface of the primordium (this is the **sympodial theory**). If this were true, the blade of the leaf would then be a totally new structure, not homologous with other parts of leaves. Kaplan (1970) was careful to map the frequency of cell divisions and concluded that the leaf tip does not arch over nor stop growing. Therefore, the expansion is due to a rather overactive but otherwise normal adaxial meristem, and the petiolar-midrib nature is confirmed.

If the leaf is to be ensiform, the growth is almost entirely radial; but, if the leaf is cylindrical, then expansion in both width and thickness occurs. In onions, as this part grows radially, the mesophyll is torn, and the leaf becomes hollow. A cross section of this part also reveals a ring, but we can notice that the lumen of this ring is not the same as the lumen of the lower, white, fleshy part, as is easy to imagine at first. The hollowness of the green lamina stops where the lamina joins the leaf base. The lamina is completely closed, the lumen having no opening anywhere; but, in the leaf base, the lumen is open at the top, as is true of all sheathing leaf bases. This opening is necessary to allow the younger leaves and the stem to emerge.

Phylogenetic Development

The term *leaf* is applied to two totally distinct structures that have had separate evolutionary origins. To distinguish between the two, the more proper, technical names of **microphyll** and **megaphyll** are used (Eames, 1936).

MICROPHYLLS   Microphylls are the more ancient structure, being present on some of the very earliest vascular plants. As their name implies, they began as small structures, simple outgrowths (**enations**) of stem tissue in such fossil plants as *Sawdonia* and *Asteroxylon*. They were just minor projections, but they in-

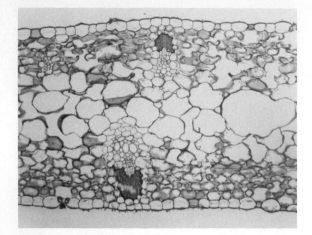

Figure 12.42
This leaf of *Iris* is unifacial; it may appear to be merely isobilateral like the *Eucalyptus* leaf of Figure 12.7, but look at the bundles carefully: in both bundles the xylem is to the inside and the phloem is to the outside. They have the appearance of being part of the same surface that has been folded over. × 160.

creased the photosynthetic surface area dramatically and must have been functional as such, because they had stomata. As these evolved to become larger, small vascular traces were formed that ran from the stele to the base of the microphyll and later actually into it. Microphylls never became large structures,

and at present, 5 to 10 mm is about their maximum length. Many believe that all leaves of the Lycopodiophyta are this type of microphyll, but Stewart (1983) has suggested that only the Lycopodiales and the fossil group Drepanophycales have them.

MEGAPHYLLS.   Megaphylls (all the leaves with which you are familiar) are so named because they originated as a large structure (an entire branch system), not just a small one like the enation precursors of microphylls. The ideas of megaphyll evolution are embodied in Zimmermann's (1952) **telome theory** (Fig. 12.45; well discussed in Stewart, 1983). They hypothesized a primitive land plant, much like *Rhynia*, composed of dichotomously branched, naked stems. The last, uppermost bifurcation in each shoot was a **telome**, and the lower ones were **mesomes**. In the hypothetical plant, branching would be in three dimensions, and all branches would be more or less equal. Many of the first land plants were like this. F. O. Bower postulated that plants would evolve such that some branches grew more rapidly ("overtopping" the others) and that the more slowly growing ones might tend to orient themselves in a plane ("planation"). Later, photosynthetic tissue might extend between the mesomes and telomes of the short side branches ("webbing"). At this stage, the plant would consist of one or several main shoots, each supporting

Figure 12.43
The formation of a unifacial, ensiform leaf. These sections represent four portions of one leaf, (A) being basalmost and (D) being most apical. Based on information provided by D. R. Kaplan.

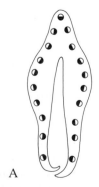

A   B   C   D

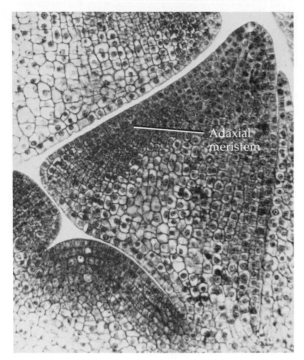

Figure 12.44
Leaf primordium of *Acorus* in which the adaxial meristem has formed and has begun causing the midregion to thicken. × 270. Micrograph generously provided by D. A. Kaplan.

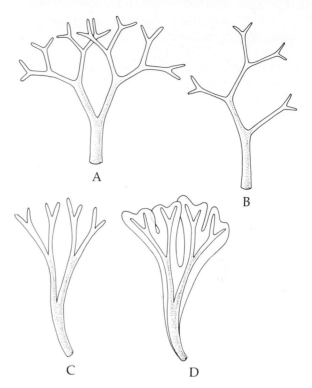

Figure 12.45
The processes by which megaphylls probably evolved. (A) The original plants had dichotomously branched, naked axes. (B) If some branches grow faster than others a main trunk is formed that has short side branches. (C) All the axes of the side branches are produced in one flat plane. (D) Webbing develops between the axes, producing an extensive photosynthetic surface.

small, flat photosynthetic side branch systems. Branches like this occur in the fossils *Pseudosporochnus nodosis* and *Cladoxylon scoparium* (Steward, 1983).

These side branches are by no means leaves, but they had the potential to continue to become megaphylls. They are sometimes called **prefronds** to indicate their status. It has been suggested (Meeuse, 1963) that the simple leaves of conifers evolved from just the last, telome portions of these prefronds but that the large compound leaves of the fern line of evolution and of the cycadophytes (and thus the angiosperms) evolved from the entire prefrond side branch. The planation of the entire prefrond was slow, and in fossils such as *Aneuro-*

phyton, Proteokalon, Protopteridium and *Triloboxylon* there is still some radial symmetry in the cauline portions. Also, the cauline portion has a stem type of stele, rather than a midrib type. Tens of millions of years were required before the branch systems were completely converted to leaves.

Some details of the telome theory are still debated, but in general it is widely accepted, and there is excellent fossil evidence in its favor. It is thought that there are three main lines of megaphyll evolution:

1. The fern megaphyll may have evolved from the prefronds of ancestors such as *Rhacophyton*.

2. The conifer needle would have come from the telomes of Upper Devonian-Lower Carboniferous *Archaeopteris* and its relatives.

3. The cycadophyte-angiosperm megaphyll originated in the prefronds of the Upper Devonian–Lower Carboniferous Aneurophytales.

According to Stewart (1983) there is increasing evidence that the Aneurophytales and the Archaeopteridales were actually closely related, being two lines of descent from the progymnosperms (Beck, 1976). If so, then the cycadophyte leaf and the gymnosperm leaf are rather closely related. The ferns, however, separated onto their own line of evolution much earlier; the very strong similarities between fern leaf anatomy and that of dicots is remarkable. In the gymnosperms and ferns, leaves have been rather static, and on a single plant are all remarkably similar, especially when compared to the enormous diversity present in the angiosperms.

At this point it is important to consider the criteria for determining whether a structure on a living plant is or is not a leaf. As mentioned earlier, some angiosperms evolved almost to the point of being leafless, especially in xeric conditions. In a variety of angiosperm groups, after certain species had so evolved, some members migrated into more moist environments in which a large leaf surface would not be selectively disadvantageous, and in fact an increased photosynthetic area would be distinctly advantageous. In some of these, mutations that allowed the expansion of the residual leaf primordium were selected, and large, true leaves resulted. Such may have happened in the formation of the monocot "phyllodial" leaves. But in others, the mutations that were selected were ones that affected the stem, causing it to become flattened, expanded, and leaflike. In many plants, this change in the stem has been mild enough that the object is still rather easily recognizable as a stem (called a cladophyll; as in *Cryptocereus anthonyanus*, *Phyllanthus*, *Opuntia*, and *Rhipsalis*), but in others the transformation is so extensive that only very close inspection shows them not to be true leaves (*Ruscus*, *Streptocarpus*, *Muehlenbeckia platyclados*, *Semele androgyna*).

The spines of cacti are composed of only a mass of fibers covered by a sclerified epidermis, yet the interpretation of this as a leaf is not questioned. In *Ruscus*, by contrast, the cladophylls have palisade and spongy parenchyma and other aspects of dorsiventral organization, yet they are not considered leaves. At present, morphologists use **position** as the overriding criterion for identifying a structure: any structure that arises in the position where a leaf should be is a leaf. This may seem arbitrary, but virtually all empirical evidence supports it: even though spines differ strongly from foliage leaves, they are produced by shoot apical meristems, their primordia look like leaf primordia, and in many species it is possible to see how leaves have become evolutionarily modified into spines. Cladophylls, however, arise where buds should arise, and, even in those species where extensive modifications have occurred, there are still significant differences between the cladophyll and true foliage leaves.

Concepts

Ideas about the role of the root system in the biology of a plant are changing rapidly at present. Just a few years ago, it was assumed that roots had only two functions of any real consequence: to anchor the plant to its substrate and to absorb water and nutrients. It is quite likely that these indeed were the main tasks for the plants in which roots originally evolved. A thalloid plant (such as the first land plants might have been) has little need for either absorptive organs or anchoring ones; but, as the land plants became taller, elevating their photosynthetic and sporogenous tissues, they became unstable, tending to topple over, even if they were a part of a spreading stoloniferous system. Moreover, these elevated tissues were no longer in contact with the soil, which is the source of water and minerals. There was obviously strong selection pressure to develop roots.

But just as is true of every plant organ, the original purposes for the evolution of the root need not remain the sole purposes. In many perennial plants, the root is now the main organ for storage of carbohydrates; in suffrutescent plants that die back to the root or a **root crown** (a root with a short piece of the stem attached), the root is the sole means of storage and of survival through the time of stress.

In all plants, annual or perennial, seed plant or cryptogam, the roots are critically important sources of hormones (Crozier and Hillman, 1984). They are a major source of the cytokinin necessary for normal shoot development and the proper functioning of apical meristems. Many gibberellins are synthesized in the roots, and others are produced in the leaves, then transported to the root to be converted to the active form.

The processes just described probably occur in all roots; in certain species, the roots have been highly modified to have specialized functions. They act as props or stilts for corn and other grasses as well as tree ferns and banyan trees. They can be a means of aeration in the stagnant marshes where bald cypress and mangroves grow (Fig. 13.1; Chapman, 1939). As holdfasts, they allow ivy and many other epiphytes to cling to their aerial substrates (Fig. 13.2). In some palms (*Crysophila, Mauritia*) the adventitious roots on the stem become spines (Tomlinson, 1961), while the adventitious roots of certain *Arum* species grow upward to form a basket that catches debris—the roots collect their own soil. In some plants the roots form adventitious shoot buds and serve as a means of asexual propagation (Peterson, 1975). Vertical **contractile roots** can pull a shoot firmly into the ground (Fig. 13.3), while horizontal ones can actually pull a plant up to a meter sideways (Jernstedt, 1984; Montenegro, 1974; Ruzin, 1979). Finally, the roots of parasitic angiosperms have become modified into **haustoria** (Calder and Bernhardt, 1983; Kuijt, 1977).

As described in Chapter 11, stems can be similarly diverse in their functions, but they

Figure 13.1
The root systems of most plants are difficult to study, but these aerial roots of *Rhizophora mangle* (mangrove) are easily accessible (except for the minor difficulty of avoiding coral snakes and alligators). The aerial portions of the root system are involved in transporting water and minerals but not absorbing them; they also act as props and as a means of channeling air to the submerged portions of the roots. See Figures 13.19 and 13.20 for cross-sectional views of these roots. × 0.25.

Figure 13.2
This strangler fig is able to climb through the dark undergrowth of a rainforest by following the trunk of a large tree. It has ordinary absorptive roots in the soil, and it also has these clasping roots. × 0.25.

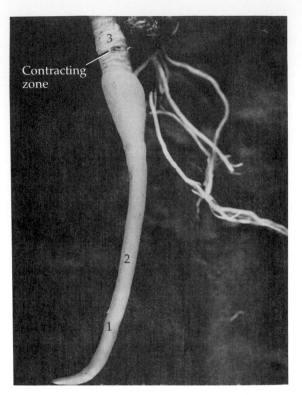

Figure 13.3
This tap root of *Freesia hybrida* was undergoing contraction at the time it was collected. The lowermost portion at 1 was elongating, at 2 it was maturing, and at 3 the tissues were contracting longitudinally, pulling the corm downward. × 0.5. Photograph generously supplied by S. E. Ruzin.

have retained a uniform organization. The same is true of roots. Only the most extreme modification, transformation into haustoria, has altered the basic organization of the root; aside from haustoria, it is always rather easy to identify a root as such on anatomical evidence (von Guttenberg, 1940; Weaver, 1926). In the primary body of a root, there is always an outermost epidermis, beneath which is the cortex, then the endodermis, pericycle, and finally the vascular tissues.

In contrast to the constancy of the internal structure of roots, the root system as a whole can be enormously variable, even within a species. There are a variety of types of roots and root systems. **Seminal roots** are those present in the seed before germination. The most important seminal root is the **radicle** which represents the basal end of the root-shoot axis in an embryo (Figs. 20.1, 20.15). In dicots and gymnosperms, the radicle usually grows out to become a **tap root**, and it is always much larger than any of the **lateral roots** that grow out of it. The tap root is known as the **primary root**, and it forms a unitary structure like the shoot, being ultimately derived from embryonic tissues. Lateral roots may be referred to as **secondary roots**. **Adventitious roots** are ones that arise on any organ that is not a root. Most commonly, they form in the lower nodes (especially in monocots, where they are then called **nodal roots**, **crown roots**, or **prop roots**), but they can form anywhere along the stem or even on leaves. Adventitious roots can originate in almost any tissue, including the epidermis, stem cortex and pericycle, ray parenchyma, immature xylem and phloem cells, and the pith. If the plant is rhizomatous, stoloniferous, or otherwise spreading, then very often the original seminal roots die when the base of the shoot dies. The root system of the surviving runners has no developmental connection with the radicle of the embryo. In monocots, the radicle dies almost immediately after germination, so a tap root system is not formed. They have a **fibrous root system** instead, consisting usually of some lateral roots from the residual base of the radicle and of adventitious crown roots.

In addition to the variability in root system structure that is caused by the presence of so many types of roots, there is the even more important factor of the root's capacity for compensatory growth in a heterogeneous environment. We tend to think of the air and the environment above ground as being heterogeneous, but it is really quite uniform. There may be important variations in sunlight, but other than a boundary effect right at the soil surface, the rest of the environment, on the scale of individual plants, has few important variations

in the concentrations of carbon dioxide, humidity, water potential, temperature, and such. By comparison, the soil is strongly heterogeneous: we find gradients of moisture (uppermost layers more moist than the lower ones immediately after a rain but drier otherwise), of temperature (upper levels hotter during summer days but cooler than lower levels at night and in the winter), of texture and porosity, of available nutrients, of oxygen and carbon dioxide, of humus and organic acids derived from the decay of fallen leaves (Russell, 1978). Roots are able to grow **compensatorily**, that is, if part of the root system is restricted, the rest will grow faster, or if one root enters a region that is especially favorable, it will greatly increase its growth rate, while the rest slow down (Fig. 13.4; Brouwer et al., 1981; Carson 1974; Crossett et al., 1975; Reynolds, 1975; Russell, 1977). The form of the root system is thus often more a result of the soil factors at that particular site, rather than any innate, predetermined habit as is true of the shoot (Kummerow et al., 1982; Reynolds, 1975). Another factor important for this is that the lateral roots form **endogenously** (in the inner root tissues) rather than **exogenously** (in surface tissues) as do buds in the shoot. Shoot buds automatically are arranged in the same phyllotaxy as the leaves that subtend them (Cutter, 1965); but root buds can apparently form almost anywhere and in almost any density, often being induced more by external factors than internal ones (Vartanian, 1981).

Structure of the Root Primary Body

The root cap and apical meristem have already been discussed in Chapter 6.

Mucigel

The apical regions of roots secrete a mucilaginous, slimy substance called **mucigel** (Fig. 13.5; Jenny and Grossenbacher, 1963). The mucigel is not a tissue but only a secretion, but

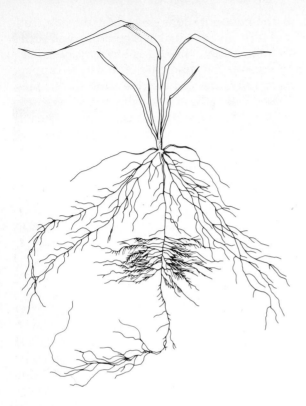

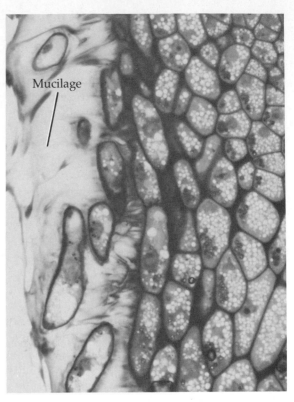

Mucilage

Figure 13.4
This diagram shows the compensatory growth of roots of barley (*Hordeum vulgare*) that occurs when one small part is given an enhanced supply of nitrate. Drawing based on information provided by M. C. Drew and L. R. Saker, 1975.

Figure 13.5
This micrograph shows the outermost cells of the root cap. They have secreted mucilage that now surrounds them almost completely. Some root cap cells are released from the root and are completely suspended in the mucigel. × 500. Micrograph generously provided by R. C. Moore.

it is so constantly present that it must be considered an important part of the root, just as fragrances are important parts of flowers and fruits. The mucigel is a complex substance. Part of it is secreted by the dictyosomes of the root cap (Juniper and Pask, 1973; Paull and Jones, 1976; see Chapter 9), and the root hairs can also secrete gel (Greaves and Darbyshire, 1972; Leppard, 1974; Werker and Kislev, 1978). In addition, both the root cap cells that are damaged by passage through the soil and those that slough off decay and contribute their debris to the mucigel. Microbes are always found in it (Foster and Rovira, 1976), and indeed, roots grown axenically (under sterile condi-

tions) have much less mucigel. This could indicate either that the microbes secrete part of it themselves or that their presence stimulates the root's production. The function of the mucigel has not been established, but it has been suggested that it serves as a culture medium for the bacteria that somehow aid the growth of the root by secreting hormones or nutrients (Bowen and Rovira, 1976; Dart and Mercer, 1964; Rovira, 1979) or that it facilitates movement of ions through the soil to the root (Greenland, 1979). The mucigel is the actual contact interface between growing roots and the soil; the cell walls do not interact directly with the soil solution (Foster et al., 1983; Greenland,

1979; Russell, 1977). Many critical ions such as phosphate and potassium are quite immobile in the soil itself, but may diffuse more rapidly through the gel. Of course, the mucigel also can act as a lubricant to aid the passage of the root apex through the soil particles, but it seems that too much gel is secreted for this to be its only function.

Epidermis

The nature of the root epidermis varies with its age. In the most apical, elongating regions of the root, the epidermal cells are small and cytoplasmic, and they do not protrude as root hairs (they would be sheared off as the root tip is pushed through the soil). Farther back from the apex, the cells begin to vacuolate and elongate, and some of them undergo an asymmetric division to produce a large ordinary epidermal cell and a small **root hair initial** (also called a **trichoblast**). As the root hair initial begins to bulge outward (Fig. 13.6), the nucleus and most of the cytoplasm migrate into it (Schnepf, 1986). The root hair grows by depositing more wall material at the tip, while regions closer to the root do not elongate. The root hair contains a giant central vacuole, and the majority of the cytoplasm forms an extremely thin layer next to the wall; the nucleus and a large aggregation of hyaloplasm remain in the tip of the root hair. If the hair runs into a small soil particle, it may grow in two directions around it, becoming forked (von Guttenberg, 1968); however, the forked region is always short, never long like many of the branched trichomes on the stem surface. A cuticle has been reported to occur on roots, even on root hairs (Scott et al., 1958; Scott, 1966), but, as Martin and Juniper (1970) point out, the stains that are used to detect cutin (especially Sudan III) are not specific and will react with many types of fatty material. Several studies with electron microscopy (Bonnett and Newcomb, 1966; Juniper and Clowes, 1965; Juniper and Roberts, 1966) could not detect a cuticle on young root tissues. When fatty de-

Figure 13.6
Germinating seed of radish. Root hairs are already abundant; they do not form on the extreme apex or in the zone of elongation. × 1.0. Photograph generously provided by J. D. Cunningham/Visuals Unlimited.

posits are found in the walls of young root epidermal cells, they are now thought to be the precursors of suberin (Martin and Juniper, 1970).

Root hairs have a greater density of plasmodesmatal connection with the adjacent cortical cells than do the ordinary hairless epidermal cells: Kurkova (1981) reported that the root hair–cortex wall has 2.06 plasmodesmata/μm^2 —10,412 in the entire wall. The ordinary cell–cortex wall has only 0.10 plasmodesmata/μm^2, which yields 630 in the entire wall. The outgrowth of root hairs is another aspect of roots that is more controlled by the environment than by the root itself. Desert plants like *Sempervivum* (Hesse, 1904) produce more hairs when the soil is dry than when it is damp, and plants produce longer root hairs when the soil is dry than when it is moist (Reid and Bowen, 1979). In moist soil or even in humid air, they can be formed in truly amazing numbers: Dittmer (1937) estimated there are 10,000,000,000 on a single mature rye plant. However, if terrestrial plants are grown in liquid culture, root hairs usually do not form (von Guttenberg, 1968), and in the aerial roots of many plants such as *Rhizophora* (Fig. 13.1, 13.19; Gill and Tomlinson, 1971), *Pandanus*, and banyan trees, the roots are completely devoid of root hairs until after they contact and penetrate the soil. The presence of root hairs is variable in aquatic angiosperms: they are often absent, but two species that are fully submerged do have them—*Posidonia australis* (Kuo and Cambridge, 1978) and *Thalassia testudinum* (Tomlinson, 1969)—as does one species of floating plant (*Hydrocharis*). Two other floating species have none: *Pistia* and *Eichhornia* (Clowes, 1985). The size, shape, and longevity of root hairs are less variable: most are unicellular, about 1,000 μm long, and 10 to 15 μm in diameter, and they last for one to three days. Multicellular root hairs are very rare, as in *Kalanchoe fedtschenkoi* (Popham and Henry, 1955), as are ones with thick walls or ones that persist longer than several days, such as those of *Gleditschia triacanthos* and some Compositae (Cormack, 1949; Dittmer, 1949; Scott et al., 1963). Older portions of roots have never been observed to produce root hairs (von Guttenberg, 1968).

Although it has always been assumed that root hairs are a means of increasing the surface area of the root so as to facilitate absorption of water and minerals, this has suddenly been thrown into some doubt by recent physiological calculations and experiments (Reid and Bowen, 1979; Russell, 1977). Although tests show that root hairs are capable of absorption

under laboratory conditions, there is no actual evidence that they can do this in real soil (Kramer, 1969). Newman (1974) concluded on theoretical grounds that, even if root hairs are produced at a density of 100 per centimeter of root length, they would still have a conductivity that is only one-twentieth that of the soil itself. Ions could diffuse through the soil solution and move by capillarity more rapidly than through the root hairs; see the theoretical treatment of movement of water through the soil given by Nobel (1983). Similarly, with the aid of micropotometers, it has been found that root hairs absorb water only in regions adjacent to concentrations of cytoplasm, not over the majority of their surface (Cailloux, 1972).

It is possible that the function of the root hairs is not to absorb material directly but rather to facilitate absorption by the regular epidermal cells; that is, the presence of the root hairs greatly alters the environment immediately adjacent to the root—the **rhizosphere** (Tinker, 1976; Weatherly, 1975). While living, their respiration produces the carbon dioxide necessary for the cation exchange that releases positively charged ions from the negatively charged soil particles (**micelles**). Also, they secrete mucigel, which may be important for changing the surface properties of the soil micelles, either by coating them with a humid film directly or by encouraging the growth of microbes that in turn can alter the soil properties and mineral availability (Rovira, 1979).

Another important consideration is that roots cannot decrease their diameter to be able to grow through small spaces and pores in the soil, nor can they increase their diameter (without secondary growth) to fill large pores (Reid and Bowen, 1979; Wiersum, 1957). Without root hairs, there would be little direct contact between root surface and soil surface in either large pores or small ones. Therefore, the root hairs may be the means by which to exploit the numerous soil pores that are not of an ideal size for the root axis itself. Whether or not root hairs absorb nutrients themselves or merely facilitate their mobility in the soil, it is known that phosphate and potassium are absorbed much more rapidly by regions with root hairs than by ones without them.

A corollary of the assumption that the root hairs were the primary means of absorption was the assumption that the epidermis over older parts of the root was impermeable and incapable of bringing in any ions. This assumption has also been shown to be incorrect; potassium and phosphate at least can be absorbed over the entire root surface (Russell and Clarkson, 1976). The rate of absorption per unit surface area definitely is much less than it is at the apex, but because the apices are just a very small portion of the root system and the older portions are so large, their uptake is significant.

If the epidermis persists for a long time without being replaced by bark, then the cells can become heavily cutinized or develop thick walls, especially the outer tangential wall (von Guttenberg, 1968). They may also become lignified. Such changes seem to have two functions: first, to combat the soil microbes that are fairly efficient at attacking the root and second, to retain the water within the older parts of the root when the soil becomes dry. The aerial roots of orchids and some aroids have a multiseriate epidermis, the velamen (Chapter 10).

Cortex

The root cortex (Fig. 13.7) is a tissue that seems to have not received all the attention it deserves. We often think mostly of the absorption across the epidermis and entry into the xylem, and we rather ignore what goes on in the cortex in between these two. With the recent increase in physiological studies, study of the cortex is becoming much more interesting.

ENDODERMIS.   The innermost layer of the cortex in roots is the **endodermis** (Figs. 13.8, 13.11, 13.12; Kroemer, 1903). Whereas an endodermis is rare in stems, it is virtually univer-

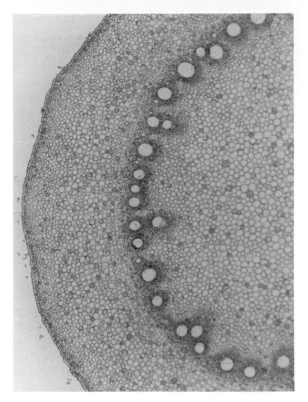

Figure 13.7
This is a cross section of a young root of corn, showing the various tissues. Note the large pith-region. × 100.

sally present in roots: von Guttenberg (1968) described only three species known not to have a root endodermis: *Canarium commune*, *Tinospora crispa*, and *Nyssa sylvatica*. As in stems, the endodermis of roots is always a unistratose cylinder of cells that line the inner boundary of the cortex and the outer boundary of the stele. The presence of the endodermis with its Casparian strips divides the root into two distinct compartments. The importance of the **continuous phase** provided by the apoplast has been discussed in regard to the movement of water in xylem, the air in leaves, and the accumulation of secretory products. The compartmentalization of the root apoplast into two **discontinuous phases** by the endodermis in

Figure 13.8
In this cross section of a root of *Smilax* (greenbriar), all tissues are easily recognizable. The exodermis and the endodermis both have extremely thickened walls on the radial sides and either the outer tangential wall (exodermis) or the inner one (endodermis). × 100.

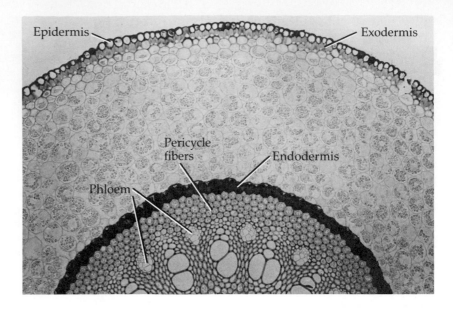

the roots is important for the selective movement of minerals and water. Any ion that is present in the soil solution can penetrate the root epidermis and cortex. Even if all of the cortical cells had plasmalemmas that were impermeable to it, it could still penetrate throughout the cortex merely by passing through the water phase of the walls and intercellular spaces. However, it cannot pass the Casparian strips; uncontrolled diffusion stops at this barrier (Brouwer, 1959; Drew, 1979; Lawton et al., 1981; Priestley, 1920, 1922; Priestley and North, 1922). For the ion to move into the stele and enter the transpiration stream of the xylem it must cross a plasmalemma of an endodermal cell (Fig. 13.9). At this point the plant can exert selection as to which ions enter and which are excluded.

On the other hand, if an ion is beneficial and required by the plant, it would be inefficient for all of the epidermal and cortical cells to absorb it actively, because it would be capable of leaking back out of the cells unless their membranes were remarkably impermeable; it is best to have the endodermal cells actively transport it into the stele. Once across this barrier, if the ion then leaks from the symplast phase to the

apoplast, it is still trapped within the stele. Thus the endodermis is the ideal structure for the location of the metabolism for actively transporting ions.

The active transport of ions into the steles has several significant consequences, the first of which is that it causes the water potential of the stele to become more negative, and water will enter, even if transpiration is not occurring, so **root pressure** builds up in the xylem (Drew, 1979). As mentioned in Chapter 7, this is important in healing small, transient embolisms. It is also partially responsible for the guttation from hydathodes (Chapter 9). This pressure in the xylem of roots necessitates the continued presence of endodermis, even in older portions of the roots that absorb minerals at lower rates. Without the endodermis, during conditions of low transpiration the pressurized water in the upper parts of the root could leave the stele and either flood the surrounding tissues or even leave the root and move out to the soil. Such flooding has been shown by growing plants that lack hydathodes in conditions that favor mineral uptake (high fertilizer) and low transpiration (high atmospheric humidity). In the stems and leaves

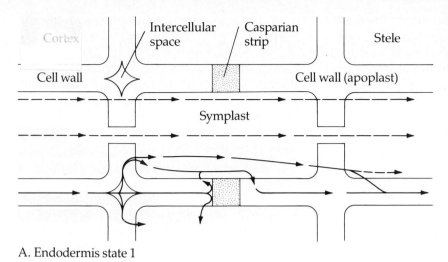

A. Endodermis state 1

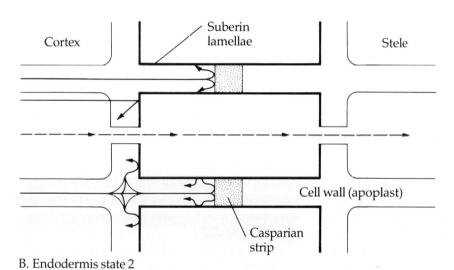

B. Endodermis state 2

Figure 13.9
The top portion of this diagram represents an endodermis in State I, having only Casparian strips on its walls. In the lower portion of the diagram, the endodermal cell is in State II, and now solutes can cross the endodermis only if they are in the cortical symplast. The endodermal plasmalemma is completely covered.

where the stele is not ensheathed by an endodermis, the tissues become waterlogged (Zimmermann, 1983).

The endodermis as just described, with only a Casparian strip, is said to be in State I (Mylius, 1913). To facilitate this pressure-resistant isolation of the stele in the older portions of the root, the walls of the endodermal cells enter a second and third phase of modification, as described in Chapter 10. In State II, suberin is deposited in all of the primary wall except where plasmodesmata occur (Robards et al., 1973; Warmbrodt, 1985). In State III of endodermal development, a thick layer of cellulose is deposited between the suberin lamella and the plasmalemma, and lignification also occurs (Figs. 13.8, 13.9, 13.12). In the angiosperms, but not the conifers (von Guttenberg, 1941;

Wilcox, 1962a), the walls can become very obviously thickened. In the grasses, the inner tangential wall and the radial walls next to it receive extra cellulose layers, but the outer tangential wall does not (Clark and Harris, 1981; Hector, 1936; Haas and Carothers, 1975). These modifications of endodermal cells do not happen simultaneously to all the cells at a particular level but rather cell by cell. In barley, at a distance of 5 cm back from the apex, about half the cells have the thickened walls with suberin lamellae; 15 to 30 cm back, about one-fourth of the cells still have just Casparian strips (Clarkson et al, 1971). The cells that remain in the State I condition are called **passage cells**, and it was thought that they regulated the movement of ions into the stele, whereas the other, secondarily modified cells did not. But now it has been established that even the thick-walled cells remain alive, with numerous cytoplasmic connections with the stele parenchyma by means of extensive fields of plasmodesmata that pass through the thickened inner tangential wall (Clarkson et al., 1971).

This type of modification to the older endodermis is truly elegant; it provides an excellent pressure barrier for the stele yet still allows ion loading into the xylem. Recall that potassium and phosphate are absorbed over the entire root surface; these can move into the stele by means of both passage cells and plasmodesmata, moving entirely symplastically. Calcium, however, can be absorbed into the cortex over the entire surface, but because it moves apoplastically, it cannot enter those regions that have an endodermis. Calcium can enter only at the extreme apex where the endodermis either has not formed or has not yet completed its Casparian strips (Robards et al., 1973).

Although the endodermis is typically analyzed with respect to its control of materials entering the stele, it also constitutes a barrier to the apoplastic unloading of sugars from the phloem (Dick and ap Rees, 1975; Giaquinta et al., 1983; Warmbrodt, 1985). Sugar unloading occurs symplastically along much of the root, but in the most active area of growth, the root apex, the endodermis is incomplete and is not a barrier.

EXODERMIS.   The internal position of the endodermis leaves all of the cortex exposed to whatever is present in the soil solution, however toxic. This problem has been overcome in some plants by the development of the **exodermis**, which is essentially an endodermis located on the exterior surface of the cortex, as a hypodermis (Fig. 13.8; von Guttenberg, 1943, 1968; van Fleet, 1950). The exodermis is usually unistratose but may be several layers thick. There have been reports that the exodermis differs from the endodermis because it deposits the suberin lamella immediately, without a Casparian band ever forming (Ferguson and Clarkson, 1975; Tippett and O'Brien, 1976; Olesen, 1978; Peterson et al., 1978). However, a reinvestigation with sensitive fluorescence methods of two species, corn and onion (Peterson et al., 1982), revealed the presence of a delicate Casparian band. The exodermis cells deposit a thin suberin lamella over the interior surface of their walls (as in Stage II development of an endodermis). Then layers of cellulose can be deposited centripetally as well, and lignin may be present (Clark and Harris, 1981). The exodermis does not mature as quickly as the endodermis, and it does not act as an exclusion barrier very near the root apex, as does the endodermis (Peterson and Pernamalla, 1984). The presence of the exodermis and endodermis separates the root into three distinct apoplastic compartments, those of the epidermis, the cortex, and the stele. The entry of ions is regulated immediately, thereby protecting the cortical cells. Final control of ion entry into the stele is still exerted by the endodermis.

In *Citrus*, the suberin lamella is reported to completely cover the wall, breaking the plasmodesmata (Walker et al., 1984) and killing the cells. Thin-walled, living passage cells do remain and allow the entrance of material into the root.

The cryptogams do not have exodermises (Fahn, 1982).

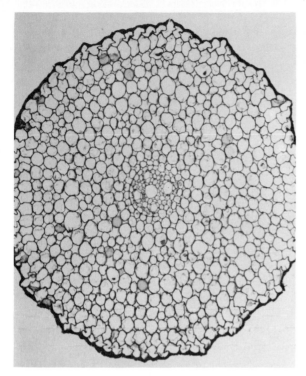

Figure 13.10
Almost the entire volume of this root of *Elodea* is composed of cortex, and a large fraction of that is intercellular space. The large apoplast permits the relatively rapid diffusion of oxygen down to the roots from the photosynthetic tissues of the shoot. × 100.

BODY OF THE CORTEX. The cortex between the exodermis and the endodermis is typically a rather uniform mass of parenchyma tissue (Figs. 13.7, 13.8, 13.10). Intercellular air spaces are more common in roots than stems, and large air canals occur in the root cortex of mangroves (Lawton et al., 1981), palms (Tomlinson, 1961), rice, and many aquatic plants (Fig. 13.10; Arikado, 1955; Bristow, 1975; Coult, 1964; Laing, 1940). Most of the types of internal secretory cells and tissues described in Chapter 9 can occur in roots, but external glands do not. Starch may be abundant (Figs. 13.8, 13.11).

The cortex of the root is typically much thinner than that of the stem, often being only three or four cells thick. If exposed to light, the roots of some plants can develop chloroplasts and turn green, but many cannot; the aerial roots of arums and most other epiphytes are white or brown, even though they grow in the presence of light.

If the root does not undergo any secondary growth, then the cortex may live as long as the rest of the root, and it may develop large masses of sclerenchyma. In some grasses and dicots, the cortex remains alive only briefly, then dies and degenerates, leaving the endodermis or pericycle as the surface of the root (Head, 1973; Waid, 1974). When this occurs, the endodermis, pericyle, and even the stelar parenchyma can become heavily lignified. Collenchyma is only rarely found in roots (von Guttenberg, 1940).

Stele

The stele of the root shows less variation than those of the stems and leaves. Root steles are almost always protosteles (Fig. 13.11); eusteles (Fig. 13.8) and atactosteles (Fig. 13.12) occur only in very wide roots (mostly in monocots). The outermost layer of the stele is a region of parenchyma called the **pericycle** (Figs. 13.8, 13.11, 13.12). This is usually just a simple, colorless parenchyma, but in some plants fibers may also occur (von Guttenberg, 1968; Warmbrodt, 1985). In most angiosperms, the pericycle is unistratose, but in some (Gramineae, Palmae, *Salix*) it can be several layers thick. The endodermis is circular in cross section, but the vascular tissues are star-shaped. Consequently the pericycle is thin or even absent where the vascular tissues project outward, and it is thicker where they do not. In grasses, the xylem extends all the way to the endodermis, and in Potamogetonaceae the phloem does (Esau, 1965); the pericycle is interrupted at those sites and exists as panels of tissue between them. Secretory cells and ducts can occur in the pericycle. In the gymnosperms and in those dicots

Figure 13.11
This cross section of *Ranunculus* (buttercup) root shows the protostele very clearly. There are four poles of protoxylem on the exterior and one central mass of metaxylem. Alternating with the protoxylem are the rods of primary phloem; in three of the phloem groups, metaphloem sieve tubes and companion cells can be identified (arrows). The endodermis is rather inconspicuous, and the Casparian strips have not stained darkly. × 200.

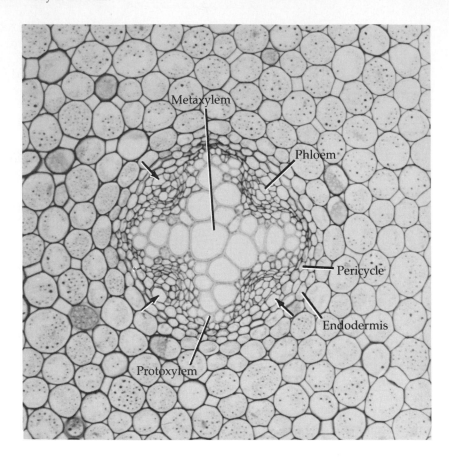

that have secondary growth, the pericycle is important in contributing to the vascular cambium. The cork cambium of the roots typically is derived from pericycle cells, also. In the seed plants, the primordia of the lateral roots arise within the pericycle (Fig. 13.15). In the monocots (which have long-lived roots without secondary growth), the pericycle cells become sclerified in older regions of the roots.

In the center of the stele is the primary xylem. As in shoots, the primary xylem is composed of the protoxylem, which differentiates while young and close to the root apex, and the metaxylem, which is older at the time of differentiation and has, consequently, much larger cells (Brouwer et al., 1981; Torrey and Clarkson, 1975; von Guttenberg, 1968). There is no reliable criterion for distinguishing unequivo-

cally between protoxylem and metaxylem (Esau, 1965); the first cells to differentiate are the most exterior and are small. After these come cells that are progressively larger and are located deeper within the stele. Whereas the bundles of all shoots are endarch, the roots of seed plants have **exarch xylem**, with the protoxylem exterior and the metaxylem forming the interior mass. The protoxylem forms as a series of rods on the outer surface of the developing xylem cylinder, and the most common numbers of rods are three (a **triarch root**) and four (**tetrarch**; Fig. 13.11). Roots with one protoxylem pole (**monarch**), two (**diarch**), or more than four (**polyarch**; Fig. 13.8) are not unusual. The number of protoxylem poles is closely correlated with the vigor of the root and also with its diameter (Brouwer et al., 1981; Jost, 1932;

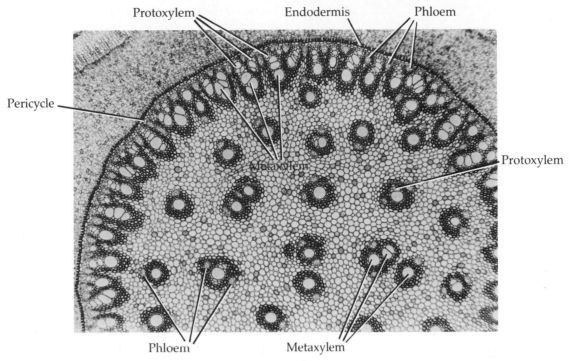

Figure 13.12
This root of *Dracaena fragrans* has an atactostele, much like the stele of its stem. A State III endodermis surrounds the stele, and, just inside, small masses of phloem can be recognized. The large metaxylem elements are easily visible and identifiable, as are the numerous protoxylem poles in the peripheral bundles. Only a few of the central bundles have recognizable protoxylem. × 100.

Torrey, 1957; Wilcox, 1962b), and it can change as the root becomes more or less robust. A single plant of *Libocedrus decurrens* can have di-, tri-, tetra-, penta-, and hexarch roots (Wilcox, 1962b).

In monocots, especially in their adventitious crown roots, which are very wide, the xylem is usually polyarch, actually broken up into distinct bundles that surround a large pith. This is interpreted, however, by some people to be still a protostele, with the central "xylary elements" converted to parenchyma. The giant adventitious roots of palms and Pandanaceae (Fig. 13.13; "screw pines," which are monocots, not conifers as their common name suggests) may have up to 100 protoxylem poles. This may be related to the fact that most monocots are often actually types of giant "herbs" that never have secondary growth and thus are not able to increase the water- and nutrient-moving capacity later when the plant is larger. This is an important difference between the

Figure 13.13
This is a large plant of *Pandanus baptisii* (screwpine) in Fairchild Tropical Gardens. Although it is a monocot that lacks secondary growth (like palms), its stems can branch and increase the total number of leaves present (unlike palms), because its stems form adventitious roots. The new roots supply more water and minerals; because they conduct directly into upper regions of the shoot, they bypass the already fully loaded vascular tissues of the lower parts of the shoot. Palms cannot do this, and they cannot increase the total number of leaves in the crown. A more common example of plants forming adventitious roots would be those on the underground rhizomes of irises and grasses.

gymnosperms and dicots versus the monocots. As plants of the former two types become larger and require larger amounts of water and nutrient transport, their roots can branch extensively. But as the distal part of a root becomes branched and increases its absorptive power, the extra water that is thus absorbed must be moved through the established proximal portions of the root, and extra conducting capacity is needed. By undergoing secondary growth, gymnosperms and dicots produce new tracheary elements to increase the capacity, and the increased load is handled. They also produce more phloem, and the larger root system can be nourished. But monocots do not have this option, because the majority of them do not have secondary growth, so the first stages of rooting must be established with an excess of capacity. The roots can branch later, increasing their absorptive capacity until they saturate the transport capacity of the proximal segments. For the root system of a monocot to supply an increased amount of water to an expanding shoot system, completely new adventitious roots must be produced, in the shoot itself, connecting directly into the shoot stele. Thus, roots with little capacity, such as monarch, diarch, and the like, would be of little use to the monocots, and only large polyarch ones with many bundles and many large metaxylem elements can meet their unusual needs.

As mentioned before, the arrangement of the large metaxylem tracheids and vessels is of great importance for the conductivity of the xylem. Apparently there are no studies that have traced individual bundles in roots (Zimmermann, 1983). However, the exarch arrangement of the root xylem allows all of the large, low-resistance, high-conductivity metaxylem elements to be clustered together. This permits easy water transfer from element to element through extensive areas of lateral wall contact. Furthermore, the metaxylem is completely surrounded by protoxylem and living parenchyma—neither of which is ever likely to contain any air—so the metaxylem channel is protected from embolisms induced by air bubbles. In those roots that have a wide stele and large pith region, wide metaxylem vessels may occur in the pith. They may occur with no obvious pattern or be arranged in two or three circles (Esau, 1965), or there may be just one single large vessel.

An almost universal observation is that the

vessel elements of roots are the widest of any within a single plant (Fahn, 1964; Zimmermann, 1983; Zimmermann and Potter, 1982). Numerous theories have been proposed concerning this, but a simple consideration of water tension offers the most acceptable explanation. At the top of the plant the tension is greatest, because there the weight and the friction of the entire water column must be supported by the leaf xylem. The tension on the base of the water column (in the roots) is very low, because only the weight of the water in the root and the small amount of friction in the root restrict the ascent of the water. The tension on the water columns in the roots is either very mild or nonexistent, usually replaced by a positive root pressure. It is logical that the widest vessels would occur in the roots, and indeed the very widest vessels ever measured (0.6–0.7 mm) do occur in roots (Jeník, 1978). Similarly, in palms that have scalariform perforation plates, Klotz (1978) found that the bars are most widely separated (and least able to block embolisms) in the roots, but are much closer together (and thus much more protective) in leaves. The conductivity of the root, in fact, seems to be restricted not by the root xylem itself, but rather by the root cortex and the shoot. Increasing the transpiration by decreasing the humidity around the leaves always leads to increased water movement through the roots, and, if the root apices are cut off from a partially immersed root such that water can enter the root xylem directly without passing through the cortex, water movement increases (Russell, 1977).

The phloem of the root occurs as bundles just exterior to the xylem, located between the protoxylem poles and equal to them in number (Figs. 13.11, 13.12). In those plants with a pith in the roots, the phloem is located just exterior to the protoxylem, forming collateral bundles; occasionally (in *Cordyline*, *Dracaena*, *Musa* and *Pandanus*) it may occur in the pith as well. The differentiation of the phloem is just as it is in the stem, with the protophloem exterior and the metaphloem closest to the xylem. Unlike

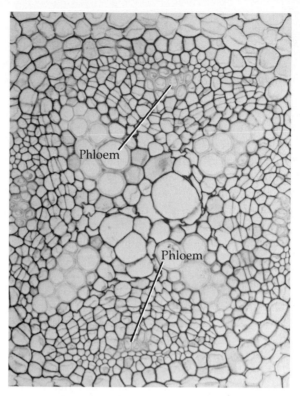

Figure 13.14
In this root of *Saja*, the phloem and the xylem are separated from each other by a considerable amount of parenchyma. A band of parenchyma cells, just exterior to the xylem and completely encircling it, has begun cell division; this is the vascular cambium forming. × 200.

the protophloem of the stem, that of the root is easy to detect, because all the surrounding cells are still meristematic and densely cytoplasmic: the protophloem sieve tubes appear empty among them (Fig. 13.12). Companion cells may be absent (Resch, 1961). In the root, the xylem and phloem are always separated by at least a small amount of parenchyma, often by rather large amounts (Fig. 13.14). If the root is persistent but does not undergo secondary growth, then much of this parenchyma may become sclerified. Whereas fibers are common in the primary phloem of stems, they occur only rarely in the primary phloem of roots (e.g., Annonaceae, Malvaceae, some legumes; von Guttenberg, 1943, 1968).

Branching

As mentioned in the introduction to this chapter, primordia for lateral roots arise endogenously, unlike the axillary buds of the shoots that are produced exogenously and automatically in the same phyllotaxy as the leaves. Both the production of the root primordia and their activation and outgrowth seem to be strongly correlated with factors in the substrate, even the presence of the roots of neighboring plants (Kummerow et al., 1982; Raper and Barber, 1970; Torrey, 1965). Recall that the soil is much more heterogeneous than the air and that root growth is compensatory, both the outgrowth of laterals and the rate of elongation of the main root axis. If roots branch at certain angles, then the soil can be optimally exploited (Russell, 1977). In monocots and ferns, which have small root systems due to the absence of secondary growth, the orientation of the lateral roots is critically important in holding the stem upright and in preventing blow down (Pinthus, 1967).

In the seed plants, the primordia of the lateral roots form in the pericycle as a locus of cells undergoing divisions with periclinal walls (Fig. 13.15; Blakely et al., 1982; McCully, 1975). In some cases, underlying parenchyma cells and occasionally even endodermal cells become mitotically active and contribute to the initial stages of growth (Bell and McCully, 1970; Esau, 1940). In most of the ferns and other cryptogams, the primordia arise in the endodermis only, without direct contribution of the pericycle (Ogura, 1938). The initiation of the primordia is related to the positions of the xylem and phloem of the stele. In roots with three or more protoxylem poles, the primordia are initiated immediately opposite the poles, whereas in roots that have two protoxylem poles, the primordia occur either opposite the

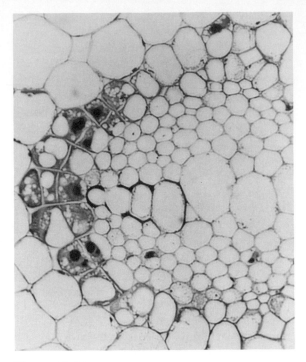

Figure 13.15
First stages in the formation of the meristem of a lateral root. Several pericycle cells have undergone one periclinal division; one cell has undergone two. Notice how extremely cytoplasmic the cells have become and how enlarged their nuclei are. The root is forming directly opposite a protoxylem pole. × 200. Micrograph generously provided by Blakely et al. (1982).

protophloem in two rows, or between the protophloem and protoxylem in either two or four rows (Knobloch, 1954; Mallory et al., 1970). This difference in positioning possibly relates to the problems of producing a hydraulically functional xylem connection between the new lateral and the existing axis. In a root that has three or more protoxylem poles, the xylem, as seen in cross section, exists in three dimensions (as a Y, an X, and so on). If a lateral root forms opposite one of the protoxylem poles, then in between the new stele and the metaxylem of the old stele is abundant parenchyma that can redifferentiate into tracheary elements to establish the connection (McCully, 1975; Seago, 1973). Furthermore, this can be done in a straight line and with minimal disturbance of the phloem. In monocots that have bundles and pith in the roots, the lateral roots are able to form xylary connections with several bundles, not just one (Fig. 13.16; von Guttenberg, 1940). The connection can extend into the deeper metaxylem vessels if the pith parenchyma cells redifferentiate into tracheary elements (Rywosch, 1909). Several studies (Riopel, 1966, 1969; Lamont, 1972a, b; Mallory et al., 1970; Yorke and Sagar, 1970) have shown clearly that the outgrowth of new roots longitudinally along the parent root is not random; instead they often occur in clumps of two, three, or more all at nearly the same level. Also, if a root is curved, the laterals form predominantly or exclusively on the convex side (McCully, 1975). The reasons or possible advantages of the clustering of lateral roots is unknown, but it seems likely that it would have a major influence on the conductivity (both xylary and phloic) at that point.

Longitudinally, the root primordia are typically initiated just behind the root hair zone, but they can occur either closer to the apex (especially in water plants) or farther back (Blakely et al., 1982; Mallory et al., 1970; Riopel, 1966). Many of the cells in this region of the root have begun to mature, and the lateral root primordia arise in partially differentiated pericycle cells that have large vacuoles (Fig. 13.15). The adjacent endodermal cells already have Casparian strips. In corn, the cells of the pericycle and endodermis have become quite thick-walled and lignified; the first visible changes associated with the initiation of the primordium is a thinning of the wall and a loss of lignin (Bell and McCully, 1970; Karas and McCully, 1973). Apparently the lateral root primordia do not arise from single cells in the pericycle but rather from groups of cells: about 30 in *Raphanus* (Fig. 13.15; Blakely et al., 1982) and 24 in *Vicia faba* (Davidson, 1972). When they begin to grow

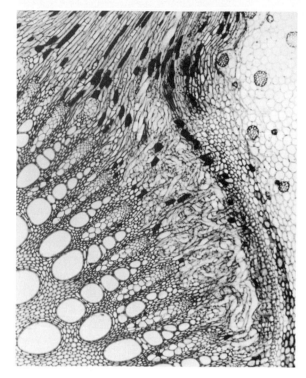

Figure 13.16
This micrograph shows a large root of the palm *Martinezia* cut in cross section at the point where a lateral root is attached. Connections with ten main bundles are visible. Notice how prominent the phloem bundles are and how large the sieve tube members are. × 300.

out, the cells become more densely cytoplasmic, then start to divide in all directions, and the primordium begins to protrude into the cortex. It is not known how the pressures are channeled such that the primordium pushes outward rather than in all directions; it has been suggested that it is capable of digesting its way through the cortex, but no convincing evidence has been presented, and this idea is not widely accepted (McCully, 1975). In all angiosperms that have been studied, the endodermis contributes to the root primordium at least in the initial stage; it usually does so only by anticlinal divisions, producing a unistratose sheath around the elongating primordium. Apparently, whatever tissue is provided by the endodermis is, in most species, just temporary; it is ruptured even before the root emerges from the cortex. This layer of endodermal cells does not become cytoplasmic and is easily distinguishable from the tissues of the new root (Byrne, 1973; Seago, 1973). It has been reported that in *Zea mays* the root cap of the new root is derived from the endodermis of the parent root (Bell and McCully, 1970; Clowes, 1978; Karas and McCulley, 1973). When the lateral root primordium has extended about one-third of the way through the cortex, the endodermal cells that cover its apex undergo periclinal divisions, thereby producing a root cap meristem. These cells develop amyloplasts that appear to be statoliths. Popham (1955) reported that the endodermis of pea contributed both to the root cap and to the root epidermis; the pericycle-derived cells formed only the inner tissues of the lateral root.

By the time the root does emerge, a well-defined apical meristem has been established, and even a root cap may be present, although it may be small (Fig. 13.17). Moore and Pasieniuk (1984) hypothesized that the small size of the newly emergent lateral root apex might be the cause for its lack of responsiveness to gravity (it grows horizontally, not downward).

The emergence of the root causes considerable damage to the cortex and epidermis, and it results in an unprotected rupture that is a danger point for penetration by bacteria and fungi. Typically, a special corky tissue is deposited around this to seal it off as much as possible. As the lateral root begins to form its own endodermis, parenchyma cells at the point of origin of the root redifferentiate as a new endodermis that connects the endodermis of the parent root with that of the new lateral (Esau, 1940; Dumbroff and Peirson, 1971).

This method of endogenous origin for roots has extremely beneficial consequences. Unlike the cortex of the shoot, that of the roots often

Figure 13.17
This is an emergent lateral root of radish. The surrounding cortical cells have been flattened, and emergence is by means of tearing through the epidermis. Micrograph generously provided by Blakely et al. (1982). × 150.

does not persist. If the soil becomes dry, the epidermis and cortex of the older portions of the roots may die. But the endodermis remains alive, as do the stele and the lateral root primordia. When moisture returns, these primordia can grow out quickly. In this case, the problem of growing through the cortex does not exist.

Dimorphic Roots

In most root systems, the primary and lateral roots are quite similar to each other; if the primary root develops into a swollen tap root, such as in beets or carrots, that process typically involves secondary growth, not the primary tissues. In some plants, the root system does consist of several types of roots, each with a distinct structure and function (Ellmore, 1981; Jeník, 1978).

An interesting adaptation to temporary water availability is the production of **root spurs**, which are analogous to short shoots (Fig. 13.18; Boke, 1979). Within the main, long root are located fully-formed root primordia that can be activated immediately, as soon as any moisture becomes available. The root primordia are unusual in having no root cap whatsoever. They grow out quickly, and they form root hairs over their entire surface, even over the apex. They are able to absorb water within hours of a rainfall, but, as soon as the soil dries, the root spurs die back almost to the parent root. Before death, however, they produce new root primordia that survive and can continue the cycle. After this process has been repeated several times, a short axis is finally built up, protruding slightly from the parent root; the root spurs are actually sympodial, not monopodial. Such a mechanism is an efficient means of harvesting water from an environment that is only periodically and briefly moist. So far root spurs have been reported only from cacti (Boke, 1979); however, they are so easily overlooked that they may be more widespread than is realized.

Roots of the members of the family Proteaceae have a special dimorphism that is thought to be related, not to the scarcity of water, but to scarcity of minerals in the soil (Dell et al., 1980; Jeffrey, 1967; Lamont, 1972a, b, 1973; Purnell, 1960). As the ordinary long root grows through the soil, at certain sites **proteoid roots** are formed. These are short lateral roots that produce an abundance of rootlets (up to 960) over their surface, opposite every protoxylem pole. The proteoid roots can form quickly, and Dell et al. (1980) calculated that these can generate new absorptive surface area at the rate of 1500 mm^2 per day and achieve a total area of 4890

Figure 13.18
Longitudinal sections of the root spurs of *Opuntia arenaria* reveal that they have no root caps but that they already have the primordium of the next root spurs that will replace these. × 100. Micrograph generously provided by N. Boke.

mm^2. It is postulated that this is a mechanism that allows a very site-specific increase in absorptive capacity as a root grows through a small area rich in nutrients.

In red mangrove (*Rhizophora mangle*), the root system consists of horizontally spreading, aerial roots that arch over the surface of the water where the plants grow, then finally descend and penetrate the mud (Fig. 13.1; Gill and Tomlinson, 1969, 1971, 1977). As the aerial portion of the root enters the soil, it undergoes striking changes (Figs. 13.19, 13.20): the long zone of elongation becomes much shorter; the number of protoxylem poles decreases, and

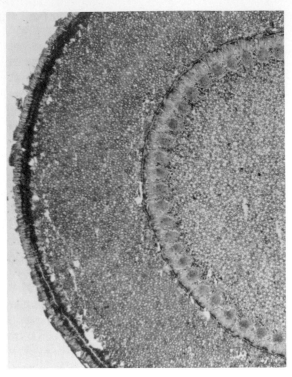

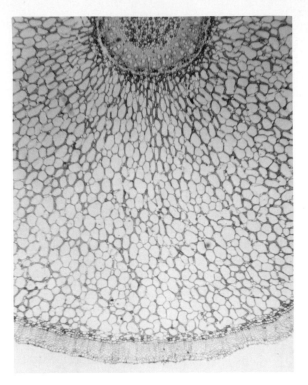

Figure 13.19
This is a portion of the aerial root of *Rhizophora mangle* (mangrove); compare its structure with that of the submerged portion in Fig. 13.20. × 10. Micrograph generously provided by A. M. Gill and P. B. Tomlinson.

Figure 13.20
The submerged root of *Rhizophora mangle*. Note that the stele is much narrower than in the aerial portion, but the aerenchymatous cortex is wider. × 10. Micrograph generously provided by A. M. Gill and P. B. Tomlinson.

the pith becomes much smaller; the cortex becomes much wider as large air-filled lacunae develop; the aerial root has a bark and lenticels; the subterranean portion retains its epidermis; the aerial root has chloroplasts, is green, and rarely branches, while the submerged root is not green and produces numerous lateral roots. The differences between the two types of roots (actually the two portions of the same root) are beautifully logical and demonstrate the adaptation to function that is possible. The aerial portion, being located in the air, is photosynthetic and has a means of gas exchange. It serves as a conduit for water back

to the trunk, but it does not collect water or minerals itself. Being out of the soil, lateral roots and root hairs would be useless. Once the root penetrates the mud of the coastal tide zone, however, chlorophyll is obviously unnecessary, and oxygen is not available for absorption by lenticels; instead it must be obtained from the atmosphere by means of cortical air canals. The aerial root grows without being attached to anything, and a long zone of elongation is feasible. Once the root penetrates the soil, then, like all other terrestrial roots, it must grow at the tip only, because of the friction of the substrate.

The Root-Shoot Interface

The interface between the root and the shoot of a plant is a problem, because the primary vascular tissues of the two organs are arranged differently. In the root there is an exarch protostele of xylem with independent phloem strands. This must be made to merge smoothly with the endarch collateral bundles of the shoot. The area where this happens is called the **transition region**, and it usually encompasses all or part of the hypocotyl.

The arrangements of tissues in the transition region are complex and not easily described; possibly for these reasons there have been few studies of it (Artschwager, 1926; Crooks, 1933; Esau, 1940, 1965; Hayward, 1938; Mauseth, et al., 1985; Pyykko, 1974; Scheirer and Hillson, 1973; Sundberg, 1983). The important conclusions to be drawn from these studies is that continuous strands of xylem and phloem are produced; there are few reports of new bundles that arise without any attachment to previously existing bundles.

The transition region of *Linum* has been described by Crooks (1933). The root is diarch, and two strands of phloem alternate with the two protoxylem poles. As the stele is followed upward into the hypocotyl, it maintains this arrangement, but then it swells as a pith appears at the center. At the lower levels, the pith consists of just a few parenchyma cells, but at higher levels it is abundant enough to divide the xylem into two exarch bundles. There are now two metaxylem groups. At about the same level, each phloem strand bifurcates, resulting in four phloem strands.

At a slightly higher level, the metaxylem elements differentiate on the sides of each group of xylem rather than adjacent to the pith. The xylem is changing from exarch to endarch, and there are now four groups of metaxylem rather than two. Furthermore, each metaxylem group is now near a phloem strand. About halfway up the hypocotyl, these have become true endarch collateral bundles; they then divide to pro-

duce a total of eight bundles. These bundles are arranged in two arcs each with four bundles; the two lateral bundles of each arc enter the cotyledons, and the two central bundles of each arc fuse into one bundle that then enters the cotyledons.

Just above the cotyledons, the vascular tissues of the epicotyl have developed as endarch collateral bundles; they begin to differentiate basipetally, downward into the hypocotyl. These bundles move toward the metaxylem and metaphloem of the transition region and link to them, forming a continuous conduit between root and stem. The bundles may initially terminate blindly in the hypocotyl parenchyma, but, as the formation of metaxylem and metaphloem continues, linkage occurs.

The gymnosperms appear to have a transition region like that of the dicots. The root stele divides into bundles that then enter the cotyledons, and the bundles of the epicotyl differentiate downward and make contact with these (Boureau, 1954). The transition is slightly more complex, however, because gymnosperms have numerous cotyledons, not just two.

Similarly, monocots are affected by having just one cotyledon. In some species the bundles that are formed in the root go into the cotyledon (as in dicots and gymnosperms). In other species, only some of the bundles enter the cotyledon; the rest enter the first foliage leaf directly. Finally, in *Triticum* (Boyd and Avery, 1936), the root vascular tissue merges into a plate called the **nodal plate**. Out of the upper side of this plate come the stem vascular bundles. Some may originate as endarch collateral bundles, but others are slightly irregular, showing features typical of bundles in transition regions.

Mycorrhizae

Until recently, it was thought that a small number of plant species formed an association with soil fungi (Powell and Bagyaraj, 1984), the association being called a **mycorrhiza** (pl.: mycorrhizae). Within just the last few years, it has become established that almost all terrestrial plants have mycorrhizal associations and that these are symbiotic relationships in which the fungus supplies the plant with nutrients, especially phosphorus, while the plant provides sugars and other organic compounds to the fungus (Daft and Okusanya, 1973). If a section of soil is treated with a fungicide that kills all the mycorrhizal fungi, the plants in the same area will be severely stunted. It can be shown that the stunting is not the result of the fungicide damaging the roots, because the plants recover immediately if a fresh inoculum of mycorrhizal fungi is added back to the soil at the base of the plants.

The mycorrhizae are divided into two broad categories: (1) **ectomycorrhizae**, in which the fungus remains predominantly on the outside of the root, with a small amount of penetration between the exterior cortical cells, and (2) **endomycorrhizae**, in which there is very little surface mycelium but an extensive network within the root.

Ectomycorrhizae

Ectomycorrhizae are rather rare, occurring on species of Betulaceae, Fagaceae, Pinaceae, and a few others, all trees or shrubs. The ectomycorrhizal roots are thicker and more branched than roots that are not associated with fungi (Gerdemann, 1974). On the surface of the root, the fungal mycelium forms a thick, dense **sheath** or **mantle** (Clowes, 1981). From the sheath, hyphae penetrate between (not through or into) the epidermal cells, then ramify in the intercellular spaces of the outer cortex. The fungus remains entirely within the apoplast of the roots. The network that they form is called a **Hartig net**, and in it the hyphae come into close contact with the plant cells. It is here that nutrient exchange is presumed to occur. Ectomycorrhizae are known to produce plant hormones (auxin and cytokinin) in culture, and

there is a possibility that they are responsible for altering the root's anatomy after infection (Gerdemann, 1974).

Endomycorrhizae

The endomycorrhizae are subdivided into those involving septate fungi and those involving nonseptate fungi. The former are rare and specialized, forming associations only with orchids and some members of the Order Ericales. The orchids are extremely dependent on their mycorrhizal associations; if not invaded by the proper fungus, the seedlings become moribund and eventually die.

The endomycorrhizae involving nonseptate fungi are the preponderant type, and possibly up to 80% of all terrestrial plants have such an association (Fig. 13.21). These are called **vesicular-arbuscular mycorrhizae** or **VA mycorrhizae** because once the fungus penetrates the host it produces characteristic **vesicles** and **arbuscules**, as described below (Powell and Bagyaraj, 1984). Within the soil, the fungus extends as a thin mycelium; when a hypha contacts a root, it flattens slightly, then penetrates either into the cell or between the cells (Scannerini and Bonfante-Fasolo, 1983). Once in the outer cortex, some hyphae penetrate the cortical cells, somehow passing through the walls. As they extend into the cell, they do not break either the plasmalemma or the tonoplast of the host. Instead the hypha is always surrounded by these membranes even though it is inside (spatially) the vacuole of the cortex cell. The hypha usually coils extensively, creating a large surface area that provides contact with the cell. Other hyphae ramify throughout the cortex, apparently pushing the cells slightly apart. Some of these hyphae swell, forming the vesicles for which they are named (note that *vesicle* is a mycological term and that it is not the same as a plant vesicle). In the inner cortex, once the hyphae invade the cells, they ramify and dichotomize, forming an arbuscule that looks like a miniature tree (Fig. 13.21; Scannerini and

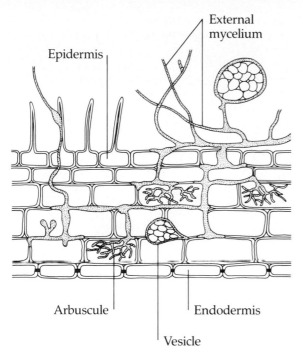

Epidermis

External mycelium

Arbuscule

Endodermis

Vesicle

Figure 13.21
This diagram illustrates the various stages of a VA mycorrhiza. Infection can occur in either a root hair or an ordinary epidermal cell; once inside, the fungus invades the outer cells to a small extent but forms large arbuscules in the inner layers of the cortex. Vesicles are also shown.

Bonfante-Fasolo, 1983). Each branch is surrounded by the plant's own membranes, so there is a large exchange surface, similar to that of the labyrinthine walls in transfer cells. The hyphae of the arbuscules fill with phosphate granules; these then dissolve, and the phosphate is absorbed by the plant (Cox and Tinker, 1976; Toth and Miller, 1984). While this happens, the plant cell responds by increasing the amount of cytoplasm it has, and the nucleus becomes deeply lobed. The development of the plastids is arrested at the proplastid stage, or they become chromoplasts. They do not accumulate starch, perhaps because the sugars are being diverted to the fungi. On a larger scale, the plant responds to an invasion of VA mycorrhizae by increasing the amount of vascular tissue in the area and increasing its content of lignin (Krishna et al., 1981). The fungus never penetrates the endodermis, never enters the stele, and does not invade the root apex (Berta and Bonfante-Fasolo, 1983). Once the arbuscule has completed its transfer of phosphate to the plant, it collapses, and the plant cell returns to a state similar to that of an uninvaded cell. Although VA mycorrhizae are essential for the healthy, rapid growth of most plants, they result in such minor changes in the anatomy of the roots that they were almost completely overlooked for years.

Root Nodules and Nitrogen Fixation

In tropical and temperate climates where water and sunlight are usually adequate for plant growth, the most severe shortage for plants is a lack of nitrogenous compounds. Although nitrogen is abundant in the air, it is in a form that plants cannot use; it must first be converted to nitrate, nitrite, or ammonium. Unfortunately, plants are unable to do this; only procaryotes (bacteria and blue-green algae) can enzymatically attack atmospheric nitrogen. In many instances, plants absorb the nitrogenous compounds that are released as the procaryotes die and their bodies decay in the soil. Other plants have formed associations with the procaryotes and are able to take up the nitrogenous compounds immediately (Becking, 1975; Bergersen, 1982; Broughton, 1983; Dart, 1975). The associations are usually symbiotic: the plant receives nitrogenous compounds and in turn supplies the procaryote with an energy source, typically sugar. These associations often result in anatomical modifications of the plant; the most dramatic and well-known examples are the root nodules in legumes (Fig. 13.22).

The bacteria that are involved in root nodules of legumes are *Rhizobium* species, and

Figure 13.22
This diagram illustrates the anatomy of a simple root nodule in a legume. The development is described in the text. The meristematic zone allows it to grow and fix nitrogen for an indefinite time.

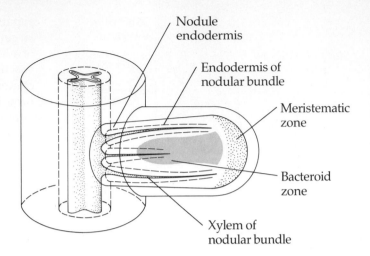

Nodule endodermis

Endodermis of nodular bundle

Meristematic zone

Bacteroid zone

Xylem of nodular bundle

they are ubiquitous in the soil (Dart, 1975). The roots exude a variety of substances into the rhizosphere, one of which apparently stimulates the proper strain of the *Rhizobium*; the bacteria in turn stimulate the root hairs to bend sharply (Broughton, 1983), and the bacterium then attaches to the root hair at the concave side. A clear spot is formed on the hair wall at the contact point; then the bacterium invades the cell. It does not break through the wall and plasmalemma, however; instead it forms a tunnel-like **infection thread** that consists of host wall material and host plasmalemma (Fig. 13.23; Baird and Webster, 1982; Bergersen, 1982). The infection thread grows toward the base of the root hair cell; then it grows through the cortex cells. As it arrives at the inner cortex, the cortical cells become meristematic, their nuclei swell, and a small, hemispherical meristem of uninvaded cells forms.

In some genera of legumes, the nodule grows in all directions and becomes spherical; in others, the meristematic region is restricted, and the nodule grows outward as a short cylinder (Fig. 13.22; Sutton, 1983; Werner and Mörschel, 1978). As the nodule continues to develop, it forms a *nodule endodermis* near its periphery; this is connected to the normal endodermis of the root stele. Interior to this is a

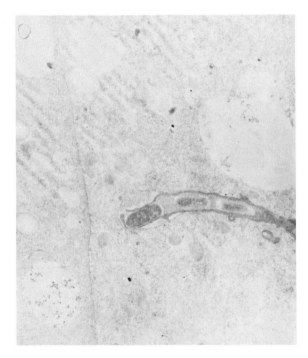

Figure 13.23
Intracellular infection thread inside a cell of *Phaseolus vulgaris*. The bean cell constitutes almost all of the picture; only three small bacteria are visible. × 26,000. Micrograph generously provided by L. M. Baird and B. D. Webster.

mass of parenchyma cells (the *nodule cortex*) and a vascular system composed of one to five bundles attached to the root stele and extending into the nodule, branching as it goes. All parts of the nodule vascular system are surrounded by a vascular bundle endodermis. The pericycle of the nodule bundle may be just parenchyma, but in some it consists of transfer cells.

As the meristematic region adds new cells to the nodule, the endodermis and vascular system extend forward. In addition, the new cortical cells become invaded by the *Rhizobium*; within the nodule the bacteria are released from the infection threads into vesicles (Bergersen, 1982; Broughton, 1983; Goodchild and Bergersen, 1966). They multiply and fill the cell with numerous procaryotic cells, now called

bacteroids. The host cell increases its cytoplasmic content, as well as the number of dictyosomes and the amounts of rough endoplasmic reticulum and vesicles. The plastids and mitochondria migrate to the wall. Each invaded cell lives for only several days; then it degenerates as all the bacteroids and cell organelles lyse (Sutton, 1983). However, new cells are being added by the nodule meristem, and these ultimately become invaded.

The anatomy of the nodule varies somewhat, depending on the particular species of legume. However, many aspects of its growth and development are influenced by the procaryotic partner; it will be fascinating to discover how two organisms together are able to control the morphogenesis of such a complex structure.

PART V

The Secondary Body of the Plant

In herbaceous plants, the primary body constitutes the entire plant; but woody species are able to initiate a new phase of growth. Different types of meristems are formed, and they produce new cells for conduction, protection, and nutrient storage. This has numerous important consequences. The continued production of conducting tissues allows the plants to outlive the functional life of their primary xylem and phloem; the extra bulk provides enough storage capacity for the plants to survive prolonged adverse conditions; and the greater height or length of the shoots and roots allows the individual plant to exploit its immediate environment more fully. Each of these consequences leads, in turn, to further consequences. The plants have the resources to become more complicated; they have more of

each type of part (a tree has more leaves than an herb), and some can become specialized for diverse functions. It is useful to think about plants such as ferns (which never have secondary growth) and compare them with the conifers and dicot trees (which do); the effects of secondary growth are extensive. However, as always, there are several anatomical and physiological solutions to the challenges that the environment presents. In several families or genera, unusual types of primary growth have evolved that provide many of the benefits of secondary growth. Also, some species have very special needs, and evolution has resulted in anomalous types of secondary growth. It is valuable to consider all of these and analyze how each type of anatomy facilitates the successful life of the individual in which it occurs.

14

Vascular Cambium

Concepts

In ferns, most monocots, and herbaceous dicots, the primary body previously described is the entire vegetative body. But in the gymnosperms and the woody dicots, a **vascular cambium** (Figs. 14.1, 14.2) produces part of the secondary body of the plant: the secondary xylem (wood) and the secondary phloem (part of the bark—the cork cambium produces the rest of the bark). Most treatments of the vascular cambium ascribe to it only one function, the production of the secondary vascular tissues, but in reality it has another function that is just as important: to survive. This is, of course, quite obvious, but it is not at all trivial, because it affects how the cambial region is organized and functions. Just as is true of the apical meristems in roots and shoots, the vascular cambium must live and operate successfully for the entire lifetime of the plant—up to 11,000 years, not just one or two seasons. Furthermore, unlike the apical meristems, the vascular cambium is irreplaceable, because there is nothing at all equivalent to the axillary buds that are activated if the main apical meristem is damaged. If the vascular cambium is destroyed, the plant cannot produce another; secondary growth, and probably life, will stop. Thus, to avoid the risks inherent in DNA replication, the vascular cambium should not be overly active; it should not be the source of all the cells for the secondary body, but instead it should be a source of initials. Relatively few cell divisions in the vas-

cular cambium are sufficient to provide new cells to the xylem mother cell region and to the phloem mother cell region, and divisions there will produce the large numbers of cells for the secondary xylem and phloem (Fig. 14.3). DNA copy errors and other somatic mutations in the xylem mother cell zone and the phloem mother cell zone are not so critical, because each cell there is only expected to produce a limited number of derivatives in its limited life (one growing season or less). Like the apical meristems, the vascular cambium must be a step meristem with an "archival" source of genetically sound cells that slowly contributes cells to zones of rapid but limited division (Klekowski, 1988).

Just as in the shoot and root apical meristems, the vascular cambium functions to establish a pattern. The subsequent growth and development of the secondary xylem and phloem are affected by this pattern. The physical arrangement of cells in the wood and the inner bark is a direct result of the organization of the vascular cambium (Fig. 14.2). Thus all three tissues are strongly interrelated, and none is free to evolve to its optimal anatomy. Certain cells in the xylem and phloem should be extremely long for efficient functioning, but if the cambial cells were that long, they would have difficulty dividing. As meristematic cells, it would be best for cambial cells to be small and isodiametric, but this would result in disastrously weak and nonfunctional wood. In considering the vascular cambium, it is neces-

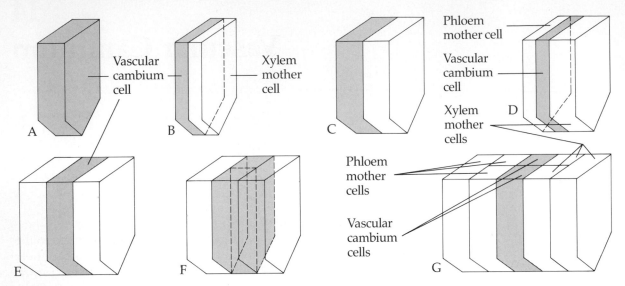

Figure 14.1
This diagram represents some of the basic features of vascular cambium cells. In (A), the bottom half of a single cell is shown; the diamond-shaped face on the left faces toward the exterior of the plant. In (B) the cell has divided; one daughter cell (in this case, the one on the left) continues to be a vascular cambium cell while the other becomes a xylem mother cell. In (C), both cells have grown; then in (D) the vascular cambium cell divides again. This time, the daughter cell on the right continues to be a vascular cambium cell, while the one on the left develops as a phloem mother cell. The xylem mother cell would have divided also, but that is not shown. In (E), all cells have grown and are ready to divide. (F) shows a division that produces two vascular cambium cells. After this, subsequent divisions as in (B) and (D) will produce two rows of cells, as in (G), not just one.

sary to think not only about its functions but also about the unavoidable influences it has on all other tissues of stems and roots.

Structure of the Vascular Cambium

In most trees and shrubs, the **cambial region** is a continuous ring-shaped, multilayered cylinder. Some of the inner layers are xylem mother cells, and some of the outer ones are phloem mother cells. During periods of activity and growth, the cambial region contains many layers (Fig. 14.2), but during dormancy it contains only a few (Fig. 14.4). How many layers comprise the cambium? This is a question that

has been difficult to answer (Larson, 1982; Philipson and Ward, 1965). Catesson (1964) considered the cambial region as the important functional unit, but Hartig (1853) thought of the vascular cambium as a double layer, a **biseriate cambium**, with the inner layer contributing to the xylem and the outer contributing to the phloem. Sanio (1873) postulated that it was a single layer, a **uniseriate cambium**, that was also **bifacial**—it could produce derivatives in both directions. After a periclinal division, either the inner cell begins differentiation and the outer cell remains active as vascular cambium, or the outer cell develops as phloem and the inner one continues as the cambial initial

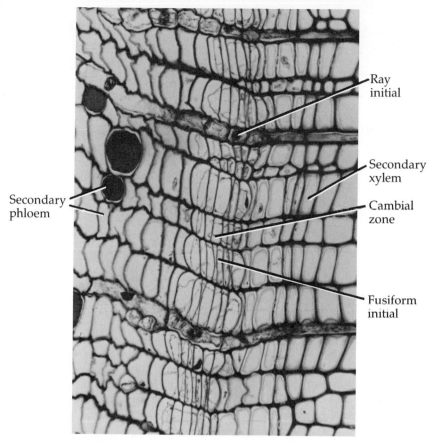

Ray
initial

Secondary
xylem

Cambial
zone

Secondary
phloem

Fusiform
initial

Figure 14.2
Cross section of the vascular cambium of *Pinus strobus*. The orientation is the
same as in Figure 14.1, with xylem on the right. The vascular cambium can be
difficult or impossible to identify among the mother cells of the xylem and
phloem. In this section, both fusiform initials and ray initials are visible. × 150.

(Fig. 14.1). This interpretation is the one most
widely accepted today. The most convincing
evidence in favor of this uniseriate, bifacial hypothesis is the fact that the secondary xylem
and phloem are more or less mirror images of
each other (Bannan, 1962). The cambium can
initiate new rays from time to time, and when
this happens the new ray is established in both
the xylem and the phloem, never in just one.
Also, as the cambium is pushed ·outward by
the maturation of the xylem derivatives interior

to it, the cambial cells divide with an anticlinal
wall and avoid becoming very wide tangentially (Fig. 14.1G). This conversion of one initial into two causes two radial files of derivatives to be produced instead of just one; when
this conversion does occur, it affects both the
xylem and the phloem. The mirror image nature of the wood and bark could be possible
with a biseriate cambium, but that would require perfect coordination between the two
layers. Bannan (1955, 1968) and Newman (1956)

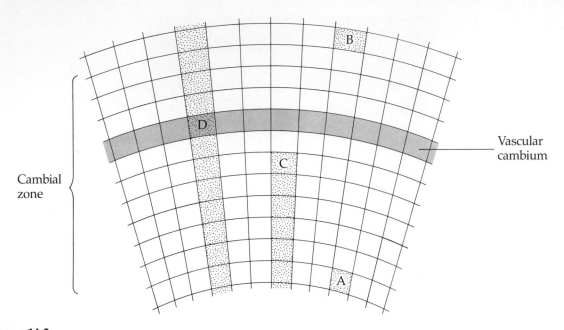

Figure 14.3
The number of cells affected by somatic mutations at various locations: if the cells at (A) or (B) mutate, they will be the only ones affected because they are outside of the cambial zone and are not dividing. A cell at (C) normally divides to produce a group of xylem cells; if it mutates, the entire group will be affected. But that cell group will be pushed out of the cambial zone by new cells coming from the cambium, so the patch of mutated cells will be limited. If a cell in the cambium itself (D) mutates, then all cells produced by it will be affected.

believed that they were in fact able to identify individual fusiform initials in radial sections because these cells were shorter than both the xylem mother cells and the phloem mother cells; the initials were present as a single layer.

It was just mentioned that the vascular cambium in many plants is a continuous ring or, in three dimensions, a cylinder. However, in many plants, especially in the succulent cacti and euphorbias and in most of the dicots that produce so little wood that they appear herbaceous, the vascular cambium exists as narrow strips that are confined to the original vascular bundles. The limited activity of the cambium causes these bundles to become either woody poles or a woody net.

The cambial cylinder or cylindrical reticulum extends without interruption from the shoot into the root and is continuous into the branches and lateral roots. The vascular cambium at the root-shoot junction is the oldest in the plant, and segments of it both above and below this area are progressively younger, until at the very tips of the shoots and roots the newest increments of vascular cambium are just forming.

Cell Types in the Vascular Cambium

The cells of the vascular cambium are of two types, the elongate **fusiform initials** that produce the axially elongated or oriented elements of the wood and inner bark, and the more isodiametric **ray initials** that produce the radially oriented rays (Figs. 14.4, 14.5, 14.6).

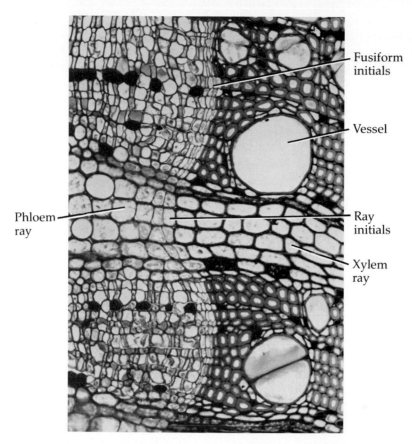

Fusiform initials

Vessel

Ray initials

Xylem ray

Phloem ray

Figure 14.4
The vascular cambium of this *Adansonia digitata* (baobab) is dormant, and the cambial zone is thin (only three layers), because most mother cells have completed differentiation into xylem or phloem by the time division has stopped. × 100.

Fusiform Initials

The fusiform initials are long cells, ranging from 140 to 462 μm in dicots and from 700 to 4500 μm in *Pinus* (Bailey, 1920), and even to 8700 μm in *Sequoia sempervirens* (Bailey, 1923). The length of fusiform initials of closely related species can be quite different: 154 μm in *Dalbergia melanoxylon* and 203 μm in *D. sissoo* (Ghouse and Yunus, 1974). Their length can vary during the year, depending on the balance between cell division and cell expansion (Sharma et al., 1979, 1980). In radial section the end walls appear straight (Fig. 14.7), but in tangential section they are pointed, tapering either gradually or abruptly to a point. In cross section they appear rectangular or slightly flattened. It would be easy to assume that they have about 8 faces, but a study of *Pinus sylvestris* has shown that 8 is the minimum; they can often have up to 32 faces with an average of 18; they have 14 contact faces shared with other cambial initials alone (Dodd, 1948).

The length of the fusiform initials has been of much interest, because it affects (at least to some degree) the length of the derivatives.

Figure 14.5
This is a tangential section through the cambial zone. The elongate, tapered cells are fusiform initials; the round cells are ray initials. × 100. Micrograph generously provided by R. Evert.

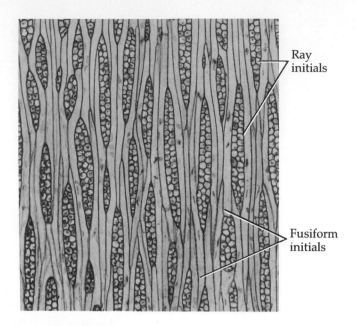

Ray initials

Fusiform initials

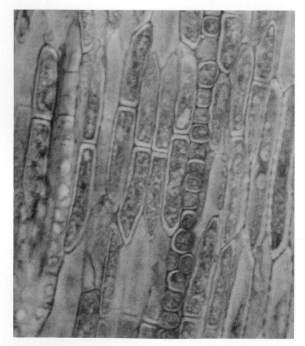

Figure 14.6
Tangential section through xylem mother cells of *Robinia,* immediately interior to the vascular cambium. The cells are imperfectly storied. × 100. Micrograph generously provided by M. Sundberg.

Although it is a simple matter to measure the length of the fusiform initials at any time, it is usually of more interest to study how the lengths of fusiform initials change with age, nutrient conditions, environmental conditions, and the general vigor of the plant. Obtaining a sample of the cambium for analysis obviously is destructive testing, and that particular cambial sample cannot be tested again later under different conditions. To follow cambial changes, most investigators have instead examined the secondary xylem, since it is a permanent record of events that occurred in the vascular cambium. The secondary phloem is not so useful, because it is not so permanent and because many fewer cambial derivatives move into the phloem. Unfortunately the xylem is not a perfect record of the cambium, because some cell elongation can occur during the maturation of the xylem elements. Bailey (1920) found that, in *Ginkgo* and conifers, the tracheids are about 5–10% longer than the fusiform initials; in dicots, the vessel elements are the same length as the fusiform initials, but the fibers are much longer. Dicot fibers are 10–950% longer than the fusiform initials that

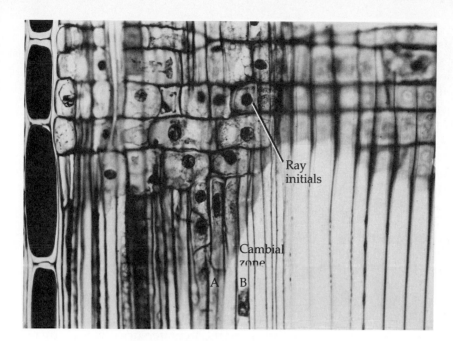

Ray
initials

Cambial
zone

A   B

Figure 14.7
Radial longitudinal section
through the cambial zone of
Pinus strobus. Mature secondary phloem is on the left; mature secondary xylem is on
the right. The cell at (A) is
probably not a fusiform initial, because its nucleus is
too close to the end of the
cell. The nucleus in (B) also
seems too high. Ray initials
are present near the top of
the picture; these cells are
wider than the fusiform initials. × 150.

produce them (Chattaway, 1936). Using this as
a measure, it has been noted that the fusiform
initials are much longer in gymnosperms than
in angiosperms and that they tend to increase
in length with age (Bailey, 1923; Bannan, 1951a,
b; Evert, 1961; Ghouse and Hashmi, 1980b;
Ghouse and Yunus, 1973). Philipson et al.
(1971) pointed out that, if the rate of height
growth of a plant is suppressed for any environmental reason, then the new fusiform initials formed in the stem tips at that time will be
shorter than normal. Furthermore, even after
they have continued to grow and function for
several years, they will continue to be shorter
than surrounding fusiform initials. Considering how variable most environments are from
year to year, and considering that the age of the
cambium varies with its position along the
shoot, root, or branch, and that one branch can
be much less vigorous than the others of the
same tree, it is obvious that the lengths of
fusiform initials will be extremely variable
within individual plants.

Even considering all of these sources of variability, it is clear that the fusiform initials of
gymnosperms are much longer than those of
most dicots. This appears to be related to the
types of cells produced. In gymnosperms the
secondary xylem contains only tracheids, and
the secondary phloem contains sieve cells. For
both of these, the optimal structure is to be as
long as possible (see Chapters 7 and 8). Even if
intrusive growth were easy (which it probably
is not), it makes sense to have the initials as
long as the derivatives and to thereby establish
a pattern. Once the pattern of the vascular cambium has been initiated, the patterns of the secondary tissues derived from it are automatically
established with no further control necessary;
there is no need for continual templating. In
the dicots, some of the most prominent elements are the vessel elements and parenchyma
cells, which need to be short. But there are also
tracheids and fibers which should be long. It
may be more efficient to have long fusiform initials for the establishment of the elongate elements and then allow the precursors of the vessel elements and parenchyma cells to divide
with transverse walls. It may be that, in very
elongate cells, nuclear control over the cyto-

plasm is poor (having the nucleus 4300 μm from the cytoplasm at the tips of the cells in *Sequoia* cannot be optimal) or that the length of time necessary for cell division becomes excessive. In *Pinus strobus*, the cell cycle lasts for 10 days, and the phragmoplast must grow for many hours at a rate of 50–100 μm per hour (Wilson, 1964).

CELL DIVISION IN THE FUSIFORM INITIALS.   Just like the apical meristems, the vascular cambium is dormant during times of stress (Aljaro et al., 1972; Avila et al., 1975; Berlyn, 1982; Perry, 1971; Riding and Little, 1984), whether the stress is imposed by heat, cold, or lack of water. As the vascular cambium enters dormancy, it stops cell division, and many of the xylem mother cells and phloem mother cells mature, causing the cambial region to become narrower (Fig. 14.4; Rao and Dave, 1984). Usually at least some of the xylem mother cells and phloem mother cells become quiescent while only partially differentiated. After overwintering, these can quickly complete differentiation, and the plant will have new conducting tissues very early. In the spring, as the cambial initials resume cell division, the first cells to become reactivated are those immediately below the swelling buds. Mitotic activity spreads downward through the cambium, presumably triggered by the basipetal movement of auxin (Berlyn, 1982; Little, 1981; Little and Wareing, 1981; Savidge, 1983). In mild climates, especially in the tropics, many trees do not become dormant, and cambial activity occurs during the entire year (Alvim, 1964; Perry, 1971).

Fusiform initials must divide with longitudinal, periclinal walls to produce new derivatives (Fig. 14.1). These are known as **proliferative divisions**. Of the cells produced by these periclinal divisions, the large majority become xylem mother cells, and very few differentiate as phloem mother cells. Furthermore, the derivatives themselves undergo cell division, usually at rates much greater than those of the fusiform initials (Bannan, 1955; Newman, 1956).

Wilson (1964) found that each derivative could divide enough to produce up to twenty cells in the xylem. By this method, large amounts of xylem and lesser amounts of phloem can be produced by relatively few divisions in the vascular cambium. Bannan (1962) calculated that, in conifers, during the most active growth the cambial initials divided only once every 4 to 6 days compared with a cell cycle of 8 to 18 hours in the apical meristems. It has been reported that the derivatives of a fusiform initial can begin dividing even before the mother cell has finished cytokinesis, and one cell will contain four nuclei and three phragmoplasts. This probably happens in the xylem mother cells, where the division rate is higher, rather than in the fusiform initials themselves.

As the xylem matures interior to the vascular cambium, it expands; the cambium is pushed outward and its circumference is constantly increased. To keep up with this increase, the cambial initials divide with anticlinal, longitudinal walls to produce more initials rather than derivatives; these are called **multiplicative divisions** (Fig. 14.1). Vascular cambia are divided into two types, and there are two types of multiplicative divisions, each related to the type of cambium in which it appears. The more primitive and more common type of vascular cambium is the **nonstoried cambium**; the **storied cambium** is rare and occurs only in some advanced dicots: *Aeschynomene, Hoheria, Robinia, Scleroxylon* (Record, 1943). In the storied cambium, the fusiform initials are aligned with each other laterally and form stories or tiers; hence the name. In nonstoried cambia there is no lateral alignment.

In the storied cambia (also called **stratified cambia**), the multiplicative divisions are by true anticlinal, longitudinal divisions in which the phragmoplast and cell plate run perfectly from tip to tip, or at least almost perfectly (Fig. 14.8A). The two daughter cells are of equal length, and the ends of the cells are aligned with the position of the original mother cell wall and therefore with the ends of the neigh-

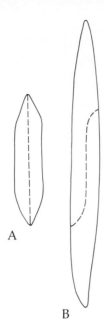

A

B

Figure 14.8
(A) Radial multiplicative division in a fusiform initial. The two daughter cells are equal in length to
each other and to the mother cell. (B) Pseudotransverse division: the new wall does not extend to
either end. Neither daughter cell is as long as the
mother cell, and, although they are equal to each
other in this drawing, they frequently are unequal.

boring cells. If the new wall misses the tip of
the original wall slightly, then a small amount
of tip growth occurs that reestablishes the
proper length (Beijer, 1927). Thus the stories
or tiers are automatically maintained by the
method of division. The fusiform initials of
storied cambia are usually rather short, often
less than 500 μm long (Bailey, 1923). The proliferative divisions will tend to maintain the
pattern of the cambium in the derivatives, either nonstoried or storied. In certain instances
(Zygophyllaceae, Bignoneaceae, and some legumes) everything in the secondary tissues,
vessel elements, tracheids, fibers, and parenchyma are storied, but typically transverse
divisions in the mother cells of the vessel ele-

ments and parenchyma cause these not to reflect the storying. Intrusive growth by the tracheids and fibers can also disrupt the pattern
such that the wood and bark do not reflect the
storied nature of the cambium.

It has been postulated by Klinken (1914) that
the storying is first established as the vascular
cambium is being initiated, because at that time
the diameter of the stem or root is very small,
and any radial growth outward is accompanied
by a relatively large increase in the circumference. Many of the divisions of the early
cambium would be multiplicative, and each
story would represent an original vascular
cambium cell. This type of development has
been found by Cumbie (1984) in *Aeschynomene*
(Fig. 14.9); however, this is a plant that produces very little wood, so it is not known if this
mechanism could be capable of maintaining
the storying in an old cambium. A nonstoried
cambium results if the developing fusiform initials undergo occasional transverse divisions
(Butterfield, 1976).

In nonstoried cambia (also called **nonstrati-
fied cambia**), the ends of adjacent fusiform
initials are not aligned, and there is less regularity to the cambium (Fig. 14.12). The multiplicative divisions are **pseudotransverse divisions** in nonstoried cambia; that is, they start
out with the appearance of a longitudinal division, but the phragmoplast and cell plate do
not reach the far ends of the cell (Fig. 14.8B).
Instead the cell plate turns toward the side
walls and fuses with them (Bannan, 1966; Harris, 1981; Hejnowicz, 1961, 1964; Hejnowicz
and Krawczszyn, 1969). The new wall is thus
shorter than the original side walls, often much
shorter (24% of the length of the original wall
in *Leitneria floridana*; Cumbie, 1967a). This results in the two daughter cells also being
shorter than the mother cell, and the original
length is regained by intrusive growth at the
tips of the daughters. This probably is not especially difficult, because the cambium is being
pushed outward and stretched apart; thus
room becomes available for the intrusion of the

Figure 14.9
Origin of the vascular cambium from the procambium and the development of storying in *Aeschynomene*. In (A), the cells are still procambium and have transverse end walls. In (B), some have begun to taper and others are undergoing transverse divisions to form rows of ray initials. In (C), the anticlinal divisions have produced prominent stories. Stippling indicates cells that will later divide and produce ray initials. Based on a diagram by B. G. Cumbie.

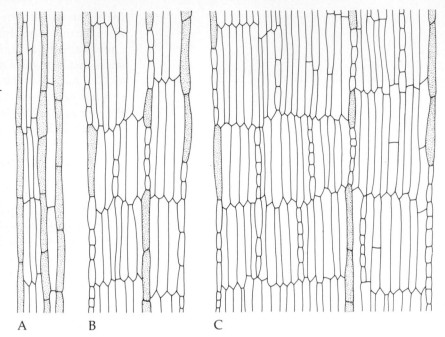

A   B   C

cell tips. Because the pseudotransverse divisions can result in cells of variable lengths that can then undergo different amounts of extension growth, the fusiform initials will quickly lose whatever arrangement they may have had, and the irregular, nonstoried condition will be maintained. The orientation of pseudotransverse divisions is not random; instead, all the fusiform initials in a large region (a **domain**) divide with pseudotransverse walls that have the same orientation (Harris, 1981; Hejnowicz and Romberger, 1973; Savidge and Farrar, 1984). One instance of pseudotransverse divisions in a storied cambium (*Tilia*) has been reported (Zagórska-Marek, 1984).

The fusiform initials of nonstoried cambia are generally much longer than those of storied cambia; lengths above 500 μm are typical, and lengths above 2000 to 4000 μm are not at all unusual (Philipson et al., 1971). It has been suggested that the cell plate does not reach the ends of the fusiform initials simply because these cells are so long, but that obviously cannot be the answer, because the cell plates in-

volved in proliferative division have no trouble at all.

The multiplicative divisions in the vascular cambium are much more dynamic than was once thought: there are many more divisions than are actually necessary to keep pace with the increase in circumference, and there is a tremendous loss of initials as they move out of the cambial region and differentiate (Bannan, 1950, 1951a, b, 1960; Evert, 1961; Whalley, 1950). Note that these are initials, not derivatives produced by periclinal divisions. In *Chamaecyparis*, after 1100 multiplicative divisions, only 162 new initials were retained in the cambial layer (Fig. 14.10; Bannan, 1950, 1960). As mentioned earlier, the amount of circumferential growth necessary with each increase in radius is geometrically determined (Fig. 14.11), and less circumferential growth is needed as the radius increases; but why there should be such an excess of anticlinal divisions at any time is just not known. It may be that it is a mechanism that "purifies" the cambium. If a detrimental somatic mutation occurred that in-

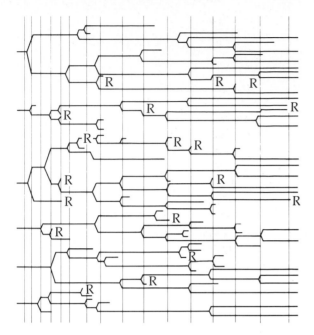

Figure 14.10
Activities of fusiform initials in a young, slowly growing stem of *Chamaecyparis thyoides* (white cedar). Each horizontal line represents the lineage of tracheids produced by a fusiform initial: where the line branches, an anticlinal division occurred. If one branch ends in an R, then one of the daughter cells was converted to a ray initial. When a line is seen simply to end, it is because the fusiform initial was eliminated from the cambium. Vertical lines represent annual rings. From Bannan, 1960.

terfered with division, expansion, or differentiation, either in the fusiform initials or in one of the recently formed xylem mother cells, it might cause that row to expand radially less rapidly than neighboring rows; and the fusiform initials would be left behind as the rest of the cambium was pushed outward at the normal rate. If the mutation occurred in a phloem mother cell, then that row in the phloem would not expand so rapidly as the ones around it, and the whole row would be carried outward due to contact with the adjacent phloem. Being unable to expand with normal speed, the fusiform initials would again be pulled out of the cambium. Only those fusiform initials and de-

rivatives that develop at maximum speed can remain in the cambium. If this hypothesis is true, then again the vascular cambium acts very much like the apical meristems.

This concept is supported by direct observations. The two initials produced by a pseudotransverse division are frequently not equal in length and do not contact equal numbers of rays. If the derivative is very short and contacts only one or two rays, this may slow its growth sufficiently to cause it to be expelled from the cambium. That this is due to ray contact and not to length is supported by a study by Bannan and Bayly (1956); they found that surviving fusiform initials had contact with 70% more ray initials than did those that did not survive. When fusiform initial length was intermediate, the shorter ones could survive if they had sufficient contact with ray initials, whereas even longer cells failed if they did not have good ray contact. The initials can be lost either abruptly or more slowly, after becoming shorter with each subsequent division. The initial is then called a **declining initial**, and the row of derivatives produced by it is a **declining tier** (Srivastava, 1963).

When an initial is lost, either slowly or abruptly, new contacts are made between the two lateral initials that fill in the vacated space. Their derivatives are able to form perfectly normal pit-pairs, and Philipson et al. (1971) report that even normal primary pit fields are formed between the two initials.

One aspect of the division of fusiform initials that has not received enough attention is the coordination of vertical files of cells. Derr and Evert (1967) found evidence that multiplicative divisions are synchronized within a file in the dicot *Robinia pseudoacacia*. If the cambium produces homogeneous tissues such as the wood of conifers, which contains only tracheids, then a high degree of coordination may not be necessary, because tracheids have long overlap areas and can probably form connections with any cell that is in the proper place. However, when a fusiform initial produces a derivative that will eventually form a

TWIG

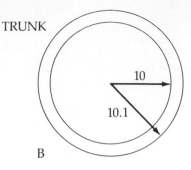

TRUNK

A

B

Circumference $= 2\pi r$

$\Delta C = 2\pi(r_2 - r_1)$

Twig:

$\Delta C = 2\pi(0.6\,\text{cm} - 0.5\,\text{cm})$
$= 2\pi(0.1\,\text{cm})$
$= 0.628\,\text{cm}$

Original circumference $= 2\pi(0.5\,\text{cm})$
$= 3.142\,\text{cm}$

Incremental circumferential growth $= \dfrac{0.628\,\text{cm}}{3.142\,\text{cm}}$
$= 0.1998$

Trunk:

$\Delta C = 2\pi(10.1\,\text{cm} - 10\,\text{cm})$
$= 2\pi(0.1\,\text{cm})$
$= 0.628\,\text{cm}$

Original circumference $= 2\pi(10\,\text{cm})$
$= 62.83\,\text{cm}$

Incremental circumferential growth $= \dfrac{0.628\,\text{cm}}{62.83\,\text{cm}}$
$= 0.0100$

Figure 14.11

The circumference of a circle is 2π times the radius. In order for the cambium to remain smooth and unwrinkled, it must grow 2π times as rapidly circumferentially than it is being pushed outward by the growth of the xylem. Any less and it would break apart; any more and initials must be crowded out. (A) When a young, small stem with a radius of 0.5 cm grows 0.1 cm radially to be 0.6 cm deep, then the circumference must increase by 0.628 cm. The original cambium was 2π (0.5 cm) = 3.142 cm in circumference, so the incremental increase in cir-cumference is 0.628 cm/3.142 cm = 0.1998. (B) If a large stem that is 10 cm in radius increases the same amount radially to 10.1 cm, then the increase in circumference is also 0.628 cm. But now the vascular cambium at the beginning was 2π (10 cm) = 62.8 cm, and the incremental increase in circumference is only 0.628 cm/62.83 cm = 0.0100. In the larger cambium, mutiplicative division can occur much more rarely than is necessary for a small cambium.

vessel element, then the cells above and below it may possibly need to be rather precisely positioned in order to form perforation plates. On the other hand, because vessel element initials usually expand greatly in diameter during differentiation (Zasada and Zahner, 1969), it may be that precise positioning is not important at first; readjustments might occur during growth.

Ray Initials

The ray initials are smaller than the fusiform initials, being short and either isodiametric or only about two to three times as tall as wide (Figs. 14.5, 14.6, 14.7). In the conifers, the ray initials are always arranged as vertical, uniseriate groups that produce vertical, uniseriate rays. The groups of ray initials may become

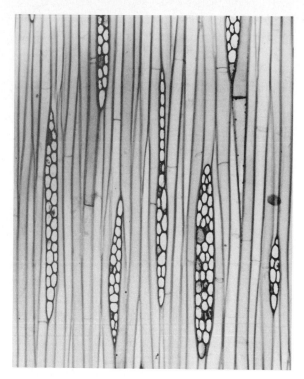

Figure 14.12
Tangential section through the young wood of
Didymopanax morototoni. The arrangement of the
fusiform initials and ray initials can be seen. Some
rays are high, others low; some are multiseriate,
others uniseriate. × 100.

taller either by the loss of fusiform initials located between two groups of ray initials, allowing them to fuse; or a fusiform initial can, by
transverse division, convert itself to a row of
ray initials (Fig. 14.13). If either one of these
mechanisms causes the group to be multiseriate, initials are quickly lost until the uniseriate
condition is reestablished. Dicots frequently
have a mixture of ray types, some uniseriate,
others wide and multiseriate, and this is reflected in the groups of ray initials (Fig. 14.12).
In either type, the group of ray initials may
contain tall initials only, isodiametric ones
only, or a mixture of the two. If both types are
present, the tall ones are almost always the up-

permost, lowermost, or both, and the isodiametric ones constitute the bulk of the group of
ray initials. The isodiametric cells may be quite
uniform in size within a group, or they may
have a diversity of sizes—from 10×10 μm to
60×65 μm in *Polyalthia longifolia* (Ghouse and
Hashmi, 1980b)—especially if the ray is large.
The multiseriate groups of ray initials often
have uniseriate "wings" extending either upward or downward, caused by the conversion
of adjacent fusiform initials into rows of ray
initials (Fig. 14.12). These wings are uniseriate
for only a short while, because ray initials, like
fusiform initials, can divide multiplicatively
(Barghoorn, 1941; Braun, 1955; Evert, 1961,
1963b), and the wing becomes multiseriate.
However, multiplicative divisions are much
rarer in ray initials than in fusiform initials.

Interconversions of Fusiform
Initials and Ray Initials

As just noted, the fusiform initials can be converted to ray initials in both gymnosperms and
dicots. This is necessary to prevent the distance between neighboring rays from becoming too great. The ratio of fusiform initials to
ray initials tends to be high. A very few dicots
have no ray initials at all, but in many species
90% of the cambium is reported to consist
of fusiform initials (Kozlowski, 1971; Wilson,
1963). Ghouse and Yunus (1974a), however,
suggested that the 90% figure is not an accurate generalization. They studied four dicot
trees and found that all had a much lower percentage of fusiform initials: *Polyalthia longifolia,*
80%; *Dillenia indica,* 75%; *Artabotrys odoratis-
simus,* 70%, and *Annona squamosa,* 68%. In
Dalbergia sissoo they constitute even less, only
60% (Ghouse and Yunus, 1973), and in a later
study of 18 trees of the family Leguminosae,
a range of 45–81% for fusiform initials was
found (Ghouse and Yunus, 1976). Ray initials
probably constitute over one-half of the volume of the vascular cambium in *Pereskiopsis*
and *Quiabentia* (Bailey, 1964).

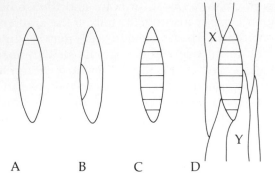

Figure 14.13
Ways in which a fusiform initial can produce ray initials. (A) An asymmetric transverse division; just the tip becomes converted, while the rest of the cell continues as a fusiform initial. (B) A small cell is cut off of the side. (C) The fusiform initial undergoes repeated transverse divisions, producing a group of ray initials. (D) If the fusiform initial at X converts to ray initials, then the existing group would become very wide; if the fusiform initial at Y converts instead, then the mass of ray initials will be multiseriate with a uniseriate wing.

The conversion of fusiform initials to ray initials can occur by any of several methods: the very tip can be cut off by a special transverse division (Fig. 14.13; Barghoorn, 1940a; Ghouse and Hashmi, 1980b); a small cell can be cut off from the side of a fusiform initial (Barghoorn, 1940a, b; Braun, 1955; Cheadle and Esau, 1964); or the entire fusiform initial can undergo repeated transverse divisions, resulting in an entire row of ray initials (Bannan, 1951a; Whalley, 1950). Several different methods may occur within a single species (Ghouse and Hashmi, 1980b). A fusiform initial next to an existing group of ray initials can convert, making the group wider or giving it a uniseriate wing (Fig. 14.13D). These are more prominent in primitive woods, because the fusiform initials are longer, so they automatically produce a longer wing. The ability of fusiform initials to convert to ray initials is important, because multiplicative divisions produce a common wall that is not in contact with a ray; a second division would then produce an entire cell with no ray contacts at all. To maintain proper horizontal conduction through the wood, bark, and cambial zone, new rays must be produced periodically.

On the other hand, in dicots, ray initials can begin to elongate and convert themselves into fusiform initials (Ghouse and Hashmi, 1980b); this does not happen in gymnosperms (Bannan and Bayly, 1956). This conversion is important in preventing rays from becoming too massive and creating large islands of weak parenchyma in the wood. Groups of ray initials can also be broken up by the intrusive growth of a fusiform initial from the periphery of the group into the mass of ray initials (Evert, 1961; Srivastava, 1963). An interesting mixture of fusiform initials and ray initials occurs in some dicots (*Fagus sylvatica*); these are **aggregate rays**, and they are large regions of the cambium that, although having the shape of a giant ray are actually mixtures of fusiform initials, uniseriate ray initials, and multiseriate ray initials (Fig. 14.14; Philipson et al., 1971).

Infected Vascular Cambia

In a variety of parasitic angiosperms (*Exocarpus, Pilostyles, Phoradendron, Tristerix*), the endophytic, absorptive tissues that penetrate the host are able to infect the vascular cambium (Calvin, 1967; Dell et al., 1982; Fineran, 1963; Mauseth et al., 1984). Somewhat surprisingly, this does not damage the cambium, which continues functioning normally. The parasite cells themselves also become cambial in nature, cutting off cells both to the interior and to the exterior. Because the growth of the parasite matches that of the host, the new secondary xylem and phloem of the host are automatically infected as they are produced.

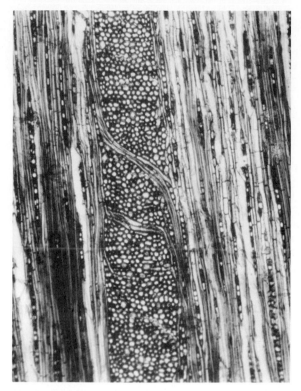

Figure 14.14
An aggregate ray in *Casuarina cunninghamiana*. It appears to be numerous rays, but basically it is a large ray that is being subdivided by the conversion of ray initials to fusiform initials and by the intrusion of nearby fusiform initials. × 50.

Development of the Vascular Cambium

Ontogenetic Development

In purely herbaceous plants, all of the procambium differentiates into vascular tissue. But in the plants that will become woody, a portion of the procambium in each strand develops into the **fascicular vascular cambium** (Philipson and Ward, 1965). There has been much conjecture as to the continuity of these two meristems: whether there is a sudden transition from procambium to vascular cambium or just a gradual change (Butterfield, 1976; Catesson, 1964; Cumbie, 1967b, 1984; Fahn et al., 1972; Larson, 1976, 1982; Philipson and Ward, 1965; Philipson et al., 1971; Sterling, 1946). Likewise, is there a slow or abrupt transition between metaxylem and secondary xylem, metaphloem and secondary phloem? There have been several claims that there are distinctive differences between procambium and vascular cambium: that is, that the procambium has transverse divisions and therefore nonpointed endwalls, and that the procambium does not have ray initials. But the vascular cambium can also have transverse divisions, and the initiation of rays is a gradual process with the first stages being quite irregular and difficult to detect (Braun, 1955; Fahn et al., 1972). Philipson et al. (1971) pointed out that the differences between primary growth and secondary growth are not as clear cut as often presented, because primary tissues are derived from divisions that take place in the subapical regions, and all lateral growth is a "continuous process, unbroken from the apex to the mature trunk." In *Aeschynomene virginica*, Cumbie (1984) found that the procambial cells with horizontal end walls gradually became longer and formed tapered walls; the transition from procambium to cambium was not abrupt (Fig. 14.9). Some of the shorter procambium cells underwent transverse divisions, gradually being converted to uniseriate files of ray initials. In *Canavalia* (Cumbie, 1967a), *Ginkgo* (Soh, 1972), *Hoheria* (Butterfield, 1976), *Phaseolus* (Enright and Cumbie, 1973), *Robinia, Syringa* (Soh, 1974b), *Sequoia* (Sterling, 1946), and *Weigela* (Soh, 1974a) the transition is similarly gradual, and fusiform initials do not acquire all of their characteristics until some time after internode elongation and primary growth have stopped. A more rapid transition occurs in *Acer* (Catesson, 1964, 1974) and *Aucuba* (Soh, 1974a), and the fusiform initials have all their characteristics by the time internode elongation has ceased. Philipson and Ward (1965)

Figure 14.15
This photograph shows all of one vascular bundle and part of a second. Fascicular cambia are well established, but the parenchyma cells of the medullary rays are just beginning to divide to produce the interfascicular cambium. × 100.

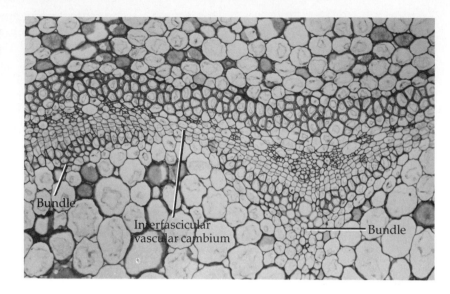

pointed out that, in some plants, the young internodes, even while still elongating, have what appears to be a cambium that produces regular files of cells. Because the internode is elongating, the xylem must undergo stretching. There are three possible conclusions that can be drawn from these observations: (1) a cambium can produce primary xylem (but the vascular cambium, by its very definition, does not produce primary tissues), (2) that secondary xylem is extensible (this also is a violation of the definition of secondary growth), or (3) there is in fact no absolute difference between a procambium and a vascular cambium. The consensus is now (keep in mind that consensus is not necessarily truth) that there is a gradual transition, usually slow but occasionally rapid, and the procambium and the vascular cambium are two developmental stages of the same meristem (Larson, 1982). Despite this, Fahn (1982) gives a very persuasive argument for at least treating and discussing the two separately: (1) in herbs, the procambium never develops into a vascular cambium, (2) the cambium can arise in regions where a procambium was not in existence before, and (3) there is a uniformity of the cambium when it is fully developed.

The second of Fahn's reasons refers to the **interfascicular vascular cambium**. As mentioned in the introduction to this chapter, in a large number of plants, only the fascicular cambium develops, and each vascular bundle enlarges. This is accompanied by very little secondary growth. Diffuse division and proliferation in the medullary rays are sufficient to keep up with the scanty wood production. The woody skeletons of these reveal the patterns of the original bundles. In trees and shrubs that form a solid wood, an interfascicular cambium arises in the parenchyma of the medullary rays, either simultaneously with the fascicular vascular cambium or shortly afterward (Figs. 14.15, 14.16). The interfascicular cambium differentiates as panels extending slowly from the edges of the fascicular cambia; the two panels of adjacent bundles meet, and a continuous interfascicular cambium is established. This also brings about the continuity of the entire cambium.

Frequently, after several months or years, the fascicular and interfascicular cambia become identical, and these terms are then not used any more for that plant; we refer only to the vascular cambium. But in some species the interfascicular cambium is distinct in the types

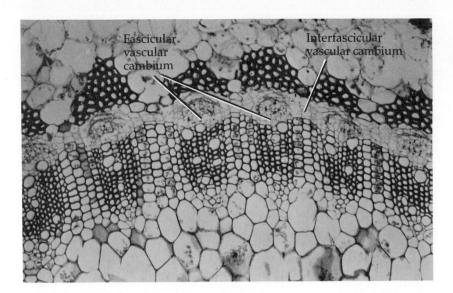

Figure 14.16
In this *Epiphyllum*, the complete cambium has formed, both the fascicular and the interfascicular portions. The interfascicular regions are producing only sclerified parenchyma on the xylem side. × 100.

of tissues that it produces. In *Casuarina*, *Epiphyllum* (Fig. 14.16), *Trichocereus* (Mauseth et al., 1984), and *Nopalea* (Mauseth, 1980a), the interfascicular cambium produces only parenchyma, lignified on the xylem side, not lignified on the phloem side. In dicots with both uniseriate and multiseriate rays, the first uniseriate rays are produced only by the fascicular cambium, and the first multiseriate rays originate in the interfascicular cambium. It is an interfascicular cambium that differentiates across a leaf gap, and this typically produces mostly parenchyma. Unlike the fascicular cambium, this one does not form until after the leaf is shed: one year for deciduous trees, several years in the evergreens. Eventually a uniform cambium is established, and all traces of the leaf gap are lost.

In many conifers (Ewers, 1982) and a small number of dicots, a fascicular cambium can develop in the petiole and midrib (see Chapter 12).

Development in the Roots

There have been few comparisons made between the vascular cambium of the shoots and roots of the same species. This is unfortunate, because the secondary tissues of roots and shoots, being in different environments and with different functions and needs, can be quite distinct. It would be possible to see how the root cambium and the shoot cambium (which are genetically identical) function to produce distinct shoot and root secondary tissues. In the roots, the vascular cambium differentiates from the procambium that lies between the xylem and phloem, just as in the shoots (Figs. 13.14, 14.17, 14.18; Esau, 1943a). An entire, undulating cylinder is formed that has differential activity: in the depressions, interior to the phloem, the cambium produces xylem more rapidly than in other areas, so the vascular cambium is pushed outward more rapidly, and a smooth ring is ultimately established (Fig. 14.18). When this is achieved, the activity becomes uniform.

As the vascular cambium forms and secondary tissues are produced, the adjacent endodermis is pushed outward. It may be stretched, then broken (Fig. 14.17; Mager, 1932), or it may undergo division and continue to act as an apoplastic barrier (Weerdenburg and Peterson, 1984).

Ultrastructurally, both fusiform initials and ray initials resemble the cells of apical meri-

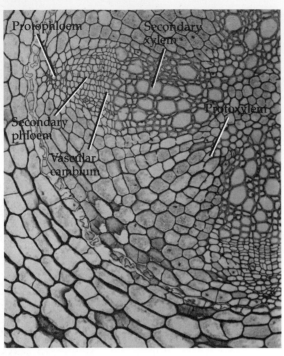

Figure 14.17
Root of *Actea alba* (baneberry). The only really well-defined vascular cambium is located interior to the protophloem. × 100.

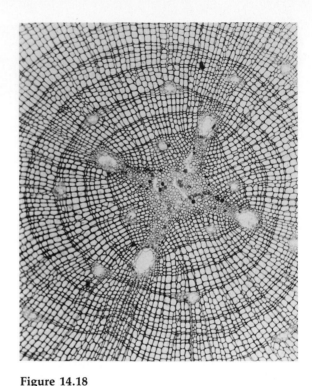

Figure 14.18
This root of *Pinus rigida* is much older than that of *Actea alba* in Figure 14.17. Because it is a conifer, its fusiform initials produce only tracheids in the xylem. The rows of tracheids can be followed to the center, indicating that a full vascular cambium arose rather uniformly. In the first season, some rows were longer than others, and at the end of the first year the cambium was square in cross section. At the end of the second year it was almost circular, and it was perfectly so by the third year. × 70.

stems, with one exception: the fusiform initials are extremely vacuolate (Buvat, 1956; Derr and Evert, 1967; Evert and Deshpande, 1970; Kidwai and Robards, 1969; Murmanis, 1971; Riding and Little, 1984; Robards and Kidwai, 1969; Srivastava, 1966; Srivastava and O'Brien, 1966). This makes it necessary for the cell plate to be suspended in a cytoplasmic bridge during the entire time of cell division (Evert and Deshpande, 1970). But other than this, all organelles are similar to those described in Chapter 6. When active in the spring, there is a large central vacuole, and the cytoplasm is all parietal; the nucleus may be either along a side wall or suspended in the middle of the cell in a bridge of cytoplasm. In ray initials the nucleus is spherical; in fusiform initials it may be elongate and spindle-shaped. During cytokinesis

of a proliferative division, the cytoplasmic bridge containing the phragmoplast and the new cell plate migrates toward the extreme ends of the cell. The bridge runs from one radial wall to the other, thus dividing the vacuole in half longitudinally. Occasionally the bridge becomes massive and connects with the tangential walls also (Evert and Deshpande, 1970). At the ultrastructural level it is usually impossible to distinguish the cambial layer from the neighboring derivatives. Also, fusiform initials and ray initials are quite similar, except for the degree of vacuolation (Riding and Little, 1984).

Phylogenetic Development

By the time of the end of the Devonian Period (about 350 million years ago), limited secondary growth existed in all major groups of terrestrial plants (Barghoorn, 1964; Banks, 1970, 1981; Cichan, 1985a). The gymnosperm and the cycadophyte-angiosperm lines of evolution can be traced back to the Progymnospermopsida (Fig. 1.1). The earliest of these are *Aneurophyton* (the wood of which had been placed in the organ genus *Callixylon*) and *Protopteridium*, both of which first appeared in the Eifelian Age of the Middle Devonian Period, 370 million years ago (Scheckler and Banks, 1971; Stewart, 1983). Both of these genera have excellent wood with little or no parenchyma, like that of modern conifers; it had to have been produced by a cambium remarkably like that which occurs in present day gymnosperms (Kräusel and Weyland, 1938; Schweitzer and Matten, 1982). Thus a well-organized vascular cambium with fusiform initials and ray initials was present at that time. Scheckler and Banks (1971) pointed out that fossil progymnosperms with highly developed secondary tissues are found in rocks that are only 10 million years more recent than those that contain fossils with the earliest known vascular cambium. Apparently once the cambium first appeared, it evolved rapidly; the wood of *Archaeopteris* is remarkably similar to that of the later gymnosperms (Beck, 1981). Within the cycadophyte-angiosperm line, the seed ferns (Lyginopteridales and Medullosales) both had secondary xylem and phloem (Taylor, 1981), but little is known of the vascular cambium.

The microphyllophytes represent a separate line of evolution that has been distinct since before the evolution of a cambium in the seed plant line (see Fig. 1.1). However, in several of these a vascular cambium did evolve independently, and the plants became woody. The vascular cambium of the Lepidodendrales is rather well known (Cichan, 1985a); it was unifacial, producing wood only and nothing to the outer side (Arnold, 1960; Eggert and Kanemoto, 1977; Frankenberg and Eggert, 1969; Pigg and Rothwell, 1983). Immediately exterior to the wood is a region of parenchyma that has been called a *post-meristematic sheath* (Cichan, 1985a). This is at least partly the result of the conversion of the vascular cambium to parenchyma; the cambium was determinate (Arnold, 1960), becoming disorganized after having produced a specific amount of wood. This may have been the result of a serious problem in the metabolism of the cells; Cichan (1985a) has concluded that the fusiform initials could not undergo multiplicative divisions. No new fusiform initials were formed as the cambium was pushed outward by the accumulation of xylem; instead, the fusiform initials became wider and wider tangentially. Certainly there must be a theoretical limit to the possible size of a fusiform initial; perhaps when they reached that size, mitosis was impossible, and meristematic activity ceased. Small nodules of narrow, irregular tracheids have been found in the wood, suggesting that the fusiform initials could occasionally undergo limited multiplicative divisions, but it seems to have been too spasmodic and too poorly controlled to have been useful.

The wood of the Lepidodendrales had rays; consequently the cambium is presumed to have had ray initials (Cichan, 1985a). The preservation of the ray cells is inadequate to reveal much about the ray initials except that they could occur singularly or in uniseriate or multiseriate groups.

Secondary growth also evolved in the sphenophytes (Cichan, 1985b; Cichan and Taylor, 1982; Eggert and Gaunt, 1973; Stewart, 1983). It had been thought that these also had a unifacial vascular cambium; the region just exterior to the xylem is usually decomposed in fossils. However, a specimen with excellent preservation showed conclusively that this vascular cambium was bifacial, although the amount of secondary phloem produced was very small (Eggert and Gaunt, 1973). The cambium must have contained both fusiform initials and ray initials. Cichan and Taylor (1982) and Cichan (1985b) found that in *Sphenophyllum* fossils there was no evidence for multiplicative divisions. However, careful measurements revealed that the length of the tracheids increased from the inner wood to the outer wood, and they hypothesized that the fusiform initials had similarly elongated. As each became longer, it would have intruded between vertically adjacent fusiform initials, resulting in an increase in the number of initials in any given cross section (Fig. 14.19). Just as in the Lepidodendrales, this must have ultimately resulted in cells so large that they could not function well; the tracheids in the wood of *Sphenophyllum* are the longest ever discovered in either living or fossil plants (Cichan and Taylor, 1982).

Limited secondary growth also occurred in some extinct ferns. This is believed to have resulted from a unifacial cambium (Lamoureux, 1961; Scheckler and Banks, 1971).

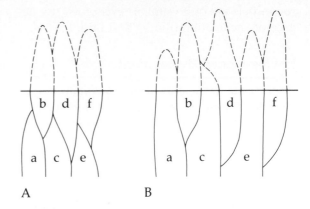

A   B

Figure 14.19
This diagram shows how intrusive growth of the fusiform initials can make it appear as though the number of initials is increasing, even though it is not. In (A) the cross section cuts through three cells: b, d, and f. In (B) after all cells have elongated and interdigitated more extensively, the same type of cross section would now reveal six cells in the cambial section. It would be easy to conclude erroneously that each of the originals had undergone a multiplicative division.

15

Secondary Xylem

Concepts

The bifacial vascular cambium described in the preceding chapter evolved in the Progymnospermopsida in the Devonian Period; the secondary tissues that it produces have been so successful that they have served as the basic organization plan of plants for about 400 million years. During this time, the cambium and its derivative tissues have undergone many modifications, some major, but none dramatic enough to obscure the underlying organizational relationships. Even the angiosperms that have evolved to be herbaceous, having lost a vascular cambium and secondary growth, have not undergone as much of a change as it may seem at first. They simply flower and die before secondary growth is necessary. Only in the long-lived monocots that have perennial primary bodies is the fundamental organization of the secondary body lost.

The subject of this chapter is the wood or secondary xylem that is produced to the interior of the vascular cambium. When it first evolved, the wood was mostly just sclerenchyma, consisting of long tracheids (Beck, 1981). But this had many consequences: it allowed the plant to have a renewed water-conducting system each year, and the plant could then live longer because it did not die when its primary xylem ceased to function. But a long-lived plant will become large and will need both strength and very long-distance conduction; fortuitously, the newly evolved secondary xylem automatically fulfilled both those needs. Because the secondary xylem persists even after it stops conducting water, each year's xylem accumulates, and the plant becomes larger and stronger. It can not only conduct more water for a more extensive photosynthetic tissue, but also can hold that tissue up to the sunlight. Being tall permits the evolution of new means of dispersal for spores and gametes, because the sporangia and gametangia can be elevated up into wind currents that allow for long-distance dispersal.

As these plants grew taller, they left the rather protected microhabitat that exists close to the ground. This put their upper parts in a more rigorous environment, which at first may have caused them much difficulty. But it acted as a selective pressure that ensured the survival of mutations that made the plants more resistant. Such adaptations allowed the plants to migrate into new environments. The evolution of secondary growth, particularly of secondary xylem, has affected virtually all aspects of the life of the plants. Here, the subject is the common types of wood found in all conifers and most woody dicots. This is not the only type of wood possible; other, anomalous types are analyzed in Chapter 18 and illustrate how well-adapted and functional common wood really is. For numerous excellent illustrations, the book by Core et al. (1979) is recommended.

Basic Structure of Wood

The cells of the secondary xylem are deposited to the inside of the vascular cambium. As these cells expand during maturation, the vascular cambium is pushed outward radially. In all woody plants the vascular cambium functions episodically: it is active during favorable times of the season and quiescent when it is too cold, too hot, or too dry. This periodicity typically results in annual increments or **annual rings** (Figs. 15.1, 15.2). In most plants, the very last two or three layers of cells in an annual ring differ markedly from the rest of the cells, so the rings are rather easy to see. In some tropical trees and in plants that produce very little wood each year (desert succulents), this difference does not occur, and annual rings cannot be seen. Where the activity and dormancy of plants are controlled by cold or heat, the rings are always annual, because these are infallible environmental signals. Where water stress controls activity, it is possible to have droughts (and cambial inactivity) that last one or several years or to have single seasons with several major moist periods, each of which activates the cambium. Thus the rings of the wood may or may not represent individual years.

The uniformity of the cambial activity and consequently the uniformity of the thickness of annual rings are interesting. In many trees and shrubs, a cross section of the trunk shows that the amount of wood produced in any area is almost identical to the amount produced in any other area. It is not known how this is controlled or if indeed it is controlled specifically.

The wood cells have only a limited period in which they live and carry out functions other than support. If they occur in stressful environments, the conducting elements may all cavitate and stop transporting water by the end of the first year; if the climate is more benign, then the tracheary elements may continue to transport water for several years, as in ebony (*Diospyros ebenum*) and letterwood (*Piratinera guianensis*). In these cases the cavitation of ves-

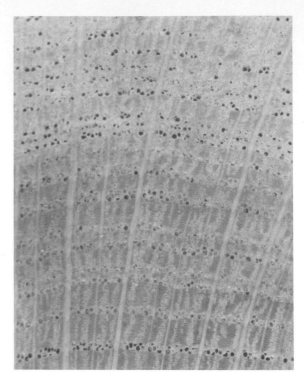

Figure 15.1
The annual rings of this wood are easily visible because the large vessels are concentrated in the spring wood. Numerous wide rays are visible. The lighter region at the top is sapwood. × 0.5.

sels and tracheids occurs sporadically such that each successively older ring has fewer active conducting elements (Zimmermann, 1982, 1983). Only when a ring is five to ten or more years old has it lost all ability to conduct axially. The outer, conducting layers constitute the **sapwood**, while the inner, nonconducting rings are the **heartwood**. Every year some sapwood is converted to heartwood, at least after the plant is several years old (all the wood of seedlings and saplings is too young to be converted). Thus every year the heartwood has a larger diameter at any given level in the stem or root. The sapwood does not become narrower, because new rings are being added by the cambium.

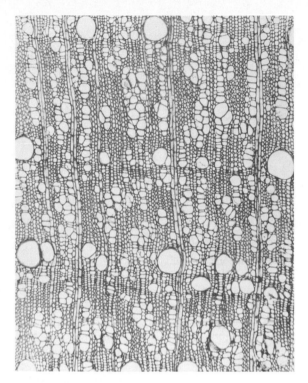

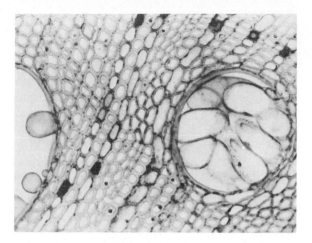

Figure 15.3
These two vessels contain tyloses; in one they are just forming, in the other they fill the vessel. Notice how thick the walls of the tyloses become. Although some spaces appear to exist between the tyloses, think about the area just above and below this region; more tyloses block fungal growth. × 400.

Figure 15.2
The annual rings of this *Pistacia mexicana* (related to pistachio) are easy to see, although the differences between springwood and summerwood are only slight. The very first layers of springwood contain the widest vessels, whereas the last layers of summerwood contain the narrowest fibers and almost no vessels at all. Notice that most vessels are narrow and occur in rather large groups, some of which are in contact with the large parenchyma cells of the rays. The vascular cambium is beyond the top of the picture. × 100.

The presence or absence of conducting tracheary elements is not the only difference between heartwood and sapwood; a vessel is an excellent channel for fungal growth. While the vessel is conducting, this is a danger that must be tolerated, but after cavitation antifungal defenses can be initiated. These take the form of filling the vessel lumen or sealing the vessel element walls. **Tyloses** constitute the first means of blocking a vessel (Fig. 15.3). A tylosis forms when a living parenchyma cell adjacent to the nonconducting element pushes part of its wall and cytoplasm through a pit into the lumen of the vessel (Chattaway, 1949; Foster, 1967; Meyer and Côté, 1968; Murmanis, 1975, 1976). The parenchyma cell may push enough material through to form a very large outgrowth that completely fills the lumen in that area, but often several tyloses enter from various parenchyma cells on all sides of the vessel. As these tyloses meet each other in the vessel lumen, they abut and form a tight-fitting seal. Variable amounts of cytoplasm may occur in the tylosis, and even the nucleus and starch may be found there (Meyer and Côté, 1968; Shibata et al., 1981). The walls often thicken secondarily, and pit-pairs become obvious in the contact walls between different tyloses. The stimulus for tylosis formation has not yet been discovered; they form both as a consequence of cavitation caused by aging as well as of cavitation caused by injury (Klein, 1923; Zimmermann, 1979).

The ability to push primary wall and protoplasm through a pit requires the pit to be suitably wide, apparently at least 10 μm; if it is any narrower the friction is too great (Chattaway, 1949). In these cases, the second method of blocking, **gummosis**, is used. The paratracheal parenchyma cells produce gums that flow through the pits and fill the lumen (Olien and Bukovac, 1982).

Other changes occur in the transition from sapwood to heartwood (Chattaway, 1952; Frey-Wyssling and Bosshard, 1959; Kramer and Kozlowski, 1979). There is often a centripetal transport of materials and a deposition of lignins and polyphenols that cause the heartwood to be dark in color and often more aromatic (Bauch et al., 1974; Hillis, 1977; Höll, 1975). Starch and water are withdrawn, and the parenchyma cells die. These changes may occur simultaneously in the rays and the axial system or in one earlier than in the other (Bhat and Patel, 1980). The primary result of these processes is that the heartwood becomes stronger, drier, and more resistant to decay. Various types of heartwood have been described by Bosshard (1966, 1967, 1968).

Secondary Xylem of Gymnosperms

The following description is based on conifers only; the wood of cycads is basically similar and is described at the end of the section.

The secondary xylem of conifers, like that of dicots and cycads, is divided into a transverse system of rays (produced by the ray initials) and a longitudinal or axial system (produced by the fusiform initials; Fig. 15.4). In the conifers, both systems are always present; no conifer wood ever lacks rays (Core et al., 1979; Panshin and de Zeeuw, 1980). The two systems work together in the functioning of the

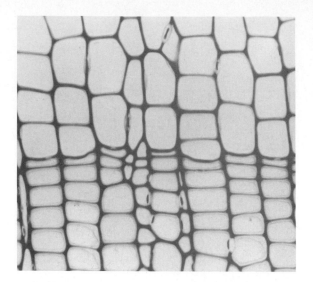

Figure 15.4

This is a cross section of *Pinus strobus* wood at the boundary between the summerwood of one year and the springwood of the next year. All the axial cells are tracheids; neither vessels nor axial parenchyma are present. Although most ray cells are parenchyma, in this section only ray tracheids are visible, recognizable because they have bordered pit-pairs on their radial walls. Notice that the axial tracheids have pit-pairs on their radial walls only, except for the last tracheids of the summerwood, which are connected to the springwood by pit-pairs in the tangential walls. × 400.

wood. Longitudinal conduction and mechanical strength are provided by the axial system. Because of the types of cells present and their orientation, however, the axial system is not well adapted for either carbohydrate or mineral storage (either long- or short-term), and it is extremely inefficient at the transverse conduction necessary to keep internal tissues in communication with the water and with the nutrient streams of the actively conducting secondary xylem and secondary phloem. This short-distance, transverse conduction and nutrient storage is the responsibility of the rays.

An important concept to bear in mind when considering the wood of conifers is that these

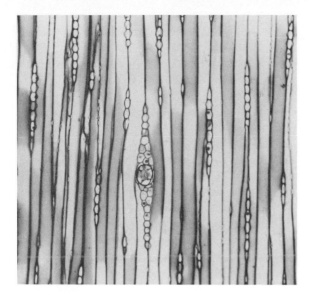

Figure 15.5
This tangential section of *Pinus strobus* shows
that the axial tracheids have extensive contact with
the rays. Although many tracheids appear to be
surrounded only by other tracheids, if the picture
were taller it would show that no tracheid is ever
without ray contact. × 200.

plants are rather homogeneous with regard to
their habit (body size and shape) and ecological requirements. None is an annual or an
herb, and they are never succulents or vines.
Taxodium grows in swamps and stream banks,
but typically conifers live in drier areas or on
thin soils with rapid drainage. This uniformity
is reflected in the wood structure. There is certainly a good possibility that this is partly the
cause of the uniformity of the wood; of course,
the uniformity of the wood could be partly responsible for their being restricted to a uniformity of habit and ecology.

Rays of Gymnosperm Wood

Conifer rays may consist of either parenchyma
cells only (if so, they are called **homocellular
rays**), or they may have a special type of cell
(called a **ray tracheid**) together with the paren-

chyma (then they are **heterocellular rays**; Figs.
15.5, 15.6, 15.8). The parenchyma of conifers
can be rather ordinary cells, being large and
with thin primary walls, and they may remain
alive for several years. In many conifers (most
Pinaceae), the parenchyma cells develop lignified secondary walls. Under a strict system
of nomenclature, it might be best to consider
these cells sclereids on the basis of their shape
and secondary wall, but they have historically
retained the designation of parenchyma.

The ray tracheids occur only in conifers,
such as *Chamaecyparis nootkatensis, Larix, Picea,
Pinus, Pseudotsuga, Tsuga* (Core et al., 1979).
They are distinguished from the sclerified ray
parenchyma cells by having bordered pits in
their secondary walls and by the loss of the
protoplast immediately after their maturation.

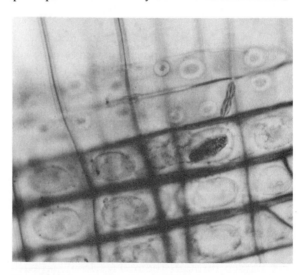

Figure 15.6
This radial section of a ray of *Pinus strobus* shows
three ray tracheids, two of which are interconnected by two circular bordered pit-pairs on their
common tangential wall. In all three, the circular
bordered pits between ray tracheids and axial tracheids are visible. Below the ray tracheids are the
ray parenchyma cells with nuclei, cytoplasm and,
for parenchyma, relatively thick walls. The large
ovals are giant pits connecting the parenchyma
cells with the axial tracheids (shown also in Figure
15.8). × 635.

The secondary walls of the ray parenchyma cells remain fairly thin, but those of the ray tracheids are typically thick and may have special thickenings (either toothlike or bandlike) that project deep into the lumen of the cell (Harada, 1965). The ray tracheids can have various distributions within the rays, occurring in rows, located at the upper or lower edges of the rays in most species, or being scattered among the ray parenchyma cells in some pines. In *Chamaecyparis nootkatensis*, some rays are composed of only ray tracheids, and other rays contain only parenchyma. Increased quantities of ray tracheids result in more rapid water movement: in *Pinus sylvestris* the ratio of ray tracheids to parenchyma is one-to-one, whereas in *Picea abies* it is one-to-four, and the velocity of radial transport is five times faster in *Pinus* (Leise and Bauch, 1967). The ray tracheids in some species do not form a continuous conduit, however; in many Pinaceae there are only parenchyma cells at the border of an annual ring (Höll, 1975; Huber, 1949a).

Because rays conduct transversely and also must interact with the longitudinally oriented tracheids, the pitting on all walls is critically important. Thin-walled parenchyma cells have only primary pit fields, of course, but the thick-walled cells have pitted walls (Fig. 15.7). On the periclinal walls these will serve for transverse conduction within the ray (Fig. 15.6); those on the radial vertical walls intercommunicate with the axial tracheids. Conifer rays are always uniseriate (except where a resin canal passes through them; Fig. 15.5), so each ray cell has contact with at least two axial elements. This pitting that interconnects the rays with the axial system is called **crossfield pitting** and usually is very different from the pitting that is confined solely within the axial system (Fig. 15.7). Both the simple pits and the bordered pits in conifer rays may be distinctive in that the border is narrow and thin, but the pit chamber is enormously wide. In many species of *Pinus* the pit chamber is so wide that the wall has only enough space available for one

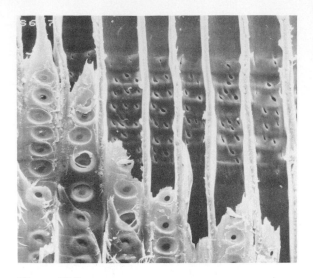

Figure 15.7
This SEM of the wood of spruce (*Picea*) shows its crossfield pitting: the small ovals are simple pits in the tracheids, and they connect the cells to the rays just behind. In the foreground is a set of radial walls with circular bordered pits. At the bottom you see the inner aperture; in the middle of the picture one secondary wall has torn away, so the outer apertures and a torn pit membrane are visible. × 500. This wood sample was taken from the Betts Stradivarius violin; micrograph generously provided by C. Ervin and S. Meyer.

single pit; in face view, this looks like a large "window," so it is called a **fenestriform pit**. Where an axial tracheid contacts a ray parenchyma cell, then the crossfield pitting probably will consist of half-bordered pit-pairs, the bordered half being in the axial element. In contacts between axial and ray tracheids, bordered pit-pairs will occur (Fig. 15.8).

In the conifers (as in the dicots) there are two types of rays in the young plant: **primary rays**, which are initiated by cells of the interfascicular region or the procambium (Fig. 15.20), and **secondary rays**, which have their origin in the conversion of fusiform initials to ray initials in the vascular cambium (see Chapter 14). The procambium cells are not extremely uniform,

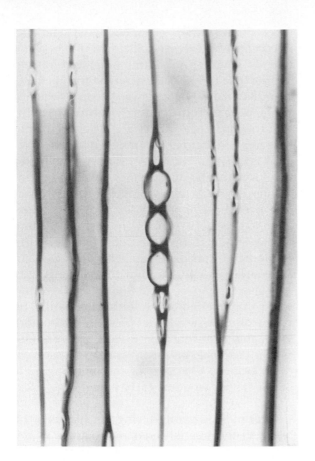
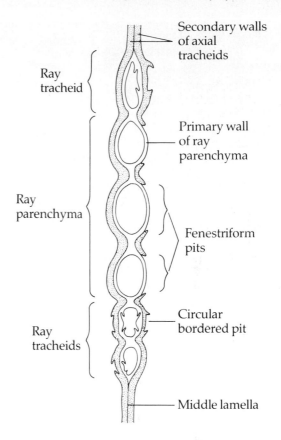

Figure 15.8
(A) This high magnification of a ray of *Pinus strobus* shows the nature of the
types of pitting. On the right are pit-pairs between tracheids. The top cell
and the two bottom cells of the ray are ray tracheids, and the three middle
cells are parenchyma cells. Try to follow the secondary walls of the tracheids, and you will see that the spherical nature of the parenchyma cells
is due to their primary walls' bulging through the fenestriform pit. × 635.
(B) This diagram shows the walls of the ray in 15.8A.

so the first cells of the primary rays, those adjacent to the pith, are similarly irregular in shape
and size, having long projections and little
contact with each other (Barghoorn, 1940).
Within a short time, however, the ray initials
begin to produce a ray of regular form. The
secondary rays arise suddenly as a fusiform
initial produces one or several ray initials.
Once established, the secondary ray can either
increase or decrease in size as more or fewer
ray initials are produced. With age, the pri-

mary rays become indistinguishable from the
secondary ones.

In a review of the research on the metabolism of rays, Höll (1975) summarized a large
body of literature that suggested that, as ray
cells age, they gradually deteriorate. However,
he pointed out that many newer studies do not
agree with this; the ray cells at the sapwood-
heartwood boundary may be extremely active,
with high levels of enzymes (Shain and Hillis,
1973; Shain and Mackay, 1973). It was con-

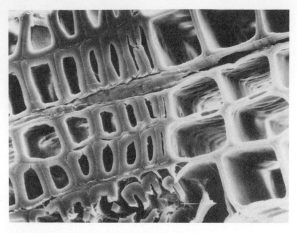

Figure 15.9
The annual ring boundary of spruce. Here the primary wall is easily distinguishable from the secondary wall. This ray shows ray parenchyma. × 500. Sample from the Betts Stradivarius violin; micrograph generously provided by C. Ervin and S. Meyer.

cluded that the ray cells do not simply die but rather play an important role in the synthesis of heartwood constituents (Höll, 1975).

Axial System of Gymnosperm Wood

In many conifers, there is only a single cell type in the axial system: the tracheid. The maximum modification that occurs is the production of fiber-tracheids at the end of the growing season (Figs. 15.4, 15.9). Due to the lack of libriform fibers, conifer wood is generally softer than dicot wood, and it is called **softwood**. Axial parenchyma is very rare in conifers, being completely absent in *Pinus* and *Taxus* and sparse in *Abies, Larix, Pseudotsuga,* and *Tsuga*. In the others (*Chamaecyparis, Cupressus, Libocedrus, Juniperus, Sequoia, Taxodium, Thuja*)—although considered more or less abundant by conifer standards (Core et al., 1979)—it still occupies only a tiny fraction of the volume of the wood.

Compared to most dicot woods, conifer wood is very simple, but it is not necessarily uni-

form. Because there are mostly just tracheids in conifer wood, these cells must function not only to conduct and store water but also to support the tree. The ideal shape for tracheids like this is long, because greater length is advantageous to both conductivity and strength. The tracheids of conifers are at least 0.5 mm long and range up to 11 mm in length. With regard to width, narrower diameter is better for strength and to minimize cavitation, but greater diameter increases conductivity. Conifers meet these opposing demands by producing two types of tracheids: in the spring, there is a great need for conduction as new needles expand and transpire heavily because their cuticles are still soft and immature. This is usually also a time of an abundance of ground water from melting snow and spring rains. The tracheids that develop at this time are quite wide and relatively thin-walled, approaching the appearance of medium-sized vessels in cross section. This part of the annual ring is called **spring wood** or **early wood** (Figs. 15.2, 15.4, 15.9).

Later in the season, during the summer, the water in the tree is under more stress as the soil begins to dry and the atmospheric temperatures increase. Furthermore, the tree is now heavier than before because of the additional weight of new leaves, twigs, and spring wood. The tree produces a stronger tracheid that is more resistant to cavitation: these **summer wood** (or **late wood**) tracheids have a narrower diameter and thicker walls (Figs. 15.2, 15.4, 15.9). At the very end of the growing season, fiber-tracheids may be produced.

The circular bordered pit-pairs that interconnect axial tracheids are located mostly near the ends of the cells and almost exclusively on the radial walls. This permits water to move primarily circumferentially within a single layer. There are occasional pit-pairs in the tangential walls, allowing some radial movement, but these are so few that most radial conduction must be by means of the rays (there is little actual data about radial movement). The presence of bordered pit-pairs on tangential walls is most common in the cells at the boundary between the late wood of one year and the early wood of the subsequent year. These can be quite prominent and have been termed **growth ring bridges**; they allow the radial movement of water between individual annual rings. Such movement is especially important at the tips of roots and shoots. Water is absorbed in regions containing primary growth and perhaps the first annual ring, but it is usually carried through the roots and trunk by a sapwood several annual rings in width, so water must be transported inward to all of these. In the twigs, the water must be moved centrifugally to the youngest ring and to the primary xylem so that it can be supplied to the leaf and branch traces.

A common feature of conifer wood is the presence of **crassulae** (sing.: crassula), also called **bars of Sanio**. These are prominent dark bands above and below most circular-bordered pits of tracheids in all conifers except the Araucariaceae. They are easily visible by light microscopy and have received considerable study; however, electron microscopy has revealed that they are merely refraction patterns. They result when the pits occur in arrangements that have a spacing that is similar in dimension to the wavelength of light used for observation. Although the crassulae are not actually a physical part of the wood, they are imporant and useful in the same way that differential staining or other analytical techniques are useful.

The wood described above is characteristic of the Coniferales and Gnetales; it is produced in large amounts and contains rather little axial parenchyma. Secondary xylem like this is termed **pycnoxylic wood**. In the Cycadales the basic characters and organization of the wood are similar, but only small amounts are produced, and it contains abundant axial parenchyma; this is **manoxylic wood** (Seward, 1917). Although some cycads can be large, up to 20 meters (Pant, 1973), most are much shorter, and the wood is always just a very thin cylinder despite the great age of the plant. A major source of the mechanical support in these stems is the outer covering of very hard, persistent leaf bases and sclerified bark. Growth rings can sometimes be distinguished, but these are related to the production of new crowns of leaves, which is not necessarily an annual event. The rays can vary from short (2 cells high) to tall (60 cells high), and from narrow to wide (2 cells to 20 cells), but they tend to be very large.

Secondary Xylem of Dicots

The discussion of wood in the angiosperms is restricted to the dicots because none of the monocots has ordinary secondary growth as occurs in gymnosperms and dicots. Those that do undergo secondary growth in addition to the primary growth do so in an unusual manner; that process is discussed in Chapter 18. Whereas the conifers are rather homogeneous with regard to body size, shape and organization, the dicots are extremely diverse. Within even a single tree it is possible to find tracheids, vessel elements, fibers, fiber-tracheids, parenchyma, and numerous types of crystals and secretory ducts (for example oaks, osage orange; Figs. 15.10, 15.11). Other dicot woods, however, are as simple as those of conifers (Fig. 15.12; *Trochodendron*). On this basis it is logical that the wood of dicots should be more complicated and should show more variations.

Rays of Dicot Wood

The rays of dicots are simpler than those of gymnosperms in that they contain only parenchyma cells (Figs. 15.13, 15.14). Ray tracheids do not occur; radially oriented vessels have been reported in the rays of several species (Botosso and Gomes, 1982; Chattaway, 1948; van Vliet, 1976a). In organization, however, the dicot rays can be much more complicated, because they are not restricted to being uniseriate. In fact, they can range from being uniseriate to being as many as fifty cells wide (Fig. 15.14), and this variability may occur

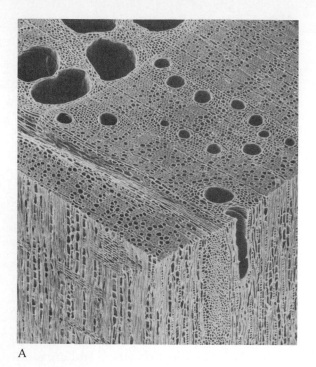

A

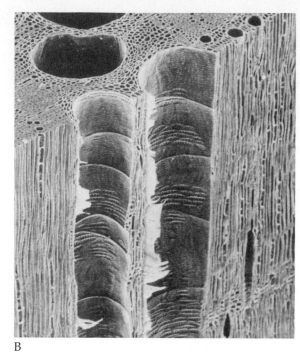

B

Figure 15.10
(A) This scanning electron micrograph shows the three-dimensional relationships of the wood elements in red oak. In cross section are visible the
vessels of springwood and summerwood, paratracheal parenchyma, and fibers. The radial longitudinal face (left) and the tangential face (right) show
the structure of the rays. × 120. (B) This magnification of the vessels shows
the regions of crossfield pitting (large pits) and intervascular pitting (small
pits). × 100. Both micrographs generously provided by W. A. Côté, N. C.
Brown Center for Ultrastructure Studies.

within one plant. In height, they can vary from
being only one cell high to being several millimeters tall. It is rare in the dicots to find species that contain just a single type of ray; ray
heterogeneity is the rule. This may be a reflection of the increased chemical and metabolic
diversity present in the dicots. With the evolution of the angiosperms came many new synthetic pathways, and the flowering plants
are much richer in the types of secretions,
idioblasts, crystals, and ergastic substances
that they employ. In the radial parenchyma
system of dicots, there are often many more
types of substances in storage (Höll, 1975;
Kramer and Kozlowski, 1979; Richter, 1980;

Ter Welle, 1976b), and the rays seem to be
more active metabolically than they are in the
gymnosperms.

With the increased size and metabolism of
rays comes an increase in the need for efficient
communication between the radial system and
the axial one. In many large deciduous trees
there is a thick, active sapwood responsible for
accumulating, storing, and then releasing large
amounts of reserve material as the roots and
crown are active in the summer, dormant in
the winter, and then put out new leaves and
root tips in the spring. In the twigs and young
sapwood of the trunk (1 to 3 years old) of apple
trees, over 10% of the dry weight of the wood

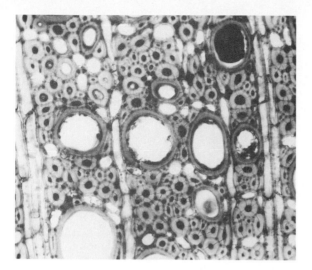

Figure 15.11
The wood of *Rochefortia grandifolia* is complex, containing vessels of several sizes, numerous fibers, a small amount of axial parenchyma, and ray parenchyma. The vessels are unusual in having walls that are even thicker than those of the fibers, but notice that, whereas the fibers have numerous simple pits, the vessels have none. × 400.

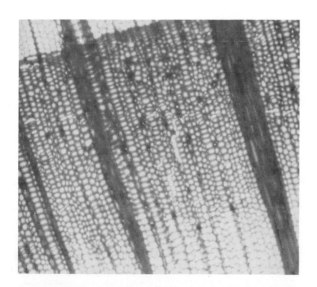

Figure 15.12
Although the absence of vessels in this wood makes it appear to have come from a gymnosperm, in fact it is from one of the primitively vesselless angiosperms, *Trochodendron aralioides*. × 100.

Figure 15.13
In cross sections, it is often difficult to distinguish the exact nature of many of the cells: tracheids may be indistinguishable from fibers, and thick-walled parenchyma cells are hard to identify if they were not properly preserved and stained. Even vessels can be narrow enough to cause confusion. In a longitudinal section, however, the nature of each is clear, and many relationships can be seen quickly. In this tangential section of *Neopringlea integrifolia*, the rays are seen to be very tall and almost completely uniseriate. There are numerous fibers, tracheids, and narrow vessels. × 160. Material prepared by D. Lemke.

is starch and sugars being stored (measured in mid-October). Up to 1% of the weight is in the form of nitrogen, and other minerals and lipid reserves are also present (Kramer and Kozlowski, 1979). Facilitating the required transfers between radial and axial conducting systems are specialized ray parenchyma cells with an upright shape (called **upright cells**) as opposed to the more common **procumbent cells**, which are wider than they are tall (Figs. 15.15, 15.16). The axial vessels and tracheids are reported to form cross-field pitting only with the upright cells, which must therefore be located on the periphery of the ray, either as one or several rows at the top and/or bottom of the ray, or as rows along the sides of the mid-

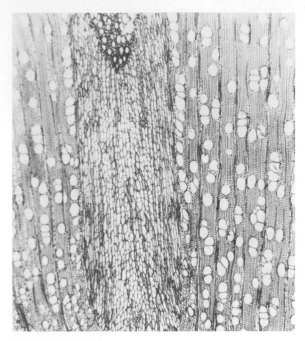

Figure 15.14
This cross section of *Didymopanax morototoni* shows the tremendous diversity of rays that may occur within a single region of one stem in dicots. Most rays are only biseriate, but the central one is over 34 cells wide; although you cannot tell from a cross section, the ray is also extremely high. × 40.

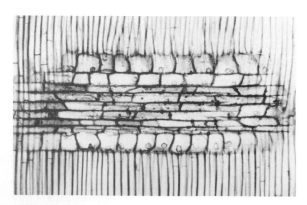

Figure 15.15
This ray of *Aucoumea klaineana* displays both the procumbent cells and the upright cells (the ray is heterocellular). In this particular area the ray is passing through a region of summerwood fibers, and there is no crossfield pitting. × 160. See Figure 15.16 for a tangential view.

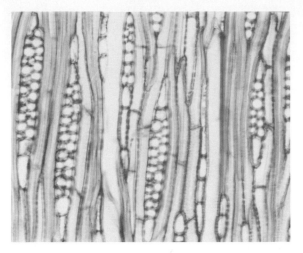

Figure 15.16
Unlike radial sections (as in Figure 15.15), tangential sections always show whether a ray is heterocellular or homocellular, and the number of rows of upright cells is easily seen. Notice that in this *Buxus sempervirens* (boxwood), the central portions of the rays (with procumbent cells) are multiseriate, whereas the margins (with upright cells) are uniseriate. This section also shows how the ends of the elongate cells overlap each other. × 400.

region (Figs. 15.15, 15.16, 15.17, 15.18, 15.19; Braun, 1970; Chattaway, 1951). It is not unusual for the upright cells to constitute extensive wings, greatly increasing the interface between the radial and the axial systems. Rays that contain both types of cells are termed **heterogeneous**, although there is a trend to use the word *heterocellular* to bring the terminology into line with that used for gymnosperms (Fahn, 1982). Likewise, rays made up only of procumbent cells are said to be **homogeneous** or *homocellular*.

It has recently been confirmed that the crossfield pitting of the upright cells is indeed the interface between the radial and the axial system. In some trees such as *Acer, Alnus, Betula,* and *Carpinus*, significant amounts of sugars have been found in the vessel sap in the spring. Apparently the newly opening buds receive a considerable portion of their carbohydrates

Figure 15.17
Crossfield pitting in *Meryta macrocarpa*; in the vessel wall, the crossfield pits are scalariform and bordered. × 800.

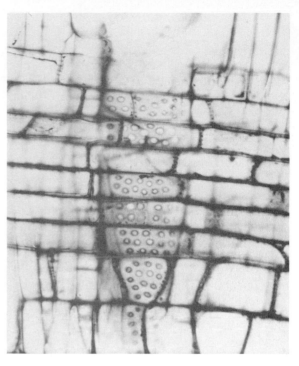

Figure 15.18
The crossfield pits of *Catalpa bignonata* are circular. Notice that here the crossfield pitting is formed against both the upright cells and the procumbent cells. × 800.

through the xylem (Höll, 1975). In two of these (*Betula populifolia, Populus tremuloides*), Sauter (1972) found that the upright cells began to lose sugar just before the buds started to swell. The rate of transfer of sugar was ten to thirty times faster than what would be expected from diffusion alone. Sugar was lost from the procumbent cells only later. It should be noted that some physiologists have begun calling upright cells *contact cells* and procumbent cells *isolation cells*.

As in the gymnosperms, young dicots have both primary and secondary rays (Fig. 15.20). The primary rays are of two types: tall uniseriate primary rays that form from procambial cells as in the gymnosperms, and large, wide multiseriate primary rays that are continua-

tions of the medullary rays. These latter are extended when cells in the interfascicular region act as ray initials. They may be quickly reduced as the fascicular cambium spreads laterally out of the bundles, forming a complete cambial ring, or they may persist indefinitely, and the wood of large plants will have several very wide rays (Figs. 15.14, 15.20). The secondary rays of dicots, like those of conifers, arise from the conversion of fusiform initials to ray initials (Chapter 14; Braun, 1970).

A number of dicots lack rays, and this seems to have come about in a variety of ways. It usually results when the cambium produces very little wood, as in woody plants that appear herbaceous, as well as in many xerophytes (*Sempervivum arboreum, Staavia glutinosa*). Carlquist

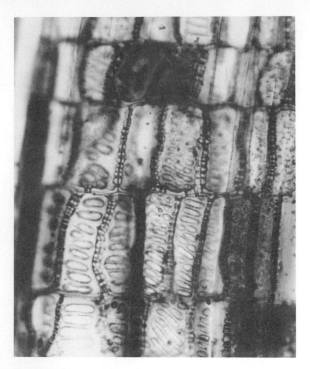

Figure 15.19
All the ray cells visible in this micrograph of *Brunellia comocladifolia* are upright; the central ones contact a vessel and have crossfield pits. These are basically scalariform, but are rather rounded. The tangential walls of the ray cells are also thick, but they have only simple, narrow pits. × 800.

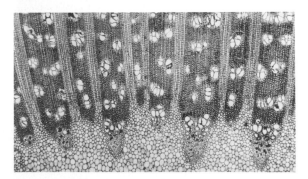

Figure 15.20
This cross section of *Coriaria* shows the pith, primary xylem, and first increments of secondary xylem. The rays here are all primary rays, being continuations of the medullary rays. The vessels are in clusters. × 40.

(1970, 1975) found that raylessness is common in species that have become more woody as they evolve from herbaceous ancestors (*Lysmachia*, *Plantago arborescens*, *P. madarensis*, *P. webbii*). This is an unusual type of evolution; it can occur when a single individual arrives on a newly formed, barren island. In the absence of normal competition and because the gene pool consists of a small number of individuals, the plants can evolve rapidly and in unusual directions.

Axial System of Dicot Wood

The wood of the more "primitive" dicots retains many gymnospermous features, the primary one being an absence of vessels. **Primitively vesselless** wood occurs in *Drimys*, *Pseudowintera*, *Tetracentron*, *Trochodendron* (Fig. 15.12) and others (Bailey, 1953; Cheadle, 1956; Lemesle, 1956). In addition, a small number of species have "lost" vessels; these **secondarily vesselless** plants are some members of Compositae (Carlquist, 1961). The Cactaceae are frequently reported to be secondarily vesselless also, but all cacti do have some vessels in their wood (Gibson, 1973, 1977); they are just few in number and difficult to detect (Fig. 15.21). Beyond these, dicot wood tends to be quite complex because of the presence of so many types of cells: vessel elements, libriform fibers, tracheids, fiber-tracheids, and axial parenchyma. Considering the possible combinations of these cell types, their possible arrangements in relation to each other, and the various modifications that each cell type can have, an almost limitless number of woods is possible. Fortunately, the arrangements tend to be quite logical. Because of the presence of fibers (Figs. 15.10, 15.11), dicot wood is called **hardwood**. The terms *hardwood* and *softwood* can be misleading, because some dicot wood (hardwood), such as balsa, is much softer than that of conifers (softwood).

Dicots usually have spring wood and summer wood; whereas these are based primarily

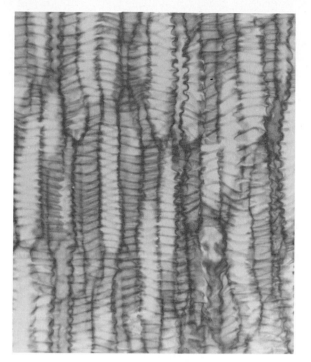

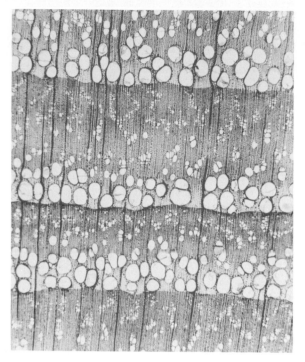

Figure 15.21
This is a rather unusual wood because it is composed almost entirely of tracheids with only annular or helical secondary walls; there are no fibers or any type of cell with an extensive secondary wall. The wood of conifers is also composed of only tracheids, but those have heavy secondary walls with bordered pits (Fig. 15.5). This is *Echinocereus knippelianus*, a cactus. × 400.

Figure 15.22
Cotinus americanus has easily recognizable ring-porous wood. The springwood is filled with large vessels and a small number of fibers, whereas the summerwood has mostly fibers and a small number of narrow tracheary elements, probably narrow vessels. × 40.

on the widths of the tracheids in the gymnosperms, in the dicots the other cell types are also involved. Because the spring wood is primarily concerned with rapid, large-volume conduction during rather nonstressful water conditions, we would expect the dicot spring wood to have large numbers of vessels (Figs. 15.10, 15.22, 15.26). Because the risk of cavitation is relatively low, these can be very long and have a large diameter. The large proportion of vessels automatically requires that there be a lower proportion of tracheids and fibers. The summer wood has the functions of me-

chanical support and sometimes also water transport under more stressful water conditions. Fibers should be abundant, and there should be some tracheids and perhaps vessels, but the vessels should be narrower to prevent cavitation and shorter to minimize the damage of whatever cavitation does occur. In climates or sites that vary from a mild spring to a stressful summer, the vessels may tend to occur very predominantly in the spring wood, or the differences in diameters between spring wood vessels and summer wood vessels would be very great. Such wood is termed **ring porous**, because distinct rings of obvious vessels are easily visible (Fig. 15.22). (*Pore* as used here is a forestry term for vessels seen in cross section.) Genera with ring porous wood include *Carya*, *Cercis*, *Fraxinus*, *Maclura*, and *Sassafras* (see Core et al., 1979, for a more complete list). In a more uniform climate, vessels may occur more uniformly through the annual ring, and this wood is **diffuse porous** (Figs. 15.2, 15.23; examples: *Acer, Cornus, Halesia, Magnolia, Populus, Salix,* and *Umbellularia*). I must emphasize here that, when relating whether or not an environment is stressful, the plants or plant parts under consideration are the only appropriate standards. Furthermore, although this climatic

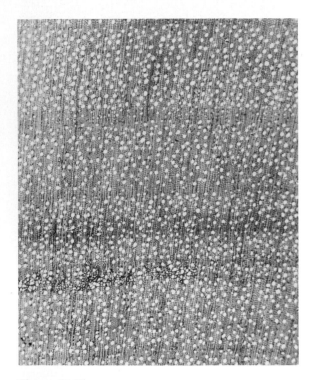

Figure 15.23
In contrast to *Cotinus americanus* (Fig. 15.22), the wood of *Buxella macowenii* is diffuse porous, with vessels distributed rather uniformly throughout each annual ring. The distribution here is so even that it is difficult to distinguish one annual ring from the next. The vessels in this wood are unusually narrow, and it would be reasonable to suspect that *B. macowenii* grows in an area of frequent water stress. × 40.

influence may be important, it is certainly not the only factor, because trees with ring porous wood may grow beside ones with diffuse porous wood. Diffuse porous wood appears to be more primitive than ring porous. Few species have ring porous wood (Gilbert, 1940), and these are considered to be advanced on the basis of other characters. Zimmermann (1983) and Handley (1936) have noted that the springwood vessels in ring porous trees are extremely long (sometimes as long as the entire trunk), whereas the vessels of diffuse porous trees are much shorter and narrower. Water movement in the very wide springwood vessels is up to ten times faster than movement in diffuse porous wood, but, because they cavitate so easily, water movement may be restricted to only the outermost annual ring (Huber, 1935; Kozlowski and Winget, 1963).

Beyond this general distribution of vessels, it is important to consider how they interconnect to each other. In a cross section of wood, vessels may appear **solitary** (Figs. 15.22, 15.23), but this is an illusion due to their length; they will always overlap two or more other vessels somewhere along their ends or sides. If vessels are grouped such that several occur in any cross section, they are said to occur in **multiples**, forming radial chains (Fig. 15.24), tangential bands (Fig. 15.20), or irregular groups (Fig. 15.25; Baas et al., 1976; Braun, 1970). In these clusters the diameter of the vessels will often not be uniform, either because some vessels really are smaller or because the section has passed through the tip of some, the middle of others. If the diameters are truly different, this is an important factor; it is not uncommon to see treatments of wood conductance in which the total cross sectional area of the vessels per square millimeter or per square centimeter is used as the basis of study. But this is a misleading figure, because it is the r^4 dimension that relates to conductance (Nobel, 1983). In a cluster of vessels, those with the largest diameters will be the most important for conduction, as described in Chapter 7. The

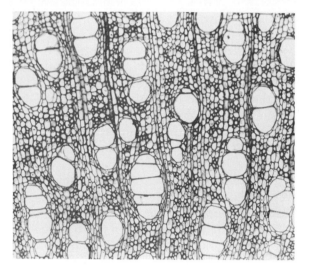

Figure 15.24
These vessels of *Adansonia digitata* occur in clusters or short chains, and each has extensive communication with its neighbors. The long areas of lateral contact allow water to pass from one vessel to another, but the pit membranes protect any vessel from embolisms that occur in other vessels of the cluster. There is also a considerable amount of axial parenchyma, both paratracheal and scattered among the fibers. × 100.

Figure 15.25
Does this *Casuarina cunninghamiana* have springwood and summerwood? A distinct layer of springwood is present with wide vessels and abundant paratracheal parenchyma. But the summerwood has large ribbons of vessels and parenchyma running through it; under low magnification the patterns can be spectacular. If the pattern that is present here were to extend longitudinally through the tree, the vessel-parenchyma bands would constitute serious panels of weakness. Probably the bands break up and form new patterns at points higher and lower in the tree, but unfortunately a three-dimensional map is not available. × 100.

narrower vessels, however, are important as vertical bridges: they cavitate less easily, and if the large-diameter vessels form an embolism, the narrower ones can then be called upon either to refill it with water or to conduct around it. Carlquist (1984b) reported that woods with clustered vessels tend not to contain tracheids, probably because they are not necessary for bypassing an embolism. However, woods that have solitary vessels do tend to have numerous tracheids that apparently can act as an alternative conducting system if the vessels cavitate.

It is also important to remember that water moves from vessel to vessel through the pit-pairs in the long, lateral-contact regions (Braun, 1959; Fig. 15.26). The narrow vessels often cover much of the surface of the wide vessels

and so can act as channels to facilitate this flow as well. In vessel elements derived from the vascular cambium, the perforation plates are always located on the radial walls (Fig. 15.26), not the periclinal ones. Therefore, vessel elements interconnect primarily tangentially through their perforation plates, but the ves-

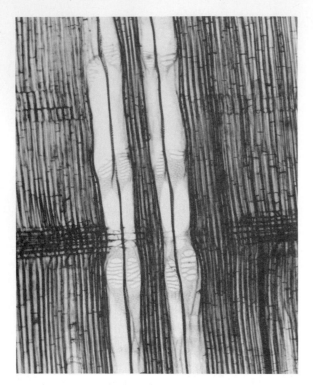

Figure 15.26
This radial longitudinal section of *Dendropanax arboreum* shows the relationship between vessels within a cluster and between other cell types. The perforation plates are radial walls, not tangential ones, so perforations are possible only between a file of cells produced by a file of fusiform initials. If perforations were to form on tangential walls, they would connect sister cells. × 100.

sels also interconnect on all sides by means of their bordered pit-pairs.

If dye is injected into the wood, it will be carried in the transpiration stream and gradually move tangentially as it passes from vessel to vessel (Braun, 1970; Zimmermann, 1983). The amount of vessel-vessel contact and intervessel pitting is important for this normal movement of water. Bosshard and Kučera (1973) found that, in beech, vessels could run side by side for as much as 2 mm, providing a very large surface area for lateral water exchange. In plants

that have clustered vessels, the amount of intervessel pitting can be very high, and both water distribution and protection from embolisms are great.

As in the gymnosperms, dicots also form interconnections across the annual ring boundaries: 30% of the vessels of the spring wood of poplar are connected to the vessels of the summer wood of the previous annual ring (Braun, 1959).

An interesting aspect of the distribution of vessel diameter is that the widest and longest vessels are located in the roots and base of the trunk, the higher portions of a tree having narrower and shorter vessels (Carlquist, 1975). This had been related to various aspects such as the greater age of the cambium at the base of the trunk or the greater phylogenetic advancement of one organ compared to another (within the same plant!), but the reason probably is much more logical. The water column at the base of the plant is under the least tension and is thus in the least danger of cavitating. The water column at the top of the plant is pulling up all of the water below it, supporting its weight and overcoming friction. This water is under the greatest tension, and vessels should be narrow to prevent embolisms and short to minimize damage if embolisms do occur.

Axial Wood Parenchyma in Dicots

Unlike the conifers, dicot wood typically contains parenchyma cells in the axial system, the cells having been derived from the fusiform initials (Fig. 15.27). Only a small number of the dicots have wood that lacks axial parenchyma (Berberidaceae, some *Bursera* species, *Aristotelia*, *Elaeocarpus*; Chalk, 1983b).

The amount and arrangement of the axial parenchyma are important considerations. The two main types are based on the relationships of the parenchyma cells to the vessels. If they are immediately adjacent to the vessels, the parenchyma cells are said to constitute **paratracheal parenchyma** (Fig. 15.24). If they are

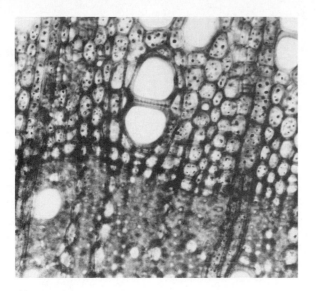

Figure 15.27
The wood used for many studies is taken from dead branches or obtained from collections of dried wood. Any parenchyma has dried, and its cytoplasm has decayed. If living wood is available, the easiest way to prepare it for microtomy is to boil it or steam it; this makes it soft but also destroys the cytoplasm. Consequently, identifying parenchyma in wood can be very difficult; if you have had trouble, don't feel bad. If the wood is preserved carefully in fixative, embedded in wax or plastic, and cut and stained properly, then the cytoplasm is retained and the parenchyma is immediately obvious, as in this sample of *Olea europa* (olive). Here, the role of the wood as a means of storage is dramatically apparent, and it is obvious that this is another function that must be considered in addition to conduction and support. × 400.

scattered among the fibers and tracheids they are **apotracheal parenchyma** (Fig. 15.11) and have no constant association with the vessels, although they may contact them occasionally. Both of these can show various modifications. If the apotracheal parenchyma cells occur as idioblasts or as isolated uniseriate strands, they are referred to as **diffuse parenchyma** (*Liquidambar, Platanus occidentalis*), but if they form large clusters or bands they are called

banded or **metatracheal parenchyma** (*Carya, Diospyros*). It is not unusual for the apotracheal parenchyma to be ordered with respect to the initiation or cessation of xylem differentiation; if there is a concentration of apotracheal parenchyma (either diffuse or banded) at the inner boundary of an annual ring, it is **initial parenchyma** (Fig. 15.28; *Ceratonia, Tilia, Zygophyllum dumosum*). If located at the outer boundary, it is **terminal parenchyma** (*Populus deltoides, Liriodendron tulipifera*).

If the paratracheal parenchyma does not form a complete sheath around the vessel or vessels, it is said to be **scanty paratracheal parenchyma** (*Acer, Betula*), and, if located on only one side of the vessels, it would be **unilaterally paratracheal parenchyma**. If the parenchyma is abundant, it will usually surround the vessels completely, and then it is **vasicentric parenchyma** (Fig. 15.24; *Fraxinus americana, Tamarix*). Such parenchyma may form a rather uniform sheath, but it may also be in the form of wings that extend out into the fibers, usually tangentially. This is **aliform parenchyma** (*Sassafras albidum*) if the parenchyma of one group of ves-

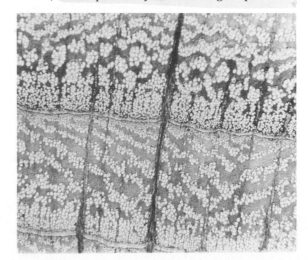

Figure 15.28
The very first layers of cells that develop in the springwood may be purely parenchyma; this is said to be an initial parenchyma. *Artemisia tridentata*; sagebrush. × 40.

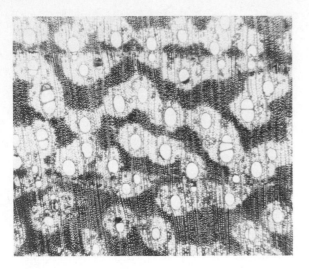

Figure 15.29
This wood of *Bombax pentandrum* (silk-cotton tree) has abundant banded parenchyma, and all vessels are located within the parenchyma regions. Although only half of the wood consists of fibers, no vessels are embedded within the fiber masses. × 40.

sels is not confluent with the parenchyma of other vessels. It is **confluent parenchyma** (Fig. 15.29) if it is and if several vessel clusters lie within a single large band of parenchyma (*Gleditsia triacanthos*).

By studying the association of various types of parenchyma with other features of plants considered to be either primitive or advanced, Kribs (1937) concluded that the earliest angiosperms contained diffuse apotracheal parenchyma. This pattern was modified to others, finally producing what he considered the most advanced organization: abundant vasicentric parenchyma. At this point, it is important to stop and consider the functions of the axial parenchyma cells. Like the radial parenchyma, they are capable of accumulating, storing, and releasing reserve materials, and in many cases they constitute an important reservoir for carbohydrates, nitrogen, and other metabolites. To fulfill a function in storage of such metabolites the arrangement of the parenchyma be-

comes very important; those cells or bands of cells that have large areas of interconnection with the vessels and rays would be able to transport material into and out of themselves most easily. Apotracheal parenchyma, especially the diffuse type, would not communicate directly with the vessels, so the ray contacts would be important. It seems reasonable to assume that terminal parenchyma might play a major role in supplying material for the differentiation of at least the first layers of the spring wood of the subsequent annual ring (Zasada and Zahner, 1969). In addition, terminal and initial parenchyma might be important in the movement of water from one annual ring to another.

Another aspect of axial parenchyma is its ability to store water, especially because the water is under, at most, only very mild tension and is free to move easily. It must be remembered that, although the vessels and tracheids conduct water, the dead fibers and fiber-tracheids, along with all of the apoplast around the axial and radial parenchyma cells, form an extensive three-dimensionally integrated capillary space. The lumens of the fibers and fiber-tracheids contain both air and water (Zimmermann, 1983). When the conducting system is under only mild tension (at night, with little transpiration), capillary action pulls water from the vessels and tracheids into the fibers and the apoplast space, and the living parenchyma cells also absorb water. When the tree begins to transpire, water tension in the vessels and tracheids increases, increasing the risk of cavitation, but the tracheary elements can draw water from the rest of the sapwood, and the danger of embolisms is reduced. The ability to pull water from this reservoir is enhanced by the elasticity of the wood. It is easy to show that trees actually become narrower as they transpire; they become wider at night when transpiration stops and the fiber and apoplast spaces are refilling.

For the mere storage of water, perhaps any of the possible arrangements of axial parenchyma would be more or less equivalent, although confluent paratracheal parenchyma would probably be most efficient in the rapid conduction of water into stressed vessels. But paratracheal parenchyma is important in another way. If a vessel is surrounded, even partially, by either intercellular spaces or dead, air-filled fibers, then—when the tension in the vessel becomes great enough—an air bubble can be drawn into it. This is called **air seeding** (Zimmermann, 1983), and cavitation will result instantly, even if the tensile strength of water has not been exceeded. However, if the vessel is completely surrounded by living parenchyma cells, all air is excluded, and, regardless of the tensions, only water can be drawn into it. The primary selective advantage of paratracheal parenchyma may be to act as an air-excluding water jacket around the vessels. This would be similar to the bundle sheaths in leaves and the endodermis in roots.

Axial Fibers in Dicot Wood

Because one of the primary functions of the wood is support, the presence of fibers is important. Conifer wood lacks fibers, as does the wood of primitive dicots, but very early in the evolution of angiosperms some of the tracheids were converted to fiber-tracheids, and libriform fibers appeared (Figs. 15.11, 15.13, 15.29; Chapter 7; Bailey, 1953). There has been some attention to the amount of fibers in wood and their relation to its strength and weight (Dinwoodie, 1976; Jeronimidis, 1976). The origin of fibers was parallel with the origin of vessel elements. The fusiform initials became shorter, apparently in relation to the needs of the vessel elements, and this automatically caused the fiber mother cells to be shorter. The mother cells undergo intrusive growth as they mature. In the development of the fiber-tracheids, they can be up to 50% longer than the fusiform initials, but if they mature into libriform fibers they can grow intrusively until they are eight times as long as the fusiform ini-

tial (Chattaway, 1936). The length of the fibers affects whether the wood is more flexible or more brittle.

The arrangement of the fibers is an important aspect of their ability to impart strength, but there appears to be a lack of study in this area. In ring porous trees, the fibers constitute only a small proportion of the springwood, which is consequently quite weak. Large amounts of banded parenchyma also cause the exclusion of fibers from certain regions, which will result in zones of weakness. It would be useful to know the three-dimensional arrangement of masses of fibers. Do they form a network of interconnections with other masses of fibers, or is each rather isolated from others by intervening masses of parenchyma or vessels? Rays also exclude fibers from regions of the wood; those trees that have very large rays typically have an abundance of heavy-walled fibers that compensate for the zones of weakness resulting from the ray parenchyma. Large aggregate rays are frequently traversed by fibers.

One of the most general trends in fiber evolution in the secondary xylem resulted in the fibers' becoming longer and having much thicker walls. In many species with heavy, dense wood, the fiber lumen is almost totally occluded by the wall as it thickens (Boraginaceae, *Lophira, Fagus*; Braun, 1970). At the same time, the selection for increased strength has resulted in these fibers' having fewer, smaller, simple pits. However, in some groups (Burseraceae, Meliaceae and others), some of the fibers have been modified in quite a different fashion, resulting in **septate fibers** (Fig. 15.26). These are most often libriform fibers (Chalk, 1983b) but occasionally fiber-tracheids also become septate (*Hypericum androsaemon, Mastixia, Nyssa*; Vestal and Vestal, 1940). These fibers undergo normal intrusive growth, and they deposit at least a thin secondary wall, but they then remain alive and undergo several subdivisions. The cross walls are usually just middle lamella and primary wall (Parameswaran and Liese, 1969), and in other features they resemble parenchyma cells (Fahn and Arnon, 1963). Septate fibers typically contain starch or crystals and thus act like storage parenchyma. Because septate fibers show so many characteristics of parenchyma, it might be expected that they evolved in wood that lacked axial parenchyma, but that is not the case; the two occur together (Frison, 1948; Spackman and Swamy, 1949).

In 1970, Braun summarized his extensive research on the patterns that the different components of wood might have. He considered the relationships between tracheids, fibers, vessels, and axial parenchyma, then divided them into 17 types. Several are described here to give you an idea of the diversity that is possible and to illustrate that the various tasks of support, conduction, and storage can be performed by several types of tissue arrangements.

1. Gymnosperm type. In this type, the ground tissue consists only of tracheids; there are no vessels. This includes most gymnosperms; among the dicots, it would encompass *Amborella trichopoda, Belliolum, Bubbia, Drimys, Exospermum, Pseudowintera,* and several others.

2. *Fagus sylvatica* type. The ground tissue also consists of tracheids, but vessels occur in it, and there is tracheid-vessel pitting. Paratracheal parenchyma is scanty, existing only as isolated strands. The only connection between annual rings is by means of pit contact between latewood tracheids and the next season's earlywood tracheids. The vessel system of one ring does not contact that of other rings.

3. *Daphne mezereum* type. This type is similar to the *Fagus sylvatica* type except that the ground tissue consists of regions of tracheids and regions of fiber-tracheids. The vessels are located only among the tracheids.

4. *Alnus glutinosa* type. In this pattern, the ground tissues consist only of fiber-tracheids; tracheids are restricted to the latewood as bridges to the next annual ring.

5. *Saurauia bracteosa/Juglans regia* type. In this type, the major innovation is the beginning of an isolation mechanism for the vessels: they are almost entirely surrounded by a sheath of axial parenchyma cells. There is much less contact with the tracheids of the ground tissue.

6. *Rhamnus cathartica* type. This type differs from the previous in two main ways. First, the ground tissue contains regions of libriform fibers and regions of tracheids, the vessels occur among the tracheids and have pit-contacts with them, and there is only scanty paratracheal parenchyma. Second, both the tracheids and the vessels of one annual ring are in contact with those of the adjacent rings.

7. *Quercus robur* type. This is similar to the *R. cathartica* type, having a ground tissue that consists of tracheid regions and libriform fiber regions, but here the vessels are isolated from the tracheids by complete sheaths of paratracheal parenchyma. As with those just described, both the tracheids and vessels are able to bridge the boundaries of annual rings.

8. *Aesculus hippocastanum* type. The ground tissue of this type consists only of libriform fibers, and the vessels occur mixed among them. There is very little pitting between the two cells types, however. Paratracheal parenchyma is scanty.

9. *Acer pseudoplatanus* type. This type contains only dead libriform fibers as ground tissue; the vessels are isolated from these fibers by complete sheaths of living fibers (not parenchyma).

10. *Bombax* type. This type is quite unusual in that the ground tissues consist of parenchyma, not tracheids or fibers.

In considering all the factors that affect water movements, tensions, cavitation, nutrient storage, and water storage, it is not surprising that dicot wood should be so complex. It must also be kept in mind that the environment provides only the first clues as to what to expect in the wood. Different parts of the plant have very different needs and constraints, and there is no reason at all to expect that the wood of the roots, trunk, and branches should be identical; they really do not even need to be similar (Kramer and Kozlowski, 1979; Zimmermann, 1982).

Reaction Wood

Although we take for granted the idea that trees grow upright, many forces of nature (wind, soil slippage on hillsides, floods) act to interfere with such growth. Also, this vertical arrangement is typical only of the main trunk; the branches grow either horizontally or at an angle between vertical and straight down. Therefore all unsupported, above-ground portions are either periodically or continuously under tension or compression. In reponse to this, the plants produce **reaction wood** (Boyd, 1977; Scurfield and Siva, 1969; Wardrop, 1964, 1965; Wilson and Archer, 1977).

Furthermore, in many cases a branch or twig is able to undergo a slow but very pronounced movement such that the orientation of existing branches is altered. Such motions are due to reaction wood formation and contraction.

In conifers, the reaction wood forms on the lower side of the branch where the tissues are compressed, so it is called **compression wood** (Wardrop and Davies, 1964; Westing, 1968). In the dicots it forms on the upper side and is called **tension wood** (Wardrop, 1956; Wardrop and Dadswell, 1948).

Compression Wood

In response to compression in conifers, there is an increased amount of wood formed in the compressed area. The branch becomes asymmetrical and has eccentric growth rings. In the normal wood of both conifers and dicots, intercellular spaces are rare and small, but the tracheids of compression wood are more rounded in cross section than those of normal wood, and intercellular spaces can form between them. However, compression wood is from 15 to 40% heavier than normal wood. The walls are slightly thicker than those of normal tracheids, and they have a higher lignin content. The innermost layer, S_3, is absent, and the inner face of S_2 is deeply grooved. These modifications cause compression wood to be heavier and more brittle than normal wood.

Tension Wood

The reaction wood of dicots is more complex than that of conifers. It is usually characterized by the presence of **gelatinous fibers**. These contain a special **G-layer (gelatinous layer)** as the innermost layer of the wall (Wardrop and Dadswell, 1948, 1955). In addition to the G-layer, the cell may produce all three of the other layers, or it may have only S_1 and S_2 or even just S_1. The G-layer is reported to have little or no lignin or hemicelluloses, so the cellulose microfibrils are only loosely interconnected (Côté et al., 1969). Overall, there is less lignin in tension wood, and there is more cellulose.

Gelatinous fibers are usually found on the upper side of the branch in the thickened part of the eccentric growth rings, but in some cases the upper side with the gelatinous fibers is the thin portion of the ring: the extra growth occurs on the lower side of the branch, and this area is free of gelatinous fibers. Gelatinous fibers may occur clustered together (**compact tension wood**) or idioblastically as single fibers or in small groups (**diffuse tension wood**). In this latter type, the gelatinous fibers occur side-by-side with normal fibers. The gelatinous fibers were completely absent from the reaction wood of 42% of the 248 species examined by Höster and Liese (1966; examples: *Aristolochia* and *Entelea*), and some dicots form no reaction wood at all (Fahn, 1982).

Taxonomic Value of Wood Anatomy

All features, both external and internal, of a plant are useful for analyzing the evolutionary relationships of one species or group with another. Flower structure and the general shape of the stems and leaves are the most commonly used characters, but wood anatomy is becoming increasingly important (Baas, 1982). It offers an abundance of characters, and in many of them the evolutionary lines are quite clear. Carlquist (1961) gives a short but very interesting summary of some of the more impressive achievements of wood anatomists in solving difficult and important taxonomic problems. Until recently, many of the wind-pollinated dicots were grouped together as the *Amentiferae*, based on the similarity of their flowers and in-florescences; however, the characters of their wood proved that this was not a natural group. Furthermore, the characters indicated how these taxa were related to other families, and how wind pollination had evolved several times. Similarly, many plants were grouped together as the *woody Ranales*, which were thought, on the basis of their flower structure, to be the most primitive angiosperms. Here, too, wood anatomy showed that this was not a homogeneous group but rather contained several species with rather advanced and specialized features (Carlquist, 1961). Similar research continues at present (Baretta-Kuipers, 1976; Core et al., 1979; Metcalfe and Chalk, 1983; Michener, 1983; Ter Welle, 1976b; van Vliet, 1976b).

16

Secondary Phloem

Concepts

Like the secondary xylem, the secondary phloem consists of a radial and an axial system (Fig. 16.1), both of which are derived from the vascular cambium (Esau, 1969, 1979). These two systems are mirror images of those in the secondary xylem: the phloem rays are produced by the same ray initials that produce the xylem rays, and the fusiform initials produce both axial systems. I emphasize this fact because it is taken for granted so often that its importance is frequently ignored. There is no a priori reason that the secondary xylem and phloem should be produced by the same meristem. It would be possible to have two unifacial cambia located back to back, one producing xylem, the other producing phloem. An important consequence of the unistratose vascular cambium is that if its initials become modified to optimize the production of xylem mother cells, the phloem mother cells that are produced will be affected, and vice versa. Therefore, the xylem mother cells and the phloem mother cells may have sizes, shapes, or organizations that are not optimal but rather are the unavoidable consequences of the dual function of the vascular cambium.

The secondary phloem, like the primary phloem, has as one of its main functions the long-distance transport of organic nutrients, but in addition it has many other roles in the plant's metabolism (Zimmermann and Brown, 1974). It is one of the main constituents of the bark and is responsible for protecting the secondary plant body from external threats such as insects and larger animals, fungi, bacteria, and abrasion by debris carried by wind and water (see Chapter 17). In the younger twigs, which have only a small amount of secondary xylem, the sclerenchyma of the secondary phloem can be an important source of support (Carlquist, 1975), especially because it is located on the periphery of the organ, which is ideal from an engineering standpoint. The sclerenchyma of the secondary phloem is often composed of fibers, so it has a flexible, elastic strength. These are important characteristics, because the support provided by the secondary phloem is most critical to the slender young twigs that carry leaves that are large in relation to the twig. If the twig were brittle or inelastic, even gentle winds acting on the leaves would cause breakage. The elasticity of the young wood and secondary phloem allows survival.

Unfortunately for the plant, the secondary phloem must not form continuous, thick layers of sclerenchyma because, although this would provide maximum protection and support, it would also lead to the asphyxiation of everything interior to it, the vascular cambium and sapwood included. Because it is one of the most exterior tissues, the secondary phloem is also important in gas exchange, and it must be permeable to oxygen.

The secondary phloem may be quite rich in secretory tissues (Fig. 16.3); this is possibly re-

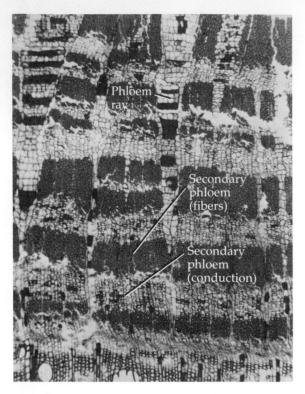

Figure 16.1

This secondary phloem of *Cochlospermum vitifolium* is fairly complex. The youngest, newest phloem is at the bottom of the photograph, nearest to the vascular cambium. Thick bands of fibers alternate with regions of sieve tube members and companion cells. The secondary phloem has both uniseriate and multiseriate rays, some of which are extremely wide near the outer parts of the phloem. These rays expand as the phloem is pushed outward. × 100.

lated to its role in protecting the plant. In addition to idioblasts and secretory chambers, the phloem may contain extensive duct systems (Ramos and Engleman, 1982; Venkaiah and Shah, 1984). It is the system of laticifers in the bark of *Hevea* that is tapped to obtain rubber (Hebant and De Fay, 1980) and the resin canals of the bark of conifers that are harvested for pitch, which is distilled to make turpentines

and resins (Hill, 1952). The secondary phloem of *Scalesia aspera* (Carlquist, 1982c) contains secretory canals that produce a resin that hardens as the phloem is converted to bark. Carlquist suggested two possible functions for the resin: (1) to deter insects and (2) to seal the plant against dessication (*S. aspera* grows on the arid lava flows in the Galapagos Islands).

The secondary phloem also has a major role in the storage of material during the dormant periods (Figs. 16.5, 16.6; Grote and Fromme, 1984; Kramer and Kozlowski, 1979). This role is important for both the conifers and the dicots. In conifers the xylem contains no axial parenchyma, so the only storage tissues available are the secondary phloem and the rays of the xylem. In dicots it is especially necessary if the plants are seasonally deciduous and must produce an entire, new set of leaves within a very short time in the spring; large amounts of carbohydrate and nitrogen are needed while the plant is incapable of photosynthesis.

Finally, the transport functions of the secondary phloem are different from those of the primary phloem. Although the bamboos, some other grasses (*Arundo*), and the palms become extremely large despite a lack of secondary growth, most herbs are quite small, usually being only 2 m or less in height. In these plants, long-distance transport in the primary phloem is not really a very long distance. In contrast, in most woody plants, the secondary phloem must be capable of moving sugars and other metabolites over truly long distances. This capability occurs not only in trees (up to 100 m tall for *Eucalyptus*) but also in various vines, especially those of tropical regions. The direction of transport is also more complicated. During the season of maximum photosynthesis, carbohydrates are transported downward into the branches, trunk, and roots for storage, respiration, and growth. But when the plants come out of dormancy (at the end of winter or after a major rainfall), carbohydrate reserves must be tapped and moved to the growing regions in the apices of roots and shoots. Thus the secondary phloem in the shoot must be capable of both upward and downward translocation. Flowers and fruits often occur on the shoot tips, but many plants are **cauliflorous** (they produce flowers on older branches or the trunk); their flowers and fruits are major sinks and may require large-volume transport through the secondary phloem.

Secondary Phloem of Conifers

Like the secondary xylem of conifers, the secondary phloem is rather simple (Figs. 14.2, 16.2). The radial system contains only paren-

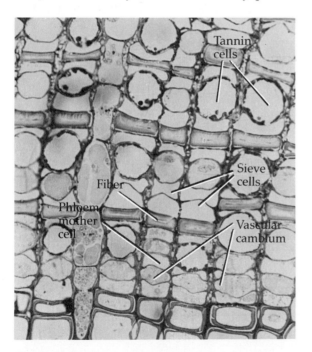

Figure 16.2
This is a cross section of the cambium and secondary phloem of the root of bald cypress (*Taxodium disticum*) collected during the winter while it was dormant. The sieve cells are somewhat difficult to identify with certainty in all cases, but they are the empty-looking cells of the axial system. Other cells are filled with tannins. The ray parenchyma cells are filled with dense, darkly stained cytoplasm. Flat, thick-walled fibers are abundant. × 400.

chyma and is in the form of tall uniseriate rays, just like the xylem rays (Fig. 16.4). The rays may contain solely storage parenchyma, or they might also have albuminous cells, tannin cells, resin canals, or other secretory structures.

The axial system always contains at least sieve cells and nonconducting parenchyma (Figs. 16.2, 16.3), and it usually also contains fibers. The sieve areas are located predominantly on the radial walls, so sieve cells of similar age are united. The sieve cells and their associated albuminous cells typically constitute a large proportion of the axial phloem while it is still actively conducting, but after the sieve cells stop transporting, they and the albuminous cells collapse and shrink to being just minor components (Fig. 16.3). The axial parenchyma ranges from sparse to abundant, as do the fibers. The arrangement of these three types of cells is constant within a species and can be an important characteristic for taxonomy; see Esau (1979) for a list of arrangements that occur in dicots. It is not unusual for the secondary phloem to consist of alternating tan-

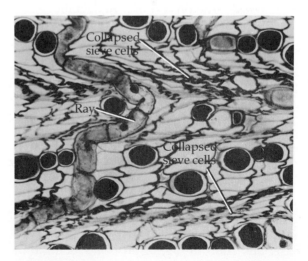

Figure 16.3
This photograph was taken from the outer, older portions of secondary phloem of *Pinus strobus*. As the sieve cells collapse and shrink, they occupy less volume; the ray cells remain alive and buckle as the surrounding cells pull them inward radially. × 400.

gential bands of conducting cells, storage cells, and fibers (Cupressaceae, Taxodiaceae; Fig. 16.2); this often gives the impression of annual rings, but typically there is no such relationship. There is no equivalent to annual rings, spring wood, or summer wood in the secondary phloem (Artschwager, 1950; Esau, 1948; Huber, 1949b). The members of the Pinaceae do not produce fibers in the secondary phloem, but the sieve cells have thick, nonlignified walls (Abbe and Crafts, 1939; Don, 1965).

The sieve cells in conifer secondary phloem usually function for only one season, then collapse. In some, however, (*Abies balsamea*, *Juniperus californica*, and *Picea mariana*) the last-formed sieve cells may overwinter, then resume conduction in the following spring (Alfieri and Evert, 1973; Kemp and Alfieri, 1974). As the sieve cells collapse, the axial phloem contracts, putting the rays under compression; the rays buckle and warp, becoming undulate in cross section (Fig. 16.3).

Even after conduction ceases, the secondary phloem continues to function for storage (Fig. 16.4) and protection. In loblolly pine, 11.2% of the nitrogen is stored in the secondary phloem; in mugo pine, sucrose constitutes as much as 5% of the dry weight of the secondary phloem, increasing during the spring. Other sugars also can reach concentrations as great as 3–4%. Such storage capacity occurs in both shoots and roots (Kramer and Kozlowski, 1979).

Secondary Phloem of Dicots

The phloem rays of dicots are composed only of parenchyma cells and are uniseriate or multiseriate, short or tall, again being identical to the xylem rays (Figs. 16.5, 16.6, 16.7). Mirror-image symmetry of xylem and phloem rays is strictly true only near the cambium; the ray derivatives can undergo slight changes as they move away from the cambium, so in older regions there will be some differences. The axial

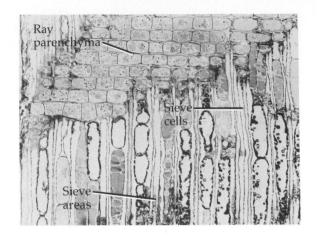

Figure 16.4
These ray cells of *Taxodium* are filled with stored nutrients; *Taxodium* is a conifer, but it requires abundant storage capacity like a dicot, because it abscises its leaves and ultimate twigs in the autumn. Expansion of new leaves and cones requires stored sugar, because there are no expanded leaves present in early spring and thus there is no photosynthesis. × 400.

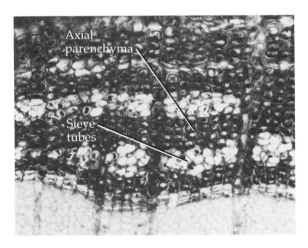

Figure 16.5
The secondary phloem of this *Eysenhardtia texana* has abundant axial parenchyma and densely cytoplasmic ray parenchyma. The sieve tube members and the companion cells are unusually easy to identify in this section. No sclerenchyma is present at all in this young secondary phloem; but see Figure 16.11 for older phloem of this plant. × 200.

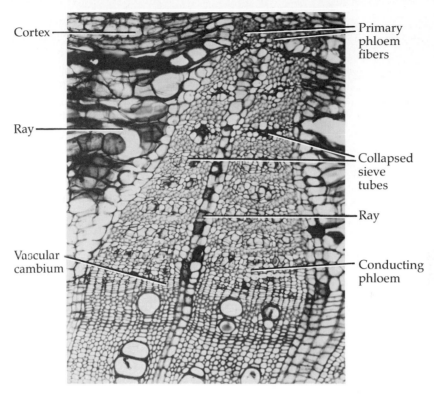

Cortex

Ray

Vascular cambium

Primary phloem fibers

Collapsed sieve tubes

Ray

Conducting phloem

Figure 16.6
This cross section of *Artabotrys odoratissima* shows a dramatic collapse of the older sieve tubes. It also shows three rays, one of which has remained narrow, the other two having expanded greatly tangentially. The primary phloem (especially the fibers) is still present and visible, showing just how little secondary phloem is produced. × 200.

system contains sieve tube members—up to 40% of the volume of the secondary phloem (Den Outer, 1982)—companion cells, storage parenchyma, fibers, and sclereids (Fig. 16.6). As in the conifers, the various cell types can have many different arrangements (Carlquist, 1980a; Datta and Chowdhury, 1982; Deshpande and Rajendrababu, 1985). In some species, the fibers are scattered, either individually (*Campsis radicans, Litsea calicaris*) or in small clusters (grapes, *Liriodendron tulipifera*) throughout the parenchyma. Secretory tissues are frequently present (Eremin, 1980; Gibson, 1981; Mustard, 1982; Ramos and Engleman, 1982). In the plants that have nonstoried cambia (the majority of

species), the secondary phloem elements are, of course, also nonstoried. Sieve tube members tend to be rather elongate with oblique endwalls; measurements for 125 species are presented by Chavan and Shah (1983) and for 444 species by Den Outer (1983). These sieve tube members may bear large compound sieve plates, while the side walls have much smaller sieve areas. Tippett and Hill (1984) found an extremely elaborate interface between adjacent sieve tube members in several genera of Myrtaceae. The interface areas are large (up to 200 μm long) and consist of thin-walled sieve areas with numerous pores; the sieve areas are supported by thick ribs. Apparently the interfaces

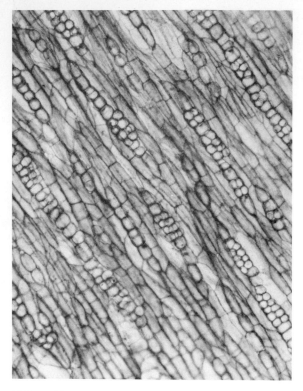

Figure 16.7
This is a tangential view of *Ligustrum sinese* (privet); it should be immediately obvious that it is in the phloem, not the xylem, even though sieve tubes are not apparent. The parenchyma is much less compact than is xylem parenchyma; the cells are more rounded and have intercellular spaces. All walls are extremely thin, and none of the walls is lignified. × 200.

are formed as intervening parenchyma cells subdivide, forming large surfaces filled with sieve areas.

In a study of five genera of the dicot family Winteraceae (which is generally considered to contain many primitive characteristics), Esau and Cheadle (1984) found no sharp differentiation between lateral sieve areas and those of the sieve plates. Also, the companion cells in these plants were short, and there was typically only one per sieve tube member. In a more advanced family (the Cactaceae), Bailey

and Srivastava (1962) and Srivastava and Bailey (1962) found that in the secondary phloem the sieve tube members were short and had transverse end walls, and that frequently there were two or three companion cells adjacent to each sieve tube member. Furthermore, they found that the derivatives of the fusiform initials divided in rather specific patterns, such that one or two axial parenchyma cells were cut off first, followed by the companion cell, which might undergo cross division (Fig. 16.8). During the course of differentiation, sieve tube members can take on unusual shapes, can develop three separate sieve plates, and can become obliquely oriented (Fig. 16.9). Thick nacreous walls may occur in many families (Fig. 16.10).

Two types of sclereids may be present in the secondary phloem of dicots. The **primary sclereids** are those that differentiate and mature at the same time as the other cells. **Secondary sclereids** appear only later, arising after conduction has stopped (Fig. 16.11). By this time continued secondary growth interior to these layers causes the phloem to begin to expand tangentially. Thus the parenchyma cells that transform themselves into secondary sclereids may be quite wide. Primary sclereids are not usually very abundant, and fibers constitute the main sclerenchyma component in active dicot phloem; but the production of secondary sclereids in the nonactive phloem can be significant with large masses resulting (Evert, 1963a). In some plants, there is no sclerenchyma in the functioning phloem, only the secondary sclereids. In *Prunus*, fibers as well as sclereids develop late. The secondary sclerification of parenchyma cells is not restricted to axial cells but may involve the rays as well, such that the inner, younger portions are parenchymatous, and the outer, older ones are sclerenchymatous (Bosshard and Stahel, 1969; Den Outer, 1982). Not all of the parenchyma may convert itself to sclereids, however, because one of the functions of the phloem parenchyma is to undergo a conversion into a cork cambium that produces the cork cells of

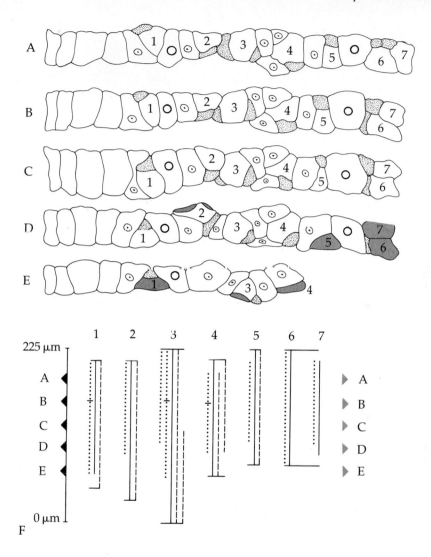

Figure 16.8

This diagram is part of an analysis of the secondary phloem of *Pereskia sacharosa:* (A) is a cross section through a fusiform initial and its derivatives, near the top of the initial (see the scale in the lower drawing). Rows (B), (C), (D), and (E) are cross sections through the same set of cells, at lower levels. Companion cells have stipples. Sieve tube members are numbered, and cells with nuclei are parenchyma cells that were sister cells to the mother cell of a sieve tube/companion cell pair. Cells with a dark circle are parenchyma cells that are not immediately related to the sieve tube members. In the lower drawing, each cell is represented as a line, to show the spatial relationships: solid lines are sieve tube members, dotted lines are companion cells, and dashed lines are related parenchyma cells. Notice that sieve tube members 1 and 4 have two companion cells each and that 3 has three. The companion cell of 6 is as long as the sieve tube member, but in all the rest the companion cells are shorter. The related parenchyma cells are often larger than the sieve tube members, indicating that they underwent intrusive growth. The dark areas in (D) and (E) are sieve plates. From Srivastava and Bailey (1962), in which there is a wealth of information.

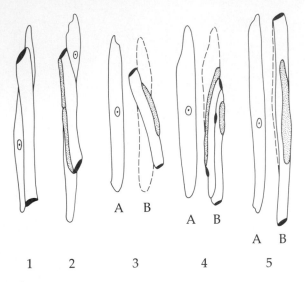

Figure 16.9
This figure represents the cell relationships in the secondary phloem of *Quiabentia* aff. *chacoensis*; the sieve plates are darkened; companion cells are stippled. In 2, there are two companion cells; in 3 the various divisions have resulted in a sieve tube member that is not aligned with its fusiform initial (shown with dashed lines). In 4, each sieve tube member has three sieve plates. The companion cells can be short (3, 4, 5) or long (4). From Srivastava and Bailey, 1962.

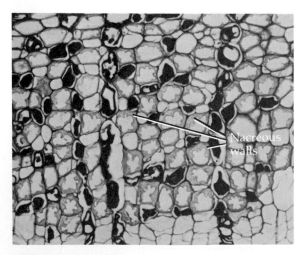

Figure 16.10
The secondary phloem of this root of *Clusia* has thick, nacreous walls. × 400.

Figure 16.11
This is an outer region of secondary phloem of *Eysenhardtia texana*, from the same section as Figure 16.5. Here all the sieve tubes have collapsed, but some of the parenchyma cells have redifferentiated into secondary sclereids. × 200.

the outer bark (this is described in Chapter 17). The presence of sclereids, either primary or secondary, and their distribution are often important characters for taxonomic purposes (Bailey, 1961; Richter, 1981).

The secondary phloem of dicots, like that of conifers, is typically active in conduction for only one year; then the sieve tube members and companion cells collapse (Fig. 16.6; Derr and Evert, 1967; Ghouse and Hashmi, 1979, 1980a; Tucker and Evert, 1969). In a few species, the phloem conducts for two years (grape, yellow poplar) or more (*Tilia*). In plants with included phloem (an anomalous type of secondary growth; see Chapter 18), the phloem functions for many years (*Bougainvillea* and the woody species of Chenopodiaceae; Fahn and Shchori, 1967). As a sieve tube member stops conducting, definitive callose is deposited. This is usually very obvious, but, if the collapse of the cells is severe, the callose may be displaced and be difficult to detect. If the sieve tube members function until late into the winter, then there may be no definitive callose formed when they collapse (Derr and Evert, 1967). In several taxa (*Antiaris, Bombax, Ricinodendron,* and *Vitis vinifera*), the nonfunctioning sieve tube members become filled with tylosoids. These are tylosis-like protrusions of protoplasm from adjacent parenchyma cells (Esau, 1948; Lawton and Lawton, 1971). In a survey of the secondary phloem in six species native to Nigeria, Lawton and Lawton (1971) were able to detect tylosoids in three genera: *Antiaris, Bombax,* and *Ricinodendron.* They pointed out that these same species also form tyloses in the xylem; furthermore, they suggested that, because tylosoids are extremely difficult to detect, they may actually be common, having simply been overlooked in most species. *Salix,* box elder, and black locust are unusual in that their sieve tube members do not collapse, nor are they filled by tylosoids.

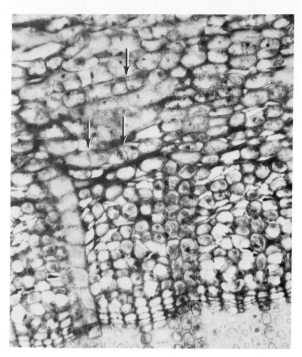

Figure 16.12
The outer portions of secondary phloem of this stem of *Vitex* have just begun producing proliferative dilatation tissue. Axial parenchyma cells (arrows) have undergone radial divisions. × 200.

Dilatation of the Secondary Phloem

As secondary growth and expansion occur near the vascular cambium, all of the peripheral regions of the secondary phloem are pushed outward. To avoid tearing, this tissue must expand circumferentially. This expansion is called **dilatation,** and it is effected by the formation of **dilatation tissue.** Two methods for the production of dilatation tissue are possible, and many plants use both. If axial parenchyma cells begin to divide and expand, they form a dilatation tissue called **proliferative tissue** (Fig. 16.12; Whitmore, 1962a). If the axial parenchyma is abundant, then the term *tissue* is fairly descriptive (Chattaway, 1955). But often the cells that resume division are mixed with

other cells that do not divide; the result is small packets of cells that can be quite difficult to detect at first. The proliferative tissue allows a uniform, balanced dilatation. The second method is by the formation of **expansion tissue;** this results from renewed cytokinesis in the ray parenchyma (Figs. 16.1, 16.6, 16.13). Because the growth is localized, it is frequently much more noticeable than proliferative tissue. The ray cells divide predominantly with walls that are parallel with the ray (anticlinal to the surface of the stem or root). As a result, the ray begins to look as if it is composed of plates of cells growing from a radial cambium (Holdheide, 1951; Schneider, 1955). Because the outer parts of the ray are older and have undergone expansion growth longer, the rays flare out-

Figure 16.13
The dilatation growth in this ray has produced striking expansion tissue; the cells are perfectly aligned in tangential rows, indicating an extremely orderly control of division. Notice that one ray is very expanded while others show little or no expansion. × 200.

ward when seen in cross section. At any given level of a stem or root, either all of the rays or only some of them will be undergoing expansion growth.

As in conifers, the secondary phloem is important in storage. The evergreen dicots are rather similar to the conifers in that they may show only one peak of accumulation of sugars (in the autumn) and then a period of release in the spring. The seasonal variation in the amounts of materials is typically low, because evergreens can use the sugar produced by existing leaves for the production of new leaves; large amounts of storage parenchyma in the secondary xylem and secondary phloem are not necessary (Dickmann and Kozlowski, 1968). In deciduous dicots, however, the entire crown of leaves must be replaced in a short period, and nutrient reserves in the secondary phloem of shoots and roots drop to low levels (Kozlowski and Keller, 1966); the large amounts of storage parenchyma that are produced by dilatation growth may be necessary. During peak storage times, carbohydrate can be up to 18% of the phloem dry weight, with sucrose and starch each accounting for about half of this (Siminovitch et al., 1953). Deciduous tropical trees are more like evergreens in showing smaller seasonal variations, but this is probably due to the fact that individual branches act independently. Some are dormant while others on the same tree are active; some produce leaves while the leaves of other branches are abscised. Thus there is a more or less continuous production of leaves balanced with photosynthetic input. Such storage needs can affect the amount of dilatation tissue needed, as well as the formation of secondary sclereids, which block conduction to exterior portions of the phloem.

17
Periderm and Bark

Concepts

The epidermis is the initial interface between the plant and its environment, and it is the first protection for the plant. In many short-lived herbaceous plants and even in some perennial woody plants, the epidermis and hypodermis are never replaced. They continue to function as the primary mechanism for the control of water loss and gas exchange, as defense against sunlight and pathogens, and as a physical barrier against abrasion. But in many plants, certainly in most of the perennials, the epidermis and hypodermis are replaced by the **periderm** (Figs. 17.1, 17.2). The periderm is actually a mixture of tissues. Certain cells of the epidermis, cortex, and secondary phloem become mitotically active and form a meristem called the **phellogen** or **cork cambium**; this produces **phellem** (**cork cells**). Thus the periderm is a mixture of layers of cork cambium and cork, and the periderm plus the remaining primary tissues and/or secondary phloem is the **bark** (Fig. 17.3). In a manner similar to that of human skin, the cork cells modify themselves to become waterproof and resistant to enzyme attack, and then die. The cork cells are extremely resistant and tough, and the actual amount of protection can be adjusted somewhat both by the nature of the chemicals deposited in the cell walls and by the amount of periderm that remains on the plant. As we can easily observe, some plants have an extremely thick bark that offers up to several centimeters of

insulation against whatever insults the outside world might present (Fig. 17.4); the bark of pines can actually enable the trees to survive forest fires very well; the heat does not penetrate to the cambium, sapwood, or inner bark, all of which remain healthy (Kramer and Kozlowski, 1979). Other plants have a much thinner, less protective bark because the individual layers of it fall off rather rapidly (Fig. 17.5). If the only aspect of the periderm were protection, then at first glance it might seem that it would be obviously advantageous to retain as much of the periderm as possible for as long as possible, thereby having a thicker, more protective bark. However, there are advantages to allowing parts of the bark to fall off. The periderm faces the same problem as the epidermis: by being waterproof it also stops the movement of CO_2 and O_2. Just as the epidermis is modified to have stomata that permit gas exchange, the periderm is modified to possess **lenticels**, patches of cork cells that have intercellular spaces that facilitate the penetration of oxygen to the living tissues beneath and the release of carbon dioxide produced by their respiration (Figs. 17.5, 17.6).

Another important function of the shedding of the bark is that the falling layers carry with them whatever might have been on or in them: fungi, bacteria, and insect larvae being the most important pathogens shed this way. In areas of high humidity, epicortical lichens can also be a problem as they grow over lenticels and inhibit gas exchange. Trees of *Carpinus car-*

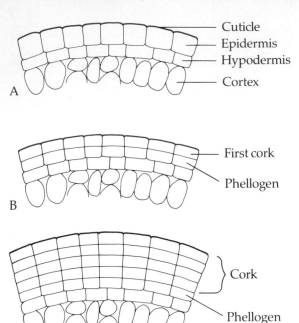

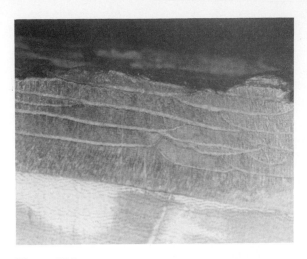

Figure 17.1
Epidermal cells undergoing divisions to produce a cork cambium and cork. (A) The epidermis before mitosis. (B) Each cell has undergone one division; the outer cell of each becomes cork, the inner cell phellogen. (C) Continued cell division in the phellogen produces a layer of cork. The cork cambium can arise in many types of tissues, and the process is similar.

Figure 17.3
The lighter areas in this outer bark are the layers of phellem. Between them are masses of secondary phloem. × 5.

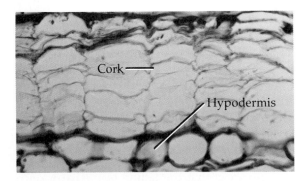

Figure 17.2
Bark of *Mammillaria nivosa*. At this stage, only thin-walled cork cells are present, and the phellogen itself has differentiated into cork also. As in Figure 17.1, the phellogen was produced from the epidermis; consequently the first layer beneath the bark is the hypodermis. × 100.

Figure 17.4
This trunk of the cork oak (*Quercus suber*) has just been harvested for its bark. The bark is allowed to grow until it is about 3 or 4 cm thick, then circular cuts are made and the bark is peeled off. The inner bark remains on the tree, so the plant does not die.

Figure 17.5
In *Betula platyphylla*, the bark is very different from that in cork oak. It peels off naturally a short time after it forms, so it never accumulates into a thick layer. Because it is so smooth, the lenticels are easily visible.

Figure 17.6
On this stem lenticels are small, circular, or oval instead of elongate as in Figure 17.5.

oliniana growing in Louisiana actually have an endoperidermal lichen (*Trypethelium eluteriae*) that grows within its bark (Lambright and Tucker, 1980). An even more severe problem of this type occurs in regions of high rainfall, in both tropical and temperate rainforests. In these areas the branches can become covered with epiphytes of all types: bromeliads, ferns, cacti, arums, orchids, and pepperomias, among others. In the desiccated state these epiphytes may not be especially heavy, but when filled with water they add a considerable weight load to the branch, often enough to cause it to break. Furthermore, their foliage can act as sails, catching the wind and, along with the leaves of the branch itself, contributing to wind damage. However, epiphytes of this nature cannot become established on trees that drop their bark

frequently, so periderm shedding for these hosts may be more protective than its retention. The shedding of bark must be considered an extremely effective method of "preening," continually providing the plant with a clean, uncluttered surface.

Rapid shedding also results in a thin bark that is translucent. The stems of *Bursera* contain chlorophyllous tissues and are the primary means of photosynthesis. Gomez-Vazques and Engelman (1984) found that in *Bursera longipes* and *B. copallifera* the periderm falls away quickly, permitting light to pass through to the chlorenchyma below. The same is true for elephant trees—*Pachycormus discolor* (Gibson, 1981)—and for *Idria columnaris* (Henrickson, 1969).

One final function should be mentioned, although it has received much more speculation than research. It is possible that, because the bark will be dropped from the plant, it is the ideal repository for waste products. Lacking anything even remotely similar to kidneys or liver, plants often must live with their waste products stored inside their cells in the vacuoles. For short-lived cells such as those of deciduous leaves, cortex and epidermis, this probably is not a major problem. But many cells are long-lived (nondeciduous leaves, sapwood parenchyma, inner bark parenchyma) or virtually immortal (apical meristem cells and vascular cambium cells). If the meristem cells of bristle cone pines (age: up to 7000 years) were required to carry along all of their wastes for such a long time, they might be expected to die eventually of the metabolic load. It seems only logical that these tissues can transport their wastes to other parts of the plant, and the bark and heartwood are reasonable repositories.

Formation of the Phellogen

Site of Initiation of the First Phellogen

The phellogen arises as living parenchyma cells resume mitotic activity and become meristematic (Fig. 17.1). This conversion can happen in almost any parenchyma cell, apparently, because phellogens have been observed to form naturally in almost all tissues: epidermis (Fig. 17.2), hypodermis, cortex (Fig. 17.7), secondary phloem, and secondary xylem (Leu and Chiang, 1981). Furthermore, it can form in all organs of the plant: shoot axis, roots, leaves, fruits, and flower parts. This diversity is based on the diversity of the functions of the bark and the requirements of that plant or plant part. The majority of bark occurs on the shoots and roots of woody plants; this is where the phellogens arise most commonly. During the first several years of secondary growth, the primary tissues (epidermis, cortex, and primary phloem) may still be present and alive, so these may be the sources for the first phellogen. Examples in the epidermis include *Malus pumlia*,

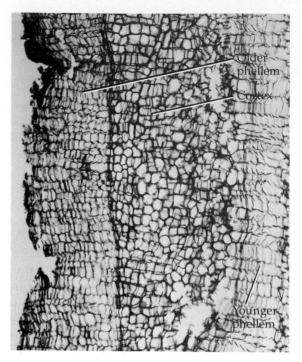

Figure 17.7
Longitudinal section through the outer bark of *Crescentia* (calabash tree); on the left is the oldest phellem. In the middle is cortex that has died after being cut off from water and nutrients by newer layers of cork (on the right). × 100.

Nerium oleander, and *Pyrus communis;* in the hypodermis, *Populus* and *Ulmus;* in the cortex, *Aristolochia* and *Pinus;* and in the phloem, *Arbutus, Punica,* and *Vitis.* Leu and Chiang (1981) surveyed more than 80 dicot species and found that the formation of the phellogen never preceded the formation of the vascular cambium. However, the first phellogen often arises soon after the vascular cambium, in the first year, especially in plants that must survive cold winters. A delay in forming a phellogen may be related to the photosynthetic activity of the stem cortex; the initiation of the first phellogen and the production of an opaque bark are delayed in plants that depend on the stem cortex to carry out a significant amount of photosynthesis. *Baccharis sarothroides* does not produce bark until its second year, and the branches of *Canotia* retain their epidermis until they are very thick (Gibson, 1983). Large trees of *Cercidium* are bright green, and many giant cacti never form any bark on their branches. Zimmermann and Brown (1974) report that in some species of *Acacia, Acer, Citrus, Eucalyptus*, and *Ilex*, a bark is never formed; the cortex and epidermis expand and continue to function as long as the tree lives.

On the other hand, in nonphotosynthetic underground stems such as stolons, the phellogen may form very quickly, even at the third internode in clover, as in *Trifolium repens* (Hay et al., 1982). By the time four more internodes have been formed by the apical meristem, suberized cork is present, and the stem is opaque.

In roots, the first phellogen usually arises in the pericycle (Chandra et al., 1982; Posluszny et al., 1984; Weerdenburg and Peterson, 1983). The exception to this is logical: fleshy roots in which the cortex functions as storage tissue. For example, in sweet potato (*Ipomoea batatas*), the phellogen arises in the layer of cells immediately interior to the exodermis (Kono and Mizoguchi, 1982). The root expands rapidly and continuously with secondary growth in the vascular tissues; the cortex also undergoes an expansion due to cell proliferation and acts as an important storage area; if the phellogen were to arise in the pericycle, this large storage component would be lost. The phellogen must arise in the epidermis or outer cortex and must also be continuously active as long as the root continues to enlarge. Otherwise the periderm would tear and become ineffective.

Formation of Subsequent Phellogens

If the first phellogen forms superficially in the epidermis or outer cortex, then the parenchyma cells of the inner cortex may produce a second phellogen later. But as secondary growth of the vascular tissues begins and then continues interior to these phellogens, they and the primary tissues in which they are embedded are pushed outward and stretched tangentially. Ultimately the existing phellogens will have to be replaced by new ones located more deeply, actually in the secondary phloem. At this point in most plants, an equilibrium is reached, and new phellogens arise at more or less the same rate as new secondary phloem is produced, and the inner bark remains rather constant in thickness (Fig. 17.3). This is a very rough equilibrium, because harsh conditions that are severe enough to inhibit the activity of the vascular cambium and thus stop the production of new secondary phloem may encourage the production of new, deeper phellogens, thus causing the inner bark to become unusually thin. The initiation and subsequent activity of the phellogen is at least partially controlled by environmental conditions in many species (Borger and Kozlowski, 1972a, b; Mogensen, 1968b), as is logical when its protective function is considered.

The formation of a new phellogen and periderm interior to an existing periderm is detrimental to the outer periderm. The new, deep layers of cork cut off all outer cells from their sources of water and nutrients in the xylem and phloem; thus all tissues exterior to the innermost periderm die (Fig. 17.7). The dead tissues divide the bark into two regions. The living tissues of the innermost periderm and all cells interior to it constitute the **inner bark**, and the dead layers exterior to this are the **outer bark** or **rhytidome** (Fig. 17.3).

In a small number of species, extremely long-lived phellogens occur. In pear, the first phellogen can last for twenty years (Evert, 1963a); in *Ceratonia*, forty years (Arzee et al., 1977). In some species of *Anabasis, Fagus, Haloxylon* (Fahn, 1982), and *Pachycormus discolor* (Gibson, 1981), one single phellogen serves for the entire life of the tree. In many cases, only a single phellogen is produced simply because of the limited lifetime of the organ being protected. Excellent examples are the bud scales of winter buds; in these, it is usually the epidermis or the hypodermis that is converted to a phellogen.

Because all growth of the scales is complete by the time this phellogen has formed, there is no subsequent expansion that necessitates the production of new cork cambia. Likewise, what at first glance appears to be the epidermis of many fruits and vegetables (apples, potatoes) is actually a thin bark produced by a single, superficial phellogen.

In the most extreme water stress of drought conditions in deserts, portions of the vascular cambium and inner bark may die. In these cases, it is necessary for a phellogen to arise in the secondary xylem itself; otherwise the sapwood would be exposed (*Achillea fragmentissima, Artemesia*; Moss, 1940; Moss and Gorham, 1953). It is not uncommon for desert shrubs to have extremely dissected wood, as the vascular cylinder cracks and breaks up into strands; these strands too must produce their own individual barks by producing phellogens from xylem parenchyma.

Activity of the Phellogen

Once formed, the phellogen begins to cut off cells and is a lateral, secondary meristem. The tissues it produces are part of the secondary plant body, just like those derived from the vascular cambium. The cork cambium is, however, simpler than the vascular cambium because it always contains only a single type of cell and usually is unifacial, producing derivatives only to the exterior, not to both sides of itself. The phellogen cells are rectangular in cross section, being wider tangentially than radially, and in face view they are polygonal in shape, without intercellular spaces except in regions that produce lenticels. In a few species, the phellogen is bifacial and produces parenchyma cells on the inner side; these constitute the **phelloderm** and are usually produced in only small amounts (Fig. 17.8).

Phellogens typically are active only once: after their formation they produce phellem, and then they become inactive. This may be simultaneous with activity in the vascular cam-

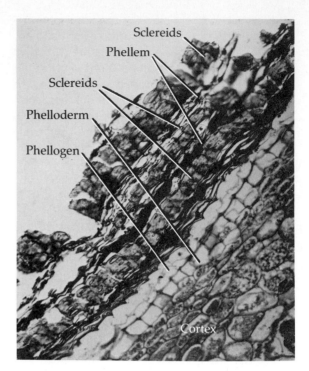

Figure 17.8
The bark on this root of *Leptocereus* is complex, containing bands of phellem and sclereids (see also Figs. 17.14 and 17.15). Toward the inside is a single layer of phelloderm. × 100.

bium, as in *Quercus ithaburensis* and *Q. infectoria* (Arzee et al., 1978), or it may not. However, some species have phellogens that are reactivated after the initial quiescence and thus have two (*Robinia pseudoacacia*) or even three (*Acacia raddiana*) periods of activity (Arzee et al., 1970; Waisel et al., 1967). The duration of activity and amount of phellem produced are also highly variable. In most species the phellogen is active only briefly (Liphschitz et al., 1984), and only a few layers are produced, but in *Quercus suber* (the commercial cork oak) each phellogen produces several millimeters of phellem (Fig. 17.4; Cooke, 1961). If a phellogen persists for several years, and if the underlying secondary xylem and phloem expand greatly, then the phellogen will have to grow circumferentially by occasional anticlinal divisions.

The phellogen can be organized in a large number of ways. It is almost always sheetlike, consisting of just one layer of cells oriented periclinally, but the size of an individual phellogen can vary from a complete cylinder that encircles an entire tree and extends for many meters up and down (Fig. 17.5), to just a tiny patch no more than one or two millimeters in diameter. Strips, bands, and irregular flakes are also possible. The nature of the phellogen (its size, shape, and activity) can vary from place to place within in a plant, especially on shoots. In *Celtis laevigata*, *Euonymus alatus*, and *Ulmus alata* (Fig. 17.9), although phellogens arise on all parts of the shoot surface, in certain strips they are very active and form large, corky wings (Bowen, 1963). In some species of the Bombacaceae, Rutaceae, and other families, the shoot is frequently covered with prominent spines. When phellogens form, those under spines are much more active than those not under spines; thus the spines become elevated on periderm bases and are even more threatening, as in *Hura crepitans* and *Zanthoxylum cubense* (Fig. 17.10). Conversely, the spines of *Euphorbia lactea* are the last areas to be cut off by bark, perhaps because they are associated with the axillary buds that must remain alive to produce branches.

Phellem

The phellem cells produced by the phellogen do not undergo any further mitotic activity, and often they grow very little. Because of this

Figure 17.9
The trunk of this small *Ulmus alata* (winged elm) has formed bark on all surfaces, but patches of cork are growing out most rapidly on the right side.

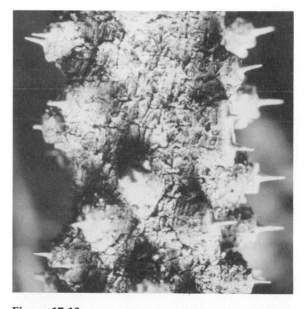

Figure 17.10
The spines on this trunk of *Zanthoxylum cubense* are raised on cushions of bark. This may help protect the tree from damage by its own spines: if an animal brushes against them, the spines cannot be pushed backward into the living phloem and vascular cambium.

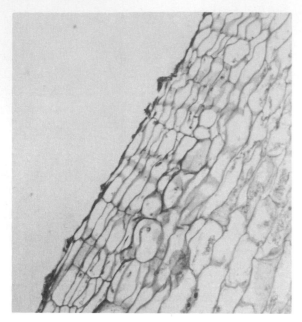

Figure 17.11
Potato bark (*Solanum tuberosum*) is simple, consisting of only phellem. It is probably the most commonly studied bark because it is pure cork, is readily obtainable, and is of tremendous economic importance. × 100.

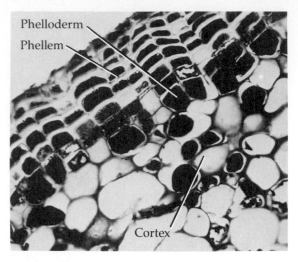

Figure 17.12
The bark on the aerial roots of *Clusia* also consists of phellem, but it is obviously quite different from potato bark (Fig. 17.11): the phellem cells fill with ergastic substances (perhaps tannins) before they die. These roots are long-lived, unlike the short-lived potato tuber, and being aerial roots they are exposed and readily visible to pests. It may be that the ergastic substances are an additional deterrent against insects. × 200.

manner of growth, they are almost always arranged in regular radial rows, and resemble the phellogen cells in being somewhat flat radially (Figs. 17.2, 17.8, 17.11, 17.12, 17.13). In face view they are polygonal like the cork cambium cells, because intrusive growth does not occur (Fig. 17.14). Because of the diversity of functions of the periderm, it is only logical to expect a variety of sizes, shapes, and wall chemistries (Cottle and Kolattukudy, 1982; Holloway, 1982c, 1983) among cork cells. Suberin is the characteristic component of the walls (up to 35% of the dry weight of the cell in *Malus pumila* cork), and it is the polymer that acts as the impervious barrier. Thus, in plants or organs that require maximum protection, the phellem cells may have very thick walls and can even become sclereids (Figs. 17.8, 17.15,

17.16). On the other hand, bud scales and fruit peels that often need just a short-term waterproofing or thermal insulation may contain phellem cells that are very thin-walled and may look rather like cortical parenchyma at first glance (Fig. 17.11).

In addition to these types, there are also cork cells that become filled with ergastic substances, tannins, resins, and very frequently crystals (Fig. 17.12). An unusual cork cell is one termed a **phelloid**; although it is produced by the cork cambium, it does not become suberized. The phelloids in *Viburnum opulus* develop walls rich in cellulose, pectin, and lignin, but suberin is absent (Wacowska and Tarkowska, 1983). In many plants, such as *Betula* and *Abies* (Mogensen, 1968a) layers of thick-walled phellem cells alternate with layers of thin-walled

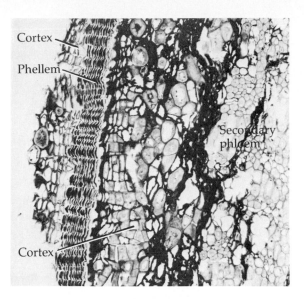

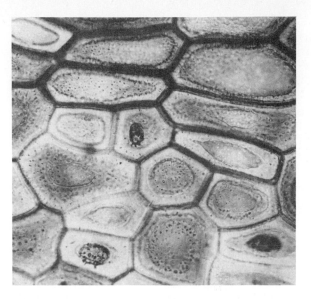

Figure 17.13

The phellem cells of this *Balanops* collapse as they mature; although there are about ten layers of cells, the entire band is less than 100 μm thick. This periderm resulted from a phellogen that arose in the cortex, and cortical cells are visible on both sides of the cork. × 100.

Figure 17.15

Face view of the sclereids in the bark of *Leptocereus*; these alternate with the cork cells (Figs. 17.8 and 17.14). Although phellem cells do not have pits, the sclereids do have them. × 150.

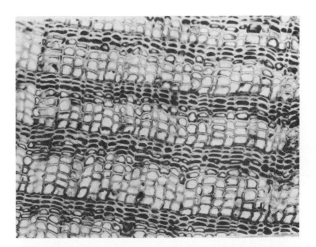

Figure 17.14

Face view of phellem cells from the bark of *Leptocereus*. The radial walls appear striated; actually what seems to be a single cell is really the entire column of cork cells, and the radial wall is the whole set of walls that are not perfectly aligned (side view in Fig. 17.8). × 150.

Figure 17.16

This periderm of *Betula* (birch) consists of alternating bands of thick-walled and thin-walled cells, although the wall thickness does not change so dramatically as it does in *Leptocereus*. × 100.

ones (Fig. 17.16). The most familiar example of phellem (the corks of wine bottles and tack boards) is not a truly representative example: most phellem is not so elastic, and it almost never occurs in such thick layers.

Considering the range of diversity among cork cells, it is possible to make only two generalizations about the differentiation of them. The first is that their primary walls are suberized (except for phelloids) and often contain a thick **suberin layer** interior to the primary wall, consisting of fine layers of suberin that alternate with layers of a substance often considered to be wax (Cottle and Kolattukudy, 1982; Knowles and Flore, 1983; Sitte, 1962; Soliday et al., 1979; Wattendorff, 1974). In the bark (skin) of potato tubers, there are ten layers in the suberin lamella (Schmidt and Schönherr, 1982). However, the alternate layers that are assumed to be wax do not dissolve when the potato phellem is soaked in wax solvents such as chloroform-methanol, and on this basis it was suggested that these layers are a special type of nonpolar suberin instead of wax. Furthermore, Vogt et al. (1983) found that, if soluble lipids are extracted from the periderm of potato, leaving the suberin in place, then the periderm lost its ability to stop water flow. They concluded that soluble lipids located in the suberin lamellae, rather than the suberin itself, were the main source of diffusion resistance. However, the periderm on potato tubers is very thin (Fig. 17.11); in a thick bark such as on birch trees (Fig. 17.16), the lipids are less important than the suberin (Schönherr and Ziegler, 1980). The suberization of cork cell walls makes it extremely resistant to chemical and enzymatic attack (Zimmermann and Seemueller, 1984). The suberization does not affect all wall layers in some species, and, if parts of the primary wall remain unsuberized, the periderm will be at least partially permeable to water (Schönherr, 1982; Schönherr and Ziegler, 1980). Interior to the suberin lamella may be another layer consisting primarily of cellulose. It has been reported that Casparian strips occur in the phellem cells of *Melaleuca* (Chiang, 1980).

The second generalization that can be made about the differentiation of cork cells is that, after the maturation of the wall, the protoplasts die: cork is always composed of dead tissues (Bowen, 1963; Mader, 1954). This is perhaps related to the fact that it is extremely difficult to maintain transport of nutrients and water to the outermost layers of phellem without many large pits or plasmodesmata in the walls of the inner layers. Such pits or plasmodesmata, of course, would greatly reduce the protective capacity of the cork, and at least in potato neither intercellular spaces nor plasmodesmata were found between phellem cells.

The cork cells function primarily as a barrier against invasion by pathogens and against water loss. However, the type of wall developed by phellem cells can act as an effective barrier for secretory spaces as well. In avocado fruits, the walls of the oil cells also have a suberin layer that presumably prevents leakage of the oil out of the cell (Platt-Aloia et al., 1983).

Phelloderm

Phelloderm is not produced by all phellogens, and when it does occur it is only in small amounts (Figs. 17.8, 17.12). One single layer is most common, and three or four are usually the maximum amount (Cucurbitaceae; Dittmer and Roser, 1963), whereas large quantities of cork may be produced to the outer side. The phelloderm, if it is unistratose, can be difficult to distinguish from the parenchyma that gave rise to the phellogen; but, if multistratose, its radial arrangement, with the cells aligned with the phellogen cells, makes it recognizable. In some cases the phelloderm can be an important tissue, containing chloroplasts and contributing to the photosynthetic capacity of the plant. In *Pachycormus discolor* (a xerophytic succulent tree), once the phellogen forms, it immediately begins to produce large quantities (more than 300 μm) of chlorenchymatous phelloderm (Gibson, 1981). Some of the phelloderm cells produce tannins, and others differentiate to produce schizogenous secretory ducts. In other species, the phelloderm stores starch (*Pinus palustris*) or becomes filled with ergastic substances; its cells may be converted to sclereids (*Libocedrus*; Chan, 1985; Den Outer, 1982). In *Fouquieria*, the phellem layers become extremely thick, and cells produced one year are radially aligned with those of previous years. Henrickson (1969) suggested that this arrangement might indicate that each year the subsequent phellogen is produced by division of the cells of the phelloderm of the previous year.

Texture of Bark

Because so many types of phellogen and phellem can occur, many types of bark occur on different trees. Furthermore, the textures of the tissues that exist between successive cork cambia also greatly affect the nature of the bark, especially if relatively little phellem is produced by each phellogen. When a plant forms successive phellogens, the bark produced by the first phellogen will usually contain epidermis and cortex, whereas subsequent phellogens will contain only secondary phloem. Thus the nature of the very first bark on a young branch is often dramatically different from the later barks on older branches, as with *Arbutus* (Fig. 17.17). Secretory tissues and crystalliferous cells are frequently part of the cortex and secondary phloem; consequently they automatically become part of the bark as phellogens arise interior to them (Ramos and Engelman, 1982). However, the phellogens also cut them off from the living phloem, causing them to die. If the cork cambia form as small patches, and if the secondary phloem contains few fibers, then the bark sloughs off as small flakes or shells, and the plant is said to have a scaly bark. These scales may be so fine as to be almost dustlike (Whitmore, 1962b), or they can be several square centimeters in size. In *Eucalyptus*, the scales fall away when thin-walled parenchyma cells break down (Chattaway,

Figure 17.17
The first bark of this *Arbutus xalapensis* (madrone) tree is visible on the right: it is flaky and sheds in large patches. The later bark is on the left; it is smooth and has a luxuriant, rich, red-grey color.

1953); because various substances are deposited in the parenchyma cells, the exfoliation reveals a patch of fresh, smooth bark that is usually bright green, red, or orange, but that later oxidizes to grey. The cork cambium of *Melaleuca* produces thick layers of phellem; the phloem fibers give this bark just enough texture to hang on the tree in large, shaggy strips.

If the cork cambium forms as a complete cylinder within a rather uniform secondary phloem, then a ring bark is formed that comes off in sheets; this can be found in *Cupressus*, *Lonicera*, *Clematis*, and *Vitis* (Fig. 17.5). In many of the Bombacaceae, especially in *Ceiba*, the bark peels off in such large, smooth, thin (actually transparent) sheets that it was used as writing paper by Mayans and Aztecs.

If the secondary phloem is sclerified, containing quantities of either fibers or sclereids (*Carya, Quercus*), the nature of the bark is affected. Sclerification often results in a much stronger bark that does not break off the tree easily, so a thick bark is built up. Because the thick bark is subjected to continued circumferential expansion from secondary growth interior to it, it is stretched tangentially and forms large, deep cracks and fissures (Fig. 17.18). If the phloem fibers are short, then the bark is shed as large chunks; if the fibers are long and abundant, then an extremely stringy bark results, as in *Juniperus*.

An interesting situation occurs in certain desert succulents: in the types that have very reduced or short-lived leaves (some cacti, euphorbias, *Cercidium floridum, Fouquieria splendens, Pachycormus discolor*), the stem is the main photosynthetic structure, but it requires more protection than an epidermis can supply (Böcher and Lyshede, 1968, 1972; Gibson, 1983; Lyshede, 1979). A bark is formed of pure phellem, and this bark is transparent (Gibson, 1981; Henrickson, 1969). When the stem becomes very old and photosynthesis ceases, a second, opaque bark may replace this first one.

The bark produced on temporary organs such as bud scales and fruits usually consists of phellem only and is not shed, because the entire organ is either abscised or eaten. Similarly, the periderm on the roots is typically different from that on the stem. Such differences can be related to two phenomena. First, the requirements for the secondary phloem may be unique in each organ, the stems perhaps requiring a more fibrous phloem, the root needing one with more storage capacity. The secondary phloem may constitute the dominant portion of the bark, so, if the phloem is distinct for the various organs, then the bark is distinct also. Second, each organ may have its own specific needs with regard to the bark. (Obviously the type of bark selectively advantageous for the trunk of an apple tree would not be advantageous on an apple fruit.) The bud scales will also have their own unique requirements.

In all plants, the nature of the bark is an important diagnostic tool and can be a key feature in plant identification (Whitmore, 1962a). Factors used in identification include texture, color, feel, and even odor; for example, *Pinus jeffreyi* is characterized by a fragrant, vanilla odor in its bark.

Figure 17.18
Bark of *Salix* (willow). It sheds slowly, so it accumulates to great thickness; it is not very elastic, so it breaks and fissures. The lenticels are not visible but are located at the deepest parts of the fissures.

Lenticels

As mentioned in the introduction, the periderm must also be capable of gas exchange to maintain the oxygen supply to the living interior tissues. The normal phellem cells, being suberized and without intercellular spaces, constitute a barrier that is gas impermeable. To circumvent this problem, regions of the phellogen produce cells (**complementary cells**) that are otherwise normal cork cells but become rounded as they differentiate, thus breaking some of the contacts with neighboring cells (Fig. 17.19). The broken contacts create a continuous, schizogenous intercellular air space. This region is a lenticel; because its cells are larger and more loosely packed than the surrounding nonlenticel cork cells, they form a protruding, spongy region in the bark. To be effective as an aerating system, the outer surface of the lenticel must be either at the surface of the plant or immediately beneath a lenticel in a more exterior periderm (Figs. 17.5, 17.6).

The phellogen cells located beneath the complementary cells also have intercellular spaces. Only a few species (*Anabasis*, *Campsis radicans*, *Haloxylon*, *Philadelphus*) are reported to have a periderm that lacks lenticels (Fahn, 1982).

The first lenticels form either simultaneously with or even shortly before the first periderm. If they form with it, then the phellogen contains two almost indistinguishable regions, one producing compact cork, the other producing complementary cells. Because the complementary cells occupy a larger volume, they push the overlying epidermis and any cortex cells outward and finally rupture them, thus permitting the entry of oxygen (Fig. 17.20). When the first lenticels arise before the first periderm, it is usually by the initiation of mitotic activity in the parenchyma below one or several stomata. The mitotic activity may occur in a rather large area, but, with time, the center of mitotic activity moves more deeply into the cortex, the orientation of the divisions becomes more precise, and a recognizable phellogen results. If the stomata are very rare, then the mitotic centers arise between them

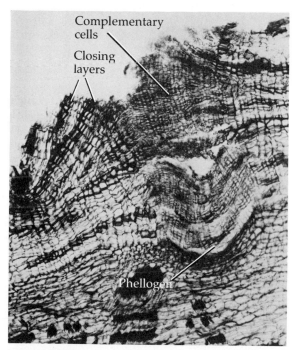

Figure 17.19
In this lenticel of *Crescentia*, the complementary cells are easily visible. Intercellular spaces are difficult to see, but the complementary cells in the lenticel are not so firmly interconnected as are the nearby phellem cells of the ordinary cork. The region of phellogen that is producing the lenticel is located deeper in the secondary phloem than is the rest of the phellogen. The layers of dark, compact cells are closing layers. × 100.

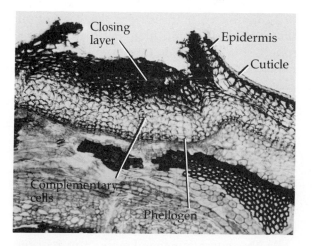

Figure 17.20
This lenticel of *Aristolochia* is smaller and thinner than that of *Crescentia* (Fig. 17.19), but the same parts are visible. The original epidermis and cuticle are still visible. × 100.

where only patches of ordinary epidermal cells occur. In roots, lenticels are usually associated with lateral roots, occurring in pairs on either side of the point where a lateral root emerges (Devaux, 1900).

Complementary cells can be suberized or nonsuberized, and they may be so loosely interconnected that they have an almost powder-like consistency. These cells can easily weather away. To hold them in place, the phellogen periodically produces a layer or two of cells that are smaller, more compact, and firmly interconnected. This is a **closing layer** (Fig. 17.19, 17.20), and it is capable of retaining the complementary cells beneath it; it has intercellular spaces that permit aeration. After the production of a closing layer, more complementary cells are produced. Ultimately the closing layers are ruptured, so their production must be cyclic. The complementary cells in some species (*Ginkgo*, *Sambucus nigra*, *Salix*) are interconnected firmly enough not to require the presence of closing layers.

In plants with smooth, thin bark, the lenticels can be quite obvious. Perhaps the most familiar example is an apple (the fruit, not the tree): the peel is really bark, and the small dots are lenticels. Another example is cherry (the tree, not the fruit): these trees have a beautiful smooth red bark interrupted by broad bands of rough, soft lenticels. In trees with thick, rough bark, the lenticels are undetectable without microscopy, but they are typically located at the bases of the cracks between individual plates or scales (Figs. 17.18, 17.21). Lenticels are usually positioned exterior to phloem rays (Wetmore, 1926), allowing a relatively easy gas exchange between the bottom of the lenticel and the more interior tissues. If lenticels were located between phloem rays, then the axial fibers of the phloem would constitute a barrier to gas movement.

As the circumference of a shoot or root increases with age, the lenticels may expand tangentially and become broad bands as the phellogen itself expands (Fig. 17.5). The tangential

Figure 17.21
As the trunk of this maple (*Acer*) increases, the bark becomes split by longitudinal furrows. At the center of each furrow is a white region where the lenticels are located.

growth of the cork cambium is a result of radial anticlinal divisions in the meristem cells. After the lenticel reaches a certain width, the center of the phellogen producing it may switch to producing normal phellem cells, so the lenticel is divided in two. If a lenticel remains very small, then its orientation is related to that of the phloem ray beneath: if the rays are tall, the lenticel is vertical; if the rays are wide or are aggregate rays, the lenticel may be horizontal. This, of course, is the logical arrangement for efficient aeration (Wetmore, 1926).

The Periderm of Monocots

Most monocots and all modern cryptogams undergo no secondary growth of any kind, yet many of them have bodies that persist for many years. In most of these, the epidermis is the only protective tissue, as in Cyperaceae (Metcalfe, 1971), Juncales (Cutler, 1969), and Marantaceae (Tomlinson, 1969a; Solereder and Meyer, 1928). To be effective for such a long time, the cells usually become sclerified.

Some monocots, such as many palms, become very large even though they do not have a vascular cambium, secondary xylem, or secondary phloem. Although very few of these have been studied (Floresta, 1905; Philipp, 1923; Tomlinson, 1961), two types of periderm have been found. The first is identical to that just described for dicots and conifers (Fig. 17.22). The second type is called **storied cork**; it is basically similar to a dicot periderm, just less regular. Bands of parenchyma cells become active as a phellogen, but they do so only

temporarily, dividing just a few times (Fig. 17.23). The daughter cells undergo little expansion, so the result is a band or bands that consist of packets of cells, each basically the shape of the original mother cell. The bands are not

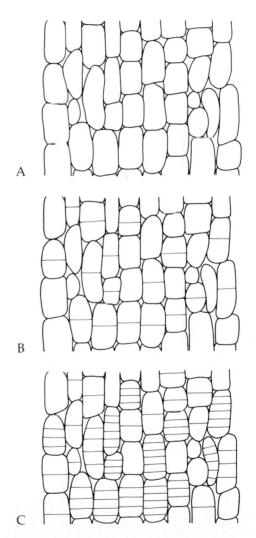

Figure 17.23
In the formation of storied cork, some of the cells in the outer cortex (A) undergo division (B). Not all cells divide, and some divide more rapidly than others, such that in (C) there are both some unaltered cells and packets of cells inside the original cell walls.

Figure 17.22
This may appear to be the trunk of a conifer such as pine or juniper, but it is a woody lily, *Yucca brevifolia* or Joshua tree.

uniform but rather may contain islands of parenchyma cells that have not undergone any divisions. All of the cells, subdivided or not, become suberized to form the storied cork, and some can develop into sclereids. This process may be repeated in deeper layers of parenchyma. Storied cork occurs in some Bromeliaceae, Commelinaceae, and Zingiberaceae (Tomlinson, 1969a). A periderm is reported to exist in Heliconiaceae, Lowiaceae, Musaceae, and Strelitziaceae, (Tomlinson, 1969a), but details about the formation or the nature of the bark were not given.

Some of the large arborescent monocots do undergo a type of secondary growth in their vascular tissues; this is described in the next chapter. These plants all form cork cambia and periderms that are similar to those of dicots and gymnosperms (Figs. 17.22, 17.24). The greatly enlarged bulbils of some dioscoreas are

reported to produce periderm by means of a phellogen (Murty and Purnima, 1983). Many dioscoreas also produce a huge, swollen tuber that has a thick, heavy bark (Ayensu, 1972). The cells for this bark are actually formed by a primary thickening meristem: the meristem produces large amounts of parenchyma to the exterior, and the parenchyma cells act as storied cork meristems. The result is a thick layer of storied cork (Martin and Ortiz, 1963).

Incidentally, in the fern *Helminthostachys ophioglossales* the outermost layers of cells of the rhizome stain with Sudan IV, and on this basis Goswami and Khandelwal (1980) stated that a periderm was present. However, the region seemed to have no other characteristics of periderm, and this stain is not specific for suberin. Ferns are mostly reported to have hardened cells in the outer cortex (Fig. 17.25), rather like most monocots.

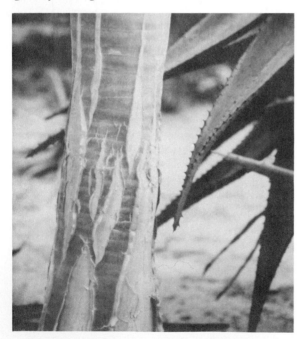

Figure 17.24
The bark of this arborescent monocot *Aloe dichotoma* is smooth and sheetlike, very different from that of Joshua tree (Fig. 17.22). Although bark is rare in monocots, it can be diverse just as in dicots.

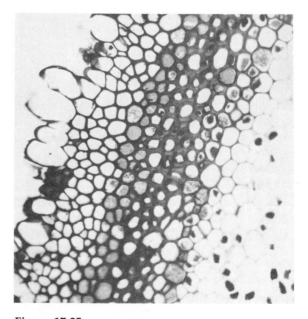

Figure 17.25
This rhizome of the fern *Matonia pectinata* shows the type of protective layers found in ferns: the remnants of epidermis and the outermost cortical cells have become thick walled and resistant, and interior to that layer is a layer of heavily lignified sclerenchyma cells. × 150.

Wound Periderm

Whenever a plant is wounded, the removal of the protective epidermis or periderm leaves the plant exposed to desiccation and attack by fungi, bacteria, and insects. In many cases, the initial damage is done by a successful fungal or bacterial invasion of the epidermis itself (Eskarous et al., 1984; de Leeuw, 1985). The wound must be sealed quickly and effectively if the plant is to survive. In most dicots and some monocots (Araceae, Liliales), the wound is closed by a two-step process: (1) the formation of a closing layer and (2) the subsequent formation of a wound periderm.

The **closing layer** is formed as the broken cells die and the adjacent cells react by depositing suberin and lignin in their walls (El Hadidi, 1969; Lipetz, 1970; Moore, 1982, 1984). The closing layer provides immediate, temporary protection. Just interior to it, other living cells are converted to a new phellogen, and this initiates the production of phellem, forming the **wound periderm** (Cline and Neely, 1983). This periderm isolates and kills the closing layer, which may soon peel away and fall off. The cork cells of a wound periderm are usually similar to those of an ordinary periderm, but they may differ in significant ways: in *Ipomoea batatas*, the cells of wound cork have only 14% as much suberin as the cork of the regular bark (Walter and Schadel, 1983). This process can occur in even very old tissues; Fisher (1981) reported that, when large branches (16 to 21 years old) were cut off trees of *Adansonia digitata*, the xylem parenchyma and even pith began to proliferate, forming a callus across the entire cut. The surface of this was covered by a wound periderm. The reactivation and redifferentiation of 21-year-old xylem parenchyma seem remarkable.

Many monocots and a small number of dicots (Jones and Lord, 1982) do not form a wound periderm; they may produce a thin, suberized (Zingiberales) or lignified (Palmae and Poaceae) closing layer instead. In onion, the suberin that is deposited in these cells is not layered as it is in ordinary phellem cells (Moon et al., 1984); instead it is amorphous and can even pass through the wall and occlude the intercellular spaces.

18
Anomalous Secondary Growth

Concepts

The type of vascular cambium and secondary growth described in the last three chapters occurs in all the gymnosperms and almost all the woody dicots. It is so ubiquitous and so unvarying that we tend to forget that many other types of secondary growth are at least theoretically possible. Furthermore, although the secondary body produced by this manner of growth has certainly shown itself to be both highly adaptable and functional, in some cases a different arrangement of tissues might be more advantageous.

Actually, several fossil groups and some extant genera do have vascular cambia and secondary growth that differ significantly from what can be termed the **common type**. In some cases, the differences are rather modest, but frequently the differences are actually startlingly exotic (Metcalfe, 1983b). It is possible to classify the types of growth in several ways, but probably the most useful and most natural classification is to divide them into (1) types that have evolved directly from the common type of vascular cambium and (2) types that represent new cambia that have evolved de novo in herbaceous ancestors (Carlquist, 1981). Although the plants with anomalous secondary growth are often studied primarily for their unusual nature (that is, they are mostly considered to be novelties), in reality they present us with the opportunity to analyze our

ideas about cambia and secondary growth. They can help us to distinguish between features that are merely common and those that are essential.

Modifications of the Common Type of Vascular Cambium

Death of the Cambium

Probably the most easily understood type of unusual growth (and the least unusual type) is that in which strips of the cambium die while adjacent areas remain alive and functional. This is exemplified by *Zygophyllum*, *Fumana thymifolia* (Fig. 18.1), and *Artemisia tridentata* (Ginzburg, 1963). These are plants of desert regions, and they grow in poor, rocky soil. During times of severe water stress, water conduction apparently stops in narrow strips of wood. This causes the adjacent vascular cambium and inner bark to die, while the neighboring regions that are still hydrated remain alive. The living regions act as though they surround a wound (which, in effect, they do) and so gradually become curved on the edges, forming more acute arcs of secondary tissues. These bands of cambium that are isolated between two strips of dead cambium can remain alive and active for years. Eventually parts of them will die also, splitting them into more cambia, which continue to live. The result is an ex-

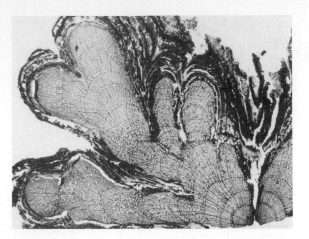

Figure 18.1
Cross section of *Fumana thymifolia* stem, showing
the split and irregular wood. At numerous points
narrow strips of cambium have died, and the adjacent regions continue as normal. The surviving
cambial areas are more sharply curved than they
would be in a normal stem of the same diameter;
this is probably due to normal growth's being slowed
along the exposed edges. × 50.

tremely rough, irregular wood. This type of
anomalous secondary growth appears to be
not so much a modification of the cambium as
simply survival under adverse conditions.

Unequal Activity of the Vascular Cambium

In several genera, the vascular cambium does
not remain uniformly active around its entire
circumference. Instead, certain regions produce tissues that are quite different from those
produced by adjacent ones. Unfortunately, we
do not know if this nonuniform growth is at all
related to the cambium or if it is due to morphogenetic factors that act on the cells after they
have been derived from the cambium. Plants
with this type of nonuniform growth present
excellent opportunities for us to test our ideas
about developmental control mechanisms.

The first example is *Aristolochia*. In plants of
this genus, regions of the cambium act like the
common type, but adjacent regions produce
pure parenchyma to both the inside and outside. It is not unusual for the interfascicular
vascular cambium to do this in many young
plants, but in this type of anomalous growth it
continues indefinitely. Furthermore, regions
that produced conducting tissues can switch to
producing just parenchyma. In effect, these
parenchymatous regions are giant rays, so this
type of growth might be based on a modification of the mechanisms that control ray formation. The presence of such large amounts of
parenchyma causes the stems (in this case,
vines) to be weak, and they often split through
these areas. A normal wound periderm then
forms and seals off the broken surfaces.

Ambrosia dumosa and *Passiflora* are the second examples of cambia with unequal activities (Fig. 18.2). In these plants, all of the cambium produces conducting xylem to the inside
and conducting phloem to the outside, but the
relative proportions vary from region to region (Ayensu and Stern, 1964; Jones and Lord,
1982). In some areas, large amounts of secondary xylem are produced, pushing the vascular
cambium rapidly outward; in adjacent regions,
very much more phloem than xylem is produced, so the cambium is pushed out slowly.
The result is a central mass of wood with an extremely irregular outline, with the cambium
following that outline. It should be remembered that root cambia in all plants have unequal activity initially, until the root becomes
round in cross section (Fig. 14.18). A cambium
such as is found here should make us more
fully appreciate the fact that it is the common
type of cambium that is remarkable for its uniform growth. Even in very large trees, the cambial cells on one side of the trunk usually grow
at the same rate as those on the other side,
thousands of cells away from each other. Unequal growth does seem to have adaptive advantages under certain conditions: in *Ambrosia
dumosa*, it results in the splitting of the stem

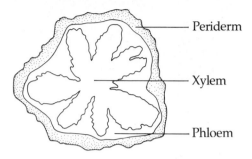

Figure 18.2
Irregular pattern of secondary xylem and phloem produced in some *Ambrosia* and *Passiflora* (passionflower) species. As certain regions of the cambium produce large amounts of xylem, they are pushed outward more rapidly than those regions producing little xylem. All parts of the vascular cambium produce roughly the same amount of tissue, and the stem is roughly circular in cross section.

into several strands, each with a number of branches. Jones (1984) found that the splitting of the stem allowed the plant to channel available water to dominant branches. There was not enough water for the entire crown, and, without splitting, all of the shoot suffered water stress equally. But with anomalous splitting, although certain branches died, others received enough water to enable them to survive.

This unequal growth happens irregularly in *Passiflora*, which causes us to infer that it might be more a nonlethal defect rather than an advantageous adaptation, but in *Bauhinia divaricata*, *B. sericella*, and *Prestonia macrocarpa* (Fig. 18.3), this type of growth occurs in a wonderfully regular pattern (Wagner, 1946). The primary growth is slightly unusual in being cruciate in form, but a normal cambium arises. By unequal growth, it produces a round shoot, just as a regular cambium does in roots. Then on two opposite sides, the cambium becomes less active, and on the alternate sides it becomes much more active. In addition, the growth within the xylem produced by the alternating sides is different: in the slowly grow-

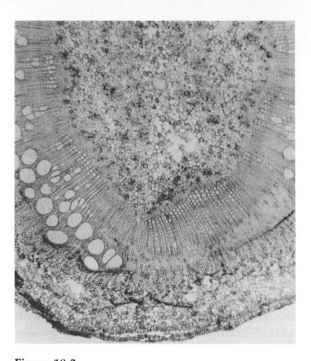

Figure 18.3
This photo was taken just after the beginning of secondary growth in *Prestonia macrocarpa*. Initially the secondary xylem consisted of rather uniform, small tracheary elements all around the pith; but now on two sides (the left is visible here) large vessels are forming, whereas on the other two sides (the bottom here) no large vessels are developing. The stem will quickly become flat and ribbonlike. × 50. See Figure 18.4 for a diagram of a mature stem of this type.

ing regions, all cells remain very small, but in the rapidly growing areas, there is a preponderance of vessels with large diameters. The result is a shoot that is flat and ribbonlike (Fig. 18.4). It is important to stop and analyze this process; it can teach us something about normal secondary growth. In the common type of secondary growth, as a stem ages it increases its diameter, resulting in both increased conductivity and increased rigidity. For many plants, both are advantageous. But these bauhinias are vines, and the ability to flex in the wind is

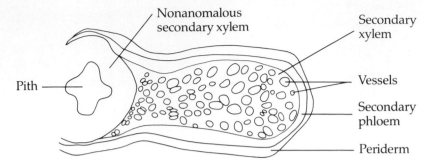

Figure 18.4
This is a cross section of a *Bauhinia* stem; it shows the result of the type of anomalous growth in which two areas of cambium produce xylem consisting mostly of large-diameter vessels, while alternate regions produce a more compact xylem. Based on a micrograph by Wagner (1946).

more advantageous than rigidity. By growing as ribbons that become wider, bauhinias can both increase conductivity and maintain their flexibility.

Anomalous Placement of the Vascular Cambium

Apparently almost any parenchyma cell can redifferentiate into a meristematic cell: parenchyma cells give rise to vascular cambia, cork cambia, dilatation tissue, and adventitious buds. In this regard, it is interesting that vascular cambia typically arise in only very precise, predictable locations. The mechanism responsible for the proper placement of vascular cambia, whatever its nature, is altered in *Ipomoea*, *Paullinia sorbilis* (Fig. 18.5) and *Serjania* (Pfeiffer, 1926). In *Ipomoea* secondary growth begins as the common type, but then additional cambia arise, each forming around an individual vascular bundle or around small groups of them. Each cambium then acts rather normally, producing xylem internally and phloem externally, although both xylem and phloem have large amounts of parenchyma (Fig. 18.6). In the secondary xylem, vessels are clustered, and more cambia can arise around them in the paratracheal parenchyma; thus this process can

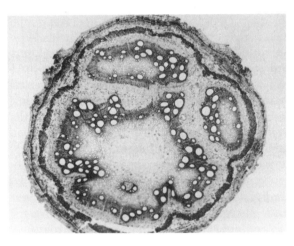

Figure 18.5
This cross section of *Paullinia sorbilis* shows that three complete cambia have arisen instead of one. Each is functioning normally, producing both xylem and phloem. × 50.

continue indefinitely. In the shoots, phellogens produce periderm in the secondary phloem. Since this phloem is produced as rings around bundles or vessel clusters, the bark that forms is also arranged as numerous rings. This causes the stem to break into numerous anastomosing, interwoven components, and it can have the appearance of many small stems that are in-

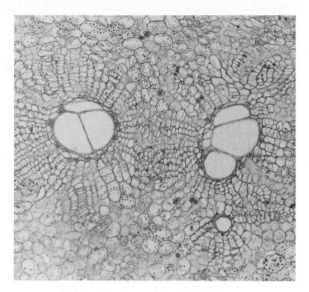

Figure 18.6
The roots of *Ipomoea batatas* (sweet potato) can produce numerous vascular cambia, most of which are centered on clusters of vessels that were produced by a previous cambium. Notice that the tissues produced by the vascular cambia are almost pure parenchyma. × 100.

tertwined. In the large storage root of *I. batatas* (sweet potato), bark does not form.

In several genera of Valerianaceae that grow high in the Andes (*Aretiastrum, Belonanthus, Phyllactis, Stangea*), fascicular cambia arise; but, instead of becoming interconnected by interfascicular cambia, each bundle of primary xylem becomes encircled as the cambia extend toward the pith instead of the adjacent bundles (Lörcher and Weberling, 1982; Weberling, 1982). These cambia then produce irregular amounts of xylem and phloem on all sides, and the bundle becomes stellate in cross section.

Successive Cambia

The common type of cambium is said to be **bidirectional**, because it produces secondary xylem to one side and secondary phloem to the other, and **continuous**, because it continues to function for the entire life of the plant. In the spring, the cambium is reactivated; it is not replaced by a new cambium. In the following plants, the cambia are **successive** because they function for a limited time (they are determinate) and then are replaced by totally new cells that differentiate from existing parenchyma.

The simplest examples to understand occur in some genera of the Amaranthaceae, Chenopodiaceae, Convolvulaceae, Menispermaceae, and Nyctaginaceae (Artschwager, 1926; Bailey, 1980; Esau and Cheadle, 1969; Fahn and Shchori, 1967; Iljin, 1950; Mennega, 1969; Miksell, 1979; Stevenson and Popham, 1973; Wheat, 1977; Zamski, 1979). In these plants, a cambium arises and functions like one of the common type. After producing normal secondary tissues, the cambium becomes inactive, and its cells differentiate, completely ceasing to exist as a cambium (Fig. 18.7). However, in the outer-

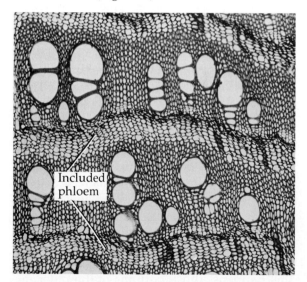

Included phloem

Figure 18.7
This photograph of *Iresine argentata* wood shows four bands of secondary xylem alternating with three bands of included phloem. One band of xylem and the phloem above it were produced by one vascular cambium; then the cambium stopped producing and a new cambium formed in the uppermost layers of that band of phloem. × 200.

most regions of the newly produced secondary phloem, a new cambium is organized. You would expect it to be a phellogen, but instead it is a normal, bidirectional vascular cambium. It too functions normally for a limited time, but then it stops and is replaced by another cambium that forms in the phloem parenchyma that it had just produced. Because the new cambia arise exterior to the majority of the secondary phloem of the previous cambium, they produce secondary xylem exterior to it, and the stem is constructed of alternating concentric rings of xylem and phloem. Any secondary phloem that is located interior to secondary xylem is said to be **included phloem**. Zamski and Azenkot (1981) found that the vascular elements produced by one cambium could have extensive radial and tangential connections with those produced by previous cambia.

Included phloem is produced in *Bougainvillea* by a process that is somewhat different from that just described (Esau and Cheadle, 1969). The vascular cambium arises normally in the root, but in the shoot it is located completely exterior to the bundles. It contains only storied fusiform initials; ray initials are absent. As the cambium begins to function, it first produces mostly just vesselless wood to the interior; then it begins to produce increased quantities of phloem to the exterior (Fig. 18.8). This phloem is largely nonconducting parenchyma. The cambium then changes activity again and produces secondary xylem that is rich in vessels. Simultaneously new cambia arise as isolated, narrow, vertical strips in the outermost parenchyma of the phloem; the new cambia begin to produce vesselless xylem. As this happens, those portions of the older, inner cambium that are located interior to these newer cambia become inactive, then differentiate. Those portions that do not become inactive continue to produce secondary xylem with vessels, so they are pushed rapidly outward; when they come into contact with the new cambial strips, they join with them to form a

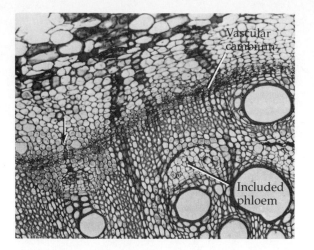

Figure 18.8
This micrograph shows the vascular cambium, secondary xylem, and included phloem of *Bougainvillea*. The young bundle of phloem at the arrow appears to have just been produced by the vascular cambium at the arrow. However, it actually was formed to the outer side of the cambium when the cambium was back at the outer edge of the vessel group. The vascular cambium just outside the bundle of phloem is just becoming organized from the phloem parenchyma; this portion of cambium is new, while the cambial regions on either side are much older. × 300.

complete cambium again (Fig. 18.8). This condition of completeness is only temporary, because the entire process is repeated. The result is that there appears to be one unitary mass of wood with isolated patches of phloem in it, but actually the wood exterior to each phloem region is distinct ontogenetically from the wood located on the other sides of the phloem. These plants have bidirectional successive cambia, but the cambium is not replaced all at once as it was in the previous examples.

A similar type of anomalous growth with included phloem occurs in *Stegnosperma* (Phytolaccaceae); successive cambia arise and produce xylem with vessels to the interior and bundles of phloem to the exterior (Horak,

1981a, b). Horak analyzed the interconnections between the vascular elements and found that the phloem of two included columns could anastomose with each other without the associated vessels' undergoing fusion. Vessels can similarly approach each other and fuse independently of the phloem. If a dye is injected into a stem, it rapidly moves into tangentially adjacent bundles. It can also move radially, although such connections are less common and the movement is slower. The functional lifetime of each set of elements is not known, but Horak (1981b) reported that the phloem in the four outermost increments appeared to be healthy, whereas the phloem in deeper layers had collapsed. It is instructive that the set of phloem bundles are able to fuse and anastomose with each other so independently of the system of vessels in the xylem. The primary body of all seed plants contains collateral bundles; this universal phenomenon would at first appear to be strong circumstantial evidence that the control of morphogenesis in the one type of vascular tissue is inextricably linked to the control of the other, but in *Stegnosperma* they are not.

An anomalous type of bidirectional continuous cambium occurs in *Azima tetracantha* (Den Outer and van Veenendaal, 1981), *Phytolacca* (Miksell, 1979), *Salvadora* (Singh, 1944) and *Thunbergia*. The primary growth is normal, and a normal vascular cambium arises, producing secondary xylem and small amounts of secondary phloem. But soon the outline of the xylem appears wavy as certain regions of the cambium begin to produce mostly parenchyma to the inside, while adjacent regions of the cambium do not. These patches of parenchyma are called **conjunctive tissue** and do not extend very far radially, because the cambium reverts to normal and produces secondary xylem, and the patch (actually a vertical rod)becomes included. Within the conjunctive tissues, small areas become mitotically active and differentiate into phloem. Several phloem bundles can occur within each patch of conjunctive tissue. Occasionally, a weak cambium

can be formed within the conjunctive tissue, but it produces only a small amount of parenchyma. Kirchoff and Fahn (1984) studied the individual traces of *Phytolacca dioica* and found that there are interconnections within a cylinder and between cylinders. Three types of connections were reported: (1) those consisting of whole vascular strands running from one cylinder to another, (2) connections made up of one or a few vessels, accompanied by a small amount of phloem, acting as a bridge, and (3) connections consisting of fibers only. Furthermore, Kirchoff and Fahn concluded that there is a correlation between leaf arrangement and the initiation of patches of included phloem.

In *Azima tetracantha*, Den Outer and van Veenendaal (1981) discovered a remarkable horizontal interconnection between rods of the included phloem and the external phloem: the central regions of the multiseriate xylem rays differentiate into sieve tube members and companion cells (Fig. 18.9). These plants have horizontally oriented phloem bundles that run through the xylem ray parenchyma. To my knowledge, this has not been reported for any other species, yet it seems like an excellent adaptation for transport through long rays, and it is rather surprising that other plants with thick sapwood do not have sieve elements in their xylem rays.

It is important to emphasize that the cambia described in this section are considered to be bidirectional and that at no time do they produce phloem to the interior. In *Salvadora* and *Thunbergia*, the phloem is produced as a result of redifferentiation of parenchyma that is produced by the cambium. There have been other interpretations that the cambia are unidirectional, that they actually produce phloem to the interior, and that they are continuous in that they do not stop and are not replaced— rather the cambium moves rapidly outward as a "wave of mitotic activity" until it reaches the outer phloem parenchyma (Balfour, 1965; Studholme and Philipson, 1966). However,

Figure 18.9
This radial, longitudinal section of *Azima tetracantha* shows two phloem strands passing horizontally through xylem rays and connecting with a vertical strand of included phloem. × 100. Micrograph generously provided by R. W. Den Outer and W. L. H. van Veenendaal.

there is no undisputed evidence for this interpretation, and all cases, when examined in close detail, have been interpreted as being bidirectional and successive.

Discontinuous, Unidirectional Cambia

In some members of the Bignoniaceae such as *Amphilophium paniculatum*, *Clytostoma* (Dobbins, 1969, 1970, 1971, 1981), *Petastoma patelliferum*, and *Pyrostegia venusta*, a normal cambium forms and functions until a thin ring of normal secondary xylem and secondary phloem has been formed. Then four small patches of cambium, located opposite each other, become

unidirectional and produce only phloem (Fig. 18.10). Because the patches of cambium are not producing any derivatives to the interior, they become stationary within the plant and push out narrow wings of secondary phloem. The cambial strips that alternate with these continue as bidirectional meristems and move outward as they deposit xylem. After a short time, the four unidirectional cambia are deep within the wood that is deposited by the bidirectional cambia. The secondary phloem that the unidirectional cambia produce must slide past the stationary wood on its sides. There will ob-

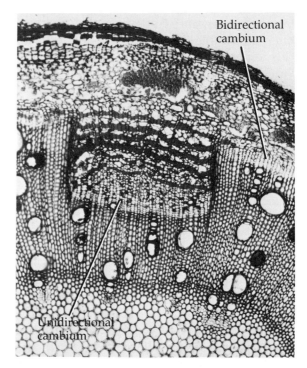

Figure 18.10
This is a young stem of *Paragonia pyramidata*; the central region of the vascular cambium has changed from bidirectional to unidirectional. Now it produces only secondary phloem to the exterior side and nothing to the interior, so it is not moving outward. The adjacent positions of cambium are still bidirectional and are being pushed outward by the deposition and maturation of xylem on their inner side. × 100.

viously be eight planes of weakness in this shoot, but consider that your own fingernails slide by the tissues beneath without being detrimentally weak.

After the shoot grows like this for a period, other portions of the bidirectional cambia convert to being unidirectional, and the process is repeated (Fig. 18.11). The new unidirectional cambia often arise adjacent to the wings of sliding phloem and are often paired on either side of it, but regions in the middle of the bidirectional cambia can also convert to unidirectional cambia. In large stems, after this process has occurred many times, there are numerous unidirectional cambia located at many sites and at many depths in the wood, each producing and pushing out its own wedge of secondary phloem.

New Cambia

Secondary Growth in Monocots

Almost all of the monocots lack secondary growth—apparently they evolved out of an herbaceous ancestor—but a few have developed the ability to produce a large, woody secondary body. This may have occurred several times, because some of the woody monocots are not at all closely related and are apparently separated from each other by herbaceous ancestors. Examples of Agavaceae include *Cordyline*, *Dasylirion*, *Dracaena*, *Sansevieria*, and *Yucca* (Fig. 18.12); of Dioscoriaceae, *Dioscorea* and *Testudinaria*; of Liliaceae, *Aloe arborescens* and *A. bainesii*; and of Xanthorrhoeaceae, *Kingia*, *Lamandra*, and *Xanthorrhoea* (Cheadle, 1937; Chouard, 1937; Fahn, 1954; Philipson et al., 1971). Unfortunately, these plants do not produce a wood of commercial value, so very little attention has been given to any of them, and absolutely no work has been done on some. As

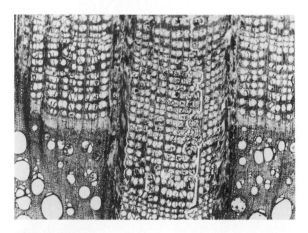

Figure 18.11
This is a portion of a stem of *Petastoma patelliferum* that was older than that in Figure 18.10. Two more regions have become unidirectional. In the entire cross section of this stem, there are four areas like this. × 150.

Figure 18.12
This is a tree of *Yucca brevifolia* (Joshua tree), one of the few arborescent monocots.

is true of the dicots with anomalous secondary growth, these unusual organizations of cambia, xylem, and phloem offer us excellent opportunities to test many of our theories of water conduction, resistance to stress and torsion, and morphogenic induction.

Although details vary from species to species, in general the vascular cambia of these plants develop as broad bands of meristematic activity just exterior to the outermost fibrovascular bundles of the shoot (Fig. 18.13). The meristem per se is not established immediately; rather, mitotic activity occurs throughout the region. When the cambium finally does become distinct as the surrounding tissues become quiescent, it may be located either centrally in the band or in the band's exterior. If located centrally, then the newly produced parenchyma on the outside is called **secondary cortex**.

In cross-sectional view the cambium has the appearance of a cambium of the common type; however, in tangential view its cells are different. In some species, they are fusiform, but they can be rectangular or half-fusiform with one end tapered and the other truncated. Ray initials are not present. The cambial cells divide by periclinal walls and produce thin-walled cells both to the inside and to the outside. The outer ones differentiate into parenchyma and contribute to the secondary cortex; the majority of the inner ones also mature into parenchyma and form conjunctive tissue. The conjunctive parenchyma may develop thick, lignified walls (Figs. 18.13, 18.14). Isolated regions (actually vertical files) of the conjunctive parenchyma cells resume mitosis and form rosettes of small cells that differentiate into fibrovascular bundles containing both secondary xylem and secondary phloem. The outermost cells of these rosettes differentiate into fibers. The cells destined to become tracheary elements undergo intrusive growth, becoming as much as 15 to 40 times longer than the cambial cells. Those cells that mature into phloem and nonconducting parenchyma do not elongate. In all species studied at present, the xylem in these **secondary bundles** contains only tracheids, not vessel elements, even though the primary bundles (of the primary body) do have large metaxylem vessels.

The result of this type of cambium is a shoot composed of a central core of primary bundles and conjunctive tissue, surrounded by a secondary body of conjunctive tissue in which are embedded secondary bundles. Exterior to this is the vascular cambium, the secondary cortex, and the primary cortex (Fig. 18.13). The epidermis usually does not persist; it is replaced by a periderm similar to that found in dicots.

Obviously, as water moves upward through

Figure 18.13
The vascular cambium of this *Cordyline terminalis* is part of a broad cambial zone. To the exterior is the secondary cortex; to the interior are the conjunctive tissue and secondary vascular bundles. At the extreme top is the original cortex. × 150.

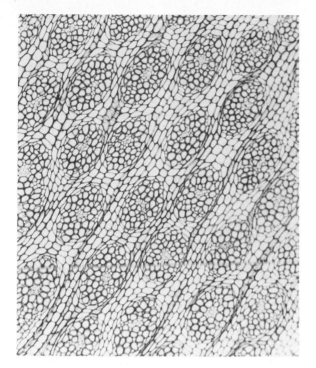

Figure 18.14
These are the secondary bundles of *Cordyline termi-nalis* embedded in conjunctive tissue, which is lig-nified. Each bundle contains just a small amount of centrally located phloem. × 150.

these secondary bundles, it must move inward to ever-younger bundles and finally move into the primary ones that are connected to the leaf and inflorescence traces. The interconnections between these bundles must be extensive, but they remain largely unstudied (Zimmermann and Tomlinson, 1970).

There has been much new work in this area, and many other examples could be discussed, but that is not necessary. You might consult the review article by Metcalfe (1983b) and the fol-lowing papers: Bailey (1980), Baird and Black-well (1980), Fahn and Zimmermann (1982), and Zamski (1979). Always try to analyze the con-sequences of the anomalous growth and con-sider what it signifies about normal secondary growth.

PART VI

Tissues and Organs of Sexual Reproduction

The various types of tissues and organs described previously in this book are vegetative tissues and organs; that is, they participate in the growth, development, and survival of the individual plant of which they are a part. Many plants are capable of asexual reproduction by means of stolons, by the production of adventitious buds on roots, or by the formation of plantlets on their leaves. Stolons, adventitious buds, and plantlets are also vegetative tissues, because each is basically identical to the tissues already discussed. Furthermore, none of the methods of asexual reproduction involves the exchange of genetic material; the new plants are genetically identical to the parental plant.

During sexual reproduction, three processes must occur: (1) the production of diverse gametes, (2) the fusion of genetically dissimilar gametes, and (3) the liberation of a new individual. A single diploid organism is able to produce numerous types of gametes by the processes of chromosomal pairing and crossing-over that occur during meiosis; cells capable of undergoing meiosis occur only in tissues concerned with sexual reproduction.

The second process, the fusion of gametes, requires that gametes be transferable from one plant to another. In the algae, transfer is most often accomplished by releasing the gametes into water currents. On land, release into water currents is not possible, and numerous mechanisms have evolved that accomplish transfer. In the simplest mechanism, the plants produce wind-borne spores that, upon germination, grow into new plants that produce gametes. Although individual gametes then travel only a very short distance, they effectively are involved in long-distance sexual reproduction. In a few gymnosperms and many angiosperms, the spore-bearing organs are capable of attracting insects that then transport the spores (pollen) from one plant to another. In other angiosperms, other animals may be involved: birds, bats, and even some rodents. The presence of diverse types of pollinators has resulted in diverse selection pressures: mutations that might be advantageous for wind pollination are usually different from those that are advantageous for insect pollination. As a consequence, numerous types of cones and flowers have evolved.

The third process, liberation of the new individuals, is simple in the cryptogams: the new individual grows directly to maturity wherever it was established. However, in the seed plants, the new individual is formed inside the tissues of one parent that is in turn still attached to its parent. The new individual is nourished and protected by its parent and grandparent, enclosed in a structure that becomes a seed. The seed can also be distributed long distances, which can be quite advantageous, because the new individuals will not then compete with their grandparents. Because a plant produces hundreds or even millions of seeds during its lifetime, long-distance dispersal permits the new individuals access to a large area in which they can grow. In many angiosperms, fruits have become the primary means of seed dispersal.

In the tissues and organs responsible for sexual reproduction, requirements are similar to those of vegetative tissues: the distribution of water, minerals, and sugars; the conservation of water; and gas exchange. But the reproductive organs have many different functions, so it is to be expected that aspects of their structure will be different. As always, two of the most important means of analyzing and understanding the structure are to consider carefully the biological role of that particular structure and to consider its ancestry.

19
Flower and Strobilus

Concepts

In order to understand the nature of flowers in angiosperms and the nature of strobili (cones) in conifers, it is essential to understand the nature of sexual reproduction in these groups. As in almost all plants, the life cycle involves two phases or generations: the diploid **sporophyte** and the haploid **gametophyte**. Sporophytes are the plants that are familiar to us as pine trees, oak trees, and ferns; as their name implies, they produce **spores** by a process of sporogenesis. The essential feature of sporogenesis is the meiotic reduction from being diploid (with two sets of chromosomes per nucleus) to being haploid (with one set per nucleus). In some algae, all the cells of the body undergo meiosis, but in the vascular plants only small groups of cells do, and these are located in the flowers and strobili. The spores are capable of germinating and growing, by mitosis, into the gametophyte generation. The gametophytes are plants, and they certainly are pines, oaks, and ferns. Containing only a few cells, however, they could not possibly be called trees or bushes. They look totally different from the sporophytes of the same species, so we say that there is an **alternation of heteromorphic generations**. It is possible for sporophytes and gametophytes to resemble each other (an alternation of **isomorphic** generations), but that occurs only in some algae, never in any land plant (Bold et al.,

1980). In the cryptogams (nonseed plants), the spores are released to the environment, and the gametophytes are free-living, independent plants. But in the seed plants (gymnosperms and angiosperms), they are retained on the parental sporophyte, and they grow heterotrophically, absorbing nutrients supplied by the sporophyte; they are never capable of an independent life. The gametophytes produce the gametes by **gametogenesis**, a mitotic process. As in animals, the gametes are oogamous; that is, the sperm cells (**microgametes**) are strikingly different from the egg cells (**megagametes**). When the sperm fertilizes the egg, the diploid condition is restored, and the cell is now a **zygote** that will grow into an embryo. The embryo is retained within the gametophyte, which is still embedded in the parental flower. In the seed plants, the embryo is encased in several special layers of cells, and the entire structure is a seed. Only when the seeds have been shed and have germinated is the embryo free of the preceding sporophyte generation.

In all the seed plants, the sperm and eggs are each produced on separate gametophytes, the male or **microgametophyte** for the sperm, the female or **megagametophyte** for the egg. These gametophytes in turn are produced from separate types of spores (**microspores** and **megaspores** respectively); the plants are said to be heterosporous. Commonly, both types of spores are produced by the same plant, in separate **microsporangia** and **megasporangia**. In dioecious species (*Ginkgo, Juniperus,* figs, marijuana), however, the microsporangia occur on one sporophyte, the megasporangia on another (often called male plants and female plants). Therefore, to analyze sexual reproduction, we need to consider two types of sporophytes, sporangia, spores, gametophytes, gametangia, and gametes and one zygote.

Secondary Sex Characteristics of the Vegetative Body

Sporophytes technically do not have a "sex" or "gender"; only the gametophytes do, and in all seed plants the microgametophytes are so different from the megagametophytes that the two types could never be confused. However, as many as one-fifth of all seed plant species are dioecious (Lloyd and Webb, 1977), with each individual having only staminate or only carpellate flowers. At the time of flowering or fruiting, the staminate plants can always be distinguished from the carpellate ones. When you think about it, though, the one type of individual has problems different from those of the other. Once pollen is released from the staminate flower, that plant's participation in sexual reproduction is completed. But for the carpellate plant, after pollen is received, the processes of fertilization, seed formation, fruit maturation, and dispersal are all still ahead. Although the new sporophyte embryo is produced by the megagametophyte, it is the carpellate sporophyte that must supply all the energy, because the megagametophyte is growing inside it. A carpellate plant must spend larger quantities of energy, carbon, and nitrogen on fruit and seed production, whereas staminate plants do not have this problem at all. Therefore, differences in the vegetative parts of the plant might be expected.

Somewhat surprisingly, it is almost impossible to distinguish between staminate and carpellate plants without actually looking at the flowers or inflorescences. Although secondary sex characters do exist, they are subtle, and the range of variation that occurs in one sex overlaps the range that occurs in the other. Ayensu (1972) mentioned that there have been instances in which staminate and carpellate individuals of *Dioscorea* were classified into separate species, but such differentiation is rare. The most striking differences occur in some hemp cultivars; the carpellate individuals have a more vigorous appearance and larger root systems, and their leaves have larger blades than do the staminate plants (Schaffner, 1919). However, Lloyd and Webb (1977) point out that this may be due to the fact that the staminate plants flower while very young, whereas the carpellate plants must be older in order to flower. Consequently small individuals are being compared with larger ones. This might be overemphasizing the differences that exist.

Studies of short-lived plants such as *Silene alba, S. dioica* (van Nigtevecht, 1966) and suffrutescent plants such as *Mercurialis* (Mukerji, 1936) have shown similar tendencies for the shoots and leaves of the carpellate plant to be slightly larger than those of the staminate plant. It has also been noted that often a population of a dioecious species contains a greater percentage of staminate individuals; examples include *Petasites japonicus* (57% staminate) and *Myrica gale* (95% staminate). This may be related to differential survival, because the ratios are more unequal as older plants are examined. Lloyd and Webb (1977) mentioned that little work has been done on comparative ecology (Vitale and Freeman, 1985) but that there seem to be cases in which the two sexes do not occupy the same environmental niche; such a difference might be expected to select for differences in secondary sex characters. Dioecious species lack strong control over their sexual expression (Lloyd and Bawa, 1984), and individuals with both types of flowers can occur sporadically in a population. Cucurbits can change from producing only staminate flowers while young to having only carpellate flowers when older. Unfortunately, almost all studies of such characters have focused on gross morphological features, not on anatomical details.

Gymnosperms: Strobili

Sporangia

The reproductive structures of conifer sporophytes are the **microsporangiate strobili** (**microstrobili** or "pollen cones") and the **megasporangiate strobili** (**megastrobili** or "seed cones"); see Allen and Owens (1972), Bold et al. (1980), Konar and Oberoi (1969). Most species are monoecious, with both types of strobili borne on the same plant, but dioecious groups also occur (Cupressaceae and Araucariaceae).

MICROSPORANGIA.   The pollen cone is a simple shoot axis whose lateral appendages are microsporophylls instead of photosynthetic leaves. The microsporophylls are variable and characteristic of each family and often even of the genus; they typically have a short stalk and an expanded, flattened region that actually bears the sporangia. There may be from two sporangia (most common) to as many as twenty (in Araucariaceae) per sporophyll (Konar and Oberoi, 1969). While still small, the sporophyll is rather homogeneous, with an epidermis, mesophyll, and a slender vascular trace (Foster and Gifford, 1974). At some point, sporogenous tissue becomes discernable (Fig. 19.1). For one group (*Cedrus deodora, Pseudotsuga menziesii*) it has been stated that these tissues develop from superficial cells in the outermost layer of the microsporophyll, but in most other species (*Cryptomeria japonica, Chamaecyparis lawsoniana, Taxus baccata*, and others) they are reported to develop from hypodermal cells. The position of these progenitor cells (**archesporial cells**) is important, because a superficial position is considered extremely primitive, so much so that its true presence in seed plants would be

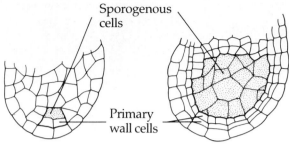

A. Microsporangium

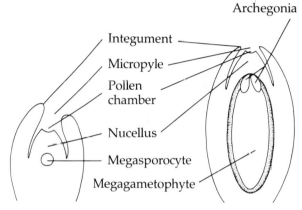

B. Megasporangium

Figure 19.1
These are diagrammatic representations of the microsporangium and the megasporangium. The microsporophyll is small and the sporagenous tissue constitutes a large fraction of its volume. The megasporangium is much more complex, having a protective integument surrounding the nucellus.

surprising. Foster and Gifford (1974) point out that the apparent superficial position of the archesporial cells in some conifers might be due to their not having a well-defined tunica and having an epidermis that is not strictly delimited until late.

Division of some of the archesporial cells results in two groups: a subepidermal **primary wall layer** and a **primary sporogenous layer**. As the microsporophyll continues to grow and expand, the primary wall layer undergoes several periclinal divisions, producing two to five

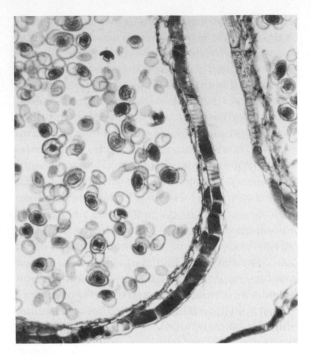

Figure 19.2
This is a microsporophyll of *Pinus*, showing the tapetum, endothecium, and developing microspores. × 160.

wall layers. The outermost frequently becomes specialized and is known as an **endothecium**, while the inner layer also becomes specialized as a nurse tissue called the **tapetum** (Figs. 19.1, 19.2; Cecich, 1984). On the inner side of the sporogenous tissue, some of the microsporophyll mesophyll parenchyma proliferates to form the inner section of the tapetum (Singh, 1978). The endothecial cells in *Pinus* have such large cytoplasmic channels between them that mitochondria can pass from cell to cell (Dickinson and Bell, 1976). The tapetum is involved in the differentiation of the sporogenous cells just interior to it; tapetal cells are glandular and frequently binucleate. The layer between the endothecium and the tapetum typically breaks down during later development.

The primary sporogenous cells increase in number by mitotic cell divisions as the spo-

rophyll enlarges. When the appropriate mass of sporogenous tissue has been achieved, the cells undergo major transformations and prepare for meiosis; at this time they are termed **microsporocytes** (pollen mother cells). The cytoplasm gradually retracts from the cell wall, and plasmodesmatal connections are broken. The cellulose walls degenerate and are replaced by thick walls of callose. Meiosis results in four microspores from each microsporocyte; these may be arranged as a tetrad or isobilaterally. The initiation of meiosis is not synchronous within a single microsporangium, and microsporocytes can be found mixed with microspores. Differentiation in *Pinus* begins in the center of the sporangium and proceeds outward (Dickinson and Bell, 1976). The events are complex, and an elaborate wall is established; an excellent review is provided by Singh (1978).

When the microspores (pollen) are ready to be released, the microsporophyll dehisces by either longitudinal or transverse (cycads) lines of weakness in the epidermis and endothecium; *Ephedra* has poricidal dehiscence. As the sporangia open, the axis of the pollen cone elongates, separating the sporophylls from each other and thereby facilitating the release of the pollen. The microspores are carried by wind to the megastrobili where they land on small drops of liquid (**pollination droplets**) exuded by the ovules, which then resorb the liquid, drawing thc pollen grains downward.

MEGASPORANGIA.   The megasporangiate strobilus (seed cone) is a compound shoot system, consisting of one stem, its leaves, and their axillary buds. The major axis bears leaves (the **bracts**; Fig. 19.3). In some extinct conifers, the axillary buds are visible and only slightly modified. In *Lebachia* each axillary bud had several small leaves and one megasporophyll, while in *Ernestiodendron* the axillary bud hád several megasporophylls (Foster and Gifford, 1974). In present-day conifers, the bracts subtend axillary buds that are highly modified.

The axis of each bud is microscopically short, and its megasporophylls have become fused through evolution such that each bud appears to be just one single **ovuliferous scale** instead of a group of sporophylls (Florin, 1944, 1950, 1951). In most Pinaceae, the scale is almost completely separate from the subtending bract, but in other conifers (Araucariaceae, Cupressaceae, Taxodiaceae) the scale and bract are also fused more or less completely to each other. The seed cone is usually analyzed in relation to the ovules that it carries and the seeds that it finally produces, but Niklas and Norstog (1984) and Niklas and Paw U (1983) have shown that the shape and arrangement of the bracts and scales are important for establishing air flow patterns that will bring wind-borne pollen to the ovules (Fig. 19.3).

Each ovuliferous scale bears from two ovules (the most common case) to as many as nine ovules (in *Sciadopitys*), each composed of an outer covering (the **integument**) and an inner mass (the **nucellus**). The two are attached to each other and to the scale at their base (the **chalaza**). The integument is not firmly attached to the nucellus on its upper portion; the gap creates a short tube (the **micropyle**) that leads down to a space (the **pollen chamber**) just above the nucellus. In gymnosperms other than conifers, the integument can be quite complex, with three distinct layers: the outer fleshy layer, the middle layer (stony layer), and the inner fleshy layer. In the conifers, the middle layer is the conspicuous layer; the inner layer is only obvious in the upper part of the integument that is not attached to the nucellus; and the outer layer is just a thin sheet of cells that eventually are destroyed as the seed matures. The vascular tissue that enters the ovule ends as a more or less extensive plate in the chalaza; very little vascular supply continues into the integuments in the conifers (Singh and Johri, 1972). Together the three layers surround and protect the nucellus, which would otherwise be completely exposed when the megastrobilus is open for pollination.

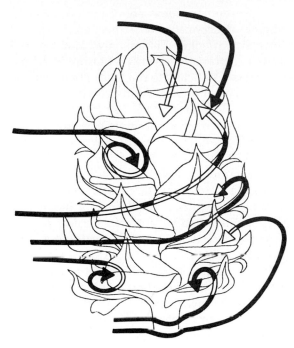

Figure 19.3
This diagram shows the flow of wind around a model of a pine seed cone. The arrows indicate the path that pollen grains might follow in falling into the ovules. Diagram generously provided by K. J. Niklas and K. T. Paw U.

As is the case of microsporangia, many investigators claim that the nucellus contains a particular hypodermal archesporial cell that divides to produce a primary wall cell and a primary sporogenous cell. However, in reviewing the literature, J. M. Coulter concluded that such claims rested on very little evidence, and he suggested that, rather than having a mass of sporogenous cells developing from one particular primary sporogenous cell, the nucellus merely grows and expands and that, at the proper size, sporogenous cells are differentiated. Konar and Banerjee (1963) did report an archesporium of six to ten cells in *Cupressus sempervirens*. This is certainly more in line with present-day ideas about morphogenesis, whereas the concepts of "archesporial cell" and "primary sporogenous cell" are based on concepts of sporogenesis in cryptogams and are also related basically to the idea of histogens.

The number of **megasporocytes** ("megaspore mother cells") is variable but typically small. One per nucellus is the most common condition; two to four occur in *Cryptomeria*, *Thuja*, and *Libocedrus*; and up to six occur in *Sequoia*. The cells immediately surrounding the megasporocytes are often large and cytoplasmic; they are sometimes called **spongy tissue** and are considered to be possibly sterile sporogenous tissue that is serving a nutritive function (see Chapter 20).

The megasporocyte is always located deep in the nucellus, separated from the pollen chamber by a thick layer of cells. The result of meiosis is a linear tetrad of potential megaspores, or if several megasporocytes are present, as many as 16 or more potential megaspores can occur. Several may begin to develop, but in all cases only one grows into a megagametophyte; the others remain quiescent or degenerate.

Gametophytes

THE MEGAGAMETOPHYTE.   The development of the megagametophyte begins while the ovule is still very small. The megaspore nucleus divides several times without concurrent cell division, resulting in a large multinucleate cell (Maheshwari and Singh, 1967). The cell expands greatly as does the central vacuole, so the cytoplasm and nuclei are pushed into a thin parietal layer. It is frequently reported that there are eight sets of divisions resulting in 256 free nuclei, but as many as 2000 nuclei occur in *Pinus* (Ferguson, 1904) and 6000 in *Cupressus sempervirens* (Mehra and Malhotra, 1947). In the second stage of development, the parietal nuclei begin to organize walls around themselves, forming cells. This stage is followed by centripetal, inward growth of cells until the entire structure becomes cellular. This can oc-

cur by having more cells formed that are completely enclosed, or it can occur by having only radial walls formed such that the new "cells" are really tube-shaped coenocytes themselves (Singh, 1978; Singh and Johri, 1972).

Simultaneous with the conversion of the megagametophyte to the cellular state from the coenocytic state, the gametangia (**archegonia**) are formed (Fig. 19.4). The archegonia may number from 4 to 100 (4 to 6 in *Sciadopitys*; 10 to 35 in *Taxodium*; 60 in *Sequoia sempervirens*; 100 in *Widdringtonia*). They may occur singly or in clusters, and they are usually located at the micropylar end of the megagametophyte but can occur along the sides (Konar and Oberoi, 1969; Singh, 1978). A superficial initial cell enlarges considerably, then divides by a periclinal wall to form an outer primary neck cell and an inner central cell. The outer cell undergoes several anticlinal divisions to form the neck. Because the surrounding gametophyte cells are growing more rapidly, the neck is soon located at the bottom of a cone-shaped depression (the archegonial chamber) in the surface of the gametophyte. The central cell begins to increase dramatically in size, and the surrounding gametophyte cells form a nutritive layer around it; this is the archegonial jacket, but it is sometimes referred to as a *tapetum*. The megagametophyte itself is frequently called *endosperm*, an unfortunate choice because the name is most commonly used for a different tissue in the angiosperms. Whereas the central cell becomes extremely large and densely cytoplasmic, the nucleus remains at the surface near the neck and divides to form a tiny **ventral canal cell** (which immediately degenerates) and a large **egg cell** (Fig. 20.2). The cytoplasm becomes dense; it contains large inclusions that are plastids that swell, then invaginate, trapping mitochondria, dictyosomes, and hyaloplasm inside themselves (Camefort, 1962; Gianordoli, 1974). The egg also has small inclusions that are regions of cytoplasm trapped inside a vesicle, but these vesicles are folded sections of endoplasmic reticulum, not plastids. The nucleus

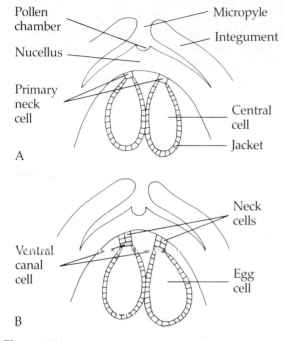

Figure 19.4
(A) A young archegonium contains only two cells, the primary neck cell and the central cell. The primary neck cells divide to form the neck (B), and the central cell divides to form a small ventral canal cell and a gigantic egg cell.

becomes gigantic; the egg nucleus in cycads is the largest nucleus known among all living organisms (Singh, 1978).

Viewed in isolation, it is difficult to understand the function of neck cells, ventral canal cells, and the archegonial jacket, but they are believed to be homologous to the structures present in the archegonia of the vascular cryptogams. In those plants, the archegonia are relatively large, flasklike structures that do have a long, tubelike neck, and a supporting layer of cells is found around the egg, which is located at the base of the neck. In seed plants, not only has the archegonium become buried in the gametophyte, but the gametophyte itself is retained and protected by the sporophyte.

Thus the supporting and protecting roles of the neck and ventral canal cells are not so important, and it is not surprising that they have become reduced evolutionarily. But it is not known why they are retained at all. It would seem to be possible to produce the egg cell directly from any one of the gametophyte cells, especially since they are all haploid, but apparently that cannot happen. A complete archegonium like those of the vascular cryptogams is complex enough that there must be numerous morphogenic interactions between them and the egg, many of which are essential for the proper development of the egg (which is, of course, much more than just any haploid cell). Almost certainly, the neck cells and the ventral canal cells are still important in contributing to the proper development of the egg.

At maturity, the ovule contains a single, large megagametophyte, completely cellular and containing several to many archegonia. It is important to remember that this megagametophyte is a complete plant that is growing parasitically on the parental sporophyte. It is separated from the sporophyte by the greatly expanded megaspore wall, which has stretched during the growth of the gametophyte within it. At maturity it is usually heavily suberized.

THE MICROGAMETOPHYTE.   The microspore begins to develop into the microgametophyte even while it is still contained within the microstrobilus. Development is quite uniform throughout the conifers and even throughout the entire set of gymnosperms (Singh, 1978). The nucleus undergoes a division to produce one small, peripheral lenticular cell and a large cell that is virtually the entire microspore. The lenticular cell degenerates immediately, but the large cell undergoes division again with the same results, and a second degenerating lenticular cell is produced. Whereas the megagametophyte becomes large with hundreds or thousands of vegetative cells, these two lenticular cells represent the vegetative body of the microgametophyte, which has been re-

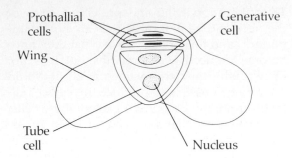

Prothallial cells

Generative cell

Wing

Tube cell

Nucleus

Figure 19.5
The microgametophyte is much smaller than the megagametophyte, and instead of having thousands of cells it has only about four.

duced almost as much as is theoretically possible. Only in the Araucariaceae do many (up to 40) vegetative cells still occur in the microgametophyte (Konar and Oberoi, 1969).

The remaining large cell represents the microgametangium (**antheridium**) and is called the **antheridial cell** (Fig. 19.5). The structures in the megagametophyte could not possibly be interpreted as archegonia without a thorough knowledge of the vascular cryptogams, and the same is true for the antheridial cell. It is an antheridium in function (it is involved in the production of the microgametes) but not in structure. It undergoes a division that results in a **generative cell** adjacent to the evanescent vegetative cells and a **tube cell**, which is the larger of the two. In many genera, the microgametophyte (pollen) remains in this condition as it is released into the wind for pollination to occur. In a few genera, it is released earlier, while still in the one-celled stage (*Cupressus, Fitzroya, Juniperus*) or two-celled stage (*Biota, Chamaecyparis, Thuja*; Konar and Oberoi, 1969).

When the microgametophyte is captured by the pollination droplet and carried into the pollen chamber, the microgametophyte begins to grow, producing a pollen tube that penetrates into the nucellus. Growth through the nucellus usually begins in the fall, stops during the

winter, then resumes in late spring and continues for about two months. The pollen tube can branch and ramify through the nucellus (Araucariaceae, *Podocarpus, Sciadopitys*), absorbing nutrients from it. It grows between the cells of the nucellus, secreting pectinases and cellulases that soften the nucellus (Willemse and Linskens, 1969). During this time, the generative nucleus divides into a **stalk cell** and a **body cell**. The body cell, the stalk cell, and the tube nucleus all pass into the pollen tube and are carried toward the archegonia. The body cell divides just before fertilization to form two sperm cells (Araucariaceae, Cupressaceae, Taxodiaceae) or two sperm nuclei located in one cell (Pinaceae). These sperm cells and those of the angiosperms are not motile, whereas those of all other vascular plants do have cilia or flagella; even cycads have motile sperm (Norstog, 1967). In the Taxodiaceae, the pollen tube arrives at the megagametophyte long before the archegonia have formed (Konar and Oberoi, 1969).

The pollen tube may break open when it reaches the archegonial chamber, releasing the sperm cells there (and also the tube nucleus and the stalk cell), or it may continue to the neck cells, crushing them, and releasing the sperm cells directly into the egg, bringing about fertilization. One of the sperm nuclei sheds its cytoplasm, migrates to the egg nucleus, and fuses with it (Chesnoy and Thomas, 1971), establishing the diploid zygote.

Angiosperms: The Flower

The ancestral group that gave rise to the flowering plants still has not been identified definitively (Stewart, 1983). Therefore the ancestral organs that evolved into the flower parts are not known, but most botanists do accept the idea that the angiosperms are a natural group, monophyletic in their origins from just one single ancestral group, not two or more.

Calyx

Calyx is the collective term for all the sepals of a flower. Sepals frequently have, at least, the role of protecting the developing flower, just as hypsophylls protect whole inflorescences. Sepals are undoubtedly modified leaves, and their anatomy can be interpreted completely as that of leaves that have adopted a protective function. A palisade parenchyma is usually absent, but the mesophyll may contain considerable chlorenchyma. The marginal growth of the sepals resembles that of the foliage leaves (Kaplan, 1968; Kaussman, 1941). Any secretory structures that occur in regular leaves may also occur here. A sepal typically receives as many vascular bundles as does a leaf on the same plant. Sepals may differ from leaves in that they are ephemeral. When they have expanded to the size at which the flower opens, sepals usually have completed their functioning and may either abscise or senesce. A photosynthetic leaf, by contrast, might be just switching from being a sink for nutrients to being a source of carbohydrates at such a size, and it might just be starting the development of its main conducting tissues of the metaxylem and metaphloem. It has sometimes been suggested that this is the only real difference between sepals and leaves, but this cannot be the case. Sepals have their own particular functional roles to fulfill; thus there are selection pressures to modify them for those purposes.

Like all flower parts, the sepals have proven to be extremely flexible evolutionarily and can vary from being entirely absent in some flowers, to being mere microscopic bristles in others, to being large and conspicuous in still others. Obviously, generalizations would be difficult and probably of little value. In *Plumbago capensis* and others, the sepals persist and form a tight, short tube around the base of the petals. The sepals are covered with sticky glands, so it is impossible for any crawling insect to get by the sepals and rob the nectaries; only flying pollinators can reach the nectar chamber (Percival, 1979). In other species (*Helleborus corsicus, Daphne, Fuchsia*), the sepals are brightly colored and petallike and may even produce fragrances and nectar. When this happens, the sepals are more delicate and are retained, as petals are, at least until the pollen is shed. In these flowers, the sepals have a double function: protection in the bud first and attraction of pollinators later.

Corolla

The petals of the flower constitute the **corolla**. In the large majority of cases, they serve as attractants for pollinators; to fulfill this function their color, texture, and fragrance are perhaps their most important features. Bees and other insects have been tested with model flowers, and the shape of the "petals" can be quite different yet will still attract the proper pollinators if color patterns and orientation of the petals are correct (Barth, 1985). Thus, the petals' most conspicuous modifications are often biochemical: they contain many pigments and textures. It is important to mention that insects do not see as we do; ultraviolet light is visible to them. Many petals have UV-absorbing flavonoid pigments in their epidermal cells (Kay and Daoud, 1981; Rieseberg and Schilling, 1985); they are invisible to us, but bees can see that they form a guide to the nectaries (Fig. 19.6). The "strategies" for pigmentation can be quite diverse. Droplets of carotenoids in the chromoplasts and betacyanins in the central vacuole are the most common, but tremendous variations of these can be achieved by other methods. If the layer of pigmented cells is just exterior to a layer of cells rich in starch grains, the color will have a creamy quality. Oils on the surface can cause the petals to glisten brilliantly, and air spaces parallel to the epidermis can cause a metallic sheen. If the petal has a smooth epidermis, the petal shines, but if it is papillose or has short trichomes, a rich, dark, velvety

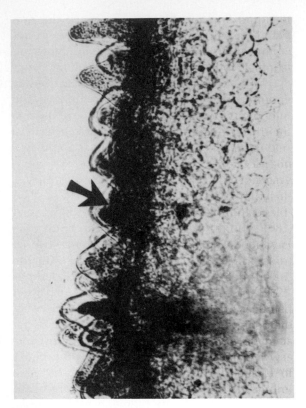

Figure 19.6
This is the petal of *Viguiera dentata*, photographed through a UV filter to show that the flavonoids (visible to insects) are located in the epidermis. Insects see a pattern different from the pattern we see. × 100. Micrograph generously provided by L. H. Riesberg and E. E. Schilling.

color is produced. In species that are wind-pollinated or self-pollinated, the petals have no attractive function and are reduced or completely absent. In the obligately self-pollinated, closed flowers in cleistogamous species, petals are reduced in size, lack pigmentation, and never expand at anthesis (Lord, 1981).

Although some petals do persist after pollination, like the sepals, they tend to be ephemeral. Some of the longest-lived petals occur in orchids. They may last up to 80 days on *Cymbidium* and *Oncidium* if the flower is not pollinated, but once the proper pollen is received

they begin to die within hours. Consequently, few resources are spent in their construction: energy and carbohydrate-rich fibers are not usually present, and even the epidermis produces little cutin or wax. Stomata are not abundant and often are incapable of movement (Fig. 19.7).

In the more primitive angiosperms, all the petals have more or less the same size and shape, and they are arranged uniformly around the receptacle; such flowers are **regular** or **actinomorphic** (*Magnolia, Nymphaea, Ranunculus*). In many families, the flowers have become modified to be bilaterally symmetrical or

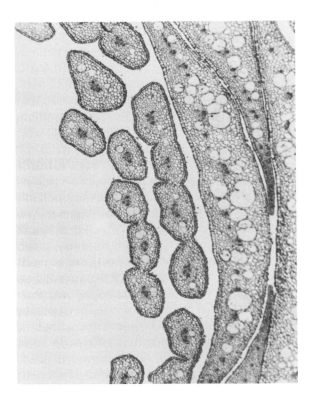

Figure 19.7
Cross section of some petals and stamen filaments in *Nopalea auberi*. The petals are thick and fleshy, without palisade parenchyma or spongy mesophyll; the filaments are slender with a single trace. The filaments are fused laterally at this level. Higher they are free; lower they are completely fused and secrete nectar. × 40.

zygomorphic. This arrangement is related to the bilateral symmetry of animal pollinators, and it tends to ensure visitation by pollinators that are specialized for that particular species. The development of zygomorphy typically results in diverse types of petals within one single flower, clover being a good example. Stamens and carpels also may be bilaterally arranged.

As flowers have become more specialized, one of the most common transformations has been the fusion of flower parts. Several aspects of fusion must be considered; the first is the type of organ undergoing fusion. It may be that similar organs unite to form one structure. For instance, all the petals may fuse into one tube-shaped structure. Any of the floral organs may undergo such a fusion. Prefixes are used to reflect whether organs are united or not: an **apo**petalous corolla consists of individual, free petals; a **sym**petalous one is a compound structure made up of fused petals (sometimes the prefix *gamo* is used: **gamo**petalous). In addition, two or more different organs may fuse, the most common example being the fusion of stamens to petals. A second important example, which is common, is the fusion of the base of the sepals, petals, and stamens to the sides of the carpels. At maturity, it is difficult to detect the fusion, and it appears as though these organs are simply produced above the ovary; this situation is called an **appendicular inferior ovary**. The clearest evidence of the fusion is the presence of the vascular bundles of the sepals, petals, and stamens in the tissues surrounding the ovary. An inferior ovary can result from a different type of fusion: in a **receptacular inferior ovary**, during development, a ring-shaped portion of the receptacle begins to grow upward more rapidly than the center. This causes the primordia of the sepals and the petals to be elevated while the carpels remain at the bottom of the concavity. The stamens line the inner side of the depression. The carpels become fused to the receptacle as they grow. The vascular tissues show the nature of this fusion, because the stele of the receptacle moves upward in the outer region and gives off traces to the sepals and petals. The stele then turns inward and downward and sends out stamen traces. In this region, because the bundles are bent over, the xylem is on the outer side of each bundle. Finally, the stele supplies traces to the ovules.

A second aspect of fusion is its extent. In ontogenetic fusion (fusion during development), the several organs are initiated as separate, distinct primordia. As these grow and swell, they crowd into each other and may merge. In such situations the individual organs may be quite visibly identifiable if they developed very much before fusing, and the contact points will be apparent as sutures. If the primordia immediately crowd into each other after formation, then it might not be at all obvious that the mature structure is compound. With phylogenetic fusion, the structure present today is the result of fusion in ancestral plants. The present structure arises as a single primordium, and it is primarily by studying related species that the fusion can be detected (Fig. 19.8). In many sympetalous species (in which the petals are united into a tube), the individual petal tips arise as small outgrowths, but, before they become distinct, meristematic activity spreads laterally on the receptacle apex until each outgrowth is sitting on top of a ring meristem. The ring meristem produces the tubular base of the corolla and is considered to be a compound structure made up of several petals, but, because it is derived from one continuous meristem, there are no sutures.

Petals are considered possibly to have been derived from two sources. In some groups they appear to be modified leaves. In others, there is good evidence that they evolved from stamens; within a flower with numerous petals and stamens, there may be a continuum of transition forms varying from true petals to true stamens, the intermediate forms being **staminodia**. These petals, like the stamens, have only one vascular trace each, whereas the

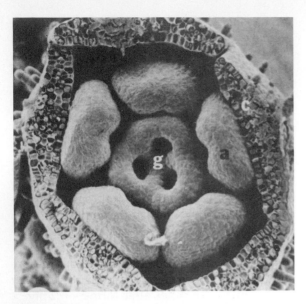

Figure 19.8
This SEM of a developing flower of *Collomia grandiflora* shows several types of fusion. The outermost layer consists of the five petals that are fused together (*c*); in the center are the three carpels (*g*). After a brief period as three separate primordia, the carpels begin growing as a single ring-shaped structure (much like a sheathing leaf base in grasses). Thus the sides of the carpels are phylogenetically fused. The tips, however, are still open; during later growth, the edges might push together and fuse—an example of ontogenetic fusion. Five stamen primordia are also visible. *c* = corolla, *a* = androecium, *g* = gynoecium. × 200. Micrograph generously provided by T. C. Mintor and E. M. Lord.

sepals and carpels will typically have more. This may be misleading evidence, however; it should be kept in mind that both petals and stamens are delicately constructed and ephemeral, not needing much conductive capacity, whereas sepals and carpels are usually more bulky and more long-lived. Sepals are formed first and are in an exposed position while all other organs are protected. The carpels may be small initially, but they expand greatly when they mature into fruits, so a large amount of vascular tissue should be present.

Much of the interest in sepals and petals has

focused on elucidating the evolutionary origin of the flower and its parts. The vascularization has been given special emphasis, and remarkably little attention has been given to the functional adaptations of the anatomy of sepals and petals (although the shape of petals receives some study). Consequently, very little information is available concerning the epidermis, mesophyll, or secretory tissues of these two floral organs.

Androecium

The microsporangia of flowers are borne in the stamens, and the collective term is **androecium**. For most stamens, production of the microspores is clearly the dominant function. But in some species, part of the stamens are converted to nectaries to attract pollinators. In other plants, part of the pollen and the stamens are eaten by the pollinators; these are both protein-rich and rather "expensive"; as a consequence, some plants (*Cavanillesia*) have two types of stamens, one type that actually produces fertile pollen and others that are feeding stamens that usually contain little or no pollen and are eaten preferentially by the pollinators. In a small number of families (Myrtaceae, some Leguminosae such as *Acacia*) the numerous stamens act like petals and supply a bright, visual display to signal pollinators. In all angiosperms except some Magnoliales, the stamen consists of two parts: the stalk (**filament**; Fig. 19.7) and the sporogenous region (the **anther**; Fig. 19.9). Stamen primordia resemble leaf primordia when initiated by the floral apical meristem (Ross, 1982).

The filament is often long, but it can be short or even absent. It almost never branches, but stamens can occur in fascicles, being grouped in bundles on the receptacle. On the filament there is an epidermis that is thin-walled and delicate, with little cutin, usually less than 1.0 μm thick, but rarely up to 5.0 μm, as in *Nerium oleander* (Schmid, 1976). Fine trichomes and papillae occur in many species. Below this is a

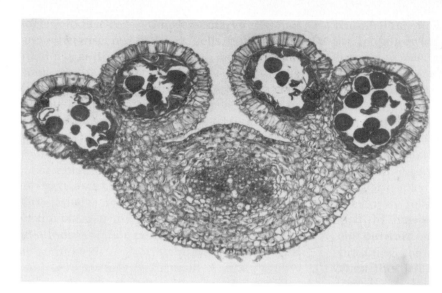

Figure 19.9
This cross section of an anther of *Asarum canadense* (wild ginger) shows the connective and the four anther sacs (microsporangia). You can see that the two anther sacs on each side are fusing with each other and that each set will have only one stomium. Notice that the vascular bundle does not branch out toward the anther sacs. × 100. Micrograph generously provided by L. D. Hufford.

narrow mesophyll, usually of parenchyma. The intercellular spaces are often quite large in filaments. Schmid (1976) reported that sclerenchyma is extremely rare in filaments, being known only in some Gentianaceae, Magnoliaceae, Nymphaeaceae, Palmae, Winteraceae, and a few others. This sclerenchyma is almost always arranged in small packets that probably do not interfere with stamen expansion during anthesis. All kinds of secretory structures are rare, although these can occur if the filament is especially large or persistent. However, nectaries on the surface of the filament or anther are not uncommon. There is one central, slender vascular bundle that contains just small amounts of conducting xylem and phloem, typically arranged as an amphicribral bundle or with the phloem almost completely encircling the xylem. In any cross section, there may be only one or two xylem elements, always with annular or spiral thickenings, never with pitted walls (Schmid, 1972). It should be kept in mind that, once the flower opens and the filaments elongate, they will live for at most only a few days—often less than a few hours—so water conduction at maturity is unneccessary. The sieve elements may occur in clusters. Schmid

(1976) reported that he had encountered no instance of secondary growth in any filament. The bundle is surrounded by a prominent sheath of large cells that may be an endodermis.

The anther is located terminally on the filament and is usually elongate, containing four tubelike microsporangia (often called pollen sacs) in which the pollen grains are located. There is an epidermis (occasionally with stomata), several layers of mesophyll cells around each microsporangium, and a rather large central mass of parenchyma (the **connective**; Fig. 19.9) that contains the vascular trace and sometimes secretory tissues (Myrtaceae). The connective is the site of attachment of the anther to the filament. The vascular tissue of the filament may branch once it enters the connective (*Asarum canadense*) and even send traces into each lobe (*Prunus virginiana*), although reports of this are rare (Hufford, 1980). Frequently, the two microsporangia on each side fuse with each other as the intervening parenchyma breaks down, and in *Arceuthobium* all the sacs fuse, resulting in a torus-shaped microsporangium. The microsporangia can be **chambered**, either by transverse (some Mimosaceae) or longitudinal (some Loranthaceae) septa of sterile parenchyma (Lersten, 1971; Maheshwari, 1950). These anthers are still interpreted as having only four microsporangia, but it is thought that during development some of the sporogenous tissue becomes septa rather than proceeding to meiosis.

Developmentally, the anther resembles the microsporophyll of the conifers. It has been interpreted as having four hypodermal archesporial regions, each containing just one row of cells (*Sansevieria*; Guérin, 1927) or many rows (most plants; Maheshwari, 1950). The cells in the archesporium undergo division by periclinal walls to produce primary parietal cells and primary sporogenous cells (Fig. 19.10). Divisions of the former result in the production of the endothecium (the exterior layer), two to four layers of parenchyma cells, and the tape-

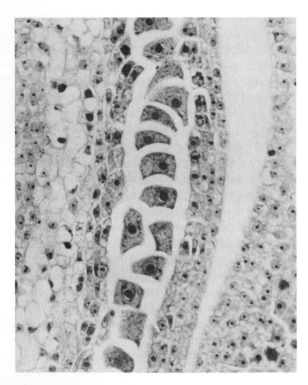

Figure 19.10
Sporogenous cells in the anther of *Opuntia*. They are preparing for meiosis and have deposited thick, white callose walls. × 120.

tum (the innermost layer), while mitosis in the primary sporogenous cells produces the sporogenous tissue. It is difficult to envision how derivatives of the primary parietal cell could form a tapetum that completely encloses the sporogenous cells. It has recently been shown (Gupta and Nanada, 1978) that the tapetum can have dual origins, the portion on the inner side arising from cells of the connective.

The endothecium is often prominent and easily identified as being located immediately below the epidermis and as having special wall thickenings in the shape of a U with the open part directed outward (Fig. 19.11; French, 1985; Noel, 1983). The epidermis of the anther is typically thin at maturity, and it may even be collapsed or torn, and the endothecium can appear to be the outermost layer (Schmid, 1976). It is thought that, as this layer dries, it aids in tearing open the anther sacs along predetermined lines of weakness, the **stomia**—sing.: stomium (Fig. 19.9). However, Percival (1979) points out that anthers dehisce even when air has a relative humidity of 90–100%, so air dryness may not be important. The stomia are typically longitudinal but can be circumscissile (transverse) or in the form of small pores. The stomium cells often contain druses of calcium oxalate, which have been considered possibly to act as a deterrent that prevents insects' feeding on the anthers. Horner and Wagner (1980) showed that the crystals form as material is removed from the wall of the stomium cells to weaken them in preparation for dehiscence. By the time the pollen grains are maturing, the walls have been greatly degraded, and the degradation products have accumulated as crystals. If dehiscence is by apical pores (for example, *Erica*), then the pore forms by the disintegration of a patch of cells and does not need to be opened by the endothecium; in these species, the U-shaped thickenings are absent (Maheshwari, 1950). There are cases known in cleistogamous flowers in which the anthers do not open; instead, the pollen tubes grow through the anther wall and

Figure 19.11
These are endothecial cells from *Scindapsus pictus*, showing the thickenings of the wall. × 570. Micrograph generously provided by J. C. French.

extend directly to the stigma nearby, as in *Viola odorata* (Mayers and Lord, 1984).

The middle layers of parenchyma located between the endothecium and the tapetum become flattened and crushed by the time meiosis occurs in the sporogenous tissue; persistent middle layers are rare (*Ranunculus, Lilium*). There are occasional reports that middle layers do not form in certain species, but reinvestigations show that these are cases in which the middle layers are established but then degenerate quickly.

The tapetum in angiosperms, as in conifers, plays an important role in the proper maturation of the microspores. All of the nutrients for the sporogenous tissues, and later for the microspores themselves, must pass through the tapetum. The tapetum is typically composed of one layer of very large, densely cytoplasmic cells. The nature of these cells can be diverse, even within a single microsporangium. In *Lathraea*, they are long, columnar, binucleate cells on the side of the microsporangium closest to the connective but are smaller, cuboidal, uninucleate cells on the side closest to the epidermis (Gates and Latter, 1927). *Lactuca* shows a similar variation, but the cells near the connective are quadrinucleate, the others binucleate (Gates and Rees, 1921). Binucleate or multinucleate tapetal cells are the rule; only 15 families have exclusively uninucleate tapeta (Buss and Lersten, 1975; Davis, 1966; Wunderlich, 1954). Endomitosis can occur (Oksala and Therman, 1977) and possibly is very common (D'Amato, 1954; Tschermak Woess, 1971). This condition of having dense cytoplasm, small vacuoles, and multiple nuclei strongly suggests that the tapetum is very active metabolically. There is a basic distinction in types of tapetum: **parietal tapetum** (formerly called "secretory" or "glandular tapetum") and **invasive tapetum** (formerly called "amoeboid tapetum." A parietal tapetum remains cellular and secretes material into the anther sac from the periphery, but in those species with an invasive tapetum the walls around the tapetal cells break down, and the protoplasts are released into the sac (Clausen, 1927; Mepham and Lane, 1969; Pacini and Juniper, 1983). The first walls to dissolve are the radial walls, which have tapetal cells on both sides; the tangential walls in contact with the microsporocytes or with the anther wall parenchyma are the last to break down, at least in *Arum italicum*. Pacini and Juniper (1983) pointed out that wall breakdown begins in the walls that have the highest density of plasmodesmata, and they suggested that, as in the formation of sieve areas in the phloem, the plasmodesmata may act as initiation sites. There plasmodesmata fuse to form a large mass of protoplasm (the **periplasmodium**) that com-

Figure 19.12
As the microsporocytes prepare for division, the cellulose wall is replaced by a callose wall with large channels (B). After meiosis (C) the young pollen grains remain in a tetrad and form the primexine (C and D). Later, probacula form, establishing the sexine (E); then the nexine and intine are laid down as the probacula are converted to bacula by the deposition of sporopollinin on them (F).

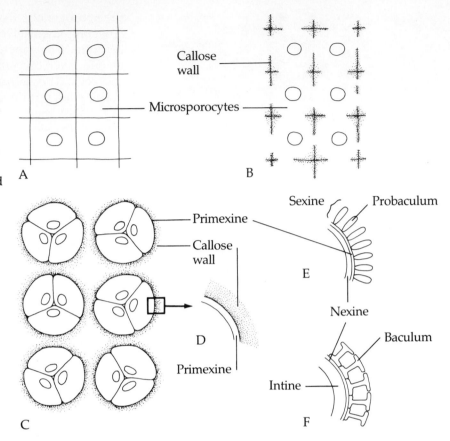

pletely encases the microsporocytes and developing microspores. As the tapetal cells move into the loculus, each microsporocyte or tetrad is enveloped in a vacuole derived from the plasmalemma of the periplasmodium. The periplasmodium may form very early, while the microsporocytes are in meiosis (some Araceae, *Potamogeton, Triglochin*) or later, when the tetrads of microspores have been fully formed (*Alisma, Sagittaria, Tradescantia*). It remains healthy for a long time, and its nuclei may undergo mitosis.

The sporogenous cells undergo mitosis and cytokinesis, growing in number as the anther enlarges. When the proper stage is reached, they are stimulated to become microsporocytes and prepare for a meiotic division. This is an extremely complex set of changes: the plasmodesmata expand greatly, becoming major cytoplasmic channels (up to 1.5 μm in diameter) between the cells, converting the entire sporogenous mass into a single syncytium. The primary, cellulosic wall becomes more difficult to detect, and a very thick callose wall replaces it; the callose wall maintains the integrity of the cytoplasmic channels (Fig. 19.12). Probably because of the syncytial nature of the microsporocytes, meiosis is either synchronous throughout the entire anther sac or begins at one site and moves as a wave throughout all the surrounding cells. At the same time that the microsporocytes are enlarging their cytoplasmic channels, they also break all plasmodesmatal connections with the tapetal cells, becoming symplastically isolated.

Two types of meiotic cytokinesis are possible: in the **successive** type, the first nuclear division of meiosis is followed by wall formation. But in the **simultaneous** type, no walls are formed until both meiotic karyokineses are completed; then both sets of walls are laid down at once. When the walls are formed simultaneously, they are not deposited by cell plates but are formed centripetally: the walls grow inward from the edges of the cell and meet at the center (Bhojwani and Bhatnagar, 1974). In the dicots, the four resulting microspores are typically arranged in a tetrad pattern; in the monocots, an isobilateral arrangement is more common. Decussate, T-shaped, and linear patterns also occur but are rare. While the microspores are still attached in tetrads, differentiation of the wall may begin, so that, even when they later separate, the inner contact faces can be distinguished from the outer free faces. In *Drimys, Drosera,* and Ericaceae, the microspores remain united in tetrads; in the Orchidaceae and Asclepiadaceae, all of the pollen grains of a single anther sac remain united as a unit: a **pollinium** (pl.: pollinia) in the orchids (Blackman and Yeung, 1983), a **massula** (pl.: massulae) in the milkweeds (Galil and Zeroni, 1969), and **polyads** in legumes (Fig. 19.13; Feuer et al, 1985).

Numerous cytological changes occur during and after meiosis (Reynolds, 1984) but a major aspect of pollen formation is the development of the most elaborate cell wall in existence (Heslop-Harrison, 1971a, 1979; Payne, 1981). The cell becomes densely cytoplasmic, and the endoplasmic reticulum is repositioned to be immediately below the wall in those areas where **germination apertures** will later form. A layer of cellulose (the **primexine**) is deposited just interior to the callose wall in all areas except those of the future apertures (Figs. 19.12, 19.14). As the cell expands, small rod-shaped **probacula** (sing.: probaculum) appear in the primexine layer. These may remain as just rods, but their bases may expand and contact the bases of adjacent probacula, forming a **foot layer**, and the tops may similarly expand to

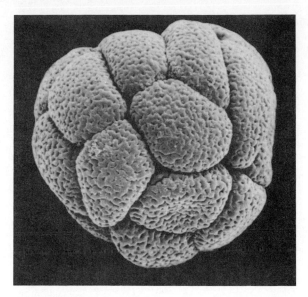

Figure 19.13
This is a polyad of *Parkia speciosa,* showing all the pollen grains clearly. All of the pollen grains will be carried to the same stigma. × 700. Micrograph generously provided by S. Feuer, C. J. Niezgoda, and L. I. Nevling.

form a **tectum** or **roof layer**. As these develop in shape, the tapetum deposits **sporopollenin** onto them, converting them from probacula to **bacula**. The wall layer formed by the bacula is the **sexine**.

The cell continues to expand, and the callose wall is lost; a smooth, uniform layer of sporopollenin is deposited between the sexine and the protoplast (Shoup et al., 1981). This layer is the **nexine**, and the sporopollenin in it is believed to be derived from the microspore itself, not the tapetum. Finally, in the regions of the future germination apertures, the innermost layer (**intine**), composed of cellulose and pectin, begins to appear beneath the nexine; it spreads laterally until it envelops the entire protoplast. The resulting wall has the following layers from inside to outside: intine, nexine, sexine, and the last two called together the **exine**. Recently, the two layers of the exine have

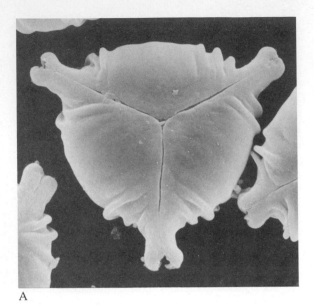

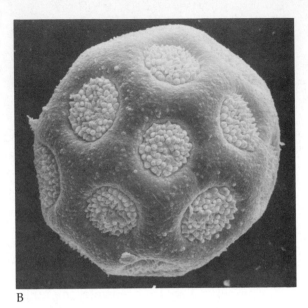

A

B

Figure 19.14
Pollen grains. (A) *Cuphea strigulosa*. × 2300. (B) *Aellenia subaphylla*. × 2500.
Both micrographs generously provided by A. Graham and J. Nowicke.

been called the **endexine** (nexine) and the **ectexine** (sexine) (Muller, 1979).

It seems logical that a tiny pollen grain could have few features, and one might expect them all to have the same basic appearance. But actually the surface details are remarkably varied (Muller, 1979), and pollen grain details are now considered as some of the most important characters for taxonomy and systematics (Fig. 19.14; Burns-Balogh, 1983; Dickison et al., 1982; Nowicke et al., 1985). Because they are very resistant, pollen grains can be found in ancient sediments, and the study of fossil pollen (**palynology**) is a rapidly expanding field (Muller, 1981).

The structure of the pollen wall has many functions, two important ones being (1) the protection of the protoplasm from desiccation during pollen transfer by the pollinator (Payne, 1981) and (2) carrying chemical agents that are involved in elaborate recognition reactions with the stigma of the receptive flower (Heslop-Harrison, 1968, 1971a, 1978). In addition, the wall must be capable of adhering (but not

too firmly) first to a pollinator and second to a stigma, and it must be capable of allowing germination.

Inside the grain, nutrients must be established to maintain the grain during the time after it is shed until it can be nourished by the style of the receptive flower. Pollen grains are classified as being either "starchy" or "starchless," the latter type storing sugars or oils instead of starch. Baker and Baker (1979) discussed the selective advantages of each type of pollen grain. In species that are pollinated by birds or certain insects, the pollination "reward" is nectar; if wind is the pollination medium, there is no reward. The pollen of these species may tend to have starch, because it is easily stored and later is easily utilized. If bees or other pollen-eating insects are the pollinators, then sugary or oily pollen is preferred by the pollinators. Another consideration is size. Starch is much bulkier than oil with the same caloric content, so, if the pollen must contain larger amounts of energy (if it is long-lived or if it must grow through a long style), it

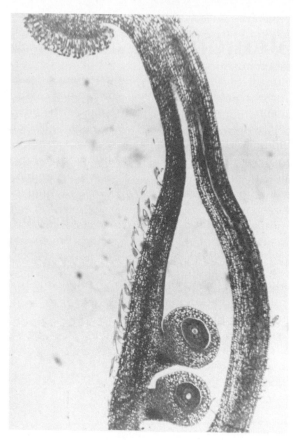

Figure 19.15
This is a longitudinal section through the carpel of *Phaseolus acutifolius*, showing the stigma, style, ovary, and ovules. The funiculi are short and broad. × 58. Micrograph generously provided by E. M. Lord and L. U. Kohorn.

may contain oil rather than starch. In a study of 990 species, Baker and Baker (1979) found that these hypotheses were reasonably accurate.

Gynoecium

The carpels are composed of three basic parts, the **stigma**, which catches and receives pollen during pollination, the **style**, which elevates the stigma and through which the pollen tubes grow, and the **ovary**, which contains the ovules

(Fig. 19.15). More than in any of the other parts, the ovaries tend to undergo fusion, most often merging with each other to produce what appears to be one single organ. When this happens, the styles and stigmas may or may not remain distinct.

At present, the carpel is most commonly interpreted as being a sporophyll (Wilson, 1942) and thus has leaflike anatomy with epidermis, mesophyll, and vascular tissue (usually with three traces). Whereas the sepals, petals, and stamens are ephemeral and therefore often rather delicate in construction, only the stigmas and styles of the carpels are temporary structures. The ovary must persist to develop into the fruit. It is therefore often more substantial, and in some cases it has more vascular tissue than one would expect if the nature of the fruit into which it would develop were unknown.

The stigma is very simple structurally; it basically consists of a small amount of mesophyll parenchyma or aerenchyma covered by a thin-walled epidermis that is papillose or has well-developed trichomes, essential for trapping pollen grains. In wind-pollinated plants, the stigmatic trichomes can be long, and the whole structure is featherlike in appearance. In many species, the stigma exudes a fluid rich in amino acids, oils, and sugars when it is ready to receive pollen (Fig. 19.16; Heslop-Harrison and Shivanna, 1977; Kristen et al., 1979; Sedgley and Scholefield, 1980; Schneider and Buchanan, 1980). The stigma may have considerable vascular tissue, especially xylem if it exudes a drop of liquid (Konar and Linskens, 1966). Other stigmas are dry (Heslop-Harrison, 1981). It is now known that in many self-incompatible species the stigmatic cells are capable of complex metabolic interactions with the pollen grains, and that if they are incompatible the stigma will deposit callose, stopping the germination of the pollen or inhibiting the growth of the pollen tube (Bawa and Beach, 1983; Heslop-Harrison, 1978; Vasil, 1974). The entrance of pollen tubes can be enhanced and guided by special channels in the stigma (*Glycine max*; Tilton et al.,

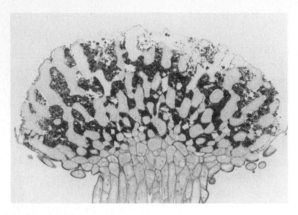

Figure 19.16
This is the stigma of soybean (*Glycine max*); it is a wet stigma, and the spaces between the stigmatic hairs are filled with exudate. Remnants of the cuticle are still present over the secretion. × 610. Micrograph generously provided by V. R. Tilton, L. W. Wilcox, R. G. Palmer, and M. C. Albertsen.

1984). After surveying almost 1000 species in about 900 genera, Heslop-Harrison and Shivanna (1977) concluded that stigma types are quite constant within most species and genera, and even within some families. The survey was extended by Heslop-Harrison (1981), and numerous excellent scanning electron micrographs were included.

The style is often similarly simple. If it is a compound style it is composed of several styles fused together; it then usually has one vascular trace for each component style. This trace is a continuation of the median trace in the carpel; rarely the lateral carpel traces also extend into the style, as in some Cunoniaceae (Dickison, 1975) and Palmae (Uhl and Moore, 1971). The traces will be located in a ring, appearing like a eustele but not being one (Fig. 19.17). The center of the style may be hollow (mostly monocots) or solid (mostly dicots). If solid, the centrally located cells are usually very cytoplasmic parenchyma called **transmitting tissue** (Arber, 1937; Clarke et al., 1977; Gawlik, 1984; Tilton and Horner, 1980b). This tissue serves as a specialized substrate through

which the pollen tubes grow (Ciampolini et al., 1978; Cresti et al., 1976; Jensen and Fisher, 1969). The pollen grain is too small to carry much nutritional reserve, so as the pollen tube elongates through the transmitting tissue it absorbs nutrients from it. The walls are soft, swollen, and gelatinous, and pollen tubes are able to grow through both the walls and the intercellular spaces. The transmitting tissue, however, is not just a passageway, because in many species it apparently interacts with the pollen tube, testing it for compatibility.

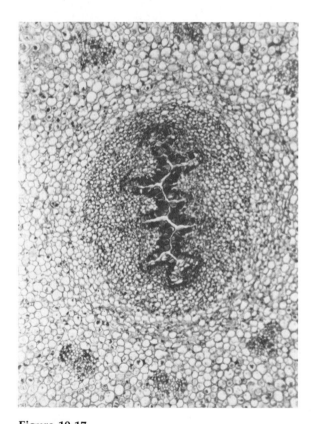

Figure 19.17
This is a cross section of a *Homalocephala texensis* (horse crippler cactus) style. The style is rather massive; it has 10 vascular bundles and rather abundant mesophyll. At the center is the transmission tissue, which is composed of the inner (adaxial) epidermis and several adjacent layers of mesophyll. × 160.

If the pollen tube is of an inappropriate type, the transmitting tissue can react to inhibit its growth (Bawa and Beach, 1983; Cresti et al., 1976; Linskens, 1975).

If the style is hollow, then transmitting tissue lines the lumen as a unistratose or multistratose, glandular layer, even projecting into the lumen as long trichomes. The pollen tubes may grow through the trichome regions or penetrate into the deeper layers. In two species of *Lilium*, the cells of this epidermal layer have been found to be transfer cells (Gawlick, 1984; Rosen and Thomas, 1970). In *Lilium leucanthum* the cells of the stigma are also transfer cells, and Gawlik referred to the stylar cells as "stigmatoid cells" because they resembled the stigma cells so greatly. In *L. longiflorum*, the stigma cells are distinctly different from the inner epidermis of the stylar canal, and Rosen and Thomas used the term "canal cells" instead of stigmatoid cells. Tilton and Horner (1980b) also found that cells of the stigma differed from those of the canal epidermis in *Ornithogalum*.

In some compound styles (**pseudosyncarpous styles**) there are as many columns of transmitting tissue as there are component carpels, and if a pollen grain lands on a particular stigma its growth is channeled by the transmitting tissue down into the ovary of the same carpel (Carr and Carr, 1961). In **eusyncarpous styles**, the transmitting tissues are fused into one mass, so a pollen tube can grow into any ovary, regardless of which stigma it germinates on.

Most styles are quite elongated. There are not many studies of this characteristic, but in *Ornithogalum* (Tilton and Horner, 1980b), *Petunia* (Linskens, 1975), and *Vinca* (Boke, 1949) it grows briefly by a basal meristem.

The ovary is often the thickest, most massive portion of the three parts of a carpel (Fig. 19.15). Its epidermis may be quite substantial and frequently has functional stomata. A hypodermis may be present, as well as considerable mesophyll and vascular tissue. The ovules are attached to the inner, ventral side. The region of attachment is called the **placenta**, but frequently this area is indistinguishable anatomically and cytologically from surrounding nonplacental areas.

The ovules may occur singly within a carpel, or there may be thousands in the ovaries of some orchids. They are attached to the placenta by a narrow stalk, the **funiculus** (Fig. 19.18), and these funiculi can be single or clustered, branched (rare) or unbranched. At the end of the funiculus is the nucellus. Various arrangements of the funiculus and nucellus are possible. If they are aligned in a straight line, the ovule is **atropous** (also called **orthotropous**), as in Cistaceae, Piperaceae, and Polygonaceae. If the funiculus is bent 180° so that the nucellus projects back toward the placenta, it is **anatropous**, the most common type, as in Buxaceae, Vitaceae, Convolvulaceae, and in general those families in which the petals are united (Maheshwari, 1950). The funiculus frequently has a narrow vascular bundle, but it almost never continues into the nucellus. *Agave* (Grove, 1941), *Casuarina* (Swamy, 1948), and *Stombasia* (Fagerlind, 1947) are among the few genera with vascularized nucelli.

From the funiculus, near the base of the nucellus, two thin layers of cells project upward and surround the nucellus. These are the **inner** and **outer integuments** (Figs. 19.18, 19.19). Many of the more advanced families have species with **unitegmic** ovules—that is, ovules with a single integument (Icacinaceae, Rafflesiaceae, Salicaceae). The single integument is produced by the fusion of the two integuments or by the suppression of either one (Bouman and Calis, 1977; Bouman and Schier, 1979). Some of the Balanophoraceae, Loranthaceae, and Olacaceae are **ategmic**, but the **bitegmic** condition is the most common. It is not known what the integuments evolved from. They may be vascularized by traces that run all the way to their tips (Betulaceae, Compositae, Convolvulaceae, Euphorbiaceae, Rhamnaceae); they occasionally have stomata and

Figure 19.18
In this diagram of an ovule, you can see the funiculus, nucellus, integuments, and micropyle.

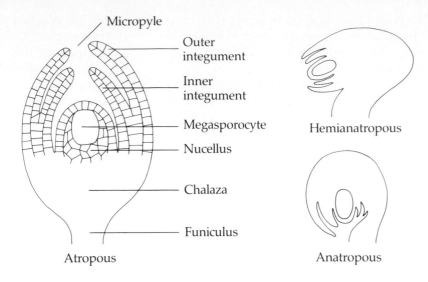

that had open sporophylls, just as is still true of the gymnosperms. With the closure of the carpel, however, they no longer offer significant protection; it may be that they function in producing the proper field for the development of the tissues inside, perhaps in creating and distributing pressures. Mogensen (1985) found evidence that the integument of tobacco can act as a transfer mechanism to pump nutrients actively to the megagametophyte and later to the embryo.

The inner integument is frequently shorter than the outer one, but it can be as long as or even project beyond it (Annonaceae, Cactaceae, Proteaceae, Trapaceae). As in the gymnosperms, the integuments do not merge on the far end of the nucellus but leave an opening, the micropyle. If both integuments contribute to the formation of the micropyle and are rather thick, then the opening of the outer integument is called the **exostome** of the micropyle, and that of the inner integument is the **endostome** (Fig. 19.20). In some taxa, about 57 families (Davis, 1966), especially those that are unitegmic and have a small nucellus, the inner epidermis of the integument can become a nutritive tissue, the **endothelium** (**integumentary tapetum**; Fig. 19.21). The endothelial cells are

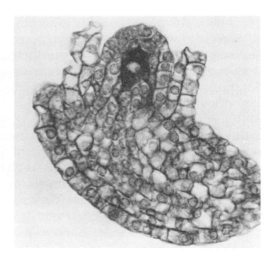

Figure 19.19
This ovule of *Frailea lepida* contains a prominent megaspore mother cell. The inner integument extends along the nucellus, but the outer integument is very short and is distinct only on one side (left). The entire top of the nucellus is exposed because the micropyle is so wide. × 500.

chlorophyll (*Amaryllis, Gladiolus, Gossypium, Hymenocallis*), and they are always covered by a cuticle (Bhatnagar and Johri, 1972). The original function of the integuments may have been to protect the nucellus in the ancestral plants

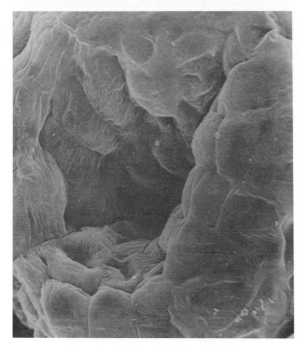

Figure 19.20
This SEM gives you a pollen-tube's-eye view of the micropyle of *Ornithogalum caudatum*. × 1273. Micrograph generously provided by V. R. Tilton.

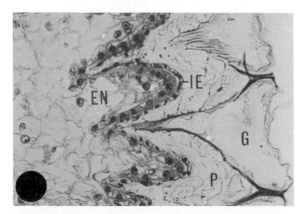

Figure 19.21
This picture shows the expanding endosperm as it encounters the inner surface of the inner integument. The integument epidermal cells have become cytoplasmic and secretory, acting as an endothelium. *G* = gynoecium, *IE* = inner epidermis, *EN* = endosperm, *P* = perisperm. × 160. Micrograph generously provided by R. Y. Berg.

densely cytoplasmic, often columnar, and frequently become multinucleate: in *Pedicularis palustris*, each cell has 32 nuclei (Bhojwani and Bhatnagar, 1974). In sunflower, the endothelium can become very thick, having as many as 10 to 12 layers. The conversion from epidermis to endothelium usually happens after fertilization of the egg, as it begins to grow into an embryo.

Within the nucellus, a hypodermal cell at the micropylar end becomes the archesporial cell. In about one hundred families, this cell functions directly as a megasporocyte, preparing for meiosis immediately (Davis, 1966). In such a case, the sporogenous cell is very close to the surface, and the ovule is said to be **tenuinucellate**. More often (in about two hundred families), the archesporial cell undergoes divisions into a primary parietal cell and a primary sporogenous cell. The primary parietal cell and surrounding nucellar cells undergo divisions; thus the nucellus grows between the epidermis and the sporogenous cell, causing the megasporocyte to become deeply buried: a **crassinucellate ovule**. In the Crossomataceae, Paeoniaceae, some Loranthaceae, and in *Sedum*, there may be several megasporocytes in each nucellus, but typically only one is fully functional.

The megasporocyte is highly asymmetric, with a stratified cytoplasm. In corn, the rough endoplasmic reticulum is most abundant in the micropylar half of the cells, but most mitochondria and plastids are either in the chalazal end or adjacent to the nucleus (Russell, 1979). The vacuole itself is also in the chalazal end.

When the megasporocyte finally is established, it undergoes meiosis (Fig. 19.19). Somewhat surprisingly, this step, which is perhaps the most critical in sporogenesis, is the most variable: numerous modifications exist and all function well (Bhatnagar and Johri, 1972). First, the nuclear divisions may or may not be accompanied by wall formation; if walls do form, then four megaspores result, but three will immediately degenerate, leaving only one functional megaspore. The four megaspores are most commonly arranged in a straight line, with the cell closest to the chalaza surviving (but in *Balanophora* and *Oenothera* the micropylar cell survives, in *Rosa* the submicropylar one survives, and in *Schizomeria serrata* it is the subchalazal one). It is not unusual for the four megaspores to be arranged in a T-shape with the top of the T near the micropyle; very rarely the top of the T is near the chalaza. Tetrahedral and isobilateral megaspores occur only as abnormalities (Bhojwani and Bhatnagar, 1974; Maheswari, 1950). This situation, in which four meioproducts are formed but only one survives, produces a **monosporic megagametophyte** (Fig. 19.22).

In *Allium*, after meiosis I, there is cytokinesis, followed by the degeneration of one cell while the other completes meiosis II and produces a binucleate cell. This cell acts as though it were formed by mitosis of a haploid spore, and it develops into a **bisporic megagametophyte** (Fig. 19.22). *Bisporic* in this instance is an unusual name, but it would be useless to try to decide if it really is a binucleate spore or two "spores" within the same cell wall and cytoplasm. Whichever is the case, the gametophyte that develops will be a chimeral mixture of two types of cells. If the gametophyte were large, this might be important; with the tiny megagametophyte formed in the angiosperms, it seems to make no difference. Furthermore, there are many plants that have **tetrasporic megagametophytes**: meiosis is not accompanied by any cell division, and the tetranucleate **coenomegaspore** that results acts like a small gametophyte that has undergone two mitotic karyokineses. The important thing seems to be that in monosporic development a morphogenic stimulus causes the megaspore to begin to develop into a megagametophyte, in bisporic development this event happens to the secondary megasporocyte, and in tetrasporic development it happens in the megasporocyte. In all events, completely functional gametophytes are formed. Klekowski (1988) suggests that the formation of bisporic and tetrasporic megagametophytes may be a mechanism that permits the healthiest, most vigorous nucleus to form the egg cell.

Gametophytes

THE MEGAGAMETOPHYTE.   In the most common type of formation of the female gametophyte (the monosporic type known as the *Polygonum* type that occurs in 81% of all angiosperms), the one surviving megaspore undergoes three rounds of mitotic nuclear division without any wall formation. During this time, the cell grows greatly in size; it has one large central vacuole but with major cytoplasmic bridges traversing it. The wall may be thickened, and it is usually without plasmodesmatal connections to the surrounding nucellar cells. During meiosis, a callose wall is often formed; the callose begins to appear before the first division of meiosis is completed (Rodkiewicz, 1970). It is not established around the entire megasporocyte uniformly, but rather first appears in the area where the surviving megaspore will be located: on the chalazal end in most, but on the micropylar end in *Oenothera*. In the monosporic and bisporic types of megasporogenesis, even the cross walls have a callose component. Rodkiewicz suggested that the callose is involved in isolating the cell that will be the one surviving megaspore, because, in tetrasporic types, where nothing degenerates, there is no formation of callose. The callose wall may persist in the megagametophyte, but it may also disappear. The cellulosic wall may become suberized.

The eight nuclei in this coenocytic megagametophyte migrate such that three occupy the chalazal end, three are in the micropylar end, and two are in the middle. Walls now become organized around the three chalazal nuclei (forming three **antipodal cells**; Fig. 19.23), around two of the micropylar nuclei (forming two **synergids**), and around the micropylar end of the other micropylar cell (forming the egg). The rest of the gametophyte is one large, binucleate **central cell** (its two nuclei are called

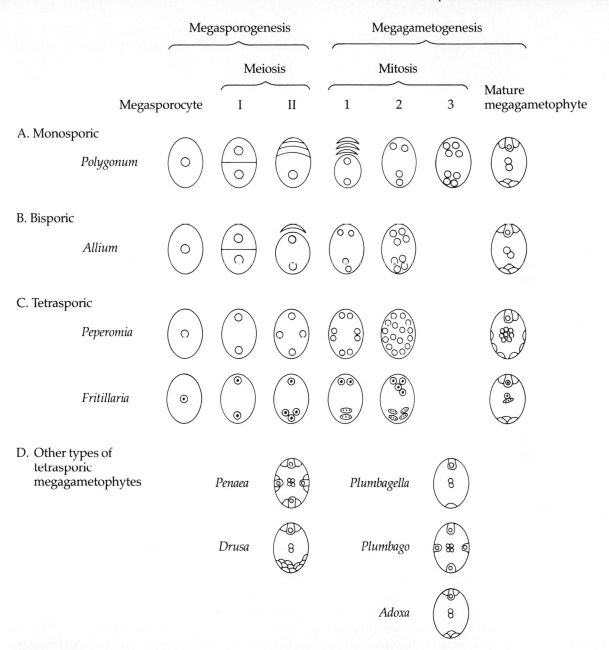

Figure 19.22
This diagram shows the various types of megagametophytes that occur in angiosperms. In the *Fritillaria* type, when three nuclei undergo karyokinesis close to each other, the spindles may merge, and the three sets of chromosomes on each side form only one nucleus, not two. Thus the result of the division of three nuclei is two nuclei, not six.

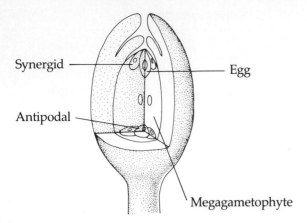

Synergid

Egg

Antipodal

Megagametophyte

Figure 19.23
This diagram should help you understand the orientation of the various cells in a megagametophyte; keep in mind that the relative sizes vary greatly from species to species, as do the shapes.

polar nuclei), and the two synergids plus the egg form the **egg apparatus** (archegonium). This type of a gametophyte is very much reduced from the type found in the conifers, and it is difficult to determine homologies. The egg is obviously homologous with all other eggs, but the synergids could be homologous with either the neck cells, the ventral canal cell, or the archegonial jacket. The antipodals might be a sterile, second egg apparatus, but they might also represent the vegetative body of the gametophyte. This is difficult to decide: in some plants (*Penaea*) there are three sets of antipodals that all resemble the egg apparatus, but in others (*Drusa, Chrysanthemum cinerariae-folium*) there is one set of antipodals, but it contains many cells. In the bamboo *Sasa paniculata*, there are up to three hundred antipodals, forming a considerable tissue (Yamaura, 1933).

One tetrasporic type (the *Fritillaria* type) is unusual, but it is important because it occurs in lilies, and these are used to prepare commercial slides for anatomy teaching labs. All four meioproducts are contained in the large coenomegaspore, but they immediately migrate

such that three are located at the chalazal end and the fourth is solitary at the micropylar end (Fig. 19.22). They all undergo a simultaneous mitotic division, but it is not possible to form three separate spindles in a space as small as the chalazal end. Instead, a single spindle results such that, as the three nuclei divide, all three sets on one side of the median plane of the spindle go to the same pole and form just one triploid nucleus, not three haploid ones. The same event happens with the other sets of chromosomes at the other pole of the spindle. After the first mitosis, then, there are still just four nuclei: two haploids at the micropylar end and two triploids at the chalaza end. A second mitosis produces eight nuclei (four haploid, four triploid). One haploid and one triploid nucleus migrate to the center of the cell, and walls are formed as described above, producing an egg apparatus with haploid cells, triploid antipodals, and a 1N + 3N central cell.

Other types of megagametophyte are presented in Fig. 19.22. It is important to note that it is not at all rare to find more than one type of megagametophyte in the same species or even in the same individual. Bhojwani and Bhatnagar (1974) mention that in *Delosperma cooperi* (Aizoaceae) 14% of the ovules are of the *Polygonum* type, 18% *Endymion* type, 32% *Drusa* type, 24% *Penaea* type, and 18% *Adoxa* type.

In the mature megagametophyte, the cytoplasm of the synergids is strongly polarized, with the majority of the organelles located at the micropylar end (Kapil and Bhatnagar, 1981). The wall in this area is a transfer wall and has numerous labyrinthine projections; it is called a **filiform apparatus** (Jensen, 1965a, b; Sarfatti and Gori, 1969; Schulz and Jensen, 1977). In the case of the synergids, the transport may be in both directions across the wall. In at least some species, the synergids produce the chemical attractant that allows the pollen tube to find the egg apparatus by chemotropism, but the synergids often are also involved in absorbing nutrients from the nucellus and passing them to the egg (Folsom and Peterson, 1984). In

Cotula australis and *Sedum sempervivoides*, the synergids grow into haustoria; in *Quinchamalium chilense*, the synergid haustoria invade the ovarian tissues and grow into the style, becoming more than 1200 μm long.

As critical as they appear, synergids are not even formed in the *Plumbagella* type or the *Plumbago* type of megagametophyte, and in most others they are ephemeral, with one degenerating almost immediately, the other dying during or just after fertilization (Schulz and Jensen, 1968a).

The egg is usually larger than the synergids and is polarized, with the bulk of its cytoplasm at the chalazal end. The few ultrastructural studies of eggs suggest a rather inactive cell. The wall is usually complete only at the base, and it either is absent or has extensive holes in it at the chalazal end (Jensen, 1965b; Kapil and Bhatnagar, 1981; Mogensen and Suthar, 1979).

The antipodals show the greatest variation: in the Sapotaceae and the Thismiaceae, the nuclei degenerate before cells can even be formed (Maheshwari, 1950). If walls do form, they can have large holes that allow the nuclei to migrate from cell to cell. Antipodal cells, if they persist, often become haustorial: in *Argemone mexicana*, they are large and are able to consume eight to ten layers of the surrounding nucellus. In *Quinchamalium chilense*, like the synergids, the antipodals are haustorial and invade through the funiculus to the placenta and attack all parts of the ovary. These antipodals are not individual cells, but rather constitute a large multinucleate coenocyte (Johri and Agarwal, 1965).

The central cell is rich in organelles and appears to be the site of much synthesis. It contains a large vacuole, and most of the cytoplasm (and the polar nuclei) is at the micropylar end, near the egg apparatus. The plastids may be large and starch rather abundant. The central cell too can be parasitic and haustorial, sending out projections that invade surrounding tissues. In *Nelsonia campestris* (Acanthaceae) the egg apparatus occurs in a part of the gameto-

phyte that invades the placenta. More striking is *Moquiniella rubra* (Loranthaceae) in which the egg apparatus is carried by the invasive megagametophyte all the way to the stigma where fertilization takes place; the developing embryo is carried back down to the ovary.

THE MICROGAMETOPHYTE. The microgametophytes are remarkably uniform in angiosperms. The nucleus of the microspore moves to the wall and divides, forming a large **vegetative cell** and a small lenticular **generative cell** (Fig. 19.24; Cresti et al., 1985; Nakamura and

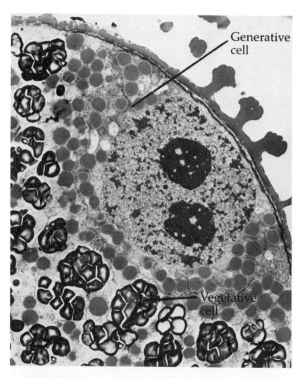

Figure 19.24
At first, the generative cell is pressed against the side of the pollen grain, as in this *Lilium longiflorum*. The nucleus is the dominant part of the generative cell, but notice also the lack of starch in it; starch is abundant in the vegetative cell. The wall that separates the generative cell from the vegetative cell is extremely thin. A portion of the pollen wall is visible. × 5000. Micrograph generously provided by S. Nakamura and H. Miki-Hirosige.

Miki-Hirosige, 1985). In some cases, the generative cell develops a cell wall (*Hyacinthus, Tradescantia*); in others, only the plasma membranes separate the two cells. Once formed, the generative cell leaves its lateral position and migrates into the protoplast of the vegetative cell. The generative cell, either within the pollen grain or in the pollen tube, will divide to produce two **sperm cells**, which are not necessarily identical (Russell, 1984). If the generative cell divides early, then the pollen grain consists of three cells; this is a condition restricted mostly to rather advanced families, especially the grasses (Brewbaker, 1967). Once the pollen has arrived at the stigma, it germinates through one of its apertures, and the pollen tube penetrates into the stigma and grows down through the transmission tissue (Figs. 19.25, 19.26).

Typically all of the pollen grain cytoplasm moves into the pollen tube, carrying the generative cell or the sperm cells with it. The name of the vegetative nucleus is now changed to **tube nucleus**, but it is not certain that it really controls the growth of the tube as its name suggests. Many species, especially wind-pollinated ones, are polysiphonous, with each pollen grain producing up to 14 separate tubes, or individual tubes may branch profusely. Only one has the tube nucleus, yet all grow well. In *Ulmus*, the vegetative nucleus degenerates even before the pollen grain germinates.

As the tube grows toward the ovules, it absorbs nutrients from the surrounding transmission tissue, but it does not produce much cytoplasm. Instead, all the protoplasm continues to move with the pollen tube tip, all the

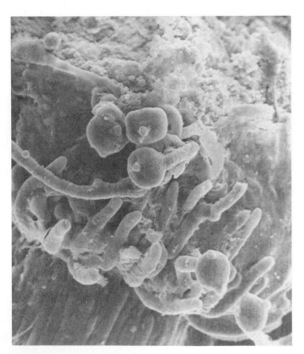

Figure 19.25
This micrograph has caught several pollen grains in the process of germinating, and their pollen tubes have just entered the stigma. × 440. Micrograph generously provided by V. R. Tilton, L. W. Wilcox, R. G. Palmer, and M. C. Albertsen.

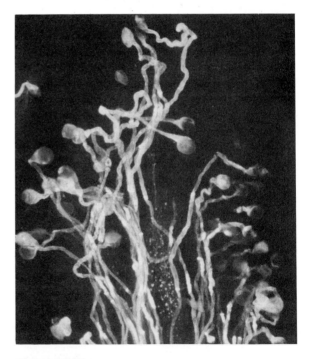

Figure 19.26
This fluorescence micrograph shows the path of the pollen tube growth in the upper portion of style and stigma of *Psychotria officinalis*. × 60. Micrograph generously provided by K. S. Bawa and J. H. Beach.

older regions being filled with vacuole. Callose plugs are deposited periodically, sealing off the empty areas. Just as in root hairs and fiber cells, the pollen tube must grow intrusively by tip growth; the lateral walls do not slide through the transmission tissue. At the extreme tip of the pollen tube, the cytoplasm is filled with organelles, especially the dictyosomes necessary for the deposition of new wall material (Cresti et al., 1977; Rosen et al., 1964).

Once inside the ovary loculus, the pollen tube may be guided to the ovule by outgrowths of surrounding placental tissue—an **obturator** (Schnarf, 1928; Tilton and Horner, 1980b)—or by a funnellike extension of the integuments (*Myriocarpa* and *Leucosyke*). Once it contacts the ovule, the tube usually grows into the micropyle (**porogamy**), but in *Casuarina* and *Rhus* it enters chalazal tissue (**chalazogamy**) and grows through the nucellus until it arrives at the egg apparatus. In *Cucurbita*, it enters the integuments (**mesogamy**). In growing through the nucellus to the egg apparatus, the pollen tube grows between cells without disturbing them. Older reports state that the pollen tube can enter the egg apparatus at any of various points, but electron microscopy (so far performed on very few species) suggests that it always enters a synergid (Cass and Jensen, 1970; Mogensen, 1978). A pore is formed in the pollen tube, either terminally as in *Epidendrum* (Cocucci and Jensen, 1969) or subterminally as in *Gossypium* (Jensen and Fisher, 1968), and the cytoplasm and nuclei are discharged. The events at this point have been difficult to study, for two reasons. First, they happen so quickly that they are difficult to catch. Second, the reaction between the pollen tube cytoplasm and the synergid cytoplasm causes both to become extremely chromophilic, staining intensely with either osmium tetroxide or the stains used for light microscopy. From what studies are available, it appears that the tube cytoplasm and the synergid cytoplasm do not mix; instead, they remain distinct but begin to degenerate.

The sperm cells migrate through this, one approaching the egg, the other approaching the central cell. As they migrate, they shed their cytoplasm, and only the nuclei cross the respective plasma membranes. The sperm nuclei move into the egg cell or central cell in areas where there is no wall. Inside the central cell, the sperm nucleus migrates to the polar nuclei (which may have already fused) and fuses with them to produce the triploid **endosperm nucleus** (this is pentaploid in lily).

Inside the egg, the sperm nucleus migrates toward the egg nucleus, and fusion may occur immediately. If so, this is a **premitotic karyogamy**, and the resulting zygote nucleus then undergoes mitosis. In *Lilium* and *Fritillaria*, there is **postmitotic karyogamy**, in which the two gamete nuclei approach and then begin mitosis. The spindles merge into one common spindle, so the result of this simultaneous mitosis is two diploid nuclei. The zygote forms a complete wall at its chalazal end (Mogensen and Suthar, 1979).

With karyogamy, the angiosperm life cycle has returned to the diploid phase in the form of a zygote. Unlike the gymnosperm zygote, this one is accompanied by an endosperm nucleus that will organize a nutritive nurse tissue that will play an integral role in the maturation of the embryo. Remember that the "endosperm" of gymnosperms is really megagametophyte tissue.

Phylogenetic Origin of the Flower

At present, the most widely accepted hypothesis about the origin of the angiosperms is that they evolved from gymnospermous ancestors about 135 million years ago during the Lower Cretaceous Period (Wilson, 1982). Many characters separate the gymnosperms from the angiosperms, so the transition must have been a slow, gradual one; it was necessary to establish vessels, libriform fibers, sieve tube members, reticulate venation, simple gametophytes, bi-

tegmic ovules, double fertilization, simple stamens, and closed carpels with a stigmatic surface. The oldest well-documented flower fossils
are from the Albian Epoch and are about 120
million years old (Dilcher, 1979). The fossils are
casts of conduplicate carpels—that is, carpels
that have a leaflike blade with the two halves of
the lamina folded together (Fig. 19.27). This
type of carpel can be found in the living
angiosperms *Degeneria*—Degeneriaceae—and
Drimys—Winteraceae (Bailey and Swamy,
1951), which are considered to have many
primitive features. It is thought, therefore, that
the carpels of *Drimys* and *Degeneria* are similar
to the original carpel: they are leaflike and bear
two rows of ovules, and the placentae are located in the center of the two halves of the lamina (Tucker and Gifford, 1966). There are three
traces to the carpel, one ventral (the midrib) and
two marginals, and all three supply bundles to
the ovules. Along the margins of the folded
lamina are rows of trichomes that form a loose
seal that protects the space (ovary loculus) inside. These trichomes act as a stigmatic surface, and pollen tubes must grow through
them. The evolution of the stigma and style
resulted as the lower and middle portions
stopped producing trichomes and the middle
region elongated.

Unfortunately, the fossil flowers mentioned
above have only carpels; below them are just
scars where other organs (petals? stamens?)
had fallen away. But all living gymnosperms
have leaflike microsporophylls, and it might be
expected that the ancestral flower also had
broad stamens (Fig. 19.27). In agreement,
Drimys and *Degeneria* (as well as two other
genera which are thought to be primitive, *Aus-
tobaileya* and *Himantandra*) have stamens that
resemble gymnosperm microsporophylls because they are flat and slightly thickened. They
do not have a filament and anther, but rather
they have four long rows of sporogenous tissues (Bailey and Nast, 1943; Bailey et al., 1943;
Bailey and Swamy, 1949; Canright, 1952; Endress, 1980; Swamy, 1948).

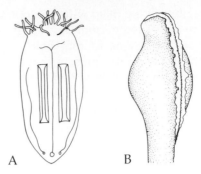

Figure 19.27
This diagram shows a stamen and carpel that are
thought to have many primitive features: the stamen
is from *Degeneria* and is somewhat leaflike, with no
well-defined filament. The carpel is from *Drimys*
and strongly resembles a young leaf whose blade
has not opened. There is no differentiation into
stigma, style, and ovary, but the margin does have
hairs that act as a stigma.

Because of the complete lack of fossils of the
early stages of flower evolution, our concepts
are based entirely on comparative morphology.
It is necessary to consider the types of structures that were available in the extinct gymnosperms, and it is necessary to consider which
extant angiosperms have the most primitive
features. Such obviously advanced and specialized groups as grasses and mistletoes will
provide little information about the characters
of the first flowers. Emphasis has been placed
on plants that are woody (no gymnosperm is
herbaceous) and that lack vessels. In carrying out the analyses, much attention has been
given to the vascular system of the flower, especially its xylem. All too often, the interpretations centered exclusively around the phylogenetic aspects of flower vasculature. But, as
Carlquist (1969) has eloquently argued, it is absolutely critical that the present-day functions,
adaptations, and biology of the flowers be considered. It is more logical to give importance
to the types of environments that the flowers
must succeed in now. This environment con-

sists not just of water scarcity, temperature extremes, and insect predators, but also of pollinators and fruit-eating animals. It is also important to consider the types of fruits that will be produced later; the receptacle and carpels are just forerunners of much greater amounts of growth and development that occur after pollination. Just as the characters of the secondary xylem and phloem are interrelated with each other and with the vascular cambium, so too are the flower and fruit interrelated with each other and with the resources that the plant can afford to spend on reproduction. If flowers did arise in the Lower Cretaceous, then the amount of diversification they have undergone in 135 million years has been quite respectable (plasticity is probably the most essential feature of angiosperms), and certain amounts of vestigial, relictual structures should be expected in such rapidly changing organs (Wilson, 1982). But, as in all other aspects of the plant's anatomy, the present biological role of the flowers, seeds, and fruits must not be ignored.

Concepts

After fertilization, the ovule and the structures within begin to develop into seeds. In almost all taxa, the requirements of the seed are the same, and consequently most seeds are remarkably uniform in structure, all having the same parts. The individual parts may vary in their composition, but usually they are easy to recognize, especially if their development is examined. In only a small number of cases (the "microspermae": Orchidaceae, Podostemonaceae, Trapaceae) has a radical departure in seed biology, and therefore also in seed anatomy, occurred. These cases will be discussed later.

In general organization, the seed contains an **embryo**, which is the young sporophyte (Fig. 20.1). It does not develop continuously but enters a state of dormancy during which it is usually resistant to many stresses: lack of water, excesses of heat or cold, crushing, and chemical attack. The embryo is always accompanied by nutrient reserves, both organic and inorganic, located either around the embryo or within its own tissues. There is a **seed coat** (**testa**) that provides a resistant and sometimes a buoyant surface.

The growth and formation of the seed occur concomitantly with those of the fructification, either the seed cone in gymnosperms or the fruit in angiosperms. As a rule, at maturity the fructification and the seed separate from each other, but in many angiosperms the fruit becomes so firmly attached to the seed coat that the two function as a unit, and it can be extremely difficult to ascertain that it is a compound structure, not just a simple seed.

The evolution of the seed had profound consequences for plants. Because of the ability of seeds to withstand extremely harsh environments, a species can thrive in poor climates by having an annual cycle that involves photosynthesis and growth by means of a plant-phase during the optimal seasons, followed by survival in the seed-phase when conditions are too severe. During the winter in the extreme northern or southern latitudes, it appears as though only woody trees exist. But beneath the snow, millions of seeds are alive, well, and quite healthy; in the following spring they will germinate and grow into new plants. The presence of trees and perennial grasses shows that it is possible for the plant-phase to become resistant enough for survival. Seeds are then not absolutely necessary, but it must be more expensive and more difficult to winterize an entire plant than merely small seeds.

Seeds have another advantage besides stress-tolerance: they are an excellent means for a species to move through the environment. Whereas animals can migrate to better locations, individual plants are rooted to their locality. Vines, runners, and rhizomes allow an individual to migrate over a very short distance through its neighborhood, and they may even permit it to encounter a patch of richer soil, more water, or brighter light. But obviously they are incapable of allowing significant mi-

minated spores. The food reserves of the seed allow the plantlet to continue this rapid initial growth, and, if the seed has landed in a mildly unfavorable microhabitat, it may be able to send its roots down to water and its leaves up to light before the nutrients run out (Fig. 21.1).

The presence of an embryo and of nutrients is so advantageous that one might expect seeds to become extremely large. Some do—some coconuts (*Lodoicea maldavica*) can weigh up to 9 kg—but many are small. The seeds of the orchid *Ilysanthes dubia* are so light that 62,000,000 barely weigh 1 kg (Kozlowski and Gunn, 1972). Why such a difference? As with all the other cells, tissues, and organs that have been discussed, there is not one single, abstract optimal structure; instead, the biology of each species determines what is best for that species. If a plant lives in a stable environment with good conditions and few seed predators, the large seeds are best (the plant must be large and perennial, of course). But some plants live in a heterogeneous environment where acceptable sites are tiny. The seeds of epiphytes and parasites must germinate on the branches of the correct host tree; weedy species grow in the disturbed areas caused by landslides, avalanches, and grass fires; many forest herbs need the bright patches of light created when a large tree blows down in a storm. If these plants have enough small seeds, although most will not land in good areas, at least a few will. If they had produced only a few large seeds, then the chances are that not even one would arrive at a proper microhabitat. An alternative is to produce seeds that can survive for several years in the soil (Kivilaan and Bandurski, 1981; Parker and Leck, 1985; Spira and Wagner, 1983), to take advantage of the fact that a site may be suboptimal one year but much better in later seasons. Also, one plant can produce several types of seeds, so more types of niches can be exploited by one parent plant (Stanton, 1984; Ungar, 1979). Any mechanism that allows the seeds to be deposited preferentially in optimal environments will be more advantageous than random dispersal. Seeds themselves have

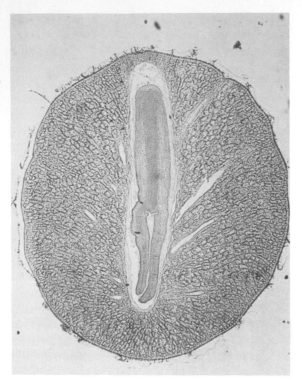

Figure 20.1
This longitudinal section of a whole seed of *Pholistoma* shows all the seed parts: the thin outermost seed coat, thick endosperm, and embryo with prominent cotyledons, hypocotyl, and radicle. × 25. Micrograph generously provided by R. Y. Berg.

gration. The spores of cryptogams are liberated into the air and can be carried long distances, permitting the plant to explore regions of millions of square kilometers. The major disadvantages of spores is that they are so small that if they land in a region that is not optimal, they will not germinate or they will die shortly after germination. The seed has a tremendous advantage here; it contains a pre-formed plant (the embryo) and food reserves. The embryo can begin growth immediately. There is no long period of organization and establishment; just by absorbing water it attains a size large enough to allow it to withstand the smaller environmental stresses that would kill newly ger-

remarkably few adaptations in regard to where they are deposited, but fruits do, and they will be discussed in the next chapter.

Nutrient Reserves External to the Embryo

Gymnosperms

In gymnosperms, the zygote is established within the archegonium located in the large megagametophyte, so nutrient reserves are already present. However, there is great proliferation of this tissue as the embryo grows, such that the very small gametophyte of the ovule produces a large seed.

Nutritive tissues are established early, even at the stage when meiosis has not yet occurred (Singh, 1978). The cells at the apex of the nucellus and even some of the sporogenous cells themselves become densely cytoplasmic, forming a **spongy tissue**. This surrounds the megaspore after meiosis occurs and then enlarges as the megagametophyte grows (Looby and Doyle, 1942; Lee, 1955; Thomson, 1905). Its cells may contain starch, as in *Pinus*, or lipids, as in *Zamia* (Pettitt, 1966). The spongy tissue is usually crushed by the later stages of expansion of the megagametophyte, but in some species, for example, *Cedrus deodara* (Roy Chowdhury, 1961), it persists and is present in the mature seed.

At the time of fertilization, the megagametophyte has become large and cellular (part may remain coenocytic), filling most of the nucellus. However, the cells of the gymnosperm megagametophyte enter a phase of meristematic activity after fertilization (Singh, 1978). This may be a diffuse activity with cell division occurring throughout the megagametophyte, or it may be restricted to certain areas (*Cephalotaxus, Ephedra, Gnetum, Torreya, Welwitschia*). This activity results in a large mass of cells that fill with nutrients; it greatly resembles the endosperm of angiosperms and is frequently called endosperm.

As the zygote begins to grow, the central portion of the megagametophyte starts to break down, and a "corrosion cavity" results (Fig. 20.2). The embryo is pushed into this nutrient-rich region and develops there. The surrounding cells of the megagametophyte fill with fats, starch, or protein (Singh and Johri, 1972). They do not undergo any further cytokinesis, but their nuclei may continue to divide. The cells of *Cephalotaxus* seeds may have up to seven nuclei each (Fujita, 1961). When the seed is mature, large amounts of megagametophyte are still present around the embryo.

Angiosperms

Within the angiosperms, the endosperm is established when the second sperm nucleus fuses with the polar nuclei, establishing the

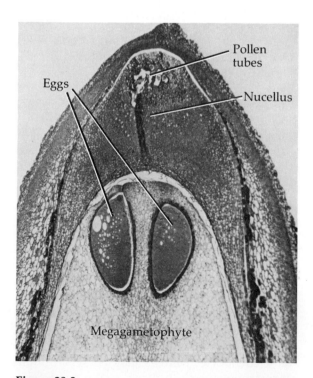

Figure 20.2
Longitudinal section of a *Pinus* ovule, showing two large eggs (each is a single cell), a thick nucellus and several pollen tubes. The megagametophyte is large, containing thousands of cells. × 50.

primary endosperm nucleus (Bhatnagar and Sawhney, 1981). This fusion results in a triploid primary endosperm nucleus in megagametophytes of the *Polygonum* type, but diploid endosperms and very high polyploids occur in bisporic and tetrasporic megagametophytes. None of the sperm cytoplasm is involved in the fusion. The primary endosperm nucleus begins one of three possible types of development.

In **cellular endosperm** development, after the first mitotic division, the central cell undergoes cytokinesis, and the first two endosperm cells are established. All subsequent nuclear divisions are accompanied by cell division, and the endosperm is cellular during all of its growth and development. There are reports (Bhatnagar and Sawhney, 1981; Sedgley, 1979) that the first wall does not form as a normal wall does but instead grows out from one side of the central cell wall and elongates until it reaches the other side. However, wall formation in endosperm is a point of controversy (see "Nuclear Endosperm"), and more studies should be performed. As cellular endosperms grow, they have a strong tendency to form haustoria, and they can invade the surrounding tissues more or less extensively, as in *Hydrocera trifora* and *Impatiens roylei* (Bhatnagar and Johri, 1972). In their review of endosperm development, Bhatnagar and Sawhney (1981) emphasized the abundance and the diversity of haustoria formed by nuclear types of endosperms; in *Euphrasia* and *Orthocarpus* they extend as tubular structures throughout the nucellus and finally contact the vascular supply of the ovule (Arekal, 1963). Cellular endosperm occurs in 72 families, all of which are dicots except for Araceae and Lemnaceae.

In **nuclear endosperm** development, nuclear division is not accompanied by cell division, at least not at first, and the endosperm grows as a coenocyte. Mitosis may be synchronous throughout the central cell, or it may be asynchronous. These multinucleate cells can become really gigantic: the center of a coconut is an endosperm coenocyte, and the liquid (coconut milk) is the protoplasm of this one cell (Fig. 21.1). In less spectacular species, the endosperm often forms a large central vacuole, and the cytoplasm and nuclei are concentrated at the two ends of the cell, with little occurring along the sides. The chalazal nuclei are often much larger and more dynamic than the micropylar ones.

The endosperm may remain free nuclear until maturity, but usually, after a certain size is attained, cell formation begins near the periphery, just as is the case in the megagametophyte of conifers. Wall formation is frequently rather irregular, and several nuclei often are trapped within a single cell; these may fuse to form large polyploid nuclei. The method of formation of these walls is still controversial: many studies (Mares et al., 1977; Morrison and O'Brien, 1976, 1978; Newcomb, 1978) report that the anticlinal walls grow inward freely, without the involvement of a mitotic spindle, perhaps in a process similar to cleavage in animal cells. As a result of this, the peripheral cytoplasm is reported to become a set of open-ended boxes (*alveoli*) that continue to grow toward the center of the coenocyte. It has been further proposed that these walls could branch and grow toward each other, fusing and establishing closed cells. Fineran et al. (1982), however, contradicted this and showed that in wheat, the peripheral nuclei divide just before wall formation, and the spindle becomes associated with the formation of the cell plate just as in all other types of plant cytokinesis (Fig. 20.3). As the new cell plate grows between the daughter nuclei, one edge meets the mother cell wall, and the other edges continue to grow. When the nuclei divide in a perpendicular plane, the other cell walls become established, and compartments are formed. A final round of nuclear division establishes a periclinal wall and closes the compartment around one daughter nucleus (Fig. 20.4). The other daughter nucleus needs to divide only once to form a complete wall around itself. The compartments continue to grow inward until those of

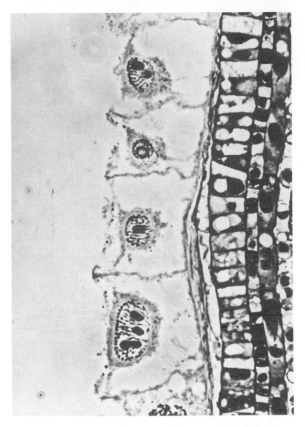

Figure 20.3
This micrograph shows the beginning of the formation of cell walls in the coenocytic endosperm of wheat. The nuclei become aligned along the edge of the endosperm, then divide; the spindles then become associated with phragmoplasts. × 1025. Micrograph generously provided by B. A. Fineran, D. J. C. Wild, and M. Ingerfeld.

one side meet those of the other, converting the entire endosperm to a cellular mass. The large cells formed by this process in wheat can then subdivide into small cells.

In many species, especially if the initial coenocyte is large, wall formation will not be uniform over the entire surface. In the Cucurbitaceae, only the upper portion becomes cellular, while the chalazal area remains coenocytic. In *Scleria foliosa* (Cyperaceae) also, the

chalazal area remains coenocytic: wall formation begins near the proembryo and then stops after only one-third of the endosperm has become cellular.

In coconut, as the endosperm enlarges, it fills with numerous free amoeboid nuclei, some as large as 90 μm in diameter, larger than most parenchyma cells (Cutter and Freeman, 1954; Cutter et al., 1955). When the fruit is about 100 mm long, large "vesicles" are seen in the endosperm; these are apparently multinucleate cells that become established as cytoplasm coalesces around groups of nuclei. Each vesicle is surrounded by its own plasmalemma, but there is no wall. As many as 40 nuclei may be in each cell. Finally, the cells settle out of the liquid cytoplasm and become attached to the endosperm wall, resulting in a peripheral gelatinous layer. The nuclei undergo mitosis, and walls are laid down, resulting in the formation of the white coconut "meat" that we eat.

Wall formation never occurs in some species, for example, *Limnanthes* or *Oxyspora* (Bhatnagar and Johri, 1972). Nuclear endosperm is definitely the most common type of development: it occurs in 161 families of both monocots and dicots. Like cellular endosperm, it frequently forms haustorial outgrowths into the surrounding tissues.

In **helobial endosperm** development, the division of the primary endosperm nucleus is accompanied by the formation of a horizontal wall that establishes a large micropylar cell and a small chalazal one. The chalazal cell may either not divide at all or may undergo a small number of divisions (either free nuclear or cellular). It contributes little to the endosperm, and it is frequently crushed by the growth of the micropylar cell (Swamy and Parameswaran, 1963). The micropylar cell grows rapidly, accompanied by free nuclear divisions at first; later, cells may form in this coenocyte.

This is a rare type of endosperm: it occurs in only 17 families, 14 of which are monocots. In a review of the literature, Swamy and Krishnamurthy (1973) concluded that the reports of

Figure 20.4
This diagram depicts the initiation and growth of the walls as the endosperm of wheat becomes compartmentalized. Diagram generously provided by B. A. Fineran, D. J. C. Wild, and M. Ingerfeld.

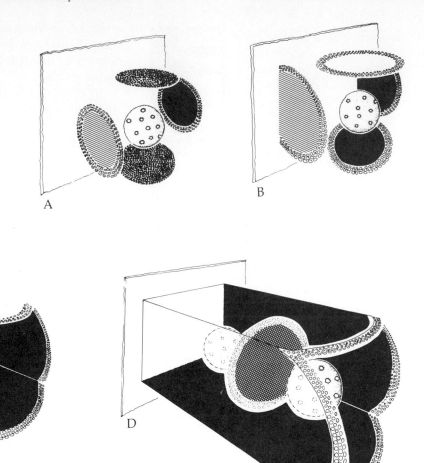

helobial endosperm in dicots may be misinterpretations; they felt that the dicot families in which it is reported to occur (Olacaceae, Saxifragaceae, and Santalaceae) have such variable and often irregular endosperm development that the studies had encountered abnormal instances of the other types of endosperm. They felt that helobial endosperm occurs only in the monocots.

It has been impossible to decide which type of endosperm is ancestral. They occur side by side in both primitive and advanced orders and can occur together in the same family.

Perhaps because the endosperm is a temporary, calluslike tissue, unusual cytological conditions are tolerated in it. In some individuals of the Trapaceae, Podostemonaceae, and Orchidaceae, the second sperm nucleus is not discharged from the pollen tube, so no endosperm is formed at all. In several species, the endosperm nucleus immediately degenerates, and in *Vanilla planifolia* and *Galeola septentrion-*

alis division stops after only about 12 to 16 nuclei are formed. In *Zea*, the sperm nuclei may not fuse with the polar nuclei, but both sets of nuclei do divide, and an endosperm results that is a mosaic of "male" and "female" cells. In *Petunia* (Ferguson, 1927) and *Lycopersicon* (Bhaduri, 1933), fusion of the sperm nucleus occurs after the polar nuclei have fused and already divided; one of these nuclei undergoes karyogamy with the sperm nucleus, and a mosaic of diploid and triploid endosperm results. Once established and growing, DNA replication becomes poorly coordinated with mitosis, and high polyploidy results, with ploidy levels greater than 9 n being common. In *Arum maculatum*, 24,576 n has been reported (Bhojwani and Bhatnagar, 1974).

After fertilization the endosperm cytoplasm becomes more dense as more organelles are formed (Briarty et al., 1979). The endoplasmic reticulum increases in extent, both the RER and the SER being involved (Bhatnagar and Sawhney, 1981). There are reports (Schulz and Jensen, 1977) that the SER is connected to the plasmalemma and may be involved in the transport and deposition of wall materials. The mitochondria increase in number, and they may also swell considerably, presumably due to greater biochemical activity (van Went, 1970). Plastids, all maternal, become numerous and may contain starch; their shape and size are variable. Storage materials accumulate in later stages of development, starch being by far the most common product (Bewley and Black, 1978). Protein bodies may also be formed (Lott, 1981), especially in the monocots that have abundant endosperm in the mature seed. As the endosperm develops, the ratios and relative abundances of the various components change: the vacuole continuously decreases, the nucleus/cytoplasm ratio increases, and finally a phase occurs in which the accumulation of nutrients predominates.

As the endosperm grows, it frequently consumes the nucellus (Fig. 20.1), and sometimes even part or all of the integuments (Boese-winkel and Bouman, 1980); in *Crinum*, it absorbs the seed coat and fruit wall and sits exposed to sunlight at maturity. More typically, endosperm just forms a rather large mass with a smooth contour that fills the seed except for the embryo, and no nucellar tissue remains when the seed is fully developed. The mechanism by which the nucellus is broken down and absorbed is not known (Norstog, 1974; Pacini et al., 1975). It has been termed autolysis, but the source of enzymes or of controlling mechanisms has not been identified. Similarly, the site for absorption of the nutrients released by nucellar degeneration is not known, but the entire surface of the endosperm seems a logical assumption; experimental studies are needed.

Although the surface of the endosperm is usually rather smooth during much of development and maturity, in some families it may be highly convoluted, projecting into any remaining nucellus or into the seed coat. It may even cause the seed coat to be rough and distorted as well. Such endosperm is said to be **ruminate endosperm** (Fig. 20.5; Bhatnagar and Sawhney, 1981), and it may develop from any of the three basic endosperm types. Rumination may be caused by localized proliferation of the peripheral cells of the endosperm (*Andrographis echioides, Elytraria acaulis*) or by generalized irregular growth of the endosperm. In many cases, it is the seed coat that causes the rumination. In Annonaceae and Aristolochiaceae the cells on the inner side of the seed coat elongate into the endosperm, whereas in Palmae and Vitaceae there are small meristems on the integuments that result in ingrowths that penetrate the endosperm.

Just as the endosperm typically consumes all of the nucellus, the embryo frequently absorbs all or most of the endosperm, and all food reserves are located inside itself. This type of seed, where the endosperm is absent or very scanty (beans, peas, peanuts), is **exalbuminous** and is much more common in the dicots than in the monocots. In other types of plants, especi-

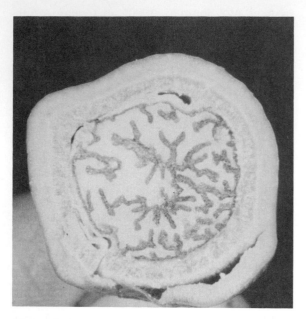

Figure 20.5
This seed of *Picrodendron macrocarpum* has ruminate endosperm. × 10.

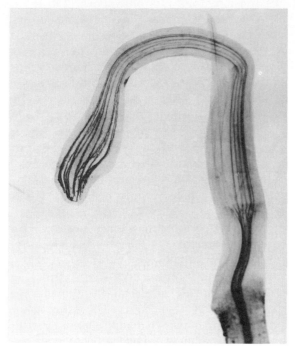

Figure 20.6
The cotyledon of *Yucca* remains in the seed, embedded in the perisperm; it absorbs the sugars and transports them to the embryo through the prominent vascular bundles that are visible here. × 5.5. Micrograph generously provided by H. J. Arnott.

ally the monocots, the seeds are **albuminous**, containing large amounts of endosperm and relatively small embryos (Fig. 20.1); for example, corn kernels are mostly just endosperm. When albuminous seeds germinate, the endosperm nutrients must be transferred to the growing seedling (Fig. 21.1). Therefore, the seedling remains attached to the seed, digesting and absorbing the endosperm by means of haustoria, usually modified cotyledons (Fig. 20.6).

An interesting observation is that even in very large masses of endosperm, no vascular tissue ever develops. In many cases, parts of the endosperm are separated from the embryo by long distances, yet special conducting tissues do not occur.

In the dicot order Caryophyllales, the endosperm often stops developing very early, and the nutritive tissue, **perisperm**, is formed by the proliferation of the nucellus instead. The nucellar cells divide and expand, filling with starch or proteins. The perisperm greatly re-

sembles endosperm, and it is difficult to distinguish the two without observing the early stages. Like the endosperm, perisperm may either persist in albuminous seeds or be completely consumed in exalbuminous ones.

It is important to point out here that there is really very little actual evidence that the endosperm is truly a nutritive tissue or that endosperm haustoria really do take up and transport nutrients to the embryo. Virtually all evidence is circumstantial. The endosperm haustoria penetrate the chalaza or the micropylar tissues of the integuments, or both, and these are typically the only routes for nutrient transfer to the embryo (Kaplan, 1969). In albuminous seeds, the germinating embryo absorbs the nutrients that result from the digestion of the endosperm by means of the cotyledon, as in *Desmaria* (Kuijt, 1985), Liliaceae (Arnott, 1962), Palmaceae (Tomlinson, 1961), Poaceae, and *Tristerix* (Mauseth et al., 1985). However, except for the grasses, virtually nothing is known about these processes, whether the cotyledons produce the digestive enzymes or whether the endosperm undergoes autolysis. Smart and O'Brien (1983) examined development in wheat and reported that the endosperm broke down more rapidly than the embryo expanded, resulting in a space between the two. But they did not know how the process is controlled or what the source of enzymes is. Bewley and Black (1978) reviewed the literature on embryo and endosperm interaction during the endosperm breakdown in albuminous seeds; they pointed out that, in grass seeds, the embryo is located to one side of the endosperm, so the two can be separated easily for experimentation. In other seeds, however, the embryo is embedded in the endosperm, and any attempt to separate the two results in significant injury to the endosperm. They concluded that experiments performed to date were inconclusive: some can be interpreted to indicate that the embryo does control endosperm breakdown, others suggest it does not. Even less is known about the processes in exalbuminous seeds.

Finally, during the very earliest stages after fertilization, it seems logical that the active, rapid growth of the endosperm would compete with the embryo for nutrients and might actually interfere with embryo nutrition. It may be that at least in exalbuminous seeds the endosperm plays a much more complex role than just accumulating, storing, and then later releasing sugars and amino acids; instead it may be a source of growth factors or complex organic compounds. The same roles may be true even of albuminous seeds while the embryo is developing: the actual accumulation of storage starch or protein occurs predominantly in the last stages of seed maturation, not continuously (Bewley and Black, 1978).

The Embryo

Gymnosperms

The embryo is not formed immediately from the egg; rather there is first a stage in which a **proembryo** and a **suspensor** are formed (Fig. 20.7; Singh, 1978). The egg nucleus divides twice mitotically; there is no cytokinesis, and the four nuclei migrate to the base of the egg, farthest from the pollen chamber. They are arranged in a flat plane, then simultaneously undergo a transverse division to form two tiers of nuclei (*Pinus*). In some species, a second round of division produces four or more tiers (*Agathis*, *Cephalotaxus*, *Podocarpus*). Cell walls are organized such that these sixteen nuclei become situated in sixteen cells; four

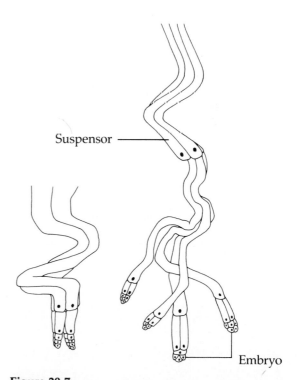

Figure 20.7
The suspensor cells of a gymnosperm embryo elongate, pushing the proembryo (the terminal cells) deep into the megagametophyte.

tiers of four each, the basal three tiers being closed cells, the top tier being cells that are open to the bulk of the egg protoplasm. The most basal tier constitutes the proembryo. The second-basalmost tier forms the suspensor; these cells will elongate, pushing the embryo cells into the corrosion cavity of the megagametophyte. As they do so, the terminal proembryo cells separate from each other and begin division. Each of the four can start to produce an embryo. Because a single megagametophyte can have several archegonia, eight, twelve, or more embryos may be initiated and thrust into the megagametophyte. Typically, only one survives in each gametophyte.

Growth of the proembryo into a mature embryo has been studied in detail only in the Pinaceae and *Welwitschia* (Martens and Waterkeyn, 1974). When the original proembryo tier splits into four separate proembryos, each becomes filamentous and automatically has an apical cell, the one farthest from the suspensor (von Guttenberg, 1961). According to Buchholz (1931), this cell acts as a pyramidal apical cell and produces derivatives just like a fern apical cell. After a short period of growth, the embryo is shaped like a ball and is in a **globular**

stage. It grows only at the two opposite ends and reaches the **torpedo stage**. The embryo then has an elongate axis with meristematic regions at both ends, and a special apical cell can no longer be detected (Fig. 20.8). At the suspensor end, a root meristem and root cap become well developed as oriented cell division results in a central columnar region and a peripheral root cap. At the opposite end, the cells proliferate in all directions. The first tissue differentiation occurs while the embryo is just slightly oblong: the central cells enlarge and become vacuolate, establishing the pith. Surrounding these cells are ones that are more elongated and narrow, but they are not designated as distinct provascular cells until the outer layers enlarge, forming the cortex. Spurr (1949) felt that the epidermis was the last tissue to be organized; although the outer layer of cells is rather prominent even from the globular stage, periclinal divisions are frequent until after the cortex forms. At this time the outer cells divide only anticlinally, and the epidermis becomes distinct. A shoot apical meristem and **cotyledons** arise last. Just below the apex, periclinal divisions in the cortex result in a swelling that is a ring-shaped foliar buttress. From it

Figure 20.8
Longitudinal section of a *Pinus strobus* embryo. The shoot apex is conical and is just beginning to produce cotyledon primordia; the root is becoming distinct from the suspensor. × 200.

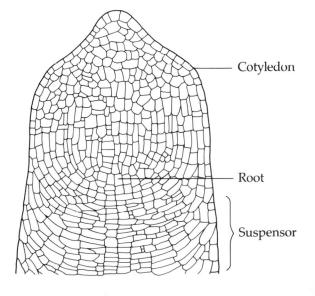

Cotyledon

Root

Suspensor

arise two to several cotyledon primordia. The number of cotyledons is variable: there are often eight or more in the conifers but only two in the other gymnosperms (Singh and Johri, 1972). In *Pinus strobus* the cotyledons initiate a prominent vein of provascular tissue before becoming dormant (Spurr, 1949). Spurr refers to the shoot apical meristem as an **epicotyl** (a shoot present within the seed), but this is not accurate because it has no leaf primordia at all; it is just an apical meristem. The portion of the embryo axis between the root apex and the cotyledons is the **hypocotyl**.

The conifer embryo stops growing while still rather small but well developed. *Cedrus* embryos are reported to have fully differentiated tracheids, and in *Pinus* the first stages of resin canals can be identified (Berlyn, 1972). Large amounts of the megagametophyte ("endosperm") remain. The embryo does not store very much nutrient reserve inside itself, but some starch and protein are deposited. During germination, these are utilized for initial growth and leaf initiation (Cecich and Horner, 1977; Kramer and Kozlowski, 1979).

Angiosperms

After fertilization, the zygote completes the wall around the chalazal end of the cell. It enters a period of apparent quiescence during which its large central vacuole is resorbed, and the amount of cytoplasm increases. The first division is almost always transverse, producing a **basal cell** and a **terminal cell**. The divisions after this may occur in a variety of patterns. Elaborate classifications have been established that contain many types based on whether the terminal cell and basal cell divide transversely or longitudinally and which cells contribute to the embryo and which to the suspensor. Charts, diagrams, and lists of these types are frequently presented, and the terms are occasionally used; the variability of development within any single species is so great that these classifications cannot be defended or

accepted, and many investigators have shown that there is no validity to these embryo typification schemes (Philip, 1972; Sachar, 1956; Wardlaw, 1955).

EMBRYO DEVELOPMENT IN DICOTS.   In many dicots, there is the formation of a suspensor (Figs. 20.9, 20.10). Suspensors are most often rather small and inconspicuous in comparison with those found in the conifers, but they can range from being virtually absent, which is the common condition, to being well developed and conspicuous, which is rare (Wardlaw, 1955). The suspensor usually consists of just a few cells that push the embryo into the endosperm (Fig. 20.11). In some species (especially in the Leguminosae), the suspensor can be large and elaborate, consisting of a row of large, flat, coin-shaped cells, as in *Lupinus pilosus*; a double

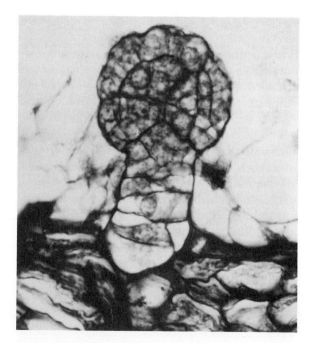

Figure 20.9
This embryo of *Cassipourea* is at the globular stage; you can see that the epidermis is already distinct and that the short suspensor is quite prominent. × 525. Micrograph generously provided by A. M. Juncosa.

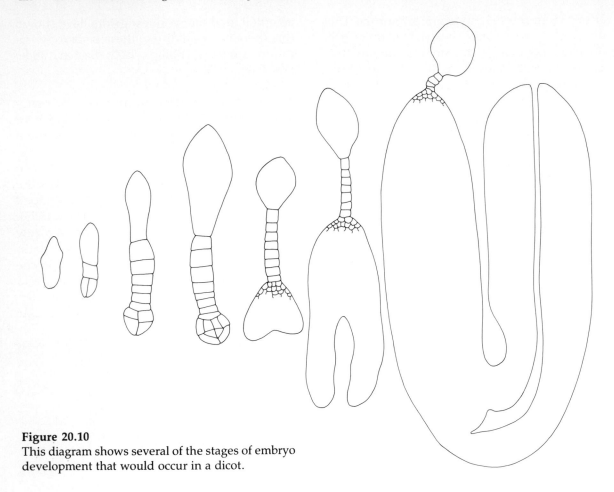

Figure 20.10
This diagram shows several of the stages of embryo development that would occur in a dicot.

row of elongate, filamentous cells, as in *Lupinus subcarnosus;* or a row of large, balloonlike cells filled with starch, as in *Ononis* (Lersten, 1983). The suspensor completes its development early, while the embryo still consists of only a few cells. By the time the cotyledons are initiated, the suspensor has usually begun to degenerate. At maturity of the seed, the suspensor has been crushed completely by the growth of the embryo.

Whereas the function of the suspensor in the gymnosperms appears to be rather easy to understand (it always elongates and pushes the proembryo into the megagametophyte), in the angiosperms the function is much less ob-

vious. As stated earlier, the suspensor is frequently absent or only rudimentary; normal embryo development in many angiosperms does not involve the suspensor at all. In those angiosperms in which it does persist, its size and structure can be variable. Lersten (1983) compiled the published descriptions of suspensors in Leguminosae, then arranged them according to a recent phylogenetic scheme for the family. He was unable to find any clear trends or correlations of suspensor structure.

There are reports that the suspensors are capable of producing growth regulators and of transporting nutrients into the proembryo (Ceccarelli et al., 1981; Hardham, 1976; Lorenzi

embryo, and each zygote produces just one embryo; four do not even begin to form as they do in the conifers. The embryo grows as a small ball of cells. Often the transition from the 8-cell stage to the 16-cell stage is by periclinal wall formation in the outer cells. This establishes the protoderm, which grows by anticlinal divisions, so the embryo is divided into two distinct populations of cells. The cells of the inner set divide in all planes, and the embryo begins to lengthen. Unlike the gymnosperm embryo, in most dicots the primordia of the cotyledons form before there is tissue differentiation in the embryo axis (Yeung and Clutter, 1978). Two areas farthest from the suspensor begin to protrude and develop into the two cotyledons, and the embryo is said to be in the **heart stage** (Fig. 20.12). With continued growth, the em-

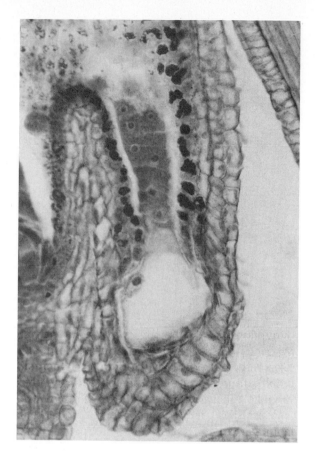

Figure 20.11
The suspensor of *Capsella* is temporarily prominent, with a large basal cell and a short uniseriate stalk. × 100.

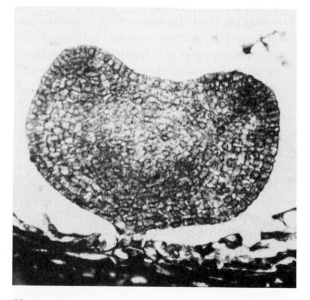

Figure 20.12
This embryo of *Cassipourea* is in the heart stage: the two cotyledons are being initiated. Notice how early this is; provascular tissue is just becoming visible, and there is no shoot or root apical meristem. The cotyledons, although called seed leaves, are not initiated by the manner common to all other leaves. × 335. Micrograph generously provided by A. M. Juncosa.

et al., 1978; Simoncioli, 1974; Yeung, 1980), but the number of such reports is too small to inspire any confidence in generalizations. Furthermore, because of technical difficulties, all of these studies have been carried out on species that have very large, well-developed suspensors.

The dicot embryo forms from the chalazal end of the zygote, either from just the terminal cell or with some contribution of the basal cell as well. Unlike the gymnosperms, angiosperms are not often considered as having a pro-

bryo becomes elongate (the torpedo stage). Finally, the root and shoot apical meristems become organized, the hypocotyl occurs between
them, and the cotyledons are attached to the
hypocotyl (Fig. 20.1). In *Phaseolus*, which has
large cotyledons, the procambium is fully differentiated in the axis before the cotyledons
have finished elongating.

At maturity the cotyledons, if they are primarily for storage, are often small, rather
spherical, and very fleshy. They may have only
a few narrow vascular bundles. After they release their nutrients during germination, the
cotyledons may either wither and die or develop chloroplasts and begin photosynthesis.
Which course they take is related to whether
the cotyledons remain underground during
germination (**hypogeal germination**) or are elevated into the light (**epigeal germination**). If
epigeal, the cotyledons are usually thin, lami-
nar, and very leaflike, examples being *Brassica*,
Galium, and *Lepidium* (Lubbock, 1896). They
begin photosynthesis immediately. Such cotyledons have vascular tissue and epidermis well
developed while still within the seed.

The root apical meristem typically grows
very little before dormancy, but at least a short
segment of root axis (the **radicle**) is produced.
In the hypocotyl, the provascular tissue becomes distinct, setting the pith off from the
cortex. Fully formed protoxylem is not unusual, especially in larger embryos. The epicotyl apex may develop full zonation before
germination, or it may remain as just a small
tunica covering a tiny, undifferentiated corpus
(Fig. 6.15; Mauseth, 1978a, 1980c). At maturity,
the embryo may be straight or curved into a
U shape.

The cells of the embryo often contain many
nutrient reserves, especially in exalbuminous
seeds, of course (Bewley and Black, 1978; Lott,
1981; Parker, 1984a, b). Most typically, the large
cotyledons are the main sites of storage, but
the hypocotyl may serve this function instead
if the cotyledons are very small, as in Cactaceae (Fig. 20.14). Starch grains and pro-

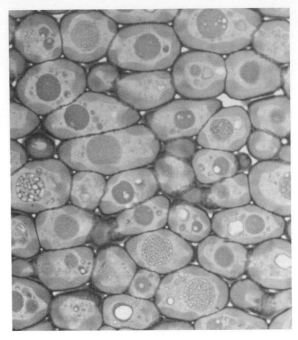

Figure 20.13
These are cotyledon cells in *Lupinus angustifolius*;
they contain large protein bodies as do many cotyledons, but they also have thickened walls that act as
a reserve of carbohydrates. × 100. Micrograph generously provided by M. L. Parker.

tein bodies are the predominant storage forms
within the embryo, just as is true of the endosperm, but other methods of sequestering nutrients are possible (Fig. 20.13; Meier and Reid,
1982). Mature seeds of many lupins contain
only traces of starch; instead, the walls of the
cotyledon mesophyll cells are massively thickened. The exact chemical composition of the
walls is not known, but they do contain large
amounts of arabinose, galactose, and xylose.
In *Lupinus angustifolius*, it was found that the
mesophyll wall increases from 0.2 μm to 20 μm
by the time the embryo is mature (Parker, 1984a,
b). The small amount of lumen that remains is
filled with protein and lipid.

Although protein is the most common storage form of nitrogen, alternatives are possible:

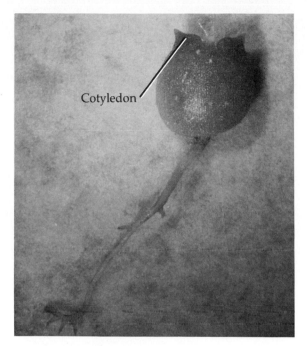

Cotyledon

Figure 20.14
This seedling of *Ferocactus acanthodes* has reduced cotyledons; the hypocotyl is greatly enlarged and acts as the nutrient reserve. × 10.

alkaloids, amides, and free amino acids occur as well. Theobromine is a major component of seeds of *Theobroma cacao* (from which we get chocolate), and caffeine is, of course, important in the seeds of *Coffea* (coffee). Certain minerals are stored as phytin, a complex of the calcium, magnesium, and potassium salts of phytic acid. The phytic acid portion of the salt itself contains a high percentage of phosphate, so phytin is an excellent reserve for four major minerals. Phytin is invariably present within protein bodies (Bewley and Black, 1978).

The method of transfer of nutrients into the embryo is not known. It is suspected (and frequently stated) that the integuments secrete material or that the suspensor (if present) is the route for secretion, but no definitive stud-

ies exist, and all evidence is rather circumstantial. However, the vascular connection from the maternal plant passes at least to the chalaza and frequently up into the integuments, so a transfer from the integuments to the endosperm to the embryo is at least logical. A set of intriguing studies found that the enzyme phosphatase (believed to be involved in the active transport of assimilates) is most highly concentrated over the entire surface of the embryo (Mogensen, 1985; Pollak and Mogensen, 1986). Therefore, the whole surface may be absorptive. Furthermore, the enzyme is abundant on the inner surface of the integument, where it could mediate secretion from the integument to the endosperm.

The types of reserves just described are usually most abundant in the endosperm, but they can be deposited in all parts of the embryo; the hypocotyl, the radicle, and the apical meristem all may become virtually packed with starch, protein, and other reserves. At this point, the embryo usually becomes dormant and dehydrates, and the seed is mature.

The development just described is the most common type, and variations occur. In the "microspermae" (Orchidaceae, Potamogetonaceae, and Trapaceae—families with seeds that are almost microscopic and dustlike) the embryo stops growing while it has less than one hundred cells. Almost all development occurs after the seed germinates and is infected by a symbiotic fungus. As a rule, the seeds of many advanced parasitic angiosperms also stop development early, before true radicles or cotyledons are produced (Okonkwo and Raghaven, 1982). In the parasite *Tristerix aphyllus*, the embryo is ephemeral and dies once the haustorium penetrates the host. The embryo consists of just the hypocotyl-haustorium and very reduced, fused cotyledons. There is no embryonic shoot, which would be superfluous anyway, because the seedling dies so quickly (Mauseth et al., 1984, 1985). In other genera, the embryo shoot apical meristem becomes active before dormancy sets in, and several foliage leaves are produced, which develop rather extensively (peanuts). The axis thus produced is the epicotyl. In corn, the epicotyl has five leaves, and only another ten will be formed after germination; one-third of all the leaves of the plant exist in the seed. In **viviparous seeds**, dormancy never occurs, and the embryo simply grows out of the seed and fruit while still on the sporophyte of the preceding generation, as in *Rhizophora* (Juncosa, 1982).

EMBRYO DEVELOPMENT IN MONOCOTS. In monocots, the initial stages of development of the embryo are similar to those described for dicots, but, at the heart-shaped stage, differences become obvious. The monocots have just one cotyledon, as their name indicates, and debate continues as to whether this is the result of the loss of one cotyledon, the fusion of the two, or something else (Fig. 20.15). At present, the most commonly accepted interpretation (Berlyn, 1972) is that the one cotyledon that does form is large and is an interface with the endosperm—most monocots have albuminous seeds (Avery, 1930; Guignard, 1961; Hayward, 1938). This cotyledon in the grasses is called a **scutellum** (Negbi and Koller, 1962), and after germination it is a source of digestive enzymes that attack the endosperm (Negbi, 1984). In the grasses, the scutellum remains in place and absorbs the hydrolyzed endosperm, but in many palms, the cotyledon proliferates rapidly and forms a large, aerenchymatous haustorium that digests and invades the endosperm. In coconuts, it becomes massive at this stage (Tomlinson, 1961). In a recent review, Negbi (1984) pointed out that even in grasses the cotyledon (scutellum) is not always so simple and uniform as it is frequently thought to be. In *Avena* it can elongate during germination and penetrate deeply into the endosperm, whereas in *Zizania aquatica* it burrows into the endosperm before the seed becomes mature. In four genera of bamboo, the scutellum digests most of the endosperm before germination; it swells and acts as the storage site, much like many dicot cotyledons.

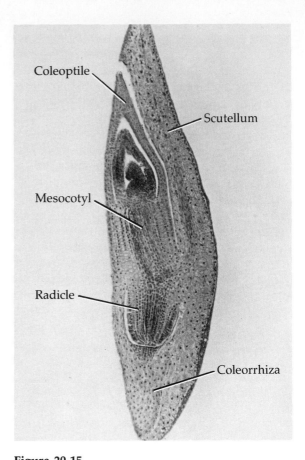

Figure 20.15
This is an embryo of *Panicum maximum*, a grass. On the right is the large scutellum; on the left are the embryo axis and leaves. The radicle and shoot apex are visible; notice that the radicle is encased by the coleorrhiza and that the large outermost leaf is the coleoptile. The point where the scutellum attaches to the rest of the embryo is the mesocotyl. The epiblast is not visible. × 78. Micrograph generously provided by C.-Y. Lu and I. K. Vasil.

The seed of the grasses has been studied in detail (Guignard, 1961; Hayward, 1938; Negbi and Koller, 1962), and shows many modifications. The cotyledon is attached to the embryo at the first node or **mesocotyl**. The term *mesocotyl* was suggested while the nature of the scutellum was unknown; now that it is considered to be a cotyledon, the mesocotyl could be more accurately called the first internode of the epicotyl. Above the mesocotyl is the epicotyl, bearing several leaves or leaflike structures. In the annual grasses that have determinate growth, almost half of the leaf primordia that will ever exist are present in the seed. The lowest of these structures is the sheathlike **coleoptile**. Its nature has been greatly debated; it has been suggested that it is either a leaf or partly cotyledon or an outgrowth of the scutellum (Worsdell, 1916). However, Negbi and Koller (1962) pointed out that the coleoptile is very leaflike. It may have an axillary bud, and in at least two unrelated grasses, *Jouvea pilosa* and *Streptochaeta* (Reeder, 1956), the coleoptiles have leaflike venation. Its present function is to extend through the soil to the surface, thus producing a smooth, protective sheath for the passage of the subsequent leaves and shoot axis. Its function is really analogous to the root cap.

Opposite the scutellum is the **epiblast**, a small, nonvascularized flap of tissue. It has been suggested that this flap is a residual second cotyledon, but that is still not settled. An objection to considering it a cotyledon is that it arises rather late in development, but as Negbi and Koller (1962) pointed out, it could be that it is biochemically initiated early but does not become physically apparent until later. It is at least a modified leaf or cotyledon, because it occurs in the proper phyllotaxy.

In the lower extremity of the axis are the radicle and root cap. Unlike the radicle of the dicots, in grasses the primordial root is protected by a sheath of tissue, the **coleorrhiza** (pl.: coleorrhizae). Negbi and Koller (1962) presented thought-provoking evidence (amassed by Pashkov) that the coleorrhiza may actually be the true radicle and that the structure that we call the radicle may actually be a giant lateral root primordium. The evidence is as follows: (1) the coleorrhiza bears root hairs; (2) in the monocots without a coleorrhiza, the radicle is attached to the suspensor, but in grasses the

coleorrhiza is attached to the suspensor; (3) the radicle differentiates after the coleorrhiza and within it. Several other families have coleorrhizae: Commelinaceae, Cycadaceae, and Lauraceae.

Not all monocot embryos are so complex: in *Yucca*, the cotyledon is not highly modified into a scutellum; instead it is large and leaflike and is deeply embedded in the abundant endosperm (Fig. 20.6). During germination, it acts as a haustorium and transfers nutrients to the embryo, just as a scutellum does. There is no coleorrhiza; the root cap is continuous with the suspensor (Arnott, 1962).

Although the root grows little before dormancy, it does produce several lateral root primordia. Adventitious root buds can also form in the mesocotyl; all of these are **seminal roots**.

The Seed Coat

The integuments are normally the source of the seed coat (testa), and the only generalization that can be made is that anything can happen. The seed coat may result from all of both integuments, all of either one, parts of both, or parts of either (Netolitzky, 1926). Occasionally, remnants of the nucellus become part of the inner surface of the testa (Euphorbiaceae, Malpighiaceae, Polygonaceae, Thymelaeaceae). The integuments often must undergo considerable cell proliferation as they expand and thicken into a seed coat. In *Dichapetalum*, the outer integument begins to grow as the inner integument is absorbed by the endosperm; although only three to four cells thick at the time of fertilization, it becomes twenty cells thick by the time the seed is mature (Fig. 20.16; Boesewinkel and Bouman, 1980). The inner several layers develop palisade characters; in the outer layers, provascular strands develop, then mature into an extensive network of vascular bundles throughout the developing testa. In the final stages of maturation, the inner palisade layers degenerate, and the tissue becomes spongy; then the vascular bundles are stretched

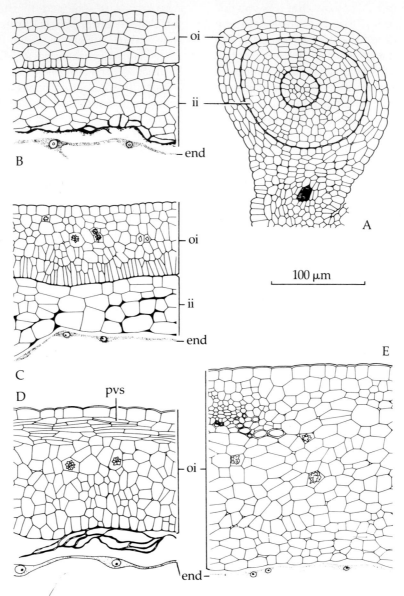

Figure 20.16
Cross section of the developing seed coat of *Dichapetalum mombut-*
tense. ii = inner integument; *oi* = outer integument; *end* = endosperm.
Diagram generously provided by F. D. Boeswinkel and F. Bouman.

and torn. The mature testa consists only of the
outer epidermis, some tannin cells, and remnants of the xylem in the bundles. In certain
species of the Boraginaceae, Convolvulaceae,
and Polemoniaceae (families that have uniteg-

mic ovules) the seed coat consists of only the
outer epidermis (Bouman and Schier, 1979).

In contrast to such thin testas, those of the
seeds of flax (*Linum usitatissimum*) are more
massive (Boesewinkel, 1980). The inner epider-

mis of the inner integument becomes a densely cytoplasmic endothelium; the middle layers of the inner integument proliferate rapidly and constitute the bulk of the entire developing seed. Initially they store much starch, but then this is lost, and the entire middle layer breaks down. The cells of the outermost layer of the inner integument elongate in the same direction that the seed does, but they keep up with the growth in circumference by rapid cell division. These cells ultimately sclerify. All cells of the outer integument enlarge greatly and fill with starch, but this is just temporary; in the final stages of maturation, the starch disappears, and mucilage precursors are formed. In the mature dry seed, these exist as layers in the outer part of the cell; when wet, they swell enormously and produce a slime sheath around the seed.

Rather remarkably, something as simple as a testa can contain a wealth of surface details and patterns such that the testa may be of critical importance in the classification and identifi-

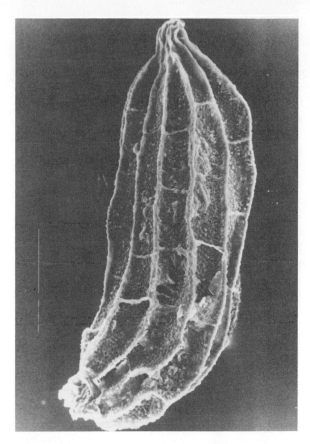

Figure 20.18
The seeds of *Stemodia* are small, and their seed coats are thin and simple. The large rectangles are the outlines of individual cells; the radial walls are thick and resistant, but the outer periclinal wall is thin and somewhat collapsed. × 187. Scanning electron micrograph generously provided by C. Cowan.

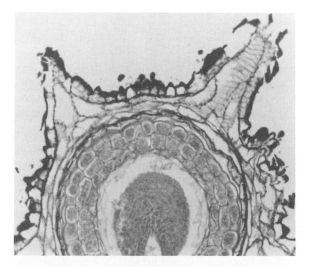

Figure 20.17
This cross section of *Galvesia fruticosa* shows a seed coat composed primarily of the epidermis and hypodermis of the integument. The elongation of the hypodermal cells produces the ornamentation on the seed coat. × 20. Micrograph generously provided by W. J. Elisens.

cation of plants (Figs. 20.17, 20.18; Benson, 1982; Chuang and Heckard, 1983; Corner, 1976; Lackey, 1981). Anyone who pays at least passing attention to food will be able to recognize instantly sesame seeds and poppy seeds, as well as those of sunflowers, pistachios, squash, and various types of beans and peas, all based primarily on testa characters. These patterns, both in color and texture, can be due to surface cells, subsurface cells, or a combination of both. Pronounced outgrowths of the testa are not common, but a few plants do produce trichomes on the seed coat: the hairs of cotton, *Fouquieria, Limonia,* and *Strychnos*; peltate hairs on *Scaphium*; and stellate hairs on *Elaeocarpus* (Corner, 1976).

One must be careful of terminology in dealing with seed coats, because the epidermis may or may not persist during maturation, and what appears to be one may not necessarily have been derived from protodermal layers. The mature testa is often layered in a complex fashion, and analyzing the ontogeny of these layers can be difficult. Furthermore, tremendous variations occur from species to species, even within a genus, and it would be difficult or impossible to develop a universal or even a broadly useful terminology. Instead, each case should be analyzed in the context of its function, requirements, and ancestors.

In the majority of seeds, the seed coat is designed to offer maximum protection, either from fruit-eating animals or from fungi and bacteria if the seed remains dormant in the soil for several years. Consequently, sclerenchyma is typically abundant as sclereids of many shapes and sizes (Fig. 20.19; Corner, 1976; Werker, 1980). Even within one testa, there may be many layers, each with its own distinct type of sclereid. Such layers also resist digestive enzymes in the alimentary tracts of the animals that eat the seeds and then distribute them. The walls may be impregnated with numerous substances such as lignin, cutin, phenols, quinones (Stuart and Loy, 1983; Werker et al., 1979), and waxes (Eames and MacDaniels, 1947).

In addition to protection, the seed coat can also be involved in maintaining the dormancy of the seed (Werker, 1980). Werker listed eight different methods by which seed coats could inhibit germination; the most common ones are by being impermeable to water or gases (excluding oxygen, retaining carbon dioxide) and by being so strong that the embryo could not break out. These three methods are often related; seeds of Chenopodiaceae, Convolvulaceae, Leguminosae, and Liliaceae have hard

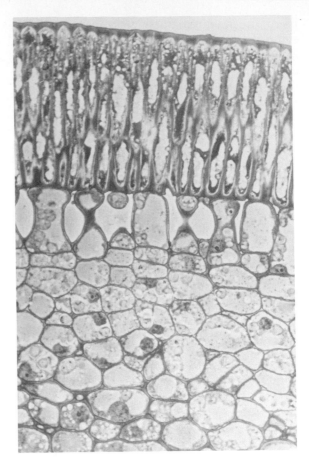

Figure 20.19
The seed coats of many legumes such as this *Pisum sativum* (pea) contain macrosclereids and osteosclereids. × 160. Micrograph generously provided by W. M. Harris.

seeds with thick walls that are cutinized, suberized, or lignified. Such a structure excludes water and gases and is also strong. However, in apples, the entry of oxygen is blocked because the seed coat contains a layer of cells filled with phenolic compounds that react with the oxygen.

Seed coats must not be totally impermeable, however, because water must enter to stimulate germination, and the embryo must be capable of breaking out of the testa. The **hilum** (pl.: hila) is the point where both of these events usually occur; it is the scar formed when the

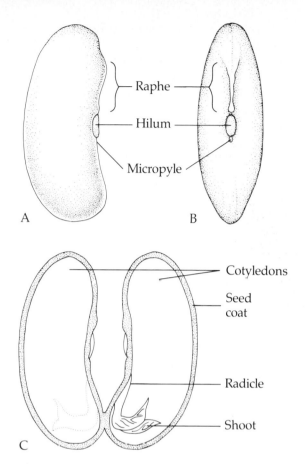

Figure 20.20
This diagram shows the various parts that are commonly present in seeds.

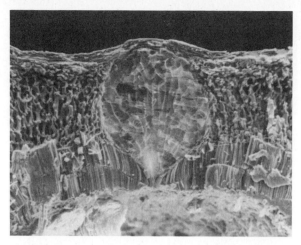

Figure 20.21
This is the tracheid bar present in the seeds of many legumes (here: *Pachyrhizus erosus*). The tracheids are believed to transfer water into the seed during germination. × 60. Scanning electron micrograph generously provided by N. R. Lersten.

seed breaks off the funiculus (Fig. 20.20). The cells of the hilum are often thin-walled and may be the only ones that are wettable. In the seeds of the Leguminosae, there is a small group of tracheidlike cells located immediately interior to the hilum (Fig. 20.21; Chowdhury and Buth, 1970; Lersten, 1982). The group of cells is called a "tracheid bar," and it is always continuous with the end of the ovular vascular bundle (Lersten, 1982). It may be involved in the absorption of water during germination. The tracheids have vestured pits, and at matu-

rity the pit membrane is completely absent, making it possible for the tracheid bar to be useful also in gas exchange. In *Albizia lophantha* (Dell, 1980) a special plug of thin-walled tissue on the residual funiculus (the "strophiolar plug" serves as the entry site for water, and in *Sida spinosa* (Fig. 20.22; Egley and Paul, 1981), a portion of the chalaza serves this purpose. Plants in regions of irregular rainfall may have a mucilaginous seed coat that absorbs and holds water long enough to permit germination (Boesewinkel, 1980; Garwood, 1985).

In addition to protection and water absorption, some seed coats are involved in attracting seed-distributing animals (Fahn and Werker, 1972). This normally is the task of the fruit, but in some cases outgrowths of the testa are responsible (Kapil et al., 1980). These can be large or small masses of cells that are rich in sugars, fats, or proteins. They are frequently brightly colored, which increases visibility and detection. In pomegranate (*Punica*), the entire outer epidermis expands and forms the red

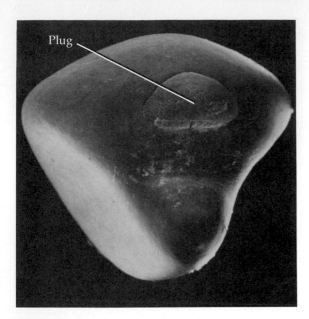

Plug

Figure 20.22
This seed of *Sida spinosa* shows a specialized method
of water uptake: the blister is a layer of cells at
the chalaza. When the blisters absorb water, they
swell and break open, permitting water to enter
the seed. × 0.5. Scanning electron micrograph
generously provided by G. H. Egley and R. N.
Paul, Jr.

edible layer. If the outgrowth is from the micropylar end of the seed, it is termed a **caruncle**;
see Fig. 21.9, *Chelidonium majus* (Szemes, 1943),
Euphorbia, Jatropha, Polygala senega, and *Ricinus
communis*. If the outgrowth is from the funiculus region, it is an **aril**, common in tropical and
subtropical plants (Fahn, 1982): *Durio zibethinus*
and *Acacia retivenea*. The spice mace is the aril
that covers the seed of *Myristica fragrans*; the
seed itself is the spice nutmeg. The seeds of the
white water lily (*Nymphaea alba*) are dispersed
by water; they have a spongy aril that provides
buoyancy.

In many cases, the testa effects distribution
not by attracting animals but rather by developing into wings for flotation on either air or
water, as in Bignoniaceae, Scrophulariaceae,
and *Campsis* (Fig. 21.11; Fahn and Werker, 1972).

In *Zanonia*, the wings are up to 10 cm wide. In
other cases, the seed coat bears bristles (*Baccharis, Leiphaimos azurens*) or soft, plumose hairs
(*Catalpa, Populus*, Asclepiadaceae), which also
are effective in seed dispersal (Schopmeyer,
1974). In orchids, the unistratose seed coat
forms a balloon around the tiny seed, and the
seeds can float on air currents (Healey et al.,
1980). Numerous further examples are described and illustrated in Fahn and Werker
(1972), Kapil et al. (1980), and van der Pijl (1982).

As mentioned before, the most common
type of ovule is the anatropous type, in which
the ovule is bent such that the funiculus is adjacent to one side of the nucellus. This often results in a marked ridge (the **raphe**) on the seed
(Fig. 20.20). In circinotropous ovules, the funiculus is wrapped completely around the nucellus and results in a raphe that similarly completely encircles the seed. In most seeds, if a
raphe is present, it is small and contains only
a narrow bundle of vascular tissue. But as
Corner (1976) pointed out, the funiculus and
raphe have not received much study, especially
in larger seeds. The funiculus is commonly
assumed to have only a single fine bundle in
all seeds, but in some tropical families with
large seeds (Clusiaceae, Myristacaceae, and
Lecythidaceae) there may be a massive bundle
in the funiculus that enlarges as the seed grows,
resulting in quite a complex bundle with
considerable xylem and phloem in the raphe.
These seeds typically also have an extensively
branched network of veins that extend into the
developing seed coat. Corner further suggested
that the big, complex funiculus/raphe may be
the more primitive condition and that the more
delicate ones are the advanced condition, because smaller seeds occur only in the more advanced families.

In conifers and other gymnosperms, the
seed coat can become especially prominent.
In many (*Cephalotaxus*, cycads, Pinaceae), the
megagametophyte pushes down into the chalazal region during seed development, so much
of the testa develops from the chalazal cells,

and the integuments contribute to only one end of the testa (Singh, 1978). As the seed develops, the outer regions of the ovule undergo cell division, and resin canals can form in the conifers, mucilage canals in the cycads (Dehgan and Yuen, 1983), and laticifers in *Gnetum* (Singh and Johri, 1972). The vascular bundles become more prominent as more xylem and phloem differentiate. The developing testa of gymnosperms initially has three layers: an outer parenchymatous **sarcotesta**, a middle sclerenchymatous **sclerotesta**, and an inner parenchymatous **endotesta**. The endotesta typically is thin and paperlike in mature seeds, and the other two layers vary in prominence. The sarcotesta is reduced or absent in *Ephedra* and Taxaceae, but in the other groups it may be thick and vascularized and have an epidermis and stomata. The sarcotesta can be extremely complex, with vascular tissue, tannin cells,

and secretory tissue. Although it is usually parenchymatous, both sclereids and fibers may occur in *Gnetum* and in several conifers such as *Cephalotaxus* and *Podocarpus*. When the sclerotesta is well developed, this usually happens only near the micropylar region, not around the entire embryo. As its name suggests, it consists predominantly of sclereids and fibers; the cells can have complex arrangements, and there may be several zones present in the sclerotesta. Singh (1978) pointed out that these are species-specific and can be helpful in taxonomy.

The seed coat of most gymnosperms (except for the conifers) is vascularized. Typically several unbranched bundles run from the chalaza to the micropyle, forming a cylinder. The bundles may branch in *Austrotaxus, Cephalotaxus,* and *Torreya.* In cycads and *Welwitschia,* there may be several sets of bundles.

Concepts

After fertilization, the zygote and ovule begin to develop into a seed, and the ovary is transformed into a fruit. The differentiation of the fruit and its final, mature form illustrate the remarkable plasticity of the angiosperms. Fruits must protect the ovule during its development into a seed, then allow its release or germination. In the more primitive of the angiosperm families the fruit opens while still on the plant; the seed itself is the unit of dispersal. But in many of the more advanced groups, this function of dispersal is transferred to the fruits, which are adapted to interact with wind, water, or animals that facilitate the distribution of the seeds located inside. This method of dispersal requires that the fruit at first remain attached to the plant and be unappealing and unpalatable to animals, while the seeds are immature, but then be capable of abscission and be attractive, when the seeds have finished growing and are ready for dispersal.

Such selection pressures alone would be expected to lead to a diversity of fruit types (van der Pijl, 1982), just as diverse pollination requirements have led to numerous types of flowers. In addition, many tissues near the ovary wall are also capable of contributing to the protection of the seeds and to their dispersal. Examples of these **accessory tissues** include: the receptacle of *Fragaria* (strawberry), *Annona cherimola* (cherimoya), and *Ochna mul-*

tiflora; the perianth of *Artocarpus* (breadfruit), *Levenhookia preissii*, and *Morus* (mulberry); the pedicel of *Anacardium occidentale* (cashew "berry"); the scales of inflorescences as in *Ananas comosus* (pineapple); and the inflorescence axis as in *Ficus* (figs). Technically, these tissues constitute a **false fruit** that complements the **true fruit** that is derived from the ovary, but this technicality is frequently overlooked, and the term *fruit* is applied to the final structure, whatever its origin. Any fruit that develops from an inferior ovary will automatically have a layer of receptacular tissue or perianth tissue, thus making it a false fruit. But apples and cucumbers (and all related fruits) show no sign of having a double structure; the carpels are perfectly fused to the accessory tissues from the time of the initiation of the primordia. It does not seem especially useful to insist on differentiating between an inner true fruit and an outer false fruit. To overcome this problem, the term **pericarp** has been introduced to be synonymous with true fruit, and **fruit wall** is used to refer to the mature structure, whether derived solely from the ovary or from the ovary plus accessory tissues. Even this terminology is not used absolutely uniformly.

The problem is compounded by the fact that during maturation, large numbers of changes occur (Roth, 1977): in many instances certain layers of cells are crushed, others greatly expanded, and dissimilar tissues changed to resemble each other and to appear to have been derived from a single layer. A casual inspection of a mature fruit is not enough to reveal the origins of many of these layers. Many mature fruits often have three layers, the **exocarp** (also called **epicarp**, "skin," and "peel"), the **mesocarp** ("flesh"), and the **endocarp** ("stone" or "pit"), but these can have very different origins depending on the species. In some, they may correspond to the outer epidermis, mesophyll, and inner epidermis of the ovary wall; in others, even accessory tissues will have contributed. Thus, in fruit structure there are often few homologous layers and many analogous ones. This has contributed to a complex and irregular terminology. Many of the schemes for fruit classification have been based on the structure of the gynoecium, but, as van der Pijl (1982) points out, the fruit is the final unit of function, and basing the classification of fruits on the structure of flowers is equivalent to classifying on the basis of primordium structure.

Another complicating factor is that in most cases the fruits are the most important part of the plant economically, so horticulturalists, pomologists, and others have developed elaborate terms with regard to individual species or even varieties. The literature may be consulted for a description of the many types and the descriptions of their anatomy (Table 21.1; Roth, 1977). Here, as in previous topics, the fruits will be discussed from a functional standpoint, because function is what they do have in common, and it is the basis for constructive analysis, interpretation, and generalizations.

General Considerations

The amount of growth and modification that can occur during fruit development is quite variable. In grasses and some other plants, each carpel contains only one ovule; when mature, the fruit has just one seed, and it is small. During maturation, very little or no cell division is necessary in the ovary and pedicel; cell expansion and sclerification are sufficient for

fruit development. In corn (*Zea mays*), maturation of the caryopsis involves very little cell division; instead there is expansion of some layers and collapse of others. The outermost layer both collapses and sclerifies, producing a dense, pitted layer (Kiesselbach and Walker, 1952). The middle layers of the carpel mesophyll disintegrate as the seed expands, and the innermost cell layers continue as thin-walled cells, but they do not grow as rapidly as the seed. This layer becomes torn and is discontinuous in the mature fruit. In a very different type of fruit, the berry of grapes, there is only enough cell division to account for a doubling of the size of the ovary, but cell expansion produces a 3000% increase in fruit size (Coombe, 1976). In general, unless a fruit becomes large or undergoes a dramatic change of shape as it matures, there is no more cell division after the flower opens (Roth, 1977).

On the other hand, fruits that do have many large seeds or that become fleshy (watermelon, squash, avocados) must have a considerable amount of cell division; they also must augment the amount of vascular tissue present (Fig. 21.1; Lilien-Kipnis and Lavee, 1971; Skene, 1966; Sterling, 1953; Tukey and Young, 1939). Obviously, the watermelon fruit needs much greater conducting capacity than the watermelon flower. In *Prunus virginiana*, Labrecque et al. (1985) found that the growth of the fruit involved three stages: an initial expansion of the cells of the original carpel wall, then a period of cell division, and finally another phase of cell expansion. The cells of avocado fruits undergo cell division continuously throughout growth and maturation (Biale and Young, 1971). The pedicel that supports a small flower is rarely strong enough to support the fruit that is formed later (grapefruit, coconuts), and it may also undergo renewed growth.

The outer epidermis may undergo periclinal divisions and give rise to a periderm, as in *Achras zapota* (sapodilla), but usually it persists as a cutinized epidermis with stomata. In quince, new stomata are formed during

Table 21.1

Dry Fruits

Indehiscent Fruits

Developing from a single carpel

1. An **achene** is simple and small, and it contains only one seed. The fruit wall usually is thin and papery, and the seed is loose inside (sunflowers).

2. A **caryopsis** is like an achene, but the pericarp and the testa of the single seed become fused (grasses: wheat, corn).

3. A **samara** is a one-seeded fruit that has winglike extensions (maple).

Developing from a compound gynoecium with several carpels

4. In a **nut**, the ovary originally contains several carpels. All but one degenerate; the mature nut contains one carpel and one seed. The pericarp is sclerenchymatous (walnut).

Dehiscent Fruits

Developing from a single carpel

5. A **follicle** is a podlike fruit that splits open on the ventral side (columbine).

6. A **legume** breaks open on both the ventral and the dorsal sides (beans, peas).

Developing from a compound gynoecium with several carpels

7. A **capsule** may open in one of many ways: splitting along fusion lines (sutures, *Hypericum*);

breaking open through the dorsal bundles (*Iris, Lilium*); splitting transversely, breaking into a top and bottom portion (*Hyocyamus, Primula*); or by small pores (*Campanula*, poppy).

Fleshy Fruits

8. A **berry** (also called a **bacca**) is a fleshy fruit in which the endocarp, mesocarp, and epicarp are easily distinguishable and are all rather soft (grape, tomato).

9. A **drupe** is like a berry except that the endocarp is especially thick and hard (peach, cherry, almond).

10. A **pome** is like a drupe except that the endocarp is papery, not stoney (apple, pear).

11. A **pepo** is like a berry except that the exocarp is hard, forming a rind (pumpkin, squash).

False Fruits

12. A **cypsela** is like an achene except that, because it develops from an inferior ovary, it contains extra tissues around the true pericarp (dandelion).

Schizocarpic Fruits

These are fruits that develop from a compound ovary but then break up into individual **mericarps** (achenes) when mature (parsley).

---

the first four to five weeks after the flower opens (Roth, 1977), but usually no new stomata are initiated, so their density decreases as the fruit becomes larger. In cucurbits, there are 13.9 stomata per square millimeter; 13.8/mm^2 in oranges (Turrel and Klotz, 1940) and 1.4/mm^2 in *Casimiroa* (white sapote; Schroeder, 1953). The stomata are frequently very large, have structural defects, and are permanently open.

The inner epidermis (the one that lines the ovary loculus) does not have a protective func-tion; it usually is not waterproof nor does it have thick walls; stomata are rare. Exceptions are the drupes, fleshy fruits with "pits" and a single seed (peaches, cherries) in which the inner epidermis and inner layers of mesophyll are converted to the sclerenchymatous endocarp. In small, hard, one-seeded fruits such as caryopses like grass grains (corn, wheat), the seed coat presses against the inner epidermis and either fuses with it or crushes it (Bradbury et al., 1956; Kiesselbach and Walker, 1952; Krauss, 1933). In fleshy, juicy fruits, the inner

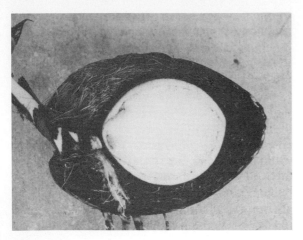

Figure 21.1
This is a cross section of a complete fruit of coconut:
the large white structure in the center is a giant
cotyledon that is digesting the endosperm (visible
as a thin layer around the cotyledon). Around this is
the hard endocarp (the "shell" of the coconut you
might see in grocery stores), then a fibrous mesocarp and a thin hard exocarp. The epicotyl and
the roots of the seedling must penetrate all fruit
layers. × 0.2.

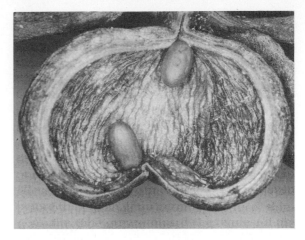

Fig. 21.2
The exocarp and endocarp of this *Sterculia foetida*
fruit are thin, and the mesocarp forms the bulk of
the fruit. Notice that the fruit loculus is much larger
than it would need to be if seed growth only were
important. × 0.3.

epidermis often produces secretory or very
cytoplasmic trichomes (Hartl, 1957; Kaniewski, 1963; Kaniewski and Wazynska, 1970).
For example, in legumes such as peas and
beans, the inner epidermis proliferates and
produces hairlike outgrowths into the fruit
locule (Kaniewski, 1963). These multicellular
chains greatly increase the area of the inner
surface, and it may be that they help maintain a
high humidity around the ovules as they develop. In many fruits that expand more than
the enclosed seeds (Fig. 21.2), the inner epidermis does not grow so rapidly as the rest of the
tissues, and it becomes torn and discontinuous
(Roth and Lindorf, 1974).

A hypodermis is common in fruits, being
collenchymatous in soft ones, sclerenchymatous in hard ones.

The mesophyll can undergo a range of transformations, just as the epidermises do. In soft,

juicy fruits, the cells enlarge, and the walls remain thin (Fig. 21.3); in the last stages of maturation they are degraded (Fig. 21.4; Pesis et al.,
1978; Platt-Aloia et al., 1980; Sobotka and Stelzig, 1974). In apple and pear, the first stages of
fruit softening are caused by a dissolution of the
middle lamella; later, in fully ripe fruits almost
all of the fibrillar structure of the cell wall is lost
(Ben-Arie et al., 1979).

In hard, dry fruits (especially nuts), most of
the mesophyll is converted to sclerenchyma, of
course. Fibers are more common, sclereids less
so. Fibers can be found in the endocarps of
drupes, in nuts, and in thin dry capsules. The
fibers in fruits are always short compared to
the fibers in other parts of the plant (Roth,
1977): in coconut they are only 0.4 to 1.0 mm
long. It might be expected that as the fruit
elongates the growing fibers would elongate in
the same direction as the fruit body, but that

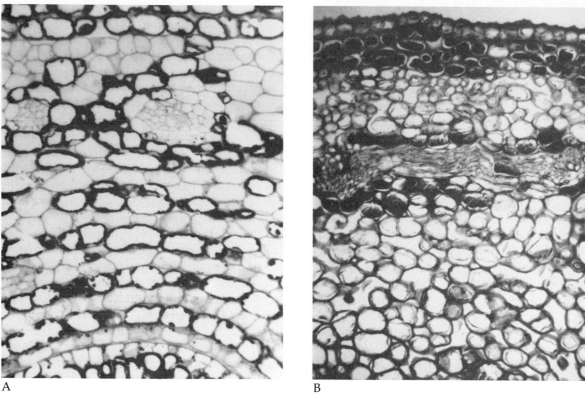

A   B

Figure 21.3
In many fruits, as in this *Amelanchier alnifolia* (service berry), there is only
a small amount of cell division during fruit maturation, but there may be
considerable cell enlargement and the formation of intercellular spaces.
(A) Ovary wall. (B) Fruit wall just before maturity. × 100. Micrograph generously provided by A. R. Olson and T. A. Steeves.

Figure 21.4
This is a cell wall of avocado as it ripens; the middle
lamella is being digested, as are the walls. Because
of the loss of material from the walls, the cellulose
fibrils are easy to see. Large oil drops fill the cytoplasm. × 10,000. Electron micrograph generously
provided by K. A. Platt-Aloia, W. W. Thomson,
and R. E. Young.

generally is not true. As the cells of individual layers develop into fibers, the orientation in each layer is different. This organization permits the fruit wall to be strong in all directions during growth and at maturity. Furthermore, once the fruit begins to dry out, each layer will contract in a different direction; this produces a tendency to twist and break the fruit open, releasing the seeds (Fig. 21.5; Monsi, 1943). If all fibers were oriented in parallel, the fruit would simply shrink as it dried, crushing the seeds instead of liberating them.

Although long-lived sclerenchyma is not unusual in wood and phloem, it seems to be very rare in fruits. This may be because most fruits are short-lived and also because fruit fibers have only a mechanical role; they do not store food reserves.

The secretory structures that normally occur in vegetative stems will often occur in fruits; these may be initiated in the ovary or only after fruit development has begun. In almond (Fig. 21.6; Morrison and Polito, 1985) and peach (Ragland, 1934; Reeve, 1959) gum ducts are initiated about 40 days after anthesis; in prune (Sterling, 1953) they form only just before fruit maturation. Oil ducts develop in the small fruits of the family Apiaceae (caraway, celery), and the opium-bearing laticifers of poppies are located in the capsules.

Carpels that mature into small fruits need little vascular tissue, but, in fruits that undergo even moderate growth, extra vascular bundles must differentiate. In large fruits, the bundles in the carpels must both increase in diameter and branch into new secondary bundles. In

Figure 21.5
This fruit of *Pithecellobium arboreum* shows the twisting of the pericarp that is necessary in order to break the sutures and cause dehiscence. The seeds remain on the fruit because of the funiculi; the seeds are black and contrast with the red fruit. × 0.2.

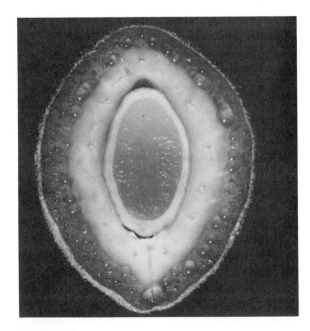

Figure 21.6
These gum ducts in an immature almond fruit show the extensive development of secretory tissues that can occur as a fruit matures. × 2.5. Photograph generously provided by J. C. Morrison and V. S. Polito.

fruits as large as pumpkins, the amount of branching is great, because all regions and layers must be vascularized. The new system of bundles can be supplied with water and sugars only if the original carpel traces increase in conducting capacity. Such an increase may occur by the addition of new xylem and phloem by a weak fascicular cambium, or the fascicular cambium can be very active, as in apple (Tetley, 1930) and citrus (Ford, 1942). In some cases, an interfascicular cambium also arises and produces a complete cylinder of secondary xylem and phloem, as in *Achras zapota* (Roth and Lindorf, 1972), *Tropaeolum pentaphyllum* (Garcin, 1890), and Proteaceae (Filla, 1926). The interfascicular cambium generally shows only moderate activity, but it can produce large amounts of wood in the fruits of such Proteaceae as *Banksia* (Filla, 1926).

Protection of Ovules and Seeds

In seeds dispersed by wind or water or other nonanimal agent, the main function of the fruit is often protection of the immature seeds against the environment as well as against attack by seed predators. This would be most easily accomplished by making the fruit as sclerenchymatous as possible, and that is indeed frequently the case, coconuts being a familiar example (remember, the coconut you see in stores is just the innermost part; in nature it is surrounded by a thick, fibrous shell). However, other factors prevent most fruits from being solid masses of sclerenchyma; before maturity they must be soft enough to grow, and after maturity they must allow water and oxygen into the embryo. They must either open to let the embryo emerge or be weak enough to allow the embryo to break out. Those fruits that open naturally as part of their development are **dehiscent fruits**, and those that do not are **indehiscent**. The latter type tends to be thinner and weaker so that embryos can crack the fruit wall by absorbing water and swelling. Alternatively, the fruit may be composed of

cells that are not especially resistant to decay; the soil bacteria and fungi attack the fruit, breaking it down until it is both permeable and weak enough to allow germination.

Even in edible fruits, the seeds must be protected from the grinding and crushing of the animal's chewing actions, and then must be isolated from digestive enzymes. This is accomplished mechanically, by a sclerenchymatous endocarp (stone fruits such as cherry and peach) or a seed coat, or chemically, by having the innermost fruit layers be so bitter that they cannot be eaten.

Mechanisms of Dehiscence

Dehiscent fruits must have a mechanism for breaking open automatically. This almost always is a two-step mechanism that includes a region of specific weakness where rupture will occur and a mechanism for generating enough force actually to tear the cells in the region of weakness. The line of weakness may be a suture where the margins of carpels meet, or it may be located in less sclerified, nonsuture regions (Roth, 1977; Stopp, 1950). The sutures may be formed by the two margins of one carpel—a **ventral suture**—or by the merger of the margins of the two adjacent carpels—a **lateral suture** (Fig. 21.7). Because a conduplicately folded carpel is primitive, a conduplicate fruit with a ventral suture is probably the original fruit type. The ventral suture is formed during development as the margins of the carpel crowd into each other; it is always a line of weakness and is rather easily broken. The "pod" (technically a legume) of leguminous plants is a simple fruit formed from a single carpel with a ventral suture. Its dehiscence has been studied carefully in *Glycine* by Monsi (1943): the exocarp is made up of the epidermis and hypodermis, both of which have thick cell walls. The endocarp consists of the inner epidermis and several layers of the mesophyll that develop into sclerenchyma cells. The outer portion of the mesophyll is the parenchyma-

Figure 21.7
This fruit of *Clusia* developed from nine carpels
that had fused. During dehiscence, the outer walls
contract and break away from the lateral walls,
which remain as septa. × 0.3.

tous mesocarp. The fibers of the exocarp are
elongated in a direction at an angle to the endocarp fibers. As the pericarp dries, the two layers
contract in cross directions, causing the wall to
twist and finally to rupture the suture.

In a syncarpous gynoecium, the individual
carpels can still be conduplicately folded; then
the carpels fuse along all of their lateral surfaces. If this lateral suture ruptures, the fruit
breaks up into separate fruitlets that then must
dehisce individually. Such fruits are called
schizocarps, and the individual fruitlets are *mericarps*; the mechanism that causes the separation
of a schizocarp is probably similar to the dehiscence mechanism just described for *Glycine*. In
more advanced gynoecia, the carpels are open
and fuse to their neighbors just along their
edges. When these sutures rupture, the locule
is opened, and the seeds are released.

If the ovary is phylogenetically fused—that
is, if individual primordia are not visible—
then sutures are not present, and a new method
of dehiscence is necessary. A narrow plate may

be formed, containing parenchymatous cells
that are especially small and thin-walled. This
region may occur anywhere, but frequently it
is located in the center of the dorsal bundle, as
a radial plate of cells that do not differentiate
into either conducting elements or vascular
sclerenchyma. This region of parenchyma is
usually accompanied by plates of especially
dense sclerenchyma on either side (Roth, 1977).
It may at first appear as though these would reinforce the region and prevent breakage. Instead, as the rest of the fruit dries and twists,
the sclerenchymatous bands focus the stresses
into the region of weakness, ensuring that all
forces are concentrated into the dehiscence
zone.

A line of dehiscence may be independent of
such carpellary features as sutures or vascular bundles (Stopp, 1950). It may be oriented
around the fruit such that the fruit breaks into
an upper and a lower half, or opens by a lid;
this is **circumscissile dehiscence**, found for
example in *Portulaca grandiflora*, *Primula*, and
Sphenoclea zeylanica (Rethke, 1946; Subraman-
yan and Raju, 1952). In some species, small
holes are formed in the wall (**poricidal dehiscence**: *Papaver, Antirrhinum*; Fig. 21.8), or flaps

Figure 21.8
This fruit of poppy has poricidal dehiscence: small
regions of pericarp break down between the vascular bundles, forming pores. As the fruit sways in
the wind, seeds are thrown out. × 0.3.

bend outward (**valvate dehiscence**). In both circumscissile and poricidal dehiscence, the rupture must cut across and break the vascular bundles.

To break open, it is necessary for the fruit wall to twist or flex. In one-seeded fruits, this may be caused by the expansion of the embryo as it swells during germination. In larger fruits or ones with many seeds, twisting and flexing are usually induced by constructing the fruit wall with two or more layers in which the cells are not oriented in parallel or in which the cellulose microfibrils of their walls are not parallel. Movement can be caused by either of two methods (von Guttenberg, 1926, 1971). In **hygroscopic movement**, as the tissues of this type dry, they tend to contract in divergent directions. Because each layer shrinks more in one direction than the surrounding layers do, warping and twisting result. This method depends on the drying of the cells, and it is the more common method, especially in temperate habitats; it is called **xerochastic opening**. The second type of movement is the result of turgor pressure caused by water uptake by living cells. This is **hygrochastic opening** and is found mostly in plants of extremely arid regions; the seeds are released only when there has been sufficient rainfall.

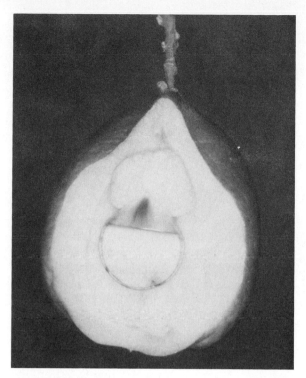

Figure 21.9
This is a fleshy fruit of *Blighia*; although "edible" because it is soft and fleshy, you would probably die before finishing even one. Certain other animals, however, are not sensitive to the poison in the pericarp. The seed is visible with a large cotyledon and a conspicuous caruncle. × 0.5.

Attraction of Animals

The primary method employed to attract animals is being edible. Fruits of this type are described as **fleshy** (remember that by *edible* we mean by some kind of animal, not just by us; see Figure 21.9), as opposed to the inedible **dry** type (Fig. 21.8). This distinction is not always appropriate, because most fleshy fruits contain a "dry" endocarp (pit) or a "dry" exocarp (rind), whereas many dry fruits are quite edible for us, either when immature (corn, green beans), with proper treatment (as in milling wheat to flour), or as is (the true fruits of strawberry—the small "seeds" that are really achenes—as opposed to the large, red false fruit).

In most cases, it is the pericarp that becomes fleshy and edible, but other tissues can be used in addition: the placenta (tomato and other Solanaceae; Fig. 21.10), the funiculi (cacti), and the seed coats (pomegranate). When ripe, the walls become soft, and the starches and acids that had been present are converted to sugars or oils, as in avocado (25% oil) and olive (20%). Pigments are produced, especially in the exocarp but frequently in the mesocarp and other tissues as well.

Fruits that are eaten must typically be more complex than dry fruits. While immature, they must actually deter consumption by being

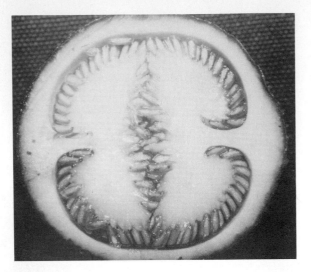

Figure 21.10
This fruit of a member of the Bignoniaceae is edible, but not because of the fruit wall; as in tomatoes, the placentae are the fleshy, edible part. × 0.5.

astringent (sour), hard, or bitter. When the embryos and seeds are mature, the fruit undergoes a rapid change in metabolism that involves the production of free sugars, the softening of walls, and a change of color and odor that signals the proper animal agent. Such changes typically occur in the outer layers, the exocarp and mesocarp if present, while the endocarp becomes especially sclerified to protect the embryo while the fruit is being chewed or while passing through a bird's crop or a digestive tract. If a sclerified endocarp is not formed, then usually the seed coat itself will be resistant. Fleshy fruits of this type may become very large (melons, apples, bananas) and thus require much more vascular tissue than the ovary does; also, the flower stalk, now a fruit stalk, must be much more active in conduction.

Fleshy fruits tend to be opened either by the animal that eats them or by the rapid rotting of the soft, fleshy, outer layers. This applies only to the outer tissues; the hardened endocarp (if one exists) must have specific dehiscence mechanisms like those of dry fruits. Fleshy fruits

also can dehisce in some cases, by the expansion of interior tissues (placentae or funiculi) or by the opening of pores or longitudinal slits.

One aspect of fruits that is often ignored and that deserves more attention is the manner in which the plant presents the fruit to the animal. In many, the fruit is abscised and allowed to fall. For birds, the fruit must be close to a branch that is stout enough to support the bird while it eats, so the pedicel must be short. Bats, however, cannot fly into trees very well, so for them the pedicels must be so elongated that the fruit hangs low, free of the foliage but not too close to the ground, as in *Kigelia* and *Mangifera indica* (mango).

Many fruits are distributed by animals without being eaten; they have claws or hooks that catch in the animal's fur or feathers (cocklebur, *Martynia*). The seeds of many mistletoes and other epiphytic parasites must be deposited on branches. Their fruits are initially eaten by birds but then regurgitated. Because of a sticky *viscid layer* the semidigested fruit sticks to the bird's beak. As the bird grooms itself, it wipes the seed onto a branch. Once attached to the host, the fruit must protect the embryo as the radicle grows out and attacks the branch (Calder and Bernhardt, 1983; Johri and Bhatnagar, 1972; Kuijt, 1969; Mauseth et al., 1985).

Dispersal by Agents Other Than Animals

Wind is commonly employed to disperse fruits (Fig. 21.11; Augspurger, 1986; Burrows, 1975; Green, 1980; Vogel, 1981). Perhaps the most familiar examples are the "parachutes" of dandelions and the plumes of *Clematis*. These are dry fruits that each enclose one seed; the parachute is formed by the dry, persistent perianth of the flower. Modifications of the fruit wall to form wings or hooks are also common, as in *Acer, Cavanillesia* (wings up to 9 cm wide), *Piscidia, Rheum,* and *Rumex* (van der Pijl, 1982). Augspurger (1986) studied 34 species of tropi-

Floater

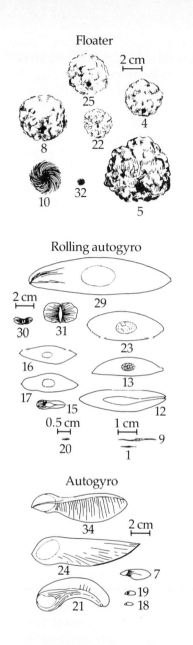

Undulator

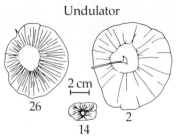

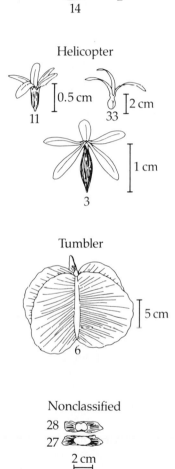

Figure 21.11
This diagram shows some of the various fruit and seed types that are dispersed by wind. The numbers refer to identification labels in the original report. Diagram generously provided by C. K. Augspurger.

cal trees that have wind-dispersed fruits and seeds and found that she could divide them into six classes: (1) "floaters" (*Bombax quinta*), (2) "rolling autogyros," which rotate on two axes simultaneously (*Lonchocarpus pentaphyllus*), (3) "autogyros," which rotate around one end (*Myroxylon balsamum*), (4) "undulators" (*Pterocarpus rohrii*), (5) "helicopters" (*Triplaris cumingiana*), and (6) "tumblers" (*Cavanillesia platanifolia*). It was found in this large sample that the mass of the fruits or seeds varied greatly, as did the maximum cross section, but that the mass divided by the area ("wing loading") varied little from species to species and that the rate of descent was similarly rather uniform.

Dispersal of fruits by water is much less common than wind dispersal; it is most important in marshes and along oceanic islands. Gunn and Dennis (1976) estimated that less than one percent of tropical seed plants produced floating fruits or seeds that could survive in seawater. The buoyancy can be provided by having a large cavity in the fruit, by having a corky or light fibrous fruit wall, by having wide, thin nonwettable wings, or by several other means. Some of the species that are distributed by floating in seawater are *Cerbera manghas*, *Barringtonia asiatica*, *Cocos nucifera* (coconut), and *Lodoicea seychellarum*. Van der Pijl (1982) pointed out that for many plants mere buoyancy does not prove or indicate water dispersal; the glumes and bracts of many dry grass fruits will allow them to float, but normally they are not distributed this way. An important feature of fruits which are dispersed by drifting on ocean currents is a resistant endocarp to protect the embryo from the toxic effects of seawater.

A small number of fruits disperse their seeds by explosive mechanisms: in *Ecballium elaterium* (squirting cucumber) pressure builds up inside the fruit until finally the pedicel breaks off and the seeds are shot out (von Guttenberg, 1926). In *Oxalis* as the fruits dry they tend to twist, but the suture is so strong that it does not break immediately. By the time that it does rupture, there is so much tension that the fruit explodes, casting the seeds in all directions. The same mechanism occurs in *Garcia*, and these fruits and seeds are large enough to cause minor injury and major surprise to the unwary.

Heterocarpy

If a plant produces either dry or fleshy fruits, the seeds are usually dispersed into the environment immediately after they mature, and all seeds are released rather synchronously. If the fruits are indehiscent and fairly resistant, requiring microbial action or mechanical weathering to weaken a protective exocarp,

then the seeds become dispersed through both space and time—some germinate relatively soon, others perhaps years later.

Some plants are able to use both strategies by producing two distinct types of fruits. This is termed **heterocarpy**, and it provides great versatility to the plant with regard to seed dispersal (Engler, 1895; Schnee and Waller, 1986; Schoen and Lloyd, 1985). In some cases, the difference may involve dehiscence and indehiscence. In *Aethionema heterocarpum* and *A. carneum* two different types of fruits are produced in the same inflorescence: dehiscent siliquas and indehiscent nuts (Zohary and Fahn, 1950). Heterocarpy may involve different types of dispersal agents. The composite *Calendula* produces three types of fruits on the same capitulum; those produced from the inner and middle florets are winged and drop off easily, and those from the outer florets are spiked and remain with the receptacle until it abscises from the plant (Becker, 1913). If one type of fruit is borne above ground and the other below, the term **amphicarpy** applies. *Amphicarpaea bracteata* (Leguminosae) produces three types of fruit from three types of flowers: on the aerial branches are cleistogamous (closed) flowers and chasmogamous (open) flowers, each of which produces its own distinct type of dehiscent legume. The subterranean, cleistogamous flowers produce fruits that have just a papery, thin pericarp around a single, large, fleshy seed (Schnee and Waller, 1986). Other examples of this are *Vicia angustifolia* and *Emex spinosa* (Evenari et al., 1977; van der Pijl, 1982; Zohary, 1937). Many species with cleistogamous flowers exhibit amphicarpy, with the closed flowers growing underground. The subterranean fruits are generally larger than those produced above ground (Engler, 1895; Koller and Roth, 1964; McNamara and Quinn, 1977; Schnee and Waller, 1986). Amphicarpy has the interesting consequence that the aerial seeds are dispersed and may or may not arrive at sites suitable for germination and growth of the plant while the subterranean seeds are automatically in a site that has proven adequate for the parental plant. At least in one species (*Cardamine chenopodifolia*), the plants that arise from the seeds of one type of fruit differ from the plants that develop from seeds of the other (Cheplick, 1983). The seeds of the subterranean fruits were six times as heavy as those of aerial fruits, and the plants they produced had more rapid root elongation and larger leaves, flowered earlier, and produced more fruits and seeds.

Abscission

An important aspect of fruit anatomy and physiology is the control of abscission (Addicott, 1983). Under normal conditions, it will be advantageous for a plant to retain its fruits until they and the seeds are mature. But if flowering and pollination are too successful, then too many fruits will begin to develop, and a process of self-thinning will be advantageous (Stephenson, 1981). Similarly, once the fruit has matured, it typically is abscised, and the scar is sealed with wound periderm. If the fruit is eaten while still on the plant, then the pedicel must be abscised.

The abscission zones of fruits are structurally very similar to those of leaves, as they fulfill the same type of function (Barnell, 1939; Stösser et al., 1969). Valdovinos and Jensen (1968) specifically pointed out how strongly the abscission zones of leaves and of fruits resembled each other. The abscission zone may become well defined early as an area with thin-walled cells of rather large size, and frequently there is a groove in the pedicel, marking the site of the abscission zone. As abscission begins, the middle lamella dissolves, and the walls may weaken from enzymatic attack. There may be some variability at this point; the weakening of the middle lamellae appears to occur across the entire diameter of the pedicel. In raspberry, *Rubus idaeus* (MacKenzie, 1979), the cells in the cortex swell, and the intercellular spaces enlarge, but the walls remain intact; only in the vascular bundles does cell breakdown occur. In tobacco and tomato, however, the cell walls are completely digested across the entire zone. Jensen and Valdovinos (1967) and Valdovinos and Jensen (1968) reported that during abscission the walls swell and become flexible. A particularly fascinating observation was made by Gilliland et al. (1976): the dictyosomes play a major role in the development of the abscission zone. While the pedicel is young, the dictyosomes transport material to the wall, participating in its construction. Later, during abscission, dictyosomes are a part of the secretory mechanism that transports digestive enzymes for the destruction of the walls that they had previously built. There may be an initial accumulation of starch before the pedicel breaks; vessels must be sealed, and a wound periderm closes the broken surface.

A fruit may have more than one abscission zone; if so, the two may appear at different times or under different circumstances. If plums fall early, it is by an abscission zone at the base of the pedicel, such that the stalk falls with the fruit; if it falls at maturity, the fruit separates from the stalk by a different abscission zone. Cherries have three abscission zones: one between the stalk and the fruit, another between the spur shoot and the stalk, and finally one that cuts off the spur shoot. *Callistemon* is remarkable in retaining the intact, unopened fruits on the tree more or less permanently, often not releasing them until a branch or the entire tree dies and decays.

Appendix A: Glossary

A-type sheath cells Cells of a bundle sheath that are actually companion cells; B-type sheath cells are transfer cells but not companion cells (Ch. 12).

Abaxial The lower surface of a leaf, also called the dorsal surface. The upper surface is adaxial or ventral (Ch. 12).

Abscission layer Synonym for **separation layer**.

Abscission zone A region in a petiole, pedicel, peduncle, or any other stalk of a deciduous organ, through which the rupture occurs that causes that organ to fall off the plant. The abscission zone typically contains a layer of weakness (the separation layer) and a layer where protective tissues will form (the protective layer) (Ch. 12).

Accessory tissue Tissues that are part of the flower or inflorescence and that contribute to the "fruit" as it matures. Accessory tissues may include the receptacle, perianth, and bracts, among others; the mature structure, because it contains more than just ovary, should be called a false fruit (Ch. 21).

Actinocytic A type of stomatal complex in which the subsidiary cells form a circle radiating outward from the guard cells (Ch. 10).

Actinomorphic flower A flower that is radially symmetrical; also called a regular flower. See **zygomorphic flower** (Ch. 19).

Actinostele A type of protostele in which the mass of xylem is undulate in cross section, not round as in a haplostele (Ch. 11).

Adaxial The upper surface of a leaf; also called the ventral surface. The lower surface is abaxial or dorsal (Ch. 12).

Adaxial meristem In a leaf primordium, cells along the adaxial surface become very active mitotically, forming an adaxial meristem. This results in a thickened midrib. Rarely called a *ventral meristem* (Ch. 12).

Adhesive cell Any cell that secretes a sticky substance that permits haustoria or holdfasts to adhere to a surface (Ch. 9).

Adult An individual plant that has matured to the point at which it is capable of sexual reproduction (Ch. 6).

Air seeding The formation of an embolism in the xylem, caused by the introduction of a small bubble of air (Ch. 15).

Albuminous cell A densely cytoplasmic cell in the phloem of gymnosperms. Albuminous cells have extensive interconnections with sieve cells and are thought to control their metabolism. Albuminous cells are analogous to the companion cells in angiosperms, but unlike companion cells they are not usually sister cells to the associated sieve elements. In the older English literature and in modern German literature, the synonym *Strasburger cell* will occur (Ch. 8).

Albuminous seed A type of seed in which there is a rather large amount of endosperm present at maturity (Ch. 20).

Alternation of generations In all vascular plants (as well as most nonvascular ones) a diploid generation, the sporophyte, alternates with a haploid generation, the gametophyte (Ch. 19).

Alternation of heteromorphic generations If the diploid generation is easily distinguishable

453

from the haploid generation, the two are hetero-morphic. All nonalgal plants have heteromorphic generations.

Amphicarpy A type of heterocarpy in which a plant produces one type of fruit above ground and a different type of fruit below ground (Ch. 21).

Amphiphloic siphonostele A siphonostele in which the phloem occurs both to the exterior of the xylem and to the interior, between the xylem and the pith. Compare with **ectophloic siphonostele** (Ch. 11).

Amyloplast A plastid that accumulates and stores starch—that is, amylose (Ch. 2).

Anastomose To fuse together, most commonly used to describe the merger of two ducts, pits, or vascular bundles (Ch. 9).

Anatropous An ovule that is bent over 180° such that the micropyle is adjacent to the funiculus (Ch. 19).

Androecium (pl.: androecia) The collective term for all the stamens of a flower (Ch. 19).

Angiosperm Any member of the Division Antho-phyta, the flowering plants. The two classes of this division are Class Magnoliopsida and Class Liliop-sida (Ch. 1).

Anisocytic A type of stomatal complex in which there are three subsidiary cells of unequal sizes. Formerly called a *cruciferous stomatal complex* (Ch. 10).

Anneau initial The hypothetical initiating ring in Plantefol's theory of apical meristems. It was postu-lated as being the region of the meristem that was capable of producing leaves when acted upon by stimuli that moved upward from pre-existing leaves (Ch. 6).

Anomocytic A type of stomatal complex in which there are no obvious subsidiary cells. Formerly called a *ranunculaceous* stomatal complex (Ch. 10).

Anther The portion of a stamen that contains the sporogenous tissue and surrounding sterile tissues (Ch. 19).

Anticlinal Perpendicular. In anatomy, used to de-scribe a wall perpendicular to the nearest important surface, either of the organ or of a duct. The syn-onym *radial* is more often used if the wall is near the center of a cylindrical structure rather than near the surface (Ch. 6). See also **periclinal.**

Antipodal cell In an angiosperm megagametophyte, the cells (usually three) located at the chalazal end (Ch. 19).

Aperture An opening. In pits, the inner aperture is nearer the protoplast; the outer aperture is closer to the primary wall (Chs. 2, 5, 7).

Apical cell The large, prominent cell in the apices of most terrestrial cryptogams. By sequential division along its inner faces, it produces all the cell lines of the organ in which it occurs. In most of these plants, the apical cells are solitary, but in some ferns they occur as rows (Ch. 6).

Apo- A prefix used to indicate that certain parts exist as individuals, not fused with anything else. An apocarpous gynoecium contains individual, free carpels (Ch. 19). See also **sym-.**

Apoplast The region of the plant not occupied by protoplasm. It consists of cell walls and intercellular spaces that form a continuum throughout the plant (Ch. 2).

Apposition The deposition of new wall microfibrils onto the inner surface of the existing wall; each suc-ceeding layer is completely interior to the previous layers (Ch. 2). Compare with **intussusception.**

Arbuscule In endomycorrhizae, the portions of the hyphae deepest in the cortex may invade cortical cells and branch profusely; the branched portion is an arbuscule (Ch. 13). See also **V/A mycorrhiza.**

Archesporial cell A cell in a sporangium that is re-puted to be the initial for the formation of all the spore mother cells and some of the surrounding sterile cell layers (Ch. 19).

Areole In leaves, the smallest portion of mesophyll that is completely surrounded by vascular bundles (Ch. 12).

Aril A fleshy outgrowth from the funiculus in some seeds; it is sometimes interpreted as an extra integu-ment (Ch. 20). Compare with **caruncle.**

Astrosclereid A sclereid that is highly branched with several arms projecting from the central body (Ch. 5).

Atactostele The stele of monocots, consisting of a complex, three-dimensional network of collateral vascular bundles (Ch. 11).

Ategmic An ovule that has no integument (Ch. 19).

Atropous An ovule that is straight, not bent, such that the micropyle is directed away from the funiculus; also called *orthotropous* (Ch. 19).

Auricles: A set of two small flaps of tissue projecting from the point at which the blade of a grass leaf attaches to the leaf sheath (Ch. 12).

Axial bundle One of the main bundles of a shoot (or root). Distinguished from leaf traces, cortical bundles, or medullary bundles. Also called a *cauline bundle* (Ch. 11).

B-type sheath cells Cells of the bundle sheath that are transfer cells but not companion cells, as are A-type sheath cells (Ch. 12).

Bacteroid Endosymbiotic nitrogen-fixing bacteria invade the roots of host plants and cause the production of a nodule. Within the cells of the host, the bacterium swells to about forty times its original size and carries out nitrogen fixation. In this stage it is a bacteroid (Ch. 13).

Baculum See **probaculum.**

Bar of sanio Synonym for **crassula.**

Bark The layers of secondary phloem and periderm on stems, roots, and rarely leaves (Ch. 17).
   Inner bark The layers of living tissue in the bark, located between the vascular cambium and the innermost cork cambium.
   Outer bark The outer, dead portions of the bark, located to the exterior of the innermost cork cambium. Also very rarely called a *rhytidome.*

Basal cell In angiosperm reproduction, one of the first two cells produced by the transverse division of the zygote. The basal cell typically contributes to the suspensor, not to the embryo (Ch. 20). In the epidermis, the lowermost cell or cells of glands or trichomes (Ch. 10).

Bifacial leaf A leaf in which the upper half (in a cross section) is different from the bottom, usually because the upper half has a palisade parenchyma and the lower half has spongy mesophyll. Also called a *dorsiventral leaf;* the alternative is an isolateral or isobilateral leaf (Ch. 12).

Bisporic megagametophyte A megagametophyte that arises from a binucleate megaspore, which is produced when two meiotic nuclei survive, not just one; see also **monosporic** and **tetrasporic** (Ch. 19).

Bitegmic Ovules that have two integuments (Ch. 19).

Border The inner aperture of a pit when it is thicker than the rest of the secondary wall (Ch. 7).

Brachysclereid A sclereid that is relatively isodiametric in shape. Also called a *stone cell* (Ch. 5).

Bract In the megasporangiate strobili (seed cones) of conifers, the leaves of the main axis. The axillary buds of the bracts develop into ovuliferous scales (Ch. 19).

Branch trace A vascular bundle that runs from the axial bundle system to an axillary bud or young axillary shoot (Ch. 11).

Bud scale A modified leaf specialized for the protection of a resting bud; also called *cataphyll* (Ch. 12).

Bulliform cell A large, thin-walled epidermal cell present in the leaves of some grasses. Debate continues about whether they are involved in the rolling and opening of leaves (Ch. 10).

Bundle sheath A layer of cells that covers a vascular bundle. The bundle sheath most frequently is composed of parenchyma cells but can be fibrous (Ch. 12).

Bundle sheath extension A panel of cells, either parenchyma or fibers, that extends from the sheath of a vascular bundle to the nearest epidermis. Bundle sheaths may be involved in the capillary movement of water from the vein to the mesophyll (Ch. 12).

C_4 photosynthesis A special metabolism in which phosphoenol pyruvate is carboxylated in the mesophyll, then transported to a bundle sheath and decarboxylated. The carbon dioxide that is liberated in the bundle sheath is then used in C_3 photosynthesis. The C_4 photosynthesis is associated with prominent bundle sheaths and a mesophyll that consists of cells arranged radially around the sheath—Krantz anatomy (Ch. 12).

Callose A polymer of glucose with β (1–3) glycosidic bonds that is present in sieve elements; when the element is damaged, the callose precipitates and contributes to sealing the wound. Callose is also deposited in other areas in response to damage and to fungal invasion and during normal growth and development of pollen (Ch. 8, 19).

Callus An irregular proliferation of parenchyma cells, either at a cut surface on a plant or as induced in pieces of tissue placed in culture medium. *Callus* is also an older synonym for *definitive callose;* this usage is obsolete (Ch. 8).

Calyptrogen The meristem of the central region of the root cap; also called the *columella mother cells.* Also designates a totally different structure in the mosses (Ch. 6).

Calyx (pl.: calyces) All the sepals of one flower (Ch. 19).

Cambial region A term that is usually employed to designate a cambium and the most recently produced derivatives; this sometimes is used to imply that a distinct, identifiable cambium does not exist (Ch. 14).

Cambium (pl.: cambia) A sheetlike fundamental type of meristem. The vascular cambium produces secondary xylem and phloem; the cork cambium (phellogen) produces the cork tissue of the bark. Both cambia are considered to produce secondary tissues (Ch. 14).

    Bifacial cambium A cambium that produces cells to both sides of itself, thus having two faces.

    Biseriate cambium A theory (generally not accepted) that the vascular cambium consists of two parallel layers of cells, one producing the secondary xylem, the other producing the secondary phloem. See **uniseriate cambium.**

    Fascicular cambium The vascular cambium that forms inside a vascular bundle, being derived from the procambium. See **interfascicular cambium.**

    Interfascicular cambium The vascular cambium that arises between vascular bundles, as the parenchyma cells of the medullary rays begin to divide.

    Nonstoried cambium Vascular cambium in which the fusiform initials are not aligned with each other as they are in a storied cambium, where the ends of one fusiform initial are even with those around it.

    Nonstratified cambium Synonym for **nonstoried cambium.**

    Storied cambium See **nonstoried cambium.**

    Stratified cambium Synonym for *storied cambium;* see **nonstoried cambium.**

Cambiumlike transition zone A region in certain shoot apical meristems in which all the cells divide with walls that are parallel with the base of the meristem. As a result, the zone does have the appearance of a diffuse cambial region (Ch. 6).

Canal Any rather long, tubular cavity, either secretory or not, either lined with an epithelium or not. Also called a *duct* (Ch. 9).

Caruncle An outgrowth of the micropylar end of the testa. It usually is brightly colored and nutritive, thus aiding in seed dispersal, often by ants. Compare with **aril** (Ch. 20).

Caryophyllaceous stomatal complex See **diacytic.**

Casparian strip (band) A layer or set of layers of suberin and sometimes also lignin in the radial walls of endodermal cells. Because the casparian strips are impermeable, the apoplast on one side of the endodermis is not continuous with the apoplast on the other side (Ch. 19).

Cataphyll A rarely used synonym for **bud scale** (Ch. 12).

Cauline bundle A synonym for **axial bundle** (Ch. 11).

Cavitation The breaking of hydrogen bonds in water such that a bubble forms in the water. When this happens within the xylem, the upper portion of the water column is not in contact with the lower portion and cannot pull it upward. This results in the loss of conducting tracheids and vessels and is a major factor in the conversion of sapwood to heartwood (Ch. 7).

Cavity Any rather isodiametric space, either secretory or not. Also called a *chamber* (Ch. 9).

Cell plate As a cell undergoes division, a set of vesicles and microtubules aggregates at the site of division and constitutes a large flat vesicle. The new cell walls and middle lamella form within, and the whole is the cell plate. The radial expansion of the cell plate divides the mother cell into two daughter cells (Ch. 2).

Cell wall Almost all plant cells are surrounded by a cell wall consisting of cellulose, hemicelluloses, pectins, and proteins. Other compounds are added to the walls of certain types of cells, which gives them special properties. The two basic types of wall are primary walls, which are usually thin, and secondary walls, which are deposited interior to the primary wall and which are often thick and lignified (Ch. 2, 4, 5).

Cellular endosperm In angiosperm reproduction, a type of endosperm in which cell division accompanies nuclear division and the endosperm consists of cells during its entire development. Compare **nuclear** and **helobial endosperm** (Ch. 20).

Cellulose One of the principal components of cell walls; it consists of glucose molecules that are interconnected by β-1,4 glycosidic bonds. The individual molecules are believed to be parallel to each other, crystalizing into microfibrils (Ch. 2).

Central cell Not a precisely defined term, it most often refers to the central, quiescent cells of a root apical meristem, but *central cells* may also be seen referring to the central mother cells of shoot apical meristems (Ch. 6). In an angiosperm megagametophyte, the large cell (usually binucleate) that will develop into endosperm after fertilization (Ch. 19).

Central cylinder meristem The portion of a root apical meristem that gives rise to the stele (Ch. 6).

Central mother cells The zone of a seed plant apical meristem that gives rise to the surrounding zones. Sometimes used for the central cells of the root (Ch. 6).

Cephalium The reproductive portions of certain genera of cacti in which the growth that occurs after sexual maturity is strikingly different from the juvenile growth (Ch. 6).

Chalaza (pl.: chalazae or chalazas) The basal portion of the nucellus, the point at which it attaches to its stalk, the funiculus (Ch. 19).

Chalazogamy The process in which a pollen tube grows into the ovule through the chalaza (Ch. 19).

Chamber See **cavity.**

Chambered pith A pith that is for the most part hollow but that has plates (septa) of parenchyma or sclerenchyma crossing it, dividing it into chambers (Ch. 11).

Chimera (pl.: chimeras) A plant or an organ whose meristematic region consists of at least two popula-

tions of cells whose genotypes are different (Ch. 6).

  Mericlinal chimera A chimera in which the two populations of cells are arranged in sectors around the axis of the meristem.

  Periclinal chimera A chimera in which the different populations of cells are arranged in layers parallel to the surface of the chimera.

  Polyploid chimera A chimera in which the two populations of cells differ in the ploidy level of their nuclei.

  Sectoral chimera Synonym for **mericlinal chimera.**

Chlorenchyma Parenchyma that is specialized for photosynthesis and that contains large numbers of chloroplasts (Ch. 3).

Chloroplast A type of plastid that is specialized for photosynthesis. Its inner membranes are extensively folded into grana, and it contains a large concentration of chlorophyll. The matrix (stroma) contains the enzymes and the intermediates for carbon fixation (Ch. 2).

Chloroplast, agranal This type of chloroplast is found in the bundle sheath of certain C_4 plants. The inner membranes do not aggregate into grana (Ch. 2).

Chromoplast A type of plastid that contains a large concentration of nonphotosynthetic pigments, especially carotenoids. These are abundant in colored petals and fruits, serving to attract pollinators and fruit-dispersing animals. They also occur in other colored plant parts, such as the roots of carrots (Ch. 2).

Chromosome The molecules of DNA within a nucleus each associate with the basic proteins, histones; these complexes are chromosomes. During cell division, the chromosomes "condense" and become visible as threadlike structures by light microscopy; during interphase they are decondensed and are too fine to be visible (Ch. 2).

Circumscissile dehiscence A method of fruit opening in which the fruit breaks into an upper and a lower portion; the upper portion may be so small as to be called a *lid* (Ch. 21).

Cladode A stem that is flattened and determinate and that resembles a leaf. Also called *phylloclades* or *cladophylls* (Ch. 11).

Cladophyll Synonym for **cladode.**

Closed meristem See **meristem.**

Closed stele A type of eustele in which the vascular bundles have such numerous, large interconnections that the appearance of a network is stronger than the appearance of long axial bundles, as is the case for an open stele (Ch. 11).

Closing layer A layer of cork cells firmly attached to each other; located in a lenticel. The closing layer holds the loose, soft cells of the lenticel in place and prevents them from being weathered away (Ch. 17).

Coated vesicle A small vesicle composed of plasmalemma and coated with clathrin; it pinches off from the plasmalemma and migrates into the cytoplasm (Ch. 9).

Coenocyte Large cells that contain many nuclei; the multinucleate condition can result from either nuclear division without cell division, or by cell fusions (Ch. 2, 21).

Coenomegaspore A term that is occasionally applied to the binucleate or tetranucleate megaspore that occurs in bisporic or tetrasporic megasporogenesis (Ch. 19).

Coleoptile In the embryos of grasses, a sheath that envelops the epicotyl and that elongates to the soil surface, forming a smooth passageway for the growth of the shoot (Ch. 20).

Coleorrhiza (pl.: coleorrhizae) In the embryos of monocots, a sheath of tissue that envelops the radicle (Ch. 20).

Collecting cell A cell that is part of a secretory structure and is (or at least appears to be) involved in transferring material from surrounding cells into the secretory structure (Ch. 9).

Collenchyma A cell and tissue type in which the primary cell walls are unevenly thickened. Collenchyma is plastically extensible and is usually the supporting mechanical tissue of growing organs such as herbaceous stems and leaves (Ch. 3, 4).
  Angular collenchyma Collenchyma in which the wall thickenings occur at the corners of the cell.
  Annular collenchyma Collenchyma in which the walls are so thickened that the lumen of the cell is round in appearance, not angular.
  Lacunar collenchyma Collenchyma that forms around an intercellular space.

Lamellar collenchyma Collenchyma in which the wall thickenings of each cell are aligned parallel to the surface of the organ, consequently appearing to form lamellae.

Columella The central region of a root cap, especially if the cells occur in extremely orderly rows (Ch. 1). In pollen, a rodlike radial component of the wall (Ch. 19).

Columella mother cells The portion of a root apical meristem that produces the cells of the central regions, the columella (Ch. 1).

Commissural bundle A short, small vascular bundle that runs transversely between the longitudinal bundles of leaves with parallel veins (Ch. 12).

Companion cell A cytoplasmic cell in the phloem of angiosperms. Companion cells are either sister cells of sieve tube members or are almost this closely related; they have extensive cytoplasmic connections to the enucleate sieve tube members and are thought both to control their metabolism and to participate in the loading and unloading of nutrients. Companion cells are analogous to the albuminous cells of the nonangiosperms (Ch. 8).

Complementary cells The rounded cork cells that are located in lenticels; unlike other cork cells, these have intercellular spaces and permit the diffusion of oxygen into the organ (Ch. 17).

Compression wood A type of reaction wood formed in gymnosperms, usually on the lower side of branches. The wood is especially dense and heavily lignified. See also **tension wood** (Ch. 15).

Conductance The capacity for a cell, tissue, or organ to conduct water through itself (Ch. 7).

Conjunctive tissue The pithlike tissue that forms the matrix surrounding the vascular bundles of a monocot root or stem (Ch. 11).

Connecting strand The strand of cytoplasm that runs through a sieve pore, connecting the protoplasm of one sieve element with that of the adjacent element (Ch. 8).

Connective The central portion of an anther, between the pollen sacs, that contains the vascular bundle (Ch. 19).

Core The inner region of a shoot apical meristem that is considered to consist of an outer mantle and the core. Many gymnosperms have this organization; a core is different from *corpus* in that it receives cells from the mantle whereas the corpus does not receive cells from the tunica (Ch. 6).

Cork cambium See **cambium** or **phellogen.**

Cork cell In the periderm, a suberized cell; see **phellem.** In the epidermis of grasses, a short cell that contains suberin in its walls; these cork cells often occur paired with another type of short cell, a silica cell (Ch. 10).

Corolla The collective term for all the petals of one flower (Ch. 19).

Corpus The inner region of a shoot apical meristem that consists of a tunica and corpus. The corpus is distinguished on the basis that its cells may divide in any direction while those of the tunica always divide with anticlinal walls. See **core** (Ch. 6).

Cortex mother cells The ring-shaped zone in a root apical meristem that produces the cortex (Ch. 6).

Cortical bundles Vascular bundles that run in the cortex outside of the stele but that are not leaf traces. Relatively few species have them (Ch. 11).

Cotyledons The first "leaves" of an embryo; in many cases, the cotyledons do resemble foliage leaves; in other cases, they are quite different, usually being storage organs. The synonym *seed leaf* is frequently mentioned, especially in introductory texts, but it is never really used (Ch. 20).

Crassinucellate A type of ovule in which the megagametophyte is located deep in the nucellus, several cell layers away from the epidermis; the alternative is tenuinucellate (Ch. 19).

Crassula (pl.: crassulae) A dark pattern located between the circular bordered pits of gymnosperms. Crassulae are often said to be thick areas of middle lamella and primary wall, but electron micrographs suggest that they are only light diffraction patterns. Also called *Bars of Sanio* (Ch. 5).

Crista (pl.: cristae) The tubular structure formed by the infolding of the inner mitochondrial membrane (Ch. 2).

Cruciferous stomatal complex See **anisocytic.**

Cuticle A term that is used to refer nonspecifically to the layers of cutin and the cutin-impregnated wall of an epidermis. See **cuticle proper** (Ch. 10).

Cuticle proper Due to the variable usage of the term *cuticle*, *cuticle proper* is used to specify just the layer of pure cutin that is exterior to the cutinized wall of the epidermis (Ch. 10).

Cuticular layer The outer portion of the outer wall of the epidermis that is impregnated with cutin (Ch. 10).

Cuticularization The deposition of cutin on the exterior surface of the outer wall of an epidermal cell, thereby forming a cuticle (Ch. 10).

Cutin A substance of variable composition, being a polymer of long-chain fatty acids and polyhydroxy derivatives of these acids. Cutin is the primary component of the cuticle (Ch. 10).

Cutinization The deposition of cutin in the walls of the epidermis and sometimes in the mesophyll cells near stomata (Ch. 10).

Cutting face One of the internal sides of an apical cell in a meristem of a cryptogam; during cell division, the cell plate is approximately parallel to a cutting face (Ch. 6).

Cyclosis The circulation or flow of protoplasm within a cell. Also called *cytoplasmic streaming* (Ch. 2).

Cystolith A crystalline aggregation of calcium carbonate on a small papilla of wall that projects into the cell. The cell is almost always an epidermal cell and is called a *lithocyst* (Ch. 2).

Cytoplasm All the components of a living cell are called protoplasm; if the cell wall, nucleus, and vacuoles are eliminated from consideration, the remainder is cytoplasm (Ch. 2).

Cytoplasmic streaming Synonym for **cyclosis** (Ch. 2).

Cytosol The liquid, non-membranous, nonfibrilar material within which all other organelles are suspended. Also called *hyaloplasm* (Ch. 2).

Declining initial A fusiform initial that becomes progressively shorter and is finally converted to a ray initial or is expelled from the cambium (Ch. 14).

Declining tier A row of axial cells in the secondary xylem or phloem that are progressively shorter because the fusiform initial is itself becoming shorter (Ch. 14).

Dehiscent fruit A fruit that opens at maturity, releasing the seeds. The alternative is indehiscent fruit (Ch. 21).

Dermatogen An older term that refers to the cells of a meristem that produce the epidermis. In its original form, *dermatogen* carried with it the concept that these cells were acting as morphogenetically unique initials. Now it is often used in general reference to any set of cells that produce epidermis, but a more accurate term for that is *protoderm* (Ch. 6).

Desmotubule The proteinaceous, tubular structure that passes through a plasmodesma. It is attached to the endoplasmic reticulum on either end, but it differs structurally from the ER (Ch. 2).

Determinate This term designates the concept of restricted potential. A plant or organ with determinate growth has a genetically limited size and cannot grow indefinitely (Ch. 6).

Diacytic A type of stomatal complex in which the long axis of the subsidiary cell is perpendicular to that of the guard cells. Formerly called a *caryophyllaceous stomatal complex* (Ch. 10).

Diarch In roots, if the xylem has two protoxylem poles, it is said to be diarch. If one pole, monarch; three, triarch; etc. (Ch. 13).

Dictyosome One of the organelles present in all cells, it consists of a stack of flat vesicles. It is involved in preparing material that is to be transported out of the cell, frequently by adding sugars onto proteins. This is synonomous with a *Golgi body* and is occasionally referred to (erroneously) as a *Golgi apparatus* (Ch. 2).

Dictyostele A type of amphiphloic siphonostele in which the leaf gaps are so numerous and so close together that the stele appears to be a network of vascular bundles (Ch. 11).

Diffuse porous A type of dicot wood in which vessels are located in both the spring wood and the summer wood and have similar diameters in each. The alternative is ring porous wood (Ch. 15).

Diffuse secondary thickening In some palms, the middle or lower portions of the trunk swell as the conjunctive tissue and the cells of the bundle sheaths begin renewed cell division (Ch. 11).

Dilatation The circumferential expansion of the secondary phloem that allows it to avoid being torn apart by the accumulation of secondary xylem interior to it. The actual cells formed are called dilatation tissue (Ch. 16).

    Expansion tissue A type of dilatation tissue formed by cell division in the phloem rays.

    Proliferation tissue A type of dilatation tissue formed by cell division in the axial parenchyma of the secondary phloem.

Disarticulate To fall to pieces; this is used of shoots that separate easily into portions, most often as a means of vegetative reproduction (Ch. 11).

Domain A region in a vascular cambium, in which all pseudotransverse divisions have the same orientation (Ch. 14).

Dorsal wall In a guard cell, the wall farthest from the stomatal pore. The closest wall is the ventral wall (Ch. 10).

Druse A crystal of calcium oxalate that has numerous acute faces. The entire crystal is approximately spherical (Ch. 2).

Dry fruit An inedible fruit. The alternative is a fleshy (edible) fruit (Ch. 21).

Duct See **canal.**

Ectexine The outermost layer of the pollen wall; also called *sexine* (Ch. 19).

Ectocarp Synonym for **epicarp.**

Ectodesma (pl.: ectodesmata) A channel in the outer wall of epidermal cells, resembling a plasmodesma. It may be that these are involved in the deposition of cutin and waxes. See also **teichode** (Ch. 10).

Ectomycorrhiza (pl.: ectomycorrhizae) A fungal association with the roots of many species of trees, in which there is a well-developed layer of hyphae covering the surface of the root. The hyphae may penetrate between the outermost root cells, forming a mycelium called a Hartig net. Also called ectophytic or ectotrophic mycorrhiza. See also **endomycorrhiza** (Ch. 13).

Ectophloic siphonostele A siphonostele in which the phloem is located only to the exterior of the xylem. See also **amphiphloic siphonostele** (Ch. 11).

Egg apparatus In an angiosperm megagametophyte, the set of synergids plus the egg cell (Ch. 19).

Elaioplast A plastid that is filled with and stores lipid (Ch. 2).

Elasticity The property of being deformable, but returning to the original size and shape once the deforming force is removed. Sclerenchyma cells are elastic (Ch. 4, 5).

Embryo The young plant that is present in a seed before germination; in the earliest stages, it may also be called a proembryo (Ch. 20).

Enation A small outgrowth from a stem or any other large structure; at present, *enation* is used almost exclusively as a synonym for *microphyll* (Ch. 12).

Endexine The inner layer of the exine of the pollen wall; also called the *nexine* (Ch. 19).

Endocarp The innermost layer of a fruit wall, often tough and papery or thickened into the "pit" or "stone" of the fruit (Ch. 21).

Endodermis A sheath of cells surrounding the vascular tissues of all roots and of some stems and leaves. The endodermal cells are characterized by having Casparian strips that prevent the diffusion of material between the cortex (or mesophyll) and the vascular tissues. Endodermal cells with Casparian strips are said to be in State I; with the deposition of more extensive layers of suberin, they are said to be in State II; they are in State III when all walls are thickened (Ch. 13).

Endomycorrhiza (pl.: endomycorrhizae) A fungal association with the roots of most plants, in which the fungal hyphae invade between the cells of the root. In the most common type, the hyphae may invade a cell and swell (a vesicle) or branch (an arbuscule). Also called an *endophytic* or *endotrophic mycorrhiza*. See also **V/A mycorrhiza** (Ch. 13).

Endoplasmic reticulum A system of tubules and small vesicles that extends through most of the cytoplasm. If covered with ribosomes, it is called rough ER (RER), and it is involved in protein synthesis and transport. If free of ribosomes, it is smooth ER (SER) and is responsible for lipid metabolism (Ch. 2).

Endostome See **exostome**.

Endotesta (pl.: endotestas) In the seeds of gymnosperms, the innermost layer of the seed coat, often being thin and paperlike (Ch. 20).

Endothecium (pl.: endothecia) In microsporophylls, the hypodermal layers formed by the primary wall layer. The endothecium often has a role in the dehiscence of the microsporophyll (Ch. 19).

Endothelium (pl.: endothelia) If the inner epidermis of the integument next to the nucellus becomes densely cytoplasmic and apparently secretory, it is called an endothelium. Also called an *integumentary tapetum* (Ch. 19).

Epiblast In the embryos of grasses, a small flap of tissue that is opposite the scutellum; its function is unknown, but it is hypothesized that it may be a residual second cotyledon (Ch. 20).

Epiblem A rarely used term for the root epidermis. Its use implies the belief that the root epidermis and the shoot epidermis are fundamentally different tissues. A synonym is *rhizodermis* (Ch. 10).

Epicarp The outermost layer of a fruit wall, exterior to the mesocarp; it is often thin and is referred to as the "peel" or "skin" of the fruit. Also frequently called the *exocarp* or *ectocarp* (Ch. 21).

Epicotyl If a shoot is present in a seed before germination, it is called an epicotyl. Many seeds have a shoot apical meristem but no leaves and no tissue that is not meristematic; these are considered to lack an epicotyl (Ch. 20).

Epidermis In the primary body, the outermost layer of cells of the stem, leaf, or root—but not root cap (Ch. 10).

 Multiple epidermis An epidermis that consists of several layers, all of which are derived by periclinal divisions in protoderm cells (Ch. 10).

Epithelium (pl.: epithelia) The cellular lining of any secretory canal or chamber; the epithelium itself may be either secretory or not, but it often is important in separating the secretion from the rest of the plant (Ch. 9).

Epithem The loose, lightly chlorophyllous or achlorophyllous cells in the central part of a hydathode (Ch. 9).

Ergastic substance Any material that is present in a cell as a storage product or crystal. An ergastic substance may be common in certain types of cells, but they are never present in all cells; they are often considered as being not part of the living protoplasm (Ch. 2).

Establishment growth A growth phase of palms and similar plants with extremely broad primary stems. This phase lasts from germination until the seedling has obtained the proper width; then height growth begins (Ch. 11).

Etioplast A plastid that has developed in the dark. The inner membrane system is in the form of a paracrystalline matrix. If exposed to light, an etioplast can be converted to a chloroplast (Ch. 2).

Euchromatin The chromatin (DNA plus histone) that is actively involved in transcribing RNA. With most stains for either light or electron microscopy, the euchromatin remains only lightly stained. Cells that are extremely active metabolically have a high proportion of euchromatin. See **heterochromatin** (Ch. 2).

Eustele A type of ectophloic siphonostele in which the stele is made up of one ring of vascular bundles that surround a pith (Ch. 11).

Eusyncarpous style A compound style in which the component styles are completely fused together and only one mass of transmitting tissue is present. See **pseudosyncarpous style** (Ch. 19).

Exalbuminous A type of seed in which there is little or no residual endosperm present at maturity. The alternative is albuminous (Ch. 20).

Exarch A stele in which the protoxylem is located on the exterior of the xylem mass and the metaxylem is located in the interior. The alternatives are endarch and mesarch (Ch. 13).

Excretion A term that was proposed to designate those secretions that consist of waste products. See **secretion** (Ch. 9).

Exine The outermost layer of the pollen wall; it is subdivided into the inner endexine, or nexine, and the outer ectexine, or sexine (Ch. 19).

Exocarp Synonym for **epicarp.**

Exodermis A layer of cells located immediately interior to the epidermis of some roots and having the characteristics of an endodermis (Ch. 13).

Exostome In ovules with thick integuments, the micropyle is canallike. The outer opening is the exostome; the inner one (adjacent to the nucellus) is the endostome (Ch. 19).

False fruit See **fruit.**

Female gametophyte Synonym for **megagametophyte** (Ch. 19).

Fenestra (pl.: fenestrae) A large depression that occurs in the walls of epidermal cells of some plants. Epidermal cells with fenestrae have very thick walls, and the fenestrae occur in the radial walls between two adjacent epidermal cells (Ch. 10).

Fiber An elongate sclerenchyma cell that does not conduct water (Ch. 5, 7).
    Extraxylary fiber Any fiber that is located in a tissue other than xylem.
    Gelatinous fiber A fiber that occurs in wood (especially tension wood) and whose walls absorb water and swell.
    Libriform fiber A fiber in the xylem that resembles a phloem fiber in being long, with thick walls and simple pits.
    Mucilaginous fiber Synonym for gelatinous fiber.
    Septate fiber A fiber that undergoes cytokinesis and forms thin cross walls, resulting in a series of cells inside the original fiber wall.
    Substitute fiber Living, elongate cells in the xylem that probably should be considered parenchyma cells, not fibers.
    Xylary fibers Fibers that occur in the xylem; they may have many forms.

Fiber bundle A group of fiber cells that at first glance resembles a vascular bundle; these are abundant in tough, leathery monocot leaves. (Ch. 12).

Fiber-tracheid A xylary cell that is intermediate between a libriform fiber and a tracheid: it has thick walls and circular bordered pits, but the pits are smaller and less conspicuous than those of tracheids (Ch. 5).

Fibril An aggregate of cellulose microfibrils, these are one of the most important structural elements of cell walls (Ch. 2).

Filament In flowers, the stalk of a stamen (Ch. 19).

Filiform apparatus In a synergid, a transfer wall that forms at the contact area between the synergid, the egg, and the nucellus (Ch. 19).

Flank meristem The lateral, ring-shaped portion of the corpus of a shoot apical meristem; it is more often called a *peripheral zone* or *peripheral meristem* (Ch. 6).

Fleshy fruit An edible fruit. The alternative is a dry (inedible) fruit (Ch. 21).

Floral bract A modified leaf that is specialized to protect an inflorescence as it develops; also called a *hypsophyll* (Ch. 12).

Foliage leaf A leaf that is specialized for photosynthesis. Rarely called a *trophophyll* (Ch. 12).

Foliar buttress The region of an apical meristem that is swelling as the first stage of leaf initation (Ch. 6).

Foliar helix The arrangement of leaves in helices on the stem (Ch. 6).

Foot cell In epidermal glands, the most proximal cell; the foot cell is usually located at the same level as the surrounding ordinary epidermal cells. Also called a *basal cell* (Ch. 10).

Foot layer The layer of a pollen wall formed by the expansion of the inner ends of the bacula (Ch. 19).

Forming face The side of a dictyosome that receives vesicles from the ER, where they fuse together to form a new vesicle of the dictyosome (see **maturing face;** Ch. 2).

Fret The portion of a thylakoid that extends from one granum to another in a chloroplast (Ch. 2).

Fruit A term that has conflicting definitions. Although the general usage in lay English is imprecise, in botanical usage it should always refer to a mature ovary of the angiosperms (Ch. 21).

 False fruit Any structure that resembles a fruit and that functions like a fruit but that contains tissues in addition to the ovary. Examples are strawberry, fig, and pineapple.
 Pericarp A strictly technical term used to specify the mature ovary wall only; this is meant to avoid the problems that accompany the term *fruit.*
 True fruit A synonym for **pericarp.**

Fruit wall A nonspecific term that is used to refer to the tissues of a fruit. It is meant to be general and to be applicable to any fruiting tissue, whether derived from the ovary (true fruit) or the ovary and accessory tissues (false fruit). The term carries no ontogenic implications (Ch. 21).

Funiculus (pl.: funiculi) The stalk of an ovule (Ch. 19).

Fusiform initial A long, tapered cell in the vascular cambium. The derivatives of fusiform initials develop into the axial elements of the secondary xylem and secondary phloem. See **ray initial** (Ch. 14).

G-layer The innermost wall layer of a gelatinous fiber; the G-layer is rich in cellulose but contains little lignin (Ch. 5).

Gametogenesis The formation of gametes (Ch. 19).

Gametophyte In the life cycle of a plant, the generation that produces the gametes. In the plants that are familiar to us, the gametophytes are the pollen and the embryo sac. In the mosses, however, the leafy plants are the gametophytes (Ch. 19).

Gas chamber A large intercellular space in aquatic plants; these provide buoyancy and facilitate the diffusion of oxygen to the lower tissues (Ch. 12).

Generative cell In a pollen grain or pollen tube, the cell that will divide to produce two sperm cells (Ch. 19).

Germination aperture A special area of the pollen wall where the pollen tube emerges during germination (Ch. 19).

Globular stage In embryo development, an early stage at which the embryo (or proembryo) consists of a small ball of cells, without any tissue differentiation (Ch. 20).

Glyoxysome An organelle of the cell that is involved in the metabolism of glyoxylic acid; these are most prominent in cells that metabolize a large amount of lipids such as in endosperm and seeds that store fats (Ch. 2).

Golgi apparatus An assemblage of golgi bodies within one cell and acting together. This term is occasionally applied to the individual members, which are properly known as *dictyosomes* or *golgi bodies* (Ch. 2).

Granular ER An older name for *rough endoplasmic reticulum;* it is not commonly used (Ch. 2).

Granum (pl: grana) A vesicle of the inner membrane of chloroplasts. The grana are associated in stacks and are connected to the grana of adjacent stacks by frets (Ch. 2).

Growth ring bridge This refers to a set of tracheids or vessels that are located at the border between two growth rings and that have pits on their tangential walls. The pits allow water to be passed from one growth ring to the other (Ch. 15).

Guard cells The pair of crescent-shaped epidermal cells that surround the stomatal pore. The swelling or shrinking of the guard cells opens or closes the stoma. Guard cells may be closely related to adjacent subsidiary cells and may form part of a stomatal complex (Ch. 10).

Gummosis A process in which the cell walls, usually of tracheary elements, are chemically altered to a gummy condition. This may occur as the result of disease or in the normal formation of heartwood (Ch. 15).

Guttation The secretion of water by hydathodes (Ch. 9).

Gymnosperm Gymnosperms are the "naked seed plants"; that is, the plants that bear seeds that are not within fruits. These are the divisions Coniferophyta, Gnetophyta, Cycadophyta, and Ginkgophyta (Ch. 1).

Hair Synonym for **trichome.**

Haplostele A type of protostele in which the mass of xylem is round in cross section (Ch. 11).

Hardwood A semitechnical term to indicate the wood of dicots; it refers to the fact that most commercial dicot wood contains libriform fibers. See **softwood** (Ch. 15).

Hartig net See **ectomycorrhiza.**

Hastula (pl.: hastulae) A rather thick flap of tissue, resembling a ligule but attached to the upper end of the petiole in palm leaves (Ch. 12).

Haustorium (pl.: haustoria) The absorptive structure of a parasite. In fungi, it is a specialized hypha capable of invading the host and absorbing nutrients. In parasitic flowering plants such as mistletoes, it is a modified root. Various parts of the gametophytes of some seed plants invade parts of the flower and absorb nutrients, passing them on to the embryo, and these are referred to as haustoria (Ch. 13).

Head In glands, the uppermost, secretory portion. If a gland has a head, then it usually also has a stalk and a foot (Ch. 10).

Heart stage In embryo development, the stage at which the cotyledons are initiated, giving the embryo a distinctive heart shape (Ch. 20).

Heartwood The wood at the center of a tree; the tracheary elements have cavitated and no longer conduct water, and the parenchyma cells have died. See **sapwood** (Ch. 15).

Helobial endosperm In angiosperm reproduction, a type of endosperm in which the division of the primary endosperm nucleus is accompanied by wall formation, resulting in a large micropylar cell and a small chalazal one. The chalazal cell contributes little to the endosperm, but the micropylar cell grows rapidly by free nuclear division. The alternatives are cellular endosperm and nuclear endosperm (Ch. 20).

Heteroblastic plant A plant that has several distinct types of leaves; this may be related to a juvenile-adult transition, to having long shoots and short shoots, to having submerged and aerial leaves, or to having sun and shade leaves (Ch. 11).

Heterocarpy The production of two or more distinct types of fruit by one plant (Ch. 21).

Heterochromatin The chromatin (DNA plus histones) that is not actively involved in transcribing RNA. With most stains for light and electron microscopy, the heterochromatin stains more darkly than the euchromatin (Ch. 2).

Heterospory A type of sexual reproduction in which two types of spores are produced—microspores and megaspores. Heterospory occurs in all seed plants and in a few cryptogams. The alternative is homospory (Ch. 19).

Hilum (pl.: hila) The nucleation center for a starch granule (Ch. 2). In seeds, the scar formed where the seed breaks away from the funiculus (Ch. 20).

Histogen An older concept, not usually accepted at present, that postulated that meristems were composed of initials, which are sets of cells that differ from each other in the types of cells and tissues that they can produce. Three histogens were postulated: dermatogen, periblem, and plerome (Ch. 6).

Homospory A type of sexual reproduction in which all spores appear to be equivalent, all having the same size and shape. Among the vascular plants, found only among some ferns and some other cryptogams (Ch. 19).

Hydathode A secretory structure usually consisting of epithem and a sheath. Hydathodes secrete almost pure water and may be a means by which minerals are transported to young tissues more rapidly than would be possible by transpiration alone (Ch. 9).

Hydrophytes Plants that grow with their stems partially or completely submerged in water (Ch. 12).

Hygrochastic opening A method of fruit opening in which the fruit walls break as a result of the stresses caused when layers of living cells absorb water and swell. Compare with **xerochastic opening** (Ch. 21).

Hygroscopic movement A movement that results from the uptake or loss of water, either by living cells or by the walls of dead cells (Ch. 21).

Hypocotyl The short portion of embryo axis that lies between the root (which has a protostele) and the shoot (which has a eustele or an atactostele; Ch. 20).

Hypodermis The layer of cells immediately interior to the epidermis, if this layer is distinct from all other cells. If it is similar to the other mesophyll or cortex cells, then it is not considered a hypodermis (Ch. 10).

Idioblast Any cell that differs significantly from the neighboring cells, usually by contents or size (Ch. 2).

Indehiscent fruit A fruit that remains closed at maturity; it does not open automatically. The alternative is dehiscent fruit (Ch. 21).

Indumentum (pl: indumenta) A collective term that refers to all the trichomes of a plant. See also **trichome complement** (Ch. 10).

Infection thread As a nitrogen-fixing bacterium invades a root hair, it passes through the hair and into the cortex, the entire time being encased in a cellulosic tube produced by the plant. The tube and the bacteria constitute an infection thread (Ch. 13).

Inferior ovary An ovary in which the other floral parts are attached to the top of the ovary (Ch. 19).

    Appendicular inferior ovary An inferior ovary in which the tissues surrounding the ovary are the fused bases of the other flower parts.

    Receptacular inferior ovary An inferior ovary in which the tissue surrounding the ovary is the receptacle.

Initiating ring See **anneau initial.**

Integument A jacket of cells that covers the central mass of cells in an ovule. The integument is always thin, and an ovule may have one integument, two, or none (Ch. 19).

Integumentary tapetum Synonym for **endothelium** (Ch. 19).

Intercellular space The space between the wall of one cell and the walls of the adjacent cells. See **apoplast** (Ch. 2).

Interfascicular region Synonym for **medullary ray.**

Intermediary cell A type of parenchyma cell in the phloem of many angiosperms. Intermediary cells are responsible for the loading and unloading of nutrients into the adjacent sieve elements (Ch. 8).

Internode The region of a stem that is located between the nodes (the areas where leaves are attached). Rosette plants, with tight whorls of leaves, do not have internodes (Ch. 11).

Intussusception A theory of cell wall deposition that postulates that the growing ends of the cellulose microfibrils are located in the wall, not in the plasmalemma. The result of intussusception would be an interweaving of the microfibrils. See **apposition** (Ch. 2).

Isolateral leaf A leaf in which the upper half (in a cross section) strongly resembles the lower half, usually because both have a layer of palisade parenchyma. Also called an isobilateral leaf; the alternative is bifacial or dorsiventral. Isolateral should not be confused with unifacial. An isolateral leaf actually does have two surfaces, which resemble each other; a unifacial leaf (in many monocots) was thought to represent an outgrowth from the abaxial surface of a leaf primordium, therefore technically having just one surface. Recent studies have thrown this interpretation into doubt (Ch. 12).

Juvenile A plant that is not yet capable of reproduction. In most plants the type of leaves and stems that are formed while the plant is juvenile differ at least slightly from those produced when it is adult (Ch. 6).

Karyokinesis (pl: karyokineses) The division of the nucleus; two methods are possible, mitosis and meiosis (Ch. 2).

Kranz anatomy In plants that have C$_4$ photosynthesis, the mesophyll consists of cells arranged radially around the bundle sheath. The term *Kranz* (wreath) was originally applied only to the mesophyll, but it is frequently used to refer to both the mesophyll and the bundle sheath (Ch. 12).

Labyrinthine wall A primary wall that has irregular ingrowths that greatly increase the area of the inner surface. Because the plasmalemma is appressed to all the contours of the ingrowths, it also has a large surface area, which presumably allows it to have more molecular pumps. Labyrinthine walls are also called *transfer walls*, and cells that contain them are *transfer cells* (Ch. 3).

Late wood A synonym for **summer wood.**

Lateral suture In fruits, the seam formed by the fusion of one margin of one carpel with the margin of an adjacent carpel. Compare with **ventral suture** (Ch. 21).

Latex A nonspecific term for any turbid, usually milky internal secretion (Ch. 9).

Laticifer Any secretory structure that produces latex (Ch. 9).
  Articulated laticifer A laticifer that is composed of individual cells, usually interconnected by holes in their common walls, much like vessels in xylem.
  Nonarticulated laticifers Laticifers that consist of single cells, usually extremely elongated. Small nonarticulated laticifers are more often called *laticiferous idioblasts.*

Leaf base The swollen nodal region, immediately below the point of attachment of a leaf. The leaf base is inconspicuous in most plants, very noticeable in a few. In a very small percentage of species it is extremely large and may be important in photosynthesis; then it may be called a *podarium* (pl.: podaria) (Ch. 12).

Leaf gap A region in a fern siphonostele, just above a leaf trace, where there is no xylem or phloem. The term has also been applied to the nodes of seed plants, but now there is evidence that they do not exist in them (Ch. 11).

Leaf sheath The basal portion of some leaves that wrap completely around the stem. The leaf sheath may envelope the entire node or just the basal portion (Ch. 12).

Closed leaf sheath A sheath that is a complete cylinder, without a longitudinal opening.
Open leaf sheath A sheath in which the two edges of the sheath touch or overlap but have not fused to each other.

Leaf trace A vascular bundle, usually collateral, that runs from the axial bundle system to a leaf (Ch. 11).

Ledge A ridge of material on either the outer or inner edge of a guard cell, on the ventral wall. When the guard cells shrink and close the stomatal pore, the ledge of one guard cell usually presses against the ledge of the other, sealing the pore (Ch. 10).

Lenticel A region in the bark that contains loose, rounded complementary cells, and that permits the diffusion of oxygen into the plant through the bark (Ch. 17).

Leucoplast Any colorless plastid in general (Ch. 2).

Lignin A complex polymer of several components, especially phenolic compounds related to coniferyl; the deposition of lignin in walls increases their strength, water impermeability, and resistance to decay (Ch. 5).

Ligule An outgrowth of the adaxial epidermis at the junction of the leaf blade and the leaf sheath (Ch. 12).

Lithocyst A cell, usually of the epidermis, that contains a cystolith (Ch. 2).

Long cell An epidermal cell in the grasses, having an elongate form. The grass epidermis also has short cells (Ch. 10).

Long shoot A term that is usually applied only if a plant also has short shoots; then, the long shoots are those with elongated internodes (almost always the main axis), and the short shoots are those with unelongated internodes and usually having a different form of leaf. Short shoots are almost always lateral shoots on a long shoot (Ch. 11).

Lumen The central space or cavity of vesicles, ducts, chambers, or virtually any other enclosed space (Ch. 2, 9).

Lysigenous An adjective that describes any structure or space that has been formed by lysis (Ch. 9).

Lysis The degeneration and breakdown of an organelle or a complete cell. When whole cells are involved, the term *lysigeny* is frequently used. Must not be confused with plasmolysis; see also **schizogeny** (Ch. 9).

Lysosome An organelle that is responsible for fusing with defective organelles and digesting them. Lysosomes are a discrete type of organelle in animals, but in plants the central vacuole acts as a lysosome (Ch. 2).

Macrosclereid A somewhat elongated sclereid, rather columnar in shape (Ch. 5).

Male gametophyte Synonym for **microgametophyte** (Ch. 19).

Manoxylic wood Wood that contains very large amounts of parenchyma; this is characteristic of cycads and some advanced dicots. The alternative is pycnoxylic (Ch. 15).

Mantle The outermost layer of cells in a shoot apical meristem that is considered to have a mantle-core organization (most of the gymnosperms). A mantle differs from a tunica in that the cells of a mantle may undergo occasional divisions with periclinal walls (Ch. 6). In ectomycorrhizae, the mycelial covering on the surface of the host root (Ch. 13).

Marginal initials In a marginal meristem of a leaf primordium, the marginal initials are the cells in the protoderm. In most species, they divide only with anticlinal walls, so they produce only the epidermis and do not contribute to the mesophyll (Ch. 12).

Marginal meristem In a leaf primordium, the lamina is produced as two rows of cells begin growing outward more rapidly than the surrounding cells. The marginal meristems contain marginal initials and submarginal initials (Ch. 12).

Margo The outer rim of the pit membrane in the circular bordered pits of gymnosperms. Many non-cellulosic components are digested out of the margo, and as a result it is extremely permeable (Ch. 7).

Massula (pl.: massulae) All the pollen grains of an anther sac of milkweed (Asclepiadaceae) stick together, forming a massula (Ch. 19).

Maturing face The side or face of a dictyosome on which the vesicles are oldest and that have finished accumulating and processing material. The vesicles of the maturing face detach and carry their contents to their destination (Ch. 2).

Medullary bundle A vascular bundle that is located in the pith of a dicot. It may be attached to an axial bundle, but it is not part of the ring of bundles (Ch. 11).

Medullary ray The regions of tissue, often parenchyma, that are located between the vascular bundles of a stem or root. Also called *interfascicular regions* (Ch. 11).

Megagamete The larger gamete in oogamy; often called an egg cell. The smaller gamete is a microgamete (Ch. 19).

Megagametophyte A gametophyte (haploid plant) that produces the megagametes (egg cells). Often called the *female gametophyte*. See **microgametophyte** (Ch. 19).

Megaphyll The leaf type of the seed plants and ferns, megaphylls evolved from branch systems (telomes). The alternative is the line of leaf evolution involving microphylls (Ch. 12).

Megasporangium (pl.: megasporangia) A structure that produces the megaspores during heterosporous reproduction. In the angiosperms, the megasporangia are the ovules. See **microsporangium** (Ch. 19).

Megaspore In plant species with heterosporous reproduction, the larger spore type. Megaspores germinate and grow to be megagametophytes. See **microspore** (Ch. 19).

Megaspore mother cell Synonym for **megasporocyte** (Ch. 19).

Megasporocyte A cell that is preparing for meiosis; one of its meiotic descendents will be the megaspore. Megasporocytes are also called megaspore mother cells (Ch. 19).

Meristem A cell or group of cells whose principal function is to divide in an organized manner. See also meristemoid (Ch. 6).
    Apical meristem A meristem that is located at the apex or most distal part of an organ.
    Axillary meristem A shoot apical meristem of a bud, located just at the upper edge of the junction of a petiole and a stem.
    Basal meristem A meristem that is located in the base or most proximal portion of an organ.
    Closed meristem A meristem in which there is a well-defined, small group of cells that produce the pattern of the tissues. See open meristem.

Ground meristem A meristem that produces relatively large amounts of uniform tissue; most frequently used to refer to that portion of the subapical meristem that produces cortex or pith.

Intercalary meristem A meristem that is located within the organ that it produces, rather than at the apex, base, or side. In practice, *intercalary meristem* perhaps most frequently is used to designate the meristematic zones of certain internodes and monocot leaves.

Lateral meristem A meristem that is located along or near the edge or margin of the tissues that are produced. The vascular cambium is often referred to as a lateral meristem to distinguish it from the shoot and root apical meristems, but it is also accurate to consider it an intercalary meristem.

Open meristem A meristem in which the patterns of the tissues cannot be traced to a small group of cells; the pattern becomes indistinct near the base of the meristem.

Primary meristem A meristem that is present in the embryo, basically the root and shoot apical meristems. See **secondary meristem.**

Secondary meristem At present, any meristem that arises within tissues produced by a primary meristem; basically the vascular cambium, cork cambium, and secondary thickening meristems.

Waiting meristem See *méristème d'attente.*

Meristematic region A zone in which cell division is frequent but which is not so well defined as to be classified as a meristem (Ch. 1).

Méristème d'attente A term that was part of a theory of meristem structure and function. The *méristème d'attente* was hypothesized as being a region that contributed little or nothing to vegetative growth but that expanded to produce the floral apex at times of reproductive growth. Also occasionally called a "waiting meristem," but the English term is rarely used (Ch. 1).

Meristemoid A cell or small group of cells that are meristematic but for only a short period. The cells that divide to produce stomatal apparatuses or trichomes are frequently referred to as meristemoids, as are small, irregular regions of cell division. The term is often used to suggest a lack of precision (Ch. 1).

Mesocarp In fruits, the fleshy middle layer of the fruit wall, if three layers are present. The mesocarp is usually the thickest layer and the most edible; in fruits with thin walls the mesocarp may be absent (Ch. 21).

Mesocotyl In monocots, that portion of the embryo axis where the scutellum is attached (Ch. 20).

Mesogamy The process in which a pollen tube grows into the ovule through the integuments (Ch. 19).

Mesogenous An adjective describing a stomatal complex in which all the subsidiary cells and the two guard cells are derived from the same initial cell. See **perigenous** (Ch. 10).

Mesome See **telome.**

Mesophyll All the internal tissue of a leaf excluding the vascular bundles. In most leaves, the mesophyll is differentiated into palisade parenchyma and spongy mesophyll (Ch. 12).

Mestome sheath In a bundle sheath, if the sheath has two layers of cells, the inner one is called a "mestome sheath," and it is basically an endodermis (Ch. 12).

Metaphloem The primary phloem that forms in a vascular bundle after that portion of the bundle has stopped elongating. The sieve elements are larger than those in the protophloem, and there will be either companion cells or albuminous cells present (Ch. 11).

Metaxylem The primary xylem that matures in a vascular bundle after the bundle has stopped elongating. The tracheary elements may be either tracheids or vessels, but they are always larger than the protoxylem elements that are formed at the same level. Because the vascular bundle has stopped elongating in the area of metaxylem maturation, the tracheary elements may have any type of secondary wall, extensible or not. See **protoxylem** (Ch. 11).

Microfibril A set of cellulose molecules that have crystalized together. The microfibrils are a major component of most cell walls (Ch. 2).

Microfilament A proteinaceous filament that is present in cells (Ch. 2).

Microgamete The smaller of the gametes in oogamy; often called a sperm cell. The larger gamete is the megagamete (Ch. 19).

Microgametophyte A gametophyte (haploid plant) that produces the microgametes (sperm cells). Sometimes called the male gametophyte. See **megaga-metophyte** (Ch. 19).

Microphyll The leaf of the lycopods and related plants, being just a small outgrowth of stem tissue; the alternative is megaphyll, a type of leaf that evolved from a branch system. The microphylls represent a distinct line of leaf evolution beginning with the earliest land plants. Also called *enations* (Ch. 12).

Micropyle The small hole in the integument at the apex of an ovule; the pollen or pollen tubes often enter through the micropyle (Ch. 19).

Microsporangium (pl.: microsporangia) A structure that produces the microspore during heterosporous reproduction. In the angiosperms, the microsporangia are the anthers. See **megasporangium** (Ch. 19).

Microspore In plant species with heterosporous reproduction, the smaller of the two types of spores. Microspores germinate and grow to become microgametophytes. See **megaspore** (Ch. 19).

Microsporocyte A cell that is preparing for meiosis and that will divide into four microspores. Also called *pollen mother cells* in the seed plants (Ch. 19).

Microtubule A proteinaceous tubule, constructed of tubulin. Microtubules are thought to act as a cytoskeleton, giving the cell a shape that is later maintained by the cell wall. Microtubules are also responsible for pulling organelles to specific sites within the cell (Ch. 2).

Mictoplasm This term is occasionally applied to the watery cytoplasm that results when the vacuolar membrane breaks down during sieve element development and the cytoplasm mixes with the contents of the vacuole (Ch. 8).

Midrib In simple leaves, the large central vascular bundle and associated sheathing tissues. Some leaves do not have an obvious midrib but rather several prominent, more or less equal-sized veins (Ch. 12).

Minor bundle The very finest veins of a leaf; it is estimated that most of the nutrient and water exchange between vascular tissue and mesophyll occurs by means of the minor veins (Ch. 12).

Mitochondrion (pl.: mitochondria) An organelle that is universally present in plant cells and that is responsible for aerobic respiration (Ch. 2).

Monarch See **diarch**.

Monocarpic A plant that reproduces only once, then dies. All annuals and biennials are monocarpic plants, and some perennials are also (Ch. 1).

Monophyletic This refers to a taxon, structure, or physiological process that has had just one evolutionary origin. Flowers are believed to be monophyletic, having evolved only once; compare with **polyphyletic** (Ch. 9).

Monopodial shoot A shoot that grows continuously from one single apical meristem. The alternative is sympodial (Ch. 11).

Monosporic megagametophyte A megagametophyte that arises from a uninucleate megaspore, produced when only one haploid nucleus survives after meiosis and the other three degenerate. See also **bisporic** and **tetrasporic** (Ch. 19).

Monostratose Consisting of just one layer (Ch. 1).

Mucigel The mucilage that is secreted by a growing root tip; it is thought to be important in lubricating the root's passage through the soil and perhaps also in affecting the properties of the soil solution and in nourishing bacteria of the rhizosphere (Ch. 13).

Multiplicative cell division A type of division in cambial initials in which the cell plate is oriented radially and longitudinally; this results in two new initials. The alternative is proliferative division (Ch. 14).

Multistratose Consisting of several layers (Ch. 6).

Mycorrhiza (pl.: mycorrhizae) An association of fungi with the roots of many plants. For many, the mycorrhiza is essential when growing in poor soil; for other plant species, the association is absolutely essential regardless of soil conditions. See **ectomycorrhiza** and **endomycorrhiza** (Ch. 13).

Nacreous layer A layer of opalescent wall material in sieve elements.

Neck In glands, the cells between the basal cells and the head; a synonym for *stalk* (Ch. 10).

Nectary A gland or trichome that secretes nectar (Ch. 9).
  Extrafloral nectary A nectary located anywhere on the plant except on the flower.
  Floral nectary A nectary located on any part of a flower.

Nexine Synonym for **endexine** (Ch. 19).

Nodal plate A transverse complex of vascular tissue located at the nodes in stems of certain plants (Ch. 13).

Node The region of the stem where a leaf is attached. The nodes are separated from each other by internodes, except in rosette plants (Ch. 11).

Nucellus (pl.: nucelli) The central mass of cells in an ovule; one of the nucellar cells is the archesporial cell (Ch. 19).

Nuclear endosperm In angiosperm reproduction, a type of endosperm in which there is only nuclear division during the first stages of development. A large multinucleate coenocyte is formed; later, cell walls are established and cells are formed. The alternatives are cellular endosperm and helobial endosperm (Ch. 20).

Nuclear envelope A set of two membranes that enclose the nucleoplasm. The nuclear envelope contains nuclear pores, and the outer membrane is continuous with the endoplasmic reticulum (Ch. 2).

Nuclear pore A hole in the nuclear envelope formed by the fusion of the two membranes at that point (Ch. 2).

Nuclear sap Synonym for **karyolymph.**

Nucleus (pl: nuclei) An organelle bounded by two membranes and containing the DNA responsible for most hereditary information. All cells have at least one nucleus when young; many cells are multinucleate (Ch. 2).

Obturator In an angiosperm carpel, an outgrowth of the placenta that acts to direct a pollen tube to the ovule (Ch. 19).

Oogamy Sexual reproduction in which one gamete is small and motile whereas the other is much larger and immobile. All vascular plants are oogamous. Compare with **anisogamy** and **isogamy** (Ch. 19).

Open stele A type of eustele in which the bundles have very little interconnection; instead each bundle maintains its identity if followed upward through the stem. Such a bundle is called a sympodium. See **closed stele** (Ch. 11).

Organ Plants are considered to consist of three organs: stems, leaves, and roots. Each is composed of characteristic arrangements of tissues. Other organs

(petals, haustoria, and the like) are considered to be modifications of these three types (Ch. 2).

Organelle The individual parts of a cell (Ch. 2).

Orthic tetrakaidecahedron A geometric body of 14 sides, frequently proposed as the shape of a parenchyma cell within a mass of pure parenchyma (Ch. 3).

Orthotropous A synonym for **atropous.**

Osmiophilic body Any body that strongly absorbs the dye osmium tetroxide; osmiophilic bodies tend to contain mostly lipids, but the term *osmiophilic body* often implies that the true chemical nature of the body is not known for certain (Ch. 2).

Osmophore A gland that secretes the fragrances of a flower (Ch. 9).

Osteosclereid A bone-shaped sclereid, being slightly elongate and having swollen ends (Ch. 5).

Ovary The lowermost portion of a carpel; the portion that contains the ovules and that develops into the fruit (Ch. 19).

Ovuliferous scale The flat, leaflike structure in a megasporangiate strobilus (seed cone) that actually bears the ovules. It is thought to represent a modified lateral branch and its sporophylls (Ch. 19).

P-protein The protein filaments that occur in sieve elements, formerly called slime. When the element is damaged and surging occurs, the P-protein forms a plug over the sieve area, formerly called a *slime plug* (Ch. 8).

P-type plastid A special type of plastid that occurs in sieve tube members; these plastids contain protein. The anatomy of P-type plastids and S-type plastids is important in the taxonomy of the dicots and especially of the Order Caryophyllales (Ch. 8).

Palisade parenchyma In the mesophyll of a leaf, a region in which the chlorenchyma cells are elongate and are arranged parallel to each other, perpendicular to the epidermis; usually located in the upper half of the leaf, but the palisade parenchyma can also be in the lower half or along both surfaces (Ch. 12).

Palynology The study of fossil pollen (Ch. 19).

Paracytic A type of stomatal complex in which the subsidiary cells are aligned parallel with the guard cells. Formerly called a *rubiaceous stomatal complex* (Ch. 10).

Parenchyma A cell and tissue type in which the cells have (usually) only thin primary walls. They are all cells that do not have the characteristic walls of collenchyma cells and that do not have the secondary walls of sclerenchyma cells (Ch. 3).

Passage cell A cell in an endodermis that is less well developed than the surrounding endodermal cells, often being in State I while the others are in State III. It was thought that the passage cells facilitated movement of material between the cortex and the stele, but now there is evidence that the presence of passage cells is just a result of unequal rates of cell differentiation (Ch. 13).

Perforation plate The end wall of a vessel element where the perforation occurs. If the perforation is large, then the perforation plate may consist of just a narrow rim (Ch. 7).

    Compound perforation plate A perforation plate that contains more than one perforation. A compound perforation plate may be reticulate, scalariform, or foraminate (ephedroid), depending on the pattern of the perforations.

    Simple perforation plate A perforation plate that contains only one perforation.

Periblem One of the histogens envisioned by Hanstein as composing the apical meristem of seed plants. The periblem was considered to give rise to the outer, nonepidermal portions of the axis. See *histogen* (Ch. 6).

Pericarp A synonym for *true fruit;* see **fruit.**

Periclinal Parallel to the surface. A periclinal cell wall is parallel to the closest surface of the organ or duct being considered. The most common alternative is anticlinal, perpendicular to the surface (Ch. 6).

Pericycle The cells that are located between the endodermis and the vascular tissues in a root. These are usually parenchyma but may also be fibers (Ch. 13).

Periderm The technical term for bark; the periderm contains the cork (phellem) that is produced by the cork cambium (phellogen) and any epidermis, cortex, and primary or secondary phloem that might be exterior to the cork cambium. Notice that unlike *epidermis* and *endodermis,* this term does not end in *is;* it is not "peridermis" (Ch. 17).

Perigenous An adjective describing a stomatal complex in which the subsidiary cells are not derived from the same initial cell as are the guard cells. See **mesogenous** (Ch. 10).

Perimedullary zone The outermost regions of the pith, those regions closest to the protoxylem. The cells of the perimedullary zone are often different from those of the rest of the pith (Ch. 11).

Perinuclear space The region that is located between the inner and the outer membranes of the nuclear envelope (Ch. 2).

Peripheral zone The outer, ring-shaped zone of the corpus in many seed plants that have a zonate corpus. The peripheral zone surrounds the central mother cells and the pith rib meristem, and it gives rise to the cortex. Also called the *flank* or *flanking meristem* (Ch. 6).

Periplasmodium (pl.: periplasmodia) The large mass of fused tapetal protoplasts that form around the developing pollen grains in an anther that has an invasive tapetum (Ch. 19).

Perisperm A nutritive tissue in the seeds of most species of the Order Caryophyllales; the perisperm is formed by the proliferation of the nucellus (Ch. 20).

Permanent initials Cells that are located within a meristem such that they are never pushed out of the meristem by the growth and division of other cells. See **initials** (Ch. 1).

Peroxisome A small organelle bounded by a single membrane, associated with chloroplasts. Peroxisomes absorb glycolic acid during photorespiration and oxidize it to carbon dioxide. This generates hydrogen peroxide, which is broken down within the peroxisome (Ch. 2).

Phanerogam These are plants that produce seeds, also known as spermatophytes. The alternative is the group of nonseed plants or cryptogams (Ch. 1).

Phellem Cork, specifically the rather large amounts produced by a phellogen. The individual cork cells that occur in grass leaves are never considered phellem (Ch. 17).

Phellogen The cork cambium of the bark. The phellogen contains only cuboidal or short columnar cells; it usually produces only cork (phellem) to its exterior, but in some species it also produces a small amount of parenchyma (phelloderm) to its interior (Ch. 17).

Phelloid A thin-walled, unsuberized cell located among the cork cells of phellem (Ch. 17).

Phragmoplast A set of microtubules that is circular in face view and that is located at the edge of the cell plate during cell division. The phragmoplast guides the direction of growth of the new cell walls (Ch. 2).

Phylloclade Synonym for **cladode**.

Phyllome A term that indicates all the leaves of a plant (Ch. 12).

Piliferous cell A young epidermal cell that will develop into a trichome. More frequently called a *trichoblast* (Ch. 10).

Pit A region in a sclerenchyma cell where there is no secondary wall but there is primary wall. An absence of both walls is a *perforation*. (Ch. 2, 5).

    Anastomosing pit Synonym for **branched pit**.

    Bordered pit A pit that has a border, a thickened margin around the inner aperture.

    Branched pit A pit that is formed as several simple pits intersect, present in the thick walls of many sclereids.

    Fenestriform pit A large "windowlike" pit that occurs in certain types of crossfield pitting.

    Half-bordered pit-pair A pit-pair in which one member is bordered and the other is simple.

    Pit canal The tubular portion of a pit in a thick wall. If the wall is thin, there is no pit canal.

    Pit cavity The enlarged portion of a pit that is adjacent to the outer aperture. Also called *pit chamber*.

    Pit membrane The primary walls and middle lamella that occur between the two pits of a pit pair.

    Simple pit A pit that is not branched or bordered.

Pith-rib meristem The lower, central portion of the corpus in the apical meristem of many seed plants, the pith-rib meristem produces the cells of the pith (Ch. 6).

Pitting The presence and nature of pits in sclerenchyma cells (Ch. 7).

    Alternate pitting The arrangement of circular bordered pits in diagonal rows on the sides of a tracheary element (Ch. 7).

    Crossfield pitting Pitting between an axial tracheary element and a ray cell.

Intervascular pitting The pitting that occurs between two adjacent tracheary elements.

Opposite pitting The arrangement of circular bordered pits in rows that are aligned transverse to the long axis of the cell (Ch. 7).

Placenta (pl.: placentae or placentas) The regions of a carpel where the ovules are attached (Ch. 19).

Plasma membrane The membrane that surrounds the entire protoplast; this term is most often used by zoologists and plant physiologists. The synonym *plasmalemma* is most often used by plant anatomists but appears to be becoming less common (Ch. 2).

Plasmalemma (pl.: plasmalemmae or plasmalemmas) See **plasma membrane**.

Plasmodesma (pl.: plasmodesmata) A small hole in a primary wall, through which the protoplasm of one cell is in contact with that of adjacent cells. Plasmodesmata may occur individually or in clusters called *primary pit fields* (Ch. 2).

Plasticity The property of being deformable and retaining the new size and shape once the deforming force is removed. Collenchyma cells are plastic (Ch. 4).

Plastid A class of organelles that are polymorphic and play numerous roles in cell metabolism and plant biology. Examples are proplastids, amyloplasts, chloroplasts, and chromoplasts (Ch. 2).

Plastoglobulus A small spherical osmiophilic body within a chloroplast. It usually consists of lipids and lipophilic compounds. The plural *plastoglobuli* is seen more often than *plastoglobulus* (Ch. 2).

Plate meristem Once the layers of a leaf lamina have been established by the marginal meristem, almost all of the cells divide anticlinally, thus causing the lamina to expand. Because the meristematic region is basically all of the plate-shaped young lamina, it is called a plate meristem (Ch. 12).

Plectostele A type of protostele in which the mass of xylem is not one single mass but rather a series of plates and small cylinders (Ch. 11).

Plerome One of the histogens envisioned by Hanstein as composing the apical meristem of seed plants. The plerome was considered to give rise to the vascular tissues of the axis. See **histogen** (Ch. 1).

Podarium (pl.: podaria) An extremely enlarged leaf base (Ch. 12).

Polar nucleus In an angiosperm megagametophyte, one of the two (or more) nuclei of the central cell; polar nuclei fuse with the second sperm nucleus and form the endosperm nucleus (Ch. 19).

Pollen chamber In conifers, the space between the apex of the nucellus and the base of the micropyle. The pollination droplet brings the pollen to the pollen chamber where it germinates (Ch. 19).

Pollen mother cell A synonym for **microsporocyte.**

Pollen sac A loculus of an anther, in which the sporogenous tissue is located (Ch. 19).

Pollination droplet A small drop of liquid that is secreted by a conifer ovule when it is fertile. After pollen lands on it, the droplet is retracted, carrying the pollen down to the ovule (Ch. 19).

Pollinium (pl.: pollinia) All the pollen grains of an anther sac of orchids stick together, forming a pollinium (Ch. 19).

Polyarch See **diarch.**

Polycyclic stele A type of stele that is present in some ferns, consisting of several concentric vascular cylinders (Ch. 11).

Polyphyletic This term signifies that a species, structure, or physiology has had multiple origins. If a genus is polyphyletic it is considered to be artificial, and studies must be undertaken to identify the natural genera involved and to determine how convergent evolution could produce several genera that are so similar that they would be mistaken for one. The same is true for structures, especially trichomes, secretory structures, and many organs or body forms. See **monophyletic** (Ch. 7).

Polyribosome Any piece of messenger RNA is long enough for several ribosomes to translate it simultaneously; this complex is a polyribosome, also called a *polysome* (Ch. 2).

Polysome See **polyribosome.**

Polysteles If more than one stele occurs within a root or shoot, they are called polysteles; this is extremely rare (Ch. 11).

Poricidal dehiscence A method of fruit opening in which small areas of the wall degenerate, resulting in holes large enough for the seed to pass through (Ch. 21).

Porogamy The process in which a pollen tube grows into the ovule through the micropyle (Ch. 19).

Prefrond A name given to the structures that were intermediate between branch systems and leaves and that occurred during the early stages in the evolution of megaphylls (Ch. 12).

Preprophase band A set of microtubules that forms in many cells during the early stages of cell division. The microtubules depolymerize before division is complete, but the cell plate fuses with the existing cell wall at the site where the preprophase band had been (Ch. 2).

Primary pit field A thin region in a primary wall, in which a high concentration of plasmodesmata occur (Ch. 2).

Primary sporogenous layer The layer of cells that are produced by division of the archesporial cell; these cells go on to become the spore mother cells (Ch. 19).

Primary thickening meristem A broad region in the subapical region of plants with extremely thickened stems. The primary thickening meristem encompasses a zone of mitotic activity that includes at least part of the pith or conjunctive tissue, the provascular tissue, and the cortex. It is present in cycads, many monocots, and some dicots with short wide stems (Ch. 11).

Primary wall See **cell wall.**

Primary wall layer The layer of subepidermal cells in a sporangium that is formed by division of the archesporial cells (Ch. 19).

Primexine One of the layers of the pollen wall, the primexine is a layer of cellulose that is deposited just interior to the original callose wall (Ch. 19).

Probaculum (pl.: probacula) Small rod-shaped structures in the pollen wall; once sporopollenin is deposited on them, they become *bacula* (Ch. 19).

Procambium (pl.: procambia) A portion of the subapical meristem that produces the cells that mature into the primary xylem and primary phloem. In woody plants, a portion of the procambium is converted into the fascicular cambium (Ch. 1).

Procumbent cell A ray cell that is longer radially than it is tall (Ch. 15).

Proembryo In gymnosperm reproduction, a complex of cells that are formed by the first several divisions of the zygote; part of the proembryo then forms the embryo proper. The term *proembryo* is also frequently used to refer to the embryo before there is any tissue differentiation. The term is occasionally also used for angiosperms (Ch. 20).

Prolepsis The outgrowth of an axillary bud after a period of quiescence. See **syllepsis** (Ch. 6).

Proliferative division A type of division in cambial initials in which the cell plate is oriented periclinally and longitudinally; this results in a new initial and a derivative cell. The alternative is multiplicative division (Ch. 14).

Promeristem A portion of a meristem that produces another portion of the same meristem. This term is most often applied to the central mother cells of shoot apical meristems because these cells flow into the peripheral zone or into the pith-rib meristem (Ch. 6).

Proplastid A type of plastid; this is the form that is found in meristematic cells. The proplastids undergo division to produce more proplastids while the cell also undergoes division (Ch. 2).

Prosenchyma A little-used term that designates the hardened parts of a plant (Ch. 5).

Protective layer A region in an abscission zone; the cells of the protective layer undergo division during abscission and form a sealing layer that is suberized (Ch. 12).

Protoderm Any set of cells that develops into an epidermis. *Protoderm* differs from *dermatogen* in that the former only describes an observable event, and carries no morphogenetic theory as *dermatogen* does (Ch. 6).

Protophloem The first primary phloem to differentiate at any particular level of a vascular bundle. It usually consists of very small sieve elements only, without companion cells or albuminous cells. The protophloem typically functions for only a short time before being replaced by metaphloem (Ch. 11).

Protoplasm All of the components of a cell; the wall has been considered historically not to be a part of the protoplasm. Ergastic substances are also excluded in the opinions of some (Ch. 2).

Protoplast The protoplasm of a single cell (Ch. 2).

Protostele A vascular cylinder in which the xylem is located at the center as a solid mass and there is no pith. See **siphonostele** (Ch. 11).

Protoxylem The first primary xylem to differentiate within a vascular bundle at any particular level of that bundle. The tracheary elements may be either tracheids or vessels, but they are always relatively small and have only extensible types of secondary wall (annular or helical). See **metaxylem** (Ch. 11).

Pseudocompound leaf A leaf that develops as a simple leaf but which, when fully expanded, is torn by the wind into a "compound" state; for example, a banana leaf (Ch. 12).

Pseudo-petiole A structure that resembles a petiole in form and function but has recently evolved from an ancestral leaf that was sessile (Ch. 12).

Pseudosyncarpous style A compound style in which each component style still has its own region of transmitting tissue; see **eusyncarpous style** (Ch. 19).

Pseudotransverse division A type of multiplicative division in fusiform cambial initials, in which the cell plate grows initially as a longitudinal radial plate but in which the ends then turn and run transversely to the lateral walls. This results in two fusiform initials that are shorter than the original and that are not aligned with each other laterally (Ch. 14).

Pulvinus (pl: pulvini) A swollen region of a petiole, either at its base or at its apex. The pulvinus usually contains motor cells, and as these absorb or lose water the leaf is raised or lowered (Ch. 12).

Pycnoxylic wood Wood that contains very little parenchyma, either axial or ray parenchyma. The alternative is manoxylic (Ch. 15).

Quiescent center A region of the root apical meristem of seed plants that is defined physiologically by a low mitotic index or even a complete lack of mitotic activity (Ch. 6).

Radicle The root portion of the embryo axis in a seed. Any lateral root primordia present on it are seminal roots; only the main axis itself is the radicle (Ch. 13).

Ranunculaceous stomatal complex See **anomocytic.**

Raphe In seeds that develop from anatropous or circinotropous ovules, the funiculus may fuse to the integument and form a ridge; the ridge is the raphe (Ch. 20).

Raphide A long, slender, needlelike crystal of calcium oxalate. Raphides occur in bundles within specialized raphide cells (Ch. 2).

Ray A set of parenchyma cells in the secondary body, which form radial masses (Ch. 15).
    Heterocellular ray A ray in gymnosperms that contains both parenchyma cells and ray tracheids.
    Heterogeneous ray A ray that contains both upright and procumbent cells.
    Homocellular ray A ray in gymnosperms that contains only parenchyma cells.
    Primary ray A ray that is a continuation of a medullary ray.
    Secondary ray A ray that is initiated by the conversion of fusiform initials to ray initials, or the conversion of procambium cells to ray initials.

Ray initial A short, rather cuboidal cell in the vascular cambium. The derivatives of ray initials develop into the cells of the rays of the secondary xylem and secondary phloem. See **fusiform initial** (Ch. 14).

Ray tracheid A ray cell that is procumbent, has lignified secondary walls, and has circular bordered pits connecting it to adjacent ray tracheids. Found only in gymnosperms (Ch. 15).

Reaction wood A specialized type of secondary xylem that is formed in response to compression (gymnosperms) or tension (dicots). See **compression wood** and **tension wood** (Ch. 15).

Recretion A term that was proposed to be used to describe the secretion of materials that had basically just passed through a plant without having been a fundamental part of its protoplasm, such as the inorganic salts of salt glands or the water of hydathodes. The term *secretion* is more often used instead. See also **excretion** and **secretion** (Ch. 9).

Refractive spherule A small round body that occurs in the sieve elements of many cryptogams (Ch. 8).

Resin duct A long canal containing the sticky resin or "pitch" of gymnosperms. The ducts are usually lined with epithelial cells and are most abundant in the wood and leaves of conifers (Ch. 9).

Rhizodermis See **epiblem**.

Rhytidome A rarely used term for the outer bark, used mostly in translations from German. See **bark** (Ch. 17).

Rib A row of cells that are aligned such that it is obvious that they have been derived from the same initial cell (Ch. 6).

Ribosome A particle consisting of ribosomal RNA and proteins; the ribosomes are responsible for translating mRNA into protein. They may occur individually or as polyribosomes. The ribosomes found in the hyaloplasm are 80s ribosomes; those in the plastids and mitochondria are 70s ribosomes (Ch. 2).

Ring porous wood Wood in which the vessels of the spring wood are wider and more numerous than those of the summer wood. The alternative is diffuse porous wood (Ch. 15).

RNA Ribonucleic acid; this is present in several forms: mRNA, or messenger RNA, which carries the genetic information from the nucleus to the cytoplasm; rRNA, or ribosomal RNA, which is a component of ribosomes; and tRNA, or transfer RNA, which carries amino acids to the ribosome to be incorporated into proteins, guided by the information in the mRNA (Ch. 2).

Root The nonshoot portion of a plant axis, usually specialized for absorption, anchorage, and nutrient storage and typically located underground or in a similar substrate (Ch. 13).
    Adventitious root Any root that is not a lateral root or a radicle; that is, it arises in tissues other than the embryo or the stele of another root.
    Contractile root A root that slowly contracts, pulling the shoot deeper into the soil.
    Fibrous root A root that is not enlarged or succulent, not a tap root.
    Lateral root any root that arises on another root.
    Nodal root A root that arises at the node of a stem, but used especially for the adventitious roots of young monocots that act as prop roots.
    Prop root The large adventitious roots produced from shoots that pass through air before striking the soil; they not only absorb nutrients but also brace the stem or branches.
    Proteoid root The specialized roots produced by species of the family Proteaceae; they are short and highly branched, being formed when a fibrous root grows into an area of rich soil.
    Root crown A short piece of shoot located on the top of a persistent root in a suffrutescent plant. Each year the shoot system dies back to the root crown, and the following year a new shoot system grows out of the axillary buds of the root crown.

Root hair initial An epidermal cell that will develop into a root hair. Also called a *trichoblast* or a *piliferous cell*.

Root pressure The hydrostatic pressure in the xylem, produced when the endodermis actively transports minerals into the stele, causing it to absorb water from the cortex.

Seminal root Any root or root primordium that is present in an ungerminated seed.

Spur root A special type of root found in many cacti and capable of growing rapidly whenever water becomes available. Spur roots have no root caps and form root hairs even on the apex when the spur root stops elongating.

Tap root If the radicle becomes prominently enlarged more than any of the laterals, it is called a tap root. Although a lateral root may become enlarged or succulent, only the radicle can be called a tap root.

Rosette A cluster of cellulose-synthesizing enzymes (cellulose synthases) that are located in the plasmalemma and that produce a cellulose microfibril to the outside of the protoplast (Ch. 2).

Rubiaceous stomatal complex See **paracytic**.

Ruminate endosperm A type of endosperm in which there is some residual nucellus and the endosperm and the nucellus interdigitate, or in which the seed coat and endosperm interdigitate (Ch. 20).

S₁ The outermost layer of a secondary wall, closest to the primary wall; the next layer is S_2, and the innermost is S_3 (Ch. 5).

S-type plastid A special type of plastid that occurs in sieve tube members; these plastids contain starch and protein. The anatomy of S-type plastids and P-type plastids is important in the taxonomy of the Caryophyllales (Ch. 8).

Salt gland An external secretory structure that secretes salts. Salt glands can be structurally simple or quite complex and are most often found on plants of saline habitats such as deserts or seashores (Ch. 9).

Sapwood The outermost portion of the wood of a tree, the wood that still contains living parenchyma and conducts water. See **heartwood** (Ch. 15).

Sarcotesta (pl.: sarcotestas) In the seeds of gymnosperms, the outermost, spongy layer of the seed coat (Ch. 20).

Schizogeny Tearing; the formation of an intercellular space or any other separation of tissues by the physical tearing of one cell away from another. Cellular breakdown is not involved. See also **schizolysigeny** and **lysigeny** (Ch. 9).

Schizolysigeny The formation of an intercellular space or any other separation of tissues by a process involving both the tearing of cells away from each other and also the actual breakdown of the cells (Ch. 9).

Sclereid One of the types of sclerenchyma cells, sclereids are not as long as fibers. They usually have very thick secondary walls, and their pits are always narrow and circular in cross section (Ch. 5).

Primary sclereid A sclereid in the secondary phloem that develops and becomes mature at the same time as the adjacent cells.

Secondary sclereid A sclereid in the secondary phloem that develops from a cell that has been a mature parenchyma cell for a period of time. The conversion from parenchyma to secondary sclereid occurs after the nearby sieve elements stop conducting.

Sclerenchyma A cell and tissue type in which the cells have both a primary cell wall and a secondary cell wall. The walls are almost always impregnated with lignin (Ch. 3, 5, 7).

Conducting sclerenchyma The conducting elements of the xylem: tracheids, fiber-tracheids, and vessel elements. Unlike the nonconducting sclerenchyma, these always have numerous prominent pits, or their secondary wall occurs as bands with large regions of uncovered primary wall that allow the movement of water (Ch. 5).

Nonconducting sclerenchyma This is any sclerenchyma other than the conducting elements of the xylem. The cells have secondary walls with pits that are too small and too few to permit rapid water conduction (Ch. 5).

Sclerophyllous An adjective that describes a leaf that is leathery and tough and often desiccation resistant; the term is also applied to plants that have leaves like this and to floras that have plants like this (Ch. 12).

Sclerotesta (pl.: sclerotestas) In the seeds of gymnosperms, the middle, sclerenchymatous layer of the seed coat (Ch. 20).

Scutellum (pl.: scutella) In monocots, the layer of

embryonic tissue that faces the endosperm; it secretes digestive enzymes, and it absorbs nutrients and passes them to the seedling. It is commonly interpreted as being a cotyledon (Ch. 20).

Secondary vein The vascular bundles of a leaf that branch off from a midrib. Branches from the secondary veins are tertiary veins (Ch. 12).

Secondary wall See **cell wall.**

Secretion The movement, either by diffusion or active transport, of material out of a plant or into a space where it can accumulate for storage (Ch. 9).

  Eccrine secretion The secretion of material by a process in which each molecule of the secretory product crosses the plasmalemma individually, not in groups.

  Endogenous secretion The secretion of material into an accumulation space that is located within the plant.

  Exogenous secretion The secretion of material to the exterior of the plant.

  Granulocrine secretion The secretion of material by a process in which it is first packed into vesicles of the ER or dictyosome, which then fuse with the plasmalemma.

  Holocrine secretion The secretion of material by a process in which the entire cell breaks down. This both liberates the secretory product and contributes cell debris to the secretion.

  Merocrine secretion Secretion by either an eccrine or a granulocrine process, as opposed to holocrine secretion.

Seed A structure that contains at least an embryo and usually a supply of stored nutrients; seeds are typically rather dry, dormant, and very resistant to environmental stress (Ch. 19).

Seed coat A synonym for **testa.**

Seed plant Any plant species that produces seeds; these are also known as spermatophytes or as phanerogams (Ch. 1).

Sepal A modified leaf that is part of most flowers, being the outermost sterile appendage; sepals usually provide protection to the flower as it develops (Ch. 12, 19).

Separation layer The region of cell weakness in which there is cell separation and rupture during the abscission of a leaf, flower, or other deciduous organ; also called an *abscission layer* (Ch. 12).

Sessile Refers to a leaf that does not have a petiole but that instead has the lamina attached directly to the stem; the term may also be applied to other structures that lack a stalk (Ch. 12).

Sexine Synonym for **ectexine.**

Sheath This term is applied in a general way to various layers of cells that surround other tissues. Vascular bundles typically have a sheath, and one separates the epithem of a hydathode from the other tissues of the leaf. The term is also applied to a layer of fungal hyphae that surround a root in an ectomycorrhizal association (Ch. 9, 15).

Shell zone A concave zone of cells that forms the base of a developing axillary meristem in many seed plants. It is detectable because the cells have thin walls parallel to the "shell" surface (Ch. 6).

Short cell The epidermis of grasses have long cells and short cells. The short cells occur in pairs of cork cells and silica cells (Ch. 10).

Sieve area Modified primary pit fields in the walls of sieve elements. The plasmodesmata have been modified into sieve pores. Sieve areas are thought to mediate the transfer of nutrients from one sieve element to the next (Ch. 8).

Sieve cell The conducting cells of the phloem of nonangiosperms. Sieve cells are characterized by not having sieve plates; all their sieve areas are relatively small and rather uniformly distributed on all walls (Ch. 8).

Sieve element The collective term used to designate either or both types of conducting cells in phloem. The two types of sieve elements are sieve cells and sieve tube members (Ch. 8).

Sieve plate The end wall of a sieve tube member, containing one or several sieve areas with sieve pores that are much larger than those on the side walls (Ch. 8).

  Compound sieve plate A sieve plate that contains more than one sieve area.

  Simple sieve plate A sieve plate that contains only a single sieve area.

Sieve pore The actual small hole in the primary wall of a sieve area, through which the cytoplasm of one sieve element is in contact with the cytoplasm of the next. Sieve pores are modified plasmodesmata (Ch. 8).

Sieve tube A long, multicellular structure composed of many sieve tube members interconnected by their sieve plates. Sieve tubes occur only in the phloem of angiosperms but are reported to be lacking in some of the most primitive ones (Ch. 8).

Silica body An amorphous aggregation of silicon dioxide within a cell. These are characteristic of certain monocot taxa, and their shapes can be important for taxonomy (Ch. 2).

Silica cell One of the two types of short cells in the epidermis of grasses, silica cells have deposits of silica in them. See also **cork cell** (Ch. 10).

Simultaneous division In reproduction, if cytokinesis and meiotic karyokinesis occur together, the division is said to be simultaneous; the alternative is successive division (Ch. 19).

Siphonostele A vascular cylinder in which there is a central pith with the xylem located around it. See **protostele** (Ch. 11).

Slime An older term, still occasionally used, for P-protein (Ch. 8).

Slime plug An older term, still occasionally used, for P-protein plug (Ch. 8).

Softwood A semitechnical term that indicates the wood of a conifer. It is based on the concept that conifer wood does not contain libriform fibers and is therefore softer than dicot wood. See **hardwood** (Ch. 15).

Solenostele A type of amphiphloic siphonostele in which the leaf gaps are widely separated (Ch. 11).

Spermatophyte A plant that reproduces with seeds; these are the phanerogams or seed plants. The alternative is reproduction without seeds, the cryptogams (Ch. 1).

Spongy mesophyll A portion of the leaf mesophyll characterized by the cells having little contact among themselves; the apoplast constitutes as much as one-half or more of the volume of the tissue (Ch. 12).

Spongy tissue In gymnosperm reproduction, the densely cytoplasmic cells at the apex of the nucellus; the spongy tissue is a nutritive tissue for the megagametophyte (Ch. 20).

Sporangium See **microsporangium** or **megasporangium.**

Spore A single cell that functions in reproduction by germinating and growing into a new plant, the gametophyte. In all the vascular plants, spores are produced by meiosis and are haploid (Ch. 19).

Sporophyte In the life cycle of a plant, the generation that produces the spores. The large plants that are familiar to us (ferns, conifers, flowering trees, shrubs, and herbs) are all sporophytes. The alternate generation, the gametophyte, produces gametes and is usually not noticed by us (Ch. 19).

Spring wood That portion of an annual ring that is formed at the beginning of a growing season. It is characterized by having wider tracheary elements than the summer wood formed later. Spring wood is also called *early wood* (Ch. 15).

Staminodium (pl.: staminodia) A sterile stamen that may remain rudimentary, may become glandular, or may become enlarged and petallike (Ch. 19).

Statolith Special starch grains located within the cells of most root caps and the nodes of some shoots. They are believed to be involved in the perception of gravity (Ch. 2).

Stegma (pl.: stegmata) A deposit of silica, similar to a silica body, but the term has been used historically for the silica deposits in families other than Poaceae and Cyperaceae (Ch. 2).

Stele The central vascular network of the shoot and roots. In the roots, the pericycle and all tissues interior to it are the stele. In shoots, the boundary is less precise, but basically everything from the outermost phloem elements inward would be part of the stele (Ch. 11).

Stereide An old term, rarely used, that indicates any individual sclerenchyma cell, regardless of the type (Ch. 5).

Stereology A method of quantifying the volumes and surface areas of the various components of a structure (Ch. 2, 6).

Sterome A term that refers to all the sclerenchyma of a plant collectively; rarely used (Ch. 5).

Stigma The uppermost portion of a carpel that receives pollen grains (Ch. 19).

Stipules Small projections of tissue at the base of a leaf. In the dicots, they may be large and vascularized and appear to be lobes of the leaf or even appear to be separate leaves. In the monocots, they are small projections from the sides of the junction between the leaf blade and the leaf sheath (Ch. 12).

Stoma (pl.: stomata) A term that is variously used. It sometimes designates the pore and guard cells of a stomatal complex, sometimes refers to just the pore, and other times indicates the entire complex (Ch. 10).

Stomatal cavity A depression in the surface of a leaf, containing one or several stomata (Ch. 10).

Stomatal complex The guard cells and subsidiary cells located around a stomatal pore (Ch. 10).

Stomatal pore The intercellular space between a pair of guard cells, the stomatal pore is the means for absorbing carbon dioxide and expelling oxygen. Frequently the stomatal pore is referred to as a *stoma*, a term that may also refer to the guard cells. *Stomatal pore* should be used for clarity (Ch. 10).

Stomium (pl.: stomia) In an anther, the part that opens and releases the pollen grains (Ch. 19).

Stone cell A small, isodiametric sclereid, also called a *brachysclereid* (Ch. 5).

Storied cork A type of periderm produced in some large perennial monocots; storied cork is formed by the limited cell division of cortical parenchyma cells, with little or no expansion of the daughter cells. The outline of the original cell is still visible (Ch. 17).

Strasburger cell An old synonym for **albuminous cell** (Ch. 8).

Strobilus (pl.: strobili) The technical term for a cone in the conifers and cycads and some other gymnosperms (Ch. 19).
  Megasporangiate strobilus The cone that produces the egg and later the seeds; much more frequently called the *seed cone* or *megastrobilus*. It consists of a short branch that bears numerous lateral branches; these lateral branches in turn bear the megasporophylls.
  Megastrobilus Synonym for **megasporangiate strobilus.**
  Microsporangiate strobilus The cone that produces the pollen. Much more often called the *pollen cone* or *microstrobilus*, it consists of a short branch that bears microsporophylls.
  Microstrobilus Synonym for **microsporangiate strobilus.**

Stroma The liquid component of plastids, especially chloroplasts (Ch. 2).

Style The portion of a carpel between the stigma and the ovary; the style elevates the stigma to a position that facilitates pollination (Ch. 19).

Styloid A large, elongate prismatic crystal of calcium oxalate. They somewhat resemble raphides but are much larger and occur singly, not in groups (Ch. 2).

Subapical meristem The region just proximal to the apical meristem in either a root or a shoot. The subapical meristems usually consist of protoderm, ground meristem, and procambium (Ch. 6).

Suberin layer A layer of the hydrophobic material suberin on the inner surface of a cell wall; this is found in cork cells (Ch. 17).

Submarginal initial In a marginal meristem of a leaf primordium, the submarginal initials are located interior to the marginal initials, and cell division in the submarginal initials establishes most of the layers of the mesophyll (Ch. 12).

Successive division In reproduction, if cytokinesis does not occur until after all stages of meiotic karyokinesis, it is said to be successive division (Ch. 19).

Summer wood That portion of an annual ring that is formed late in a growing season. It is characterized by having narrower tracheary elements than the spring wood. Summer wood is also called *late wood* (Ch. 15).

Suspensor A cell or set of cells that are derived from the zygote but that do not become part of the embryo. The suspensor cells push the embryo into the nutritive tissues (Ch. 20).

Syllepsis The outgrowth of an axillary bud immediately after formation, without a period of dormancy. See **prolepsis** (Ch. 6).

Sym- A prefix used to indicate that certain parts exist fused to other parts, not as individuals. A sympetalous corolla is one structure, consisting of several petals fused together. See **apo-** (Ch. 19).

Symplasm All the protoplasm of the plant. The important concept is that the protoplasm of individual cells is connected to that of adjacent cells by means of plasmodesmata; therefore, there is actually only one mass of protoplasm, regardless of the number of cells (Ch. 2).

Symplast Like *symplasm*, all the protoplasm of a plant. The term *symplast* is usually used with relation to the concept of apoplast (all the walls and intercellular spaces of a plant; Ch. 2).

Sympodial shoot A shoot that grows for a short period by one apical meristem. That apex is then replaced by one of the uppermost axillary buds. This may result in a coiled or zig-zag shoot, but often the shoot is so straight that only a careful developmental study shows that it is sympodial rather than monopodial (Ch. 11).

Sympodium (pl.: sympodia) In an open eustele, an axial bundle and its leaf and branch traces. See **open stele** (Ch. 11).

Synergid In an angiosperm megagametophyte, a cell located immediately adjacent to the egg and involved in the reception of the pollen tube. The sperm cells are discharged into a synergid (Ch. 19).

Tail The short, narrow section of a vessel element that extends beyond the perforation plate (Ch. 7).

Tapetum (pl.: tapeta) The innermost layer of sterile cells that surrounds the sporogenous tissue in a microsporophyll. It seems to be involved in nourishing the microspores and perhaps also has considerable effect on their morphogenesis (Ch. 19).
   Amoeboid tapetum Synonym for **invasive tapetum.**
   Glandular tapetum Synonym for **parietal tapetum.**
   Invasive tapetum A tapetum in which the walls break down and the tapetal protoplasts surround the microspore mother cells.
   Parietal tapetum A tapetum in which the cells remain on the periphery of the anther loculus.
   Secretory tapetum Synonym for **parietal tapetum.**

Tectum (pl.: tecta) The layer of a pollen wall formed by the expansion of the tops of the bacula; also called a roof layer (Ch. 19).

Teichode A channel in the outer wall of epidermal cells, resembling a plasmodesma. These are usually called *ectodesmata*, but the term *teichode* is used if it is believed that these are fundamentally different from plasmodesmata (Ch. 10).

Telome In the telome theory, the telome is the upper-most bifurcation of a dichotomously branched system. The lower dichotomies are mesomes (Ch. 12).

Tension wood A type of reaction wood that is formed in dicots, usually on the upper side of branches. It is characterized by the presence of gelatinous fibers.
   Compact tension wood Tension wood in which the gelatinous fibers form continuous regions.
   Diffuse tension wood Tension wood in which the gelatinous fibers are scattered among the normal fibers.

Tenuinucellate A type of ovule in which the megagametophyte is located immediately adjacent to the epidermis of the nucellus; the alternative is crassinucellate (Ch. 19).

Terminal cell In angiosperm reproduction, one of the first two cells formed by the transverse division of the zygote. The terminal cell produces all or most of the embryo (Ch. 20).

Tertiary vein A vascular bundle in a leaf that branches off from the secondary veins (Ch. 12).

Testa (pl.: testas or, rarely, testae) The seed coat, formed from the integument(s) of the ovule (Ch. 20).

Tetrarch See **diarch.**

Tetrasporic megagametophyte A megagametophyte that arises from a tetranucleate megaspore, produced when all four haploid nuclei survive after meiosis. See also **bisporic** and **monosporic** (Ch. 19).

Thallophyte These are the plants that do not contain any vascular tissue in their bodies; they include all the algae, mosses, and liverworts. The alternative is the group of vascular plants, also known as the *tracheophytes* (Ch. 1).

Thylakoids The membranes of the interior of a chloroplast. Thylakoids occur either stacked as grana or individually as frets (Ch. 2).

Tissue A set of cells that function together. In a simple tissue, all the cells are similar; in a complex tissue, there are various distinct types of cells (Ch. 2).

Tonoplast The membrane that surrounds a vacuole; the term *vacuolar membrane* is perhaps more common (Ch. 2).

Torpedo stage In embryo development, the stage at which the embryo is shaped like a cylinder (Ch. 20).

Torus (pl.: tori) The thickened central part of a pit membrane of the circular bordered pits in gymnosperms (Ch. 7).

Tracheary element The two types of conducting cells in xylem: tracheids and vessel members (Ch. 7).

Tracheid One of the types of conducting cells that occur in xylem. Tracheids are characterized by having a complete primary wall without any perforations (Ch. 7).

   Terminal tracheid A tracheid at the end of a minor vein in the leaf. The terminal tracheids are often large and irregular, and they may not actually be in contact with each other; it may be that they store water rather than conduct it (Ch. 12).

Tracheoid idioblast Water storage cells in the mesophyll of a leaf that have the appearance of being tracheids. However, they are usually not in direct contact with the tracheids of the vascular bundles (Ch. 12).

Transfer cell A cell that has labyrinthine ingrowths on one or several walls and that apparently functions in high-volume transfer of solutes over short distances. See **labyrinthine wall** (Ch. 3).

Transfer wall See **labyrinthine wall** and **transfer cell.**

Transfusion tissue A special type of vascular tissue that occurs in the leaves of gymnosperms; it consists of transfusion tracheids and transfusion parenchyma. The transfusion tracheids are short with blunt ends, but they have circular bordered pits (Ch. 12).

   Accessory transfusion tissue A set of horizontally elongate transfusion tracheids that run laterally away from the vascular bundles to the mesophyll in leaves of cycads.

Transmitting tissue The tissue in the style through which pollen tubes grow (Ch. 19).

Triarch See **diarch.**

Trichoblast A young epidermal cell that will develop into a trichome. Occasionally referred to as a *piliferous cell* (Ch. 10, 13).

Trichome A plant hair; basically any outgrowth from the epidermis. Trichomes can be glandular or nonglandular, uniseriate, multiseriate, branched, or unbranched (Ch. 10).

Trichome complement A term that refers to all the types of trichomes on a plant. This differs from *indumentum*, which refers to all the trichomes themselves (Ch. 10).

Trichosclereid A long, almost hairlike sclereid, somewhat resembling a fiber (Ch. 5).

Trophophyll A rarely used synonym for **foliage leaf.**

True fruit A synonym for *pericarp;* see **fruit.**

Tube nucleus In a pollen tube, the nucleus that is not the sperm nuclei; it is usually located near the tip of the pollen tube and may be involved in controlling its metabolism (Ch. 19).

Tunica (pl.: tunicas) The portion of the protoderm that lies within the shoot apical meristem. A tunica is distinguished from a mantle by not having periclinal divisions (Ch. 6).

Ultrastructure This term was developed to describe the anatomy that can be studied only by electron microscopy. Currently it is used to indicate any structure that is studied by electron microscopy even if the structure can also be studied by light microscopy (Ch. 2).

Unitegmic An adjective that describes ovules that have one integument (Ch. 19).

Upright cell A ray cell that is taller than it is long radially. Upright cells, when they are present, appear to be especially important in the transfer of material between the rays and the axial xylary elements (Ch. 15).

V/A mycorrhiza (pl.: V/A mycorrhizae) The most common type of endomycorrhiza, in which the fungal hyphae are able to penetrate the host cells and form either vesicles (swellings) or arbuscles (highly branched hyphae; Ch. 13).

Vacuole An organelle that is bounded by a single membrane (the tonoplast) and whose contents are aqueous. Vacuoles are similar to vesicles, but are larger (Ch. 2).

Vacuome A collective term that refers to all of the vacuoles of a cell (Ch. 2).

Valvate dehiscence A method of fruit opening in which flaps (valves) of fruit wall break open, then bend outward, creating large openings for seed release (Ch. 21).

Vascular bundle A set of primary xylem and primary phloem running side by side in the plant (Ch. 11).

    Amphicribral bundle A vascular bundle in which the phloem surrounds the xylem.

    Amphivasal bundle A vascular bundle in which the xylem completely surrounds the phloem.

    Bicollateral vascular bundle A vascular bundle that contains one mass of xylem and two of phloem, one located exterior to the xylem and the other (the internal phloem) located interior to the xylem (Ch. 11).

    Collateral vascular bundle A vascular bundle that contains one mass of xylem and one of phloem.

Vascular cryptogam Any species of plant that has vascular tissue but that does not produce seeds (Ch. 1).

Vascular plant Any species of plant that contains vascular tissue (xylem and phloem; Ch. 1).

Vascular tracheid A small, short, wide tracheid that perhaps has evolved from vessel elements that failed to form perforations (Ch. 7).

Vegetative cell In angiosperm reproduction, the larger of the two cells in a young pollen grain, the other cell being a generative cell (Ch. 19).

Velamen The multiseriate epidermis on the roots of orchids and aroids (Ch. 10).

Ventral meristem A synonym for **adaxial meristem.**

Ventral suture In fruits, the seam formed by the fusion of the two margins of one carpel. Compare with **lateral suture** (Ch. 21).

Ventral wall On a guard cell, the wall closest to the stomatal pore. See also **ledge** and **dorsal wall** (Ch. 10).

Vesicle A small portion of liquid material surrounded by a single membrane (Ch. 2). In endomycorrhizae, a swollen portion of the hyphae (Ch. 13).

Vesicular-arbuscular mycorrhiza See **V/A mycorrhiza.**

Vessel A long, multicellular structure in the xylem of most angiosperms and a few nonangiosperms. Vessels are essentially long tubes composed of vessel elements that are interconnected by perforations (holes) in their common walls (Ch. 7).

Vessel element An individual cell of a vessel. The necessary aspect of a vessel element is the presence of one or two relatively large holes (perforations) in its wall. See **tracheid** (Ch. 7).

Vesselless A plant that does not have vessels in its xylem (Ch. 7).

    Primitively vesselless Any plant that does not have vessels and that evolved from ancestors that also lacked vessels. Basically, most nonangiosperms and the most primitive angiosperms.

    Secondarily vesselless Any plant that lacks vessels but evolved from ancestors that did contain them.

Vestured pit A pit in which there are short outgrowths from the inner surface of the border (Ch. 7).

Viviparous seed A seed that germinates while it is still attached to the maternal plant (Ch. 20).

Water pore A modified stoma located over a hydathode. Water pores are often larger than stomata and usually do not close (Ch. 9).

Wax A substance of variable composition, being a polymer of very long-chain fatty acids (Ch. 10).

    Epicuticular wax The wax that is located as a layer exterior to the cuticle proper.

    Intracuticular wax The wax that occurs as plates and patches inside of the cuticle.

Wound periderm The layers of cork cells that are induced to form when most plant tissues are cut or otherwise damaged (Ch. 17).

Xerochastic opening A method of fruit opening in which the fruit wall breaks as a result of the stresses caused by the differential shrinkage experienced by various layers as they dry. Compare with **hydrochastic opening** (Ch. 21).

Xerophyte A plant of arid regions having xeromorphic characters such as hard leaves, spines, and succulence (Ch. 12).

Zygomorphic flower A flower that is bilaterally symmetrical (Ch. 19).

Zygote The cell that is produced after an egg is fertilized. In oogamous species, because the egg is so much larger than the sperm, the zygote is basically still the egg, with the important difference being that it now contains a paternal genome (Ch. 19).

Appendix B: References

Abbe, L. B., and A. S. Crafts. 1939. Phloem of white pine and other coniferous species. Bot. Gaz. 100:695–722.

Addicott, F. T. 1983. Abscisic acid in abscission. In *Abscisic Acid,* ed. F. T. Addicott. Praeger Press. New York.

Agashe, S. N. 1968. Phloem studies in the pteridophytes. Part I, *Equisetum.* Amer. Fern J. 58:74–77.

Aist, J. R. 1976. Papillae and related wound plugs of plant cells. Ann. Rev. Phytopathol. 14:145–163.

Alberts, B., D. Bray, J. Lewis, M. Raff, K. Roberts, and J. D. Watson. 1983. *Molecular Biology of the Cell.* Garland Publishing: New York.

Aldaba, V. C. 1927. The structure and development of the cell wall in plants: I. Bast fibers of *Boehmeria* and *Linum.* Amer. J. Bot. 14:16–24.

Alexandrov, W. G., and I. I. Djaparidze. 1927. Über das Entholzen und Verholzen der Zellhaut. Planta 4.467–475.

Alfieri, F. J., and R. F. Evert. 1973. Structure and seasonal development of the secondary phloem in the Pinaceae. Bot. Gaz. 134:17–25.

Aljaro, M. E., G. Avila, A. Hoffmann, and J. Kummerow. 1972. The annual rhythm of cambial activity in two woody species of the Chilean "Matorral." Amer. J. Bot. 59:879–885.

Allen, G. S., and J. N. Owens. 1972. *The Life History of Douglas-Fir.* Information Canada: Ottawa.

Allen, R. D., and C. L. Nessler. 1984. Cytochemical localization of pectinase activity in laticifers of *Nerium oleander* L. Protoplasma 119:74–78.

Allsopp, A. 1965. Heteroblastic development in cormophytes. Handb. Pflanzenphysiol. 15:1172–1221.

Aloni, R., and T. Sachs. 1973. The three-dimensional structure of primary phloem systems. Planta 113:345–353.

Alosi, M. C., and F. J. Alfieri. 1972. Ontogeny and structure of the secondary phloem in *Ephedra.* Amer. J. Bot. 59:818–827.

Alvim, P. de T. 1964. Tree growth and periodicity in tropical climates. In *The Formation of Wood in Forest Trees,* ed. M. H. Zimmermann. Academic Press: New York.

Ambronn, H. 1881. Über die Entwicklungsgeschichte und die mechanischen Eigenschaften des Collenchyms. Ein Beitrag zur Kenntnis des mechanischen Gewebesystems. Jb. wiss. Bot. 12:473–541.

Antonova, T. S., and A. V. Golovin. 1983. Pathological changes in cell walls of sunflower torus infected with gray mold (*Botrytis cinerea*). S-KH. Biol. 0:88–92.

Arber, A. 1918. The phyllode theory of the monocotyledonous leaf, with special reference to anatomical evidence. Ann. Bot. 32:465–501.

Arber, A. 1925. *Monocotyledons. A Morphological Study.* Cambridge University Press: Cambridge.

Arber, A. 1937. The interpretation of the flower: a study of some aspects of morphological thought. Biol. Rev. 12:157–184.

Arber, A. 1950. *The Natural Philosophy of Plant Form.* Cambridge University Press: Cambridge.

Arekal, G. D. 1963. Embryological studies in Canadian representatives of the tribe Rhinantheae, Scrophulariaceae. Can. J. Bot. 41:267–302.

Arikado, H. 1955. The ventilating pressure of rice plants growing under paddy field conditions. Bull. Fac. Agr. Mie. Univ., no. 11, Japan.

Armstrong, J. E., and C. Heimsch. 1976. Ontogenetic reorganization of the root meristem in the Compositae. Amer. J. Bot. 63:212–219.

Arnholt, P. J. 1969. A comparative study of the ontogeny of macroscereids and osteoscereids in the integument of *Cassia fasiculata* and *Desmodium canadense.* Proc. Iowa Acad. Sci. 76:21–31.

Arnold, C. A. 1960. A lepidodendrid stem from Kansas and its bearing on the problem of cambium and phloem in Paleozoic lycopods. Contrib. Mus. Paleontol. Univ. Mich. 10:249–267.

Arnott, H. J. 1962. *The seed, germination, and seedling of Yucca.* University of California Press: Berkeley, California.

Arnott, H. J. 1976. Calcification in higher plants. In *The Mechanisms of Mineralization in the Invertebrates and Plants,* ed. N. Watanabe and K. M. Wilbur. University of South Carolina Press: Columbia, South Carolina.

Arnott, H. J., and F. G. E. Pautard. 1965. Development of raphide idioblasts in *Lemna*. Amer. J. Bot. 52:618.

Artschwager, E. 1924. Studies on the potato tuber. J. Agric. Res. 27:809–835.

Artschwager, E. 1925. Anatomy of vegetative organs of sugar cane. J. Agric. Res. 30:197–221.

Artschwager, E. 1926. Anatomy of the vegetative organs of the sugar beet. J. Agric. Res. 33:143–176.

Artschwager, E. 1943. Contribution to the morphology and anatomy of Guayule (*Parthenium argentatum*). Tech. Bull. 842. U.S. Department of Agriculture: Washington, D.C.

Artschwager, E. 1950. The time factor in the differentiation of the secondary xylem and phloem in pecan. Amer. J. Bot. 37:15–24.

Arzee, T. 1953. Morphology and ontogeny of foliar sclereids in *Olea europaea*: II: Ontogeny. Amer. J. Bot. 40:745–752.

Arzee, T., E. Arbel, and L. Cohen. 1977. Ontogeny of periderm and phellogen activity in *Ceratonia siliqua* L. Bot. Gaz. 138:329–333.

Arzee, T., D. Kamir, and L. Cohen. 1978. On the relationship of hairs to periderm development in *Quercus ithaburensis* and *Q. infectoria*. Bot. Gaz. 139:95–101.

Arzee, T., Y. Waisel, and N. Liphschitz. 1970. Periderm development and phellogen activity in shoots of *Acacia raddiana* Savi. New Phytol. 69:395–398.

Atkinson, A. W., Jr., P. C. L. John, and B. E. S. Gunning. 1974. The growth and division of the single mitochondrion and other organelles during the cell cycle of *Chlorella*, studied by quantitative stereology and three dimensional reconstruction. Protoplasma 81:77–109.

Augspurger, C. K. 1986. Morphology and dispersal potential of wind-dispersed diaspores of neotropical trees. Amer. J. Bot. 73:353–363.

Avery, G. S., Jr. 1930. Comparative anatomy and morphology of embryos and seedlings of maize, oats, and wheat. Bot. Gaz. 89:1–39.

Avila, G., M. E. Aljaro, S. Araya, G. Montenegro, and J. Kummerow. 1975. The seasonal cambium activity of Chilean and Californian shrubs. Amer. J. Bot. 62:473–478.

Ayensu, E. S. 1972. *Anatomy of the Monocotyledons. VI. Dioscoreales*. Clarendon Press: Oxford.

Ayensu, E. S., and W. L. Stern. 1964. Systematic anatomy and ontogeny of the stem in Passifloraceae. Contrib. U.S. Natl. Herb. 34:45–73

Aylor, D. E., J.-Y. Parlange, and A. D. Krikorian. 1973. Stomatal mechanics. Amer. J. Bot. 60:163–171.

Baas, P. 1970. Anatomical contributions to plant taxonomy. I. Floral and vegetative anatomy of *Eliaea* from Madagascar and *Cratoxylum* from Indo-Malesia (Guttiferae). Blumea 18:369–391.

Baas, P. 1973. The wood anatomical range in *Ilex* and its ecological and phylogenetic significance. Blumea 21:193–258.

Baas, P. 1982. Systematic, phylogenetic, and ecological wood anatomy—History and perspectives. In *New Perspectives in Wood Anatomy*, ed. P. Baas. Martinus Nijhoff/Dr W. Junk Publishers: The Hague.

Baas, P., A. J. Bolton, and D. M. Catling. 1976. *Wood Structure in Biological and Technological Research*. Leiden University Press: Leiden.

Baas, P., and R. C. V. J. Zweypfenning. 1979. Wood anatomy of the Lythraceae. Acta Bot. Neerl. 28:117–155.

Bailey, D. C. 1980. Anomalous growth and vegetative anatomy of *Simmondsia chinensis*. Amer. J. Bot. 67:147–161.

Bailey, I. W. 1920a. The cambium and its derivative tissues. II. Size variations of cambial initials in gymnosperms and angiosperms. Amer. J. Bot. 7:355–367.

Bailey, I. W. 1920b. The cambium and its derivative tissues. III. A reconnaissance of cytological phenomena in the cambium. Amer. J. Bot. 7:417–434.

Bailey, I. W. 1923. The cambium and its derivative tissues: IV. The increase in girth of the cambium. Amer. J. Bot. 10:499–509.

Bailey, I. W. 1944. The development of vessels in angiosperms and its significance in morphological research. Amer. J. Bot. 31:421–428.

Bailey, I. W. 1953. Evolution of the tracheary tissue of land plants. Amer. J. Bot. 40:4–8.

Bailey, I. W. 1956. Nodal anatomy in retrospect. J. Arnold Arbor. 37:269–287.

Bailey, I. W. 1957a. The potentialities and limitations of wood anatomy in the study of the phylogeny and classification of angiosperms. J. Arnold Arbor. 38:243–254.

Bailey, I. W. 1957b. Die Struktur der Tüpfelmembranen bei den Tracheiden der Koniferen. Holz als Roh-und Werkstoff 15:210–213.

Bailey, I. W. 1961. Comparative anatomy of the leaf-bearing Cactaceae, II. Structure and distribution of sclerenchyma in the phloem of *Pereskia, Pereskiopsis* and *Quiabentia*. J. Arnold Arbor. 42:144–150.

Bailey, I. W. 1963. Comparative anatomy of the leaf-bearing Cactaceae, VIII. The xylem of pereskias from southern Mexico and Central America. J. Arnold Arbor. 44:211–216.

Bailey, I. W. 1966. The significance of the reduction of vessels in the Cactaceae. J. Arnold Arbor. 47:288–292.

Bailey, I. W., and C. G. Nast. 1943. The comparative morphology of the Winteraceae. I. Pollen and stamens. J. Arnold Arbor. 24:340–346.

Bailey, I. W., C. G. Nast, and A. C. Smith. 1943. The family Himantandraceae. J. Arnold Arbor. 24:190–206.

Bailey, I. W., and L. M. Srivastava. 1962. Comparative anatomy of the leaf-bearing Cactaceae, IV. The fusiform initials of the cambium and the form and structure of their derivatives. J. Arnold Arbor. 43:187–197.

Bailey, I. W., and B. G. L. Swamy. 1949. The morphology and relationships of *Autrobaileya*. J. Arnold Arbor. 30: 211–226.

Bailey, I. W., and B. G. L. Swamy. 1951. The conduplicate carpel of dicotyledons and its initial trends of specialization. Amer. J. Bot. 38:373–379.

Baird, L. M., and B. D. Webster. 1982. Morphogenesis of effective and ineffective root nodules in *Phaseolus vulgaris* L. Bot. Gaz. 143:41–51.

Baird, W. V., and W. H. Blackwell. 1980. Secondary growth in the axis of *Halogeton glomeratus* (Bieb.) Meyer (Chenopodiaceae). Bot. Gaz. 141:269–276.

Baker, E. A. 1982. Chemistry and morphology of plant epicuticular waxes. In *The Plant Cuticle*, ed. D. F. Cutler, K. L. Alvin, and C. E. Price. Academic Press: London.

Baker, E. A., and P. J. Holloway. 1970. The constituent acids of angiosperm cutins. Phytochem. 9:1557–1562.

Baker, E. A., and J. Procopiou. 1975. The cuticles of *Citrus* species. Composition of the intracellular lipids of leaves and fruits. J. Sci. Food and Agri. 26:1347–1352.

Baker, H. G., and I. Baker. 1973a. Amino-acids in nectar and their evolutionary significance. Nature, London 241: 543–545.

Baker, H. G., and I. Baker. 1973b. Some authecological aspects of the evolution of nectar-producing flowers, particularly amino acid production in nectar. In *Taxonomy and Ecology*, ed. V. H. Heywood. Academic Press: London.

Baker, H. G., and I. Baker. 1975. Studies of nectar-constitution and pollinator-plant coevolution. In *Coevolution of Animals and Plants*, ed. L. E. Gilbert and P. H. Raven. University of Texas Press: Austin, Texas.

Baker, H. G., and I. Baker. 1979. Starch in angiosperm pollen grains and its evolutionary significance. Amer. J. Bot. 66:591–600.

Balfour, E. 1965. Anomalous secondary thickening in Chenopodiaceae, Nyctaginaceae and Amaranthaceae. Phytomorph. 15:111–122.

Balfour, E. E., and W. R. Philipson. 1962. The development of the primary vascular system of certain dicotyledons. Phytomorph. 12:110–143.

Ball, E. 1941. The development of the shoot apex and the primary thickening meristem in *Phoenix canariensis* Chaub., with comparisons to *Washingtonia filifera* Wats. and *Trachycarpus excelsa* Wendl. Amer. J. Bot. 28:820–832.

Ball, E. 1956. Growth of the embryo of *Ginkgo biloba* under experimental conditions. II. Effects of a longitudinal split in the tip of the hypocotyl. Amer. J. Bot. 43:802–810.

Banks, H. P. 1970. Major evolutionary events and the geological record of plants. Biol. Rev. 47:451–454.

Banks, H. P. 1981. Time of appearance of some plant biocharacters during Siluro-Devonian time. Can. J. Bot. 59: 1292–1296.

Bannan, M. W. 1950. The frequency of anticlinal divisions in fusiform cambial cells of *Chamaecyparis*. Amer. J. Bot. 37:511–519.

Bannan, M. W. 1951a. The reduction of fusiform cambial cells in *Chamaecyparis* and *Thuja*. Can. J. Bot. 29:57–67.

Bannan, M. W. 1951b. The annual cycle of size changes in the fusiform cambial cells of *Chamaecyparis* and *Thuja*. Can. J. Bot. 29:421–437.

Bannan, M. W. 1955. The vascular cambium and radial growth in *Thuja occidentalis* L. Can. J. Bot. 33:113–138.

Bannan, M. W. 1960. Ontogenetic trends in conifer cambium with respect to frequency of anticlinal division and cell length. Can. J. Bot. 38:795–802.

Bannan, M. W. 1962. The vascular cambium and tree-ring development. In *Tree Growth*, ed. T. T. Kozlowski. Ronald Press: New York.

Bannan, M. W. 1966. Spiral grain and anticlinal divisions in the cambium of conifers. Can. J. Bot. 44:1515–1538.

Bannan, M. W. 1968. Polarity in the survival and elongation of fusiform initials in conifer cambium. Can. J. Bot. 46:1005–1008.

Bannan, M. W., and I. L. Bayly. 1956. Cell size and survival in conifer cambium. Can. J. Bot. 34:769–776.

Barber, D. A., M. Ebert, and N. T. S. Evans. 1962. The movement of $^{15}O_2$ through barley and rice plants. J. Exp. Bot. 13:397–403.

Barckhausen, R. 1978. Ultrastructural changes in wounded plant storage tissue cells. In *Biochemistry of Wounded Plant Tissues*, ed. G. Kahl. Walter de Gruyter: Berlin.

Baretta-Kuipers, T. 1976. Comparative wood anatomy of Bonnetiaceae, Theaceae and Guttiferae. In *Wood Structure in Biological and Technological Research*, ed. P. Baas, A. J. Bolton, and D. M. Catling. Leiden University Press: Leiden.

Barghoorn, E. S., Jr. 1940a. Origin and development of the uniseriate ray in the Coniferae. Bull. Torrey Bot. Club 67:303–328.

Barghoorn, E. S., Jr. 1940b. The ontogenetic development and phylogenetic specialization of rays in the xylem of dicotyledons. I. The primitive ray structure. Amer. J. Bot. 27:918–928.

Barghoorn, E. S., Jr. 1941. The ontogenetic development and phylogenetic specialization of rays in the xylem of dicotyledons. II. Modification of the multiseriate and uniseriate rays. Amer. J. Bot. 28:273–282.

Barghoorn, E. S., Jr. 1963. The cambium and seasonal development of the phloem of *Pyrus malus*. Amer. J. Bot. 50:149–159.

Barghoorn, E. S., Jr. 1964. Evolution of cambium in geologic time. In *The Formation of Wood in Forest Trees*, ed. M. H. Zimmermann. Academic Press: New York.

Barlow, P. 1970. Vacuoles in the nucleoli of *Zea mays* root apices and their possible significance in nucleolar physiology. Caryologia 23:61–70.

Barlow, P. W. 1974. Regeneration of the cap of primary roots of *Zea mays*. New Phytol. 73:937–954.

Barlow, P. W. 1985. The nuclear endoreduplication cycle in metaxylem cells of primary roots of *Zea mays* L. Ann. Bot. 55:445–457.

Barnell, E. 1939. Studies in tropical fruits. V. Some anatomical aspects of fruit-fall in two tropical arboreal plants. Ann. Bot. N.S. 3:77–89.

Barnett, J. R. 1981. Secondary xylem cell development. In *Xylem Cell Development,* ed. J. R. Barnett. Castle House Publications, Ltd.: Kent, England.

Barnett, J. R., and J. M. Harris. 1975. Early stages of bordered pit formation in radiata pine. Wood Sci. Technol. 9:233–241.

Barth, F. G. 1985. *Insects and Flowers; the Biology of a Partnership*. Princeton University Press: Princeton, New Jersey.

Bauch, J., W. Schweers, and H. Berndt. 1974. Lignification during heartwood formation: comparative study of rays and bordered pit membranes in coniferous woods. Holzforsch. 28:86–91.

Bawa, K. S., and J. H. Beach. 1983. Self-incompatibility systems in the Rubiaceae of a tropical lowland wet forest. Amer. J. Bot. 70:1281–1288.

Beck, C. B. 1976. Current status of the Progymnospermopsida. Rev. of Paleobot. and Paly. 21:5–23.

Beck, C. B. 1981. *Archaeopteris* and its role in vascular plant evolution. In *Paleobotany, Paleoecology, and Evolution,* ed. K. J. Niklas, vol. 1. Praeger Special Studies: New York.

Becker, H. 1913. Über die Keimung verschiedartiger Früchte und Samen bei derselben Species. Beih. Bot. Zentralbl. 21–143.

Becking, J. H. 1975. Root nodules in non-legumes. In *The Development and Function of Roots,* ed. J. G. Torrey and D. T. Clarkson. Academic Press: London.

Beer, M., and G. Setterfield. 1958. Fine structure in thickened primary walls of collenchyma cells of celery petioles. Amer. J. Bot. 45:571–580.

Beevers, H. 1979. Microbodies in higher plants. Ann. Rev. Plant Physiol. 30:159–193.

Behnke, H.-D. 1965. Über das Phloem der Dioscoreaceen unter besonderer Berücksichtigung ihrer Phloembecken. Z. Pflanzenphysiol. 53:214–244.

Behnke, H.-D. 1971. Über den Feinbau verdickter (nacré) Wänden und der Plastiden in den Siebröhren von *Annona* und *Myristica*. Protoplasma 72:69–78.

Behnke, H.-D. 1975. P-type sieve-element plastids: A correlative ultrastructural and ultrahistochemical study on the diversity and uniformity of a new reliable character in seed plant systematics. Protoplasma 83:91–101.

Behnke, H.-D. 1976. Ultrastructure of sieve-element plastids in Caryophyllales (Centrospermae), evidence for the delimitation and classification of the order. Plant Syst. Evol. 126:31–54.

Beijer, J. J. 1927. Die Vermehrung der radialen Reihen im Cambium. Rec. trav. Bot. Neerl. 24:631–786.

Bell, A. 1980. The vascular pattern of a rhizomatous ginger (*Alpinia speciosa* L., Zingiberaceae). Ann. Bot. 46:203–220.

Bell, J. K., and M. E. McCully. 1970. A histological study of lateral root initiation and development in *Zea mays*. Protoplasma 70:179–205.

Bell, P. R. 1978. A microtubule–nuclear envelope complex in the spermatozoid of *Pteridium*. J. Cell Sci. 29:189–195.

Ben-Arie, R., N. Kislev, and C. Frenkel. 1979. Ultrastructural changes in the cell walls of ripening apple and pear fruit. Plant Physiol. 64:197–202.

Ben-Shaul, Y., and Y. Naftali. 1969. The development and ultrastructure of lycopene bodies in chromoplasts of *Lycopersicum esculentum*. Protoplasma 67:333–344.

Ben-Shaul, Y., T. Treffry, and S. Klein. 1968. Fine structure studies of carotene body development. J. Microscopie 7:265–274.

Benayoun, J. 1977. *The ultrastructure of resin duct cells and resin secretion in* Pinus halepensis. (In Hebrew.) Ph.D. Thesis of the Hebrew University of Jerusalem.

Benayoun, J., A. M. Catesson, and Y. Czaninski. 1981. A cytochemical study of differentiation and breakdown of vessel end walls. Ann. Bot. 47:687–698.

Benson, L. 1982. *The Cacti of the United States and Canada*. Stanford University Press: Stanford, California.

Bentley, B., and T. Elias. 1983. *The Biology of Nectaries*. Columbia University Press: New York.

Benzing, D. H. 1980. *The Biology of the Bromeliads*. Mad River Press: Eureka, California.

Benzing, D. H., W. E. Friedman, G. Peterson, and A. Renfrow. 1983. Shootlessness, velamentous roots, and the preeminence of Orchidaceae in the epiphytic biotope. Amer. J. Bot. 70:121–133.

Benzing, D. H., D. W. Ott, and W. E. Friedman. 1982. Roots of *Sobralia macrantha* (Orchidaceae): structure and function of the velamen-exodermis complex. Amer. J. Bot. 69:608–614.

Benzing, D. H., and A. M. Pridgeon. 1983. Foliar trichomes of Pleurothallidinae (Orchidaceae): functional significance. Amer. J. Bot. 70:173–180.

Bergersen, F. J. 1982. *Root Nodules of Legumes: Structure and Functions*. Research Studies Press: New York.

Bergfeld, R., Y. Hong, T. Kühnl, and P. Schopfer. 1978. Formation of oleosomes (storage lipid bodies) during embryogenesis and their breakdown during seedling development in cotyledons of *Sinapis alba* L. Planta 143:297–307.

Berlin, J., J. E. Quisenberry, F. Bailey, M. Woodworth, and B. L. McMichael. 1982. Effect of water stress on cotton leaves. I. An electron microscopic stereological study of the palisade cells. Plant Physiol. 70:238–243.

Berlyn, G. P. 1972. Seed germination and morphogenesis. In *Seed Biology*, ed. T. T. Kozlowski, vol. 1. Academic Press: New York.

Berlyn, G. P. 1982. Morphogenetic factors in wood formation and differentiation. In *New Perspectives in Wood Anatomy*, ed. P. Baas. Martinus Nijhoff/Dr W. Junk Publishers: The Hague.

Bernstein, Z., and A. Fahn. 1960. The effect of annual and bi-annual pruning on the seasonal changes in xylem formation in the grapevine. Ann. Bot. 24:159–171.

Berta, G., and P. Bonfante-Fasolo. 1983. Apical meristems in mycorrhizal and uninfected roots of *Calluna vulgaris*. Plant Soil 71:285–291.

Bewley, J. D., and M. Black. 1978. *Physiology and Biochemistry of Seeds in Relation to Germination*. Springer-Verlag: Berlin.

Bhaduri, P. N. 1933. A note on the "new type of fertilization" in plants. Current Sci. (India) 2:95.

Bhat, K. V., and J. D. Patel. 1980. Histology and histochemistry of sapwood and heartwood in *Garuga pinnata* Roxb. (Burseraceae) and *Ougeinia oojeinensis* (Roxb.) Hochr. (Leguminosae). Flora 170:144–157.

Bhatnagar, S. P., and B. M. Johri. 1972. Development of angiosperm seeds. In *Seed Biology*, ed. T. T. Kozlowski. Academic Press: New York.

Bhatnagar, S. P., and V. Sawhney. 1981. Endosperm—Its morphology, ultrastructure, and histochemistry. Int. Rev. Cytol. 73:55–102.

Bhojwani, S. S., and S. P. Bhatnagar. 1974. *The Embryology of Angiosperms*. Vikas Publishing House: Delhi.

Biale, J. B., and R. E. Young. 1971. The avocado pear. In *The Biochemistry of Fruits and their Products*, ed. A. C. Hulme. Academic Press: London.

Bieniek, M. E., and W. F. Millington. 1967. Differentiation of lateral shoots as thorns in *Ulex europeus*. Amer. J. Bot. 54:61–70.

Bierhorst, D. W. 1958. Vessels in *Equisetum*. Amer. J. Bot. 45:534–537.

Bierhorst, D. W. 1960. Observations on tracheary elements. Phytomorph. 10:249–305.

Bierhorst, D. W. 1971. *Morphology of Vascular Plants*. Macmillan, Inc.: New York.

Bierhorst, D. W., and P. M. Zamora. 1965. Primary xylem elements and element associations of angiosperms. Amer. J. Bot. 52:657–710.

Biggs, A. R., and J. Northover. 1985. Formation of the primary protective layer and phellogen after leaf abscission in peach. Can. J. Bot. 63:1547–1550.

Björkman, R. 1976. Properties and function of plant myrosinases. In *The Biology and Chemistry of the Cruciferae*, eds. J. G. Vaughan, A. J. MacLeod, and B. M. G. Jones. Academic Press: London.

Black, R. F. 1954. The leaf anatomy of Australian members of the genus *Atriplex*. Aust. J. Bot. 2:269–286.

Blackman, S. J., and E. C. Yeung. 1983. Comparative anatomy of pollinia and caudicles of an orchid (*Epidendrum*). Bot. Gaz. 144:331–337.

Blake, S. T. 1972. *Idiospermum* (Idiospermaceae), a new genus and family for *Calycanthus australiensis*. Contrib. Queensland Herb. 12:1–37.

Blakely, L. M., M. Durham, T. A. Evans, and R. M. Blakely. 1982. Experimental studies on lateral root formation in radish seedling roots. I. General methods, developmental stages, and spontaneous formation of laterals. Bot. Gaz. 143:341–352.

Blaser, H. W. 1945. Anatomy of *Cryptostegia grandiflora* with special reference to the latex system. Amer. J. Bot. 32:135–141.

Blaser, H. W. 1956. Morphology of the determinate thorn-shoots of *Gleditsia*. Amer. J. Bot. 43:22–28.

Bleckmann, C. A., H. M. Hull, and R. W. Hoshaw. 1980. Cuticular ultrastructure of *Prosopis velutina* and *Acacia greggii* leaflets. Bot. Gaz. 141:1–8.

Bloch, R. 1946. Differentiation and pattern in *Monstera deliciosa*. The idioblastic development of the trichosclereids in the air root. Amer. J. Bot. 33:544–551.

Blyth, A. 1958. Origin of primary extraxylary stem fibers in dicotyledons. Calif. Univ. Publs., Bot. 30:145–232.

Böcher, T. W., and O. B. Lyshede. 1968. Anatomical studies in xerophytic apophyllous plants. I. *Monttea aphylla*, *Bulnesia retama* and *Bredemeyera colletioides*. Biol. Skrifter Danske Videnskabernes Selskab 16:1–44.

Böcher, T. W., and O. B. Lyshede. 1972. Anatomical studies in xerophytic apophyllous plants. II. Additional species from South American shrub steppes. Biol. Skrifter Danske Videnskabernes Selskab. 18:1–137.

Boesewinkel, F. D. 1980. Development of ovule and testa of *Linum usitatissimum* L. Acta Bot. Neerl. 29:17–32.

Boesewinkel, F. D., and F. Bouman. 1980. Development of ovule and seed-coat of *Dichapetalum mombuttense* Engl. with notes on other species. Acta Bot. Neerl. 29:103–115.

Boke, N. H. 1941. Zonation in the shoot apices of *Trichocereus spachianus* and *Opuntia cylindrica*. Amer. J. Bot. 28:656–664.

Boke, N. H. 1944. Histogenesis of the leaf and areole in *Opuntia cylindrica*. Amer. J. Bot. 31:299–316.

Boke, N. H. 1949. Development of the stamens and carpels in *Vinca rosea* L. Amer. J. Bot. 36:535–547.

Boke, N. H. 1951. Histogenesis of the vegetative shoot in *Echinocereus*. Amer. J. Bot. 38:23–38.

Boke, N. H. 1953. Tubercle development in *Mammillaria heyderi*. Amer. J. Bot. 40:239–247.

Boke, N.H. 1954. Organogenesis of the vegetative shoot in *Pereskia*. Amer. J. Bot. 41:619–637.

Boke, N. H. 1961. Structure and development of the shoot in *Dolichothele*. Amer. J. Bot. 48:316–321.

Boke, N. H. 1976. Dichotomous branching in *Mammillaria* (Cactaceae). Amer. J. Bot. 63:1380–1384.

Boke, N. H. 1979. Root glochids and root spurs of *Opuntia arenaria* (Cactaceae). Amer. J. Bot. 66:1085–1092.

Boke, N. H. 1980. Developmental morphology and anatomy in Cactaceae. BioScience 30:605–610.

Boke, N. H., and R. G. Ross. 1978. Fasciation and dichotomous branching in *Echinocereus* (Cactaceae). Amer. J. Bot. 65:522–530.

Bold, H. C., C. J. Alexopoulos, and T. Delevoryas. 1980. *Morphology of Plants and Fungi*. Harper and Row: New York.

Bold, H. C., and M. J. Wynne. 1978. *Introduction to the Algae. Structure and Reproduction*. Prentice-Hall: Englewood Cliffs, New Jersey.

Bonde, S. D., and N. V. Biradar. 1981. On 2 palm woods from the Deccan Inter-Trappean beds of Dongargaon district, Chandrapur Maharashtra, India. J. Univ. Poona Sci. Technol. 0:247–257.

Bonner, J., and A. W. Galston. 1947. The physiology and biochemistry of rubber formation in plants. Bot. Rev. 13:543–596.

Bonnett, H. T., Jr. 1968. The root endodermis: fine structure and function. J. Cell Biol. 37:199–205.

Bonnett, H. T., Jr., and E. H. Newcomb. 1966. Coated vesicles and other cytoplasmic components of growing root hairs of radish. Protoplasma 62:59–75.

Bonnett, O. T. 1961. The oat plant: its histology and development. Bull. Ill. Agr. Exp. Sta., no. 672.

Bonnier, G. 1879. *Les Nectaires*. Theses Fac. Sci. Paris.

Booker, C. E., and R. S. Dwivedi. 1973. Ultrastructure of meristematic cells of dormant and released buds in *Tradescantia paludosa*. Exptl. Cell Res. 82:255–261.

Borger, G. A., and T. T. Kozlowski. 1972a. Effects of water deficits on first periderm and xylem development in *Fraxinus pennsylvanica*. Can. J. For. Res. 2:144–151.

Borger, G. A., and T. T. Kozlowski. 1972b. Effects of light intensity on first periderm and xylem development in *Pinus resinosa*, *Fraxinus pennsylvanica*, and *Robinia pseudoacacia*. Can. J. For. Res. 2:190–197.

Bosabalidis, A. M., and W. W. Thomson. 1984. Ultrastructural differentiation of an unusual structure lining the anticlinal walls of the inner secretory cells of *Tamarix* salt glands. Bot. Gaz. 145:427–435.

Bosshard, H. H. 1966. Notes on the biology of heartwood formation. News Bull. Int. Assoc. Wood Anat. 1966:11–14.

Bosshard, H. H. 1967. Über die fakultative Farbkernbildung. Holz Roh-u. Werkstoff 25:409–416.

Bosshard, H. H. 1968. On the formation of facultatively colored heartwood in *Beilschmiedia tawa*. Wood Sci. Tech. 2:1–12.

Bosshard, H. H., and L. Kučera. 1973. Die dreidimensionale Strukturanalyse des Holzes. I. Die Vernetzung des Gefäßsystems in *Fagus sylvatica* L. Holz Roh-u. Werkstoff 31:437–445.

Bosshard, H. H., and J. Stahel. 1969. Modifikationen in der sekundären Rinde von *Populus robusta*. Holzforsch. 5:1–5.

Botosso, P. C., and A. V. Gomes. 1982. Radial vessels and series of perforated ray cells in Annonaceae. IAWA Bull. 3:39–44.

Bouchet, P., and G. Deysson. 1971. Aspects ultrastructuraux de la différenciation des cellules à mucilage de la rose trémière, *Althaea rosea* Cav. C. R. Acad. Sci. Paris. 272:819–822.

Bouchet, P., and G. Deysson. 1974. Les canaux à mucilage des angiospermes. Etude morphologique et ultrastructurale des cellules constituant les canaux à mucilage du *Sterculia bidwilli* Hook. Revue gén. Bot. 81:369–402.

Bouman, F, and J. I. M. Calis. 1977. Integumentary shifting—a third way to unitegmy. Ber. Dt. Bot. Ges. 90:15–28.

Bouman, F. and S. Schier. 1979. Ovule ontogeny and seed coat development in *Gentiana*, with a discussion on the evolutionary origin of the single integument. Acta Bot. Neerl. 28:467–478.

Boureau, E. 1954. *Anatomie végétale, vol. 1*. Paris: Presses Universitaires de France.

Bowen, G. D., and A. D. Rovira. 1976. Microbial colonization of plant roots. Ann. Rev. Phytopath. 14:121–144.

Bowen, W. R. 1963. Origin and development of winged cork in *Euonymus alatus*. Bot. Gaz. 124:256–261.

Boyd, J. D. 1977. Basic cause of differentiation of tension wood and compression wood. Aust. For. Res. 7:121–143.

Boyd, L., and G. S. Avery, Jr. 1936. Grass seedling anatomy: the first internode of *Avena* and *Triticum*. Bot. Gaz. 97:765–779.

Boyd, W. D., W. M. Harris, and L. E. Murry. 1982. Sclereid development in *Camellia* petioles. Amer. J. Bot. 69:339–347.

Bradbury, D., M. M. MacMasters, and J. M. Cull. 1956. Structure of mature wheat kernel: II. Microscopic structure of pericarp, seed coat, and other coverings of the endosperm and germ of hard red winter wheat. Cereal Chem. 33:342–360.

Bradley, M. O. 1973. Microfilaments and cytoplasmic streaming: inhibition of streaming with cytochalasin. J. Cell Sci. 12:327–343.

Braun, H. J. 1955. Beiträge zur Entwicklungsgeschichte der Markstrahlen. Bot. Stud. Jena 4:73–131.

Braun, H. J. 1959. Die Vernetzung der Gefäße bei *Populus*. Z. Bot. 47:421–434.

Braun, H. J. 1970. Funktionelle Histologie der sekundären Sprossachse. I. Das Holz. In *Encyclopedia of Plant Anatomy*, 2nd ed. Gbdr. Bornträger: Berlin.

Brewbaker, J. L. 1967. The distribution and phylogenetic significance of binucleate and trinucleate pollen grains in the angiosperms. Amer. J. Bot. 54:1069–1083.

Briarty, L. G. 1980. Stereological analysis of cotyledon cell development in *Phaseolus*. II. The developing cotyledon. J. Exptl. Bot. 31:1387–1398.

Briarty, L. G., C. E. Hughes, and A. D. Evers. 1979. The developing endosperm of wheat—A stereological analysis. Ann. Bot. 44:641–658.

Bristow, J. M. 1975. The structure and function of roots in aquatic vascular plants. In *The Development and Function of Roots*, ed. J. G. Torrey and D. T. Clarkson. Academic Press: London.

Broughton, W. J. 1983. *Nitrogen Fixation. Vol. 3. Legumes.* Clarendon Press: Oxford.

Brouwer, R. 1953. The arrangement of the vascular bundles in the nodes of the Dioscoreaceae. Acta Bot. Neerl. 2:66–73.

Brouwer, R. 1959. Diffusible and exchangeable Rb ions in pea roots. Acta bot. Neerl. 8:68–76.

Brouwer, R., O. Gašpariková, J. Kolek, and B. C. Loughman. 1981. *Structure and Function of Plant Roots.* Martinus Nijhoff/Dr W. Junk Publishers: The Hague.

Brown, W. V. 1958. Leaf anatomy in grass systematics. Bot. Gaz. 119:170–178.

Brown, W. V. 1975. Variations in anatomy, associations, and origins of Kranz tissue. Amer. J. Bot. 62:395–402.

Brown, W. V. 1977. The Kranz syndrome and its subtypes in grass systematics. Mem. Torrey Bot. Club 23:1–97.

Brown, W. V., and S. C. Johnson. 1962. The fine structure of the grass guard cell. Amer. J. Bot. 49:110–115.

Bruck, D. K., and D. R. Kaplan. 1980. Heterophyllic development in *Muehlenbeckia* (Polygonaceae). Amer. J. Bot. 67:337–346.

Bruck, D. K., and D. J. Paolillo, Jr. 1984. Anatomy of nodes vs. internodes in *Coleus:* the longitudinal course of xylem differentiation. Amer. J. Bot. 71:151–157.

Brunner, C. 1909. Beiträge zur vergleichenden Anatomie der Tamaricaceae. Jb. wiss. Anst. Hamburg 89–162.

Buchholz, J. T. 1931. The pine embryo and the embryos of related genera. Trans. Illinois Acad. Sci. 23:117–125.

Buckhout, T. J., B. M. Gripshover, and D. J. Morré. 1981. Endoplasmic reticulum formation during germination of wheat seeds. A quantitative electron microscope study. Plant Physiol. 68:1319–1322.

Bugnon, P. 1925. Origine, évolution et valeur des concepts de protoxylème et de metaxylème. Soc. Linn. de Normandie, Bull. 7:123–151.

Bünning, E., and F. Biegert. 1953. Die Bildung der Spaltöffnungsinitialen bei *Allium cepa*. Z. wiss. Bot. 41:17–39.

Burgess, J. 1985. *An Introduction to Plant Cell Development.* Cambridge University Press: Cambridge.

Burns-Balogh, P. 1983. A theory on the evolution of the exine in Orchidaceae. Amer. J. Bot. 70:1304–1312.

Burrows, F. M. 1975. Wind-borne seed and fruit movement. New Phytol. 75:405–418.

Buss, P. A., Jr., and N. R. Lersten. 1975. Survey of tapetal nuclear number as a taxonomic character in Leguminosae. Bot. Gaz. 136:388–395.

Butterfield, B. G. 1974. Cambial cells. In *Dynamic Aspects of Plant Ultrastructure*, ed. A. W. Robarbs. McGraw-Hill: London.

Butterfield, B. G. 1976. The ontogeny of the vascular cambium in *Hoheria angustifolia* Raoul. New Phytol. 77:409–420.

Butterfield, B. G., and B. A. Meylan. 1975. Simple to scalariform combination perforation plates in *Vitex lucens* Kirk and *Brachyglottis repanda* J. R. and G. Forst. IAWA Bull. 3:39–42.

Butterfield, B. G., and B. A. Meylan. 1980. *The Three Dimensional Structure of Wood: An Ultrastructural Approach.* Chapman and Hall: London.

Buttrose, M. S. 1962. The influence of environment on the shell structure of starch granules. J. Cell Biol. 14:159–167.

Buvat, R. 1952. Structure, évolution et fonctionnement du méristème apical de quelques dicotylédones. Ann. Sci. Nat. Bot. Ser. II. 13:199–300.

Buvat, R. 1955. Le méristème apical de la tige. Ann. Biol. 31:595–656.

Buvat, R. 1956. Variations saisonnières du chondriome dans le cambium de *Robinia pseudoacacia*. C. R. Acad. Sci. Paris. 243:1908–1911.

Buvat, R., and O. Roger-Liard. 1954. La proliferation cellulaire dans le méristème radiculaire d'*Equisetum arvense*. C. R. Acad. Sci. Paris 238:1257–1258.

Buxbaum, F. 1950. *Morphology of Cacti.* Abbey Garden Press: Pasadena, California.

Byott, G. S. 1976. Leaf air space systems in C_3 and C_4 species. New Phytol. 76:295–299.

Byrne, J. M. 1973. The root apex of *Malva sylvestris*. III. Lateral root development and the quiescent center. Amer. J. Bot. 60:657–662.

Cailloux, M. 1972. Metabolism and the absorption of water by root hairs. Can. J. Bot. 50:557–573.

Calder, M., and P. Bernhardt. 1983. *The Biology of Mistletoes.* Academic Press: Sydney.

Calvin, C. L. 1967. Anatomy of the endophytic system of the mistletoe, *Phoradendron flavescens*. Bot. Gaz. 128:117–137.

Calvin, C. L., and R. L. Null. 1977. On the development of collenchyma in carrot. Phytomorph. 27:323–331.

Camefort, H. 1962. L'organisation du cytoplasme dans l'oosphère et la cellule centrale du *Pinus laricio* Poir. (var. *austriaca*). Ann. Sci. Nat. Bot. Biol. Veg. 12e Ser. 3:269–291.

Campbell, C. J., and J. E. Strong. 1964. Salt gland anatomy in *Tamarix pentandra* (Tamaricaceae). SWest. Nat. 9: 232–238.

Canright, J. E. 1952. The comparative morphology and relationships of the Magnoliaceae. I. Trends of specialization in the stamens. Amer. J. Bot. 31:484–497.

Capesius, I., and W. Barthlott. 1975. Isotopen-Markierungen und rasterelektronenmikroskopische Untersuchungen des Velamen radicum der Orchideen. Z. Pflanzenphysiol. 75:436–448.

Carde, J. P. 1978. Ultrastructural studies of *Pinus pinaster* needles: The endodermis. Amer. J. Bot. 65:1041–1054.

Carlquist, S. 1957. Leaf anatomy and ontogeny in *Argyroxiphium* and *Wilkesia* (Compositae). Amer. J. Bot. 44: 696–705.

Carlquist, S. 1958. Structure and ontogeny of glandular trichomes of Madiinae (Compositae). Amer. J. Bot. 45: 675–682.

Carlquist, S. 1961. *Comparative Plant Anatomy.* Holt, Rinehart & Winston: New York.

Carlquist, S. 1962. A theory of paedomorphosis in dicotyledonous woods. Phytomorph. 12:30–45.

Carlquist, S. 1966. Wood anatomy of Compositae: a summary, with comments on factors controlling wood evolution. Aliso 6:25–44.

Carlquist, S. 1969. Toward acceptable evolutionary interpretations of floral anatomy. Phytomorph. 19:332–362.

Carlquist, S. 1970. Wood anatomy of insular species of *Plantago* and the problem of raylessness. Bull. Torrey Bot. Club 97:353–361.

Carlquist, S. 1975. *Ecological Strategies of Xylem Evolution.* University of California Press: Berkeley, California.

Carlquist, S. 1980a. Anatomy and systematics of Balanopaceae. Allertonia 2:191–246.

Carlquist, S. 1980b. Further concepts in ecological wood anatomy, with comments on recent work in wood anatomy and evolution. Aliso 9:499–553.

Carlquist, S. 1981. Types of cambial activity and wood anatomy of *Stylidium* (Stylidiaceae). Amer. J. Bot. 68: 778–785.

Carlquist, S. 1982a. Wood anatomy of *Illicium* (Illiciaceae): Phylogenetic, ecological, and functional interpretations. Amer. J. Bot. 69:1587–1598.

Carlquist, S. 1982b. Wood anatomy of Onagraceae: Further species; root anatomy; significance of vestured pits and allied structures in dicotyledons. Ann. Missouri Bot. Gard. 69:755–769.

Carlquist, S. 1982c. Wood and bark anatomy of *Scalesia* (Asteraceae). Aliso 10:301–312.

Carlquist, S. 1983. Wood anatomy of *Bubbia* (Winteraceae), with comments on origin of vessels in dicotyledons. Amer. J. Bot. 70:578–590.

Carlquist, S. 1984a. Wood and stem anatomy of *Bergia suffruticosa*: relationships of Elatinaceae and broader significance of vascular tracheids, vasicentric tracheids, and fibriform vessel elements. Ann. Missouri Bot. Gard. 71: 232–242.

Carlquist, S. 1984b. Vessel grouping in dicotyledon wood: significance and relationship to imperforate tracheary elements. Aliso 10:505–525.

Carlquist, S. 1985. Wood anatomy of Begoniaceae, with comments on raylessness, paedomorphosis, relationships, vessel diameter, and ecology. Bull. Torrey Bot. Club 112: 59–69.

Carolin, R. C. 1971. The trichomes of the Goodeniaceae. Proc. Linn. Soc. N. S. W. 96:8–22.

Carolin, R. C., S. W. L. Jacobs, and M. Vesk. 1973. The structure of the cells of the mesophyll and parenchymatous bundle sheath of the Gramineae. J. Linn. Soc. Bot. 66: 259–275.

Carolin, R. C., S. W. L. Jacobs, and M. Vesk. 1975. Leaf structure in Chenopodiaceae. Bot. Jahrb. Syst. 95:226–255.

Carpenter, C. H., and L. Leney. 1952. 382 *Photomicrographs of 91 Papermaking Fibers.* State University of New York, College of Forestry at Syracuse: Syracuse, New York.

Carr, D. J. 1976. Plasmodesmata in growth and development. In *Intercellular Communication in Plants: Studies on Plasmodesmata,* eds. B. E. S. Gunning and A. W. Robards. Springer-Verlag: Berlin.

Carr, S. G. M., and D. J. Carr. 1961. The functional significance of syncarpy. Phytomorphology 11:249–256.

Carr, S. G. M., L. Milkovits, and D. J. Carr. 1971. Eucalypt phytoglyphs: the microanatomical features of the epidermis in relation to taxonomy. Aust. J. Bot. 19:173–190.

Carson, E. W. 1974. *The Plant Root and its Environment.* University Press of Virginia: Charlottesville, Virginia.

Cartwright, S. C., W. M. Lush, and M. J. Canny. 1977. A comparison of translocation of labelled assimilate by normal and lignified sieve elements in wheat leaves. Planta 134:207–208.

Cass, D. D., and W. A. Jensen. 1970. Fertilization in barley. Amer. J. Bot. 57:62–70.

Cassens, D. L. 1980. Vestured pits in the New World *Pithecellobium* (sensu lato). IAWA Bull., n. s. 1:59–64.

Catesson, A. M. 1964. Origine, fonctionnement et variations cytologiques saisonnières du cambium de l'*Acer pseudoplatanus* L. (Aceracées). Annls. Sci. nat. Bot. ser. 12, 5:229–498.

Catesson, A. M. 1974. Cambial cells. In *Dynamic Aspects of Plant Ultrastructure,* ed. A. W. Robards. McGraw-Hill: New York.

Catling, D., and J. Grayson. 1982. *Identification of Vegetable Fibers.* Chapman and Hall: London.

Ceccarelli, N., R. Lorenzi, and A. Alpi. 1981. Gibberellin biosynthesis in *Phaseolus coccineus* suspensor. Z. Pflanzenphysiol. 102:37–44.

Cecich, R. A. 1977. An electron microscope evaluation of cytohistological zonation in the shoot apical meristem of *Pinus banksiana*. Amer. J. Bot. 64:1263–1271.

Cecich, R. A. 1984. The histochemistry and ultrastructure of jack pine microsporangia during winter. Amer. J. Bot. 71:851–864.

Cecich, R. A., and H. T. Horner, Jr. 1977. An ultrastructural and microspectrophotometric study of the shoot apex during the initiation of the first leaf in germinating *Pinus banksiana*. Amer. J. Bot. 64:207–222.

Chafe, S. C. 1970. The fine structure of the collenchyma cell wall. Planta 90:12–21.

Chafe, S. C., and A. B. Wardrop. 1970. Microfibril orientation in plant cell walls. Planta 92:13–24.

Chaffey, N. J. 1982. Presence of stomata-like structures in the ligule of *Agrostis gigantea* Roth. Ann. Bot. 50:717–720.

Chaffey, N. J. 1985. Structure and function in the grass ligule: optical and electron microscopy of the membranous ligule of *Lolium temulentum* L. Ann. Bot. 55:65–75.

Chalk, L. 1933. Multiperforate plates in vessels with special reference to the Bignoniaceae. Forestry 7:16–25.

Chalk, L. 1983a. Vessels. In *Anatomy of the Dicotyledons*, vol. 2, eds. C. R. Metcalfe and L. Chalk. Clarendon Press: Oxford.

Chalk, L. 1983b. The Wood. In *Anatomy of the Dicotyledons*, vol. 2, eds. C. R. Metcalfe and L. Chalk. Clarendon Press: Oxford.

Chalk, L., and M. M. Chattaway. 1934. Measuring the length of vessel members. Trop. Woods 40:19–26.

Chan, L–L. 1985. The anatomy of the bark of *Libocedrus* in New Zealand. IAWA Bull. 6:23–24.

Chandra, A., R. Singh, and V. S. Rathore. 1982. Anatomical changes associated with suberization and secondary growth in *Citrus limon* Burm. roots. Indian J. Bot. 5: 128–130.

Chandra Sekhar, K. N., and V. K. Sawhney. 1985. Ultrastructure of the shoot apex of tomato (Lycopersicon esculentum). Amer. J. Bot. 72:1813–1822.

Chapman, V. J. 1939. Cambridge University Expedition to Jamaica. Part 3. The morphology of *Avicennia nitida* Jacq. and the function of its pneumatophores. Bot. J. Linn. Soc. 52:487–533.

Chapman, V. J. 1944. The morphology of *Avicennia nitida*. J. Linn. Soc. Bot. 52:487–533.

Chattaway, M. M. 1936. Relation between fibre and cambial initial length in dicotyledonous woods. Trop. Woods 46:16–20.

Chattaway, M. M. 1948. Note on the vascular tissue in the rays of *Banksia*. J. Council Sci. Ind. Res. 21:275–278.

Chattaway, M. M. 1949. The development of tyloses and secretion of gum in heartwood formation. Aust. J. Sci. Res. Ser. B, Biol. Sci. 2:227–240.

Chattaway, M. M. 1951. Morphological and functional variations in the rays of pored timbers. Aust. J. Sci. Res. 4:12–27.

Chattaway, M. M. 1952. The sapwood-heartwood transition. Aust. For. 16:25–34.

Chattaway, M. M. 1953. The anatomy of bark: I, The genus *Eucalyptus*. Aust. J. Bot. 1:402–433.

Chattaway, M. M. 1955. The anatomy of bark. VI. Peppermints, boxes, ironwoods, and other eucalypts with cracked and furrowed bark. Aust. J. Bot. 3:170–176.

Chavan, R. R., and J. J. Shah. 1983. Statistical approach for the understanding of secondary phloem in 125 tropical dicotyledons. Proc. Indian Natl. Sci. Acad. Part B, Biol. Sci. 49:28–36.

Chazdon, R. L. 1985. Leaf display, canopy structure, and light interception of two understory palm species. Amer. J. Bot. 72:1493–1502.

Cheadle, V. I. 1937. Secondary growth by means of a thickening ring in certain monocotyledons. Bot. Gaz. 98: 535–555.

Cheadle, V. I. 1943. The origin and certain trends of specialization of the vessel in the Monocotyledonae. Amer. J. Bot. 30:11–17.

Cheadle, V. I. 1944. Specialization of vessels within the xylem of each organ in the Monocotyledonae. Amer. J. Bot. 31:81–92.

Cheadle, V. I. 1948. Observations on the phloem in the Monocotyledoneae. II, Additional data on the occurrence and phylogenetic specialization in structure of the sieve tubes in the metaphloem. Amer. J. Bot. 35:129–131.

Cheadle, V. I. 1953. Independent origin of vessels in the monocotyledons and dicotyledons. Phytomorph. 3: 23–44.

Cheadle, V. I. 1956. Research on xylem and phloem— progress in fifty years. Amer. J. Bot. 43:719–731.

Cheadle, V. I., and K. Esau. 1964. Secondary phloem of *Liriodendron tulipifera*. Univ. Calif. Publ. Bot. 36:143–252.

Cheadle, V. I., and N. W. Uhl. 1948. The relation of metaphloem to the types of vascular bundles in the Monocotyledonae. Amer. J. Bot. 35:578–583.

Cheadle, V. I., and N. B. Whitford. 1941. Observations on the phloem in the Monocotyledoneae: I. The occurrence and phylogenetic specialization in structure of the sieve tubes in the metaphloem. Amer. J. Bot. 28:623–627.

Cheplick, G. P. 1983. Differences between plants arising from aerial and subterranean seeds in the amphicarpic annual *Cardamine chenopodifolia* (Cruciferae). Bull. Torrey Bot. Club 110:442–448.

Chesnoy, L., and M. J. Thomas. 1971. Electron microscopy studies on gametogenesis and fertilization in gymnosperms. Phytomorphology 21:50–63.

Chiang, S-H. T. 1980. Casparian strips in the lattice work phellem of *Melaleuca leucadendra*. Taiwania 25:48–56.

Chlyah, A., and M. Tran Thanh Van. 1984. Histological changes in epidermal and subepidermal cell layers of *Begonia rex* induced to form de novo unicellular hairs, buds and roots. Bot. Gaz. 145:55–59.

Chouard, P. 1937. La nature et le rôle des formations dites "secondaires" dans l'édification de la tige des monocotylédones. Bull. Soc. Bot. Fr. 83:819–836.

Chowdhury, K. A. 1941. The formation of growth rings in Indian trees. III. A study of the effect of locality. Indian For. Rec., n. s., Util. 2:59–75.

Chowdhury, K. A., and G. M. Buth. 1970. Seed coat structure and anatomy of Indian pulses. In *New Research in Plant Anatomy*, eds. N. K. B. Robson, D. F. Cutler, and M. Gregory. J. Linn. Soc. Bot. 63, Suppl. 1. Academic Press: London.

Chrispeels, M. J. 1980. The endoplasmic reticulum. In *The Biochemistry of Plants. A Comprehensive Treatise*, vol. 1, ed. N. E. Tolbert. Academic Press: New York.

Chuang, T. I., and L. R. Heckard. 1983. Systematic significance of seed-surface features in *Orthocarpus* (Scrophulariaceae—subtribe Castillejinae). Amer. J. Bot. 70:877–890.

Ciampolini, F., M. Cresti, and E. Pacini. 1978. Caratteristiche ultrastrutturali ed istochimiche del tessuto trasmittente stilare de melo. Seminario sulla "Fertilita delle piante da frutto." Progetto finalizzato CNR "Biologia della riproduzione." Bologna, 15 dicembre 1978.

Cichan, M. A. 1985a. Vascular cambium and wood development in carboniferous plants. I. Lepidodendrales. Amer. J. Bot. 72:1163–1176.

Cichan, M. A. 1985b. Vascular cambium and wood development in carboniferous plants. II. *Sphenophyllum plurifoliatum* Williamson and Scott (Sphenophyllales). Bot. Gaz. 146:395–403.

Cichan, M. A., and T. N. Taylor. 1982. Vascular cambium development in *Sphenophyllum*, a carboniferous arthrophyte. IAWA 3:155–160.

Clark, L. H., and W. H. Harris. 1981. Observations on the root anatomy of rice (*Oryza sativa* L.). Amer. J. Bot. 68:154–161.

Clarke, A. E., J. A. Considine, R. Ward, and R. B. Knox. 1977. Mechanism of pollination in *Gladiolus*: roles of the stigma and pollen tube guide. Ann. Bot. 41:14–20.

Clarkson, D. T., and A. W. Robards. 1975. The endodermis, its structural development and physiological role. In *The Development and Function of Roots*, eds. J. G. Torrey and D. T. Clarkson. Academic Press: London.

Clarkson, D. T., A. W. Robards, and J. Sanderson. 1971. The tertiary endodermis in barley roots: fine structure in relation to radial transport of ions and water. Planta 96:292–305.

Claussen, P. 1927. Über das Verhalten des Antherentapetums bei einigen Monocotylen und Ranales. Bot. Arch. 18:1–27.

Cline, M. N., and D. Neely. 1983. The histology and histochemistry of the wound healing process in geranium (*Pelargonium hortorum* cultivar Yours Truly) cuttings. J. Am. Soc. Hort. Sci. 108:496–502.

Clowes, F. A. L. 1961. *Apical Meristems*. Blackwell: Oxford.

Clowes, F. A. L. 1978a. Development of the shoot apex in *Zea mays*. New Phytol. 81:663–669.

Clowes, F. A. L. 1978b. Chimeras and the origin of lateral root primordia in *Zea mays*. Ann. Bot. 42:801–807.

Clowes, F. A. L. 1981a. Cell proliferation in ectotrophic mycorrhizas of *Fagus sylvatica*. L. New Phytol. 87:547–555.

Clowes, F. A. L. 1981b. The difference between open and closed meristems. Ann. Bot. 48:761–767.

Clowes, F. A. L. 1985. Origin of epidermis and development of root primordia in *Pistia, Hydrocharis* and *Eichhornia*. Ann. Bot. 55:849–857.

Clowes, F. A. L., and B. E. Juniper. 1964. The fine structure of the quiescent centre and neighbouring tissues in root meristems. J. Exptl. Bot. 15:622–630.

Clowes, F. A. L., and B. E. Juniper. 1968. *Plant Cells*. Blackwell: Glasgow.

Cocucci, A. E., and W. A. Jensen. 1969. Orchid embryology: Megagametophyte of *Epidendrum scutella* following fertilization. Amer. J. Bot. 56:629–640.

Cohen, L., and T. Arzee. 1980. Twofold pathways of apical determination in the thorn system of *Carissa grandiflora*. Bot. Gaz. 141:258–263.

Colvin, J. R. 1981. Ultrastructure of the plant cell wall; biophysical viewpoint. In *Encyclopedia of Plant Physiology*, new series, vol. 13B, eds. W. Tanner and R. A. Loewus. Springer-Verlag: Berlin.

Conrad, P. A., G. L. Steucek, and P. K. Hepler. In press (Planta).

Considine, J. A. 1982. Physical aspects of fruit growth, cuticular fracture and fracture patterns in relation to fruit structure in *Vitis vinifera* cultivar *sultana*. J. Hortic. Sci. 57:79–92.

Cook, P. R., and R. A. Laskey. 1984. *Higher Order Structure in the Nucleus*. Supplement 1, J. Cell Sci. The Company of Biologists, Limited: Cambridge.

Cooke, G. B. 1961. *Cork and the Cork Tree*. Pergamon Press: New York.

Coombe, B. G. 1976. The development of fleshy fruits. Ann. Rev. Plant Physiol. 27:507–528.

Cooper, T. G., and H. Beevers. 1969. Mitochondria and glyoxysomes from castor bean endosperm. J. Biol. Chem. 244:3507–3513.

Core, H. A., W. A. Côté, and A. C. Day. 1979. *Wood Structure and Identification*. Syracuse University Press: Syracuse, New York.

Cormack, R. G. H. 1949. The development of root hairs in angiosperms. Bot. Rev. 15:583–612.

Corner, E. J. H. 1976. *The Seeds of Dicotyledons*. 2 vols. Cambridge University Press: Cambridge.

Corson, G. E., Jr. 1969. Cell division studies of the shoot apex of *Datura stramonium* during transition to flowering. Amer. J. Bot. 56:1127–1134.

Côté, W. A., Jr., A. C. Day, and T. E. Timell. 1969. A contribution to the ultrastructure of tension wood fibres. Wood Sci. Tech. 3:257–271.

Cottle, W., and P. E. Kolattukudy. 1982. Biosynthesis, deposition and partial characterization of potato (*Solanum tuberosum* cultivar White Rose) suberin phenolics. Plant Physiol. 69:393–399.

Coult, D. A. 1964. Observations on gas movement in the rhizome of *Menyanthes trifoliata* L., with comments on the role of the endodermis. J. Exp. Bot. 15:205–218.

Cowan, J. M. 1950. *The Rhododendron Leaf, a Study of the Epidermal Appendages*. Oliver and Boyd: Edinburgh.

Cox, G., and P. B. Tinker. 1976. Translocation and transfer of nutrients in vesicular-arbuscular mycorrhizas. I. The arbuscule and phosphorus transfer: a quantitative ultrastructural study. New Phytol. 77:371–378.

Crafts, A. S., and C. E. Crisp. 1971. *Phloem Transport in Plants*. W. H. Freeman and Co.: San Francisco.

Cragg, F. J., and J. H. M. Willison. 1980. The ultrastructure of quiescent buds of *Tilia europaea*. Can. J. Bot. 58:1804–1813.

Cresti, M., F. Ciampolini, D. L. M. Mulcahy, and G. Mulcahy. 1985. Ultrastructure of *Nictotiana alata* pollen, its germination and early tube formation. Amer. J. Bot. 72:719–727.

Cresti, M., E. Pacini, F. Ciampolini, and G. Sarfatti. 1977. Germination and early tube development in vitro of *Lycopersicum peruvianum* pollen: ultrastructural features. Planta 136:239–247.

Cresti, M., J. L. van Went, E. Pacini, and M. T. M. Willemse. 1976. Ultrastructure of transmitting tissue of *Lycopersicon peruvianum* style: development and histochemistry. Planta 132:305–312.

Critchfield, W. B. 1970. Shoot growth and heterophylly in *Ginkgo biloba*. Bot. Gaz. 131:150–162.

Cronquist, A. 1981. *An Integrated System of Classification of Flowering Plants*. Columbia University Press: New York.

Cronshaw, J., and K. Esau. 1967. Tubular and fibrillar components of mature and differentiating sieve elements. J. Cell Biol. 34:801–816.

Cronshaw, J., and K. Esau. 1968. Cell division in leaves of *Nicotiana*. Protoplasma 65:1–24.

Crooks, D. M. 1933. Histological and regenerative studies on the flax seedling. Bot. Gaz. 95:209–239.

Cross, G. L. 1936. The structure of the growing point and the development of the bud scales of *Morus alba* L. Bull. Torrey Bot. Club 63:451–465.

Cross, G. L. 1938. A comparative histogenetic study of the bud scales and foliage leaves of *Viburnum opulus* Amer. J. Bot. 25:246–258.

Cross, G. L. 1941. Some histogenetic features of the shoot of *Cryptomeria japonica*. Amer. J. Bot. 28:573–582.

Cross, G. L. 1942. Structure of the apical meristem and development of the foliage leaves of *Cunninghamia lanceolata*. Amer. J. Bot. 29:288–301.

Crossett, R. N., D. J. Campbell, and H. E. Stewart. 1975. Compensatory growth in cereal root systems. Pl. Soil 42:673–683.

Croxdale, J. G. 1976. Origin and early morphogenesis of lateral buds in the fern *Davallia*. Amer. J. Bot. 63:226–238.

Crozier, A., and J. R. Hillman. 1984. *The Biosynthesis and Metabolism of Plant Hormones*. Cambridge University Press: London.

Cumbie, B. G. 1967a. Developmental changes in the vascular cambium of *Leitneria floridana*. Amer. J. Bot. 54:414–424.

Cumbie, B. G. 1967b. Development and structure of the cambium of *Canavalia*. Bull. Torrey Bot. Club 94:162–175.

Cumbie, B. G. 1984. Origin and development of the vascular cambium in *Aeschynomene virginica*. Bull. Torrey Bot. Club 3:42–50.

Curtis, J. D., and N. R. Lersten. 1974. Morphology, seasonal variation, and function of resin glands on buds and leaves of *Populus deltoides* (Salicaceae). Amer. J. Bot. 61:835–845.

Cutler, D. F. 1969. *Anatomy of the Monocotyledons. IV. Juncales*. Clarendon Press: Oxford.

Cutler, D. F., K. L. Alvin, and C. E. Price. 1982. *The Plant Cuticle*. Academic Press: London.

Cutler, E. G. 1965. Recent experimental studies of the shoot apex and shoot morphogenesis. Bot. Rev. 31:7–113.

Cutter, E. G., and L. J. Feldman. 1970a. Trichoblasts in *Hydrocharis*: I. Origin, differentiation, dimensions and growth. Amer. J. Bot. 57:190–201.

Cutter, E. G., and L. J. Feldman. 1970b. Trichoblasts in *Hydrocharis*: II. Nucleic acids, proteins and a consideration of cell growth in relation to endopolyploidy. Amer. J. Bot. 57:202–211.

Cutter, V. M. Jr., and B. Freeman. 1954. Development of the syncytial endosperm of *Cocos nucifera*. Nature 173:827–828.

Cutter, V. M. Jr., K. S. Wilson, and B. Freeman. 1955. Nuclear behavior and cell formation in the developing endosperm of *Cocos nucifera*. Amer. J. Bot. 42:109–115.

Czaninski, Y. 1964. Variations saisonnières du chondriome et de l'amidon dans les fibres libriforms du xylème du *Robinia pseudo-acacia*. C. R. Acad. Sci. (Paris) 258:5945–5948.

Czaninski, Y. 1968. Étude cytologique de la différentiation cellulaire du bois de Robinier. J. Microscopie 7:1051–1068.

Dadswell, H. E., and A. B. Wardrop. 1955. The structure and properties of tension wood. Holzforschung 9:97–104.

Daft, M. J., and B. O. Okusanya. 1973. Effect of *Endogone* mycorrhiza on plant growth. VI. Influence of infection on the anatomy and reproductive development in four hosts. New Phytol. 72:1333–1339.

Dahlgren, R. M. T., and H. T. Clifford. 1982. *The Monocotyledons: A Comparative Study*. Academic Press: New York.

D'Amato, F. 1954. A brief discussion of "endomitosis." Caryologia 6:341–344.

Dart, P. J. 1975. Legume root nodule initiation and development. In *The Development and Function of Roots,* eds. J. G. Torrey and D. T. Clarkson, Academic Press: London.

Dart, P. J., and F. V. Mercer. 1964. The legume rhizosphere. Arch. Mikrobiol. 47:344–378.

Darvill, A., M. McNeil, P. Albersheim, and D. P. Delmer. 1980. The primary cell walls of flowering plants. In *The Biochemistry of Plants: A Comprehensive Treatise,* vol. 1, ed. N. E. Tolbert. Academic Press: New York.

Datta, P. C., and A. Chowdhury. 1982. Fiber forming sequence in secondary phloem of *Hibiscus surattensis.* Bangladesh J. Bot. 11:24–32.

Daumann, E. 1970. Das Blütennektarium der Monocotyledonen unter besonderer Berücksichtigung seiner systematischen und phylogenetischen Bedeutung. Feddes Reperotorium. 80:463–590.

Davidson, D. 1972. Morphogenesis of primordia of lateral roots. In *The Dynamics of Meristem Cell Populations,* eds. M. Miller and C. Kuehnert. Plenum Press: New York.

Davies, E., and B. A. Larkins. 1980. Ribosomes. In *The Biochemistry of Plants: A Comprehensive Treatise,* vol. 1, ed. N. E. Tolbert. Academic Press: New York.

Davis, E. L. 1961. Medullary bundles in the genus *Dahlia* and their possible origin. Amer. J. Bot. 48:108–113.

Davis, G. L. 1966. Systematic embryology of the angiosperms. Wiley: New York.

Davis, R. W. 1984. A comparison of cystolith structure in some members of the Moraceae. J. Texas Soc. Electron Micro. 15:31.

Dayanandan, P., P. B. Kaufman, and C. I. Franklin. 1983. Detection of silica in plants. Amer. J. Bot. 70:1079–1084.

De Bary, A. 1877. *Vergleichende Anatomie der Vegetationsorgane der Phanerogamen und Farne.* W. Engelmann: Leipzig.

De Bary, A. 1884. *Comparative Anatomy of the Vegetative Organs of the Phanerogams and Ferns.* Clarendon Press: Oxford.

Dehgan, B. 1982. Comparative anatomy of the petiole and infrageneric relationships in *Jatropha* (Euphorbiaceae). Amer. J. Bot. 69:1283–1295.

Dehgan, B., and M. E. Craig. 1978. Types of laticifers and crystals in *Jatropha* and their taxonomic implications. Amer. J. Bot. 65:345–352.

Dehgan, B., and C. K. K. H. Yuen. 1983. Seed morphology in relation to dispersal, evolution, and propagation of *Cycas* L. Bot. Gaz. 144:412–418.

De Leeuw, G. T. N. 1985. Deposition of lignin, suberin and callose in relation to the restriction of infection by *Botrytis cinerea* in ghost spots of tomato fruits. Phytopathol Z. 112:143–152.

Dell, B. 1980. Structure and function of the strophiolar plug in seeds of *Albizia lophantha.* Amer. J. Bot. 67:556–563.

Dell, B., J. Kuo, and A. H. Burbidge. 1982. Anatomy of *Pilostyles hamiltonii* (Rafflesiaceae) in stems of *Daviesia.* Aust. J. Bot. 30:1–10.

Dell, B., J. Kuo, and G. J. Thomson. 1980. Development of proteoid roots in *Hakea obliqua* R. Br. (Proteaceae) grown in water culture. Aust. J. Bot. 28:27–37.

Dell, B., and A. J. McComb. 1978. Plant resins—their formation, secretion and possible functions. In *Advances in Botanical Research,* ed. E. W. Woolhouse. Academic Press: London.

DeMaggio, A. E., and C. L. Wilson. 1986. Floral structure and organogenesis in *Podophyllum peltatum* (Berberidaceae). Amer. J. Bot. 73:21–32.

DeMason, D. A. 1979a. Function and development of the primary thickening meristem in the monocotyledon *Allium cepa* L. Bot. Gaz. 140:51–66.

DeMason, D. A. 1979b. Histochemistry of the primary thickening meristem in the vegetative stem of *Allium cepa* L. Amer. J. Bot. 66:347–350.

DeMason, D. A. 1980. Localization of cell division activity in the primary thickening meristem in *Allium cepa* L. Amer. J. Bot. 67:393–399.

DeMason, D. A. 1983. The primary thickening meristem: definition and function in monocotyledons. Amer. J. Bot. 70:955–962.

Dengler, N. G., R. E. Dengler, and P. W. Hattersley. 1985. Differing ontogenetic origins of PCR ("Kranz") sheaths in leaf blades of C_4 grasses (Poaceae). Amer. J. Bot. 72:284–302.

Den Outer, R. W. 1982. Bark anatomy of *Tambourissa* (Monimiaceae) from Madagascar. Acta Bot. Neerl. 31:275–280.

Den Outer, R. W. 1983. Comparative study of the secondary phloem of some woody dicotyledons. Acta Bot. Neerl. 32:29–38.

Den Outer, R. W., and W. L. H. van Veenendaal. 1981. Wood and bark anatomy of *Azima tetracantha* Lam. (Salvadoraceae) with description of its included phloem. Acta Bot. Neerl. 30:199–207.

Dermen, H. 1945. The mechanism of colchicine-induced cytohistological changes in cranberry. Amer. J. Bot. 32:387–394.

Dermen, H. 1947. Periclinal cytochimeras and histogenesis in cranberry. Amer. J. Bot. 34:32–43.

Dermen, H. 1951. Ontogeny of tissues in stems and leaf of cytochimeral apples. Amer. J. Bot. 38:753–760.

De Roon, S. C. 1967. Foliar sclereids in the Marcgraviaceae. Acta Bot. Neerl. 15:585–623.

Derr, W. F., and R. F. Evert. 1967. The cambium and seasonal development of the phloem in *Robinia pseudoacacia.* Amer. J. Bot. 54:147–153.

Deschamp, P. A., and T. J. Cooke. 1985. Leaf dimorphism in the aquatic angiosperm *Callitriche heterophylla.* Amer. J. Bot. 72:1377–1387.

Deshpande, B. P. 1976. Observations on the fine structure of plant cell walls. III. The sieve-tube wall in *Cucurbita.* Ann. Bot. 40:443–446.

Deshpande, B. P., and T. Rajendrababu. 1985. Seasonal

changes in the structure of the secondary phloem of *Grewia tiliaefolia,* a deciduous tree from India. Ann. Bot. 56: 61–71.

Devaux, M. H. 1900. Recherches sur les lenticelles. Ann. Sci. Nat. Bot. VIII. 12:1–240.

Dick, P. S., and T. ap Rees. 1975. The pathway of sugar transport in roots of *Pisum sativum.* J. Exp. Bot. 26: 305–314.

Dickenson, P. B., and J. W. Fairbairn. 1975. The ultra-structure of the alkaloidal vesicles of *Papaver somniferum* latex. Ann. Bot. 39:707–712.

Dickinson, H. G., and P. R. Bell. 1976. Development of the tapetum in *Pinus banksiana* preceding sporogenesis. Ann. Bot. 40:103–113.

Dickison, W. C. 1974. Trichomes. In *Vascular Plant Systematics,* eds. A. E. Radford, W. C. Dickison, J. R. Massey, and C. R. Bell. Harper and Row: New York.

Dickison, W. C. 1975. Studies on the floral anatomy of the Cunoniaceae. Amer. J. Bot. 62:433–447.

Dickison, W. C., J. W. Nowicke, and J. J. Skvarla. 1982. Pollen morphology of the Dilleniaceae and Actinidiaceae. Amer. J. Bot. 69:1055–1073.

Dickmann, D. I., and T. T. Kozlowski. 1968. Mobilization by *Pinus resinosa* cones and shoots of C^{14}-photosynthate from needles of different ages. Amer. J. Bot. 55:900–906.

Dilcher, D. L. 1979. Early angiosperm reproduction: an introductory report. Review of Palaeobotany and Palynology 27:291–328.

Dinwoodie, J. M. 1976. Causes of brashness in timber. In *Wood Structure in Biological and Technological Research,* eds. P. Baas, A. J. Bolton, and D. M. Catling. Leiden University Press: Leiden.

Dittmer, H. J. 1937. A quantitative study of the roots and root hairs of a winter rye plant (*Secale cereale*). Amer. J. Bot. 24:417–420.

Dittmer, H. J. 1949. Root hair variations in plant species. Amer. J. Bot. 36:152–155.

Dittmer, H. J., and M. L. Roser. 1963. The periderm of certain members of the Cucurbitaceae. Southwest Nat. 8:1–9.

Doak, C. C. 1935. Evolution of foliar types, dwarf shoots, and cone scales in *Pinus.* Illinois Biol. Monogr. 13:1–106.

Dobbins, D. R. 1969. Studies on the anomalous cambial activity in *Doxantha unguis-cati* (Bignoniaceae). I. Development of the vascular pattern. Can. J. Bot. 47:2101–2106.

Dobbins, D. R. 1970. The influence of leaves on cambial activity in the cross vine. Amer. J. Bot. 57:735.

Dobbins, D. R. 1971. Studies on the anomalous cambial activity in *Doxantha unguis-cati* (Bignoniaceae). II. A case of differential production of secondary tissues. Amer. J. Bot. 58:697–705.

Dobbins, D. R. 1981. Anomalous secondary growth in lianas of the Bignoniaceae is correlated with the vascular pattern. Amer. J. Bot. 68:142–144.

Dobbins, D. R., H. Alden, and D. Marvel. 1983. Develop-

mental anatomy of juvenile and adult shoots of *Marcgravia rectifolia* L. Amer. J. Bot. 70:1263–1271.

Dodd, J. D. 1948. On the shapes of cells in the cambial zone of *Pinus sylvestris* L. Amer. J. Bot. 35:666–682.

Don, D. 1965. Secondary walls in phloem of *Pinus radiata.* Nature 207:657–658.

Dormer, K. J. 1945. An investigation of the taxonomic value of shoot structure in angiosperms with especial reference to Leguminosae. Ann. Bot. N. S. 9:141–153.

Dormer, K. J. 1972. *Shoot Organization in Vascular Plants.* Chapman and Hall: London.

Dormer, K. J. 1980. *Fundamental Tissue Geometry for Biologists.* Cambridge University Press: Cambridge.

Dosier, L. W., and J. L. Riopel. 1978. Origin, development, and growth of differentiating trichoblasts in *Elodea canadensis.* Amer. J. Bot. 65:813–822.

Douce, R. 1985. *Mitochondria in Higher Plants.* Academic Press: Orlando, Florida.

Doyle, M. H., and J. Doyle. 1948. Pith structure in conifers. I. Taxodiaceae. Roy. Irish Acad., Proc. Sect. B. 52: 15–39.

Dressler, R. L. 1981. *The Orchids.* Smithsonian Institution: Washington, D. C.

Drew, M. C. 1979. Properties of roots which influence rates of absorption. In *The Soil-Root Interface,* eds. J. L. Harley and R. S. Russell. Academic Press: New York.

Duchaigne, A. 1955. Les divers types de collenchymes chez les Dicotylédones; leur ontogénie et leur lignification. Annls. Sci. nat. Bot. 16:455–479.

Duckett, J. G., and R. Toth. 1977. Giant mitochondria in a periclinal chimera, *Ficus elastica* Roxb. forma variegata. Ann. Bot. 41:903–912.

Duerden, H. 1934. On the occurrence of vessels in *Selaginella.* Ann. Bot. 48:459–465.

Dumbroff, E. B., and H. W. Elmore. 1977. Living fibers are a principal feature of the xylem in seedlings of *Acer saccharum* Marsh. Ann. Bot. 41:471–472.

Dumbroff, E. B., and D. R. Peirson. 1971. Possible sites for a passive movement of ions across the endodermis. Can. J. Bot. 49:35–38.

Dute, R. R. 1983. Phloem of primitive angiosperms. I. Sieve-element ontogeny in the petiole of *Liriodendron tulipifera* L. (Magnoliaceae). Amer. J. Bot. 70:64–73.

Dute, R. R., and R. F. Evert. 1978. Sieve-element ontogeny in the aerial shoot of *Equisetum hyemale* L. Ann. Bot. 42: 23–32.

Eames, A. J. 1936. *Morphology of Vascular Plants: Lower Groups.* McGraw-Hill: New York.

Eames, A. J., and L. H. MacDaniels. 1947. *An Introduction to Plant Anatomy.* 2nd ed. McGraw-Hill: New York.

Eastman, A., and R. L. Peterson. 1985. Root apex structure in *Regnellidium diphyllum* (Marsileaceae). Bot. Gaz. 146: 44–55.

Edwards, D., D. S. Edwards, and R. Rayner. 1982. The cuticle of early vascular plants and its evolutionary significance. In *The Plant Cuticle*, eds. D. F. Cutler, K. L. Alvin, and C. E. Price. Academic Press: London.

Edwards, M., and H. Meidner. 1979. Direct measurements of turgor pressure potentials. IV. Naturally occurring pressures in guard cells and their relation to solute and matric potentials in the epidermis. J. Exptl. Bot. 30:829–837.

Eggert, D. A., and D. D. Gaunt. 1973. Phloem of *Sphenophyllum*. Amer. J. Bot. 60:755–770.

Eggert, D. A., and N. Y. Kanemoto. 1977. Stem phloem of a Middle Pennsylvanian *Lepidodendron*. Bot. Gaz. 138:102–111.

Eggli, U. 1984. Stomatal types of Cactaceae. Pl. Syst. Evol. 146:197–214.

Egley, G. H., and R. N. Paul, Jr. 1981. Morphological observations on the early imbibition of water by *Sida spinosa* (Malvaceae) seed. Amer. J. Bot. 68:1056–1065.

Eglinton, G., and R. J. Hamilton. 1967. Leaf epicuticular waxes. Science 156:1322–1335.

El Hadidi, M. N. 1969. Observations on the wound-healing process in some flowering plants. Mikroskopie 25:54–69.

Elisens, W. J. 1985. The systematic significance of seed coat anatomy among New World species of Tribe Antirrhineae (Scrophulariaceae). Syst. Bot. 10:282–299.

Eller, B. M., and A. Nipkow. 1983. Diurnal course of the temperature in a *Lithops* sp. (Mesembryanthemaceae Fenzl) and its surrounding soil. Plant, Cell and Environ. 6:559–565.

Elliott, J. H. 1937. The development of the vascular system in evergreen leaves more than one year old. Ann. Bot. N. S. 1:107–127.

Ellis, R. P. 1976. A procedure for standardizing comparative leaf anatomy in the Poaceae. I. The leaf-blade as viewed in transverse section. Bothalia 12:65–109.

Ellmore, G. S. 1981. Root dimorphism in *Ludwigia peploides* (Onagraceae): development of two root types from similar primordia. Bot. Gaz. 142:525–533.

Endress, P. K. 1980. The reproductive structures and systematic position of the Austrobaileyaceae. Bot. Jahrb. Syst. 101:393–433.

Engler, A. 1895. Über die Amphicarpie bei *Fleurya podocarpa* Wedd., nebst einigen allgemeinen Bemerkungen über Erscheinung der Amphicarpie und Geocarpie. Sitzungsber. König. Preuss. Akad. Wiss. 5:1–10.

Enright, A. M., and B. G. Cumbie. 1973. Stem anatomy and internodal development in *Phaseolus vulgaris*. Amer. J. Bot. 60:915–922.

Eremin, V. M. 1980. Anatomy of *Pseudolarix kaempferi* bark. Biol. Nauki. (Mosc.) 0:74–79.

Esau, K. 1936. Ontogeny and structure of collenchyma and of vascular tissues in celery petioles. Hilgardia 10:431–476.

Esau, K. 1940. Developmental anatomy of the fleshy storage organ of *Daucus carota*. Hilgardia 13:175–226.

Esau, K. 1943a. Vascular differentiation in the pear root. Hilgardia 15:299–324.

Esau, K. 1943b. Ontogeny of the vascular bundle in *Zea mays*. Hilgardia 15:327–368.

Esau, K. 1943c. Origin and development of primary vascular tissues in seed plants. Bot. Rev. 9:125–206.

Esau, K. 1948. Phloem structure in the grapevine, and its seasonal changes. Hilgardia 18:217–296.

Esau, K. 1965a. *Plant Anatomy*. John Wiley and Sons: New York.

Esau, K. 1965b. *Vascular Differentiation in Plants*. Holt, Rinehart & Winston: New York.

Esau, K. 1967. Minor veins in *Beta* leaves: Structure related to function. Proc. Am. Phil. Soc. 111:219–233.

Esau, K. 1969. The Phloem. In *Encyclopedia of Plant Anatomy*. ed. W. Zimmermann, P. Ozenda, H. D. Wulff. vol. 5, part 2. Gbdr. Borntraeger. Berlin.

Esau, K. 1979. Phloem. In *Anatomy of the Dicotyledons*, eds. C. R. Metcalfe and L. Chalk, vol. I. Clarendon Press: Oxford.

Esau, K., and I. Charvat. 1978. On vessel member differentiation in the bean (*Phaseolus vulgaris* L.). Ann. Bot. 42:665–677.

Esau, K., and V. I. Cheadle. 1959. Size of pores and their contents in sieve elements of dicotyledons. Proc. Natl. Acad. Sci. 45:156–162.

Esau, K., and V. I. Cheadle. 1969. Secondary growth in *Bougainvillea*. Ann. Bot. 33:807–819.

Esau, K., and V. I. Cheadle. 1984. Anatomy of the secondary phloem in Winteraceae. Int. Assoc. Wood Anat. Bull. 5:13–43.

Esau, K., V. I. Cheadle, and E. B. Risley. 1962. Development of sieve-plate pores. Bot. Gaz. 123:233–243.

Esau, K., and J. Cronshaw. 1967. Tubular components in cells of healthy and tobacco mosaic virus-infected *Nicotiana*. Virology 33:26–35.

Esau, K., and R. H. Gill. 1971. Aggregation of endoplasmic reticulum and its relation to the nucleus in a differentiating sieve element. J. Ultrastruct. Res. 34:144–158.

Esau, K., and W. M. B. Hewitt. 1940. Structure of end wall in differentiating vessels. Hilgardia 13:229–244.

Esau, K., and J. Thorsch. 1984. The sieve plate of *Echium* (Boraginaceae). Developmental aspects and response of P-protein to protein digestion. J. Ultrastruct. Res. 86:31–45.

Esau, K., and J. Thorsch. 1985. Sieve plate pores and plasmodesmata, the communication channels of the symplast: Ultrastructural aspects and developmental relations. Amer. J. Bot. 72:1641–1653.

Eschrich, W. 1963. Beziehung zwischen dem Auftreten

von Callose und der Feinstruktur des primären Phloems bei *Cucurbita ficifolia*. Planta 59:243–261.

Eskarous, J. K., H. M. Habib, and H. E. Sweiha. 1984. The histology of the necrotic lesions induced by tomato mosaic virus on tobacco (*Nicotiana tabacum* var. *angustifolia*) and *Datura metel* leaves. Z. Pflanzenkr. Pflanzenschutz. 91: 12–19.

Espinosa, R. 1932. Ökologische Studien über Kordillerenpflanzen (morphologisch und anatomisch dargestellt). Bot. Jb. 65:120–211.

Evenari, M., A. Kadouri, and Y. Gutterman. 1977. Ecophysiological investigations on the amphicarpy of *Emex spinosa* (L.) Campd. Flora 166:223–238.

Evert, R. F. 1961. Some aspects of cambial development in *Pyrus communis*. Amer. J. Bot. 48:479–488.

Evert, R. F. 1963a. Ontogeny and structure of the secondary phloem in *Pyrus malus*. Amer. J. Bot. 50:8–37.

Evert, R. F. 1963b. The cambium and seasonal development of the phloem of *Pyrus malus*. Amer. J. Bot. 50:149–159.

Evert, R. F. 1977. Phloem structure and histochemistry. Ann. Rev. Plant Physiol. 28:199–222.

Evert, R. F. 1978. Leaf structure in relation to solute transport and phloem loading in *Zea mays* L. Planta 138: 279–294.

Evert, R. F. 1982. Sieve-tube structure in relation to function. BioScience 32:789–795.

Evert, R. F., J. T. Davis, C. M. Tucker, and F. J. Alfieri. 1970. On the occurrence of nuclei in mature sieve elements. Planta 95:281–296.

Evert, R. F., and W. F. Derr. 1964. Callose substance in sieve elements. Amer. J. Bot. 51:552–559.

Evert, R. F., and B. P. Deshpande. 1970a. Nuclear P-protein in sieve elements of *Tilia americana*. J. Cell Biol. 44:463–466.

Evert, R. F., and B. P. Deshpande. 1970b. An ultrastructural study of cell division in the cambium. Amer. J. Bot. 57:942–961.

Evert, R. F., B. P. Deshpande, and S. E. Eichhorn. 1971. Lateral sieve-area in woody dicotyledons. Can. J. Bot. 49:1509–1515.

Evert, R. F., and S. E. Eichhorn. 1974. Sieve-element ultrastructure in *Platycerium bifurcatum* and some other polypodiaceous ferns: The refractive spherules. Planta 119: 319–334.

Evert, R. F., W. Eschrich, and S. E. Eichhorn. 1973. P-protein distribution in mature sieve elements of *Cucurbita maxima*. Planta 109:193–210.

Evert, R. F., W. Eschrich, and W. Heyser. 1977. Distribution and structure of the plasmodesmata in mesophyll and bundle-sheath cells of *Zea mays* L. Planta 136:77–89.

Ewers, F. W. 1982. Secondary growth in needle leaves of *Pinus longaeva* (bristlecone pine) and other conifers: Quantitative data. Amer. J. Bot. 69:1552–1559

Ewers, F. W., and R. Schmid. 1985. The fate of the dwarf shoot apex in bristlecone pine (*Pinus longaeva*). Amer. J. Bot. 72:509–513.

Fabbri, F., and L. M. Bonzi. 1975. Observations on nuclear pores of freeze-etched spore mother cells of *Psilotum nudum* (L.) Beauv. during early stages of meiotic prophase. I. Preliminary report. Caryologia 28:549–559.

Facey, V. 1950. Abscission of leaves in *Fraxinus americana* L. New Phytol. 49:103–116.

Fagerberg, W. R. 1980. Stereology; quantitative electron micrograph analysis. In *The Handbook of Phycological Techniques*, vol. 3, ed. E. Gantt. Cambridge University Press: Cambridge.

Fagerberg, W. R., and G. Culpepper. 1984. A morphometric study of anatomical changes during sunflower leaf development under low light. Bot. Gaz. 145:346–350.

Fagerlind, F. 1947. Gynöceummorphologische und embryologische Studien in der Familie Olaeaceae. Bot. Notiser 1947:207–230.

Fahn, A. 1953. Metaxylem elements in some families of the Monocotyledonae. New Phytol. 53:530–540.

Fahn, A. 1954. The anatomical structure of the Xanthorrhoeaceae Dumort. Bot. J. Linn. Soc. 55:158–184.

Fahn, A. 1979. *Secretory Tissues in Plants*. Academic Press: New York.

Fahn, A. 1982. *Plant Anatomy*. Pergamon Press: Oxford.

Fahn, A., and N. Arnon. 1963. The living wood fibres of *Tamarix aphylla* and the changes occurring in them in transition from sapwood to heartwood. New Phytol. 62: 99–104.

Fahn, A., and T. Arzee. 1959. Vascularization of articulated Chenopodiaceae and the nature of their fleshy cortex. Amer. J. Bot. 46:330–338.

Fahn, A., and I. W. Bailey. 1957. The nodal anatomy and the primary vascular cylinder of the Calycanthaceae. J. Arnold Arbor. 38:107–117.

Fahn, A., and P. Benouaiche. 1979. Ultrastructure of the nectary of *Musa paradisiaca* L. var. *sapientum* Kuntze and its relation to nectar secretion. Ann. Bot. 44:85–93.

Fahn, A., R. Ben-Sasson, and T. Sachs. 1972. The relation between the procambium and the cambium. In *Research Trends in Plant Anatomy*, eds. A. K. M. Ghouse and M. Yunus. Tata McGraw-Hill: New Delhi.

Fahn, A., and N. Dembo. 1964. Structure and development of the epidermis in articulated Chenopodiaceae. Israel J. Bot. 13:177–192.

Fahn, A., and R. F. Evert. 1974. Ultrastructure of the secretory ducts of *Rhus glabra* L. Amer. J. Bot. 61:1–14.

Fahn, A., and B. Leshem. 1963. Wood fibres with living protoplasts. New Phytol. 62:91–98.

Fahn, A., and T. Rachmilevitz. 1970. Ultrastructure and nectar secretion in *Lonicera japonica*. In *New Research in Plant Anatomy*, eds. N. K. B. Robson, D. F. Cutler, and

M. Gregory. Bot. J. Linn. Soc. 63:51–56 and Academic Press: London.

Fahn, A., and T. Rachmilevitz. 1975. An autoradiological study of nectar secretion in *Lonicera japonica* Thumb. Ann. Bot. 39:975–976.

Fahn, A., and Y. Shchori. 1967. The organization of the secondary conducting tissues in some species of the Chenopodiaceae. Phytomorphology 17:147–154.

Fahn, A., and E. Werker. 1972. Anatomical mechanisms of seed dispersal. In *Seed Biology*, ed. T. T. Kozlowski. 1: 151–221. Academic Press: New York.

Fahn, A., and M. H. Zimmermann. 1982. Development of the successive cambia in *Atriplex halimus* (Chenopodiaceae). Bot. Gaz. 143:353–357.

Fairbairn, J. W., and L. D. Kapoor. 1960. The laticiferous vessels of *Papaver somniferum*. Planta Med. 8:49–61.

Farmer, J. B., and T. G. Hill. 1902. On the arrangement and structure of the vascular strands in *Angiopteris evecta*, and some other Marattiaceae. Ann. Bot. 16:371–402.

Feldhofen, E. 1933. Beiträge zur physiologischen Anatomie der nuptialen Nektarien aus den Reihen der Dikotylen. Beih. bot. Zbl. I. Abt. 50:459–634.

Feldman, L. J., and J. G. Torrey. 1976. The isolation and culture in vitro of the quiescent center of *Zea mays*. Amer. J. Bot. 63:345–355.

Ferguson, I. B., and D. T. Clarkson. 1975. Ion transport and endodermal suberization in the roots of *Zea mays*. New Phytol. 75:69–79.

Ferguson, M. C. 1904. Contributions to the knowledge of the life history of *Pinus* with special reference to sporogenesis, the development of the gametophytes, and fertilization. Proc. Wash. Acad. Sci. 6:1–202.

Ferguson, M. C. 1927. A cytological and genetical study of *Petunia*. I. Bull. Torrey Bot. Club 54:657–664.

Feuer, S., C. J. Niezgoda, and L. I. Nevling. 1985. Ultrastructure of *Parkia* polyads (Mimosoideae: Leguminosae). Amer. J. Bot. 72:1871–1890.

Feustel, H. 1921. Anatomie und Biologie der Gymnospermenblätter. Bot. Centbl. Beihefte 38:177–257.

Filla, F. 1926. Das Perikarp der Proteaceae. Flora 120: 99–142.

Fincher, G. B., and B. A. Stone. 1981. Metabolism of noncellulosic polysaccharides. In *Encyclopedia of Plant Physiology*, new series, vol. 13B, eds. W. Tanner and R. A. Loewus. Springer-Verlag: Berlin.

Fineran, B. A. 1963. Studies on the root parasitism of *Exocarpus bidwillii* Hook.f. IV. Structure of the mature haustorium. Phytomorph. 13:249–267.

Fineran, B. A., D. J. C. Wild, and M. Ingerfeld. 1982. Initial wall formation in the endosperm of wheat, *Triticum aestivum*: A reevaluation. Can. J. Bot. 60:1776–1795.

Fink, S. 1983. The occurrence of adventitious and preventitious buds within the bark of some temperate and tropical trees. Amer. J. Bot. 70:532–542.

Fink, S. 1984. Some cases of delayed or induced development of axillary buds from persisting detached meristems in conifers. Amer. J. Bot. 71:44–51.

Fisher, D. B. 1970. Kinetics of C-14 translocation in soybean. III. Theoretical considerations. Plant Physiol. 45: 119–125.

Fisher, D. G. 1985. Morphology and anatomy of the leaf of *Coleus blumei* (Lamiaceae). Amer. J. Bot. 72:392–406.

Fisher, D. G., and R. F. Evert. 1979. Endoplasmic reticulum-dictyosome involvement in the origin of refractive spherules in sieve elements of *Davallia fijiensis* Hook. Ann. Bot. 43:255–258.

Fisher, D. G., and R. F. Evert. 1982a. Studies on the leaf of *Amaranthus retroflexus* (Amaranthaceae): Quantitative aspects, and solute concentration in the phloem. Amer. J. Bot. 69:1375–1388.

Fisher, D. G., and R. F. Evert. 1982b. Studies on the leaf of *Amaranthus retroflexus* (Amaranthaceae): Ultrastructure, plasmodesmatal frequency, and solute concentration in relation to phloem loading. Planta 155:377–387.

Fisher, D. G., and P. R. Larson. 1983. Structure of leaf/branch gap parenchyma and associated vascular tissues in *Populus deltoides*. Bot. Gaz. 144:73–85.

Fisher, J. B. 1974. Axillary and dichotomous branching in the palm *Chamaedorea*. Amer. J. Bot. 61:1046–1056.

Fisher, J. B. 1976. Development of dichotomous branching and axillary buds in *Strelitzia* (Monocotyledoneae). Can. J. Bot. 54:578–592.

Fisher, J. B. 1981. Wound healing by exposed secondary xylem in *Adansonia digitata* (Bombacaceae). IAWA Bull. 2:193–199.

Fisher, J. B., and P. B. Tomlinson. 1972. Morphological studies in *Cordyline* (Agavaceae). II. Vegetative morphology of *Cordyline terminalis*. J. Arnold Arbor. 53:113–127.

Floresta, P. la. 1905. Ricerche sul periderma delle Palme. Cont. Biol. Veg. Palermo 3:333–354.

Florin, R. 1931. Untersuchungen zur Stammesgeschichte der Coniferales und Cordaitales. K. svenska Vetensk. Acad. Handl. ser. 5. 10:1–588.

Florin, R. 1944. Die Koniferen des Oberkarbons und des unteren Perms. Paleontographica 85B:365–654.

Florin, R. 1950. Upper Carboniferous and Lower Permian Conifers. Bot. Rev. 16:258–282.

Florin, R. 1951. Evolution in Cordaites and conifers. Acta Horti Bergiani 15:285–388.

Folsom, M. W., and C. M. Peterson. 1984. Ultrastructural aspects of the mature embryo sac of soybean, *Glycine max* (L.) Merr. Bot. Gaz. 145:1–10.

Ford, E. S. 1942. Anatomy and histology of the Eureka lemon. Bot. Gaz. 104:288–305.

Forde, B. J. 1965. Differentiation and continuity of the phloem in the leaf intercalary meristem of *Lolium perenne*. Amer. J. Bot. 52:953–961.

Forer, A. 1982. Possible roles of cytoskeletal elements in mitosis. In *The Cytoskeleton in Plant Growth and Development*, ed. C. W. Lloyd. Academic Press: London.

Forsaith, C. C. 1926. The technology of New York State timbers. Tech. Pub. 18, vol. 26. Syracuse, New York: N. Y. State College of Forestry, Syracuse University.

Foster, A. S. 1928. Salient features of the problem of bud-scale morphology. Biol Rev. 3:123–164.

Foster, A. S. 1932. Investigations on the morphology and comparative history of development of foliar organs. III. Cataphyll and foliage-leaf ontogeny in the black hickory (*Carya buckleyi* var. *arkansana*). Amer. J. Bot. 19:75–99.

Foster, A. S. 1935a. A histogenetic study of foliar determination in *Carya buckleyi* var. *arkansana*. Amer. J. Bot. 22:88–147.

Foster, A. S. 1935b. Comparative histogenesis of foliar transition forms in *Carya*. Univ. California Publ. Bot. 19: 159–186.

Foster, A. S. 1936. Leaf differentiation in angiosperms. Bot. Rev. 2:349–372.

Foster, A. S. 1937. Structure and behavior of the marginal meristems in the bud scales of *Rhododendron*. Amer. J. Bot. 24:304–316.

Foster, A.S. 1938. Structure and growth of the shoot apex in *Ginkgo biloba*. Bull. Torrey Bot. Club 65:531–556.

Foster, A. S. 1940. Further studies on zonal structure and growth of the shoot apex of *Cycas revoluta* Thunb. Amer. J. Bot. 27:487–501.

Foster, A. S. 1941. Comparative studies on the structure of the shoot apex in seed plants. Bull. Torrey Bot. Club 68: 339–350.

Foster, A. S. 1945a. Origin and development of sclereids in the foliage leaf of *Trochodendron aralioides* Sieb. and Zucc. Amer. J. Bot. 32:456–468.

Foster, A. S. 1945b. The foliar sclereids of *Trochodendron aralioides* Sieb. and Zucc. J. Arnold Arbor. 26:155–162.

Foster, A. S. 1946. Comparative morphology of the foliar sclereids in the genus *Mouriria* Aubl. J. Arnold Arbor. 27:253–271.

Foster, A. S. 1950. *Practical plant anatomy*. van Nostrand: New York.

Foster, A. S. 1956. Plant idioblasts: Remarkable examples of cell specialization. Protoplasma 46:184–193.

Foster, A. S., and E. M. Gifford, Jr. 1959. *Comparative Morphology of Vascular Plants*. W. H. Freeman: San Francisco.

Foster, A. S., and E. M. Gifford, Jr. 1974. *Comparative Morphology of Vascular Plants*. 2nd. ed. W. H. Freeman: San Francisco.

Foster, R. C. 1967. Fine structure of tyloses in three species of Myrtaceae. Aust. J. Bot. 15:25–34.

Foster, R. C., and A. D. Rovira. 1976. Ultrastructure of wheat rhizosphere. New Phytol. 76:343–352.

Foster, R. C., A. D. Rovira, and T. W. Cock. 1983. *Ultrastructure of the Root-Soil Interface*. The Americal Phytopathological Society. St. Paul, Minnesota.

Franceschi, V.R., and R. T. Giaquinta. 1983a. The paraveinal mesophyll of soybean leaves in relation to assimilate transfer and compartmentation. I. Ultrastructure and histochemistry during vegetative development. Planta 157:411–421.

Franceschi, V. R., and R. T. Giaquinta. 1983b. The paraveinal mesophyll of soybean leaves in relation to assimilate transfer and compartmentation. II. Structural, metabolic, and compartmental changes during seed filling. Planta 157:422–431.

Francheschi, V. R., and H. T. Horner, Jr. 1980. Calcium oxalate crystals in plants. Bot. Rev. 46:361–427.

Franck, D. H. 1979. Development of vein pattern in leaves of *Ostrya virginiana* (Betulaceae). Bot. Gaz. 140:77–83.

Franke, W. 1971a. The entry of residues into plants via ectodesmata (ectocythodes). Residue Rev. 38:81–115.

Franke, W. 1971b. Uber die Natur der Ectodesmen und einen Vorschlag zur Terminologie. Ber. Dtsch. Bot. Ges. 84:533–537.

Franke, W. W., W. Herth, W. J. Van Der Woude, and D. J. Morré. 1972. Tubular and filamentous structures in pollen tubes: Possible involvement as guide elements in protoplasmic streaming and vectorial migration of secretory vesicles. Planta 105:317–341.

Frankenberg, J. M., and D. A. Eggert. 1969. Petrified *Stigmaria* from North America: Part 1. *Stigmaria ficoides*, the underground portions of the Lepidodendraceae. Palaeontographica 128B:1–47.

Frei, E. 1955. Die Innervierung der floralen Nektarien dikotyler Pflanzenfamilien. Berschweiz. bot. Ges. 65: 60–114.

French, J. C. 1985. Patterns of endothecial wall thickenings in Araceae: Subfamilies Pothoideae and Monsteroideae. Amer. J. Bot. 72:472–486.

French, J. C., K. Clancy, and P. B. Tomlinson. 1983. Vascular patterns in stems of the Cyclanthaceae. Amer. J. Bot. 70:1386–1400.

French, J. C., and P. B. Tomlinson. 1981a. Vascular patterns in stems of Araceae: Subfamilies Calloideae and Lasioideae. Bot. Gaz. 142:366–381.

French, J. C., and P. B. Tomlinson. 1981b. Vascular patterns in stems of Araceae: Subfamily Pothoideae. Amer. J. Bot. 68:713–729.

French, J. C., and P. B. Tomlinson. 1981c. Vascular patterns in stems of Araceae: Subfamily Monsteroideae. Amer. J. Bot. 68:1115–1129.

French, J. C., and P. B. Tomlinson. 1981d. Vascular patterns in stems of Araceae: Subfamily Philodendroideae. Bot. Gaz. 142:550–563.

Frey-Wyssling, A. 1935. Die Stoffausscheidung der hoheren Pflanzen. Julius Springer Verlag: Berlin.

Frey-Wyssling, A. 1941. Die Guttation als allgemeine Erscheinung. Ber. schweiz. bot. Ges. 51:321–325.

Frey-Wyssling, A. 1959. *Die Pflanzliche Zellwand*. Springer-Verlag: Berlin.

Frey-Wyssling, A. 1972. Elimination processes in higher plants. Saussurea. 3:79–90.

Frey-Wyssling, A. 1976. The plant cell wall. In *Encyclopedia of Plant Anatomy*, vol. III, part 4. Gbdr. Borntraeger: Berlin.

Frey-Wyssling, A., and H. H. Bosshard. 1959. Cytology of ray cells in sapwood and heartwood. Holzforschung 13: 129–137.

Frey-Wyssling, A., and K. Mühlethaler. 1959. Über das submikroskopische Geschehen bei der Kutinisierung pflanzlicher Zellwände. Vierteljahrsschrift Naturforsch. Gesell. Zurich 104:294–299.

Frison, E. 1948. De la présence dámidon dans le lumen des fibres du bois. Bull. agric. congo belge. 39:869–874.

Fritz, F. 1935. Über Kutikula von Aloë-und Gasteriaarten. Jb. wiss. Bot. 81:718–746.

Fritz, F. 1937. Untersuchungen über die Kutinisierung der Zellmembranen und den rhythmischen Verlauf dieses Vorganges. Planta 26:693–704.

Frydman, V., and P. F. Wareing. 1974. Phase change in *Hedera helix* L. III. The effects of gibberellins, abscisic acid and growth retardants on juvenile and adult ivy. J. Expt. Bot. 25:420–429.

Fryns-Claessens, E., and W. van Cotthem. 1973. A new classification of the ontogenetic types of stomata. Bot. Rev. 39:71–138.

Fuchs, C. 1975. Ontogenèse foliaire et acquisition de la forme chez le *Tropaeolum peregrinum* L. Annls. Sci. nat. Bot. ser. 12, 16:321–389.

Fuhrman, M. H., and W. L. Koukkari. 1981. Anatomical and physiological characteristics of the petiole of *Abutilon theophrasti* in relation to circadian leaf movements. Physiol. Plant. 51:309–313.

Fujita, T. 1961. Multinucleate endosperm cells in *Cephalotaxus drupacea* Siebold et Zuccarini. J. Jap. Bot. 36:29.

Gahan, P. B. 1981. An early cytochemical marker of commitment to stelar differentiation in meristems from dicotyledonous plants. Ann. Bot. 48:769–775.

Galatis, B., and K. Mitrakos. 1980. The ultrastructural cytology of the differentiating guard cells of *Vigna sinensis*. Amer. J. Bot. 67:1243–1261.

Galil, J., and M. Zeroni. 1969. On the organization of the pollinium in *Asclepias curassavica*. Bot. Gaz. 130:1–4.

Gambles, R. L., and R. E. Dengler. 1982a. The anatomy of the leaf of red pine, *Pinus resinosa*. I. Nonvascular tissues. Can. J. Bot. 60:2788–2803.

Gambles, R. L., and R. E. Dengler. 1982b. The anatomy of the leaf of red pine, *Pinus resinosa*. II. Vascular tissues. Can. J. Bot. 60:2804–2824.

Garcin, A. G. 1890. Recherches sur l'histogénèse des péricarpes charnus. Thesis, Paris (G. Masson Edit.)

Gardiner, W. 1883. On the physiological significance of water-glands and nectaries. Phil. Soc. Proc. 5:35–50.

Gardner, J. S., W. M. Hess, and E. J. Trione. 1985. Development of the young wheat spike: A SEM study of Chinese spring wheat. Amer. J. Bot. 72:548–559.

Garrison, R., and R. H. Wetmore. 1961. Studies in shoot-tip abortion: *Syringa vulgaris*. Amer. J. Bot. 48:789–795.

Garwood, N. C. 1985. The role of mucilage in the germination of cuipo, *Cavanillesia platanifolia* (H. and B.) H. B. K. (Bombacaceae), a tropical tree. Amer. J. Bot. 72:1095–1105.

Gates, D. M. 1980. *Biophysical Ecology*. Springer-Verlag: Heidelberg.

Gates, R. R., and J. Latter. 1927. Observations on the pollen development of two species of *Lathraea*. J. Roy. Micros. Soc. 1927:209–224.

Gates, R. R., and E. M. Rees. 1921. A cytological study of pollen development in *Lactuca*. Ann. Bot. 16:123–148.

Gaudet, J. 1960. Ontogeny of foliar sclereids in *Nymphaea odorata*. Amer. J. Bot. 47:525–532.

Gawlik, S. R. 1984. An ultrastructural study of transmitting tissue development in the pistil of *Lilium leucanthum*. Amer. J. Bot. 71:512–521.

Geesteranus, R. A. M. 1941. On the development of the stellate form of the pith cells of *Juncus* species. Proc. Nederl. Akad. Wetensch. 44:489–501, 648–653.

Gerdemann, J. W. 1974. Mycorrhizae. In *The Plant Root and Its Environment*, ed. E. W. Carson. University Press of Virginia: Charlottesville, Virginia.

Ghouse, A. K. M. 1974. Transfusion tissue in the leaves of *Taxus baccata* L. Cellule 70:159–162.

Ghouse, A. K. M., and S. Hashmi. 1979. Longevity of phloem in *Polyalthia longifolia* Benth. and Hook. Bull. Torrey Bot. Club 106:182–184.

Ghouse, A. K. M., and S. Hashmi. 1980a. Longevity of secondary phloem in *Delonix regia*. Proc. Indian Acad. Sci., Plant Sci. 89:67–72.

Ghouse, A. K. M., and S. Hashmi. 1980b. Changes in the vascular cambium of *Polyalthia longifolia* Benth. and Hook. (Annonaceae) in relation to the girth of the tree. Flora 170:135–143.

Ghouse, A. K. M., and M. Yunus. 1973. Some aspects of cambial development in the shoots of *Dalbergia sissoo* Roxb. Flora 162:549–558.

Ghouse, A. K. M., and M. Yunus. 1974a. The ratio of ray and fusiform initials in some woody species of the Ranalian complex. Bull. Torrey Bot. Club 101:363–366.

Ghouse, A. K. M., and M. Yunus. 1974b. Cambial structure in *Dalbergia*. Phytomorph. 24:152–158.

Ghouse, A. K. M., and M. Yunus. 1974c. Transfusion tissue in the leaves of *Cunninghamia lanceolata* (Lambert) Hooker (Taxodiaceae). J. Linn. Soc. Bot. 69:147–151.

Ghouse, A. K. M., and M. Yunus. 1975. Transfusion tissue in the leaves of *Thuja orientalis*. L. Ann. Bot. 39:225–227.

Ghouse, A. K. M., and M. Yunus. 1976. Ratio of ray and fusiform initials in the vascular cambium of certain leguminous trees. Flora 165:23–28.

Gianordoli, M. 1974. A cytological investigation on gametes and fecundation among *Cephalotaxus drupacea*. In *Fertilization in Higher Plants,* ed. H. F. Linskens. North Holland: Amsterdam.

Giaquinta, R. 1980. Mechanism and control of phloem loading of sucrose. Ber. Dtsch. Bot. Ges. 93:187–201.

Giaquinta, R. T., W. Lin, N. Sadler, and V. R. Franceschi. 1983. Pathway of phloem unloading of sucrose in corn roots. Plant Physiol. 72:362–367.

Gibson, A. C. 1973. Comparative anatomy of secondary xylem in Cactoideae (Cactaceae). Biotropica 5:29–65.

Gibson, A. C. 1976. Vascular organization in shoots of Cactaceae. I. Development and morphology of primary vasculature in Pereskioideae and Opuntioideae. Amer. J. Bot. 63:414–426.

Gibson, A. C. 1977. Wood anatomy of opuntias with cylindrical to globular stems. Bot. Gaz. 138:334–351.

Gibson, A. C. 1978a. Wood anatomy of platyopuntias. Aliso 9:279–307.

Gibson, A. C. 1978b. Dimorphism of secondary xylem in two species of cacti. Flora 167:403–408.

Gibson, A. C. 1981. Vegetative anatomy of *Pachycormus* (Anacardiaceae). Bot. J. Linn. Soc. 83:273–284.

Gibson, A. C. 1983. Anatomy of photosynthetic old stems of nonsucculent dicotyledons from North American deserts. Bot. Gaz. 144:347–362.

Gibson, A. C., H. W. Calkin, and P. S. Nobel. 1984. Xylem anatomy, water flow, and hydraulic conductance in the fern *Cyrtomium falcatum*. Amer. J. Bot. 71:564–574.

Gibson, A. C., and P. S. Nobel. 1986. *The Cactus Primer.* Harvard University Press: Cambridge, Massachusetts.

Giddings, T. H., Jr., D. L. Brower, and L. A. Staehelin. 1980. Visualization of particle complexes in the plasma membrane of *Micrasterias denticulata* associated with the formation of cellulose fibrils in primary and secondary cell walls. J. Cell Biol. 84:327–339.

Gifford, E. M., Jr. 1954. The shoot apex in angiosperms. Bot. Rev. 20:477–529.

Gifford, E. M., Jr. 1983. Concept of apical cells in bryophytes and pteridophytes. Ann. Rev. Plant Physiol. 34:419–440.

Gifford, E. M., Jr., and E. Kurth. 1982. Quantitative studies of the root apical meristem of *Equisetum scirpoides*. Amer. J. Bot. 69:464–473.

Gifford, E. M., Jr., and V. S. Polito. 1981. Mitotic activity at the shoot apex of *Azolla filiculoides*. Amer. J. Bot. 68:1050–1055.

Gifford, E. M., Jr., and K. D. Stewart. 1967. Ultrastructure of the shoot apex of *Chenopodium album* and certain other seed plants. J. Cell Biol. 33:131–142.

Gilbert, S. G. 1940. Evolutionary significance of ring porosity in woody angiosperms. Bot. Gaz. 102:105–120.

Gilchrist, A. J., and B. E. Juniper. 1974. An excitable membrane in the stalked glands of *Drosera capensis* L. Planta 119:143–147.

Gill, A. M., and P. B. Tomlinson. 1969. Studies on the growth of red mangrove (*Rhizophora mangle* L.). 1. Habit and general morphology. Biotropica 1:1–9.

Gill, A. M., and P. B. Tomlinson. 1971. Studies on the growth of red mangrove (*Rhizophora mangle* L.). 2. Growth and differentiation of aerial roots. Biotropica 3:63–77.

Gill, A. M., and P. B. Tomlinson. 1977. Studies on the growth of red mangrove (*Rhizophora mangle* L.). 4. The adult root system. Biotropica 9:145–155.

Gilliland, M. G., C. H. Bornman, and F. T. Addicott. 1976. Ultrastructure and acid phosphatase in pedicel abscission of *Hibiscus*. Amer. J. Bot. 63:925–935.

Ginzburg, C. 1963. Some anatomic features of splitting of desert shrubs. Phytomorph. 13:92–97.

Goebel, K. 1928. *Organographie der Pflanzen.* Teil 1. Allgemeine Organographie. Gustav Fischer: Jena, Germany.

Goffinet, M. C., and P. R. Larson. 1982a. Xylary union between the new shoot and old stem during terminal bud break in *Populus deltoides*. Amer. J. Bot. 69:432–446.

Goffinet, M. C., and P. R. Larson. 1982b. Lamina abortion in terminal bud-scale leaves of *Populus deltoides* during dormancy induction. Bot. Gaz. 143:331–340.

Gómez, L. D. 1974. Biology of the potato-fern *Solanopteris brunei*. Brenesia 4:37–61.

Gómez-Vazquez, B. G., and E. M. Engleman. 1984. Bark anatomy of *Bursera longipes* and *Bursera copallifera*. IAWA Bull. 5:335–340.

Good, C. W., and T. N. Taylor. 1972. The ontogeny of Carboniferous articulates: The apex of *Sphenophyllum*. Amer. J. Bot. 59:617–626.

Goodchild, D. G., and F. J. Bergersen. 1966. Electron microscopy of the infection and subsequent development of soybean nodule cells. J. Bacteriol. 92:204–213.

Goodwin, T. W., and E. I. Mercer. 1983. *Introduction to Plant Biochemistry.* Pergamon Press: Oxford.

Goosen-De Roo, L. 1973a. The relationship between cell organelles and cell wall thickenings in primary tracheary elements of the cucumber. I. Morphological aspects. Acta Bot. Neerl. 22:279–300.

Goosen-De Roo, L. 1973b. The relationship between cell organelles and cell wall thickenings in primary tracheary elements of the cucumber. II. Quantitative aspects. Acta Bot. Neerl. 22:301–320.

Goswami, H. K., and S. Khandelwal. 1980. Observations on *Helminthostachys ophioglossales*. Acta Bot Neerl. 29:199–202.

Gottwald, H., and N. Parameswaran. 1964. Vielfache Gefassdurchbrechungen in der Familie Dipterocarpaceae. Z. Bot. 52:321–334.

Graham, L. E. 1982. The occurrence, evolution, and phylogenetic significance of parenchyma in *Coleochaete* Bréb. (Chlorophyta). Amer. J. Bot. 69:447–454.

Grassmann, P. 1884. Die Septaldrüsen. Ihre Verbreitung, Entstehung und Verrichtung. Flora. 67:113–136.

Gray, M. W., and W. F. Doolittle. 1982. Has the endosymbiont hypothesis been proven? Microbiol. Rev. 46:1–42.

Greaves, M. P., and J. F. Darbyshire. 1972. The ultrastructure of the mucilaginous layer on plant roots. Soil Biol. Biochem. 4:443–449.

Green, D. S. 1980. The terminal velocity and dispersal of spinning samaras. Amer. J. Bot. 67:1218–1224.

Green, P. B. 1980. Organogenesis—A biophysical review. Ann. Rev. Plant Physiol. 31:51–82.

Greenland, D. J. 1979. The physics and chemistry of the soil-root interface: some comments. In *The Soil-Root Interface*, eds. J. L. Harley and R. S. Russell. Academic Press: New York.

Gregory, R. A. 1978. Living elements of the conducting secondary xylem of sugar maple (*Acer saccharum* Marsh.). IAWA Bull. 1978:65–70.

Gregory, R. A., and J. A. Romberger. 1972. The shoot apical ontogeny of the *Picea abies* seedling. I. Anatomy, apical dome diameter,and plastochron duration. Amer. J. Bot. 59:587–597.

Grob, K., and P. Matile. 1979. Vacuolar location of glucosinolates in horseradish root cells. Plant Sci. Lett. 14:327–335.

Grote, M., and H. G. Fromme. 1984. Electron microscopic investigations of the cell structure in fresh and processed vegetables and green bean pods. Food Microstruct. 3:55–64.

Grove, A. R., 1941. Morphological study of *Agave lechuguilla*. Bot. Gaz. 103:354–365.

Guérin, P. 1927. Le développement de l'anthère et du pollen chez les Liliacées (*Sansevieria, Ophiopogon, Peliosanthes*). Bull. Soc. Bot. de France 74:102–107.

Guignard, J. L. 1961. Recherches sur l'embryogénie des Graminées; rapports des Graminées avec l'autres Monocotylédones. Ann. Sci. Nat.: Bot. Biol. Veg. 2:491–610.

Gunn, C. R., and J. V. Dennis. 1976. *World Guide to Tropical Drift Seeds and Fruits*. Demeter Press: New York.

Gunning, B. E. S. 1978. Age-related and origin-related control of the numbers of plasmodesmata in cell walls of developing *Azolla* roots. Planta 143:181–190.

Gunning, B. E. S. 1982. The cytokinetic apparatus: Its development and spatial regulation. In *The Cytoskeleton in Plant Growth and Development*, ed. C. W. Lloyd. Academic Press: London.

Gunning, B. E. S., and A. R. Hardham. 1982. Microtubules. Ann. Rev. Plant Physiol. 33: 651–698.

Gunning, B. E. S., and J. S. Pate. 1969. "Transfer cells"—Plant cells with wall ingrowths, specialized in relation to short distance transport of solutes: Their occurrence, structure and development. Protoplasma 68:107–33.

Gunning, B. E. S., and J. S. Pate. 1974. Transfer cells. In *Dynamic Aspects of Plant Ultrastructure*, ed. A. W. Robards. McGraw-Hill: London.

Gunning, B. E. S., J. S. Pate, and L. G. Briarty. 1968. Specialized "transfer cells" in minor veins of leaves and their possible significance in phloem translocation. J. Cell Biol. 37: C7-C12.

Gunning, B. E. S., J. S. Pate, and L. W. Green. 1970. Transfer cells in the vascular system of stems: Taxonomy, association with nodes, and structure. Protoplasma 71: 147–171.

Gunning, B. E. S., J. S. Pate, F. R. Minchin, and I. Marks. 1974. Quantitative aspects of transfer cell structure in relation to vein loading in leaves and solute transport in legume nodules. Symp. Soc. Exp. Biol. 28:87–126.

Gunning, B. E. S., and A. W. Robards. 1976. *Intercellular Communication in Plants: Studies on Plasmodesmata*. Springer-Verlag: Berlin.

Gunning, B. E. S., and M. W. Steer. 1975. *Ultrastructure and the Biology of Plant Cells*. Edward Arnold, Ltd.: London.

Gupta, S. C., and K. Nanda. 1978. Studies in the Bignoniaceae. I. Ontogeny of dimorphic anther tapetum in *Pyrostegia*. Amer. J. Bot. 65:395–399.

Haas, D. L., and Z. B. Carothers. 1975. Some ultrastructural observations on endodermal cell development in *Zea mays* roots. Amer. J. Bot. 62:336–348.

Haberlandt, G. 1886. Zur Anatomie und Physiologie der pflanzlichen Brennhaare. Sitzber. Akad. Wien. 93: 122–145.

Haberlandt, G. 1894. Über Bau and Funktion der Hydathoden. Ber. Dtsch. Bot. Ges. 12:367–378.

Haberlandt, G. 1914. *Physiological plant anatomy*. Translated from the 4th German edition by M. Drummond. MacMillan: London.

Haberlandt, G. 1918. *Physiologische Pflanzenanatomie*. 5th ed. W. Engelmann: Leipzig.

Hajibagheri, M. A., A. R. Yeo, and T. J. Flowers. 1985. Salt tolerance in *Suaeda maritima* (L.) Dum. Fine structure and ion concentrations in the apical region of roots. New Phytol. 99:331–343.

Hall, D. M. 1967a. The ultrastructure of wax deposits on plant leaf surfaces. II. Cuticular pores and wax formation. J. Ultrastruct. Res. 17:34–44.

Hall, D. M. 1967b. Wax microchannels in the epidermis of white clover. Science 158:505–506.

Hallam, N. D. 1970. Leaf wax fine structure and ontogeny in *Eucalyptus* demonstrated by means of a specialized fixation technique. J. Microscopy 92:137–144.

Hallé, F., and R. A. A. Oldemann. 1970. *Essai sur l'Architecture et Dynamique de Croissance des Arbres Tropicaux*. Masson et Cie.: Paris

Hallé, F., R. A. A. Oldemann, and P. B. Tomlinson. 1978. *Tropical Trees and Forests: An Architectural Analysis*. Springer-Verlag: Heidelberg.

Halperin, W. 1978. Organogenesis at the shoot apex. Ann. Rev. Plant Physiol. 29:239–262.

Halperin, W. 1986. Attainment and retention of morphogenetic capacity in vitro. In *Cell Culture and Somatic Cell Genetics of Plants*, vol. 3, ed. I. K. Vasil. Academic Press: Orlando, Florida.

Hammond, C. T., and P. G. Mahlberg. 1977. Morphogenesis of capitate glandular hairs of *Cannabis sativa* (Cannabaceae). Amer. J. Bot. 64:1023–1031.

Handley, W. R. C. 1936. Some observations on the problem of vessel length determination in woody dicotyledons. New Phytol. 35:456–471.

Hanson, J. B., and D. A. Day. 1980. Plant mitochondria. In *The Biochemistry of Plants: A Comprehensive Treatise*, vol. 1, ed. N. E. Tolbert. Academic Press: New York.

Hanstein, J. 1868. Die Scheitelzellgruppe im Vegetationspunkt der Phanerogamen. Fetschr. Niederrhein. Ges. Natur-und Heilkinde 1868:109–134.

Harada, H. 1965. Ultrastructure of angiosperm vessels and ray parenchyma. In *Cellular Ultrastructure of Woody Plants*, ed. W. A. Côté. Syracuse University Press: Syracuse, New York.

Harborne, J. B. 1982. *Introduction to Ecological Biochemistry*. Academic Press: London.

Hardham, A. R. 1976. Structural aspects of the pathways of nutrient flow to the developing embryo and cotyledons of *Pisum sativum* L. Aust. J. Bot. 24:711–721.

Hardham, A. R., and B. E. S. Gunning. 1979. Interpolation of microtubules into cortical arrays during cell elongation and differentiation in roots of *Azolla pinnata*. J. Cell Sci. 37:411–442.

Harper, J. L., and R. A. Benton. 1966. The behavior of seeds in soil. II. The germination of seeds on the surface of a water supplying substrate. J. Ecol. 54:151–166.

Harris, J. M. 1954. Heartwood formation in *Pinus radiata* (D. Don). New Phytol. 53:517–524.

Harris, J. M. 1981. Spiral grain formation. In *Xylem Cell Development*, ed. J. R. Barnett. Castle House Publishers: Tunbridge Wells, Kent, England.

Harris, M. 1954. *Handbook of Textile Fibers*. Harris Research Laboratories: Washington, D.C.

Harris, W. M. 1983. On the development of macroscereids in seed coats of *Pisum sativum* L. Amer. J. Bot. 70:1528–1535.

Harris, W. M. 1984. On the development of osteoscereids in seed coats of *Pisum sativum* L. New Phytol. 98:135–141.

Hartig, T. 1837. Vergleichende Untersuchungen über die Organisation des Stammes der einheimischen Waldbäume. Jahresber. Forsch. Fortwissensch. und Forstl. Naturkunde 1:125–168.

Hartig, T. 1853. Ueber die Entwicklung des Jahrringes der Holzpflanzen. Bot. Zeit. 11:533–566, 569–579.

Hartl, D. 1957. Struktur und Herkunft des Endokarps der Rutaceen. Beitr. Biol. Pfl. 34:35–49.

Hartmann, H. 1979. Surface structures of leaves: Their ecological and taxonomic significance in members of the subfamily Ruschioideae Schw. (Mesembryanthemaceae Fenzl). In *Electron Microscope Studies of the Leaf Epidermis in Some Succulents*, eds. D F. Cutler and H. Hartmann. Akad. d. Wiss. u. lit: Mainz, Germany.

Hasselberg, G. B. F. 1937. Zur Morphologie des vegetativen Sprosses der Loganiaceen. Symb. bot. Upsal. II, 3:1–170.

Hattersley, P. W., and L. Watson. 1976. C_4 grasses: An anatomical criterion for distinguishing between NADP-malic enzyme species and PCK or NAD-malic enzyme species. Aust. J. Bot. 24:297–308.

Hauri, H. 1917. Anatomische Untersuchungen an Polsterpflanzen nebst morphologischen und ökologischen Notizen. Beih. bot. Zbl. 33:257–293.

Havelange, A. 1980. The quantitative ultrastructure of the meristematic cells of *Xanthium strumarium* during the transition to flowering. Amer. J. Bot. 67:1171–1178.

Havelange, A., and G. Bernier. 1974. Descriptive and quantitative study of ultrastructural changes in the apical meristem of mustard in transition to flowering. I. The cell and nucleus. J. Cell Sci. 15:633–644.

Havelange, A., G. Bernier, and A. Jacqmard. 1974. Descriptive and quantitative study of ultrastructural changes in the apical meristem of mustard in transition to flowering. II. The cytoplasm, mitochondria and proplastids. J. Cell Sci. 16:421–432.

Hay, M. J. M., J. Dunlop, and D. H. Hopcroft. 1982. Anatomy and development of the superficial layers in stolons of white clover (*Trifolium repens* L.). N. Z. J. Bot. 20:315–324.

Hayward, H. E. 1938. *The Structure of Economic Plants*. Macmillan: New York.

Head, G. C. 1973. Shedding of roots. In *Shedding of Plant Parts*, ed. T. T. Kozlowski. Academic Press: New York.

Healy, P. L., J. D. Michaud, and J. Arditti. 1980. Morphometry of orchid seeds. III. Native California and related species of *Goodyera*, *Piperia*, *Platanthera* and *Spiranthes*. Amer. J. Bot. 67:508–518.

Heath, I. B., and R. W. Seagull. 1982. Oriented cellulose fibrils and the cytoskeleton: A critical comparison of models. In *The Cytoskeleton in Plant Growth and Development*, ed. C. W. Lloyd. Academic Press: London.

Hebant, C., and E. De Fay. 1980. Functional organization of the bark of *Hevea brasilensis* (rubber tree). A structural and histo-enzymological study. Z. Pflanzenphysiol. 97:391–398.

Hector, J. M. 1936. Introduction to the botany of field crops. Vol. 1. Cereals. South Afr. Agric. Series 16. Central News Agency Ltd.: Johannesburg.

Heide-Jørgensen, H. S. 1978. The xeromorphic leaves of *Hakea suaveolens* R. Br. II. Structure of epidermal cells, cuticle development and ectodesmata. Bot. Tidsskr. 72:227–244.

Heide-Jørgensen, H. S. 1980. The xeromorphic leaves of *Hakea suaveolens* R. Br. III. Ontogeny, structure and function of the T-shaped trichomes. Bot. Tidsskrift. 75: 181–198.

Heinrich, G. 1967. Licht-und elektronenmikroskopische Untersuchungen der Milchröhren von *Taraxacum bicorne*. Flora A 158:413–420.

Heinricher, E. 1884. Über Eiweisstoffe führende Idioblasten bei einigen Cruciferen. Ber. Dtsch. Bot. Ges. 2:463–466.

Heitzelman, C. E., and R. E. Howard. 1948. The comparative morphology of the Icacinaceae. V. The pubescence and the crystals. Amer. J. Bot. 35:42–52.

Hejnowicz, Z. 1961. Anticlinal divisions, intrusive growth, and loss of fusiform initials in nonstoried cambium. Acta Soc. Bot. Pol. 30:729–752.

Hejnowicz, Z. 1964. Orientation of the partition in pseudo-transverse division in cambia of some conifers. Can. J. Bot. 42:1685–1691.

Hejnowicz, Z., and J. Krawczyszyn. 1969. Oriented morphogenetic phenomena in cambium of broadleaved trees. Acta Soc. Bot. Pol. 38:547–560.

Hejnowicz, Z., and J. A. Romberger. 1973. Migrating cambial domains and the origin of wavy grain in xylem of broad-leaved trees. Amer. J. Bot. 60:209–222.

Henke, R. R., K. W. Hughes, M. J. Constantin, and A. Hollaender. 1985. *Tissue Culture in Forestry and Agriculture*. Plenum Press: New York.

Henrickson, J. 1969. Anatomy of periderm and cortex of Fouquieriaceae. Aliso 7:97–126.

Hensel, W., and A. Sievers. 1980. Effects of prolonged omnilateral gravistimulation on the ultrastructure of statocytes and on the graviresponse of roots. Planta 150: 338–346.

Hepler, P. K. 1982. Endoplasmic reticulum in the formation of the cell plate and plasmodesmata. Protoplasma 111:121–133.

Hepler, P. K., and D. E. Fosket. 1971. The role of microtubules in vessel member differentiation in *Coleus*. Protoplasma 72:213–236.

Hepler, P. K., and E. H. Newcomb. 1964. Microtubules and fibrils in the cytoplasm of *Coleus* cells undergoing secondary wall deposition. J. Cell Biol. 20:529–533.

Hepton, C. E. L., and R. D. Preston. 1960. Electron microscopic observations of the structure of sieve-connections in the phloem of angiosperms and gymnosperms. J. Exp. Bot. 11:381–394.

Herth, W. 1984. Oriented "rosette" alignment during cellulose formation in mung bean hypocotyl. Naturwissenschaften 71:216–217.

Herth, W. 1985. Plasma-membrane rosettes involved in localized wall thickening during xylem vessel formation of *Lepidium sativum* L. Planta 164:12–21.

Herth, W., and G. Weber. 1984. Occurrence of the putative cellulose-synthesizing "rosettes" in the plasma membrane

of *Glycine max* suspension culture cells. Naturwissenschaften 71:153–154.

Heslop-Harrison, J. 1968. Tapetal origin of pollen coat substances in *Lilium*. New Phytol. 67:779–786.

Heslop-Harrison, J. 1971a. *Pollen: Development and Physiology*. Butterworths: London.

Heslop-Harrison, J. 1971b. Wall pattern formation in angiosperm microsporogenesis. Symp. Soc. Exp. Biol. 25: 277–300.

Heslop-Harrison, J. 1978. *Cellular Recognition Systems in Plants*. Studies in Biology, no. 100. Edward Arnold: London.

Heslop-Harrison, J. 1979. Pollen walls as adaptive systems. Ann. Missouri Bot. Gard. 66:813–829.

Heslop-Harrison, J., and Y. Heslop-Harrison. 1982. The specialized cuticles of the receptive surfaces of angiosperm stigmas. In *The Plant Cuticle*, eds. D. F. Cutler, K. L. Alvin, and C. E. Price. Academic Press: London.

Heslop-Harrison, Y. 1976. Carnivorous plants a century after Darwin. Endeavour 35:114–122.

Heslop-Harrison, Y. 1981. Stigma characteristics and angiosperm taxonomy. Nord. J. Bot. 1:401–420.

Heslop-Harrison, Y., and R. B. Knox. 1971. A cytochemical study of the leaf-gland enzymes of insectivorous plants of the genus *Pinguicula*. Planta 96:183–211.

Heslop-Harrison, Y., and K. R. Shivanna. 1977. The receptive surface of the angiosperm stigma. Ann. Bot. 41: 1233–1258.

Hess, D. 1983. *Die Blüte*. Verlag Eugen Ulmer: Stuttgart, Germany.

Hesse, H. 1904. *Beiträge zur Morphologie und Biologie der Wurzelhaare*. Dissertation Jena.

Hickey, L. J. 1973. Classification of the architecture of dicotyledonous leaves. Amer. J. Bot. 60:17–33.

Hickey, L. J. 1979. A revised classification of the architecture of dicotyledonous leaves. In *Anatomy of the Dicotyledons*, 2nd ed., eds. C. R. Metcalfe and L. Chalk. Clarendon Press: Oxford.

Hicks, G. S. 1980. Patterns of organ development in plant tissue culture and the problem of organ determination. Bot. Rev. 46:1–23.

Higinbotham, N. 1942. The three-dimensional shape of undifferentiated cells in the petiole of *Angiopteris evecta*. Amer. J. Bot. 29:851–858.

Hill, A. F. 1952. *Economic Botany*. McGraw-Hill: New York.

Hill, A. W. 1931. A hybrid Daphne (*D. petraea* Leybold X *D. cneorum* L.). Ann. Bot. 45:229–231.

Hillis, W. E. 1977. Secondary changes in wood. In *The Structure, Biosynthesis, and Degradation of Wood*, eds. F. A. Loewus and V. C. Runeckles. Plenum Press: New York.

Hiltz, M. P. 1950. Contribution à l'étude des lithocystes et des cystolithes de *Ficus elastica*. Rev. Gén. Bot. 57: 453–477.

Hirsch, A. M., and D. R. Kaplan. 1974. Organography, branching, and the problem of leaf versus bud differentiation in the vining epiphytic fern genus *Microgramma*. Amer. J. Bot. 61:217–229.

Hitch, P. A., and B. C. Sharman. 1971. The vascular pattern of festucoid grass axes, with particular reference to nodal plexi. Bot. Gaz. 132:38–56.

Hodson, M. J., A. G. Sangster, and D. W. Parry. 1985. An ultrastructural study on the developmental phases and silicification of the glumes of *Phalaris canariensis* L. Ann. Bot. 55:649–665.

Hoefert, L. L. 1979. Ultrastructure of developing sieve elements in *Thlaspi arvense* L. I. The immature state. Amer. J. Bot. 66:925–932.

Hoefert, L. L. 1980. Ultrastructure of developing sieve elements in *Thlaspi arvense* L. II. Maturation. Amer. J. Bot. 67:194–201.

Höhn, K. 1950. Untersuchungen über Hydathoden und Funktion. Akad. Wiss. Lit. Mainz. Abh. Math.-Nat. Kl. 2:9–42.

Höhn, K. 1951. Beziehungen zwischen Blutung und Guttation bei *Zea mays*. Planta 39:65–74.

Holdheide, W. 1951. Anatomie mitteleuropaischer Geholzrinden (mit mikrophotographischem Atlas). In *Handbuch der Mikroskopie in der Technik*, Band 5, Heft 1, ed. H. Freund. Umschau Verlag: Frankfurt, Germany.

Höll, W. 1975. Radial transport in rays. In *Transport in plants. I. Encyclopedia of Plant Physiology*, eds. M. H. Zimmermann and J. A. Milburn, new ser. vol. 1. Springer-Verlag: Berlin.

Holloway, P. J. 1982a. Structure and histochemistry of plant cuticular membranes: An overview. In *The Plant Cuticle*, eds. D. F. Cutler, K. L. Alvin, and C. E. Price. Academic Press: London.

Holloway, P. J. 1982b. The chemical constitution of plant cutins. In *The Plant Cuticle*, eds. D. F. Cutler, K. L. Alvin, and C. E. Price. Academic Press: London.

Holloway, P. J. 1982c. Suberins of *Malus pumila* stem and root corks. Phytochemistry 21:2517–2522.

Holloway, P. J. 1983. Some variations in the composition of suberin from the cork layers of higher plants. Phytochemistry 22:495–502.

Holloway, P. J., and E. A Baker. 1970. The cuticles of some angiosperm leaves and fruits. Ann. Applied Biol. 66:145–154.

Holm, T. 1921. Morphological study of *Carya alba* and *Juglans nigra*. Bot. Gaz. 72:375–389.

Hoober, J. K. 1984. *Chloroplasts*. Plenum Press: New York.

Hooker, J. D. 1875. Address to the Department of Botany and Zoology. Rep. 44th Meeting Brit. As. Adv. Sci. Belfast 1874:102–116.

Horak, K. E. 1981a. Anomalous secondary thickening in *Stegnosperma* (Phytolaccaceae). Bull. Torrey Bot. Club 108:189–197.

Horak, K. E. 1981b. The three-dimensional structure of vascular tissues in *Stegnosperma* (Phytolaccaceae). Bot. Gaz. 142:545–549.

Horn, H. S. 1971. *The Adaptive Geometry of Trees*. Princeton University Press: Princeton, New Jersey.

Horner, H. T., Jr., A. P. Kausch, and B. L. Wagner. 1981. Growth and change in shape of raphide and druse calcium oxalate crystals as a function of intracellular development in *Typha angustifolia* L. (Typhaceae) and *Capsicum annuum* L. (Solanaceae). Scanning Electron Microscopy 1981:251–262.

Horner, H. T., Jr., and N. R. Lersten. 1967. Development, structure and function of dendroid secretory trichomes in *Psychotria bacteriophila* (Rubiaceae). Amer. J. Bot. 54:638 (Abstr.).

Horner, H. T., Jr., and N. R. Lersten. 1968. Development, structure and function of secretory trichomes in *Psychotria bacteriophila* (Rubiaceae). Amer. J. Bot. 55:1089–1099.

Horner, H. T., Jr., and B. L. Wagner. 1980. The association of druse crystals with the developing stomium of *Capsicum annuum* (Solanaceae) anthers. Amer. J. Bot. 67:1347–1360.

Höster, H. R., and W. Liese. 1966. Über das Vorkommen von Reaktionsgewebe in Wurzeln und Ästen der Dikotyledonen. Holzforsch. 20:80–90.

Howard, R. A. 1962. The vascular structure of the petiole as a taxonomic character. Proc. 15th Inter. Hort. Cong. Nice 1958:7–13.

Howard, R. A. 1969. The ecology of an elfin forest in Puerto Rico. 8. Studies of stem growth and form and of leaf structure. J. Arnold Arbor. 50:225–267.

Howard, R. A. 1970. Some observations on the nodes of woody plants with special reference to the problem of the "split lateral" versus the "common gap." In *New Research in Plant Anatomy*, eds. N. K. B. Robson, D. F. Cutler, and M. Gregory. Academic Press: London.

Howard, R. A. 1979a. The stem-node-leaf continuum of the Dicotyledonae. In *Anatomy of the Dicotyledons*, 2nd ed. eds. C. R. Metcalfe and L. Chalk. Clarendon Press: Oxford.

Howard, R. A. 1979b. The petiole. In *Anatomy of the Dicotyledons*, 2nd ed., eds. C. R. Metcalfe and L. Chalk. Clarendon Press: Oxford.

Huang, Shiu-Mei, and C. Sterling. 1970. Laticifers in the bulb scales of *Allium*. Amer. J. Bot. 57:1000–1003.

Huber, B. 1935. Die physiologische Bedeutung der Ring- und Zerstreutporigkeit. Ber. Dt. Bot. Ges. 53:711–719.

Huber, B. 1949a. Zur Frage der anatomischen Unterscheidbarkeit des Holzes von *Pinus sylvestris* L. und *Pinus nicricans* Host. mit Betrachtungen über heterogene Markstrahlen. Forstwiss. Zentr. 68:456–468.

Huber, B. 1949b. Zur Phylogenie des Jahrringbaues der Rinde. Svensk. bot. Tidskr. 43:376–382.

Hufford, L. D. 1980. Staminal vascular architecture in five dicotyledonous angiosperms. Proc. Iowa Acad. Sci. 87:96–102.

Hulbary, R. L. 1944. The influence of air spaces on the three-dimensional shapes of cells in *Elodea* stems, and a comparison with pith cells of *Ailanthus*. Amer. J. Bot. 31:561–580.

Hulbary, R. L. 1948. Three-dimensional cell shape in the tuberous roots of *Asparagus* and in the leaf of *Rhoeo*. Amer. J. Bot. 35:558–566.

Hunt, D. R. 1971. Schumann and Buxbaum reconciled. Cact. Succ. J. Gr. Brit. 33:53–72.

Huxley, C. 1980. Symbiosis between ants and epiphytes. Biol. Rev. 55:321–340.

Hyde, B. B. 1970. Mucilage-producing cells in the seed coat of *Plantago ovata*: Developmental fine structure. Amer. J. Bot. 57:1197–1206.

Ihlenfeldt, H.-D., and H. E. K. Hartmann. 1982. Leaf surfaces in Mesembryanthemaceae. In *The Plant Cuticle*, eds. D. F. Cutler, K. L. Alvin, and C. E. Price. Academic Press: London.

Iljin, M. M. 1950. Polykambialnost' i evoliutsya. Problemy Bot. 1:232–249.

Innamorati, M. 1966. Transformazioni strutturali e formazione degli strati nei granuli di amido *Triticum* coltivato in condizioni constanti e in ambiente naturale. Caryologia 19:343–367.

Isebrands, J. G., and P. R. Larson. 1977a. Organization and ontogeny of the vascular system in the petiole of eastern cottonwood. Amer. J. Bot. 64:65–77.

Isebrands, J. G., and P. R. Larson. 1977b. Vascular anatomy of the nodal region in *Populus deltoides* Bartr. Amer. J. Bot. 64:1066–1077.

Jaccard, M., and P. E. Pilet. 1975. Extensibility and rheology of collenchyma cells. I. Creep relaxation and viscoelasticity of young and senescent cells. Plant and Cell Physiol. 16:113–120.

Jaccard, M., and P. E. Pilet. 1977. Extensibility and rheology of collenchyma cells. II. Low-pH effect on the extension of collocytes isolated from high- and low-growing material. Plant and Cell Physiol. 18:883–891.

Jaccard, M., and P. E. Pilet. 1979. Growth and rheological changes of collenchyma cells. The fusicoccin effect. Plant and Cell Physiol. 20:1–7.

Jarvis, M. C., A. S. Logan, and H. J. Duncan. 1984. Tensile characteristics of collenchyma cell walls at different calcium contents. Physiol. Plant. 61:81–86.

Jeffree, C. E., M. M. Yeoman, and D. C. Kilpatrick. 1982. Immunofluorescence studies on plant cells. Int. Rev. Cytol. 80:231–265.

Jeffrey, D. W. 1967. Phosphate nutrition of Australian heath plants. I. The importance of proteoid roots in *Banksia* (Proteaceae). Aust. J. Bot. 15:403–411.

Jeník, J. 1978. Roots and root systems in tropical trees: Morphologic and ecologic aspects. In *Tropical Trees as Living Systems*, eds. P. B. Tomlinson and M. H. Zimmermann. Cambridge University Press: New York.

Jenny, H., and K. Grossenbacher. 1963. Root soil bound-

ary zones as seen in the electron microscope. Proc. Soc. Soil Sci. Am. 27:273–277.

Jensen, R. G. 1980. Biochemistry of the chloroplast. In *The Biochemistry of Plants: A Comprehensive Treatise*, vol. 1, ed. N. E. Tolbert. Academic Press: New York.

Jensen, T. E., and J. G. Valdovinos. 1967. Fine structure of abscission zones. I. Abscission zones of the pedicels of tobacco and tomato flowers at anthesis. Planta 77:298–318.

Jensen, W. A. 1965a. The ultrastructure and histochemistry of the synergids of cotton. Amer. J. Bot. 52:238–256.

Jensen, W. A. 1965b. The ultrastructure and composition of the egg and central cell of cotton. Amer. J. Bot. 52:781–798.

Jensen, W. A., and D. B. Fisher. 1968. Cotton embryogenesis: The entrance and discharge of the pollen tube into the embryo sac. Planta 58:158–183.

Jensen, W. A., and D. B. Fisher. 1969. Cotton embryogenesis: The tissues of the stigma and style and their relation to the pollen tube. Planta 84:97–121.

Jernstedt, J. A. 1984. Seedling growth and root contraction in the soap plant, *Chlorogalum pomeridianum* (Liliaceae). Amer. J. Bot. 71:69–75.

Jernstedt, J. A., and C. Clark. 1979. Stomata on the fruits and seeds of *Eschscholzia* (Papaveraceae). Amer. J. Bot. 66:586–590.

Jeronimides, G. 1976. The fracture behavior of wood in relation to its structure. In *Wood Structure in Biological and Technological Research*, eds. P. Baas, A. J. Bolton, and D. M. Catling. Leiden University Press: Leiden, The Netherlands.

Joel, D. M, and B. E. Juniper. 1982. Cuticular gaps in carnivorous plant glands. In *The Plant Cuticle*, eds. D. F. Cutler, K. L. Alvin, and C. E. Price. Academic Press: London.

Johansen, D. A. 1940. *Plant Microtechnique*. McGraw-Hill: New York.

Johnson, H. B. 1975. Plant pubescence: An ecological perspective. Bot. Rev. 41:233–258.

Johnson, J. M., and L. E. Jones. 1967. Behavior of nucleoli and contracting nucleolar vacuoles in tobacco cells growing in microculture. Amer. J. Bot. 54:189–198.

Johnson, M. A. 1937. Hydathodes in the genus *Equisetum*. Bot. Gaz. 98:598–608.

Johnson, M. A., and F. H. Truscott. 1956. On the anatomy of *Serjania*. I. Path of the bundles. Amer. J. Bot. 43:509–518.

Johnson, R. W., and R. T. Riding. 1981. Structure and ontogeny of the stomatal complex in *Pinus strobus* L. and *Pinus banksiana* Lamb. Amer. J. Bot. 68:260–268.

Johri, B. M., and S. Agarwal. 1965. Morphological and embryological studies in the family Santalaceae. VIII. *Quinchamalium chilense* Lam. Phytomorphology 15:360–372.

Johri, B. M., and S. P. Bhatnagar. 1972. *Loranthaceae*. Botanical Monograph No. 8. Council of Science and Industrial Research: New Delhi.

Jones, C. S. 1984. The effect of axis splitting on xylem pressure potentials and water movement in the desert shrub *Ambrosia dumosa* (Gray) Payne (Asteraceae). Bot. Gaz. 145:125–131.

Jones, C. S., and E. M. Lord. 1982. The development of split axes in *Ambrosia dumosa* (Gray) Payne (Asteraceae). Bot. Gaz. 143:446–453.

Jones, M. G. K. 1976. The origin and development of plasmodesmata. In *Intercellular Communication in Plants: Studies on Plasmodesmata*, eds. B. E. S. Gunning and A. W. Robards. Springer-Verlag: Berlin.

Jones, M. G. K., and H. L. Payne. 1978. Cytokinesis in *Impatiens balsamina* and the effect of caffeine. Cytobios 20: 79–91.

Jordan, E. G., J. N. Timmis, and A. J. Trewavas. 1980. The plant nucleus. In *The Biochemistry of Plants: A Comprehensive Treatise*, vol. 1, ed. N. E. Tolbert. Academic Press: New York.

Jordan, L. S., J. L. Jordan, and C. M. Jordan. 1985. Changes induced by water on *Euphorbia supina* seed coat structures. Amer. J. Bot. 72:1530–1536.

Jørgensen, L. B., J. D. Møller, and P. Wagner. 1975. Secondary phloem of *Trochodendron aralioides*. Bot. Tidskr. 69:217–238.

Jost, L. 1932. Die Determinierung der Wurzelstruktur. Z. Bot. 25:481–522.

Joubert, A. M., R. L. Verhoeven, and H. J. T. Venter. 1984. An anatomical investigation of the stem and leaf of the South African species of *Lycium* (Solanaceae). S. Afr. J. Bot. 3:219–230.

Juncosa, A. M. 1982. Developmental morphology of the embryo and seedling of *Rhizophora mangle* L. (Rhizophoraceae). Amer. J. Bot. 69:1599–1611.

Juniper, B. 1976. Geotropism. Ann. Rev. Plant Physiol. 27:385–406.

Juniper, B. E., and F. A. L. Clowes. 1965. Cytoplasmic organelles and cell growth in root caps. Nature 208: 864–865.

Juniper, B. E., and A. French. 1973. The distribution and redistribution of endoplasmic reticulum (ER) in geoperceptive cells. Planta 109:211–224.

Juniper, B. E., and A. J. Gilchrist. 1976. Absorption and transport of calcium in the stalked glands of *Drosera capensis* L. In *Perspectives in Experimental Biology*, vol. 2, Botany, ed. N. Sunderland. Pergamon Press: Oxford.

Juniper, B. E., J. R. Lawton, and P. J. Harris. 1981. Cellular organelles and cell-wall formation in fibres from the flowering stem of *Lolium temulentum* L. New Phytol. 89:609–619.

Juniper, B. E., and G. Pask. 1973. Directional secretion by Golgi bodies in maize root cells. Planta 109:225–231.

Juniper, B. E., and R. M. Roberts. 1966. Polysaccharide synthesis and the fine structure of root cells. J. Royal Micro. Soc. 85:63–72.

Junttila, O. 1976. Apical growth cessation and shoot tip abscission in *Salix*. Physiol. Plant. 38:278–286.

Kallarackal, J., and J. A. Milburn. 1983. Studies on the phloem sealing mechanism in *Ricinus communis* var. *gibsonii* fruit stalks. Aust. J. Plant Physiol. 10:561–568.

Kallen, F. 1882. Verhalten des Protoplasma in den Geweben von *Urtica urens* entwicklungsgeschichtlich dargestellt. Flora 65:65–80, 81–96, 97–105.

Kaniewski, K. 1963. Hairs in the loculus of the broad-bean (*Vicia faba* L.) fruit. Bull. Acad. Polon. Sci. Cl. V. Sér. Sci. Biol. 16:585–594.

Kaniewski, K., and Z. Wazynska. 1970. Sclerenchymatous endocarp with hairs in the fruit of *Acer pseudoplatanus* L. Bull. Acad. Polon. Sci. II. 18:413–420.

Kapil, R. N., and A. K. Bhatnagar. 1981. Ultrastructure and biology of female gametophyte in flowering plants. Int. Rev. Cytol. 70:291–341.

Kapil, R. N., J. Bor, and F. Bouman. 1980. Seed appendages in Angiosperms. Bot. Jahrb. Syst. 101:555–573.

Kaplan, D. R. 1968. Structure and development of the perianth in *Downingia bacigalupii*. Amer. J. Bot. 55:406–420.

Kaplan, D. R. 1969. Seed development in *Downingia*. Phytomorphology 19:253–378.

Kaplan, D. R. 1970. Comparative foliar histogenesis in *Acorus calamus* and its bearings on the phyllode theory of monocotyledonous leaves. Amer. J. Bot. 57:331–361.

Kaplan, D. R. 1973. Comparative developmental analysis of the heteroblastic leaf series of axillary shoots of *Acorus calamus* L. (Araceae). Cellule 69:253–290.

Kaplan, D. R. 1975. Comparative developmental evaluation of the morphology of unifacial leaves in the monocotyledons. Bot. Jahrb. Syst. 95:1–105.

Karas, I., and M. E. McCully. 1973. Further studies of the histology of lateral root development in *Zea mays*. Protoplasma 77:243–269.

Kaufman, P. B. 1959. Development of the shoot of *Oryza sativa* L. II. Leaf histogenesis. Phytomorph. 9:277–311.

Kaufman, P. B., L. B. Petering, C. S. Yocum, and D. Baic. 1970. Ultrastructural studies on stomata development in internodes of *Avena sativa*. Amer. J. Bot. 57:33–49.

Kausch, A. P., and H. T. Horner. 1983. The development of mucilaginous raphide crystal idioblasts in young leaves of *Typha angustifolia* L. (Typhaceae). Amer. J. Bot. 70: 691–705.

Kaussmann, B. 1941. Vergleichende Untersuchungen über die Blattnatur der Kelch-, Blumen-und Staubblätter. Bot. Arch. 42:503–572.

Kay, Q. O. N., and H. S. Daoud. 1981. Pigment distribution, light reflection and cell structure in petals. Bot. J. Linn. Soc. 83:57–84.

Kelly, G. J., and E. Latzko. 1980. The cytosol. In *The Biochemistry of Plants: A Comprehensive Treatise*, vol. 1, ed. N. E. Tolbert. Academic Press: New York.

Kemp, R., and F. J. Alfieri. 1974. Seasonal phloem development in *Juniperus californica*. Amer. J. Bot. 61:58.

Kenda, G. 1952. Stomata in Antheren. I. Anatomischer Teil. Phyton, Horn 4:83–96.

Kerr, T., and I. W. Bailey. 1934. The cambium and its derivative tissues. X. Structure, optical properties and chemical composition of the so-called middle lamella. J. Arnold Arbor. 15:327–349.

Kidwai, P., and A. W. Robards. 1969. On the ultrastructure of resting cambium of *Fagus sylvatica* L. Planta 89:361–368.

Kiesselbach, T. A., and E. R. Walker. 1952. Structure of certain specialized tissues in the kernel of corn. Amer. J. Bot. 39:561–569.

Kirby, E. J. M., and J. L. Rymer. 1975. The vascular anatomy of the barley spikelet. Ann. Bot. 39:205–211.

Kirby, R. H. 1963. *Vegetable Fibers*. Leonard Hill (Books) Ltd.: London.

Kirchoff, B. K., and A. Fahn. 1984. Initiation and structure of the secondary vascular system in *Phytolacca dioica* (Phytolaccaceae). Can. J. Bot. 62:2580–2586.

Kirk, J. T. O., and R. A. E. Tilney-Bassett. 1978. *The Plastids: Their Chemistry, Structure, Growth and Inheritance*. Elsevier/North-Holland: Amsterdam.

Kisser, J. 1958. Der Stoffwechsel sekundärer Pflanzenstoffe. In *Handbuch der Pflanzenphysiologie*, ed. W. Ruhland. Springer-Verlag: Berlin.

Kivilaan, A., and R. S. Bandurski. 1981. The one-hundred-year period for Dr. Beal's seed viability experiment. Amer. J. Bot. 68:1290–1292.

Klein, G. 1923. Zur Aetiologie der Thyllen. Z. Bot. 15:418–439.

Klekowski, E. J., Jr. 1988. *Mutation, Developmental Selection and Plant Evolution*. Columbia University Press: New York.

Klekowski, E. J., Jr., and N. Kazarinova-Fukshansky. 1984a. Shoot apical meristems and mutation: Fixation of selectively neutral cell genotypes. Amer. J. Bot. 71:22–27.

Klekowski, E. J., Jr., and N. Kazarinova-Fukshansky. 1984b. Shoot apical meristems and mutation: Selective loss of disadvantageous cell genotypes. Amer. J. Bot. 71:28–34.

Klepper, B, and M. R. Kaufmann. 1966. Removal of salt from xylem sap by leaves and stems of guttating plants. Plant Physiol. 41:1743–1747.

Klinken, J. 1914. Ueber das gleitende Wachstum der Initialen im Kambium der Koniferen und der Markstrahlenverlauf in ihrer sekundaren Rinde. Bibl. Bot. 19:1–37.

Klotz, L. H. 1978. Form of the perforation plates in the wide vessels of metaxylem in palms. J. Arnold Arbor. 59:105–128.

Knobloch, I. W. 1954. Developmental anatomy of chicory. The root. Phytomorph. 4:47–54.

Knowles, L. O., and J. A. Flore. 1983. Quantitative and qualitative characterization of carrot (*Daucus carota*) root periderm during development. J. Am. Soc. Hort. Sci. 108:923–928.

Koch, A. 1884. Ueber den Verlauf und die Endigungen der Siebröhren in den Blättern. Bot. Z. 42:401–427.

Kolattukudy, P. E. 1980a. Biopolyester membranes of plants: Cutin and suberin. Science 208:990–999.

Kolattukudy, P. E. 1980b. Cutin, suberin, and waxes. In *The Biochemistry of Plants*, vol. 4, ed. P. K. Stumpf. Academic Press: New York.

Kolattukudy, P. E. 1981. Structure, biosynthesis and biodegradation of cutin and suberin. Ann. Rev. Plant Physiol. 32:539–567.

Koller, D., and N. Roth. 1964. Studies of the ecological and physiological significance of amphicarpy in *Gymnarrhena micrantha* (Comp.). Amer. J. Bot. 51:26–35.

Kollmann, R., and W. Schumacher. 1961. Über die Feinstruktur des Phloems von *Metasequoia glyptostroboides* und seine jahreszeitlichen Veränderungen. I. Das Ruhephloem. Planta 57:583–607.

Konar, R. N., and S. K. Banerjee. 1963. The morphology and embryology of *Cupressus funebris* Endl. Phytomorphology 13:321–338.

Konar, R. N., and H. F. Linskens. 1966. The morphology and anatomy of the stigma of *Petunia hybrida*. Planta 71:356–371.

Konar, R. N., and Y. P. Oberoi. 1969. Recent work on reproductive structures of living conifers and taxads—a review. Bot. Rev. 35:89–116.

Kono, Y., and T. Mizoguchi. 1982. The origin of root periderm in the sweet potato plant (*Ipomoea batatas*). Japan. J. Crop Sci. 51:535–541.

Kozlowski, T. T. 1971. *Growth and Development of Trees*, vol. 2. Academic Press: New York.

Kozlowski, T. T., and C. R. Gunn. 1972. Importance and characteristics of seeds. In *Seed Biology*, ed. T. T. Kozlowski, vol. 1. Academic Press: New York.

Kozlowski, T. T., and T. Keller. 1966. Food relations of woody plants. Bot. Rev. 32:294–382.

Kozlowski, T. T., and C. H. Winget. 1963. Patterns of water movement in forest trees. Bot. Gaz. 124:301–311.

Kramer, P. J. 1956. Physical and physiological aspects of water absorption. In *Encyclopedia of Plant Physiology*, ed. W. Ruhland, 3:124–159. Springer-Verlag: Berlin.

Kramer, P. J. 1969. *Plant and Soil Water Relationships: A Modern Synthesis*. McGraw-Hill: New York.

Kramer, P. J., and T. T. Kozlowski. 1979. *Physiology of Woody Plants*. Academic Press: New York.

Kräusel, R., and H. Weyland. 1938. Neue Pflanzenfunde im Mitteldevon von Elberfeld. Palaeontographica 83B:172–195.

Krauss, L. 1933. Entwicklungsgeschichte der Früchte von *Hordeum, Triticum, Bromus*, und *Poa* mit besonderer Berücksichtigung ihrer Samenschalen. Jb. wiss. Bot. 77:773–808.

Kribs, D. A. 1937. Salient lines of structural specialization in the wood parenchyma of dicotyledons. Bull. Torrey Bot. Club 64:177–188.

Krishna, K. R., H. M. Suresh, J. Syamsunder, and D. J. Bagyaraj. 1981. Changes in the leaves of finger millet due to VA mycorrhizal infection. New Phytol. 87:717–722.

Kristen, U., M. Biedermann, G. Liebezeit, R. Dawson, and L. Böhm. 1979. The composition of stigmatic exudate and the ultrastructure of the stigma papillae in *Aptenia cordifolia*. Europ. J. Cell Biol. 19:281–287.

Kroemer, K. 1903. Wurzelhaube, Hypodermis und Endodermis der Angiospermenwurzel. Bibliotheca Bot. 12:1–159.

Kruatrachue, M., and R. F. Evert. 1978. Structure and development of sieve elements in the root of *Isoetes muricata* Dur. Ann. Bot. 42:15–21.

Kuijt, J. 1969. *The Biology of Parasitic Flowering Plants*. University of California Press: Berkeley, California.

Kuijt, J. 1977. Haustoria of phanerogamic parasites. Ann. Rev. Phytopathol. 17:91–118.

Kuijt, J. 1982. Epicortical roots and vegetative reproduction in Loranthaceae (s. s.) of the New World. Beitr. Biol. Pflanzen 56:307–316.

Kuijt, J. 1985. Morphology, biology, and systematic relationships of *Desmaria* (Loranthaceae). Pl. Syst. Evol. 151:121–130.

Kummerow, J., G. Avila, M.-E. Aljaro, S. Araya, and G. Montenegro. 1982. Effect of fertilizer on fine root density and shoot growth in Chilean matorral. Bot. Gaz. 143:498–504.

Kundu, B. C., and A. De. 1968. Taxonomic position of the genus *Nyctanthes*. Bull. bot. Surv. India 10:397–408.

Kuo, J. 1983. The nacreous walls of sieve elements in seagrasses. Amer. J. Bot. 70:159–164.

Kuo, J., and M. L. Cambridge. 1978. Morphology, anatomy and histochemistry of the Australian seagrasses of the genus *Posidonia* Konig (Posidoniaceae). II. Rhizome and root of *Posidonia australis* Hook. F. Aquatic Bot. 5:191–206.

Kuo, J., and T. P. O'Brien. 1974. Lignified sieve elements in the wheat leaf. Planta 117:349–353.

Kuo, J., T. P. O'Brien, and M. J. Canny. 1974. Pit-field distribution, plasmodesmatal frequency, and assimilate flux in the mestome sheath cells of wheat leaves. Planta 121:97–118.

Kuo, J., T. P. O'Brien, and S-Y. Zee. 1972. The transverse veins of the wheat leaf. Aust. J. Biol. Sci. 25:721–737.

Kurkova, E. B. 1981. Distribution of plasmodesmata in root epidermis. In *Structure and Function of Plant Roots*, eds. R. Brouwer, O. Gašpaříková, J. Kolek, and B. C. Loughman. Martinus Nijhoff/Dr W. Junk Publishers: The Hague.

Kursanov, A. L. 1984. *Assimilate Transport in Plants*. Elsevier: Amsterdam.

Kurt, J. 1930. Über die Hydathoden der Saxifrageae. Beih. bot. Zbl. 46:203–246.

Kurth, E. 1981. Mitotic activity in the root apex of the water fern *Marsilea vestita* Hook. and Grev. Amer. J. Bot. 68:881–896.

Küster, E. 1956. *Die Pflanzenzelle*. 3rd ed. Gustav Fischer: Jena, Germany.

Labavitch, J. M. 1981. Cell wall turnover in plant development. Ann. Rev. Plant Physiol. 32:385–406.

Labouriau, L. G. 1952. On the latex of *Regnellidum* (sic) *diphyllum* Lindm. Phyton 2:57–74.

Labrecque, M., D. Barabé, and J. Vièth. 1985. Développement du fruit de *Prunus virginiana* (Rosaceae). Can. J. Bot. 63:242–251.

Lackey, J. A. 1981. Systematic significance of the epihilum in Phaseoleae (Fabaceae, Faboideae). Bot. Gaz. 142:160–164.

Laetsch, W. M. 1974. The C_4 syndrome: A structural analysis. Ann. Rev. Plant Physiol. 25:27–52.

LaFrankie, J. V., Jr. 1985. Morphology, growth, and vasculature of the sympodial rhizome of *Smilacina racemosa* (Liliaceae). Bot. Gaz. 146:534–544.

Lai, V., and L. M. Srivastava. 1976. Nuclear changes during differentiation of xylem vessel elements. Cytobiologie 12:220–243.

Laing, H. E. 1940a. Respiration of the rhizomes of *Nuphar advenum* and other water plants. Amer. J. Bot. 27:574–581.

Laing, H. E. 1940b. The composition of the internal atmosphere of *Nuphar advenum* and other water plants. Amer. J. Bot. 27:861–868.

Lambright, D. D., and S. C. Tucker. 1980. Observations on the ultrastructure of *Trypethelium eluteriae* Spreng. Bryologist 83:170–178.

Lamont, B. 1972a. The effect of soil nutrients on the production of proteoid roots by *Hakea* species. Aust. J. Bot. 20:27–40.

Lamont, B. 1972b. The morphology and anatomy of proteoid roots in the genus *Hakea*. Aust. J. Bot. 20:155–174.

Lamont, B. 1973. Factors affecting the distribution of proteoid roots within the root system of two *Hakea* species. Aust. J. Bot. 21:165–187.

Lamont, B. N. 1980. Tissue longevity of the arborescent monocotyledon, *Kingia australis* (Xanthorrhoeaceae). Amer. J. Bot. 67:1262–1264.

Lamoureux, C. H. 1961. *Comparative studies on phloem of vascular cryptogams*. Ph.D. Dissertation. University of California, Davis, California.

Landré, P. 1972. Origine et développement des epidermes cotyledonaires et foliaires de la moutarde (*Sinapis alba* L.). Différentiation ultrastructurale des stomates. Ann. Sci. Nat. (Bot.) 12:247–322.

Larson, P. R. 1975. Development and organization of the primary vascular system in *Populus deltoides* according to phyllotaxy. Amer. J. Bot. 62:1084–1099.

Larson, P. R. 1976. Procambium vs. cambium and protoxylem vs. metaxylem in *Populus deltoides* seedlings. Amer. J. Bot. 63:1332–1348.

Larson, P. R. 1982. The concept of cambium. In *New Perspectives in Wood Anatomy*, ed. P. Baas. Martinus Nijhoff/Dr W. Junk Publishers: The Hague.

Larson, P. R., and J. G. Isebrands. 1978. Functional significance of the nodal constricted zone in *Populus deltoides*. Can. J. Bot. 56:801–804.

Larson, P. R., and T. D. Pizzolato. 1977. Axillary bud development in *Populus deltoides*. I. Origin and early ontogeny. Amer. J. Bot. 64:835–848.

Laubenfels, D. J. de. 1953. The external morphology of coniferous leaves. Phytomorphology 3:1–20.

Laval-Martin, D. 1974. La maturation du fruit de tomate cerise: mise en évidence, par cryodécapage, de l'évolution des chloroplastes en deux types de chromoplastes. Protoplasma 82:33–59.

Lawton, J. R. 1980. Observations on the structure of epidermal cells, particularly the cork and silica cells, from the flowering stem internode of *Lolium temulentum* L. (Gramineae). J. Linn. Soc. Bot. 80:161–77.

Lawton, J. R., P. J. Harris, and B. E. Juniper. 1979. Ultrastructural aspects of the development of fibres from the flowering stem of *Lolium temulentum* L. New Phytol. 82:529–536.

Lawton, J. R., and J. R. S. Lawton. 1971. Seasonal variations in the secondary phloem of some forest trees from Nigeria. New Phytol. 70:187–196.

Lawton, J. R., A. Todd, and D. K. Naidoo. 1981. Preliminary investigations into the structure of the roots of the mangroves, *Avicennia marina* and *Bruguiera gymnorrhiza*, in relation to ion uptake. New Phytol. 88:713–722.

Lawton, J. R. S., and M. J. Canny. 1970. The proportion of sieve elements in the phloem of some tropical trees. Planta 95:351–354.

Leach, J. E., M. A. Cantrell, and L. Sequeira. 1982. A hydroxyproline-rich bacterial agglutinin from potato (*Solanum tuberosum* cultivar *Katahdin*.) Its location by immunofluorescence. Physiol. Plant Pathol. 21:319–326.

Lederer, B. 1955. Vergleichende Untersuchungen über das Transfusionsgewebe einiger rezenter Gymnospermen. Bot. Stud. 4:1–42.

Lee, C. L. 1955. Fertilization in *Ginkgo*. Bot. Gaz. 117:79–100.

Lefebvre, D. D. 1985. Stomata on the primary root of *Pisum sativum* L. Ann. Bot. 55:337–341.

Leinfellner, W. 1937. Beiträge zur Kenntniss der Cactaceen-Areolen. Österr. Bot. Zeitschr. 86:1–60.

Leitão, M. M. N. 1984. Comparative study of the subgenus *Pharmacosycea* in Brazil. 1. *Ficus obtusiuscula*. Bradea 4:31–40.

Lemesle, R. 1956. Les éléments du xyléme dans les Angiospermes à caractères primitifs. Soc. Bot. de France Bul. 103:629–677.

Leppard, G. G. 1974. Rhizoplane fibrils in wheat: Demonstration and derivation. Science, NY 185:1066–1067.

Lersten, N. R. 1971. A review of septate microsporangia in vascular plants. Iowa St. J. Sci. 45:487–497.

Lersten, N. R. 1974a. Morphology and distribution of colleters and crystals in relation to the taxonomy and bacterial leaf nodule symbiosis of *Psychotria* (Rubiaceae). Amer. J. Bot. 61:973–981.

Lersten, N. R. 1974b. Colleter morphology in *Pavetta, Neorosea* and *Tricalysia* (Rubiaceae) and its relationship to the bacterial leaf nodule symbiosis. Bot. J. Linn. Soc. 69:125–136.

Lersten, N. R. 1982. Tracheid bar and vestured pits in legume seeds (Leguminosae: Papilionoideae). Amer. J. Bot. 69:98–107.

Lersten, N. R. 1983. Suspensors in Leguminosae. Bot. Rev. 49:233–257.

Lersten, N. R., and J. D. Curtis. 1985. Distribution and anatomy of hydathodes in Asteraceae. Bot. Gaz. 146:106–114.

Leu, S-W., and S-H. T. Chiang. 1981. The site of the first periderm in the stem. Taiwania 26:12–21.

Levin, D. A. 1973. The role of trichomes in plant defence. Q. Rev. Biol. 48:3–15.

Lewin, J., and B. E. F. Reimann. 1969. Silicon and plant growth. Ann. Rev. Plant Physiol. 20:289–304.

Lieberman, S. J., J. G. Valsovinos, and T. E. Jensen. 1982. Ultrastructural localization of cellulase in abscission cells on tobacco flower pedicels. Bot. Gaz. 143:32–40.

Liese, W., and J. Bausch. 1967. On the anatomical causes of the refractory behavior of spruce and douglas fir. J. Inst. Wood Sci. 19:3–14.

Lignier, M. O. 1887. Recherches sur l'anatomie comparée des Calycanthées, des Mélastomacées et des Myrtacées. Arch. Sci. Nord. de la France 4:455.

Lilien-Kipnis, H., and S. Lavee. 1971. Anatomical changes during development of "Ventura" peach fruits. J. Hortic. Sci. 46:103–110.

Lin, J., and E. M. Gifford, Jr. 1976. The distribution of ribosomes in the vegetative and floral apices of *Adonis aestivalis*. Can. J. Bot. 54:2478–2483.

Linsbauer, K. 1930. Die Epidermis. In *Encyclopedia of Plant Anatomy*, vol. 4, part 2. Gbdr. Borntraeger: Berlin.

Linskens, H. F. 1975. Incompatibilty in *Petunia*. Proc. Roy. Soc. London. B. 188:299–311.

Lipetz, J. 1970. Wound healing in higher plants. Int. Rev. Cytol. 27:1–28.

Liphschitz, N., S. Lev-Yadun, E. Rosen, and Y. Waisel. 1984. The annual rhythm of activity of the lateral meristems, cambium and phellogen, in *Pinus halepensis* and *Pinus pinea*. IAWA Bull. 5:263–274.

List, A., Jr. 1963. Some observations on DNA content and cell and nuclear volume growth in the developing xylem cells of certain higher plants. Amer. J. Bot. 50:320–329.

Little, C. H. A. 1981. Effect of cambial dormancy state on

the transport of [1-¹⁴C]indol-3-ylacetic acid in *Abies balsamea* shoots. Can. J. Bot. 59:342–348.

Little, C. H. A., and P. F. Wareing. 1981. Control of cambial activity and dormancy in *Picea sitchensis* by indol-3-ylacetic and abscisic acids. Can. J. Bot. 59:1480–1493.

Lloyd, C. W. 1982. *The Cytoskeleton in Plant Growth and Development.* Academic Press: London.

Lloyd, D. G., and K. S. Bawa. 1984. Modification of the gender of seed plants in varying conditions. Evol. Biol. 17:255–338.

Lloyd, D. G., and C. J. Webb. 1977. Secondary sex characters in plants. Bot. Rev. 43:177–216.

Lloyd, F. E. 1942. *The Carnivorous Plants.* Ronald Press Co.: Waltham, Massachusetts.

Lloyd, F. E. 1976. *The Carnivorous Plants.* Dover Publications: New York.

Looby, W. J., and J. Doyle. 1942. Formation of gynospore, female gametophyte and archegonia in *Sequoia.* Sci. Proc. Roy. Dublin Soc. 23:35.

Lörcher, H., and F. Weberling. 1982. Zur Achsenverdickung hochandiner Valerianaceen. Ber. Deutsch Bot. Ges. 95:57–74.

Lord, E. M. 1981. Cleistogamy: a tool for the study of floral morphogenesis, function and evolution. Bot. Rev. 47:421–449.

Lord, E. M., and K. J. Eckard. 1985. Shoot development in *Citrus sinensis* L. (Washington naval orange). I. Floral and inflorescence ontogeny. Bot. Gaz. 146:320–326.

Lorenzi, R., A. Bennici, P. G. Cionini, A. Alpi, and F. D'Amato. 1978. Embryo-suspensor relations in *Phaseolus coccineus:* Cytokinins during seed development. Planta 143:59–62.

Lott, J. N. A. 1980. Protein bodies. In *The Biochemistry of Plants: A Comprehensive Treatise,* vol. 1, ed. N. E. Tolbert. Academic Press: New York.

Lott, J. N. A. 1981. Protein bodies in seeds. Nord. J. Bot. 1:421–432.

Lubbock, J. 1896. *A Contribution to our Knowledge of Seedlings.* Kegan Paul, Trench, Trübner and Co.: London.

Lütge, U., and G. Krapf. 1969. Die Ultrastruktur der *Nymphaea*—Hydropoten in Zusammenhang mit ihrer Funktion als Salztransportierende Drüsen. Cytobiologie 1:121–131.

Lynch, D. V., and E. R. Rivera. 1981. Ultrastructure of cells in the overwintering dormant shoot apex of *Rhododendron maximum* L. Bot. Gaz. 142:63–72.

Lyndon, R. F., and E. F. Robertson. 1976. The quantitative ultrastructure of the pea shoot apex in relation to leaf initiation. Protoplasma 87:387–402.

Lyshede, O. B. 1978. Studies on outer epidermal cell walls with microchannels in a xerophytic species. New Phytol. 80:421–426.

Lyshede, O. B. 1979. Xeromorphic features of three stem assimilants in relation to their ecology. J. Linn. Soc. Bot. 78:85–98.

Lyshede, O. B. 1982. Structure of the outer epidermal wall in xerophytes. In *The Plant Cuticle,* eds. D. F. Cutler, K. L. Alvin, and C. E. Price. Academic Press: London.

Mabry, T. J., J. H. Hunziker, and D. R. Difeo, Jr. 1977. *Creosote Bush, Biology and Chemistry of Larrea in New World Deserts.* Dowden, Hutchinson and Ross: Stroudsburg, Pennsylvania.

McAlpin, B. W., and R. A. White. 1974. Shoot organization in the Filicales: The promeristem. Amer. J. Bot. 61:562–579.

McClendon, J. H. 1984. The micro-optics of leaves. I. Patterns of reflection from the epidermis. Amer. J. Bot. 71:1391–1397.

McCully, M. E. 1975. The development of lateral roots. In *The Development and Function of Roots,* eds. J. G. Torrey and D. T. Clarkson. Academic Press: London.

MacFarlane, W. V. 1963. The stinging properties of *Laportea.* Econ. Bot. 17:303–311.

MacKenzie, K. A. D. 1979. The structure of the fruit of the red raspberry (*Rubus idaeus* L.) in relation to abscission. Ann. Bot. 43:355–362.

McNamara, J., and J. A. Quinn. 1977. Resource allocation and reproduction in populations of *Amphicarpum purshii* (Gramineae). Amer. J. Bot. 64:17–23.

Mader, H. 1954. Kork. Handb. der Pflanzenphysiol. 10:282–299.

Maercker, U. 1965. Über das Vorkommen von Stomata in der Epidermis bunter Perianthblätter. Z. Pflanzenphysiol. 53:422–428.

Mager, H. 1932. Beiträge zur Kenntnis der primaren Wurzelrinde. Planta 16:666–708.

Maheshwari, P. 1950. *An Introduction to the Embryology of Angiosperms.* McGraw-Hill: New York.

Maheshwari, P., and H. Singh. 1967. The female gametophyte of gymnosperms. Biol. Rev. 42:88–130.

Mahlberg, P. G. 1959a. Karyokinesis in non-articulated laticifers of *Nerium oleander* L. Phytomorphology 9:110–118.

Mahlberg, P. G. 1959b. Development of the non-articulated laticifer in proliferated embryos of *Euphorbia marginata* Pursh. Phytomorphology 9:156–162.

Mahlberg, P. G. 1961. Embryogeny and histogenesis in *Nerium oleander* L. II. Origin and development of the non-articulated laticifer. Amer. J. Bot. 48:90–99.

Mahlberg, P. G. 1975. Evolution of the laticifer in *Euphorbia* as interpreted from starch grain morphology. Amer. J. Bot. 62:577–583.

Mahlberg, P. G. 1982. Comparative morphology of starch grains in latex from varieties of *Poinsettia, Euphorbia pulcherrima* Willd. (Euphorbiaceae). Bot. Gaz. 143:206–209.

Mahlberg, P., and J. Pleszczynska. 1983. Phylogeny of *Euphorbia* interpreted from sterol composition of the laticifer.

In *Numerical Taxonomy*, ed. J. Felsenstein. Springer-Verlag: Berlin.

Mahlberg, P., J. Pleszczynska, E. Schnepf, and W. Rauh. 1983. Evolution of succulent *Euphorbia* as interpreted from latex composition. Bothalia 14:533–539.

Mahlberg, P. G., and P. Sabharwal. 1966. Mitotic waves in laticifers of *Euphorbia marginata*. Science 152:518–519.

Mahlberg, P. G., and P. S. Sabharwal. 1967. Mitosis in the non-articulated laticifer of *Euphorbia marginata*. Amer. J. Bot. 54:465–472.

Mahlberg, P. G., and P. Sabharwal. 1968. Origin and early development of non-articulated laticifers in embryos of *Euphorbia marginata*. Amer. J. Bot. 55:375–381.

Maiti, R. K. 1980. *Plant Fibers*. Bishen Singh Mahendra Pal Singh: Dehra Dun-248001, India.

Majumdar, G. P., and R. D. Preston. 1941. The fine structure of collenchyma cells in *Heracleum sphondylium* L. Proc. Roy. Soc. London B. 130:201–217.

Maksymowych, R., and R. O. Erickson. 1960. Development of the lamina in *Xanthium italicaum* represented by the plastochron index. Amer. J. Bot. 47:451–459.

Maksymowych, R., A. B. Maksymowych, and J. A. Orkwiszewski. 1985. Stem elongation of *Xanthium* plants presented in terms of relative elemental rates. Amer. J. Bot. 72:1114–1119.

Mallory, T. E., S. Chiang, E. G. Cutter, and E. M. Gifford, Jr. 1970. Sequence and pattern of lateral root formation in five selected species. Amer. J. Bot. 57:800–809.

Mangenot, S. 1968. Sur la presence de leucoplastes chez les végétaux vasculaires mycotrophes ou parasites. C. r. hebd. Séanc. Acad. Paris 267:1193–1195.

Marc, J., and J. H. Palmer. 1982. Changes in mitotic activity and cell size in the apical meristem of *Helianthus annuus* L. during transition to flowering. Amer. J. Bot. 69:768–775.

Marchant, H. J. 1982. The establishment and maintenance of plant cell shape by microtubules. In *The Cytoskeleton in Plant Growth and Development*, ed. C. W. Lloyd. Academic Press: London.

Marco, H. F. 1939. The anatomy of spruce needles. J. Agr. Res. 58:357–368.

Mares, D. J., B. A. Stone, C. Jeffery, and K. Norstog. 1977. Early stages in development of wheat endosperm. II. Ultrastructural observations on cell wall formation. Aust. J. Bot. 25:599–613.

Marloth, R. 1909. Die Schutzmittel der Pflanzen gegen übermässige Insolation. Ber. Dt. Bot. Ges. 27:362–371.

Marsden, M. P. F., and I. W. Bailey. 1955. A fourth type of nodal anatomy in dicotyledons, illustrated by *Clerodendron trichotomum* Thunb. J. Arnold Arbor. 36:1–51.

Martens, P. 1937. L'origine des espaces intercellulaires. Cellule 46:357–388.

Martens, P. 1938. Nouvelles recherches sur l'origine des espaces intercellulaires. Beih. Bot. Zbl. 58:349–364.

Martens, P. 1971. Les Gnétophytes. *Handbook of Plant Anatomy*, vol. 12. Gbdr. Borntraeger: Berlin.

Martens, P., and L. Waterkeyn. 1974. Étude sur les Gnétales. XIII. Recherches sur *Welwitschia mirabilis*. Cellule 70:163–258.

Martin, F. W., and S. Ortiz. 1963. Origin and anatomy of tubers of *Dioscorea floribunda* and *D. spiculiflora*. Bot. Gaz. 124:416–421.

Martin, J. T., and B. E. Juniper. 1970. *The Cuticles of Plants*. St. Martin's Press: New York.

Marty, F. 1968. Infrastructures des organes sécréteurs de la feuille d'*Urtica urens*. C. R. Acad. Sci. Paris 266:1712–1714.

Marty, F. 1978. Cytochemical studies on GERL, provacuoles, and vacuoles in root meristematic cells of *Euphorbia*. Proc. Natl. Acad. Sci. 75:852–856.

Marty, F., D. Branton, and R. A. Leigh. 1980. Plant vacuoles. In *The Biochemistry of Plants: A Comprehensive Treatise*, vol. 1, ed. N. E. Tolbert. Academic Press: New York.

Marvin, J. W. 1944. Cell shape and cell volume relations in the pith of *Eupatorium perfoliatum* L. Amer. J. Bot. 31:208–218.

Matile, P., and H. Moor. 1968. Vacuolation: Origin and development of the lysosomal apparatus in root-tip cells. Planta 80:159–175.

Matzke, E. B. 1939. Volume-shape relationships in lead shot and their bearing on cell shapes. Amer. J. Bot. 26:288–295.

Matzke, E. B. 1945. The three-dimensional shapes of bubbles in foams. Proc. Natl. Acad. Sci. 31:281–289.

Matzke, E. B. 1946. The three-dimensional shape of bubbles in foam—an analysis of the role of surface forces in three-dimensional cell shape determination. Amer. J. Bot. 33:58–80.

Mauseth, J. D. 1976. Cytokinin- and gibberellic acid-induced effects on the structure and metabolism of shoot apical meristems in *Opuntia polyacantha* (Cactaceae). Amer. J. Bot. 63:1295–1301.

Mauseth, J. D. 1977. Cytokinin- and gibberellic acid-induced effects on the determination and morphogenesis of leaf primordia in *Opuntia polyacantha* (Cactaceae). Amer. J. Bot. 64:337–346.

Mauseth, J. D. 1978a. An investigation of the morphogenetic mechanisms which control the development of zonation in seedling shoot apical meristems. Amer. J. Bot. 65:158–167.

Mauseth, J. D. 1978b. An investigation of the phylogenetic and ontogenetic variability of shoot apical meristems in the Cactaceae. Amer. J. Bot. 65:326–333.

Mauseth, J. D. 1978c. The structure and development of an unusual type of articulated laticifer in *Mammillaria* (Cactaceae). Amer. J. Bot. 65:415–420.

Mauseth, J. D. 1978d. Further studies of the unusual type of laticiferous canals in *Mammillaria* (Cactaceae): Structure and development of the semi-milky type. Amer. J. Bot. 65:1098–1102.

Mauseth, J. D. 1980a. Release of whole cells of *Nopalea* (Cactaceae) into secretory canals. Bot. Gaz. 141:15–18.

Mauseth, J. D. 1980b. A stereological morphometric study of the ultrastructure of mucilage cells in *Opuntia polyacantha* (Cactaceae). Bot. Gaz. 141:374–378.

Mauseth, J. D. 1980c. A morphometric study of the ultrastructure of *Echinocereus engelmannii* (Cactaceae). I. Shoot apical meristems at germination. Amer. J. Bot. 67:173–181.

Mauseth, J. D. 1981a. A morphometric study of the ultrastructure of *Echinocereus engelmannii* (Cactaceae). II. The mature, zonate shoot apical meristem. Amer. J. Bot. 68:96–100.

Mauseth, J. D. 1981b. A morphometric study of the ultrastructure of *Echinocereus engelmannii* (Cactaceae). III. Subapical and mature tissues. Amer. J. Bot. 68:531–534.

Mauseth, J. D. 1982a. A morphometric study of the ultrastructure of *Echinocereus engelmannii* (Cactaceae). IV. Leaf and spine primordia. Amer. J. Bot. 69:546–550.

Mauseth, J. D. 1982b. A morphometric study of the ultrastructure of *Echinocereus engelmannii* (Cactaceae). V. Comparison with the shoot apical meristems of *Trichocereus pachanoi* (Cactaceae). Amer. J. Bot. 69:551–555.

Mauseth, J. D. 1982c. A morphometric study of the ultrastructure of *Echinocereus englemannii* (Cactaceae). VI. The individualized ultrastructures of diverse types of meristems. Amer. J. Bot. 69:1524–1526.

Mauseth, J. D. 1982d. Development and ultrastructure of extrafloral nectaries in *Ancistrocactus scheeri* (Cactaceae). Bot. Gaz. 143:273–277.

Mauseth, J. D. 1984a. Effect of growth rate, morphogenetic activity, and phylogeny on shoot apical ultrastructure in *Opuntia polyacantha* (Cactaceae). Amer. J. Bot. 71:1283–1292.

Mauseth, J. D. 1984b. Introduction to cactus anatomy. Part 7, Epidermis. Cactus and Succ. J. (US) 56:33–37.

Mauseth, J. D. 1988. Systematic anatomy of the primitive cereoid cactus *Leptocereus quadricostatus*. Bradleya. In press.

Mauseth, J. D., and W. Halperin. 1975. Hormonal control of organogenesis in *Opuntia polyacantha* (Cactaceae). Amer. J. Bot. 62:869–877.

Mauseth, J. D., G. Montenegro, and A. M. Walckowiak. 1984. Studies of the holoparasite *Tristerix aphyllus* (Loranthaceae) infecting *Trichocereus chilensis* (Cactaceae). Can. J. Bot. 62:847–857.

Mauseth, J. D., G. Montenegro, and A. M. Walckowiak. 1985. Host infection and flower formation by the parasite *Tristerix aphyllus* (Loranthaceae). Can. J. Bot. 63:567–581.

Mauseth, J. D., and K. J. Niklas. 1979. Constancy of relative volumes of zones in shoot apical meristems in Cactaceae: Implications concerning meristem size, shape, and metabolism. Amer. J. Bot. 66:933–939.

Maximov, N. A. 1931. The physiological significance of the xeromorphic structure of plants. J. Ecol. 19:272–282.

Mayers, A. M., and E. M. Lord. 1984. Comparative flower development in the cleistogamous species *Viola odorata*. III. A histological study. Bot. Gaz. 145:83–91.

Meeuse, A. D. 1963. From ovule to ovary: a contribution to the phylogeny of the megasporangium. Acta Biotheoretica 16:127–182.

Meeuse, B., and S. Morris. 1984. *The Sex Life of Flowers*. Facts on File Publications: New York.

Mehra, P. N., and R. K. Malhotra. 1947. Stages in the embryogeny of *Cupressus sempervirens* with particular reference to the occurrence of multiple male cells in the male gametophyte. Proc. Nat. Acad. Sci. India 17B:129–153.

Mehra, P. N., and S. L. Soni. 1971. Morphology of tracheary elements in *Marsilea* and *Pteridium*. Phytomorph. 21:68–71.

Meidner, H., and D. W. Sheriff. 1976. *Water and Plants*. Blackie and Son: Glasgow.

Meier, H., and J. S. G. Reid. 1982. Reserve polysaccharides other than starch in higher plants. In *Plant Carbohydrates. I. Intracellular carbohydrates. Encyclopedia of Plant Physiology*, vol. 13A, eds. F. A. Loewus and W. Tanner. Springer-Verlag: Berlin.

Mennega, A. M. W. 1969. The wood structure of *Dicranostyles* (Convolvulaceae). Acta Bot. Neerl. 18:173–179.

Mepham, R. H., and G. R. Lane. 1970. Observations on the fine structure of developing microspores of *Tradescantia braceata*. Protoplasma 70:1–20.

Mersky, M. L. 1973. Lower Cretaceous (Potamic group) angiosperm cuticles. Amer. J. Bot. 60 (supplement):17–18.

Metcalfe, C. R. 1960. *Anatomy of the Monocotyledons. I. Gramineae*. Clarendon Press: Oxford.

Metcalfe, C. R. 1961. The anatomical approach to systematics. In *Recent Advances in Botany*. Toronto University Press: Toronto.

Metcalfe, C. R. 1963. Comparative anatomy as a modern botanical discipline. Adv. Bot. Res. 1:101–147.

Metcalfe, C. R. 1967. Distribution of latex in the plant kingdom. Econ. Bot. 21:115–125.

Metcalfe, C. R. 1971. *Anatomy of the Monocotyledons. V. Cyperaceae*. Clarendon Press. Oxford.

Metcalfe, C. R. 1979a. The leaf: General topography and ontogeny of the tissues. In *Anatomy of the Dicotyledons*, vol. 1, eds. C. R. Metcalfe and L. Chalk. Clarendon Press: Oxford.

Metcalfe, C. R. 1979b. Secreted mineral substances. In *Anatomy of the Dicotyledons*, vol. 2, eds. C. R. Metcalfe and L. Chalk. Clarendon Press: Oxford.

Metcalfe, C. R. 1979c. Some basic types of cells and tissues. In *Anatomy of the Dicotyledons*, second edition, vol. 1, eds. C. R. Metcalfe and L. Chalk. Clarendon Press: Oxford.

Metcalfe, C. R. 1979d. The stem. In *Anatomy of the Dicotyledons*, vol. 1, eds. C. R. Metcalfe and L. Chalk. Clarendon Press: Oxford.

Metcalfe, C. R. 1983a. Ecological anatomy and morphology.

General survey. In *Anatomy of the Dicotyledons*, vol. 2, eds. C. R. Metcalfe and L. Chalk. Clarendon Press: Oxford.

Metcalfe, C. R. 1983b. Anomalous structure. In *Anatomy of the Dicotyledons*, vol. 2, eds. C. R. Metcalfe and L. Chalk. Clarendon Press: Oxford.

Metcalfe, C. R. 1983c. Secreted mineral substances. In *Anatomy of the Dicotyledons*, vol. 2. eds. C. R. Metcalfe and L. Chalk. Clarendon Press: Oxford.

Metcalfe, C. R., and L. Chalk. 1950. *Anatomy of the Dicotyledons*, vols. 1 and 2. Clarendon Press: Oxford.

Metcalfe, C. R., and L. Chalk. 1979. *Anatomy of the Dicotyledons*, vol. 1. Clarendon Press: Oxford.

Metcalfe, C. R., and L. Chalk. 1983. *Anatomy of the Dicotyledons*, vol. 2, *Wood Structure and Conclusion of the General Introduction*. Clarendon Press: Oxford.

Mettenius, G. H. 1865. Über die Hymenophyllaceae. Abh. sächs Akad. Wiss. (Math.-Phys. Kl) 7:403–504.

Meyer, F. J. 1928. Die Begriffe "stammeigene Bündel" und "Blattspurbündel" in Lichte unserer heutigen Kenntnisse von Aufbau und der physiologischen Wirkungsweise der Leitbundel. Jb. Wiss. Bot. 69: 237–263.

Meyer, F. J. 1962. Das tropische Parenchym: A, Assimilationsgewebe. In *Encyclopedia of Plant Anatomy*, vol. 4, part 7A. Gbdr. Borntraeger. Berlin.

Meyer, R. W., and W. A. Côté, Jr. 1968. Formation of the protective layer and its role in tylosis development. Wood Sci. Technol. 2:84–94.

Meylan, B. A., and B. G. Butterfield. 1972. Perforation plate development in *Knightia excelsa* R. Br.: A scanning electron microscope study. Aust. J. Bot. 20:79–86.

Meylan, B. A., and B. G. Butterfield. 1973. Unusual perforation plates: Observations using scanning electron microscopy. Micron 4:47–59.

Meylan, B. A., and B. G. Butterfield. 1975. Occurrence of simple, multiple and combination perforation plates in the vessels of New Zealand woods. N. Z. J. Bot. 13:1–18.

Meylan, B. A., and B. G. Butterfield. 1981. Perforation plate development in the vessels of hardwoods. In *Xylem Cell Development*, ed. J. R. Barnett. Castle House Publications, Ltd.: Kent, England.

Michener, D. C. 1983. Systematic and ecological wood anatomy of Californian Scrophulariaceae. I. *Antirrhinum, Castilleja, Galvezia,* and *Mimulus* sect. *Diplacus.* Aliso 10: 471–487.

Mikesell, J. E. 1979. Anomalous secondary thickening in *Phytolacca americana* L. (Phytolaccaceae). Amer. J. Bot. 66:997-1005.

Mikesell, J. E., and A. C. Schroeder. 1980. Development of chambered pith in stems of *Phytolacca americana* L. (Phytolaccaceae). Amer. J. Bot. 67:111–118.

Miller, R. H. 1985. The prevalence of pores and canals in leaf cuticular membranes. Ann. Bot. 55:459–471.

Millet, M. A., A. J. Baker, and L. D. Satter. 1975. Pretreatments to enhance chemical, enzymatic, and microbiological attack of cellulosic materials. Biotechnol. Bioeng. Symp. 5:193–219.

Mitchison, J. M. 1971. *The Biology of the Cell Cycle.* Cambridge University Press: Cambridge.

Mogensen, H. L. 1968a. Studies on the bark of the cork bark fir: *Abies lasiocarpa* var. *arizonica* (Merriam) Lemmon. I. Periderm ontogeny. Ariz. Acad. Sci. J. 5:36–40.

Mogensen, H. L. 1968b. Studies on the bark of the cork bark fir: *Abies lasiocarpa* var. *arizonica* (Merriam) Lemmon. II. The effect of exposure on the time of initial rhytidome formation. Ariz. Acad. Sci. J. 5:108–109.

Mogensen, H. L. 1978. Pollen tube-synergid interactions in *Proboscidea louisianica* (Martineaceae). Amer. J. Bot. 65: 953–964.

Mogensen, H. L. 1985. Ultracytochemical localization of plasma membrane-associated phosphatase activity in developing tobacco seeds. Amer. J. Bot. 72:741–754.

Mogensen, H. L., and H. K. Suthar. 1979. Ultrastructure of the egg apparatus of *Nicotiana tabacum* (Solanaceae) before and after fertilization. Bot. Gaz. 140:168–179.

Moline, H. E., and J. M. Bostrack. 1972. Abscission of leaves and leaflets in *Acer negundo* and *Fraxinus americana.* Amer. J. Bot. 59:83–88.

Mollenhauer, H. H. 1967. The fine structure of mucilage secreting cells of *Hibiscus esculentus* pods. Protoplasma 63:353–362.

Mollenhauer, H. H., and D. J. Morré. 1966. Golgi apparatus and plant secretion. Ann. Rev. Plant Physiol. 17: 27–46.

Mollenhauer, H. H., and D. J. Morré. 1980. The golgi apparatus. In *The Biochemistry of Plants: A Comprehensive Treatise*, vol. 1, ed. N. E. Tolbert. Academic Press: New York.

Monsi, M. 1943. Untersuchungen über den Mechanismus der Schleuderbewegung der Sojabohnen-Hülse. Jap. J. Bot. 12:437–474.

Montenegro, G. 1974. Desarrollo de raíces contráctiles en *Hippeastrum chilense,* geofita del matorral chileno. Acta Cient. Venez. 25:82–86.

Montenegro, G. 1984. *Atlas de Anatomía de Especies Vegetales Autótonas de la Zona Central.* Ediciones Universidad Católica de Chile: Santiago, Chile.

Montenegro, G., G. Avila, and P. Schatte. 1983. Presence and development of lignotubers in shrubs of the Chilean matorral. Can. J. Bot. 61:1804–1808.

Moon, G. J., C. A. Peterson, and R. L. Peterson. 1984. Structural, chemical and permeability changes following wounding in onion roots. Can. J. Bot. 62:2253–2259.

Mooney, H. A., P. J. Weisser, and S. L. Gulmon. 1977. Environmental adaptations of the Atacaman desert cactus *Copiapoa haseltoniana.* Flora 166:117–124.

Moore, R. 1982. Studies of vegetative compatibility-incompatibility in higher plants. V. A morphometric analysis of the development of a compatible and an incompatible graft. Can. J. Bot. 60:2780–2787.

Moore, R. 1983. A morphometric analysis of the ultrastructure of columella statocytes in primary roots of *Zea mays*. L. Ann. Bot. 51:771–778.

Moore, R. 1984. Cellular interactions during the formation of approach grafts in *Sedum telephoides* (Crassulaceae). Can. J. Bot. 62:2476–2484.

Moore, R., and C. E. McClelen. 1983a. A morphometric analysis of cellular differentiation in the root cap of *Zea mays*. Amer. J. Bot. 70:611–617.

Moore, R., and C. E. McClelen. 1983b. Ultrastructural aspects of cellular differentiation in the root cap of *Zea mays*. Can. J. Bot. 61:1566–1572.

Moore, R., and J. Pasieniuk. 1984. Structure of columella cells in primary and lateral roots of *Ricinus communis* (Euphorbiaceae). Ann. Bot. 53:715–726.

Morrison, I. N., and T. P. O'Brien. 1976. Cytokinesis in the developing wheat grain; division with and without a phragmoplast. Planta 130:57–67.

Morrison, I. N., and T. P. O'Brien. 1978. Initial cellularization and differentiation of the aleurone cells in the ventral region of the developing wheat grain. Planta 140:19–30.

Morrison, J. C., and V. S. Polito. 1985. Gum duct development in almond fruit, *Prunus dulcis* (Mill.) D. A. Webb. Bot. Gaz. 146:15–25.

Morton, A. G. 1981. *History of Botanical Science*. Academic Press: London.

Moss, E. H. 1940. Interxylary cork in *Atemisia* with a reference to its taxonomic significance. Amer. J. Bot. 27:762–768.

Moss, E. H., and A. I. Gorham. 1953. Interxylary cork and fission of stems and roots. Phytomorph. 3:285–294.

Mueller, R. J. 1982. Shoot morphology of the climbing fern *Lygodium* (Schizaeaceae): General organography, leaf initiation, and branching. Bot. Gaz. 143:319–330.

Mueller, R. J. 1985. Determinate branch development in *Alstonia scholaris* (Apocynaceae): The plagiotropic module. Amer. J. Bot. 72:1435–1444.

Muhammad, A. F., and R. Sattler. 1982. Vessel structure of *Gnetum* and the origin of angiosperms. Amer. J. Bot. 69:1004–1021.

Mukerji, S. K. 1936. Contributions to the autecology of *Mercurialis perennis* L. J. Ecol. 24:38–81.

Müller, C. 1890. Ein Beitrag zur Kenntnis der Formen des Collenchyms. Ber. dt. bot. Ges. 8:150–166.

Muller, J. 1979. Form and function in angiosperm pollen. Ann. Missouri Bot. Gard. 66:593–632.

Muller, J. 1981. Fossil pollen records of extant angiosperms. Bot. Rev. 47:1–142.

Muller-Stoll, W., and H. Suss. 1969. Uber aussergewöhnlich gestaltete vielfache Gefassdurchbrechungen bei *Aeschynomene virginica* (L.) B. S. P. (*Papili naceae*). Ber. Dt. Bot. Ges. 82:613–619.

Murmanis, L. 1971. Structural changes in vascular cambium of *Pinus strobus* L. during an annual cycle. Ann. Bot. 35:133–141.

Murmanis, L. 1975. Formation of tyloses in felled *Quercus rubra* L. Wood Sci. Technol. 9:3–14.

Murmanis, L. 1976. The protective layer in xylem parenchyma cells of *Quercus rubra* L. App. Pol. Symp. 28:1283–1292.

Murmanis, L. 1978. Breakdown of end walls in differentiating vessels of secondary xylem in *Quercus rubra* L. Ann. Bot. 42:679–682.

Murmanis, L., and R. F. Evert. 1966. Some aspects of sieve cell ultrastructure in *Pinus strobus*. Amer. J. Bot. 53:1065–1078.

Murty, Y. S. 1983. Morphology, anatomy and development of bulbil in some *Dioscorea*. Proc. Indian Acad. Sci., Plant Sci. 92:443–450.

Mustard, M. J. 1982. Origin and distribution of secondary articulated anastomosing laticifers in *Manilkara zapota* (Sapotaceae). J. Am. Soc. Hort. Sci. 107:355–360.

Mylius, G. 1913. Das Polyderm. Bibl. Bot. 1:1–119.

Nägeli, C. W. 1858. Das Wachsthum des Stammes und der Wurzel bei den Gefässpflanzen und die Anordnung der Gefässträge in Stengel. Beitr. z. Wiss. Bot. 1:1–156.

Nakamura, S., and H. Miki-Hirosige. 1985. Fine-structural study on the formation of the generative cell wall and intine-3 layer in a growing pollen grain of *Lilium longiflorum*. Amer. J. Bot. 72:365–375.

Nakamura, Y., and T. Higuchi. 1976. Ester linkage of *p*-coumaric acid in bamboo lignin. Holzforschung 30:187–191.

Namboodiri, K. K., and C. B. Beck. 1968a. A comparative study of primary vascular system in conifers. I. Genera with helical phyllotaxis. Amer. J. Bot. 55:447–457.

Namboodiri, K. K., and C. B. Beck. 1968b. A comparative study of primary vascular system in conifers. III. Stelar evolution in gymnosperms. Amer. J. Bot. 55:461–472.

Napp-Zinn, K. 1973, 1974. Anatomie des Blattes. II. Blattanatomie der Angiospermen. A. Entwicklungsgeschichtliche und topographische Anatome des Angiospermenblattes, 2 vols. In *Encyclopedia of Plant Anatomy*, vol. 8, part 2A, Gbdr. Borntraeger: Berlin.

Negbi, M. 1984. The structure and function of the scutellum of the Gramineae. Bot. J. Linn. Soc. 88:205–222.

Negbi, M., and D. Koller. 1962. Homologies in the grass embryo—A re-evaluation. Phytomorph. 12:289–295.

Nessler, C. L., and P. G. Mahlberg. 1977. Ontogeny and cytochemistry of alkaloidal vesicles in laticifers of *Papaver somniferum* L. (Papaveraceae). Amer. J. Bot. 64:541–551.

Nessler, C. L., and P. G. Mahlberg. 1978. Laticifer ultrastructure and differentiation in seedlings of *Papaver bracteatum* Lindl., population Arya II (Papaveraceae). Amer. J. Bot. 65:978–983.

Netolitzky, F. 1926. *Anatomie der Angiospermen-Samen*. Gbdr. Borntraeger: Berlin.

Neumann, D., and E. Müller. 1972. Beiträge zur Physiologie der Alkaloide. III. *Chelidonium majus* L. und *Sanguinaria canadensis* L: Ultrastruktur der Alkaloidbehälter Alkaloidaufnahme und Verteilung. Biochem. Physiol. Pflanzen 163:375–391.

Newcomb, W. 1978. The development of cells in the coenocytic endosperm of the African blood lily *Haemanthus*. Can. J. Bot. 56:483–501.

Newman, E. I. 1974. Root and soil water relations. In *The Plant Root and Its Environment*, ed. E. W. Carson. University Press of Virginia: Charlottesville, Virginia.

Newman, I. V. 1956. Pattern in the meristems of vascular plants. I. Cell partition in living apices and in the cambial zone in relation to the concepts of initial cells and apical cells. Phytomorph. 6:1–19.

Nicharat, S., and G. W. Gillett. 1970. A review of the taxonomy of hawaiian *Pipturus* (Urticaceae) by anatomical and cytological evidence. Brittonia 22:191–206.

Niklas, K. J., and V. Kerchner. 1984. Mechanical and photosynthetic constraints on the evolution of plant shape. Paleobiology 10:79–101.

Niklas, K. J., and J. D. Mauseth. 1981. Relationships among shoot apical meristem ontogenic features in *Trichocereus pachanoi* and *Melocactus matanzanus* (Cactaceae). Amer. J. Bot. 68:101–106.

Niklas, K. J., and K. Norstog. 1984. Aerodynamics and pollen grain depositional patterns on cycad megastrobili: Implications on the reproduction of three cycad genera (*Cycas, Dioon,* and *Zamia*). Bot. Gaz. 145:92–104.

Niklas, K. J., and K. T. Paw U. 1983. Conifer ovulate cone morphology: Implications on pollen impaction patterns. Amer. J. Bot. 70:568–577.

Nobel, P. S. 1978. Microhabitat, water relations, and photosynthesis of a desert fern, *Notholaena parryi*. Oecologia 31:293–309.

Nobel, P. S. 1983. *Biophysical Plant Physiology and Ecology*. W. H. Freeman and Company: San Francisco.

Noel, A. R. A. 1983. The endothecium—a neglected criterion in taxonomy and phylogeny? Bothalia 14:833–838.

Nolan, J. R. 1969. Bifurcation of the stem apex in *Asclepias syriaca*. Amer. J. Bot. 56:603–609.

Norstog, K. 1967. Fine structure of the spermatozoids of *Zamia* with special reference to the flagellar apparatus. Amer. J. Bot. 54:831–840.

Norstog, K. 1974. Nucellus during early embryogeny in barley: Fine structure. Bot. Gaz. 135:97–103.

Northcote, D. H. 1969. Fine structure of cytoplasm in relation to synthesis and secretion in plant cells. Proc. R. Soc. B. 173:21–30.

Northcote, D. H., and J. D. Pickett-Heaps. 1966. A function of the Golgi apparatus in polysaccharide synthesis and transport in the root-cap cells of wheat. Biochem. J. 98:159–167.

Nougarède, A. 1965. Organisation et fonctionnement du méristème apical des végétaux vasculaires. In *Travaux dédiés au Lucien Plantefol*. Masson et Cie: Paris.

Nowicke, J. W., V. Patel, and J. J. Skvarla. 1985. Pollen morphology and the relationships of *Aëtoxylon, Amyxa* and *Gonystylus* to the Thymelaeaceae. Amer. J. Bot. 72:1106–1113.

O'Brien, T. P. 1981. The primary xylem. In *Xylem Cell Development*, ed. J. R. Barnett. Castle House Publications, Ltd.: Kent, England.

O'Brien, T. P., and D. J. Carr. 1970. A suberized layer in the cell walls of the bundle sheath of grasses. Aust. J. Biol. Sci. 23:275–287.

O'Brien, T. P., and K. V. Thimann. 1967. Observations on the fine structure of the oat coleoptile. III. Correlated light and electron microscopy of vascular tissues. Protoplasma 63:443–478.

Ogura, Y. 1937. Disarticulation of the branches in *Bladhia* (Myrsinaceae). Bot. Mag. Tokyo 51:158–167.

Ogura, Y. 1938. Anatomie der Vegetationsorgane der Pteridophyten. In *Handbuch der Pflanzenanatomie*, Bd. 7, Lief. 36, ed. K. Linsbauer. Gbdr. Borntraeger: Berlin.

Ogura, Y. 1972. Comparative anatomy of vegetative organs of the pteridophytes. In *Encyclopedia of Plant Anatomy*, vol. 7, pt. 3, ed. K. Linsbauer. Gbdr. Borntraeger: Berlin.

Ohtani, J., and S. Ishida. 1973. An observation of the sculptures of the vessel wall of *Fagus crenata* Bl. using scanning electron microscopy. Res. Bull. College Exp. For. Hokkaido Univ. 33:115–126.

Okonkwo, S. N. C., and V. Raghavan. 1982. Studies on the germination of seeds of the root parasites, *Alectra vogelii* and *Striga gesnerioides*. I. Anatomical changes in the embryos. Amer. J. Bot. 69:1636–1645.

Oksala, T., and E. Therman. 1977. Endomitosis in tapetal cells of *Eremurus* (Liliaceae). Amer. J. Bot. 64:866–872.

Olesen, P. 1978. Studies on the physiological sheaths in roots. I. Ultrastructure of the exodermis in *Hoya carnosa* L. Protoplasma 94:325–340.

Olien, W. C., and M. J. Bukovac. 1982. Ethephon-induced gummosis in sour cherry (*Prunus cerasus* L.). Plant Physiol. 70:547–555, 556–559.

Olson, J. M. 1981. Evolution of photosynthetic and respiratory prokaryotes and organelles. Ann. N. Y. Acad. Sci. 361:8–17.

Öpik, H. 1974. Mitochondria. In *Dynamic Aspects of Plant Ultrastructure*, ed. A. W. Robards. McGraw-Hill: London.

Oppenheimer, H. R. 1960. Adaptation to drought: Xerophytism. Plant-water relationships in arid and semiarid conditions. Reviews of research, Unesco. Arid Zone Res. 15:105–138.

Orr, A. R. 1981. A quantitative study of cellular events in the shoot apical meristem of *Brassica campestris* (Cruciferae) during transition from vegetative to reproductive condition. Amer. J. Bot. 68:17–23.

Orr, A. R. 1985. Histochemical study of enzyme activity in

the shoot apical meristem of *Brassica campestris* L. during transition to flowering. III. Glucose-6-phosphate dehydrogenase and 6-phosphogluconate dehydrogenase. Bot. Gaz. 146:477–482.

Osmond, C. B., U. Lüttge, K. R. West, C. K. Pallaghy, and B. Shachar-Hill. 1969. Ion absorption in *Atriplex* leaf tissue. II. Secretion of ions to epidermal bladders. Aust. J. Bibl. Sci. 22:797–814.

Ozenda, P. 1949. Rescherches sur les dicotylédones apocarpiques. Contribution à l'étude des angiospermes dites primitives. Publ. Lab. Biol. Ecole norm supér. Masson and Cir: Paris.

Pacini, E., and B. E. Juniper. 1983. The ultrastructure of the formation and development of the amoeboid tapetum in *Arum italicum* Miller. Protoplasma 117:116–129.

Pacini, E., C. Simoncioli, and M. Cresti. 1975. Ultrastructure of nucellus and endosperm of *Diplotaxis erucoides* during embryogenesis. Caryologia 28:525–538.

Palevitz, B. A., and P. K. Hepler. 1976. Cellulose microfibril orientation and cell shaping in developing guard cells of *Allium:* The role of microtubules and ion accumulation. Planta 132:71–93.

Panshin, A. J., and C. de Zeeuw. 1980. *Textbook of Wood Technology.* McGraw-Hill: New York.

Pant, D. D. 1965. On the ontogeny of stomata and other homologous structures. Plant Sci. Ser. (Allahabad, India) 1:1–24.

Pant, D. D. 1973. Cycas *and the Cycadales.* Central Book Deposit Publications in Botany: Allahabad, India.

Pant, D. D., and R. Banerji. 1965. Epidermal structure and development of stomata in Convolvulaceae. Senckenberg. Biol. 46:155–173.

Pant, D. D., and P. F. Kidwai. 1967. Development of stomata in some Cruciferae. Ann. Bot. 31:513–521.

Pant, D. D., and B. Mehra. 1964. Nodal anatomy in retrospect. Phytomorph. 14:384–387.

Parameswaran, N., and W. Liese. 1969. On the formation and fine structure of septate wood fibres of *Ribes sanguineum.* Wood Sci. Tech. 3:272–286.

Parameswaran, N., and W. Liese. 1973. Scanning electron microscopy of multiperforate perforation plates. Holzforsch. 27:181–186.

Parker, M. L. 1984a. Cell wall storage polysaccharides in cotyledons of *Lupinus angustifolius* L. I. Deposition during seed development. Protoplasma 120:224–232.

Parker, M. L. 1984b. Cell wall storage polysaccharides in cotyledons of *Lupinus angustifolius* L. II. Mobilization during germination and seedling development. Protoplasma 120:233–241.

Parker, V. T., and M. A. Leck. 1985. Relationships of seed banks to plant distribution patterns in a freshwater tidal wetland. Amer. J. Bot. 72:161–174.

Parkhurst, D. F. 1982. Stereological methods for measuring internal leaf structure variables. Amer. J. Bot. 69:31–39.

Parthasarathy, M. V., and K. Mühlethaler. 1969. Ultastructure of protein tubules in differentiating sieve elements. Cytobiologie 1:17–36.

Parthasarathy, M. V., and P. B. Tomlinson. 1967. Anatomical features of metaphloem in stems of *Sabal, Cocos,* and two other palms. Amer. J. Bot. 54:1143–1151.

Pate, J. S., and B. E. S. Gunning. 1969. Vascular transfer cells in angiosperm leaves. A taxonomic and morphological survey. Protoplasma 68:135–156.

Pate, J. S., and B. E. S. Gunning. 1972. Transfer cells. Ann. Rev. Pl. Physiol. 23:173–196.

Patel, J. D. 1978. How should we interpret and distinguish subsidiary cells? J. Linn. Soc. Bot. 77:65–72.

Patrick, J. W. 1972. Vascular system of the stem of the wheat plant. Aust. J. Bot. 20:65–78.

Patterson, D. T. 1975. Nutrient return in the stem flow and throughfall of individual trees in the piedmont deciduous forest. In *Mineral Cycling in Southeastern Ecosystems,* eds. F. G. Howell, J. B. Gentry, and M. H. Smith. ERDA Symp. Ser. (Conf. 740513). National Technical Information Service: Springfield, Virginia.

Paull, R. E., and R. L. Jones. 1976. Studies on the secretion of maize root cap slime. IV. Evidence for the involvement of dictyosomes. Plant Physiol. 57:249–256.

Payne, W. W. 1970. Helicocytic and allelocytic stomata: Unrecognized patterns in the Dicotyledonae. Amer. J. Bot. 57:140–147.

Payne, W. W. 1978. A glossary of hair terminology. Brittonia 30:239–255.

Payne, W. W. 1979. Stomatal patterns in embryophytes: Their evolution, ontogeny and interpretation. Taxon 28:117–132.

Payne, W. W. 1981. Structure and function in angiosperm pollen wall evolution. Rev. Palaeobotany and Palynology 35:39–59.

Pearl, I. A. 1967. *The Chemistry of Lignin.* Marcel Dekker, Inc.: New York.

Pentecost, A. 1980. Calcification in plants. Int. Rev. Cytol. 62:1–27.

Percival, M. 1979. *Floral Biology.* Pergamon Press: Oxford.

Percy, K. E. 1985. Effects of acid rain on forest vegetation: Morphological and non-mensurational growth effects. Proceedings of the Acid Rain and Forest Resources Conference: Quebec, Canada.

Percy, K. E., and R. T. Riding. 1978. The epicuticular waxes of *Pinus strobus* subjected to air pollutants. Can. J. Forest Res. 8:474–477.

Perrin, A. 1971. Présence de "cellules de transfert" au sein de l'épithème de quelques hydathodes. Z. Pflanzenphysiol. 65:39–51.

Perry, T. O. 1971. Dormancy of trees in winter. Science 171:29–36.

Pesis, E., Y. Fuchs, and G. Zauberman. 1978. Cellulase and softening in avocado. Plant Physiol. 61:416–419.

Peterson, C. A., M. E. Emanuel, and C. Wilson. 1982. Identification of a Casparian band in the hypodermis of onion and corn roots. Can. J. Bot. 60:1529–1535.

Peterson, C. A., and C. J. Perumalla. 1984. Development of the hypodermal Casparian band in corn and onion roots. J. Exp. Bot. 35:51–57.

Peterson, C. A., R. L. Peterson, and A. W. Robards. 1978. A correlated histochemical and ultrastructural study of the epidermis and hypodermis of onion roots. Protoplasma 96:1–21.

Peterson, R. L. 1975. The initiation and development of root buds. In *The Development and Function of Roots*, eds. J. G. Torrey and D. T. Clarkson. Academic Press: London.

Peterson, R. L., and S. Hambleton. 1978. Guard cell ontogeny in leaf stomata of the fern *Ophioglossum petiolatum*. Can. J. Bot. 56:2836–2852.

Peterson, R. L., and J. Vermeer. 1980. Root apex structure in *Ephedra monosperma* and *Ephedra chilensis* (Ephedraceae). Amer. J. Bot. 67:815–823.

Pettitt, J. M. 1966. A new interpretation of the structure of the megaspore membrane in some gymnospermous ovules. J. Linn. Soc. London Bot. 59:253.

Pfeiffer, H. 1926. Das abnorme Dickenwachstum. In *Handbuch der Pflanzenanatomie*, Bd. 9, Lief. 15, ed. K. Linsbauer. Gbdr. Borntraeger: Berlin.

Pfeiffer, H. 1928. Die pflanzlichen Trennungsgewebe. In *Encyclopedia of Plant Anatomy*, ed. K. Linsbauer. Gbdr. Borntraeger: Berlin.

Philip, V. J. 1972. Embryogenesis and seedling anatomy of *Catharanthus roseus* (Linn.) G. Don. I. Embryogeny and procambialization. Cellule 69:155–172.

Philipp, M. 1923. Über die verkorkten Abschlussgewebe der Monokotylen. Bibliotheca bot. 92:1–28.

Philipson, W. R., and J. M. Ward. 1965. The ontogeny of the vascular cambium in the stem of seed plants. Biol. Rev. 40:534–579.

Philipson, W. R., J. M. Ward, and B. G. Butterfield. 1971. *The Vascular Cambium*. Chapman and Hall: London.

Pickett-Heaps, J. D. 1967a. The effects of colchicine on the ultrastructure of dividing plant cells, xylem, wall differentiation and distribution of cytoplasmic microtubules. Devl. Biol. 15:206–236.

Pickett-Heaps, J. D. 1967b. Further observations on Golgi apparatus and its functions in cells of the wheat seedling. J. Ultrastruct. Res. 18:287–303.

Pickett-Heaps, J. D. 1968. Xylem wall deposition: Radioautographic investigations using lignin precursors. Protoplasma 65:181–205.

Pickett-Heaps, J. D., and D. H. Northcote. 1966a. Organization of microtubules and endoplasmic reticulum during mitosis and cytokinesis in wheat meristems. J. Cell Sci. 1:109–120.

Pickett-Heaps, J. D., and D. H. Northcote. 1966b. Cell division in the formation of the stomatal complex of the young leaves of wheat. J. Cell Sci. 1:121–128.

Pierre, L. 1896. Plantes du Gabon. Bull. Mens. Soc. Linn. Paris 2:1249–1256.

Pigg, K. B., and G. W. Rothwell. 1983. *Chaloneria* gen. nov., heterosporous lycophytes from the Pennsylvanian of North America. Bot. Gaz. 144:132–147.

Pilet, P. E., and J. C. Roland. 1974. Growth and extensibility of collenchyma cells. Plant Sci. Let. 2:203–207.

Pinthus, M. J. 1967. Spread of the root system as indicator for evaluating lodging resistance of wheat. Crop Sci. 7:107–110.

Pireyre, N. 1961. Contribution à l'étude morphologique, histologique et physiologique des cystolithes. Rev. Cyt. et Biol. Végét. 23:93–320.

Pirwitz, K. 1931. Physiologische und anatomische Untersuchungen an Speichertracheiden und Velamina. Planta 14:19–76.

Pizzolato, T. D. 1983. Vascular system of lodicules of *Dactylis glomerata*. Amer. J. Bot. 70:17–29.

Pizzolato, T. D., J. L. Burbano, J. D. Berlin, P. R. Morey, and R. W. Pease. 1976. An electron microscope study of the path of water movement in transpiring leaves of cotton (*Gossypium hirsutum* L.). J. Exp. Bot. 27:145–161.

Pizzolato, T. D., and C. Heimsch. 1975. Ontogeny of the protophloem fibers and secondary xylem fibers within the stem of *Coleus*. I. A light microscope study. Can. J. Bot. 53:1658–1671.

Plantefol, L. 1947. Hélices foliares, point végétatif et stèle chez les Dicotylédones. La notion d'anneau initial. Rev. gén. Bot. 54:49–80.

Platt-Aloia, K. A., J. W. Oross, and W. W. Thomson. 1983. Ultrastructural study of the development of oil cells in the mesocarp of avocado (*Persea americana*) fruit. Bot. Gaz. 144:49–55.

Platt-Aloia, K. A., W. W. Thomson, and R. E. Young. 1980. Ultrastructural changes in the walls of ripening avocados: Transmission, scanning, and freeze fracture microscopy. Bot. Gaz. 141:366–373.

Plymale, E. L., and R. B. Wylie. 1944. The major veins of mesomorphic leaves. Amer. J. Bot. 31:99–106.

Pobeguin, T. 1951. Précipitation du carbonate de calcium chez quelques végétaux: existence in vivo et in vitro du calcaire amorph. Ann. Sci. nat. bot. sér. 11, 12:219–225.

Pobeguin, T. 1954. Contribution à l'étude des carbonates de calcium, précipitation du calcaire par les végétaux, comparaisons avec le monde animal. Ann. Sci. nat. bot. sér. 11, 15:29–109.

Pollak, P. E., and H. L. Mogensen. 1986. Ultracytochemical and biochemical studies of plasma membrane ATPase in the individual parts of the mature, dormant seed of *Pisum sativum*. Amer. J. Bot. 73:48–59.

Poovaiah, B. W. 1974. Formation of callose and lignin during leaf abscission. Amer. J. Bot. 61:829–834.

Popham, R. A. 1952. *Developmental Plant Anatomy*. Long: Columbus, Ohio.

Popham, R. A. 1955. Zonation of primary and lateral root apices of *Pisum sativum.* Amer. J. Bot. 42:267–273.

Popham, R. A. 1960. Variability among vegetative shoot apices. Bull. Torrey Bot. Club 87:139–150.

Popham, R. A., and A. P. Chan. 1950. Zonation in the vegetative stem tip of *Chrysanthemum morifolium* Bailey. Amer. J. Bot. 37:476–484.

Popham, R. A., and R. D. Henry. 1955. Multicellular root hairs on adventitious roots of *Kalanchoë fedtschenkoi.* Ohio J. Sci. 55:301–307.

Posluszny, U., M. J. Sharp, and P. A. Keddy. 1984. Vegetative propagation in *Rhexia virginica* (Melastomataceae): Some morphological and ecological considerations. Can. J. Bot. 62:2118–2121.

Possingham, J. V. 1980. Plastid replication and development in the life cycle of higher plants. Ann. Rev. Plant Physiol. 31:113–129.

Possingham, J. V., and M. E. Lawrence. 1983. Controls to plastid division. Int. Rev. Cytol. 84:1–56.

Possingham, J. V., and W. Saurer. 1969. Changes in chloroplast number per cell during leaf development in spinach. Planta 86:186–194.

Postek, M. T. 1981. The occurrence of silica in the leaves of *Magnolia grandiflora* L. Bot. Gaz. 142:124–134.

Postek, M. T., and S. C. Tucker. 1982. Foliar ontogeny and histogenesis in *Magnolia grandiflora* L. I. Apical organization and early development. Amer. J. Bot. 69:556–569.

Powell, C. L., and D. J. Bagyaraj. 1984. *VA Mycorrhiza.* CRC Press: Boca Raton, Florida.

Prat, R., B. Vian, D. Reis, and J. Roland. 1977. Evolution of internal pressure, vacuolation and membrane flow during cell growth in mung bean hypocotyl. Biol. Cellulaire 28:269–280.

Pray, T. R. 1957. Marginal growth of leaves of monocotyledons: *Hosta, Maranta* and *Philodendron.* Phytomorph. 7:381–387.

Pray, T. R. 1959. Pattern and ontogeny of the foliar venation of *Bobea elatior* (Rubiaceae). Pacific Sci. 13:3–13.

Preston, R. D. 1982. The case for multinet growth in growing walls of plant cells. Planta 155:356–363.

Price, C. E. 1982. A review of the factors influencing the penetration of pesticides through plant leaves. In *The Plant Cuticle,* eds. D. F. Cutler, K. L. Alvin, and C. E. Price. Academic Press: London.

Pridgeon, A. M. 1981. Absorbing trichomes in the Pleurothallidinae (Orchidaceae). Amer. J. Bot. 68:64–71.

Pridgeon, A. M. 1982. Diagnostic anatomical characters in the Pleurothallidinae (Orchidaceae). Amer. J. Bot. 69:921–938.

Priestley, J. H. 1920. The mechanism of root pressure. New Phytol. 19:189–200.

Priestley, J. H. 1922. The mechanism of root pressure. New Phytol. 21:41–47.

Priestley, J. H. 1943. The cuticle in angiosperms. Bot. Rev. 9:593–616.

Priestley, J. H., and E. K. North. 1922. The structure of the endodermis in relation to its function. New Phytol. 21:111–139.

Purnell, H. M. 1960. Studies of the family Proteaceae. I. Anatomy and morphology of the roots of some Victorian species. Aust. J. Bot. 8:38–50.

Pyykko, M. 1974. Developmental anatomy of the seedling of *Honkenya peploides.* Ann. Bot. Fenn. 11:253–261.

Quézel, P. 1966. A propos des xérophytes épineus en coussinet. Mém. Soc. bot. fr. 113:109–120.

Quézel, P. 1967. A propos des xérophytes épineus en coussinet du pourtour Mediterranéen. Ann. Fac. Sci. Marseille 39:173–181.

Rachmilevitz, T., and A. Fahn. 1973. Ultrastructure of nectaries of *Vinca rosea* L., *Vinca major* L. and *Citrus sinensis* Osbeck cv. *Valencia* and its relation to the mechanism of nectar secretion. Ann. Bot. 37:1–9.

Ragetli, H. W. J., M. Weintraub, and E. Lo. 1972. Characteristics of *Drosera* tentacles. I. Anatomical and cytological detail. Can. J. Bot. 50:159–168.

Ragland, C. H. 1934. The development of the peach fruit, with special reference to split-pit and gumming. Proc. Am. Soc. Hort. Sci. 31:1–21.

Ramayya, N. 1964. Morphology of the emergences. Curr. Sci. 33:577–580.

Ramos, G. S., and E. M. Engleman. 1982. Resin canals in the bark of *Bursera copallifera* and *Bursera grandifolia.* Bol. Soc. Bot. Mex. 0:41–54.

Rao, K. S., and Y. S. Dave. 1980. Development of the pericarp septa and zone of dehiscence in the fruit of *Cassia tora.* Acta Soc. Bot. Pol. 49:409–414.

Rao, K. S., and Y. S. Dave. 1984. Seasonal variations in the vascular cambium of *Holoptelea integrifolia* (Ulmaceae). Beitr. Biol. Pflanz. 59:321–332.

Rao, T. A. 1951a. Studies on foliar sclereids. A preliminary survey. J. Indian Bot. Soc. 30:28–39.

Rao, T. A. 1951b. Studies on foliar sclereids in Dicotyledons. I. Structure and ontogeny of sclereids in the leaf of *Diospyros discolor* Willd. Proc. Indian Acad. Sci. B. 34:92–98.

Rao, T. A. 1951c. Studies on foliar sclereids in Dicotyledons. V. Structure of the terminal sclereids in the leaf of *Memecylon heyneanum.* Benth. Proc. Indian Acad. Sci. B. 34:329–334.

Rao, T. A. 1957. Comparative morphology and ontogeny of foliar sclereids in seed plants. I. *Memecylon* L. Phytomorph. 7:306–330.

Rao, T. A., and S. Das. 1979. Leaf sclereids—occurrence and distribution in the angiosperms. Bot. Notiser. 132:319–324.

Raper, C. D., and S. A. Barber. 1970. Rooting systems of

soybeans. I. Differences in root morphology among varieties. Agron. J. 62:581–584.

Raschke, K. 1979. Movements of stomata. In *Physiology of Movements: Encyclopedia of Plant Physiology*, N. S., eds. W. Haupt and M. E. Feinleib. Springer-Verlag: Berlin. 7:383–441.

Rasmussen, H. 1981. Terminology and classification of stomata and stomatal development—a critical survey. J. Linn. Soc. Bot. 83:199–212.

Rauh, W. 1979a. *Kakteen an ihren Standorten*. Verlag Paul Parey: Berlin.

Rauh, W. 1979b. *Die großartige Welt der Sukkulenten*. Verlag Paul Parey: Berlin.

Rauh, W., and F. Rappert. 1954. Über das Vorkommen und die Histogenese von Scheitelgruben bei Krautigen Dikotylen, mit besonderer Berucksichtigung der ganzend Halbrosettenpflanzen. Planta 43:325–360.

Record, S. J. 1943. Key to the American woods. X. Woods with storied structure. Trop. Woods 76:32–47.

Reed, T. 1910. On the anatomy of some tubers. Ann. Bot. 24:537–548.

Reeder, J. R. 1956. The embryo of *Jouvea pilosa* as further evidence for the foliar nature of the coleoptile. Bull. Torrey Bot. Club 83:1–4.

Reeve, R. M. 1959. Histological and histochemical changes in developing and ripening peaches. II. The cell walls and pectins. Amer. J. Bot. 46:241–248.

Reid, C. P. P., and G. D. Bowen. 1979. Effects of soil moisture on V/A mycorrhiza formation and root development in *Medicago*. In *The Soil-Root Interface*, eds. J. L. Harley and R. S. Russell. Academic Press: New York.

Renaudin, S., and M. Capdepon. 1977. Sur la structure des parois des glandes pédicellées et des glandes en bouclier de *Tozzia alpina* L. Bull. Soc. bot. Fr. 124:29–43.

Renaudin, S., and R. Garrigues. 1966. Sur l'ultrastructure des glandes en bouclier de *Lathraea clandestina* L. et leur rôle physiologique. C. R. Acad. Sci. Paris 264:1984–1987.

Resch, A. 1961. Zur Frage nach den Geleitzellen im Protophloem der Wurzel. Ztschr. f. Bot. 49:82–95.

Rethke, R. V. 1946. The anatomy of circumscissile dehiscence. Amer. J. Bot. 33:677–683.

Reynolds, E. R. C. 1975. Tree rootlets and their distribution. In *The Development and Function of Roots*, eds. J. G. Torrey and D. T. Clarkson. Academic Press: London.

Reynolds, M. E. 1942. Development of the node in *Ricinus communis*. Bot. Gaz. 104:167–170.

Reynolds, T. L. 1984. An ultrastructural and stereological analysis of pollen grains of *Hyocyamus niger* during normal ontogeny and induced embryogenic development. Amer. J. Bot. 71:490–504.

Richter, H. G. 1980. Occurrence, morphology and taxonomic implications of crystalline and siliceous inclusions in the secondary xylem of the Lauraceae and related families. Wood Sci. Tech. 14:35–44.

Richter, H. G. 1981. Wood and bark anatomy of Lauraceae. I. *Aniba*. IAWA Bull. 2:79–87.

Ride, J. P., and R. B. Pearce. 1979. Lignification and papilla formation at sites of attempted penetration of wheat leaves by nonpathogenic fungi. Physiol. Plant Pathol. 15:79–92.

Riding, R. T., and E. M. Gifford, Jr. 1973. Histochemical changes occurring at the seedling shoot apex of *Pinus radiata*. Can. J. Bot. 51:501–512.

Riding, R. T., and C. H. A. Little. 1984. Anatomy and histochemistry of *Abies balsamea* cambial zone cells during the onset and breaking of dormancy. Can. J. Bot. 62:2570–2579.

Riding, R. T., and K. E. Percy. 1985. Effects of SO_2 and other air pollutants on the morphology of epicuticular waxes on needles of *Pinus strobus* and *Pinus banksiana*. New Phytol. 99:555–563.

Rieseberg, L. H., and E. E. Schilling. 1985. Floral flavonoids and ultraviolet patterns in *Viguiera* (Compositae). Amer. J. Bot. 72:999–1004.

Riopel, J. L. 1966. The distribution of lateral roots in *Musa acuminata* "Gros Michel." Amer. J. Bot. 53"403–406.

Riopel, J. L. 1969. Regulation of lateral root positions. Bot. Gaz. 130:80–83.

Robards, A. W. 1976. Plasmodesmata in higher plants. In *Intercellular Communication in Plants: Studies on Plasmodesmata*, eds. B. E. S. Gunning and A. W. Robards. Springer-Verlag: Berlin.

Robards, A. W., S. M. Jackson, D. T. Clarkson, and J. Sanderson. 1973. The structure of barley roots in relation to the transport of ions into the stele. Protoplasma 77:291–311.

Robards, A. W., and P. Kidwai. 1969. A comparative study of the ultrastructure of resting and active cambium of *Salix fragilis* L. Planta 84:239–249.

Roberts, K., and D. H. Northcote. 1971. Ultrastructure of the nuclear envelope; structural aspects of the interphase nucleus of sycamore suspension culture cells. Microscopia Acta 71:102–120.

Robinson, D. G., and U. Kristen. 1982. Membrane flow via the Golgi apparatus of higher plant cells. Int. Rev. Cytol. 77:89–127.

Robinson, D. G., and H. Quader. 1982. The microtubule–microfibril syndrome. In *The Cytoskeleton in Plant Growth and Development*, ed. C. W. Lloyd. Academic Press: London.

Rodkiewicz, B. 1970. Callose in cell walls during megasporogenesis. Planta 93:39–47.

Rodriguez, E., P. L. Healey, and I. Mehta. 1984. *Biology and Chemistry of Plant Trichomes*. Plenum Press: New York.

Roelofsen, P. A. 1952. On the submicroscopic structure of cuticular cell walls. Acta bot. neerl. 1:99–114.

Roelofsen, P. A. 1959. The plant cell-wall. In *Handbuch der Pflanzenanatomie*, Bd. 3. T. 4, ed. K. Linsbauer. Gbdr. Borntraeger: Berlin.

Roland, J. C. 1967. Recherches en microscopie photonique et en microscopie électronique sur l'origine et la différenciation des cellules du collenchyme. Annls. Sci. nat. Bot. (ser. 12) 8:141–214.

Rosen, W. G. 1968. Ultrastructure and physiology of pollen. Ann. Rev. Plant Physiol. 19:435–462.

Rosen, W. G., S. R. Gawlik, W. V. Dashek, and K. A. Siegesmund. 1964. Fine structure and cytochemistry of *Lilium* pollen tubes. Amer. J. Bot. 51:61–71.

Rosen, W. G., and H. R. Thomas. 1970. Secretory cells of lily pistils. I. Fine structure and function. Amer. J. Bot. 57:1108–1114.

Rosowski, J. R. 1968. Laticifer morphology in the mature stem and leaf of *Euphorbia supina*. Bot. Gaz. 129:113–120.

Ross, R. 1982. Initiation of stamens, carpels, and receptacle in the Cactaceae. Amer. J. Bot. 69:369–379.

Rost, T. L. 1969. Vascular pattern and hydathodes in leaves of *Crassula argentea* (Crassulaceae). Bot. Gaz. 130:267–270.

Roth, I. 1977. Fruits of angiosperms. In *Handbuch der Pflanzenanatomie*, Spez. Teil, Bd. 10, T. 1, ed. K. Linsbauer. Gbdr. Borntraeger: Berlin.

Roth, I., and I. Clausnitzer. 1972. Desarrollo de los hidatodos en *Sedum argenteum*. Acta Bot. Venez. 7:207–217.

Roth, I., and H. Lindorf. 1972. Anatomía y desarrollo del fruto y de la semilla de *Achras zapota* L. (Níspero). Acta Bot. Venez. 7:121–141.

Roth, I., and H. Lindorf. 1974. Desarrollo y anatomía del fruto y de la semilla de *Myristica fragrans* (van Houtt.). Acta Bot. Venez. 9:149–176.

Rovira, A. D. 1979. Biology of the soil-root interface. In *The Soil-Root Interface*, eds. J. L. Harley and R. S. Russell. Academic Press: New York.

Roy Chowdhury, C. 1961. The morphology and embryology of *Cedrus deodara* Loud. Phytomorph. 11:283–304.

Rugenstein, S. R., and N. R. Lersten. 1981. Stomata on seeds and fruits of *Bauhinia* (Leguminosae: Caesalpinioideae). Amer. J. Bot. 68:873–876.

Russell, E. W., and E. J. Russell. 1978. *Soil Conditions and Plant Growth*. Longman: London.

Russell, R. S. 1977. *Plant Root Systems*. McGraw-Hill: Maidenhead, U.K.

Russell, R. S., and D. T. Clarkson. 1976. Ion transport in root systems. In *Perspectives in Experimental Biology*, vol. 2, *Botany*, ed. N. Sunderland. Pergamon Press: Oxford.

Russell, S. D. 1979. Fine structure of megagametophye development in *Zea mays*. Can. J. Bot. 57:1093–1110.

Russell, S. D. 1984. Ultrastructure of the sperm of *Plumbago zeylanica*. II. Quantitative cytology and three-dimensional organization. Planta 162:385–391.

Russin, W. A., and R. F. Evert. 1985. Studies on the leaf of *Populus deltoides* (Salicaceae): Ultrastructure, plasmodesmatal frequency, and solute concentrations. Amer. J. Bot. 72:1232–1247.

Ruth, J., E. J. Klekowski, Jr., and O. L. Stein. 1985. Impermanent initials of the shoot apex and diplontic selection in a *Juniper* chimera. Amer. J. Bot. 72:1127–1135.

Rutherford, R. J. 1970. The anatomy and cytology of *Pilostyles thurberi* Gray (Rafflesiaceae). Aliso 7:263–288.

Ruzin, S. E. 1979. Root contraction in *Freesia* (Iridaceae). Amer. J. Bot. 66:522–531.

Rywosch, S. 1909. Untersuchungen über die Entwicklungsgeschichte der Seitenwurzeln der Monokotylen. Ztschr. f. Bot. 1:253–283.

Sachar, R. C. 1956. The embryogeny of *Argemone mexicana* L.—A criticism. Phytomorphology 6:148–151.

Sacher, J. A. 1955. Cataphyll ontogeny in *Pinus lambertiana*. Amer. J. Bot. 42:82–91.

Salisbury, F. B., and C. W. Ross. 1985. *Plant Physiology*. Wadsworth Publishing Company: Belmont, California.

Samantarai, B., and T. Kabi. 1953. Secondary growth in petiole and the partial shoot theory of the leaf. Nature 172:37.

Sangster, A. G. 1985. Silicon distribution and anatomy of the grass rhizome, with special reference to *Miscanthus sacchariflorus* (Maxim.) Hackel. Ann. Bot. 55:621–634.

Sanio, K. 1873. Anatomie der gemeinen Kiefer (*Pinus sylvestris* L.). Jahrb. wiss. Bot. 9:50–126.

Sarfatti, G., and P. Gori. 1969. Embryo sac of *Euphorbia dulcis* L., and ultrastructural study. G. Bot. Ital. 103:631–632.

Sargent, C. 1976. *Studies on the Ultrastructure and Development of the Plant Cuticle*. Ph.D. Thesis. London University.

Sargent, C., and J. L. Gay. 1977. Barley epidermal apoplast structure and modifications by powdery mildew contact. Physiol. Plant Pathol. 11:195–205.

Sassen, M. M. A. 1965. Breakdown of plant cell wall during the cell-fusion process. Acta bot. neerl. 14:165–196.

Satina, S., and A. F. Blakeslee. 1941. Periclinal chimeras in *Datura stramonium* in relation to development of leaf and flower. Amer. J. Bot. 28:862–871.

Satina, S., A. F. Blakeslee, and A. G. Avery. 1940. Demonstration of the three germ layers in the shoot apex of *Datura* by means of induced polyploidy in periclinal chimeras. Amer. J. Bot. 27:895–905.

Satter, R. L., R. C. Garber, L. Khairallah, and Yi-Shan Cheng. 1982. Elemental analysis of freeze-dried thin sections of *Samanea* motor organs: Barriers to ion diffusion through the apoplast. J. Cell Biol. 95:893–902.

Satterthwait, D. F., and J. W. Schopf. 1972. Structurally preserved phloem zone tissue in *Rhynia*. Amer. J. Bot. 59:373–376.

Sauter, J. J. 1972. Respiratory and phosphatase activities in contact cells of wood rays and their possible role in sugar secretion. Z. Pflanzenphysiol. 67:135–145.

Sauter, J. J. 1974. Structure and physiology of Strasburger cells. Ber. Dtsch. Bot. Ges. 87:327–336.

Sauter, J. J. 1980. The strasburger cells—equivalents of companion cells. Ber. Dtsch. Bot. Ges. 93:29–42.

Savidge, R. A. 1983. The role of plant hormones in higher plant cellular differentiation. II. Experiments with the vascular cambium, and sclereid and tracheid differentiation in the pine, Pinus contorta. Histochemical Journal 15:447–466.

Savidge, R. A., and J. L. Farrar. 1984. Cellular adjustments in the vascular cambium leading to spiral grain formation in conifers. Can. J. Bot. 62:2872–2879.

Sawhney, V. K., P. J. Rennie and T. A. Steeves. 1981. The ultrastructure of the central zone cells of the shoot apex of Helianthus annuus. Can. J. Bot. 59:2009–2015.

Saxena, P. R., M. C. Pant, K. Kishor, and K. P. Bhargave. 1965. Pharmacologically active constituents of Urtica parviflora (Rosb.). Can. Jour. Physiol. Pharmacol. 43:869–876.

Scagel, R. F., R. J. Bandoni, J. R. Maze, G. E. Rouse, W. B. Schofield, and J. R. Stein. 1982. Nonvascular Plants: An Evolutionary Survey. Wadsworth Publishing Company: Belmont, California.

Scannerini, S., and P. Bonfante-Fasolo. 1983. Comparative ultrastructural analysis of mycorrhizal associations. Can. J. Bot. 61:917–943.

Schaffner, J. H. 1919. Complete reversal of sex in hemp. Science 50:311–312.

Schaffstein, G. 1932. Untersuchungen an ungegliederten Milchröhren. Beih. bot. Zbl. 49:197–220.

Scheckler, S. E., and H. P. Banks. 1971. Anatomy and relationships of some Devonian progymnosperms from New York. Amer. J. Bot. 58:737–751.

Scheirer, D. C., and C. J. Hillson. 1973. The vascular transition region of Helianthus annuus. I. Bilateral and unilateral patterns of differentiation. Amer. J. Bot. 60:242–246.

Schieferstein, R. H., and W. E. Loomis. 1956. Wax deposits on leaf surfaces. Plant Physiol. 31:240–247.

Schiff, J. A. 1980. Development, inheritance, and evolution of plastids and mitochondria. In The Biochemistry of Plants: A Comprehensive Treatise, vol. 1, ed. N. E. Tolbert. Academic Press: New York.

Schmid, R. 1972. A resolution of the Eugenia-Syzygium controversy (Myrtaceae). Amer. J. Bot. 59:423–436.

Schmid, R. 1976. Filament histology and anther dehiscence. Bot. J. Linn. Soc. 73:303–315.

Schmidt, A. 1924. Histologische Studien an Phanerogamen Vegetationspunkten. Bot. Arch. 8:345–404.

Schmidt, H. W., and J. Schönherr. 1982. Fine structure of isolated and non-isolated potato tuber periderm. Planta 154:76–80.

Schnarf, K. 1928. Embryologie der Angiospermen. In Handbuch der Pflanzenanatomie, ed. K. Linsbauer, Bd. 10, Lief. 23. Gbr. Borntraeger: Berlin.

Schnee, B. K., and D. M. Waller. 1986. Reproductive behavior of Amphicarpaea bracteata (Leguminosae), an amphicarpic annual. Amer. J. Bot. 73:376–386.

Schneider, A., and R. Dargent. 1977. Localisation et compartement du mycelium de Taphrina deformans dans le mésophylle et sans le cuticule des feuilles de pêcher (Prunus persica). Can. J. Bot. 55:2485–2495.

Schneider, E. L., and J. D. Buchanan. 1980. Morphological studies of the Nymphaeaceae. XI. The floral biology of Nelumbo pentapetala. Amer. J. Bot. 67:182–193.

Schneider, H. 1955. Ontogeny of the lemon tree bark. Amer. J. Bot. 42:893–905.

Schneider, W. 1913. Vergleichend-morphologische Untersuchung ueber die Kurztriebe einige Arten von Pinus. Flora 105:385–446.

Schnepf, E. 1959. Untersuchungen über Darstellung und Bau der Ektodesmen und ihre Beeinflussbarkeit durch stoffliche Faktoren. Planta 52:644–708.

Schnepf, E. 1961. Licht- und elektronenmikroskopische Beobachtungen an Insektivorendrüsen über die Sekretion des Fangschleims. Flora 151:73–87.

Schnepf, E. 1963. Zur Cytologie und Physiologie pflanzlicher Drüsen. I. Teil. Über den Fangschleim der Insektivoren. Flora 153:1–22.

Schnepf, E. 1968. Zur Feinstruktur der schleimsezernierenden Drüsenhaare auf der Ochrea von Rumex and Rheum. Planta 79:22–34.

Schnepf, E. 1969a. Über den Feinbau von Öldrüsen. I. Die Drüsenhaare von Arctium lappa. Protoplasma 67:185–194.

Schnepf, E. 1969b. Sekretion und Exkretion bei Pflanzen. Protoplasmatologie VIII/8. Springer-Verlag: Vienna.

Schnepf, E. 1986. Cellular polarity. Ann. Rev. Plant Physiol. 37:23–47.

Schoen, D. J., and D. G. Lloyd. 1985. The selection of cleistogamy and heteromorphic diaspores. Biol. J. Linn. Soc. 23:303–322.

Scholander, P. F., L. Van Dam, and S. I. Scholander. 1955. Gas exchange in the roots of mangroves. Amer. J. Bot. 42:92–98.

Schönbohm, E. 1973. Kontraktile Fibrillen als aktive Elemente bei der Mechanik der Chloroplastenverlagerung. Ber. Dt. Bot. Ges. 86:407–422.

Schönherr, J. 1982. Resistances of plant surfaces to water loss: Transport properties of cutin, suberin and associated lipids. In Encyclopedia of plant physiology, n.s. 12B: Physiological plant ecology, eds. O. L. Lange, P. S. Nobel, C. B. Osmund, and H. Ziegler. Springer-Verlag: Berlin.

Schönherr, J., and H. Ziegler. 1980. Water permeability of Betula periderm. Planta 147:345–354.

Schopmeyer, C. S. 1974. Seeds of Woody Plants in the United States. Agricultural Handbook No. 450. Forest Service, U.S. Department of Agriculture: Washington, D.C.

Schroeder, C. A. 1953. Fruit morphology and anatomy in the white sapote. Bot. Gaz. 115:248–254.

Schulz, P., and W. A. Jensen. 1977. Cotton embryogenesis: The early development of the free nuclear endosperm. Amer. J. Bot. 64:384–394.

Schulz, S. R., and W. A. Jensen. 1968. *Capsella* embryogenesis: The synergids before and after fertilization. Amer. J. Bot. 55:541–552.

Schumacher, A. 1948. Beitrag zur Kenntnis des Stofftransportes in dem Siebröhrensystem höherer Pflanzen. Planta 35:642–700.

Schweitzer, H. J., and L. Matten. 1982. *Aneurophyton germanicum* and *Protopteridium thompsonii* from the Middle Devonian of Germany. Palaeontographica 184B:65–106.

Scott, D. H. 1891. Origin of polysteley in dicotyledons. Ann. Bot. 5:514–517.

Scott, F. M. 1966. Cell wall surface of the higher plants. Nature 210:1015–1017.

Scott, F. M., and B. G. Bystrom. 1970. Mucilaginous idioblasts in okra, *Hibiscus esculentus* L. In *New Research in Plant Anatomy*, eds. N. K. B. Robson, D. F. Cutler, and M. Gregory. Bot. J. Linn. Soc. 63:15–24 and Academic Press: London.

Scott, F. M., B. G. Bystrom, and E. Bowler. 1963. Root hairs, cuticle and pits. Science 140:63–64.

Scott, F. M., K. C. Hamner, E. Baker, and E. Bowler. 1957. Ultrasonic and electron microscope study on onion epidermal wall. Science 125:399–400.

Scott, F. M., K. C. Hamner, E. Baker, and E. Bowler. 1958. Electron microscope studies of the epidermis of *Allium cepa*. Amer. J. Bot. 45:449–461.

Scott, P. C., L. W. Miller, B. D. Webster, and A. C. Leopold. 1967. Structural changes during bean leaf abscission. Amer. J. Bot. 56:730–734.

Scurfield, G., and S. Silva. 1969. The structure of reaction wood as indicated by scanning electron microscopy. Aust. J. Bot. 17:391–402.

Seago, J. L. 1973. Developmental anatomy in roots of *Ipomoea purpurea*. II. Initiation and development of secondary roots. Amer. J. Bot. 60:607–618.

Sedgley, M. 1979. Light microscope study of pollen tube growth, fertilization and early embryo and endosperm development in the avocado varieties Fuerte and Hass. Ann. Bot. 44:353–359.

Sedgley, M., and P. B. Scholefield. 1980. Stigma secretion in the watermelon before and after pollination. Bot. Gaz. 141:428–434.

Setia, R. E., M. V. Parthasarathy, and J. J. Shah. 1977. Development, histochemistry and ultrastructure of gum-resin ducts in *Commiphora mukul* Engl. Ann. Bot. 41:999–1004.

Severs, N. J. 1976. Nuclear envelope inclusions demonstrated by freeze-fracture. Cytobios 16:125–132.

Seward, A. C. 1917. *Fossil Plants*. Vol. III. Cambridge University Press: London.

Sexton, R., and A. J. Redshaw. 1981. The role of cell expansion in the abscission of *Impatiens sultani* leaves. Ann. Bot. 48:745–756.

Shah, J. J., and R. L. Fotedar. 1974. Sieve-tube members in the stem of *Cyathea gigantea*. Amer. Fern J. 64:27–28.

Shah, J. J., and R. Jacob. 1969. Development and structure of phloem in the petiole of *Lagenaria siceraria* (Mol.) Standl. and *Momordica charantia* L. Ann. Bot. 33:855–863.

Shah, J. J., and M. R. James. 1968. Sieve tube elements in the stem of *Neptunia oleracea* Lour. Aust. J. Bot. 16:433–444.

Shah, J. J., and J. D. Patel. 1972. The shell zone: Its differentiation and probable function in some dicotyledons. Amer. J. Bot. 59:683–690.

Shah, J. J., and K. Unnikrishnan. 1968. The shoot apex and the ontogeny of axillary buds in *Cuminum cyminum* L. Aust. J. Bot. 17:241–253.

Shain, L., and W. E. Hillis. 1973. Ethylene production in xylem of *Pinus radiata* in relation to heartwood formation. Can. J. Bot. 51:1331–1335.

Shain, L., and J. F. G. Mackay. 1973. Seasonal fluctuation in respiration of aging xylem in relation to heartwood formation in *Pinus radiata*. Can. J. Bot. 51:737–741.

Sharma, D. D., H. K. Sharma, and G. S. Paliwal. 1980. Annual rhythm of size variations in the cambial initials of *Chorisa speciosa* St. Hil. Science and Culture 45:96–97.

Sharma, H. K., D. D. Sharma, and G. S. Paliwal. 1979. Annual rhythm of size variations in cambial initials of *Azadirachta indica* A. Juss. Geobios 6:127–129.

Sharma, M. 1971. Ontogenetic studies of the myrosin idioblasts in *Brassica napus* and *Brassica montana*. Bot. Tidsskr. 66:51–59.

Sharman, B. C. 1942. Developmental anatomy of the shoot of *Zea mays* L. Ann. Bot. n. s. 6:245–282.

Sheriff, D. W. 1982. The hydraulic pathways in *Nicotiana glauca* (Grah.) and *Tradescantia virginiana* (L.) leaves, and water potentials in leaf epidermis. Ann. Bot. 50:535–548.

Shibata, N., H. Harada, and H. Saiki. 1981. Difference in the development of incubated tyloses within the sapwood of *Castanea crenata* Sieb. et Zucc. Bull. Kyoto Univ. For. 53:231–240.

Shields, L. M. 1951. The involution mechanism in leaves of certain xeric grasses. Phytomorphology 1:225–241.

Shimony, C., and A. Fahn. 1968. Light-and electron-microscopical studies on the structure of salt glands of *Tamarix aphylla* L. J. Linn. Soc. Bot. 60:283–288.

Shimony, C., A. Fahn, and L. Reinhold. 1973. Ultrastructure and ion gradients in the salt glands of *Avicennia marina* (Forssk.) Vierh. New Phytol. 72:27–36.

Shininger, T. L. 1979. The control of vascular development. Ann. Rev. Plant Physiol. 30:313–337.

Shoup, J. R., J. Overton, and M. Ruddat. 1981. Ultrastructure and development of the nexine and intine in the pollen wall of *Silene alba* (Caryophyllaceae). Amer. J. Bot. 68:1090–1095.

Sievers, A. 1959. Untersuchungen über die Darstelbarkeit

der Ektodesmen und ihre Beeinflussung durch physikalische Faktoren. Flora 147:263–316.

Sievers, A. 1963a. Beteiligung des Golgi-Apparatus bei der Bildung der Zellwand von Wurzelhaaren. Protoplasma 56:188–192.

Sievers, A. 1963b. Über die Feinstruktur des Plasmas wachsender Wurzelhaare. Z. Naturf. 18b:830–836.

Sifton, H. B. 1945. Air space tissues in plants. Bot. Rev. 11:108–143.

Sifton, H. B. 1957. Air space tissues in plants. Bot. Rev. 23:303–312.

Siminovitch, D., C. M. Wilson, and D. R. Briggs. 1953. Studies on the chemistry of the living bark of the black locust in relation to frost hardiness. V. Seasonal transformations and variations in the carbohydrates: Starch-sucrose interconversions. Plant Physiol. 28:383–400.

Simoncioli, C. 1974. Ultrastructural characteristics of *Diplotaxis erucoides* (L.) DC. suspensor. Giorn. Bot. Ital. 108:175–189.

Simpson, W. L., and R. Moore. 1984. Leaf structure and light absorption in *Frithia pulchra* (Mesembryanthemaceae). Ann. Bot. 53:413–420.

Singh, A. P., and L. M. Srivastava. 1973. The fine structure of pea stomata. Protoplasma 76:61–82.

Singh, B. 1944. A contribution to the anatomy of *Salvadora persica* L. with special reference to the origin of the included phloem. J. Indian Bot. Soc. 23:71–78.

Singh, H. 1978. *Embryology of Gymnosperms.* Gbdr. Borntraeger: Berlin.

Singh, H., and B. M. Johri. 1972. Development in gymnosperm seeds. In *Seed Biology,* ed. T. T. Kozlowski, vol. 1. Academic Press: New York.

Singh, R., D. R. Bohra, and B. D. Sharma. 1978. Vessels in the rhizome of *Actiniopteris radiata*. Phytomorph. 28:455–457.

Sinnott, E. W. 1914. The anatomy of the node as an aid in the classification of angiosperms. Amer. J. Bot. 1:303–322.

Sinnott, E. W., and R. Bloch. 1945. The cytoplasmic basis of intercellular patterns in vascular differentiation. Amer. J. Bot. 32:151–156.

Sitte, P. 1962. Zum Feinbau der Suberinschichten im Flaschenkork. Protoplasma 54:555–559.

Skene, D. S. 1966. The distribution of growth and cell division in the fruit of Cox's Orange Pippin. Ann. Bot. 30:494–512.

Skutch, A. F. 1932. Anatomy of the axis of the banana. Bot. Gaz. 93:233–258.

Slade, B. F. 1971. Stelar evolution in vascular plants. New Phytol. 70:879–884.

Slatyer, R. O. 1967. *Plant-Water Relationships.* Academic Press: London.

Smaoui, M. A. 1971. Differentiation des trichomes chez *Atriplex halimus* L. C. R. Acad. Sci. Paris 273:1268–1271.

Smart, M. G., and T. P. O'Brien. 1983. The development of the wheat embryo in relation to the neighboring tissues. Protoplasma 114:1–13.

Smith, B. N., and B. J. Meeuse. 1966. Production of volatile amines and skatole at anthesis in some *Arum* Lily species. Plant Physiol. 41:343–347.

Smith, F. H. 1958. Anatomical development of the hypocotyl of Douglas-fir. Forest Sci. 4:61–70.

Smith, G. M. 1955. *Cryptogamic Botany. Vol. II, Bryophytes and Pteridophytes.* 2nd ed. McGraw-Hill: New York.

Smith, M., and R. D. Butler. 1971. Ultastructural aspects of petal development in *Cucumis sativus* with particular reference to the chromoplasts. Protoplasma 73:1–13.

Sobotka, F. E., and D. A. Stelzig. 1974. An apparent cellulase complex in tomato (*Lycopersicon esculentum* L.) fruit. Plant Physiol. 53:759–763.

Soh, W. Y. 1972. Early ontogeny of vascular cambium. I. *Ginkgo biloba.* Bot. Mag. Tokyo 85:111–124.

Soh, W. Y. 1974a. Early ontogeny of vascular cambium. II. *Aucuba japonica* and *Weigela coraeensis.* Bot. Mag. Tokyo 87:17–32.

Soh, W. Y. 1974b. Early ontogeny of vascular cambium. III. *Robinia pseudo-acacia* and *Syringa oblata.* Bot. Mag. Tokyo 87:99–112.

Sokolova, S. V. 1968. Fine structure of the phloem cells of *Beta vulgaris* L. petioles. Fiziol. Rast. 15:757–763.

Solereder, H. 1908. *Systematic Anatomy of the Dicotyledons* (Translated by L. A. Boodle and F. E. Fritsch, revised by D. H. Scott). 2 vols. Clarendon Press: Oxford.

Solereder, H., and F. J. Meyer. 1928. *Systematische Anatomie der Monokotyledonen.* Heft III. Gebrüder Borntraeger: Berlin.

Soliday, C. L., P. E. Kolattukudy, and R. W. Davis. 1979. Chemical and ultrastructural evidence that waxes associated with the suberin polymer constitute the major diffusion barrier to water vapor in potato tuber (*Solanum tuberosum* L.). Planta 146:607–614.

Soper, K., and K. J. Mitchell. 1956. The developmental anatomy of perennial ryegrass (*Lolium perenne* L.). N. Z. Jour. Sci. Tech. A. 37:484–504.

Southorn, W. A. 1960. Complex particles in *Hevea* latex. Nature, London 188:165–166.

Southorn, W. A. 1964. A complex sub-cellular component of widespread occurrence in plant latices. J. Exp. Bot. 15:616–621.

Spackman, W., and B. G. L. Swamy. 1949. The nature and occurrence of septate fibres in dicotyledons. Amer. J. Bot. 36:804.

Spira, T. P., and L. K. Wagner. 1983. Viability of seeds up to 211 years old extracted from adobe brick buildings of California and Northern Mexico. Amer. J. Bot. 70:303–307.

Sporne, K. R. 1962. *The Morphology of Pteridophytes.* Hutchinson University Library: London.

Spurr, A. R. 1949. Histogenesis and organization of the embryo in *Pinus strobus* L. Amer. J. Bot. 36:629–641.

Spurr, A. R. 1957. The effect of boron on cell wall structure in celery. Amer. J. Bot. 44:637–650.

Spurr, A. R., and W. M. Harris. 1968. Ultrastructure of chloroplasts and chromoplasts in *Capsicum annuum*. I. Thylakoid membrane changes during fruit ripening. Amer. J. Bot. 55:1210–1224.

Srivastava, L. M. 1963. Secondary phloem in the Pinaceae. Univ. Calif. Publ. Bot. 36:1–42.

Srivastava, L. M. 1966. On the fine structure of the cambium of *Fraxinus americana* L. J. Cell Biol. 31:79–93.

Srivastava, L. M. 1970. The secondary phloem of *Austrobaileya scandens*. Can. J. Bot. 48:341–359.

Srivastava, L. M., and I. W. Bailey. 1962. Comparative anatomy of the leaf-bearing Cactaceae. V. The secondary phloem. J. Arnold Arbor. 43:234–272.

Srivastava, L. M., and T. P. O'Brien. 1966. On the ultrastructure of cambium and its vascular derivatives. I. Cambium of *Pinus strobus* L. Protoplasma 61:257–276.

Srivastava, L. M., and A. P. Singh. 1972. Stomatal structure in corn leaves. J. Ultrastr. Res. 39:345–363.

Stace, C. A. 1965. Cuticular studies as an aid to plant taxonomy. Bull. Br. Mus. (Nat. Hist.) Bot. 4:1–78.

Staff, I. A. 1968. A study of the apex and growth pattern in the shoot of *Xanthorrhoea media* R. Br. Phytomorph. 18:153–166.

Stålfelt, M. G. 1956. Morphologie und Anatomie des Blattes als Transpirationsorgan. Handb. der Pflanzenphysiol. 3:324–341.

Stant, M. Y. 1964. Anatomy of the Alismataceae. Bot. J. Linn. Soc. 59:1–42.

Stant, M. Y. 1967. Anatomy of the Butomaceae. Bot. J. Linn. Soc. 60:31–60.

Stanton, M. L. 1984. Developmental and genetic sources of seed weight variation in *Raphanus raphanistrum* L. (Brassicaceae). Amer. J. Bot. 71:1090–1098.

Stebbins, G. L., and S. K. Jain. 1960. Developmental studies of cell differentiation in the epidermis of monocotyledons. I. *Allium*, *Rhoeo* and *Commelina*. Devl. Biol. 2:409–426.

Stebbins, G. L., and G. S. Khush. 1961. Variation in the organization of the stomatal complex in the leaf epidermis of monocotyledons and its bearing on their phylogeny. Amer. J. Bot. 48:51–59.

Steer, M. W. 1974. The development of tapetal cells in *Avena sativa* L. 8th Int. Congress Electron Microscopy, Canberra II:594–595.

Steer, M. W. 1981. *Understanding Cell Structure*. Cambridge University Press: Cambridge.

Steeves, T. A., M. A. Hicks, J. M. Naylor, and P. Rennie. 1969. Analytical studies on the shoot apex of *Helianthus annuus*. Can. J. Bot. 47:1367–1375.

Steeves, T. A., and I. M. Sussex. 1972. *Patterns in Plant Development*. Prentice-Hall. New Jersey.

Stein, O. L., and E. B. Fosket. 1969. Comparative developmental anatomy of shoots of juvenile and adult *Hedera helix*. Amer. J. Bot. 56:546–551.

Steinberger-Hart, A. L. 1922. Über Regulation des osmotischen Wertes in den Schliesszellen von Luft-und Wasserspalten. Biol. Zbl. 42:405–419.

Stephenson, A. G. 1981. Flower and fruit abortion: Proximate causes and ultimate functions. Ann. Rev. Ecol. Syst. 12:253–279.

Sterling, C. 1946. Cytological aspects of vascularization in *Sequoia*. Amer. J. Bot. 33:35–45.

Sterling, C. 1968. The structure of the starch grain. In *Starch and its Derivatives*, 4th ed., ed. J. A. Radley. Chapman and Hall: London.

Sterling, G. 1953. Developmental anatomy of the fruit of *Prunus domestica* L. Bull. Torrey Bot. Club 80:457–477.

Stevens, R. A., and E. S. Martin. 1978a. A new ontogenetic classification of stomatal types. J. Linn. Soc. Bot. 77:53–64.

Stevens, R. A., and E. S. Martin. 1978b. Structural and functional aspects of stomata. I. Developmental studies in *Polypodium vulgare*. Planta 142:307–316.

Stevenson, D. W. 1980. Radial growth in the Cycadales. Amer. J. Bot. 67:465–475.

Stevenson, D. W., and J. B. Fisher. 1980. The developmental relationship between primary and secondary thickening growth in *Cordyline* (Agavaceae). Bot. Gaz. 141:264–268.

Stevenson, D. W., and R. A. Popham. 1973. Ontogeny of the primary thickening meristem in seedlings of *Bougainvillea spectabilis*. Amer. J. Bot. 60:1–9.

Stewart, J. M. 1975. Fiber initiation on the cotton ovule (*Gossypium hirsutum*). Amer. J. Bot. 62:723–730.

Stewart, K. D., and K. R. Mattox. 1975. Comparative cytology, evolution and classification of the green algae with some consideration of the origin of other organisms with chlorophylls a and b. Bot. Rev. 41:104–135.

Stewart, R. N., and H. Dermen. 1979. Ontogeny in the monocotyledons as revealed by studies of the developmental anatomy of periclinal chloroplast chimeras. Amer. J. Bot. 66:47–58.

Stewart, W. N. 1983. *Paleobotany and the Evolution of Plants*. Cambridge University Press: Cambridge.

Stodola, J. 1967. *Encyclopedia of Water Plants*. TFH Publications: Neptune City, New Jersey.

Stopp, K. 1950. Karpologische Studien I und II. Abh. Akad. Wiss. Lit. Mainz, Math.-Nat. Kl. 7:165–218.

Stösser, R., H. P. Rasmussen, and M. J. Bukovac. 1969. Histochemical changes in the developing abscission layer in fruits of *Prunus cerasus* L. Planta 86:151–164.

Strain, R. W. 1933. A study of vein endings in leaves. Am. Midland. Natur. 14:367–373.

Strelis, I. and R. W. Kennedy. 1967. *Identification of North*

American Commercial Pulpwoods and Pulp Fibers. University of Toronto Press: Toronto.

Stuart, S. G., and J. B. Loy. 1983. Comparison of testa development in normal and hull-less seeded strains of *Cucurbita pepo* L. Bot. Gaz. 144:491–500.

Studholme, W. P., and W. R. Philipson. 1966. A comparison of the cambium in two woods with included phloem: *Heimerliodendron brunonianum* and *Avicennia resinifera.* N. Z. J. Bot. 4:355–365.

Subramanyam, K., and M. V. S. Raju. 1952. Circumscissile dehiscence in *Sphenoclea zeylanica* Gaertn. Curr. Sci. 21:139–140.

Sundberg, M. D. 1983. Vascular development in the transition region of *Populus deltoides* Bartr. ex Marsh. seedlings. Amer. J. Bot. 70:735–743.

Sunell, L. A., and P. L. Healey. 1979. Distribution of calcium oxalate crystal idioblasts in corms of taro (*Colocasia esculenta*). Amer. J. Bot. 66:1029–1032.

Sunell, L. A., and P. L. Healey. 1985. Distribution of calcium oxalate crystal idioblasts in leaves of taro (*Colocasia esculenta*). Amer. J. Bot. 72:1854–1860.

Sutherland, M. 1933. A microscopical study of the structure of leaves of the genus *Pinus.* Trans. R. Soc. N. Z. 63:517–568.

Sutton, W. D. 1983. Nodule development and senescence. In *Nitrogen Fixation, vol. 3, Legumes,* ed. W. J. Broughton. Clarendon Press: Oxford.

Swamy, B. G. L. 1948. A contribution to the life history of *Casuarina.* Proc. Amer. Acad. Arts and Sci. 77:1–32.

Swamy, B. G. L., and K. V. Krishnamurthy. 1973. The helobial endosperm: A decennial review. Phytomorph. 23:74–79.

Swamy, B. G. L., and N. Parameswaran. 1963. The helobial endosperm. Biol Rev. 38:1–50.

Swiecki, T. J., A. G. Endress, and O. C. Taylor. 1982. The role of surface wax in susceptibility of plants to air pollutant injury. Can. J. Bot. 60:316–319.

Szemes, G. 1943. Zur Entwicklung des Elaiosoms von *Chelidonium majus.* Wien. Bot. Z. 92:215–219.

Taylor, T. N. 1981. *Paleobotany: An Introduction to Fossil Plant Biology.* McGraw-Hill: New York.

Teal, J. M., and J. W. Kanwisher. 1966. Gas transport in the marsh grass, *Spartina alterniflora.* J. Exptl. Bot. 17:355–361.

Ter Welle, B. J. H. 1976a. On the occurrence of silica grains in the secondary xylem of the Chrysobalanaceae. IAWA Bull. 1976:19–29.

Ter Welle, B. J. H. 1976b. Silica grains in woody plants of the neotropics especially Surinam. In *Wood Structure in Biological and Technological Research,* eds. P. Baas, A. J. Bolton, and D. M. Catling. Leiden Bot. Ser. 3. Leiden University Press: Leiden.

Ter Welle, B. J. H., and A. M. W. Mennega. 1977. On the presence of large styloids in the secondary xylem of *Henriettea* (Melastomataceae). IAWA Bull. 1977:31–35.

Tetley, U. 1930. A study of the anatomical development of the apple and some observations on the pectic constituents of the cell walls. J. Pom. Hort Sci. 8:153–172.

Tewari, R. B. 1975. Structure of vessels and tracheids of *Regnellidium diphyllum* Lindman (Marsileaceae). Ann. Bot. 39:229–231.

Theobald, W. L., J. L. Krahulik, and R. C. Rollins. 1979. Trichome description and classification. In *Anatomy of the Dicotyledons,* vol. 1, eds. C. R. Metcalfe and L. Chalk. Clarendon Press: Oxford.

Thielke, C. 1957. Über Differenzierungsvorgänge bei Cyperaceen. II. Entstehung von epidermalen Faserbündeln in der Scheide von *Carex.* Planta 49:33–46.

Thomas, B. A., and D. L. Masarati. 1982. Cuticular and epidermal studies in fossil and living lycophytes. In *The Plant Cuticle,* eds. D. F. Cutler, K. L. Alvin, and C. E. Price. Academic Press: London.

Thompson, W. P. 1923. The relationship of the different types of angiospermic vessels. Ann. Bot. 37:183–192.

Thomson, R. B. 1905. The megaspore membrane of the gymnosperms. Univ. Toronto Stud., Biol. Ser. 4:85.

Thomson, W. W. 1975. The structure and function of salt glands. In *Plants in Saline Environments,* eds. A. Poljakoff-Mayber and J. Gale. Springer-Verlag: Berlin.

Thomson, W. W., and L. L. Liu. 1967. Ultrastructural features of the salt gland of *Tamarix aphylla* L. Planta 73:201–220.

Thomson, W. W, and K. Platt-Aloia. 1976. Ultrastructure of the epidermis of developing, ripening, and senescing navel oranges. Hilgardia 44:61–82.

Thorpe, N. O. 1984. *Cell Biology.* John Wiley and Sons: New York.

Thureson-Klein, A. 1970. Observations on the development and fine structure of the articulated laticifers of *Papaver somniferum.* Ann. Bot. 34:751–759.

Thurston, E. L. 1974. Morphology, fine structure and ontogeny of the stinging emergence of *Urtica dioica.* Amer. J. Bot. 61:809–817.

Thurston, E. L. 1976. Morphology, fine structure and ontogeny of the stinging emergence of *Tragia ramosa* and *T. saxicola* (Euphorbiaceae). Amer. J. Bot. 63:710–718.

Thurston, E. L., and N. R. Lersten. 1969. The morphology and toxicology of plant stinging hairs. Bot. Rev. 35:393–412.

Tilton, V. R., and H. T. Horner, Jr. 1980a. Calcium oxalate raphide crystals and crystalliferous idioblasts in the carpels of *Ornithogalum caudatum.* Ann. Bot. 46:533–539.

Tilton, V. R., and H. T. Horner, Jr. 1980b. Stigma, style and obturator of *Ornithogalum caudatum* (Liliaceae) and their function in the reproductive process. Amer. J. Bot. 67:1113–1131.

Tilton, V. R., L. W. Wilcox, R. G. Palmer, and M. C. Albertsen. 1984. Stigma, style and obturator of soybean, *Glycine max* (L.) Merr. (Leguminosae) and their function in the reproductive process. Amer. J. Bot. 71:676–686.

Ting, I. P. 1982. *Plant Physiology*. Addison-Wesley: Reading, Massachusetts.

Tinker, P. B. 1976. Roots and water. Transport of water to plant roots in soil. Phil. Trans. R. Soc. London B. 273: 445–461.

Tippett, J. T., and T. C. Hill. 1984. Junction complexes between sieve tubes in the secondary phloem of Myrtaceae. Ann. Bot. 53:421–429.

Tippett, J. T., and T. P. O'Brien. 1976. The structure of eucalypt roots. Aust. J. Bot. 24:619–632.

Tolbert, N. E. 1980a. *The Biochemistry of Plants: A Comprehensive Treatise*. Vol. 1. Academic Press: New York.

Tolbert, N. E. 1980b. Microbodies—peroxisomes and glyoxysomes. In *The Biochemistry of Plants: A Comprehensive Treatise*, vol. 1, ed. N. E. Tolbert. Academic Press: New York.

Tomlinson, P. B. 1959. An anatomical approach to the classification of the Musaceae. J. Linn. Soc. Bot. 55:779–809.

Tomlinson, P. B. 1961. *Anatomy of the Monocotyledons. II. Palmae*. Claredon Press: Oxford.

Tomlinson, P. B. 1969a. *Anatomy of the Monocotyledons. III. Commelinales-Zingiberales*. Clarendon Press: Oxford.

Tomlinson, P. B. 1969b. On the morphology and anatomy of turtle grass, *Thalassia testudinum* (Hydrocharitaceae). II. Anatomy and development of the root in relation to function. Bull. Marine Sci. 19:57–71.

Tomlinson, P. B. 1970a. Monocotyledons—towards an understanding of their morphology and anatomy. In *Advances in Botanical Research*, vol. 3, ed. R. D. Preston. Academic Press: New York.

Tomlinson, P. B. 1970b. Dichotomous branching in *Flagellaria indica* (Monocotyledones). In *New Research in Plant Anatomy*, eds. N. K. Robson, D. F. Cutler, and M. Gregory. Bot. J. Linn. Soc. London, vol. 63, suppl. 1.

Tomlinson, P. B. 1971. The shoot apex and its dichotomous branching in the *Nypa* palm. Ann. Bot. 35:865–879.

Tomlinson, P. B. 1974. Development of the stomatal complex as a taxonomic character in the monocotyledons. Taxon 23:109–128.

Tomlinson, P. B., and A. E. Esler. 1973. Establishment growth in woody monocotyledons native to New Zealand. N. Z. J. Bot. 11:627–644.

Tomlinson, P. B., and U. Posluszny. 1977. Features of dichotomizing apices in *Flagellaria indica* (Monocotyledones). Amer. J. Bot. 64:1057–1065.

Torrey, J. G. 1957. Auxin control of vascular pattern formation in regenerating pea root meristems grown in vitro. Amer. J. Bot. 44:859–870.

Torrey, J. G. 1965. Physiological bases of organization and development in the root. In *Encyclopedia of Plant Physiology* 15:1256–1327.

Torrey, J. G. and D. T. Clarkson. 1975. *The Development and Function of Roots*. Academic Press: London.

Torrey, J. G., and L. J. Feldman. 1977. The organization and function of the root apex. Am. Sci. 65:334–344.

Toth, R. 1982. An introduction to morphometric cytology and its application to botanical research. Amer. J. Bot. 69:1694–1706.

Toth, R., and R. M. Miller. 1984. Dynamics of arbuscule development and degeneration in a *Zea mays* mycorrhiza. Amer. J. Bot. 71:449–460.

Trachtenberg, S., and A. Fahn. 1981. The mucilage cells of *Opuntia ficus-indica* (L.) Mill.—Development, ultrastructure, and mucilage secretion. Bot. Gaz. 142:206–213.

Trachtenberg, S., and A. M. Mayer. 1981. Composition and properties of *Opuntia ficus-indica* mucilage. Phytochemistry 20:2665–2668.

Trachtenberg, S., and A. M. Mayer. 1982. Quantitative autoradiography of mucilage secretion in *Opuntia ficus-indica* (L.) Mill. Biol. Cell 44:69–76.

Tran, T.-T. H. 1963. Sur l'existence de deux catégories histologiques de ligules nerviées chez les Graminées. Bull. Soc. Bot. France 110:204–209.

Troll, W. 1935. *Vergleichende Morphologie der höheren Pflanzen*. Bd. 1, Lief. 2. Gbdr. Borntraeger: Berlin.

Troll, W. 1937. *Vergleichende Morphologie der höheren Pflanzen*. Bd. 1, Erster Teil. Gbdr. Borntraeger: Berlin.

Troll, W. 1939. *Vergleichende Morphologie der höheren Pflanzen. Band 1. Vegetationsorgane*. Heft 2. Gbdr. Borntraeger: Berlin.

Troll, W., and W. Rauh. 1950. *Das Erstarkungswachstum krantiger Dikotylen, mit besonderer Berucksichtigung der primären Verdickungsorgane*. Sitzungeber. Heidel. Akad. Wiss. 1 Abh.

Tschermak-Woess, E. 1971. Endomitose. Handb. allgem. Pathol. 2:569–625.

Tschirch, A. 1889. *Angewandte Pflanzenanatomie*. Urban and Schwarzenberg: Vienna.

Tucker, C. M., and R. F. Evert. 1969. Seasonal development of the secondary phloem in *Acer negundo*. Amer. J. Bot. 56:275–284.

Tucker, S. C. 1960. Ontogeny of the floral apex of *Michelia fuscata*. Amer. J. Bot. 47:266–277.

Tucker, S. C. 1964. The terminal idioblasts in magnoliaceous leaves. Amer. J. Bot. 51:1051–1062

Tucker, S. C. 1985. Initiation and development of inflorescence and flower in *Anemopsis californica* (Saururaceae). Amer. J. Bot. 72:20–31.

Tucker, S. C., and E. M. Gifford, Jr. 1966. Carpel development in *Drimys lanceolata*. Amer. J. Bot. 53:671–678.

Tukey, H. B., Jr. 1970. The leaching of substances from plants. Ann. Rev. Plant Physiol. 21:305–324.

Tukey, H. B., and J. O. Young. 1939. Histological study of the developing fruit of the sour cherry. Bot. Gaz. 100: 723–749.

Turrell, F. M. 1936. The area of internal exposed surface of dicotyledon leaves. Amer. J. Bot. 23:255–264.

Turrell, F. M., and L. J. Klotz. 1940. Density of stomata and oil glands and incidence of water spot in the rind of Washington navel orange. Bot. Gaz. 101:862–871.

Uhl, N. W., and H. E. Moore, Jr. 1971. The palm gynoecium. Amer. J. Bot. 58:945–992.

Ungar, I. A. 1979. Seed dimorphism in *Salicornia europaea* L. Bot. Gaz. 140:102–108.

Uphof, J. C. T. 1962. Plant hairs. In *Handbook of Plant Anatomy*, vol. 4, pt. 5. bdr. Borntraeger: Berlin.

Urschler, I. 1956. Untersuchungen mit dem Phycomyces-Test. Protoplasma 46:794–797.

Valdovinos, J. G., and T. E. Jensen. 1968. Fine structure of abscission zones. II. Cell-wall changes in abscising pedicels of tobacco and tomato flowers. Planta 83:295–302.

Vallade, J. 1983. Les racines ont elles un épiderme? Bull. Sci. de Bourgogne 36:7–17.

van Bel, A. J. E., and A. J. Koops. 1985. Uptake of [^{14}C]sucrose in isolated minor-vein networks of *Commelina benghalensis* L. Planta 164:362–369.

van Cotthem, W. R. J. 1970. A classification of stomatal types. Bot. J. Linn. Soc. 63:235–246.

van Cotthem, W. R. J. 1971. Vergleichende morphologische Studien über Stomata und eine neue Klassifikation ihrer Typen. Ber. Dtsch. Bot. Ges. 84:141–168.

van der Pijl, L. 1982. *Principles of Dispersal in Higher Plants.* 3rd ed. Springer-Verlag: Berlin.

van Fleet, D. S. 1942. The development and distribution of the endodermis and an associated oxidase system in monocotyledonous plants. Amer. J. Bot. 29:1–15.

van Nigtevecht, G. 1966. Genetic studies in dioecious *Melandrium*. I. Sex-linked and sex-influenced inheritance in *Melandrium album* and *Melandrium dioicum*. Genetica 37:281–306.

van Staveren, M. G. C., and P. Baas. 1973. Epidermal leaf characters of the Malesian Icacinaceae. Acta Bot. Neerl. 22:329–359.

van Tieghem, P., and H. Douliot. 1886. Sur la polystélie. Annls. Sci. nat. Bot. 3:275–322.

van Vliet, G. J. C. M. 1976a. Radial vessels in rays. IAWA Bull. 1976:35–37.

van Vliet, G. J. C. M. 1976b. Wood anatomy of the Rhizophoraceae. In *Wood Structure in Biological and Technological Research*, eds. P. Baas, A. J. Bolton, and D. M. Catling. Leiden University Press: Leiden.

van Went, J. L. 1970. The ultrastructure of the egg and central cell of *Petunia*. Acta Bot. Neerl. 19:313–322.

Vance, C. P., T. K. Kirk, and R. T. Sherwood. 1980. Lignification as a mechanism of disease resistance. Ann. Rev. Phytopathol. 18:259–288.

Vartanian, N. 1981. Some aspects of structural and functional modifications induced by drought in root systems. In *Structure and Function of Plant Roots*, eds. R. Brouwer, O. Gašparíková, J. Kolek, and B. C. Loughman. Martinus Nijhoff/Dr W. Junk Publishers: The Hague.

Vasil, I. K. 1974. The histology and physiology of pollen germination and pollen tube growth on the stigma and in the style. In *Fertilization in Higher Plants*, ed. H. F. Linskens. North Holland: Amsterdam.

Vasil, I. K. 1984. *Cell Culture and Somatic Cell Genetics of Plants. Vol. I. Laboratory Procedures and Their Applications.* Academic Press: New York.

Venkaiah, K., and J. J. Shah. 1984. Distribution, development and structure of gum ducts in *Lannea coromandelica* (Houtt.) Merril. Ann. Bot. 54:175–186.

Veres, J. S., and G. J. Williams, III. 1985. Leaf cavity size differentiation and water relations in *Carex eleocharis*. Amer. J. Bot. 72:1074–1077.

Vertrees, G. L., and P. G. Mahlberg. 1978. Structure and ontogeny of laticifers in *Cichorium intybus* (Compositae). Amer. J. Bot. 65:764–771.

Vesque, J. 1889. De l'emploi des caractères anatomiques dans la classification des végétaux. Bull. Soc. Bot. Fr. 36, XLI–LXXVII.

Vestal, P. A., and M. R. Vestal. 1940. The formation of septa in the fiber-tracheids of *Hypericum androsaemum* L. Leafl. Bot. Mus. Harv. Univ. 8:169–188.

Vitale, J. J., and D. C. Freeman. 1985. Secondary sex characteristics in *Spinacia oleracea* L.: Quantitative evidence for the existence of at least three sexual morphs. Amer. J. Bot. 72:1061–1066.

Vogel, A. 1960. Zur Feinstruktur der Drüsen von *Pinguicula*. Beih. Zn. schweiz. Forstver. 30:113–122.

Vogel, S. 1955. Niedere "Fensterpflanzen" in der südafrikanischen Wüste. Beit. Biol. Pflanzen 31:45–135.

Vogel, S. 1962. Duftdrüsen im Dienst der Bestaübung. Über Bau und Funktion der Osmophoren. Akad. Wiss. Lit. Mainz, Abh. Math.-Nat. Kl. 10:598–763.

Vogel, S. 1981. *Life in Moving Fluids*. Willard Grant Press: Boston.

Vogt, E., J. Schönherr, and H. W. Schmidt. 1983. Water permeability of periderm membranes isolated enzymatically from potato tubers (*Solanum tuberosum* L.). Planta 158:294–301.

Volkens, G. 1887. *Die Flora der Aegyptisch-Arabischen Wuste*. Gbdr. Borntraeger: Berlin.

von Guttenberg, H. 1926. Die Bewegungsgewebe. In *Handbuch der Pflanzenanatomie*, Bd. 5, Lief 18, ed. K. Linsbauer. Gbdr. Borntraeger: Berlin.

von Guttenberg, H. 1940. Der primäre Bau der Angiospermenwurzel. In *Handbuch der Pflanzenanatomie*, Bd. 8, Lief. 39, ed. K. Linsbauer. Gbdr. Borntraeger: Berlin.

von Guttenberg, H. 1941. Der primäre Bau der Gymnospermenwurzel. In *Handbuch der Pflanzenanatomie*, Bd. 8, Lief. 41, ed. K. Linsbauer. Gbdr. Borntraeger: Berlin.

von Guttenberg, H. 1943. Die physiologischen Scheiden.

In *Handbuch der Pflanzenanatomie*, Bd. 5, Lief. 42, ed. K. Linsbauer. Gbdr. Borntraeger: Berlin.

von Guttenberg, H. 1961. Grundzüge der Histogenese höherer Pflanzen. II. Gymnospermen. Gbdr. Borntraeger. Berlin.

von Guttenberg, H. 1968. Der Primäre Bau der Angiospermenwurzel. In *Encyclopedia of Plant Anatomy*, vol. 8, part 5. Gbdr. Borntraeger: Berlin.

von Guttenberg, H. 1971. Bewegunsgewebe und Perzeptionsorgane. In *Handbuch der Pflanzenanatomie*, Bd. 6, Teil 1, ed. K. Linsbauer. Gbdr. Borntraeger: Berlin.

Wacowska, M., and J. A. Tarkowska. 1983. Ontogenesis and structure of phelloid in *Viburnum opulus*. Acta Soc. Bot. Pol. 52:107–114.

Wagner, G., W. Haupt, and A. Laux. 1972. Reversible inhibition of chloroplast movement by cytochalasin B in the green alga *Mougeotia*. Science, N. Y. 176:808–809.

Wagner, K. A. 1946. Notes on the anomalous stem structure of a species of *Bauhinia*. American Midland Naturalist 36:251–256.

Wagner, W. H., Jr. 1972. *Solanopteris brunei*, a little-known fern epiphyte with dimorphic stems. Amer. Fern J. 62: 33–43.

Wagner, W. H., Jr. 1979. Reticulate veins in the systematics of modern ferns. Taxon 28:87–95.

Waid, J. S. 1974. Decomposition of roots. In *Biology of Plant Litter Decomposition*, eds. C. H. Dickinson and G. J. F. Pugh, vol. 1. Academic Press: London.

Waisel, Y., N. Liphschitz, and T. Arzee. 1967. Phellogen activity in *Robinia pseudoacacia* L. New Phytol. 66:331–335.

Walker, R. R., M. Sedgley, M. A. Blesing, and T. J. Douglas. 1984. Anatomy, ultrastructure and assimilate concentrations of roots of *Citrus* genotypes differing in ability for salt exclusion. J. Exp. Bot. 35:1481–1494.

Walker, W. S. 1957. The effect of mechanical stimulation on the collenchyma of *Apium graveolens* L. Proc. Iowa Acad. Sci. 64:177–186.

Walker, W. S. 1960. The effect of mechanical stimulation and etiolation of the collenchyma of *Datura stramonium*. Amer. J. Bot. 47:717–724.

Walles, B., B. Nyman, and T. Alden. 1973. On the ultrastructure of needles of *Pinus silvestris* L. Stud. Forest. Suec. 106:1–26.

Walsh, M. A., and R. F. Evert. 1975. Ultrastructure of metaphloem sieve elements in *Zea mays*. Protoplasma 83:365–388.

Walsh, M. A., and J. E. Melaragno. 1981. Structural evidence for plastid inclusions as a possible sealing mechanism in the phloem of monocotyledons. J. Exp. Bot. 32: 311–320.

Walter, H., and M. Steiner. 1937. Die Okologie der Ost-Afrikanischen Mangroven. Z. Bot. 30:65–193.

Walter, W. M., Jr., and W. E. Schadel. 1983. Structure and composition of normal skin periderm and wound tissue from cured sweet-potatoes. J. Am. Hort. Sci. 108:909–914.

Wardlaw, C. W. 1955. *Embryogenesis in Plants*. John Wiley: New York.

Wardrop, A. B. 1956. The nature of reaction wood. V. The distribution and formation of tension wood in some species of *Eucalyptus*. Aust. J. Bot. 4:152–166.

Wardrop, A. B. 1964. The reaction anatomy of arborescent angiosperms. In *The Formation of Wood in Forest Trees*, ed. M. H. Zimmermann. Academic Press: New York.

Wardrop, A. B. 1965. The formation and function of reaction wood. In *Cellular Ultrastructure of Woody Plants*, ed. W. A. Côté, Jr. Syracuse University Press: Syracuse, New York.

Wardrop, A. B. 1969. The structure of the cell wall in lignified collenchyma of *Eryngium* sp. (Umbelliferae). Aust. J. Bot. 17:229–240.

Wardrop, A. B. 1971. Occurrence and formation in plants. In *Lignins: Occurrence, Formation, Structure, and Reactions*, eds. K. V. Sarkanen and C. H. Ludwig. Wiley-Interscience: New York.

Wardrop, A. B. 1981. Lignification and xylogenesis. In *Xylem Cell Development*, ed. J. R. Barnett. Castle House Publications, Ltd: Kent, England.

Wardrop, A. B., and H. E. Dadswell. 1948. The nature of reaction wood. I. The structure and properties of tension wood fibres. Aust. J. Sci. Res. B1:3–16.

Wardrop, A. B., and H. E. Dadswell. 1955. The nature of reaction wood. IV. Variations in cell wall organization of tension wood fibres. Aust. J. Bot. 3:177–189.

Wardrop, A. B., and G. W. Davies. 1964. The nature of reaction wood. VIII. The structure and differentiation of compression wood. Aust. J. Bot. 12:24–36.

Warmbrodt, R. D. 1980. Characteristics of structure and differentiation in the sieve element of lower vascular plants. Ber. dtsch. bot. Ges. 93:13–28.

Warmbrodt, R. D. 1985. Studies on the root of *Hordeum vulgare* L. Ultastructure of the seminal root with special reference to the phloem. Amer. J. Bot. 72:414–432.

Warmbrodt, R. D., and W. Eschrich. 1985. Studies on the mycorrhizas of *Pinus sylvestris* L. produced in vitro with the basidiomycete *Suillus variegatus* (Sw. ex Fr.) O. Kuntze. II. Ultrastructural aspects of the endodermis and vascular cylinder of the mycorrhizal rootlets. New Phytol. 100: 403–418.

Warmbrodt, R. D., and R. F. Evert. 1974. Structure and development of the sieve element in the stem of *Lycopodium lucidulum*. Amer. J. Bot. 61:267–277.

Warmbrodt, R. D., and R. F. Evert. 1978. Comparative leaf structure of six species of heterosporous ferns. Bot. Gaz. 139:393–429.

Warmbrodt, R. D., and R. F. Evert. 1979a. Comparative leaf structure of six species of eusporangiate and protoleptosporangiate ferns. Bot. Gaz. 140:153–167.

Warmbrodt, R. D., and R. F. Evert. 1979b. Comparative leaf structure of several species of homosporous leptosporangiate ferns. Amer. J. Bot. 66:412–440.

Watson, R. W. 1942. The effect of cuticular hardening on the form of epidermal cells. New Phytol. 41:223–229.

Wattendorff, J. 1974. The formation of cork cells in the periderm of *Acacia senegal* Willd. and their ultrastructure during suberin deposition. Z. Pflanzenphysiol. 72:119–134.

Wattendorff, J., and P. J. Holloway. 1980. Studies on the ultrastructure and histochemistry of plant cuticles: The cuticular membrane of *Agave americana* L. in situ. Ann. Bot. 46:13–28.

Weatherly, P. E. 1975. Water relations of the root system. In *The Development and Function of Roots*, eds. J. G. Torrey and D. T. Clarkson. Academic Press: London.

Weaver, J. E. 1926. *Root Development of Field Crops*. McGraw-Hill: New York.

Weberling, F. 1982. *Stangea, Belonanthus* und *Phyllactis*— ein Vergleich. Ber. Deutsch Bot. Ges. 95:165–179.

Webster, B. D. 1968. Anatomical aspects of abscission. Plant Physiol. 43:1512–1544.

Weerdenburg, C. A., and C. A. Peterson. 1983. Structural changes in phi thickenings during primary and secondary growth in roots. 1. Apple (*Pyrus malus*) Rosaceae. Can. J. Bot. 61:2570–2576.

Weerdenburg, C. A., and C. A. Peterson. 1984. Effect of secondary growth on the conformation and permeability of the endodermis of broad bean (*Vicia faba*), sunflower (*Helianthus annuus*), and garden balsam (*Impatiens balsamina*). Can. J. Bot. 62:907–910.

Weibel, E. R. 1979. *Stereological Methods. Vol. 1. Practical Methods for Biological Morphometry*. Academic Press: New York.

Werker, E. 1980. Review. Seed dormancy as explained by the anatomy of embryo envelopes. Israel J. Bot. 29:22–44.

Werker, E., and M. Kislev. 1978. Mucilage on the root surface and root hairs of *Sorghum*: Heterogeneity in structure, manner of production and site of accumulation. Ann. Bot. 42:809–816.

Werker, E., I. Marbach, and A. M. Mayer. 1979. Relation between the anatomy of the testa, water permeability and the presence of phenolics in the genus *Pisum*. Ann. Bot. 43:765–771.

Werker, E., and J. G. Vaughan. 1974. Anatomical and ultrastructural changes in aleurone and myrosin cells of *Sinapis alba* during germination. Planta 116:243–255.

Werker, E., and J. G. Vaughan. 1976. Ontogeny and distribution of myrosin cells in the shoot of *Sinapis alba* L. A light- and electron-microscope study. Israel J. Bot. 25:140–151.

Werner, D., and E. Mörschel. 1978. Differentiation of nodules of *Glycine max*. Planta 141:169–177.

West, C. 1917. A contribution to the study of the Marattiaceae. Ann. Bot. 31:361–414.

Westing, A. H. 1968. Formation and function of compression wood in gymnosperms. II. Bot. Rev. 34:51–78.

Wetmore, R. H. 1926. Organization and significance of lenticels in dicotyledons. I. Lenticels in relation to aggregate and compound storage rays in woody stems. Lenticels and roots. Bot. Gaz. 82:71–88.

Whaley, W. G., and J. H. Leech. 1961. A function of Golgi apparatus in outer root cap cells. J. Ultrastruct. Res. 5:193–200.

Whalley, B. E. 1950. Increase in girth of the cambium in *Thuja occidentalis* L. Can. J. Res. Sect. C. Bot. Sci. 28:331–340.

Whatley, J. M. 1980. Plastid growth and division in *Phaseolus vulgaris*. New Phytol. 86:1–16.

Wheat, D. 1977. Successive cambia in the stem of *Phytolacca dioica*. Amer. J. Bot. 64:1209–1217.

Wheat, D. 1980. Sylleptic branching in *Myrsine floridana* (Myrsinaceae). Amer. J. Bot. 67:490–499.

Wheeler, G. E. 1979. Raphide files in vegetative organs of *Zebrina*. Bot. Gaz. 140:189–198.

Whitmore, T. C. 1962a. Studies in systematic bark morphology. I. Bark morphology in Dipterocarpaceae. II. General features of bark construction in Dipterocarpaceae. New Phytol. 61:191–220.

Whitmore, T. C. 1962b. Why do trees have different sorts of bark? New Scientist 16:330–331.

Wiersum, L. K. 1957. The relationship of the size and structural rigidity of pores to their penetration by roots. Pl. Soil 9:75–85.

Wiesner, J., and K. Linsbauer. 1920. *Anatomie und Physiologie der Pflanzen*. 6th ed. Vienna.

Wilcox, H. 1962a. Growth studies of the root of incense cedar *Libocedrus decurrens*. I. The origin and development of primary tissues. Amer. J. Bot. 49:221–236.

Wilcox, H. 1962b. Growth studies of the root of incense cedar *Libocedrus decurrens*. II. Morphological features of the root system and growth behaviour. Amer. J. Bot. 49:237–245.

Wilder, G. J. 1985. Anatomy of noncostal portions of lamina in the Cyclanthaceae (Monocotyledoneae). I. Epidermis. Bot. Gaz. 146:82–105.

Wilder, G. J., and D. H. Harris. 1982. Laticifers in *Cyclanthus bipartitus* Poit. (Cyclanthaceae). Bot. Gaz. 143:84–93.

Wilkinson, H. P. 1971. *Leaf anatomy of various Anacardiaceae with special reference to the epidermis and some contributions to the taxonomy of the genus* Dracontomelon *Blume*. Thesis. University of London.

Wilkinson, H. P. 1979. The plant surface (mainly leaf). In *Anatomy of the Dicotyledons*, vol. 1, ed. C. R. Metcalfe and L. Chalk. Clarendon Press: Oxford.

Willemse, M. T. M., and H. F. Linskens. 1969. Développement du microgamétophyte chez le *Pinus sylvestris* entre la meiose et la fécondation. Rev. Cytol. Biol. vég. 32:121–128.

Williams, S. E., and B. G. Pickard. 1974. Connections and barriers between cells of *Drosera* tentacles in relation to their electrophysiology. Planta 116:1–16.

Williams, W. T., and D. A. Barber. 1961. The functional significance of aerenchyma in plants. Soc. Exp. Biol. Symp. 15:132–144.

Willison, J. H. M., and F. J. Cragg. 1980. Nuclear pore structure in quiescent buds of *Tilia europaea*. Can. J. Bot. 58:1814–1819.

Willmer, C. M., and R. Sexton. 1979. Stomata and plasmodesmata. Protoplasma 100:113–124.

Wilson, B. F. 1963. The fusiform cells contribute more than 90% by volume of the cambium and its derivatives. Amer. J. Bot. 50:95–102.

Wilson, B. F. 1964. A model for cell production by the cambium of conifers. In *The Formation of Wood in Forest Trees*, ed. M. H. Zimmermann. Academic Press: New York.

Wilson, B. F., and R. R. Archer. 1977. Reaction wood: Induction and mechanical action. Ann. Rev. Plant Physiol. 28:23–43.

Wilson, C. L. 1924. Medullary bundle in relation to primary vascular system in Chenopodiaceae and Amaranthaceae. Bot. Gaz. 78:175–199.

Wilson, C. L. 1942. The telome theory and the origin of the stamen. Amer. J. Bot. 21:759–764.

Wilson, C. L. 1982. Vestigial structures and the flower. Amer. J. Bot. 69:1356–1365.

Wilson, K. J., and P. G. Mahlberg. 1978. Ultrastructure of non-articulated laticifers in mature embryos and seedlings of *Asclepia syriaca* L. (Asclepiadaceae). Amer. J. Bot. 65:98–109.

Wilson, K. J., C. L. Nessler, and P. G. Mahlberg. 1976. Pectinase in *Asclepias* latex and its possible role in laticifer growth and development. Amer. J. Bot. 63:1140–1144.

Winkler, H. 1931. Über die eigenartige Stellung der Blüten bei der Rubiacee *Stichianthus minutiflorus* Valenton. Planta 13:85–101.

Withner, C. L. *The Orchids: Scientific Studies*. John Wiley and Sons: New York.

Wittler, G. H., and J. D. Mauseth. 1984a. The ultrastructure of developing latex ducts in *Mammillaria heyderi* (Cactaceae). Amer. J. Bot. 71:100–110.

Wittler, G. H., and J. D. Mauseth. 1984b. Schizogeny and ultrastructure of developing latex ducts in *Mammillaria guerreronis* (Cactaceae). Amer. J. Bot. 71:1128–1138.

Wodzicki, T. J., and C. L. Brown. 1973. Organization and breakdown of the protoplast during maturation of pine tracheids. Amer. J. Bot. 60:631–640.

Wodzicki, T. J., and W. J. Humphreys. 1972. Cytodifferentiation of maturing pine tracheids: The final stage. Tissue and Cell 4:525–528.

Wolkinger, F. 1969. Morphologie und systematische Verbreitung der lebenden Holzfasern bei Sträucher und Bäumen. I. Zur Morphologie und Zytologie. Holzforsch. 23:135–144.

Wolkinger, F. 1970a. Morphologie und systematische Verbreitung der lebenden Holzfasern bei Sträucher und Bäumen. II. Zur Histologie. Holzforsch. 24:141–151.

Wolkinger, F. 1970b. Das Vorkommen lebender Holzfasern in Sträuchern und Bäumen. Phyton 14:55–67.

Woodhouse, R. M., and P. S. Nobel. 1982. Stipe anatomy, water potentials, and xylem conductances in seven species of ferns (Filicopsida). Amer. J. Bot. 69:135–140.

Worsdell, W. C. 1916. The morphology of the monocotyledonous embryo and of that of the grass in particular. Ann. Bot. 30:509–524.

Wright, M., and D. J. Osborne. 1974. Abscission in *Phaseolus vulgaris*. The positional differentiation and ethylene induced expansion growth of specialized cells. Planta 120:163–170.

Wunderlich, R. 1954. Über das Antherentapetum mit besonderer Berücksichtigung seiner Kernzahl. Österreich. Bot. Z. 101:1–63.

Wylie, R. B. 1939. Relations between tissue organization and vein distribution in dicotyledon leaves. Amer. J. Bot. 26:219–225.

Wylie, R. B. 1943. The role of the epidermis in foliar organization and its relations to the minor venation. Amer. J. Bot. 30:273–280.

Wylie, R. B. 1947. Conduction in dicotyledon leaves. Proc. Iowa Acad. Sci. 53:195–202.

Wylie, R. B. 1948. The dominant role of the epidermis in leaves of *Adiantum*. Amer. J. Bot. 35:465–473.

Wylie, R. B. 1949. Differences in foliar organization among leaves from four locations in the crown of an isolated tree (*Acer platanoides*). Proc. Iowa Acad. Sci. 56:189–198.

Wylie, R. B. 1951. Principles of foliar organization shown by sun-shade leaves from ten species of deciduous dicotyledonous trees. Amer. J. Bot. 38:355–361.

Yamaura, A. 1933. Karyologische und embryologische Studien über einige *Bambusa*-Arten. (Vorläufige Mitteilung). Bot. Mag. (Tokyo) 47:551–555.

Yarrow, G. L., and R. A. Popham. 1981. The ontogeny of the primary thickening meristem of *Atriplex hortensis* L. (Chenopodiaceae). Amer. J. Bot. 68:1042–1049.

Yeoman, M. M., A. J. Tulett, and V. Bagshaw. 1970. Nuclear extension in dividing vacuolated plant cells. Nature, London 226:557–558.

Yeung, E. C. 1980. Embryogeny of *Phaseolus*: The role of the suspensor. Z. Pflanzenphysiol. 96:17–28.

Yeung, E. C., and M. E. Clutter. 1978. Embryogeny of *Phaseolus coccineus*: Growth and microanatomy. Protoplasma 94:19–40.

Yorke, J. S., and G. R. Sagar. 1970. Distribution of secondary root growth potential in the root system of *Pisum sativum*. Can. J. Bot. 48:699–704.

Young, D. A. 1981. Are the angiosperms primitively vesselless? Systematic Bot. 6:313–330.

Zagórska-Marek, B. 1984. Pseudotransverse divisions and intrusive elongation of fusiform initials in the storeyed cambium of *Tilia*. Can. J. Bot. 62:20–27.

Zahur, M. S. 1959. Comparative study of secondary phloem of 423 species of woody dicotyledons belonging to 85 families. Mem. Cornell Univ. Agric. Exp. Sta., no. 358. Cornell University: Ithaca, New York.

Zamski, E. 1979. The mode of secondary growth and the three-dimensional structure of the phloem in *Avicennia*. Bot. Gaz. 140:67–76.

Zamski, E., and A. Azenkot. 1981. Sugarbeet vasculature. I. Cambial development and the three-dimensional structure of the vascular system. Bot. Gaz. 142:334–343.

Zasada, J. C., and R. Zahner. 1969. Vessel element development in the earlywood of red oak (*Quercus rubra*). Can. J. Bot. 47:1965–1971.

Zee, S. Y., and T. P. O'Brien. 1970. A special type of tracheary element associated with "xylem discontinuity" in the floral axis of wheat. Aust. J. Biol. Sci. 23:783–791.

Zheng-Hai, H. 1963. Studies on the structure and the ontogeny of laticiferous canals of *Decaisnea fargesii* Franch. Acta Bot. Sinica 11:129–140.

Ziegenspeck, H. 1944. Vergleichende Untersuchungen der Entwicklung der Spaltöffnungen von Monokotyledonen und Dikotyledonen im Lichte der Polariskopie und Dichroskopie. Protoplasma 38:197–224.

Ziegler, H., E. Shmueli, and G. Lange. 1974. Structure and function of the stomata of *Zea mays*. I. The development. Cytobiologie 9:162–168.

Zimmermann, A. 1922. Die Cucurbitaceen, Heft 1. Beiträge zur Anatomie und Physiologie. Gustav Fischer: Jena, Germany.

Zimmermann, J. G. 1932. Über die extrafloralen Nektarien der Angiospermen. Beih. Bot. Zbl. 49:99–196.

Zimmermann, M. H. 1979. The discovery of tylose formation by a Viennese Lady in 1845. IAWA Bull. 1979:51–56.

Zimmermann, M. H. 1982. Functional xylem anatomy of angiosperm trees. In *New Perspectives in Wood Anatomy*, ed. P. Baas. Martinus Nijhoff/Dr W. Junk Publishers: The Hague.

Zimmermann, M. H. 1983. *Xylem Structure and the Ascent of Sap.* Sringer-Verlag: Berlin.

Zimmermann, M. H., and C. L. Brown. 1974. *Tree Structure and Function.* Springer-Verlag: Berlin.

Zimmermann, M. H., and A. A. Jeje. 1981. Vessel-length distribution in stems of some American woody plants. Can. J. Bot. 59:1882–1892.

Zimmermann, M. H., K. F. McCue, and J. S. Sperry. 1982. Anatomy of the palm *Rhapis excelsa*. VIII. Vessel network and vessel-length distribution in the stem. J. Arnold Arbor. 63:83–95.

Zimmermann, M. H., and J. S. Sperry. 1983. Anatomy of the palm *Rhapis excelsa*. IX. Xylem structure of the leaf insertion. J. Arnold Arbor. 64:599–609.

Zimmermann, M. H., and P. B. Tomlinson. 1965. Anatomy of the palm *Rhapis excelsa*. I. Mature vegetative axis. J. Arnold Arbor. 46:160–180.

Zimmermann, M. H., and P. B. Tomlinson. 1967. Anatomy of the palm *Rhapis excelsa*. IV. Vascular development in apex of vegetative aerial axis and rhizome. J. Arnold Arbor. 48:122–142.

Zimmermann, M. H., and P. B. Tomlinson. 1968. Vascular construction and development in the aerial stem of *Prionium* (Juncaceae). Amer. J. Bot. 55:1100–1109.

Zimmermann, M. H., and P. B. Tomlinson. 1969. The vascular system of the axis of *Dracaena fragrans* (Agavaceae). 1. Distribution and development of primary strands. J. Arnold Arbor. 50:370–383.

Zimmermann, M. H., and P. B. Tomlinson. 1970. The vascular system in the axis of *Dracaena fragrans* (Agavaceae). 2. Distribution and development of secondary vascular tissue. J. Arnold Arbor. 51:478–491.

Zimmermann, M. H., and P. B. Tomlinson. 1974. Vascular patterns in palm stems: Variations of the *Rhapis* principle. J. Arnold Arbor. 55:402–424.

Zimmermann, M. H., P. B. Tomlinson, and J. LeClaire. 1974. Vascular construction and development in the stems of certain Pandanaceae. Bot. J. Linn. Soc. 68:21–41.

Zimmermann, W. 1952. The main results of the "telome theory." The Palaeobotanist 1:456–470.

Zimmermann, W., and E. Seemueller. 1984. Degradation of raspberry suberin by *Fusarium solani* f. sp. *pisi* and *Armillaria mellea*. Phytopathol. Z. 110:192–199.

Zobel, A. M. 1985. Ontogenesis of tannin coenocytes in *Sambucus racemosa* L. II. Mother tannin cells. Ann. Bot. 56:91–104.

Zohary, M. 1937. Die verbreitungsökologischen Verhältnisse der Pflanzen Palästinas. I. Die antitelechorischen Erscheinungen. Beih. Bot. Zbl. Abt. A. 56:1–155.

Zohary, M., and A. Fahn. 1950. On the heterocarpy of *Aethionema*. Palestine J. Bot. Jerusalem ser. 5:28–31.

Zweypfenning, R. C. V. J. 1978. A hypothesis on the function of vestured pits. IAWA Bull. 1978:13–15.

Index